Unsteady Aerodynamics

Aerospace Series

David Allerton. *Principles of Flight Simulation*
Dora Musielak. *Scramjet Propulsion: A Practical Introduction*
Allan Seabridge, Mohammad Radaei. *Aircraft Systems Classifications: A Handbook of Characteristics and Design Guidelines*
Douglas M. Marshall. *UAS Integration into Civil Airspace: Policy, Regulations and Strategy*
Paul G. Fahlstrom, Thomas J. Gleason, Mohammad H. Sadraey. *Introduction to UAV Systems, 5th Edition*
James W. Gregory, Tianshu Liu. *Introduction to Flight Testing*
Ashish Tewari. *Foundations of Space Dynamics*
Egbert Torenbeek. *Essentials of Supersonic Commercial Aircraft Conceptual Design*
Mohammad H. Sadraey. *Design of Unmanned Aerial Systems*
Saeed Farokhi. *Future Propulsion Systems and Energy Sources in Sustainable Aviation*
Rama K. Yedavalli. *Flight Dynamics and Control of Aero and Space Vehicles*
Allan Seabridge, Ian Moir. *Design and Development of Aircraft Systems, 3rd Edition*
Gareth D. Padfield. *Helicopter Flight Dynamics: Including a Treatment of Tiltrotor Aircraft, 3rd Edition*
Craig A. Kluever. *Space Flight Dynamics, 2nd Edition*
Trevor M. Young. *Performance of the Jet Transport Airplane: Analysis Methods, Flight Operations, and Regulations*
Andrew J. Keane, Andros Sobester, James P. Scanlan. *Small Unmanned Fixed-wing Aircraft Design: A Practical Approach*
Pascual Marques, Andrea Da Ronch. *Advanced UAV Aerodynamics, Flight Stability and Control: Novel Concepts, Theory and Applications*
Farhan A. Faruqi. *Differential Game Theory with Applications to Missiles and Autonomous Systems Guidance*
Grigorios Dimitriadis. *Introduction to Nonlinear Aeroelasticity*
Nancy J. Cooke, Leah J. Rowe, Winston Bennett Jr., DeForest Q. Joralmon. *Remotely Piloted Aircraft Systems: A Human Systems Integration Perspective*
Stephen Corda. *Introduction to Aerospace Engineering with a Flight Test Perspective*
Wayne Durham, Kenneth A. Bordignon, Roger Beck. *Aircraft Control Allocation*
Ashish Tewari. *Adaptive Aeroservoelastic Control*
Ajoy Kumar Kundu, Mark A. Price, David Riordan. *Theory and Practice of Aircraft Performance*
Peter Belobaba, Amedeo Odoni, Cynthia Barnhart, Christos Kassapoglou. *The Global Airline Industry, 2nd Edition*
Jan R. Wright, Jonathan Edward Cooper. *Introduction to Aircraft Aeroelasticity and Loads, 2nd Edition*
Tapan K. Sengupta. *Theoretical and Computational Aerodynamics*
Andros Sobester, Alexander I.J. Forrester. *Aircraft Aerodynamic Design: Geometry and Optimization*
Roy Langton. *Stability and Control of Aircraft Systems: Introduction to Classical Feedback Control*
T. W. Lee. *Aerospace Propulsion*
Ian Moir, Allan Seabridge, Malcolm Jukes. *Civil Avionics Systems, 2nd Edition*
Wayne Durham. *Aircraft Flight Dynamics and Control*
Konstantinos Zografos, Giovanni Andreatta, Amedeo Odoni. *Modelling and Managing Airport Performance*
Egbert Torenbeek. *Advanced Aircraft Design: Conceptual Design, Analysis and Optimization of Subsonic Civil Airplanes*
Christos Kassapoglou. *Design and Analysis of Composite Structures: With Applications to Aerospace Structures, 2nd Edition*
Keith A. Rigby. *Aircraft Systems Integration of Air-Launched Weapons*
Doug McLean. *Understanding Aerodynamics: Arguing from the Real Physics*
Mohammad H. Sadraey. *Aircraft Design: A Systems Engineering Approach*
G.D. McBain. *Theory of Lift: Introductory Computational Aerodynamics in MATLAB/Octave*
Plamen Angelov. *Sense and Avoid in UAS: Research and Applications*
John Valasek. *Morphing Aerospace Vehicles and Structures*
Reg Austin. *Unmanned Aircraft Systems: UAVS Design, Development and Deployment*
Visit www.wiley.com to view more titles in the Aerospace Series.

Unsteady Aerodynamics

Potential and Vortex Methods

Grigorios Dimitriadis
University of Liège
Belgium

WILEY

This edition first published 2024.

Registered Offices
John Wiley & Sons, Inc., 111 River Street, Hoboken, NJ 07030, USA
John Wiley & Sons Ltd, The Atrium, Southern Gate, Chichester, West Sussex, PO19 8SQ, UK

For details of our global editorial offices, customer services, and more information about Wiley products visit us at www.wiley.com.

Wiley also publishes its books in a variety of electronic formats and by print-on-demand. Some content that appears in standard print versions of this book may not be available in other formats.

Library of Congress Cataloging-in-Publication Data

Names: Dimitriadis, Grigorios, 1972- author.
Title: Unsteady aerodynamics : potential and vortex methods / Grigorios Dimitriadis.
Description: Hoboken, NJ : Wiley, 2024. | Series: Aerospace series | Includes index.
Identifiers: LCCN 2023030217 (print) | LCCN 2023030218 (ebook) | ISBN 9781119762478 (cloth) | ISBN 9781119762539 (adobe pdf) | ISBN 9781119762553 (epub)
Subjects: LCSH: Unsteady flow (Aerodynamics)
Classification: LCC TL574.U5 D56 2024 (print) | LCC TL574.U5 (ebook) | DDC 629.132/3–dc23/eng/20230719
LC record available at https://lccn.loc.gov/2023030217
LC ebook record available at https://lccn.loc.gov/2023030218

Cover Design: Wiley
Cover Image: Image provided by G. Dimitriadis

Set in 9.5/12.5pt STIXTwoText by Straive, Chennai, India
Printed and bound by CPI Group (UK) Ltd, Croydon, CR0 4YY

C9781119762478_171123

Contents

Preface

The term 'unsteady aerodynamics' is used to denote fluid flow problems whereby either a body moves in a fluid in a time-varying fashion or the flow is time-varying in itself. The reader should also take note of the subtitle: potential and vortex methods. Although physical phenomena related to unsteady fluid motion are discussed in detail, the focus of this book is on modelling methods. Furthermore, the subtitle makes it clear that there will be no discussion in this book of what is commonly referred to as Computational Fluid Dynamics. Even though many numerical approaches will be presented, none of them will rely on discretising the entire flowfield around the body. All numerical solutions will be obtained by discretising the surface of the body and shedding vorticity in its wake, propagating the latter using a Lagrangian approach.

The book is called 'Unsteady Aerodynamics' and not Unsteady Fluid Dynamics because the main focus is on flow over wings. Nevertheless, many of the methodologies presented here could be applied to hydrodynamic problems, and the validation data used for some of the theories were obtained in water tunnels. Even though unsteadiness is the main subject area, several steady aerodynamic problems are presented and discussed in detail. However, the book is not an introduction to all aerodynamics or fluid dynamics in general. This means that the reader should have good background knowledge of basic aerodynamics.

The emphasis of this book is on application so that all theories are accompanied by practical examples solved by means of Matlab and C codes. These codes are available to the reader on the Wiley website; they have been tested on Matlab version 2020a but could also be compatible with other versions. The C codes are written as Matlab mex functions and should be compiled appropriately. It is the reader's responsibility to do so, neither Wiley nor the author will provide technical support. The Mathworks website includes a list of compatible compilers for different architectures here: https://www.mathworks.com/support/requirements/supported-compilers.html. The reader should note that the purpose of the codes is to illustrate the examples and the underlying theories. They solve the particular problems for which they were written, but they should not be seen as general unsteady aerodynamic analysis codes that can be directly applied to different problems.

Chapter 1 is a brief introduction to steady and unsteady aerodynamics and provides a more detailed outline of the book. Chapter 2 presents the fundamentals of unsteady flow, focusing on the concepts and equations that will be of use throughout the book. Classical 2D unsteady potential theory is presented in Chapter 3, with a focus on analytical solutions. Chapter 4 discusses numerical 2D potential flow solutions and concentrates on interesting physical phenomena such as thrust production and propulsive efficiency. Chapter 5 introduces the aerodynamic analysis of finite wings in incompressible flows, initially by means of analytical methods and then using numerical techniques. Compressible unsteady flows are treated in Chapter 6, which addresses subsonic, supersonic and transonic aerodynamic problems. Finally, Chapter 7 introduces viscous unsteady flows featuring significant flow separation.

I would like to take this opportunity to thank my colleagues Thomas Andrianne, Adrien Crovato, Thierry Magin and Ludovic Noels who took over my teaching during my sabbatical year. A particularly warm thank you goes to Kyros Iakinthos and Pericles Panagiotou, who welcomed me into their research group at the Aristotle University of Thessaloniki during that year. I would also like to thank Adrien Crovato for spotting a negative steady drag issue in the source and doublet panel code I developed for this book, Thomas Lambert for our exchanges on the Vatistas model, Mariano Sánchez Martínez for carrying out Euler simulations on the LANN wing, Johan Boutet for our collaboration on unsteady lifting line methods and Vincent Terrapon and Boyan Mihaylov for a late-night Zoom session discussing compound rotations (among other things) during the lockdown.

August 2023

Grigorios Dimitriadis
Liège and Thessaloniki

About the Companion Website

This book is accompanied by a companion website:

www.wiley.com/go/dimitriadis/unsteady_aerodynamics

The website includes sample programs with MATLAB codes and solutions.

1

Introduction

Unsteady aerodynamics refers to flow of air over bodies whose velocity field is changing in time. The causes of unsteadiness can be

- Translational and rotational acceleration of the body relative to the fluid. The vast majority of this book is devoted to this type of unsteadiness.
- Upstream or free stream unsteadiness. In the atmosphere, this phenomenon is referred to as atmospheric turbulence and is caused by a variety of meteorological and geographical phenomena. In the laboratory, we encounter wind tunnel turbulence. There will be no discussion of this type of unsteadiness in this book.
- Turbulence in boundary layers, which is a ubiquitous source of unsteadiness in most practical flows. A short discussion of laminar and turbulent boundary layers is included in Chapter 7.
- Flow that separates from the surface of the body, either instantaneously or permanently. This type of flow is inherently unsteady, even when the motion of the body is steady. We will discuss this type of unsteadiness in Chapter 7.

Other sources of unsteadiness, such as acoustics, jet impingement, wake interaction, and thermal effects, lie beyond the scope of this book.

The vast majority of practical airflows will feature some degree of turbulence upstream in the boundary layer and in the wake of the body. As a consequence, nearly all aerodynamics is unsteady. However, aircraft, rotors, wind turbines and other engineering structures are generally designed to operate under attached flow conditions, so that the turbulence is confined to a thin layer of fluid in contact with the surface. Under such conditions, the effect of turbulence is averaged and therefore the flow can be treated as steady. Then, the major source of unsteadiness becomes the motion of the body itself. Conversely, civil engineering structures are mostly aerodynamically bluff bodies; even though they seldom move, they are subjected to significant unsteadiness due to separated flow and upstream turbulence.

This book deals mostly with wings and therefore the source of unsteadiness it will address most of the time is body motion. Our ancestral prototype of flight is bird flight, which involves flapping wings. Yet the first man-made flying objects were kites which in their simplest form do not flap or deform in any way. The first gliders and aircraft also had fixed wings, and flapping blades were introduced in helicopter rotors much later. From a practical point of view, it is clearly easier to work with steady aerodynamics. This is also the case

Unsteady Aerodynamics: Potential and Vortex Methods, First Edition. Grigorios Dimitriadis.

Companion website: www.wiley.com/go/dimitriadis/unsteady_aerodynamics

from a mathematical point of view; the flow equations are simpler and easier to solve. From the experimental point of view too, setting up and measuring a steady flow is more straightforward.

Even fixed-wing aircraft undergo unsteady motion, both rigid and flexible. Rigid aircraft motion is the field of study of flight dynamics; aircraft have both oscillatory and non-oscillatory rigid body eigenmodes that cannot be predicted adequately using purely steady aerodynamic analysis. We will give an example of the calculation of aerodynamic stability derivatives in Chapter 5. Furthermore, aircraft structures are flexible and are becoming increasingly so. The study of vibrating structures in an airflow is the subject area of aeroelasticity. Again, a steady or quasi-steady aerodynamic analysis is insufficient to predict aeroelastic phenomena. Chapter 3 includes one example of a direct application of unsteady aerodynamics to flutter prediction. Nevertheless, all of the methods presented in this book can be used for flight dynamic, aeroelastic or combined aeroservoelastic analysis.

1.1 Why Potential and Vortex Methods?

The equations of fluid flow are notorious for being unsolvable. The Millennium Prize (Clay Mathematics Institute, 2000) for proving the existence and smoothness of solutions to the 3D Navier–Stokes equations was still unclaimed at the time of writing of this book and the original US$1 million prize money had already depreciated to US$575,000 due to inflation. Numerical solutions of these equations are possible, but turbulence renders them impractical. In order to capture all the spatial scales of turbulence at a Reynolds number encountered in aeronautical practice, the computational requirements of a direct numerical simulation of the Navier–Stokes equations exceed the capabilities of even the fastest and biggest modern computers. Therefore, in order to model practical problems, we resort to solving easier equations. These can be averaged or filtered versions of the original Navier–Stokes relations or simpler equations that are developed after making assumptions about the physics of the flow.

The fastest solutions are obtained for potential flow equations, whereby the flow is assumed to be inviscid, irrotational and isentropic, if not incompressible. Even though a significant amount of the physics of fluid flow is discarded in order to obtain such solutions, their range of validity can include many aeronautical applications under nominal operating conditions. For example, potential flow methods are the industrial standard for aircraft aeroelastic calculations. As long as the flow remains attached to the surface, its Reynolds number is high and there are no strong shock waves, potential methods can provide fast and reliable solutions to practical engineering problems. Their main advantage is that they do not require the calculation of the solution in the entire flowfield; calculations on the surface of the body and in its wake are sufficient, and the computational cost of such solutions is very low. Even separated flows can be approximated in this manner, by shedding vortices from the separated flow region of the body into the wake.

Potential flow approaches for steady aerodynamics are presented in detail in many textbooks, notably Katz and Plotkin (2001). Gülçat (2016) discusses many potential flow methods for unsteady aerodynamics for various flow conditions, from incompressible to hypersonic. Potential flow techniques for unsteady transonic flows are also presented in Landahl (1961) and Nixon (1989). The present book focuses on application; each method is presented in detail and applied to practical, usually experimental test cases. Furthermore, the book is accompanied by computer codes in the Matlab programming environment that can be used to solve these test cases. The text and computer codes should be studied in parallel. It is hoped that this application-based approach will help the reader to develop a deeper understanding of the various methodologies.

The focus on potential and vortex methods means that the reader will not find any information in this book on what is commonly referred to as Computational Fluid Dynamics (CFD). The latter requires the numerical solution of the flow in a very wide region around the body, usually by means of finite volume, finite element or finite difference discretization. Even though most of the methods discussed in this book are numerical, they only require the discretization of the surface of the body; the wake is treated in a Lagrangian manner so that its vorticity propagates at the local flow velocity. Readers interested in unsteady CFD can consult alternative texts, such as Tucker (2014).

1.2 Outline of This Book

Chapter 2 constitutes an introduction to the mathematics of unsteady flow. The full flow equations are presented, and their simplifications using specific flow assumptions are derived. Both compressible and incompressible flow equations are presented, and their boundary conditions are discussed. Solutions to these equations are developed, and their implementation by means of Green's theorem is described in detail. The chapter finishes with a discussion of vorticity and the viscous flow equations.

Chapter 3 introduces the classical unsteady aerodynamic theories for 2D incompressible inviscid flow. The modelling of a flat plate airfoil oscillating in a flow is presented, and analytical equations for the resulting aerodynamic loads are derived. The example of impulsive airfoil motion is used in order to introduce the Wagner function while oscillating motion is used for the definition of Theodorsen's function. It is shown how general small amplitude motion can be represented using these theories and the generation of thrust or drag due to unsteady phenomena is explained. Finally, finite state theory is derived in detail.

Chapter 4 introduces numerical methods that can be used to model 2D inviscid unsteady flow with higher fidelity, for example by modelling more accurately the wake behind an oscillating airfoil or by representing the geometry of airfoils with non-negligible thickness. Three such methods are presented, all with their advantages and disadvantages. They are used in order to demonstrate the physics of the wake behind oscillating airfoils and the mechanisms of thrust generation. Comparison of the predicted aerodynamic loads to experimental results demonstrates how the higher fidelity of numerical methods can represent more of the physics of the real phenomenon. Furthermore, it is shown that all numerical panel methods can be linearized and transformed to the frequency domain in order to obtain faster aerodynamic load predictions for harmonically oscillating wings.

Chapter 5 presents unsteady aerodynamic theories for 3D finite wings. It starts with a description of finite wing geometry. Then, analytical solutions are developed, starting with an impulsively started elliptical wing. Unsteady lifting line theories are discussed before detailing two numerical panel methods, the Vortex Lattice Method and the Source and

Doublet Panel Method. These approaches are applied to several practical problems, such as the calculation of aerodynamic stability derivatives and prediction of the unsteady pressure distribution on a flexible wing.

Chapter 6 treats 3D compressible unsteady flows. Subsonic flow is treated first, with the presentation of the Doublet Lattice Method and the subsonic Source and Doublet Panel Method. Supersonic flow is modelled by means of the Mach box and Mach panel techniques. Then, unsteady transonic flows are discussed, and their modelling by means of field panel techniques is outlined. Finally, steady and unsteady corrections that allow subsonic approaches to model transonic flows are presented.

Chapter 7, which is the last chapter of this book, addresses viscous flows. It starts with a brief presentation of the boundary layer and its separation and then proceeds to discuss leading edge separation and dynamic stall. Finally, the Discrete Vortex Method is used to model highly separated flows around bluff bodies.

References

Clay Mathematics Institute (2000). Millennium problems. https://www.claymath.org/millennium-problems (accessed 14 March 2023).

Gülçat, U. (2016). *Fundamentals of Modern Unsteady Aerodynamics*, 2e. Springer.

Katz, J. and Plotkin, A. (2001). *Low Speed Aerodynamics*. Cambridge University Press.

Landahl, M.T. (1961). *Unsteady Transonic Flow*. Dover Publications, Inc.

Nixon, D. (ed.) (1989). *Unsteady Transonic Aerodynamics, Progress in Astronautics and Aeronautics*, vol. 120. AIAA.

Tucker, P.G. (2014). *Unsteady Computational Fluid Dynamics in Aeronautics, Fluid Mechanics and Its Applications*, vol. 104. Springer.

2

Unsteady Flow Fundamentals

2.1 Introduction

This chapter introduces the concepts and equations that will be used throughout the rest of the book. It is not intended as an introduction to all fluid dynamics and many results will be taken for granted. We will not introduce the continuum assumption or constitutive fluid models and we will not derive the flow equations; there are several good textbooks that do. The main focus lies in deriving the compressible and incompressible potential flow equations, discussing their boundary conditions and developing fundamental solutions for these equations.

2.2 From Navier–Stokes to Unsteady Incompressible Potential Flow

For a Newtonian fluid, the flow equations (see for example Anderson Jr. (1985) or Kuethe and Chow (1986)) are given by the continuity equation

$$\frac{\partial \rho}{\partial t} + \frac{\partial(\rho u)}{\partial x} + \frac{\partial(\rho v)}{\partial y} + \frac{\partial(\rho w)}{\partial z} = 0 \tag{2.1}$$

and the momentum equations, also known as the Navier–Stokes equations,

$$\frac{\partial(\rho u)}{\partial t} + \frac{\partial(\rho u^2)}{\partial x} + \frac{\partial(\rho uv)}{\partial y} + \frac{\partial(\rho uw)}{\partial z} = -\frac{\partial p}{\partial x} + \mu\left(\frac{\partial^2 u}{\partial x^2} + \frac{\partial^2 u}{\partial y^2} + \frac{\partial^2 u}{\partial z^2}\right) \tag{2.2}$$

$$\frac{\partial(\rho v)}{\partial t} + \frac{\partial(\rho uv)}{\partial x} + \frac{\partial(\rho v^2)}{\partial y} + \frac{\partial(\rho vw)}{\partial z} = -\frac{\partial p}{\partial y} + \mu\left(\frac{\partial^2 v}{\partial x^2} + \frac{\partial^2 v}{\partial y^2} + \frac{\partial^2 v}{\partial z^2}\right) \tag{2.3}$$

$$\frac{\partial(\rho w)}{\partial t} + \frac{\partial(\rho uw)}{\partial x} + \frac{\partial(\rho vw)}{\partial y} + \frac{\partial(\rho w^2)}{\partial z} = -\frac{\partial p}{\partial z} + \mu\left(\frac{\partial^2 w}{\partial x^2} + \frac{\partial^2 w}{\partial y^2} + \frac{\partial^2 w}{\partial z^2}\right) \tag{2.4}$$

where ρ is the fluid density, u, v, w the flow velocities in the x, y and z directions, p the pressure and μ the dynamic viscosity. The Navier–Stokes equations reduce to the Euler equations for inviscid flow by setting $\mu = 0$, such that

$$\frac{\partial(\rho u)}{\partial t} + \frac{\partial(\rho u^2)}{\partial x} + \frac{\partial(\rho uv)}{\partial y} + \frac{\partial(\rho uw)}{\partial z} = -\frac{\partial p}{\partial x} \tag{2.5}$$

Unsteady Aerodynamics: Potential and Vortex Methods, First Edition. Grigorios Dimitriadis.

Companion website: www.wiley.com/go/dimitriadis/unsteady_aerodynamics

$$\frac{\partial(\rho v)}{\partial t} + \frac{\partial(\rho uv)}{\partial x} + \frac{\partial(\rho v^2)}{\partial y} + \frac{\partial(\rho vw)}{\partial z} = -\frac{\partial p}{\partial y} \tag{2.6}$$

$$\frac{\partial(\rho w)}{\partial t} + \frac{\partial(\rho uw)}{\partial x} + \frac{\partial(\rho vw)}{\partial y} + \frac{\partial(\rho w^2)}{\partial z} = -\frac{\partial p}{\partial z} \tag{2.7}$$

We can write the Euler equations in a different form that does not contain derivatives of the density. We first multiply the continuity equation by u to obtain

$$u\frac{\partial \rho}{\partial t} + u\frac{\partial(\rho u)}{\partial x} + u\frac{\partial(\rho v)}{\partial y} + u\frac{\partial(\rho w)}{\partial z} = 0 \tag{2.8}$$

and then we write Eq. (2.5) as

$$u\frac{\partial \rho}{\partial t} + \rho\frac{\partial u}{\partial t} + u\frac{\partial(\rho u)}{\partial x} + \rho u\frac{\partial u}{\partial x} + u\frac{\partial(\rho v)}{\partial y} + \rho v\frac{\partial u}{\partial y} + u\frac{\partial(\rho w)}{\partial z} + \rho w\frac{\partial u}{\partial z} = -\frac{\partial p}{\partial x}$$

Subtracting Eq. (2.8) from this latest expression yields

$$\rho\frac{\partial u}{\partial t} + \rho u\frac{\partial u}{\partial x} + \rho v\frac{\partial u}{\partial y} + \rho w\frac{\partial u}{\partial z} = -\frac{\partial p}{\partial x}$$

Carrying out similar operations to Eqs. (2.6) and (2.7), we obtain

$$\frac{\partial u}{\partial t} + u\frac{\partial u}{\partial x} + v\frac{\partial u}{\partial y} + w\frac{\partial u}{\partial z} = -\frac{1}{\rho}\frac{\partial p}{\partial x} \tag{2.9}$$

$$\frac{\partial v}{\partial t} + u\frac{\partial v}{\partial x} + v\frac{\partial v}{\partial y} + w\frac{\partial v}{\partial z} = -\frac{1}{\rho}\frac{\partial p}{\partial y} \tag{2.10}$$

$$\frac{\partial w}{\partial t} + u\frac{\partial w}{\partial x} + v\frac{\partial w}{\partial y} + w\frac{\partial w}{\partial z} = -\frac{1}{\rho}\frac{\partial p}{\partial z} \tag{2.11}$$

Despite the lack of density derivatives, Eqs. (2.9)–(2.11) can still describe compressible flow if ρ is not constant; the only simplification we have imposed is to ignore viscosity.

2.2.1 Irrotational Flow

We can apply a further simplification by assuming that the flow is irrotational, that is

$$\nabla \times \mathbf{u} = \mathbf{0} \tag{2.12}$$

where $\mathbf{u} = (u, v, w)$ and

$$\nabla = \left(\frac{\partial}{\partial x}, \frac{\partial}{\partial y}, \frac{\partial}{\partial z}\right)$$

from which we obtain the three irrotationality relationships:

$$\frac{\partial w}{\partial y} - \frac{\partial v}{\partial z} = 0, \quad \frac{\partial u}{\partial z} - \frac{\partial w}{\partial x} = 0, \quad \frac{\partial v}{\partial x} - \frac{\partial u}{\partial y} = 0 \tag{2.13}$$

Equation (2.9) can be rewritten as

$$\frac{\partial u}{\partial t} + \frac{1}{2}\frac{\partial}{\partial x}\left(u^2 + v^2 + w^2\right) - v\frac{\partial v}{\partial x} - w\frac{\partial w}{\partial x} + v\frac{\partial u}{\partial y} + w\frac{\partial u}{\partial z} = -\frac{1}{\rho}\frac{\partial p}{\partial x}$$

and re-arranged in the form

$$\frac{\partial u}{\partial t} + \frac{1}{2}\frac{\partial}{\partial x}\left(u^2 + v^2 + w^2\right) - v\left(\frac{\partial v}{\partial x} - \frac{\partial u}{\partial y}\right) + w\left(\frac{\partial u}{\partial z} - \frac{\partial w}{\partial x}\right) = -\frac{1}{\rho}\frac{\partial p}{\partial x}$$

Substituting from the irrotationality relationships of Eq. (2.13), the x momentum equation simplifies to

$$\rho\frac{\partial u}{\partial t}+\frac{1}{2}\rho\frac{\partial}{\partial x}\left(u^2+v^2+w^2\right)=-\frac{\partial p}{\partial x}$$

Carrying out similar operations to Eqs. (2.10) and (2.11), we obtain the irrotational Euler equations

$$\frac{\partial u}{\partial t}+\frac{1}{2}\frac{\partial}{\partial x}\left(u^2+v^2+w^2\right)=-\frac{1}{\rho}\frac{\partial p}{\partial x} \tag{2.14}$$

$$\frac{\partial v}{\partial t}+\frac{1}{2}\frac{\partial}{\partial y}\left(u^2+v^2+w^2\right)=-\frac{1}{\rho}\frac{\partial p}{\partial y} \tag{2.15}$$

$$\frac{\partial w}{\partial t}+\frac{1}{2}\frac{\partial}{\partial z}\left(u^2+v^2+w^2\right)=-\frac{1}{\rho}\frac{\partial p}{\partial z} \tag{2.16}$$

Furthermore, we define the velocity potential function $\Phi(x, y, z, t)$ such that

$$u=\frac{\partial \Phi}{\partial x},\ v=\frac{\partial \Phi}{\partial y},\ w=\frac{\partial \Phi}{\partial z} \tag{2.17}$$

Substituting these definitions into Eqs. (2.14)–(2.16) leads to

$$\frac{\partial}{\partial t}\left(\frac{\partial \Phi}{\partial x}\right)+\frac{1}{2}\frac{\partial}{\partial x}\left(u^2+v^2+w^2\right)=-\frac{1}{\rho}\frac{\partial p}{\partial x} \tag{2.18}$$

$$\frac{\partial}{\partial t}\left(\frac{\partial \Phi}{\partial y}\right)+\frac{1}{2}\frac{\partial}{\partial y}\left(u^2+v^2+w^2\right)=-\frac{1}{\rho}\frac{\partial p}{\partial y} \tag{2.19}$$

$$\frac{\partial}{\partial t}\left(\frac{\partial \Phi}{\partial z}\right)+\frac{1}{2}\frac{\partial}{\partial z}\left(u^2+v^2+w^2\right)=-\frac{1}{\rho}\frac{\partial p}{\partial z} \tag{2.20}$$

2.2.2 Laplace's and Bernoulli's Equations

We now apply the final simplification by assuming that the flow is incompressible so that the density is constant everywhere in the flowfield and at all times. The continuity Eq. (2.1) becomes

$$\frac{\partial u}{\partial x}+\frac{\partial v}{\partial y}+\frac{w}{\partial z}=0 \tag{2.21}$$

Substituting from the definition of the potential in expressions (2.17) results in

$$\frac{\partial^2 \Phi}{\partial x^2}+\frac{\partial^2 \Phi}{\partial y^2}+\frac{\partial^2 \Phi}{\partial z^2}=0 \tag{2.22}$$

which is known as Laplace's equation and can also be expressed as

$$\nabla^2\Phi=0 \tag{2.23}$$

where $\nabla^2=\nabla\cdot\nabla$.

The inviscid, irrotational and incompressible assumptions have simplified the continuity equation from the form (2.1) to the form (2.22), which is a linear partial differential equation with a single unknown, the potential $\Phi(x, y, z)$. The irrotational Euler equations can also be simplified using the incompressible assumption; we can multiply Eqs. (2.18)–(2.20) by their respective spatial differentials to obtain

$$\frac{\partial}{\partial t}\left(\frac{\partial \Phi}{\partial x}\right) \mathrm{d}x + \frac{1}{2}\frac{\partial}{\partial x}\left(u^2 + v^2 + w^2\right) \mathrm{d}x = -\frac{1}{\rho}\frac{\partial p}{\partial x}\mathrm{d}x \tag{2.24}$$

$$\frac{\partial}{\partial t}\left(\frac{\partial \Phi}{\partial y}\right) \mathrm{d}y + \frac{1}{2}\frac{\partial}{\partial y}\left(u^2 + v^2 + w^2\right) \mathrm{d}y = -\frac{1}{\rho}\frac{\partial p}{\partial y}\mathrm{d}y \tag{2.25}$$

$$\frac{\partial}{\partial t}\left(\frac{\partial \Phi}{\partial z}\right) \mathrm{d}z + \frac{1}{2}\frac{\partial}{\partial z}\left(u^2 + v^2 + w^2\right) \mathrm{d}z = -\frac{1}{\rho}\frac{\partial p}{\partial z}\mathrm{d}z \tag{2.26}$$

Since ρ is constant, exchanging the order of the time and space derivatives in the first terms of each of the equations and integrating in space results in

$$\frac{\partial \Phi}{\partial t} + \frac{1}{2}\left(u^2 + v^2 + w^2\right) = -\frac{p}{\rho} + \text{constant}$$

which is known as the unsteady Bernoulli equation. Note that all three momentum equations lead to the same Bernoulli expression. The constant of integration can be written as p_{ref}/ρ without loss of generality, where p_{ref} is a reference pressure. Then the unsteady Bernoulli equation becomes

$$\rho\left(\frac{1}{2}\left(u^2 + v^2 + w^2\right) + \frac{\partial \Phi}{\partial t}\right) + p = p_{\text{ref}} \tag{2.27}$$

Using the definition of the potential, it can also be written as

$$\rho\left(\frac{1}{2}(\nabla\Phi)^2 + \frac{\partial \Phi}{\partial t}\right) + p = p_{\text{ref}} \tag{2.28}$$

where $(\nabla\Phi)^2 = \nabla\Phi \cdot \nabla\Phi$. Bernoulli's equation is valid everywhere in an incompressible, inviscid and irrotational flowfield but the reference pressure needs to be specified. As it stands, Eq. (2.27) relates conditions at any point x, y, z to a point in space where the pressure is equal to p_{ref}, the total flow speed is zero and the time derivative of the potential is also equal to zero. Bernoulli's equation can be written more intuitively by relating conditions at point x, y, z to a faraway point where the pressure is p_∞, the total flow speed is Q_∞ and the potential is constant or zero, that is

$$\rho\left(\frac{1}{2}\left(u^2 + v^2 + w^2\right) + \frac{\partial \Phi}{\partial t}\right) + p = \frac{1}{2}\rho Q_\infty^2 + p_\infty \tag{2.29}$$

The quantity $1/2\rho\left(u^2 + v^2 + w^2\right)$ is known as the dynamic pressure and $1/2\rho Q_\infty^2$ is the far-field dynamic pressure. The pressure coefficient is defined as

$$c_p = \frac{p - p_\infty}{\frac{1}{2}\rho Q_\infty^2} \tag{2.30}$$

and, using Eq. (2.29)

$$c_p(x, y, z, t) = 1 - \frac{\left(u^2 + v^2 + w^2\right)}{Q_\infty^2} - \frac{2}{Q_\infty^2}\frac{\partial \Phi}{\partial t} \tag{2.31}$$

Equations (2.23) and (2.31) will be used to solve all incompressible flow problems presented in this book, for which viscous phenomena are not important. Laplace's equation is a second-order linear partial differential equation with a single unknown, the potential $\Phi(x, y, z, t)$. Once it is solved, the potential can be substituted into Bernoulli's equation in order to calculate the pressure anywhere in the flow. Since Laplace's equation is of second order, it requires two boundary conditions.

2.2.3 Motion in an Incompressible, Inviscid, Irrotational Fluid

Consider a body in unsteady motion in still air. Points on the body's surface are denoted by vector $\mathbf{x}_s(t) = (x_s(t), y_s(t), z_s(t))$ with respect to a static origin; they are defined as the solutions of the equation $S(\mathbf{x}_s, t) = 0$. The body is rotating, translating and deforming so that the velocity of the points on the surface is denoted by $\mathbf{V}_S(\mathbf{x}_s, t)$. Consequently, if the equation of the surface is $S(\mathbf{x}_s(t), t) = 0$ at time t, at time $t + \Delta t$, it becomes

$$S(\mathbf{x}_s + \Delta\mathbf{x}, t + \Delta t) = 0$$

Expanding $S(\mathbf{x}_s + \Delta\mathbf{x}, t + \Delta t)$ as a Taylor series around $S(\mathbf{x}_s, t) = 0$ gives

$$S(\mathbf{x}_s + \Delta\mathbf{x}, t + \Delta t) = S(\mathbf{x}_s, t) + \left.\frac{\partial S}{\partial \mathbf{x}}\right|_{\mathbf{x}_s, t} \cdot \Delta\mathbf{x} + \left.\frac{\partial S}{\partial t}\right|_{\mathbf{x}_s, t} \Delta t + \cdots = 0$$

where $\partial S/\partial \mathbf{x}$ is the vector $(\partial S/\partial x, \partial S/\partial y, \partial S/\partial z) = \nabla S$. Therefore, recalling that $S(\mathbf{x}_s, t) = 0$,

$$\nabla S(\mathbf{x}_s, t) \cdot \Delta\mathbf{x} + \left.\frac{\partial S}{\partial t}\right|_{\mathbf{x}_s, t} \Delta t + \cdots = 0 \tag{2.32}$$

where the notation $\nabla S(\mathbf{x}_s, t)$ signifies 'evaluate ∇S at $\mathbf{x}_s$ and t'. Now, as the surface of the body is moving with velocity $\mathbf{V}_S(\mathbf{x}_s, t)$, we can approximate $\Delta\mathbf{x}$ as

$$\Delta\mathbf{x} = \mathbf{V}_S(\mathbf{x}_s, t)\Delta t$$

Substituting back into Eq. (2.32) yields

$$\nabla S(\mathbf{x}_s, t) \cdot \mathbf{V}_S(\mathbf{x}_s, t)\Delta t + \left.\frac{\partial S}{\partial t}\right|_{\mathbf{x}_s, t} \Delta t + O(\Delta t^2) = 0$$

or dividing throughout by Δt

$$\nabla S(\mathbf{x}_s, t) \cdot \mathbf{V}_S(\mathbf{x}_s, t) + \left.\frac{\partial S}{\partial t}\right|_{\mathbf{x}_s, t} + O(\Delta t) = 0$$

Finally, taking the limit of this latest expression as $\Delta t \to 0$,

$$\nabla S(\mathbf{x}_s, t) \cdot \mathbf{V}_S(\mathbf{x}_s, t) + \left.\frac{\partial S}{\partial t}\right|_{\mathbf{x}_s, t} = 0 \tag{2.33}$$

Equation (2.33) describes how the surface of the body deforms with respect to its own velocity $\mathbf{V}_S(\mathbf{x}_s, t)$. However, we are also interested in how the fluid deforms due to the motion of the body's surface.

We denote the coordinates of any point in the fluid by $\mathbf{x} = (x, y, z)$. Assuming that the air is still, far from the body its velocity will be zero. This is known as the far-field boundary condition, which states that the flow disturbance caused by the body decays to zero as $r \rightarrow \infty$, where

$$r = ||\mathbf{x} - \mathbf{x}_s|| = \sqrt{(x - x_s)^2 + (y - y_s)^2 + (z - z_s)^2}$$

As the flow velocity is given by $\nabla\Phi = (u, v, w)$, the far-field condition can be formulated mathematically as

$$\lim_{r\rightarrow\infty} \nabla\Phi(\mathbf{x}, t) = 0 \tag{2.34}$$

Far from the body, the potential must be constant so that its derivatives in all directions are equal to zero. This constant value of the potential may be chosen to be equal to zero.

Close to the body, the flow will be disturbed by the body's motion and therefore the flow velocity and potential will be non-zero. The objective of unsteady potential flow modelling is to calculate the flow velocities $u(\mathbf{x}_s, t)$, $v(\mathbf{x}_s, t)$, $w(\mathbf{x}_s, t)$ on the surface of the body. From these, the pressure around the body, $p(\mathbf{x}_s, t)$, can be evaluated from Eq. (2.29); the total aerodynamic force acting on the body is then given by

$$\mathbf{F} = \int_S p(\mathbf{x}_s, t)\mathbf{n}(\mathbf{x}_s, t)\mathrm{d}S \tag{2.35}$$

where $\mathbf{n}(\mathbf{x}_s, t)$ is a unit vector normal to the surface at point $\mathbf{x}_s$ and time t, while $\int_S$ denotes an integral over the entire surface of the body and $\mathrm{d}S$ is an infinitesimal element of this surface. The fundamental flow equation to be solved is Laplace's equation (2.22), which requires two boundary conditions. One of them is the far-field condition but we still need to define the second. Assuming that the surface of the body is impermeable, the layer of fluid in contact with the surface will also obey $S(\mathbf{x}_s, t) = 0$ and, hence, Eq. (2.32). The velocity of the fluid on the surface is given by $\nabla\Phi(\mathbf{x}_s, t)$ so that for the fluid,

$$\Delta\mathbf{x} = \nabla\Phi(\mathbf{x}_s, t)\Delta t$$

Substituting in Eq. (2.32), dividing by Δt and taking the limit as $\Delta t \rightarrow 0$ leads to

$$\nabla S(\mathbf{x}_s, t) \cdot \nabla\Phi(\mathbf{x}_s, t) + \left.\frac{\partial S}{\partial t}\right|_{\mathbf{x}_s, t} = 0$$

Furthermore, dividing throughout by $||\nabla S(\mathbf{x}_s, t)||$ gives

$$\frac{\nabla S(\mathbf{x}_s, t)}{||\nabla S(\mathbf{x}_s, t)||} \cdot \nabla\Phi(\mathbf{x}_s, t) + \frac{1}{||\nabla S(\mathbf{x}_s, t)||}\left.\frac{\partial S}{\partial t}\right|_{\mathbf{x}_s, t} = 0$$

The quantity $\nabla S(\mathbf{x}_s, t)/||\nabla S(\mathbf{x}_s, t)||$ is in fact the unit vector normal to the surface $\mathbf{n}(\mathbf{x}_s, t)$, so that

$$\mathbf{n}(\mathbf{x}_s, t) \cdot \nabla\Phi(\mathbf{x}_s, t) + \frac{1}{||\nabla S(\mathbf{x}_s, t)||}\left.\frac{\partial S}{\partial t}\right|_{\mathbf{x}_s, t} = 0 \tag{2.36}$$

Finally, we solve Eq. (2.33) for $\partial S/\partial t|_{\mathbf{x}_s, t}$ and substitute the result into Eq. (2.36) to obtain

$$\left(\nabla\Phi(\mathbf{x}_s, t) - \mathbf{V}_S(\mathbf{x}_s, t)\right) \cdot \mathbf{n}(\mathbf{x}_s, t) = 0 \tag{2.37}$$

Equation (2.37) is known as the impermeability boundary condition, or zero normal flow condition. It states that the relative velocity between the fluid and the surface in a direction normal to the surface must be equal to zero so that no flow can cross the solid boundary.

The far-field equation (2.34) and the impermeability equation (2.37) are the two boundary conditions that are required to obtain solutions of Laplace's equation.

Example 2.1 *Determine the impermeability boundary condition for a sphere that translates with speed* $\mathbf{V}_0(t)$ *and whose radius* $R(t)$ *is a function of time.*

We will select the centre of the sphere as the body datum. Since we are measuring all distances from the centre of the sphere, the equation of the surface of the sphere is

$$S(\mathbf{x}_s, t) = x_s(t)^2 + y_s(t)^2 + z_s(t)^2 - R(t)^2 = 0$$

The solutions of this equation are the points on the surface $\mathbf{x}_s(t)$ *such that* $\mathbf{x}_s(t) \cdot \mathbf{x}_s(t) = R(t)^2$. *The quantity* ∇S *evaluated at* $\mathbf{x}_s, t$ *becomes*

$$\nabla S(\mathbf{x}_s, t) = (2x_s(t),\ 2y_s(t),\ 2z_s(t)) = 2\mathbf{x}_s(t)$$

and its magnitude is

$$||\nabla S(\mathbf{x}_s, t)|| = \sqrt{2\mathbf{x}_s(t) \cdot 2\mathbf{x}_s(t)} = \sqrt{4x_s(t)^2 + 4y_s(t)^2 + 4z_s(t)^2} = 2R(t)$$

so that the normal vector on the surface is given by

$$\mathbf{n}(\mathbf{x}_s, t) = \frac{\nabla S(\mathbf{x}_s, t)}{|\nabla S(\mathbf{x}_s, t)|} = \frac{1}{R(t)}\mathbf{x}_s(t) \tag{2.38}$$

The deformation velocity of the surface of the sphere has magnitude $\dot{R}$ *and its direction is radial, that is parallel to* $\mathbf{n}(\mathbf{x}_s, t)$. *Recalling that the sphere also translates with speed* $\mathbf{V}_0(t)$, *the total velocity of the surface in the normal direction is given by*

$$\mathbf{V}_S(\mathbf{x}_s, t) \cdot \mathbf{n}(\mathbf{x}_s, t) = \dot{R}(t) + \mathbf{V}_0(t) \cdot \mathbf{n}(\mathbf{x}_s, t)$$

Substituting this latest result in the boundary condition of Eq. (2.37) gives

$$\nabla\Phi(\mathbf{x}_s, t) \cdot \mathbf{n}(\mathbf{x}_s, t) = \dot{R}(t) + \mathbf{V}_0(t) \cdot \mathbf{n}(\mathbf{x}_s, t)$$

or after substituting for $\mathbf{n}(\mathbf{x}_s, t)$ *from expression (2.38)*

$$\nabla\Phi(\mathbf{x}_s, t) \cdot \mathbf{x}_s(t) = R(t)\dot{R}(t) + \mathbf{V}_0(t) \cdot \mathbf{x}_s(t)$$

As expected, this equation states that the flow velocity in a direction normal to the surface of the sphere must be equal to the normal component of the motion of the surface, if the flow is not to penetrate inside the sphere.

As the flow is inviscid, the fluid velocity tangent to the surface is non-zero and contributes directly to the surface pressure. If $\boldsymbol{\tau}(\mathbf{x}_s, t)$ is a unit vector tangent to the surface and parallel to the local flow direction, then the tangential flow velocity is given by $\nabla\Phi(\mathbf{x}_s, t) \cdot \boldsymbol{\tau}(\mathbf{x}_s, t)$ and the pressure on the surface is obtained from Eq. (2.28) as

$$\rho\left(\frac{1}{2}\left(\left(\nabla\Phi(\mathbf{x}_s, t) - \mathbf{V}_S(\mathbf{x}_s, t)\right) \cdot \boldsymbol{\tau}(\mathbf{x}_s, t)\right)^2 + \frac{\partial\Phi(\mathbf{x}_s, t)}{\partial t}\right) + p(\mathbf{x}_s, t) = p_{\mathrm{ref}} \tag{2.39}$$

where $\left(\nabla\Phi(\mathbf{x}_s, t) - \mathbf{V}_S(\mathbf{x}_s, t)\right) \cdot \boldsymbol{\tau}(\mathbf{x}_s, t)$ is the relative velocity between the fluid and the body's surface in the tangential direction.

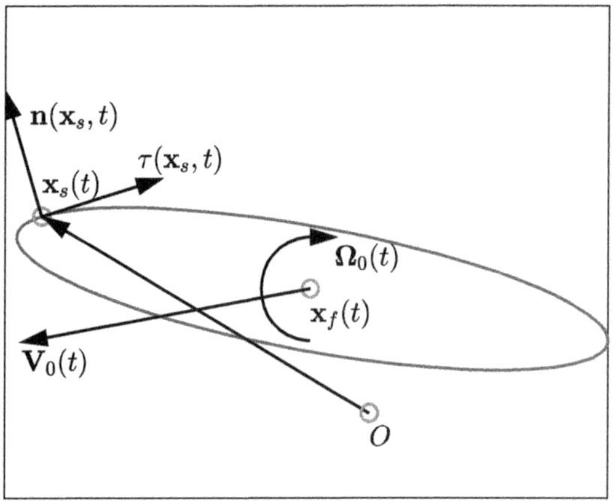

Figure 2.1 A body translating and rotating in a still fluid.

Now, let us consider the special case where the body does not deform, it only translates with velocity $\mathbf{V}_0(t)$ and rotates rigidly around a point $\mathbf{x}_f(t)$ with rotational velocity $\mathbf{\Omega}_0(t)$. The surface velocity is given by

$$\mathbf{V}_S(t) = \mathbf{V}_0(t) + \mathbf{\Omega}_0(t) \times (\mathbf{x}_s(t) - \mathbf{x}_f(t))$$

The situation is depicted in Figure 2.1; $\mathbf{x}_f(t)$ is the instantaneous position of the rotation centre with respect to the origin O and $\mathbf{x}_s(t)$ is the instantaneous position of a point on the surface, again with respect to O. Both $\mathbf{x}_f(t)$ and $\mathbf{x}_s(t)$ translate with velocity $\mathbf{V}_0(t)$, while $\mathbf{x}_s(t)$ also rotates with velocity $\mathbf{\Omega}_0(t)$ around $\mathbf{x}_f(t)$. The figure also draws the unit vectors normal and tangent to the surface at point $\mathbf{x}_s(t)$, $\mathbf{n}(\mathbf{x}_s, t)$ and $\boldsymbol{\tau}(\mathbf{x}_s, t)$, respectively. For this rigid case, the far-field boundary condition is still given by Eq. (2.34). The impermeability boundary condition of Eq. (2.37) becomes

$$\left(\nabla\Phi(\mathbf{x}_s, t) - \mathbf{V}_0(t) - \mathbf{\Omega}_0(t) \times (\mathbf{x}_s(t) - \mathbf{x}_f(t))\right) \cdot \mathbf{n}(\mathbf{x}_s, t) = 0 \tag{2.40}$$

where $\mathbf{V}_0(t) + \mathbf{\Omega}_0(t) \times (\mathbf{x}_s(t) - \mathbf{x}_f(t))$ is the velocity of point $\mathbf{x}_s(t)$ on the surface of the body.

Next, we assume that the translation velocity $\mathbf{V}_0(t)$ is composed of a steady component, which we will call $\mathbf{V}_0$ and an unsteady component $\mathbf{v}_0(t)$. In such cases, it is customary to set the fluid in motion with velocity $\mathbf{Q}_\infty = -\mathbf{V}_0$, known as the free stream velocity, and to only attribute the unsteady component $\mathbf{v}_0(t)$ to the body, as seen in Figure 2.2. The free stream has components $\mathbf{Q}_\infty = (U_\infty, V_\infty, W_\infty)$; usually, they are chosen such that $V_\infty \ll U_\infty$ and $W_\infty \ll U_\infty$. The angle $\tan^{-1}(W_\infty/U_\infty)$ is the free stream angle of attack α_∞, while the angle $\tan^{-1}(V_\infty/U_\infty)$ is the free stream sideslip angle β_∞. If $Q_\infty = ||\mathbf{Q}_\infty||$ is the magnitude of the free stream velocity and $\beta_\infty = 0$, then

$$U_\infty = Q_\infty \cos\alpha_\infty,\ V_\infty = 0,\ W_\infty = Q_\infty \sin\alpha_\infty$$

The relative velocity between the fluid and the body does not change, but as the steady velocity component is attributed to the flow, the flow's potential will increase. The total flow potential, $\Phi(\mathbf{x}, t)$, is the sum of the potential due to the free stream and the perturbation potential due to the shape of the body and its unsteady motion, $\phi(\mathbf{x}, t)$. It can be written as

$$\Phi(\mathbf{x}, t) = \phi_\infty(\mathbf{x}) + \phi(\mathbf{x}, t) \tag{2.41}$$

Figure 2.2 A body translating and rotating in a steady free stream.

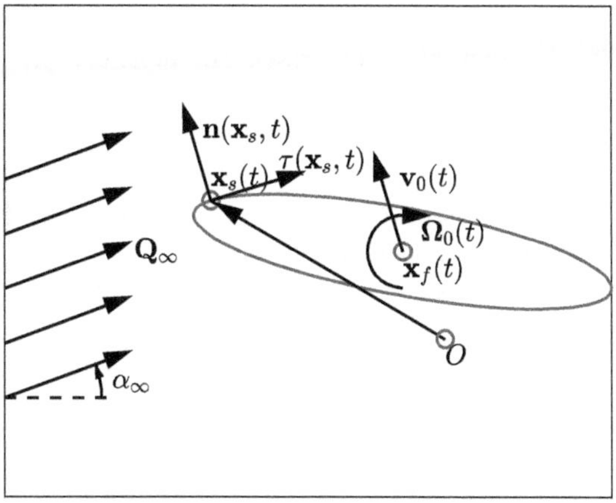

where $\phi_\infty(\mathbf{x})$ is the free stream potential, defined as $\nabla\phi_\infty(\mathbf{x}) = \mathbf{Q}_\infty$. Consequently, the boundary conditions must be adapted to

- Far-field. The total potential far from the body is equal to $\phi_\infty(\mathbf{x})$ while the perturbation velocity tends to zero, that is

$$\lim_{r\to\infty} \Phi(\mathbf{x}, t) = \phi_\infty(\mathbf{x}), \ \lim_{r\to\infty} \nabla\phi(\mathbf{x}, t) = 0 \tag{2.42}$$

  This means that the total potential is not a perturbation quantity.
- Impermeability. The velocity anywhere in the flow is now $\nabla\Phi(\mathbf{x}, t) = \mathbf{Q}_\infty + \nabla\phi(\mathbf{x}, t)$, so that Eq. (2.40) becomes

$$\left(\nabla\phi(\mathbf{x}_s, t) + \mathbf{Q}_\infty - \mathbf{v}_0(t) - \boldsymbol{\Omega}_0(t) \times (\mathbf{x}_s(t) - \mathbf{x}_f(t))\right) \cdot \mathbf{n}(\mathbf{x}_s, t) = 0 \tag{2.43}$$

Finally, the motion can also be described in a quasi-fixed form, by keeping the body static and assigning all the steady and unsteady velocity components to the fluid, as seen in Figure 2.3. The coordinate system is such that the body is horizontal and the origin lies somewhere on the body, e.g. on its geometric centre. The centre of rotation lies at position $\mathbf{x}_f = (x_f, y_f, z_f)$, which is not necessarily coincident with the origin. The points $\mathbf{x}_s$

Figure 2.3 A static body in an unsteady free stream.

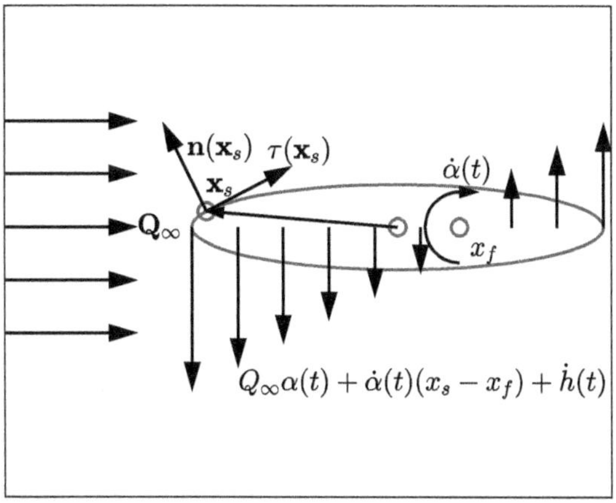

on the surface are no longer functions of time and neither are the normal and tangential unit vectors, $\mathbf{n}(\mathbf{x}_s)$ and $\boldsymbol{\tau}(\mathbf{x}_s)$. The most classical example of the quasi-fixed description is pitching and plunging motion, whereby the body is immersed in a steady free stream $\mathbf{Q}_\infty = (U_\infty, 0, 0)$, is changing its pitch angle with respect to the horizontal, $\alpha(t)$, and is translating in the vertical direction with plunge velocity $\dot{h}(t)$, defined positive downwards. Consequently, $\boldsymbol{\Omega}_0(t) = \dot{\alpha}(t)(0, 1, 0)$ and $\mathbf{v}_0(t) = (0, 0, -\dot{h}(t))$. However, these unsteady motions are not applied to the body, they are attributed instead to the relative velocity between the body and the flow $\mathbf{u}_m(\mathbf{x}_s, t)$ with components

$$\begin{aligned} u_m(\mathbf{x}_s, t) &= Q_\infty \cos\alpha(t) - \dot{\alpha}(t)(z_s - z_f) \\ v_m(\mathbf{x}_s, t) &= 0 \\ w_m(\mathbf{x}_s, t) &= Q_\infty \sin\alpha(t) + \dot{\alpha}(t)(x_s - x_f) + \dot{h}(t) \end{aligned} \tag{2.44}$$

The vertical velocity $w_m(\mathbf{x}_s, t)$ is usually referred to as the upwash. The total potential is still given by $\Phi(x, y, z, t) = \phi_\infty(x, y, z) + \phi(x, y, z, t)$.

In this quasi-fixed case, the boundary conditions become

- Far-field. The total potential far from the body is equal to $\phi_\infty(x, y, z)$, while the perturbation velocity far from the body is equal to zero, as in Eq. (2.42).
- Impermeability. Equation (2.43) becomes

$$\left(\nabla\phi(\mathbf{x}_s, t) + \mathbf{u}_m(\mathbf{x}_s, t)\right) \cdot \mathbf{n}(\mathbf{x}_s) = 0 \tag{2.45}$$

Assuming that $\alpha(t) \ll 1$ and the body is very thin so that the x and y components of $\mathbf{n}(\mathbf{x}_s)$ are negligible, while the z component is close to unity, this boundary condition can be simplified further to

$$\phi_z(x_s, t) + U_\infty\alpha(t) + \dot{\alpha}(t)(x_s - x_f) + \dot{h}(t) = 0 \tag{2.46}$$

The relative flow scheme of Figure 2.2 and the quasi-fixed scheme of Figure 2.3 will be used most often in this book. The two schemes give similar results for small motion amplitudes and frequencies.

2.3 Incompressible Potential Flow Solutions

Solving for incompressible potential flow around a solid body amounts to solving Laplace's equation (2.23)

$$\nabla^2\Phi = 0 \tag{2.47}$$

for the potential $\Phi(x, y, z)$ everywhere in the flow, given boundary conditions (2.34) and (2.40), and then using the unsteady Bernoulli equation (2.28) to calculate the pressure distribution on the surface.

Laplace's equation is a linear partial differential equation of the second order; as it is linear, its solutions can be superimposed. If $\phi_1(x, y, z)$ and $\phi_2(x, y, z)$ are both solutions to Laplace's equation, then $\phi_1(x, y, z) + \phi_2(x, y, z)$ will also be a solution. This means that the desired flow can be constructed using super-positions of elementary solutions to Laplace's equation. Several such solutions exist, the simplest of which is the free stream: assume that there is no body in the flow, only a free stream with velocity $\mathbf{Q}_\infty = (U_\infty, V_\infty, W_\infty)$. The potential of this flow is given by

$$\phi_\infty(x, y, z) = U_\infty x + V_\infty y + W_\infty z \tag{2.48}$$

Substituting this potential into Eq. (2.47), we can see that the latter is indeed satisfied. Furthermore, substituting this potential into the definitions of Eq. (2.17), we can see that the flow speed is indeed $\nabla\phi_\infty = (U_\infty, V_\infty, W_\infty) = \mathbf{Q}_\infty$ everywhere. As stated in Section 2.2.3, the free stream solution is used very regularly in practice, since it can represent the steady part of the relative flow velocity between a body and the surrounding fluid. In such cases, the total potential in the flow is given by Eq. (2.41), where $\phi(\mathbf{x}, t)$ is the perturbation potential due to the shape and unsteady motion of the body.

The perturbation potential is also modelled using solutions of Laplace's equations. These solutions are usually singularities, which mean that a point $\mathbf{x}_0 = (x_0, y_0, z_0)$ in space generates a flow around it, but this flow is singular at exactly $\mathbf{x}_0$. The fundamental singularity is the source, defined as

$$\phi(\mathbf{x}, \mathbf{x}_0) = -\frac{1}{\sqrt{(x - x_0)^2 + (y - y_0)^2 + (z - z_0)^2}} = -\frac{1}{r(\mathbf{x}, \mathbf{x}_0)} \tag{2.49}$$

where $\phi(\mathbf{x}, \mathbf{x}_0)$ is the potential induced at a general point $\mathbf{x} = (x, y, z)$ in space by a source lying at $\mathbf{x}_0$ and $r(\mathbf{x}, \mathbf{x}_0) = \sqrt{(x - x_0)^2 + (y - y_0)^2 + (z - z_0)^2}$ is the cartesian distance between $\mathbf{x}$ and $\mathbf{x}_0$. Substituting expression (2.49) into Eq. (2.17) yields

$$\begin{aligned}
u &= \frac{\partial\phi}{\partial x} = \frac{x - x_0}{\left((x - x_0)^2 + (y - y_0)^2 + (z - z_0)^2\right)^{3/2}} \\
v &= \frac{\partial\phi}{\partial y} = \frac{y - y_0}{\left((x - x_0)^2 + (y - y_0)^2 + (z - z_0)^2\right)^{3/2}} \\
w &= \frac{\partial\phi}{\partial z} = \frac{z - z_0}{\left((x - x_0)^2 + (y - y_0)^2 + (z - z_0)^2\right)^{3/2}}
\end{aligned} \tag{2.50}$$

Differentiating u, v, w once more results in

$$\begin{aligned}
\frac{\partial^2\phi}{\partial x^2} &= \frac{-2(x - x_0)^2 + (y - y_0)^2 + (z - z_0)^2}{\left((x - x_0)^2 + (y - y_0)^2 + (z - z_0)^2\right)^{5/2}} \\
\frac{\partial^2\phi}{\partial y^2} &= \frac{(x - x_0)^2 - 2(y - y_0)^2 + (z - z_0)^2}{\left((x - x_0)^2 + (y - y_0)^2 + (z - z_0)^2\right)^{5/2}} \\
\frac{\partial^2\phi}{\partial z^2} &= \frac{(x - x_0)^2 + (y - y_0)^2 - 2(z - z_0)^2}{\left((x - x_0)^2 + (y - y_0)^2 + (z - z_0)^2\right)^{5/2}}
\end{aligned}$$

It can be clearly seen that adding up these last three expressions yields

$$\frac{\partial^2\phi}{\partial x^2} + \frac{\partial^2\phi}{\partial y^2} + \frac{\partial^2\phi}{\partial z^2} = 0$$

Equation (2.47) is indeed satisfied, but there is a difficulty when $\mathbf{x} = \mathbf{x}_0$ that will be discussed later.

Figure 2.4a plots the flowfield defined by Eq. (2.50), demonstrating that the flow moves radially away from the origin. As the magnitude of the velocity depends on the inverse of

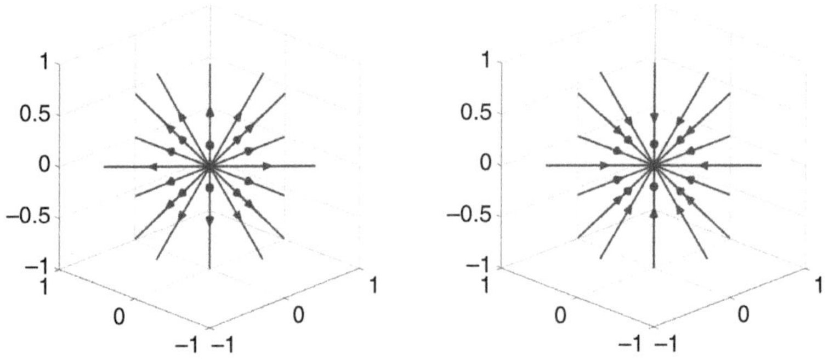

(a) (b)

Figure 2.4 Flow induced by a 3D point (a) source and (b) sink.

the distance from the origin, it decreases quickly as this distance increases. Figure 2.4a justifies the fact that this singularity is known as a source; flow appears at the origin and radiates outwards. On the contrary, the singularity $1/r$ is known as a sink. The flow induced by the sink is plotted in Figure 2.4b, which shows that the fluid moves radially towards the origin.

Figures 2.4a and 2.4b introduce another useful flow concept, that of the streamline. A streamline is a line in 3D space that is always parallel to the local flow velocity. If we use s to denote the arc length of a streamline, then the points $\mathbf{x}(s)$ on the streamline obey

$$\frac{\mathrm{d}\mathbf{x}}{\mathrm{d}s} \times \mathbf{u}(\mathbf{x}(s)) = 0 \tag{2.51}$$

everywhere on the streamline. This also means that, on a streamline,

$$\frac{u}{\mathrm{d}x} = \frac{v}{\mathrm{d}y} = \frac{w}{\mathrm{d}z} \tag{2.52}$$

For the source and sink, the streamlines are straight lines, as plotted in Figure 2.4.

For both the source and the sink, Eq. (2.50) show that the flow velocity is singular when $x = x_0, y = y_0, z = z_0$. In Appendix A.8 it is shown that when Eq. (2.49) is substituted into Laplace's equation (2.47), the latter becomes

$$\nabla^2 \left(-\frac{1}{r(\mathbf{x}, \mathbf{x}_0)} \right) = \delta(\mathbf{x} - \mathbf{x}_0) \tag{2.53}$$

where $\delta(\mathbf{x} - \mathbf{x}_0)$ is the 3D Dirac delta function

$$\delta(\mathbf{x} - \mathbf{x}_0) = \begin{cases} \infty & \text{if } \mathbf{x} - \mathbf{x}_0 = 0 \\ 0 & \text{if } \mathbf{x} - \mathbf{x}_0 \neq 0 \end{cases}$$

This function has the sifting property

$$\int_V F(\mathbf{x})\delta(\mathbf{x} - \mathbf{x}_0)\mathrm{d}V = F(\mathbf{x}_0), \text{ or } \int_V F(\mathbf{x}_0)\delta(\mathbf{x} - \mathbf{x}_0)\mathrm{d}V = F(\mathbf{x}) \tag{2.54}$$

for any other function $F(\mathbf{x})$, where V is an infinitely big volume and $\mathrm{d}V$ an infinitesimal volume. Equation (2.53) states that using the source as a solution, Laplace's equation is

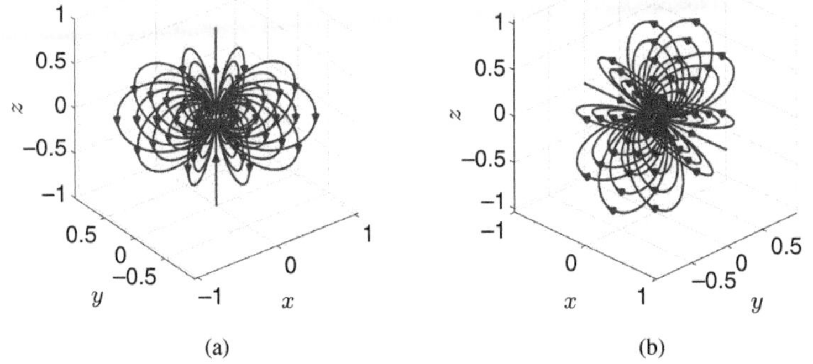

Figure 2.5 Flow induced by a 3D point doublet. (a) $\mathbf{n} = (0\,0\,1)$ and (b) $\mathbf{n} = (1\,0\,0)$.

satisfied everywhere in the flow except at the location of the source. This means that we need to be careful about where we place the source and where we evaluate the potential.

If the source is a solution to Laplace's equation, it stands to reason that any of its spatial derivatives will also be such solutions. Consider for example the derivative of the source in the x direction, given in Eq. (2.50),

$$\frac{\partial}{\partial x}\left(-\frac{1}{r}\right) = \frac{x - x_0}{\left((x-x_0)^2 + (y-y_0)^2 + (z-z_0)^2\right)^{3/2}}$$

If we take the Laplacian of this function, we obtain

$$\nabla^2\left(\frac{\partial}{\partial x}\left(-\frac{1}{r}\right)\right) = \frac{\partial}{\partial x}\left(\nabla^2\left(-\frac{1}{r}\right)\right) = 0$$

which means that this derivative is also a solution of Laplace's equation, except at $\mathbf{x}_0$. This is the case for the derivatives with respect to y and z and any linear combination of these derivatives. We therefore define a function known as the doublet, expressed as

$$\phi(\mathbf{x}, \mathbf{x}_0) = \mathbf{n} \cdot \nabla \frac{1}{r(\mathbf{x}, \mathbf{x}_0)} \tag{2.55}$$

where $\mathbf{n}$ is any unit direction vector. Note that the doublet is defined as the derivative of the sink in direction $\mathbf{n}$, not of the source. As an example, the flowfield induced by a doublet for $\mathbf{n} = (0, 0, 1)$ is plotted in Figure 2.5a; the fluid moves upwards through the centre of the doublet and downwards away from it. The similarity to the magnetic field of a dipole justifies the choice of the name doublet for this function. Figure 2.5b plots the flowfield induced by a doublet aligned with the x direction, obtained by choosing $\mathbf{n} = (1, 0, 0)$. It should be noted that the doublet is also a singularity when $\mathbf{x} = \mathbf{x}_0$. The flow induced by the doublet looks spherical; this means that the doublet may be a good candidate for developing a potential flow solution around a sphere immersed in a free stream. This will be our first flow solution, which we will derive in the next example.

Example 2.2 *Use the doublet to derive an exact solution for the potential flow around an impermeable motionless and undeforming sphere immersed in a steady free stream.*

We consider a sphere of radius R centred around the origin and immersed in a free stream with potential $\phi_\infty = U_\infty x$. The far-field boundary condition is automatically satisfied by the doublet; the impermeability boundary condition states that the total flow velocity on the surface of the sphere must be equal to zero. As the sphere does not move or deform and the free stream is steady, we remove all time-dependency from Eq. (2.43) so that impermeably states that

$$\left(\nabla\phi(\mathbf{x}_s) + \mathbf{Q}_\infty\right)\cdot\mathbf{n}(\mathbf{x}_s) = 0 \tag{2.56}$$

where $\mathbf{x}_s$ is any point on the surface of the sphere, $\phi(\mathbf{x}_s)$ is the perturbation potential at that point and $\mathbf{n}(\mathbf{x}_s)$ is the unit vector normal to the sphere at the same point, given by Eq. (2.38). We can choose to enforce this boundary condition anywhere on the surface, but enforcing it at $\mathbf{x}_s = (-R, 0, 0)$ makes sense because it renders the boundary condition scalar, that is

$$\frac{\partial\phi}{\partial x} + U_\infty = 0 \tag{2.57}$$

If we are to model the flow using a doublet, the first step is to choose the direction of this doublet. Figure 2.5a shows that a doublet aligned with the z direction induces zero velocity in the x direction in the z = 0 plane; this is also the case for a doublet aligned with the y direction but not for one aligned with the x direction. As we want the velocity induced by the doublet at z = 0 to be equal and opposite to U_∞, we must choose a doublet aligned with the x direction, that is

$$\phi(\mathbf{x},\mathbf{x}_0) = \frac{\partial}{\partial x}\left(\frac{1}{(\mathbf{x},\mathbf{x}_0)}\right) = -\frac{x-x_0}{\left((x-x_0)^2+(y-y_0)^2+(z-z_0)^2\right)^{3/2}}$$

However, a doublet of unit strength cannot necessarily satisfy the boundary condition, which depends on U_∞. We therefore introduce the doublet strength, μ, such that

$$\phi(\mathbf{x},\mathbf{x}_0) = \mu\frac{\partial}{\partial x}\left(\frac{1}{(\mathbf{x},\mathbf{x}_0)}\right) = -\mu\frac{x-x_0}{\left((x-x_0)^2+(y-y_0)^2+(z-z_0)^2\right)^{3/2}}$$

The velocities induced by this doublet are then given by

$$\frac{\partial\phi}{\partial x} = -\mu\frac{-2(x-x_0)^2+(y-y_0)^2+(z-z_0)^2}{\left((x-x_0)^2+(y-y_0)^2+(z-z_0)^2\right)^{5/2}}$$
$$\frac{\partial\phi}{\partial y} = 3\mu\frac{(x-x_0)(y-y_0)}{\left((x-x_0)^2+(y-y_0)^2+(z-z_0)^2\right)^{5/2}}$$
$$\frac{\partial\phi}{\partial z} = 3\mu\frac{(x-x_0)(z-z_0)}{\left((x-x_0)^2+(y-y_0)^2+(z-z_0)^2\right)^{5/2}}$$

We select $\mathbf{x}_0 = \mathbf{0}$ so that the doublet lies at the origin, evaluate $\partial\phi/\partial x$ at $\mathbf{x} = (-R, 0, 0)$ and insert the result in Eq. (2.57) to obtain the boundary condition:

$$\frac{2\mu}{R^3} + U_\infty = 0$$

from which the necessary doublet strength is $\mu = -U_\infty R^3/2$.

We can now assemble the total potential anywhere in the flow

$$\Phi(\mathbf{x}) = \phi_\infty(\mathbf{x}) + \phi(\mathbf{x}) = U_\infty x + \frac{U_\infty R^3}{2}\frac{x}{\left(x^2+y^2+z^2\right)^{3/2}} \tag{2.58}$$

and the total velocities

$$\begin{aligned}
\frac{\partial \Phi}{\partial x} &= U_\infty + \frac{U_\infty R^3}{2}\frac{-2x^2+y^2+z^2}{\left(x^2+y^2+x^2\right)^{5/2}}\\
\frac{\partial \Phi}{\partial y} &= -\frac{3U_\infty R^3}{2}\frac{xy}{\left(x^2+y^2+x^2\right)^{5/2}}\\
\frac{\partial \Phi}{\partial z} &= -\frac{3U_\infty R^3}{2}\frac{xz}{\left(x^2+y^2+x^2\right)^{5/2}}
\end{aligned} \tag{2.59}$$

These expressions for the potential and velocities are valid everywhere in the flow, but we are particularly interested in their values on the surface of the sphere, $\mathbf{x} = \mathbf{x}_s$. *There*

$$\begin{aligned}
\Phi(\mathbf{x}_s) &= U_\infty x_s + \frac{U_\infty}{2}x_s\\
\left.\frac{\partial \Phi}{\partial x}\right|_{\mathbf{x}_s} &= U_\infty + \frac{U_\infty}{2}\frac{-2x_s^2+y_s^2+z_s^2}{R^2}\\
\left.\frac{\partial \Phi}{\partial y}\right|_{\mathbf{x}_s} &= -\frac{3U_\infty}{2}\frac{x_s y_s}{R^2}\\
\left.\frac{\partial \Phi}{\partial z}\right|_{\mathbf{x}_s} &= -\frac{3U_\infty}{2}\frac{x_s z_s}{R^2}
\end{aligned} \tag{2.60}$$

Note that $\nabla\Phi(\mathbf{x}) = \nabla\phi(\mathbf{x}_s) + \mathbf{Q}_\infty$ *and*

$$\nabla\Phi(\mathbf{x})\cdot\mathbf{n}(\mathbf{x}_s) = U_\infty\frac{x_s}{R} + \frac{U_\infty}{2R^2}\left(-2x_s^3 + x_s y_s^2 + x_s z_s^2 - x_s y_s^2 - 3x_s z_2^2\right) = 0$$

so that the sphere is indeed impermeable everywhere, despite the fact that we imposed impermeability at one point only.

The impermeability of the sphere is demonstrated graphically in Figure 2.6a, which plots the velocity vectors on the surface of a sphere of radius $R = 2$ *m immersed in a free stream of speed* $U_\infty = 10$ *m/s. Figure 2.6b plots the flowfield around and inside the sphere in the plane* $y = 0$. *It shows that there is both an external flow, which goes around the sphere without ever penetrating it, and an internal flow, which is very similar to the flow induced by a doublet (see*

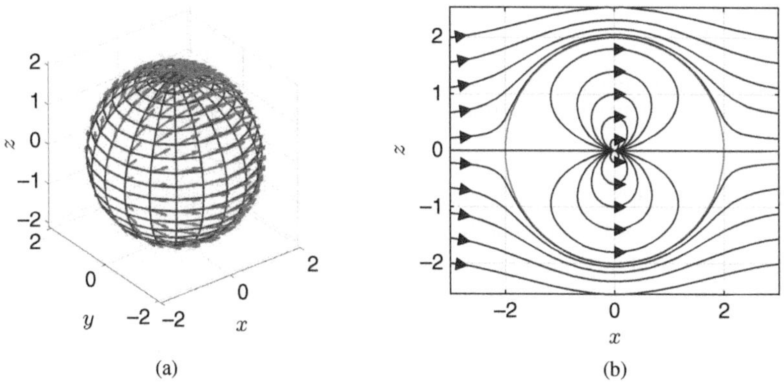

Figure 2.6 Potential flow on and around an impermeable sphere immersed in a free stream. (a) Flow on surface and (b) Flow streamlines in plane $y = 0$.

Figure 2.5b). Only the external flow is of practical interest, but the fact that there is an internal flow is important. The streamline starting at $x = -3, z = 0$ remains always horizontal and decelerates to zero velocity as it reaches the surface of the sphere. The flow stops there and the point $x = -R$, $z = 0$ is referred to as a stagnation point. The flow streamline that ends at a stagnation point is known as the stagnation streamline. At $x = R$, $z = 0$, there is another stagnation point; the velocity of the corresponding streamline is zero there but increases as we move further downstream to reach U_∞ as $z \to \infty$. The reader can substitute the points $(\pm R, 0, 0)$ in Eq. (2.60) in order to verify that the total flow velocity is indeed zero there.

Figure 2.7a plots the variation of the potential induced by the doublet on the surface of the sphere. This variation is linear with x_s and does not depend on y_s or z_s, as shown in the first of Eq. (2.60). The pressure coefficient on the surface of the sphere can be calculated using the steady version of Eq. (2.31)

$$c_p(\mathbf{x}_s) = 1 - \frac{\left(\left.\frac{\partial\Phi}{\partial x}\right|_{\mathbf{x}_s}\right)^2 + \left(\left.\frac{\partial\Phi}{\partial y}\right|_{\mathbf{x}_s}\right)^2 + \left(\left.\frac{\partial\Phi}{\partial z}\right|_{\mathbf{x}_s}\right)^2}{U_\infty^2}$$

As the total flow velocity is equal to zero at the two stagnation points, the value of c_p there is equal to 1 and is known as the stagnation pressure. This phenomenon is clearly seen in Figure 2.7b that plots the pressure coefficient variation on the surface against x_s and y_s. Note that $c_p(x_s, y_s, z_s) = c_p(x_s, y_s, -z_s)$ due to the symmetry in the flow.

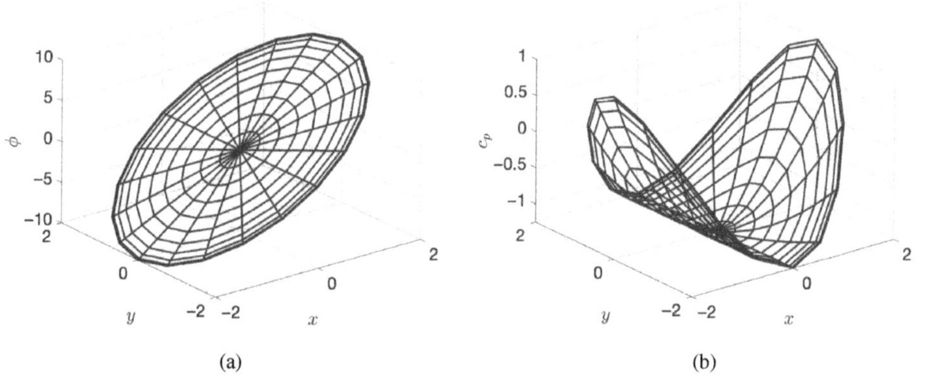

Figure 2.7 Potential and pressure distribution on the surface of the impermeable sphere. (a) Potential on surface and (b) pressure on surface.

Example 2.2 shows that it is possible to superimpose solutions of Laplace's equation in order to model the flow around a closed body. The sphere is axisymmetric in all directions and satisfying the impermeability boundary condition on one point on its surface results in the automatic satisfaction of the condition everywhere. The same procedure cannot be used for most practical bodies, such as aircraft wings and fuselages. For general shapes, the usual approach is to place a continuous distribution of solutions to Laplace's equation over the entire surface of the body. Then, the impermeability boundary condition must be explicitly satisfied on every point on the surface.

As a final thought on the sphere example, it must be stated that the flowfield of Figure 2.6 is not realistic, at least not everywhere. The sphere is a bluff body and flow cannot remain attached everywhere on its surface; viscous phenomena are very important over the rear part of the sphere and therefore such flows cannot be modelled using potential methods. Nevertheless, for $x < 0$, viscous phenomena are less important and the flow of Figure 2.6 is a reasonable approximation of reality; the major modelling errors occur for $x > 0$.

2.3.1 Green's Third Identity

The free stream, source and doublet are sufficient to solve a large class of aerodynamic problems. The free stream velocity is prescribed and the number, positions and strengths of singularities are calculated such that the impermeability boundary condition is satisfied. In order to simplify and speed up the calculations, we would like to keep the number of singularities to a minimum. This means that, ideally, we do not want to place singularities everywhere in the flowfield. We could choose to place singularities only on the surface of the body, but then the question becomes the following: is determining the value of the potential on the surface sufficient to describe the entire flowfield?

In order to start answering this question, we will first study a very simple flow case. Consider the flow drawn in Figure 2.8a, whose potential is $\Phi(\mathbf{x})$ and is a solution of Laplace's equation. This potential is causing flow velocities denoted by the arrow field in the figure, whose components are $u(\mathbf{x}) = \partial\Phi/\partial x$, $v(\mathbf{x}) = \partial\Phi/\partial y$. The flow is drawn in 2D for clarity, but the reader should imagine it as a 3D flow. The figure also plots a surface boundary S, which delimits a flow volume V and is completely permeable; it has no effect on the flow. We will now repose the question we asked earlier: if we know the value of the potential on the surface S, can we calculate the value of the potential anywhere inside the volume V?

In order to answer this question, we will make use of the divergence theorem, which states that for every smooth vector function $\mathbf{F}(\mathbf{x})$,

$$\int_V \nabla \cdot \mathbf{F}(\mathbf{x}) \mathrm{d}V = \int_S \mathbf{n}(\mathbf{x}_s) \cdot \mathbf{F}(\mathbf{x}_s) \mathrm{d}S \tag{2.61}$$

where $\mathbf{x}_s$ is any point on surface S, $\mathbf{n}(\mathbf{x}_s)$ is a unit vector normal to S at $\mathbf{x}_s$ and pointing away from V, $\int_V$ denotes a volume integral over V, $\mathrm{d}V$ is an infinitesimal element of V, $\int_S$ denotes a surface integral over S and $\mathrm{d}S$ is an infinitesimal element of S. The next step is to place a single sink at position $\mathbf{x}_0$ inside the volume, as shown in Figure 2.8a. This sink induces an additional potential $1/r(\mathbf{x}, \mathbf{x}_0)$ everywhere in the flow so that the total potential is now $\Phi' = \Phi(\mathbf{x}) + 1/r(\mathbf{x}, \mathbf{x}_0)$. For the vector function $\mathbf{F}$, we select the form

$$\mathbf{F}(\mathbf{x}) = \frac{1}{r(\mathbf{x}, \mathbf{x}_0)} \nabla\Phi' - \Phi' \nabla \left(\frac{1}{r(\mathbf{x}, \mathbf{x}_0)} \right)$$

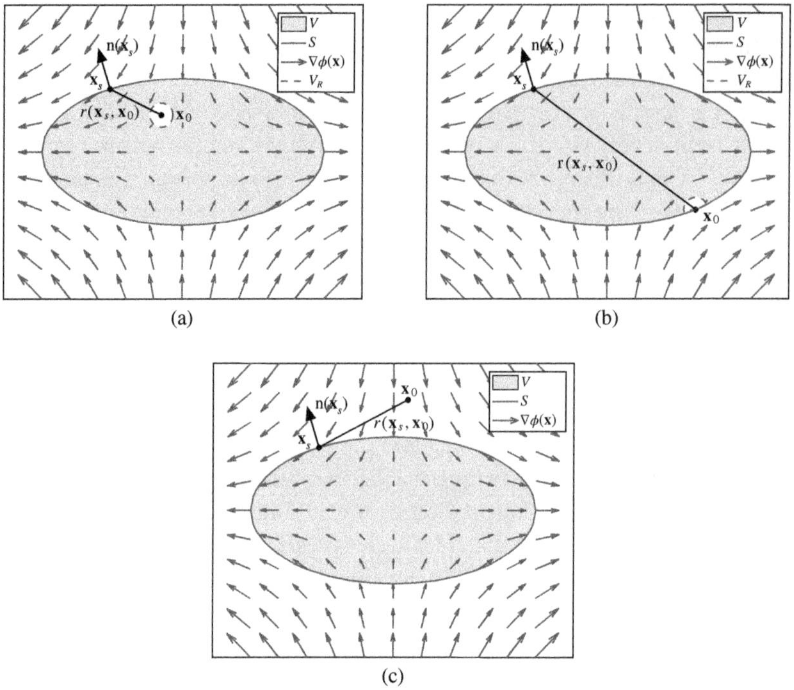

Figure 2.8 Definition of control volume V with bounding surface S in a flowfield and placement of point source at $\mathbf{x}_0$. (a) $\mathbf{x}_0$ in V, (b) $\mathbf{x}_0$ on S, and (c) $\mathbf{x}_0$ not in V and not on S.

Substituting for Φ' yields

$$\begin{aligned}\mathbf{F}(\mathbf{x}) &= \frac{1}{r(\mathbf{x},\mathbf{x}_0)}\nabla\left(\Phi(\mathbf{x})+\frac{1}{r(\mathbf{x},\mathbf{x}_0)}\right)-\left(\Phi(\mathbf{x})+\frac{1}{r(\mathbf{x},\mathbf{x}_0)}\right)\nabla\left(\frac{1}{r(\mathbf{x},\mathbf{x}_0)}\right)\\ &= \frac{1}{r(\mathbf{x},\mathbf{x}_0)}\nabla\Phi(\mathbf{x})-\Phi(\mathbf{x})\nabla\left(\frac{1}{r(\mathbf{x},\mathbf{x}_0)}\right)\end{aligned} \tag{2.62}$$

The divergence of this function is

$$\begin{aligned}\nabla\cdot\mathbf{F}(\mathbf{x}) &= \nabla\left(\frac{1}{r(\mathbf{x},\mathbf{x}_0)}\right)\cdot\nabla\Phi(\mathbf{x})+\frac{1}{r(\mathbf{x},\mathbf{x}_0)}\nabla^2\Phi(\mathbf{x})\\ &\quad-\nabla\Phi(\mathbf{x})\cdot\nabla\left(\frac{1}{r(\mathbf{x},\mathbf{x}_0)}\right)-\Phi(\mathbf{x})\nabla^2\left(\frac{1}{r(\mathbf{x},\mathbf{x}_0)}\right)\\ &= \frac{1}{r(\mathbf{x},\mathbf{x}_0)}\nabla^2\Phi(\mathbf{x})-\Phi(\mathbf{x})\nabla^2\left(\frac{1}{r(\mathbf{x},\mathbf{x}_0)}\right)\end{aligned}$$

We now recall that $\Phi(\mathbf{x})$ is a solution to Laplace's equation so that $\nabla^2\Phi(\mathbf{x}) = 0$ everywhere in the flow. Therefore, the divergence of $\mathbf{F}(\mathbf{x})$ becomes simply

$$\nabla\cdot\mathbf{F}(\mathbf{x}) = -\Phi(\mathbf{x})\nabla^2\left(\frac{1}{r(\mathbf{x},\mathbf{x}_0)}\right)$$

We substitute the function of expression (2.62) into the divergence theorem 2.61 to obtain

$$\int_V -\Phi(\mathbf{x})\nabla^2\left(\frac{1}{r(\mathbf{x},\mathbf{x}_0)}\right)\mathrm{d}V$$
$$= \int_S \mathbf{n}(\mathbf{x}_s)\cdot\left(\frac{1}{r(\mathbf{x}_s,\mathbf{x}_0)}\nabla\Phi(\mathbf{x}_s) - \Phi(\mathbf{x}_s)\nabla\left(\frac{1}{r(\mathbf{x},\mathbf{x}_0)}\right)\bigg|_{\mathbf{x}_s}\right)\mathrm{d}S \tag{2.63}$$

where $\nabla\Phi(\mathbf{x}_s)$ is used to denote the vector

$$\nabla\Phi(\mathbf{x}_s) = \left(\frac{\partial\Phi(\mathbf{x})}{\partial x}, \frac{\partial\Phi(\mathbf{x})}{\partial y}, \frac{\partial\Phi(\mathbf{x})}{\partial z}\right)\bigg|_{\mathbf{x}=\mathbf{x}_s}$$

and $\nabla(1/r(\mathbf{x},\mathbf{x}_0))|_{\mathbf{x}_s}$ represents the vector

$$\nabla\left(\frac{1}{r(\mathbf{x},\mathbf{x}_0)}\right)\bigg|_{\mathbf{x}_s} = \left(\frac{\partial}{\partial x}\left(\frac{1}{r(\mathbf{x},\mathbf{x}_0)}\right), \frac{\partial}{\partial y}\left(\frac{1}{r(\mathbf{x},\mathbf{x}_0)}\right), \frac{\partial}{\partial z}\left(\frac{1}{r(\mathbf{x},\mathbf{x}_0)}\right)\right)\bigg|_{\mathbf{x}=\mathbf{x}_s}$$

In both cases, the derivatives of $\Phi(\mathbf{x})$ and $1/r(\mathbf{x},\mathbf{x}_0)$ are taken with respect to x, y, z and then we substitute $\mathbf{x} = \mathbf{x}_s$.

The integrand on the left-hand side of Eq. (2.63) is proportional to the Laplacian of the sink function $\nabla^2(1/r(\mathbf{x},\mathbf{x}_0))$. Equation (2.53) states that the Laplacian of the source is zero everywhere except at $\mathbf{x} = \mathbf{x}_0$, where it is infinite; the same applies to the sink. We can therefore calculate this integral inside a small sphere centred at $\mathbf{x}_0$ with infinitesimal radius $R \ll 1$ and volume V_R. We assume that $\Phi(\mathbf{x})$ is continuous inside the sphere so that its value is nearly constant and equal to $\Phi(\mathbf{x}_0)$ over such small distances. Consequently,

$$\int_V -\Phi(\mathbf{x})\nabla^2\left(\frac{1}{r(\mathbf{x},\mathbf{x}_0)}\right)\mathrm{d}V = -\Phi(\mathbf{x}_0)\int_{V_R}\nabla^2\left(\frac{1}{r(\mathbf{x},\mathbf{x}_0)}\right)\mathrm{d}V$$

In Appendix A.8, we show that this latest integral is equal to -4π (see Eq. (A.50)). Consequently,

$$\int_V -\Phi(\mathbf{x})\nabla^2\left(\frac{1}{r(\mathbf{x},\mathbf{x}_0)}\right)\mathrm{d}V = 4\pi\Phi(\mathbf{x}_0) \tag{2.64}$$

and Eq. (2.63) becomes

$$4\pi\Phi(\mathbf{x}_0) = \int_S \mathbf{n}(\mathbf{x}_s)\cdot\left(\frac{1}{r(\mathbf{x}_s,\mathbf{x}_0)}\nabla\Phi(\mathbf{x}_s) - \Phi(\mathbf{x}_s)\nabla\left(\frac{1}{r(\mathbf{x},\mathbf{x}_0)}\right)\bigg|_{\mathbf{x}_s}\right)\mathrm{d}S \tag{2.65}$$

This result is known as Green's third identity and is the basis of potential flow aerodynamics for wings and bodies. In essence, the sink we have added at $\mathbf{x}_0$ is an observation instrument, usually referred to as a Green's function. If we know the values of $\Phi(\mathbf{x}_s)$ and $\nabla\Phi(\mathbf{x}_s)$ on the surface, then we can integrate them over S in order to obtain the potential at $\mathbf{x}_0$. If we then move the sink, we can calculate the potential at this new position or anywhere else inside the volume.

Figure 2.8b draws the special case where $\mathbf{x}_0$ is placed exactly on the surface S. Assuming that the curvature of V_R is much higher than that of S, only half of V_R lies inside V.

Therefore, the integral on the left-hand side of Eq. (2.63) gives

$$\int_V -\Phi(\mathbf{x})\nabla^2\left(\frac{1}{r(\mathbf{x},\mathbf{x}_0)}\right)\mathrm{d}V = -\Phi(\mathbf{x}_0)\int_{V_R/2}\nabla^2\left(\frac{1}{r(\mathbf{x},\mathbf{x}_0)}\right)\mathrm{d}V$$
$$= -\frac{\Phi(\mathbf{x}_0)}{2}\int_{V_R}\nabla^2\left(\frac{1}{r(\mathbf{x},\mathbf{x}_0)}\right)\mathrm{d}V = 2\pi\Phi(\mathbf{x}_0)$$

As point $\mathbf{x}_0$ lies on the surface, it must be excluded from the surface integral on the right-hand side of Eq. (2.65) to avoid the singularity. Consequently, Eq. (2.65) becomes

$$2\pi\Phi(\mathbf{x}_0) = \int_{S-\mathbf{x}_0}\mathbf{n}(\mathbf{x}_s)\cdot\left(\frac{1}{r(\mathbf{x}_s,\mathbf{x}_0)}\nabla\Phi(\mathbf{x}_s) - \Phi(\mathbf{x}_s)\nabla\left(\frac{1}{r(\mathbf{x},\mathbf{x}_0)}\right)\right)\Bigg|_{\mathbf{x}_s}\mathrm{d}S \tag{2.66}$$

where the notation $\int_{S-\mathbf{x}_0}$ states that point $\mathbf{x}_0$ is excluded from the surface integral.

Finally, Figure 2.8c depicts the case where $\mathbf{x}_0$ lies outside V. In this case, $1/r(\mathbf{x},\mathbf{x}_0)$ is finite in V and therefore

$$\nabla^2\left(\frac{1}{r(\mathbf{x},\mathbf{x}_0)}\right) = 0$$

everywhere in V. Equation (2.65) reduces to

$$0 = \int_S \mathbf{n}(\mathbf{x}_s)\cdot\left(\frac{1}{r(\mathbf{x}_s,\mathbf{x}_0)}\nabla\Phi(\mathbf{x}_s) - \Phi(\mathbf{x}_s)\nabla\left(\frac{1}{r(\mathbf{x},\mathbf{x}_0)}\right)\Bigg|_{\mathbf{x}_s}\right)\mathrm{d}S \tag{2.67}$$

Equations (2.65), (2.66), and (2.67) can be written compactly as (Morino 1974)

$$E(\mathbf{x}_0)\Phi(\mathbf{x}_0) = \frac{1}{4\pi}\int_S \mathbf{n}(\mathbf{x}_s)\cdot\left(\frac{1}{r(\mathbf{x}_s,\mathbf{x}_0)}\nabla\Phi(\mathbf{x}_s) - \Phi(\mathbf{x}_s)\nabla\left(\frac{1}{r(\mathbf{x},\mathbf{x}_0)}\right)\Bigg|_{\mathbf{x}_s}\right)\mathrm{d}S \tag{2.68}$$

where

$$E(\mathbf{x}_0) = \begin{cases} 1 & \text{if } \mathbf{x}_0 \text{ lies in } V \text{ but not on } S \\ 1/2 & \text{if } \mathbf{x}_0 \text{ lies on } S \\ 0 & \text{if } \mathbf{x}_0 \text{ lies outside } V \end{cases} \tag{2.69}$$

although it must be recalled that point $\mathbf{x}_0$ must be excluded from the surface integral if $E(\mathbf{x}_0) = 1/2$.

Example 2.3 *Consider a flow that features only a free stream so that its total potential is $\phi_\infty = U_\infty x + V_\infty y + W_\infty z$. Define a permeable sphere of radius R centred at the origin and use Green's third identity to calculate the potential at any point $\mathbf{x}_0$ inside or on that sphere. Compare to the exact result.*

We select values for the velocities $U_\infty = 4$ m/s, $V_\infty = -2$ m/s, $W_\infty = 2$ m/s. We already know that the free stream

$$\phi_\infty = 4x - 2y + 2z \tag{2.70}$$

is a solution of Laplace's equation so that we can indeed apply Green's third identity. We create a sphere of radius $R = 2$ m centred at the origin $\mathbf{x} = 0$, which is the surface S that delimits

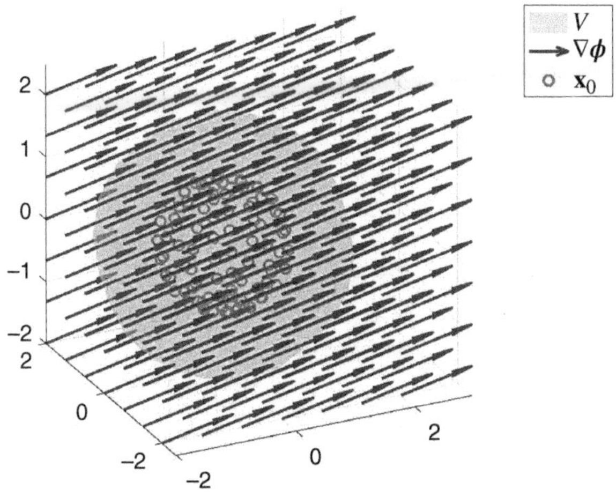

Figure 2.9 Control volume definition in a flow with potential $\phi_\infty(\mathbf{x}) = 4x - 2y + 2z$.

the control volume V; we also define points $\mathbf{x}_0$ at which we want to calculate the potential. Figure 2.9 plots the flow's velocity field $\nabla\phi_\infty(\mathbf{x})$, the control volume V of radius R and points $\mathbf{x}_0$ that lie on the surface of a sphere with smaller radius R_2. It can be seen that many flow vectors enter or exit S; the surface is not impermeable.

Recall from Example (2.1) that the unit vector normal to the surface of a sphere of radius R centred around the origin and pointing outwards (hence, away from V, as required by the divergence theorem) is given by

$$\mathbf{n}(\mathbf{x}_s) = \frac{1}{R}\mathbf{x}_s$$

Therefore,

$$\mathbf{n}(\mathbf{x}_s) \cdot \nabla\phi_\infty(\mathbf{x}_s) = \frac{1}{R}\left(4x_s - 2y_s + 2z_s\right) = \frac{\phi_\infty(\mathbf{x}_s)}{R}$$

We now turn our attention to the sink lying at $\mathbf{x}_0$ and calculate

$$\frac{1}{r(\mathbf{x}_s, \mathbf{x}_0)}\mathbf{n}(\mathbf{x}_s) \cdot \nabla\phi_\infty(\mathbf{x}_s) = \frac{\phi_\infty(\mathbf{x}_s)}{Rr(\mathbf{x}_s, \mathbf{x}_0)}$$

Furthermore,

$$\nabla\left(\frac{1}{r(\mathbf{x}_s, \mathbf{x}_0)}\right)\Bigg|_{\mathbf{x}_s} = -\frac{1}{r(\mathbf{x}_s, \mathbf{x}_0)^3}\left((x_s - x_0), (y_s - y_0), (z_s - z_0)\right) = -\frac{(\mathbf{x}_s - \mathbf{x}_0)}{r(\mathbf{x}_s, \mathbf{x}_0)^3}$$

so that

$$\mathbf{n}(\mathbf{x}_s) \cdot \nabla\left(\frac{1}{r(\mathbf{x}_s, \mathbf{x}_0)}\right)\Bigg|_{\mathbf{x}_s} = -\frac{\mathbf{x}_s \cdot (\mathbf{x}_s - \mathbf{x}_0)}{Rr(\mathbf{x}_s, \mathbf{x}_0)^3} \tag{2.71}$$

Consequently, Eq. (2.68) becomes

$$E(\mathbf{x}_0)\phi_\infty(\mathbf{x}_0) = \frac{1}{4\pi}\int_S \phi_\infty(\mathbf{x}_s)\left(\frac{1}{Rr(\mathbf{x}_s, \mathbf{x}_0)} + \frac{\mathbf{x}_s \cdot (\mathbf{x}_s - \mathbf{x}_0)}{Rr(\mathbf{x}_s, \mathbf{x}_0)^3}\right)\mathrm{d}S \tag{2.72}$$

This integral is not easy to evaluate, except for the special case $\mathbf{x}_0 = \mathbf{0}$*, that is the case where we place the sink at the origin and look for the value of* $\phi_\infty(\mathbf{0})$*. Substituting* $\mathbf{x}_0 = \mathbf{0}$ *in Eq. (2.72) and noting that* $r(\mathbf{x}_s, \mathbf{0}) = R$, $\mathbf{x}_s \cdot \mathbf{x}_s = R$ *and* $E(\mathbf{0}) = 1$*, we obtain*

$$\phi_\infty(\mathbf{0}) = \frac{1}{2\pi R^2} \int_S \phi_\infty(\mathbf{x}_s) \mathrm{d}S$$

We can calculate the integral using spherical coordinates

$$x_s = R\cos\psi \sin\theta,\ y_s = R\sin\psi \sin\theta,\ z_s = R\cos\theta$$

where the azimuthal angle ψ *takes values between* 0 *and* 2π *and the polar angle* θ *between* 0 *and* π*. Consequently,* $\mathrm{d}S = R^2 \sin\theta \mathrm{d}\psi \mathrm{d}\theta$ *and Eq. (2.72) yields*

$$\phi_\infty(\mathbf{0}) = \frac{R}{2\pi} \int_0^{2\pi} \int_0^{\pi} (4\cos\psi \sin\theta - 2\sin\psi \sin\theta + 2\cos\theta) \sin\theta \mathrm{d}\psi \mathrm{d}\theta = 0$$

We obtain the same result when substituting $\mathbf{x}_0 = \mathbf{0}$ *in Eq. (2.70), so Green's identity is verified for the case where the sink is placed at the origin.*

For the general case where $\mathbf{x}_0 \neq \mathbf{0}$*, there is no obvious analytical expression for the integral of Eq. (2.72), so we have to resort to numerical integration. Using the Matlab function* `sphere.m`*, we can create three* $(n+1) \times (n+1)$ *matrices* $x_{p_{i,j}}, y_{p_{i,j}}, z_{p_{i,j}}$*, for* $i = 1, \dots, n+1$*,* $j = 1, \dots, n+1$ *that contain the coordinates of* $(n+1)^2$ *regularly spaced points on the surface of a sphere of radius 1. We can then create two spheres:*

- *The spherical surface S of radius R that delimits V and on which we know* $\phi_\infty(\mathbf{x}_s)$*. The coordinates of the points on this surface are denoted by* $x_{s_{i,j}} = Rx_{p_{i,j}}$, $y_{s_{i,j}} = Ry_{p_{i,j}}$, $z_{s_{i,j}} = Rz_{p_{i,j}}$*, such that* $\mathbf{x}_{s_{i,j}} = (x_{s_{i,j}}, y_{s_{i,j}}, z_{s_{i,j}})$*, for* $i = 1, \dots, n+1, j = 1, \dots, n+1$.
- *A spherical surface of radius* $R_2 \leq R$ *on which we would like to calculate the potential* $\phi_\infty(\mathbf{x}_0)$*. The coordinates of the points on this surface are denoted by* $x_{0_{k,l}} = R_2 x_{p_{k,l}}$, $y_{0_{k,l}} = R_2 y_{p_{k,l}}$, $z_{0_{k,l}} = R_2 z_{p_{k,l}}$*, such that* $\mathbf{x}_{0_{k,l}} = (x_{0_{k,l}}, y_{0_{k,l}}, z_{0_{k,l}})$*, for* $k = 1, \dots, n+1$, $l = 1, \dots, n+1$.

The objective is to calculate $\phi_\infty(\mathbf{x}_{0_{k,l}})$ *knowing* $\phi_\infty(\mathbf{x}_{s_{i,j}})$ *from a discrete version of Eq. (2.72).*

Some of the points $\mathbf{x}_{s_{i,j}}$ *on the sphere with radius R are plotted in Figure 2.10. Every four neighbouring points define a quadrilateral panel so that there is a total of* n^2 *panels. The panel i, j is delimited by vertices* $\mathbf{x}_{s_{i,j}}, \mathbf{x}_{s_{i,j+1}}, \mathbf{x}_{s_{i+1,j+1}}, \mathbf{x}_{s_{i+1,j}}$*, for* $i = 1, \dots, n, j = 1, \dots, n$*. We define a control point on each panel, denoted by* $\mathbf{x}_{c_{i,j}}$*, which is generally calculated from the mean of the vertices*

$$\mathbf{x}_{c_{i,j}} = (\mathbf{x}_{s_{i,j}} + \mathbf{x}_{s_{i,j+1}} + \mathbf{x}_{s_{i+1,j+1}} + \mathbf{x}_{s_{i+1,j}})/4$$

Furthermore, panel i, j has surface area $\Delta s_{i,j}$ *that can be calculated from*

$$\Delta s_{i,j} = \frac{1}{2}\left(\mathbf{x}_{s_{i+1,j+1}} - \mathbf{x}_{s_{i,j}}\right) \times \left(\mathbf{x}_{s_{i,j+1}} - \mathbf{x}_{s_{i,j}}\right) + \frac{1}{2}\left(\mathbf{x}_{s_{i+1,j+1}} - \mathbf{x}_{s_{i,j}}\right) \times \left(\mathbf{x}_{s_{i+1,j}} - \mathbf{x}_{s_{i,j}}\right) \tag{2.73}$$

Figure 2.10 Discretisation of a surface into quadrilateral panels.

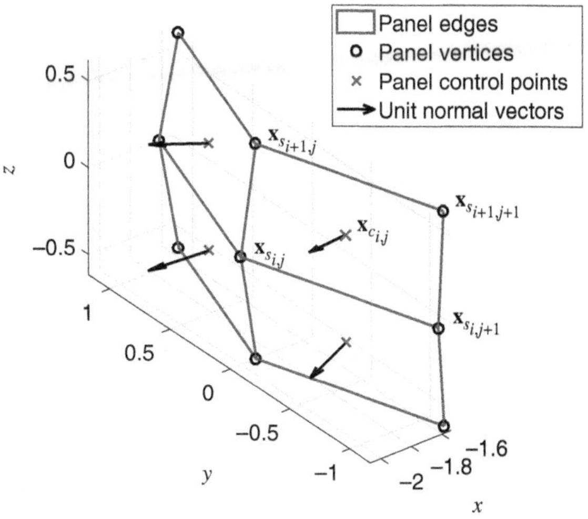

Since we have replaced the surface of the sphere by a series of flat quadrilateral panels, Eq. (2.72) becomes a summation of integrals over each panel, such that

$$E(\mathbf{x}_0)\phi_\infty(\mathbf{x}_0) = \frac{1}{4\pi}\sum_{J=1}^{N}\int_{\Delta s_J}\phi_\infty(\mathbf{x}_s)\left(\frac{1}{Rr(\mathbf{x}_s,\mathbf{x}_0)} + \frac{\mathbf{x}_s\cdot(\mathbf{x}_s-\mathbf{x}_0)}{Rr(\mathbf{x}_s,\mathbf{x}_0)^3}\right)\mathrm{d}S$$

where J is a single counter $J = ni + j$ that cycles through all the panels and takes values $J = 1, \ldots, N$ for $N = n^2$, while $\mathbf{x}_s$ denotes any point lying on panel J. The situation has not improved significantly; the remaining integral is still difficult to evaluate analytically. The next step is to assume that each of the panels is so small that the integrand is approximately constant over $\Delta s_{i,j}$ and equal to its value at the control point. Then the integrand can be removed from the integral,

$$E(\mathbf{x}_0)\phi_\infty(\mathbf{x}_0) = \frac{1}{4\pi}\sum_{J=1}^{N}\phi_\infty(\mathbf{x}_{c_J})\left(\frac{1}{Rr(\mathbf{x}_{c_J},\mathbf{x}_0)} + \frac{\mathbf{x}_{c_J}\cdot(\mathbf{x}_{c_J}-\mathbf{x}_0)}{Rr(\mathbf{x}_{c_J},\mathbf{x}_0)^3}\right)\int_{\Delta s_J}\mathrm{d}S$$

The remaining integral is simply equal to $\Delta s_{i,j}$. Applying this procedure for each of the $(n+1)^2$ points $\mathbf{x}_{0_I}$, where $I = ki + l$ takes values $I = 1, \ldots, (n+1)^2$, leads to

$$\phi_\infty(\mathbf{x}_{0_I}) \approx \frac{1}{4\pi}\sum_{J=1}^{N}\phi_\infty(\mathbf{x}_{c_J})\left(\frac{1}{Rr(\mathbf{x}_{c_J},\mathbf{x}_{0_I})} + \frac{\mathbf{x}_{c_J}\cdot\left(\mathbf{x}_{c_J}-\mathbf{x}_{0_I}\right)}{Rr(\mathbf{x}_{c_J},\mathbf{x}_{0_I})^3}\right)\Delta s_J \tag{2.74}$$

for $R_2 < R$, that is for $E(\mathbf{x}_{0_I}) = 1$. The assumption is that the panels are so small that the error in Eq. (2.74) is not significant. The sum in Eq. (2.74) can be readily calculated as all the quantities inside it are known.

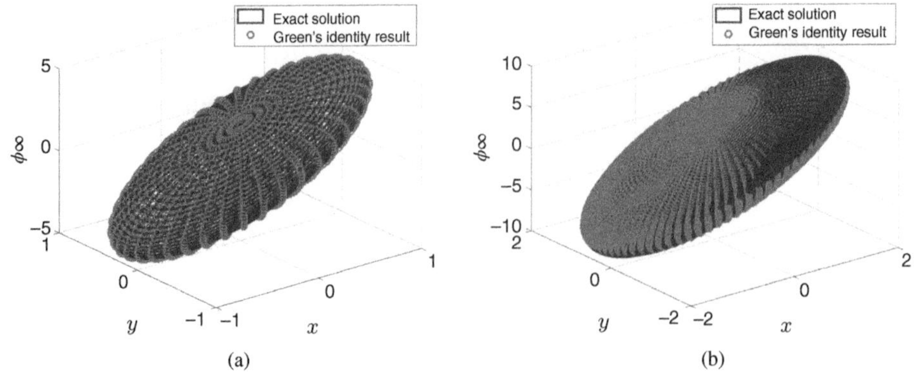

Figure 2.11 Potential distribution in flowfield with potential $\phi_\infty(\mathbf{x}) = 4x - 2y + 2z$. (a) $R_2 = R/2$ and (b) $R_2 = R$.

Figure 2.11a plots the values of $\phi_\infty(\mathbf{x}_{0_{k,l}})$ calculated using Eq. (2.74) on a sphere of radius $R_2 = R/2$ against x and y and compares them to the exact values from Eq. (2.70). We have chosen $n = 40$ such that there are $N = 800$ panels; clearly this discretisation results in accurate estimates for the potential at $R/2$. Next, we set $R_2 = R$, such that $\mathbf{x}_{0_I} = \mathbf{x}_{s_I}$ to calculate the potential on surface S itself. In this case, $E(\mathbf{x}_{0_I}) = 1/2$ and Eq. (2.72) becomes

$$\frac{1}{2}\phi_\infty(\mathbf{x}_{s_I}) \approx \frac{1}{4\pi}\sum_{J=1}^{N}\phi_\infty(\mathbf{x}_{c_J})\left(\frac{1}{Rr(\mathbf{x}_{c_J}, \mathbf{x}_{s_I})} + \frac{\mathbf{x}_{c_J}\cdot\left(\mathbf{x}_{c_J} - \mathbf{x}_{s_I}\right)}{Rr(\mathbf{x}_{c_J}, \mathbf{x}_{s_I})^3}\right)\Delta s_J \tag{2.75}$$

In other words, we are calculating the potential on the vertices of the panels on S, knowing the value of the potential on the control points of the same panels. Note that these control points do not actually lie on the surface of the sphere; they only do so if the size of the panels becomes infinitesimal. In order to obtain accurate estimates of the potential on the vertices, we need to increase the number of panels, by setting $n = 80$ for instance. Figure 2.11b compares the numerical prediction to the exact value of Eq. (2.70); the agreement between the two sets of results is good and becomes better if the number of panels is increased even further.

Finally, we apply Green's identity to calculate the potential on points lying on a sphere with radius $R_2 = 2R$. In this case, all the points on circle R_2 lie outside V and $E(\mathbf{x}_{0_I}) = 0$. Equation (2.72) becomes

$$0 \approx \frac{1}{4\pi}\sum_{J=1}^{N}\phi_\infty(\mathbf{x}_{c_J})\left(\frac{1}{Rr(\mathbf{x}_{c_J}, \mathbf{x}_{0_I})} + \frac{\mathbf{x}_{c_J}\cdot\left(\mathbf{x}_{c_J} - \mathbf{x}_{0_I}\right)}{Rr(\mathbf{x}_{c_J}, \mathbf{x}_{0_I})^3}\right)\Delta s_J \tag{2.76}$$

for all I. Calculating the right-hand side yields a left-hand side that is equal to zero to within numerical error. This does not mean that the potential on these points is really equal to zero, Eq. (2.70) shows otherwise; it simply means that the left-hand side of Green's identity is equal *to zero when the points $\mathbf{x}_0$ lie outside the surface S. This example is solved by Matlab code* `flow_through_sphere.m`.

In the previous example, the sphere was not impermeable; we only used it to verify the accuracy of Green's third identity. In practical aerodynamic problems, the surface of the body is impermeable, and we are interested in calculating the potential of the flow around it. There may be flow inside the body, but it does not communicate with the external flow and, in any case, we are not interested in it. Figure 2.12 plots the domain in which flow of interest occurs around a wing. It defines two surfaces, the body surface S and the far-field boundary surface S_∞. Flow occurs at all points $\mathbf{x}_0$ that lie in the volume defined between these surfaces. Consequently, we would like to calculate the potential $\Phi(\mathbf{x}_0)$ anywhere in V in terms of the potential $\Phi(\mathbf{x}_s)$ everywhere on S and S_∞.

If we compare the flow plotted in Figure 2.12 to that in Figure 2.8, we can see two major differences:

- The flow volume in Figure 2.12 is bounded by both S and S_∞
- The unit vector normal to S_∞ still points outside V, as in Figure 2.8. Conversely, the unit vector normal to S now points towards V.

Consequently, Green's third identity (2.68) must now be written such that the effect of both boundaries is included and that the signs of the normal vectors are appropriate to each boundary,

$$E(\mathbf{x}_0)\Phi(\mathbf{x}_0) = \frac{1}{4\pi}\int_S \mathbf{n}(\mathbf{x}_s)\cdot\left(-\frac{1}{r(\mathbf{x}_s,\mathbf{x}_0)}\nabla\Phi(\mathbf{x}_s) + \Phi(\mathbf{x}_s)\nabla\left(\frac{1}{r(\mathbf{x},\mathbf{x}_0)}\right)\bigg|_{\mathbf{x}_s}\right)\mathrm{d}S$$
$$+\frac{1}{4\pi}\int_{S_\infty} \mathbf{n}(\mathbf{x}_s)\cdot\left(\frac{1}{r(\mathbf{x}_s,\mathbf{x}_0)}\nabla\Phi(\mathbf{x}_s) - \Phi(\mathbf{x}_s)\nabla\left(\frac{1}{r(\mathbf{x},\mathbf{x}_0)}\right)\bigg|_{\mathbf{x}_s}\right)\mathrm{d}S \quad (2.77)$$

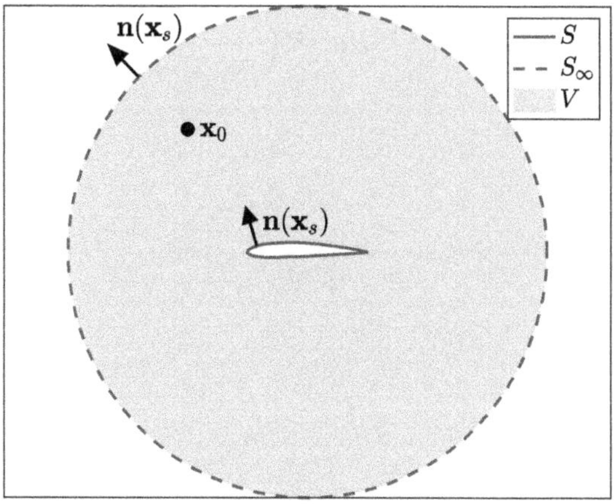

Figure 2.12 Flow domain around a wing.

where $E(\mathbf{x}_0)$ is defined in Eq. (2.69), $\mathbf{x}_s$ are points on S or S_∞ and $\mathbf{x}_0$ is any point inside V, on either of the two surfaces or outside V. If the motion is described by Figure 2.2, then the total flow potential is given by

$$\Phi(\mathbf{x}) = \phi_\infty(\mathbf{x}) + \phi(\mathbf{x})$$

where $\phi_\infty(\mathbf{x})$ is the free stream potential and $\phi(\mathbf{x})$ is the perturbation potential due to the flow disturbance caused by the presence of the body. This potential decays to zero on boundary S_∞ so that Eq. (2.77) becomes

$$\begin{aligned} E(\mathbf{x}_0)\Phi(\mathbf{x}_0) &= \frac{1}{4\pi}\int_S \mathbf{n}(\mathbf{x}_s)\cdot\left(-\frac{1}{r(\mathbf{x}_s,\mathbf{x}_0)}\nabla\phi(\mathbf{x}_s) + \phi(\mathbf{x}_s)\nabla\left(\frac{1}{r(\mathbf{x},\mathbf{x}_0)}\right)\bigg|_{\mathbf{x}_s}\right)\mathrm{d}S \\ &+ \frac{1}{4\pi}\int_S \mathbf{n}(\mathbf{x}_s)\cdot\left(-\frac{1}{r(\mathbf{x}_s,\mathbf{x}_0)}\nabla\phi_\infty(\mathbf{x}_s) + \phi_\infty(\mathbf{x}_s)\nabla\left(\frac{1}{r(\mathbf{x},\mathbf{x}_0)}\right)\bigg|_{\mathbf{x}_s}\right)\mathrm{d}S \\ &+ \frac{1}{4\pi}\int_{S_\infty} \mathbf{n}(\mathbf{x}_s)\cdot\left(\frac{1}{r(\mathbf{x}_s,\mathbf{x}_0)}\nabla\phi_\infty(\mathbf{x}_s) - \phi_\infty(\mathbf{x}_s)\nabla\left(\frac{1}{r(\mathbf{x},\mathbf{x}_0)}\right)\bigg|_{\mathbf{x}_s}\right)\mathrm{d}S \end{aligned} \tag{2.78}$$

Recall example 2.3; the flow of interest occurs inside boundary S_∞ so that

$$\frac{1}{4\pi}\int_{S_\infty} \mathbf{n}(\mathbf{x}_s)\cdot\left(\frac{1}{r(\mathbf{x}_s,\mathbf{x}_0)}\nabla\phi_\infty(\mathbf{x}_s) - \phi_\infty(\mathbf{x}_s)\nabla\left(\frac{1}{r(\mathbf{x},\mathbf{x}_0)}\right)\bigg|_{\mathbf{x}_s}\right)\mathrm{d}S = \phi_\infty(\mathbf{x}_s)$$

(see Eq. (2.74) for a discretised spherical surface, but the result holds for any shape of surface). Conversely, the flow of interest occurs outside boundary S, so that

$$\frac{1}{4\pi}\int_S \mathbf{n}(\mathbf{x}_s)\cdot\left(-\frac{1}{r(\mathbf{x}_s,\mathbf{x}_0)}\nabla\phi_\infty(\mathbf{x}_s) + \phi_\infty(\mathbf{x}_s)\nabla\left(\frac{1}{r(\mathbf{x},\mathbf{x}_0)}\right)\bigg|_{\mathbf{x}_s}\right)\mathrm{d}S = 0$$

(see Eq. (2.76)).[1] Consequently, if point $\mathbf{x}_0$ lies in V but not on S, Eq. (2.78) becomes

$$\begin{aligned} \phi(\mathbf{x}_0) + \phi_\infty(\mathbf{x}_0) &= \frac{1}{4\pi}\int_S \mathbf{n}(\mathbf{x}_s)\cdot\left(-\frac{1}{r(\mathbf{x}_s,\mathbf{x}_0)}\nabla\phi(\mathbf{x}_s) + \phi(\mathbf{x}_s)\nabla\left(\frac{1}{r(\mathbf{x},\mathbf{x}_0)}\right)\bigg|_{\mathbf{x}_s}\right)\mathrm{d}S \\ &+ \phi_\infty(\mathbf{x}_0) \end{aligned}$$

or, after simplification

$$\phi(\mathbf{x}_0) = \frac{1}{4\pi}\int_S \mathbf{n}(\mathbf{x}_s)\cdot\left(-\frac{1}{r(\mathbf{x}_s,\mathbf{x}_0)}\nabla\phi(\mathbf{x}_s) + \phi(\mathbf{x}_s)\nabla\left(\frac{1}{r(\mathbf{x}_s,\mathbf{x}_0)}\right)\bigg|_{\mathbf{x}_s}\right)\mathrm{d}S \tag{2.79}$$

1 The right-hand side of Green's identity is always equal to zero when evaluated outside a region that does not contain singularities. Conversely, the right-hand side of Green's identity is always equal to zero when evaluated inside a region that contains singularities. These statements are true irrespective of the direction of the normal vector.

If point $\mathbf{x}_0$ lies on S, Eq. (2.77) becomes

$$\frac{1}{2}\Phi(\mathbf{x}_0) = \frac{1}{4\pi}\int_S \mathbf{n}(\mathbf{x}_s)\cdot\left(-\frac{1}{r(\mathbf{x}_s,\mathbf{x}_0)}\nabla\phi(\mathbf{x}_s) + \phi(\mathbf{x}_s)\nabla\left(\frac{1}{r(\mathbf{x},\mathbf{x}_0)}\right)\bigg|_{\mathbf{x}_s}\right)\mathrm{d}S$$
$$+\frac{1}{4\pi}\int_S \mathbf{n}(\mathbf{x}_s)\cdot\left(-\frac{1}{r(\mathbf{x}_s,\mathbf{x}_0)}\nabla\phi_\infty(\mathbf{x}_s) + \phi_\infty(\mathbf{x}_s)\nabla\left(\frac{1}{r(\mathbf{x},\mathbf{x}_0)}\right)\bigg|_{\mathbf{x}_s}\right)\mathrm{d}S \quad (2.80)$$
$$+\frac{1}{4\pi}\int_{S_\infty} \mathbf{n}(\mathbf{x}_s)\cdot\left(\frac{1}{r(\mathbf{x}_s,\mathbf{x}_0)}\nabla\phi_\infty(\mathbf{x}_s) - \phi_\infty(\mathbf{x}_s)\nabla\left(\frac{1}{r(\mathbf{x},\mathbf{x}_0)}\right)\bigg|_{\mathbf{x}_s}\right)\mathrm{d}S$$

Using the generalisation of Eq. (2.75) we write

$$\frac{1}{4\pi}\int_S \mathbf{n}(\mathbf{x}_s)\cdot\left(-\frac{1}{r(\mathbf{x}_s,\mathbf{x}_0)}\nabla\phi_\infty(\mathbf{x}_s) + \phi_\infty(\mathbf{x}_s)\nabla\left(\frac{1}{r(\mathbf{x},\mathbf{x}_0)}\right)\bigg|_{\mathbf{x}_s}\right)\mathrm{d}S = -\frac{1}{2}\phi_\infty(\mathbf{x}_0)$$

because the sign of the argument of the integral has changed. The last integral in Eq. (2.80) is still equal to $\phi_\infty(\mathbf{x}_0)$ since any point on S lies inside S_∞. Consequently, Eq. (2.80) becomes

$$\frac{1}{2}\left(\phi(\mathbf{x}_0) + \phi_\infty(\mathbf{x}_0)\right) = \frac{1}{4\pi}\int_S \mathbf{n}(\mathbf{x}_s)\cdot\left(-\frac{1}{r(\mathbf{x}_s,\mathbf{x}_0)}\nabla\phi(\mathbf{x}_s) + \phi(\mathbf{x}_s)\nabla\left(\frac{1}{r(\mathbf{x},\mathbf{x}_0)}\right)\bigg|_{\mathbf{x}_s}\right)\mathrm{d}S$$
$$-\frac{1}{2}\phi_\infty(\mathbf{x}_0) + \phi_\infty(\mathbf{x}_0)$$

or after simplification

$$\frac{1}{2}\phi(\mathbf{x}_0) = \frac{1}{4\pi}\int_S \mathbf{n}(\mathbf{x}_s)\cdot\left(-\frac{1}{r(\mathbf{x}_s,\mathbf{x}_0)}\nabla\phi(\mathbf{x}_s) + \phi(\mathbf{x}_s)\nabla\left(\frac{1}{r(\mathbf{x}_s,\mathbf{x}_0)}\right)\bigg|_{\mathbf{x}_s}\right)\mathrm{d}S \quad (2.81)$$

Finally, Eqs. (2.79) and (2.81) can be combined into

$$E(\mathbf{x}_0)\phi(\mathbf{x}_0) = \frac{1}{4\pi}\int_S \mathbf{n}(\mathbf{x}_s)\cdot\left(-\frac{1}{r(\mathbf{x}_s,\mathbf{x}_0)}\nabla\phi(\mathbf{x}_s) + \phi(\mathbf{x}_s)\nabla\left(\frac{1}{r(\mathbf{x}_s,\mathbf{x}_0)}\right)\bigg|_{\mathbf{x}_s}\right)\mathrm{d}S \quad (2.82)$$

This latest version of Green's third identity involves only the perturbation potential and is independent of the free stream potential and the far-field boundary. If the motion is described by Figure 2.1, there is no free stream and $\Phi(\mathbf{x}_s)$ is negligible on S_∞ due to the far-field boundary condition. Consequently, Eq. (2.77) simplifies to the same form as Eq. (2.82), except that it is written in terms of $\Phi(\mathbf{x}_0)$ and $\Phi(\mathbf{x}_s)$ instead of $\phi(\mathbf{x}_0)$ and $\phi(\mathbf{x}_s)$.

Example 2.4 *Verify Green's third identity (2.82) for the flow around an impermeable sphere immersed in a free stream.*

We have already developed an exact solution for the flow around a sphere immersed in a free stream in Example 2.2. The potential induced by the doublet anywhere in the flow is given by Eq. (2.58) as

$$\phi(\mathbf{x}) = \frac{U_\infty R^3}{2}\frac{x}{\left(x^2+y^2+z^2\right)^{3/2}} \quad (2.83)$$

assuming that the sphere has radius R and is centred around the origin. On the surface of the sphere,

$$\phi(\mathbf{x}_s) = \frac{U_\infty}{2} x_s \tag{2.84}$$

The total flow velocities on the surface are given by Eq. (2.59) so that the velocities on S due only to the doublet are

$$\left.\frac{\partial \phi}{\partial x}\right|_{\mathbf{x}_s} = \frac{U_\infty}{2} \frac{-2x_s^2 + y_s^2 + z_s^2}{R^2}$$

$$\left.\frac{\partial \phi}{\partial y}\right|_{\mathbf{x}_s} = -\frac{3U_\infty}{2} \frac{x_s y_s}{R^2}$$

$$\left.\frac{\partial \phi}{\partial z}\right|_{\mathbf{x}_s} = -\frac{3U_\infty}{2} \frac{x_s z_s}{R^2}$$

The unit vector normal to the surface of the sphere is given by Eq. (2.38)

$$\mathbf{n}(\mathbf{x}_s) = \frac{1}{R} \mathbf{x}_s$$

It points outwards from S and into V so that it is compatible with Figure 2.12 and Eq. (2.82). Consequently, the product $\mathbf{n}(\mathbf{x}_s) \cdot \nabla(\mathbf{x}_s)$ *is given by*

$$\mathbf{n}(\mathbf{x}_s) \cdot \nabla \phi(\mathbf{x}_s) = -\frac{U_\infty}{R^3} \left(x_s^3 + x_s y_s^2 + x_s z_s^2\right) = -\frac{U_\infty}{R} x_s \tag{2.85}$$

Furthermore, Eq. (2.71) is still valid here, that is

$$\mathbf{n}(\mathbf{x}_s) \cdot \nabla \left.\left(\frac{1}{r(\mathbf{x}, \mathbf{x}_0)}\right)\right|_{\mathbf{x}_s} = -\frac{\mathbf{x}_s \cdot (\mathbf{x}_s - \mathbf{x}_0)}{R r(\mathbf{x}_s, \mathbf{x}_0)^3}$$

We can now assemble Eq. (2.82)

$$E(\mathbf{x}_0)\phi(\mathbf{x}_0) = \frac{1}{4\pi} \int_S \frac{U_\infty}{R r(\mathbf{x}_s, \mathbf{x}_0)} x_s \mathrm{d}S - \frac{1}{4\pi} \int_S \frac{U_\infty}{2} x_s \frac{\mathbf{x}_s \cdot (\mathbf{x}_s - \mathbf{x}_0)}{R r(\mathbf{x}_s, \mathbf{x}_0)^3} \mathrm{d}S \tag{2.86}$$

As in Example 2.3, there is no obvious analytical expression for the integrals in the Eq. (2.86) and, again, we need to resort to numerical integration. We create two spheres, one with radius R and coordinates $\mathbf{x}_{s_{i,j}}$, *for* $i = 1, \ldots, n+1, j = 1, \ldots, n+1$, *and one with radius* R_2 *and coordinates* $\mathbf{x}_{0_{k,l}}$, *for* $k = 1, \ldots, n+1$, $l = 1, \ldots, n+1$. *As before, there are* $N = n^2$ *panels that can be counted with single counters* $I = nk + l$, $J = ni + j$. *The only difference is that, in the present case,* $R_2 \geq R$ *since we are interested in flow outside the impermeable sphere. The panels on the latter have control points lying at* $\mathbf{x}_{c_{i,j}}$ *and surface areas* $\Delta s_{i,j}$, *for* $i = 1, \ldots, n, j = 1, \ldots, n$. *Consequently, Eq. (2.86) can be approximated by*

$$\phi(\mathbf{x}_{0_I}) \approx \frac{U_\infty}{4\pi R} \sum_{J=1}^{N} x_{c_J} \left(\frac{1}{r(\mathbf{x}_{c_J}, \mathbf{x}_{0_I})} - \frac{\mathbf{x}_{c_J} \cdot (\mathbf{x}_{c_J} - \mathbf{x}_{0_I})}{2r(\mathbf{x}_{c_J}, \mathbf{x}_{0_I})^3} \right) \Delta s_J$$

for $R_2 > R$, *having again made the assumption that the integrands of Eq. (2.86) can be evaluated at the panel control points and are constant over each panel. If* $R_2 = R$, *then* $\mathbf{x}_{0_I} = \mathbf{x}_{s_I}$, *while* $E(\mathbf{x}_{0_I}) = 1/2$ *so that the approximation to Eq. (2.86) becomes*

$$\phi(\mathbf{x}_{s_I}) \approx \frac{U_\infty}{2\pi R} \sum_{J=1}^{N} x_{c_J} \left(\frac{1}{r(\mathbf{x}_{c_J}, \mathbf{x}_{s_I})} - \frac{\mathbf{x}_{c_J} \cdot (\mathbf{x}_{c_J} - \mathbf{x}_{s_I})}{2r(\mathbf{x}_{c_J}, \mathbf{x}_{s_I})^3} \right) \Delta s_J \tag{2.87}$$

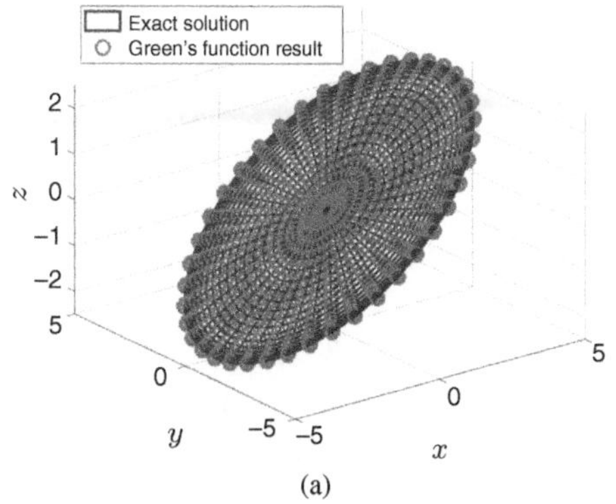

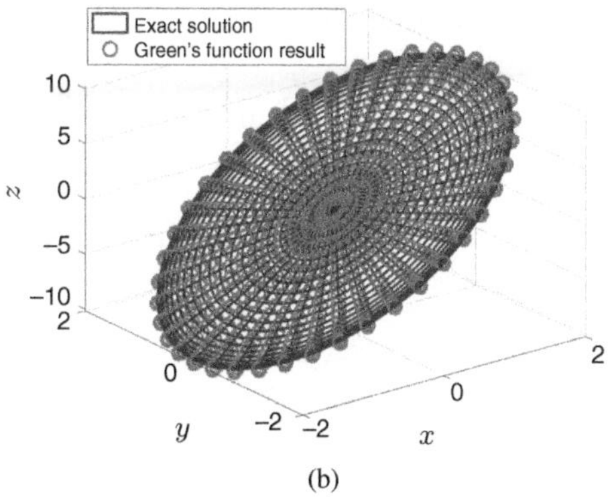

Figure 2.13 Potential distribution in flowfield around an impermeable sphere of radius R immersed in a free stream. (a) $R_2 = 2R$ and (b) $R_2 = R$.

Figure 2.13 plots the potential estimated using the numerical approximation of Green's third identity, $\phi(\mathbf{x}_{0_{i,j}})$, for $n = 40$ on spheres with radius $R_2 = 2R$ and $R_2 = R$. The comparison with the exact result of Eq. (2.83) is excellent in both cases. Finally, we set $R = R/2$ so that all points $\mathbf{x}_{0_I}$ lie inside S. We find that

$$0 = \frac{U_\infty}{4\pi R}\sum_{J=1}^{N} x_{c_J}\left(\frac{1}{r(\mathbf{x}_{c_J}, \mathbf{x}_{0_I})} - \frac{\mathbf{x}_{c_J}\cdot(\mathbf{x}_{c_J} - \mathbf{x}_{0_I})}{2r(\mathbf{x}_{c_J}, \mathbf{x}_{0_I})^3}\right)\Delta s_J$$

to within numerical error for all I. This result is compatible with the fact that $E(\mathbf{x}_0) = 0$ inside S. This example is solved by Matlab code `doublet_sphere_Green.m`.

The example has served to verify Green's third identity on a flow for which we have an exact solution and, therefore, we know already the potential on the surface. In practical aerodynamic applications, we do not have an exact solution and therefore we do not know the potential on the surface. Consider a general closed surface discretised into N panels with vertices $\mathbf{x}_{s_J}$ and control points $\mathbf{x}_{c_J}$. In Eq. (2.87), we applied Green's identity in order to calculate the potential on the panel vertices $\mathbf{x}_{s_I}$ from the known values of the potential on the control points $\mathbf{x}_{c_J}$. It would have made little sense to try to calculate the potential on the control points from the known values of the potential on the control points. However, such a calculation would make a lot of sense if we did not know the values of the potential on the control points in the first place.

Let us discretise Green's third identity (2.82) for the case where we evaluate the potential on the panel control points from unknown values of the potential on the control points,

$$\begin{aligned}\phi(\mathbf{x}_{c_I}) = &-\frac{1}{2\pi}\sum_{J=1}^{N}\mathbf{n}(\mathbf{x}_{c_J})\cdot\nabla\phi(\mathbf{x}_{c_J})\frac{1}{r(\mathbf{x}_{c_J}, \mathbf{x}_{c_I})}\Delta s_J \\ &+\frac{1}{2\pi}\sum_{J=1}^{N}\phi(\mathbf{x}_{c_J})\mathbf{n}(\mathbf{x}_{c_J})\cdot\nabla\left(\frac{1}{r(\mathbf{x}_{c_J}, \mathbf{x}_{c_I})}\right)\Delta s_J\end{aligned}$$

Recall that, in developing Green's third identity, we excluded from the integrals the point $\mathbf{x}_0$ when it lies on the surface. In Eq. (2.88), $\mathbf{x}_0 = \mathbf{x}_{c_{k,l}}$, which means that we must exclude the case $J = I$ from the summations, that is

$$\phi(\mathbf{x}_{c_I}) = -\frac{1}{2\pi}\sum_{J=1,J\neq I}^{N} \mathbf{n}(\mathbf{x}_{c_J}) \cdot \nabla\phi(\mathbf{x}_{c_J})\frac{1}{r(\mathbf{x}_{c_J},\mathbf{x}_{c_I})}\Delta s_J + \frac{1}{2\pi}\sum_{J=1,J\neq I}^{N} \phi(\mathbf{x}_{c_J})\mathbf{n}(\mathbf{x}_{c_J}) \cdot \nabla\left(\frac{1}{r(\mathbf{x}_{c_J},\mathbf{x}_{c_I})}\right)\Delta s_J \tag{2.88}$$

The terms $1/r(\mathbf{x}_{c_J},\mathbf{x}_{c_I})$ and $\mathbf{n}(\mathbf{x}_{c_J}) \cdot \nabla\left(1/r(\mathbf{x}_{c_J},\mathbf{x}_{c_I})\right)$ depend uniquely on the geometry of the surface S and are known for any value of I, J. The term $\mathbf{n}(\mathbf{x}_{c_J}) \cdot \nabla\phi(\mathbf{x}_{c_J})$ is not known directly but can be calculated from the impermeability boundary condition (2.43). For steady flow, this condition is given in Eq. (2.56) and can be re-written as

$$\mathbf{n}(\mathbf{x}_s) \cdot \nabla\phi(\mathbf{x}_s) = -\mathbf{n}(\mathbf{x}_s) \cdot \mathbf{Q}_\infty \tag{2.89}$$

Applying this expression to the panel control points yields $\mathbf{n}(\mathbf{x}_{c_J}) \cdot \nabla\phi(\mathbf{x}_{c_J}) = -\mathbf{n}(\mathbf{x}_{c_J}) \cdot \mathbf{Q}_\infty$. Consequently, Eq. (2.88) is a single equation with N unknowns, the values of $\phi(\mathbf{x}_{c_J})$. Writing the same equation for all $I = 1, \ldots, N$ results in a system of N linear algebraic equations with N unknowns that is straightforward to solve.

Example 2.5 *Use the discrete version of Green's third identity in order to calculate the potential on the control points of an impermeable sphere immersed in a free stream.*

Here, we are essentially repeating Example 2.4 but without making use of the exact solution. We set up a single sphere with radius R, discretised into $N = n^2$ panels with control points lying at $\mathbf{x}_{c_{i,j}}$, for $i = 1, \ldots, n$ and $j = 1, \ldots, n$. The vertices of the panels are denoted by $\mathbf{x}_{s_J}$ for $J = ni + j$ but are only used in order to calculate the areas of the panels, Δs_J, from Eq. (2.73). The unit vectors normal to the surface at the control points are given by

$$\mathbf{n}(\mathbf{x}_{c_J}) = \frac{1}{R}\mathbf{x}_{c_J}$$

The free stream velocity is $\mathbf{Q}_\infty = (U_\infty, 0, 0)$, so that using Eq. (2.89)

$$\mathbf{n}(\mathbf{x}_{c_J}) \cdot \nabla\phi(\mathbf{x}_{c_J}) = -\frac{U_\infty}{R}x_{c_J}$$

Note that the expression above is identical to Eq. (2.85) that was derived knowing the exact solution for this flow. Here we have assumed no such knowledge, and we have just applied the impermeability boundary condition. Furthermore, using Eq. (2.71),

$$\mathbf{n}(\mathbf{x}_{c_J}) \cdot \nabla\left(\frac{1}{r(\mathbf{x}_{c_J},\mathbf{x}_{c_I})}\right) = -\frac{\mathbf{x}_{c_J} \cdot (\mathbf{x}_{c_J} - \mathbf{x}_{c_I})}{Rr(\mathbf{x}_{c_J},\mathbf{x}_{c_I})^3}$$

We are now ready to apply Eq. (2.88),

$$\phi(\mathbf{x}_{c_I}) = -\frac{1}{2\pi}\sum_{J=1,J\neq I}^{N} \mathbf{n}(\mathbf{x}_{c_J}) \cdot \nabla\phi(\mathbf{x}_{c_J})\frac{1}{r(\mathbf{x}_{c_J},\mathbf{x}_{c_I})}\Delta s_J + \frac{1}{2\pi}\sum_{J=1,J\neq I}^{N} \phi(\mathbf{x}_{c_J})\mathbf{n}(\mathbf{x}_{c_J}) \cdot \nabla\left(\frac{1}{r(\mathbf{x}_{c_J},\mathbf{x}_{c_I})}\right)\Delta s_J \tag{2.90}$$

Substituting for $\mathbf{n}(\mathbf{x}_{c_J}) \cdot \nabla\phi(\mathbf{x}_{c_J})$ *and* $\mathbf{n}(\mathbf{x}_{c_J}) \cdot \nabla(1/r(\mathbf{x}_{c_J}, \mathbf{x}_{c_I}))$ *from the expressions above,*

$$\phi(\mathbf{x}_{c_I}) = \frac{U_\infty}{2\pi R} \sum_{J=1,J\neq I}^{N} x_{c_I} \frac{1}{r(\mathbf{x}_{c_J}, \mathbf{x}_{c_I})} \Delta s_J - \frac{U_\infty}{2\pi R} \sum_{J=1,J\neq I}^{N} \phi(\mathbf{x}_{c_J}) \frac{\mathbf{x}_{c_J} \cdot (\mathbf{x}_{c_J} - \mathbf{x}_{c_I})}{r(\mathbf{x}_{c_J}, \mathbf{x}_{c_I})^3} \Delta s_J \tag{2.91}$$

This latest equation is very similar to Eq. (2.87), except that we are evaluating the potential at control points instead of panel vertices and we have not used the exact solution for $\phi(\mathbf{x}_{c_J})$ *to derive it. Writing out the equation for all* $I = 1, \dots, N$ *leads to the matrix equation*

$$\boldsymbol{\phi} = \boldsymbol{A}\boldsymbol{x}_c + \boldsymbol{B}\boldsymbol{\phi} \tag{2.92}$$

where $\boldsymbol{\phi}$ *and* $\boldsymbol{x}_c$[2] *are* $N \times 1$ *column vectors defined as* $\boldsymbol{\phi} = (\phi(\mathbf{x}_{c_1}), \dots, \phi(\mathbf{x}_{c_N}))^T$, $\boldsymbol{x}_c = (x_{c_1}, \dots, x_{c_N})^T$, *while* $\boldsymbol{A}, \boldsymbol{B}$ *are* $N \times N$ *matrices with elements*

$$A_{I,J} = \frac{U_\infty}{2\pi R} \begin{cases} \frac{1}{r(\mathbf{x}_{c_J}, \mathbf{x}_{c_I})} \Delta s_J & \text{if } I \neq J \\ 0 & \text{if } I = J \end{cases}$$

$$B_{I,J} = -\frac{U_\infty}{2\pi R} \begin{cases} \frac{\mathbf{x}_{c_J} \cdot (\mathbf{x}_{c_J} - \mathbf{x}_{c_I})}{r(\mathbf{x}_{c_J}, \mathbf{x}_{c_I})^3} \Delta s_J & \text{if } I \neq J \\ 0 & \text{if } I = J \end{cases}$$

Equation (2.92) can be rewritten as $(\boldsymbol{I} - \boldsymbol{B})\boldsymbol{\phi} = \boldsymbol{A}\boldsymbol{x}_c$ *and solved to obtain*

$$\boldsymbol{\phi} = (\boldsymbol{I} - \boldsymbol{B})^{-1}\boldsymbol{A}\boldsymbol{x}_c$$

Consequently, the values of the potential on all the control points have been obtained without using any a priori knowledge. Nevertheless, the control points do not actually lie on the surface of the sphere. We can now re-apply Green's third identity in order to calculate the potential on the panel vertices (which do lie on the surface) from the potential on the control points that we have just evaluated,

$$\phi(\mathbf{x}_{s_I}) = \frac{U_\infty}{2\pi R} \sum_{J=1}^{N} x_{c_I} \frac{1}{r(\mathbf{x}_{c_J}, \mathbf{x}_{s_I})} \Delta s_J - \frac{U_\infty}{2\pi R} \sum_{J=1}^{N} \phi(\mathbf{x}_{c_J}) \frac{\mathbf{x}_{c_J} \cdot (\mathbf{x}_{c_J} - \mathbf{x}_{s_I})}{r(\mathbf{x}_{c_J}, \mathbf{x}_{s_I})^3} \Delta s_J \tag{2.93}$$

for $I = 1, \dots, (n+1)^2$, *since there are* $(n+1)^2$ *panel vertices; J still takes values between 1 and* $N = n^2$ *since there are* n^2 *panels. Note that we do not need to exclude* $J = I$ *from the summations as the panel vertices never coincide with the panel control points.*

We have now calculated the potential on the surface of the sphere, and we can compare it to the values predicted by the exact solution of Eq. (2.84). Figure 2.14 plots this comparison for a sphere with $R = 2$ *and* $n = 40$; *it can be seen that the agreement between these two sets of results is excellent. It is important to understand exactly what is stipulated by Eq. (2.91): the potential at the control points (which lie inside the surface S) is calculated in terms of the potential at the same control points. There are three conclusions to be drawn*

- *The true surface of the impermeable sphere has been replaced by a collection of flat panels that approximate S with increasing accuracy as n increases.*

---

2 In this book, we will use $\mathbf{x} = (x, y, z)$, $\mathbf{u} = (u, v, w)$ to denote vectors of position and velocity in three dimensions, while $\boldsymbol{x} = (x_1, \dots, x_N)^T$, $\boldsymbol{y} = (y_1, \dots, y_N)^T$, $\boldsymbol{z} = (z_1, \dots, z_N)^T$ and $\boldsymbol{u} = (u_1, \dots, u_N)^T$, $\boldsymbol{v} = (v_1, \dots, v_N)^T$, $\boldsymbol{w} = (w_1, \dots, w_N)^T$ will be used to denote column vectors that contain all the values of x, y, z, u, v, w on the panels of a body or wake.

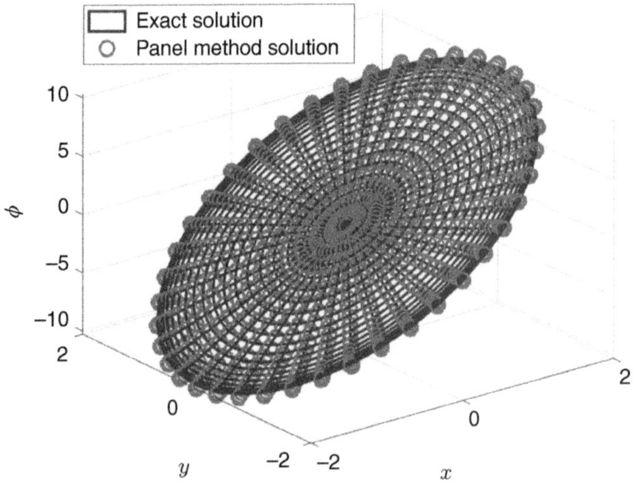

Figure 2.14 Comparison of numerical and exact solutions for the potential around an impermeable sphere in a free stream.

- *Both the panel control points and the panel vertices lie on this new approximate surface; therefore, $E = 1/2$ on both of these types of points. In reality, the control points lie inside S, while the panel vertices lie on it.*
- *As we are using an approximate surface and an approximate version of Green's identity, E does not obey the definition of Eq. (2.69); it varies gradually between 0, 0.5 and 1.*

The gradual nature of E can be demonstrated by calculating the potential on points off the surface of the sphere and comparing with the prediction of the exact solution. As an example, we can calculate the potential on a sphere with radius $R_2 \neq R$. If we denote the panel vertices on this new sphere by $\mathbf{x}_{s_I}(R_2)$, then Eq. (2.93) becomes

$$\phi(\mathbf{x}_{s_I}(R_2)) = \frac{U_\infty}{4\pi R}\sum_{J=1}^{N} x_{c_J}\frac{1}{r(\mathbf{x}_{c_J}, \mathbf{x}_{s_I}(R_2))}\Delta s_J - \frac{U_\infty}{4\pi R}\sum_{J=1}^{N}\phi(\mathbf{x}_{c_J})\frac{\mathbf{x}_{c_J}\cdot(\mathbf{x}_{c_J} - \mathbf{x}_{s_I}(R_2))}{r(\mathbf{x}_{c_J}, \mathbf{x}_{s_I}(R_2))^3}\Delta s_J$$

having enforced $E(\mathbf{x}_{s_I}(R_2)) = 1$. Figure 2.15 plots the variation of the potential with x for $y = 0$ and for three different values of R_2. For all figures, the impermeable sphere S has radius $R = 2$ and the grid size is defined by $n = 40$. Figures 2.15a and 2.15b plot the potential inside S; at $R_2 = 0.9R$ this potential is nearly zero, as expected from Green's theory. However, at $R_2 = 0.98R$, the potential is not only non-zero but also it is not even linear; its maximum value is slightly less than half the maximum value of the exact solution. Similar behaviour can be observed in Figure 2.15c, which plots the potential at $R_2 = 1.02R$; here, the maximum value of the potential is slightly more than half the exact maximum. Finally, at $R_2 = 1.1R$ the numerical and exact solutions are very similar. If we take the ratio of the maximum numerical and exact potential as an estimate of E, it can be stated that E increases gradually from nearly zero to 1 as the surface of the sphere is crossed.

A clearer picture of the behaviour of E can be observed in Figure 2.16, which plots the ratios of the maximum potential values against $(R_2 - R)/R$ for different values of the grid size. As n

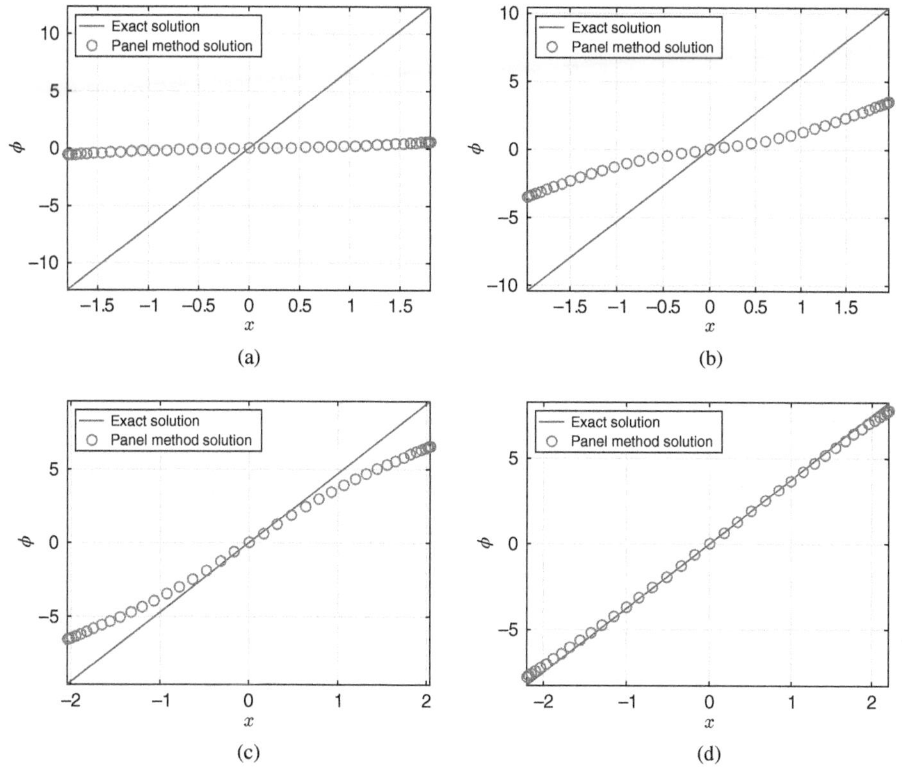

Figure 2.15 Potential variation with x for $y = 0$ on spheres of different radii around impermeable sphere of radius R. (a) $R_2 = 0.9R$, (b) $R_2 = 0.98R$, (c) $R_2 = 1.02R$, and (d) $R_2 = 1.1R$.

increases, the behaviour of E approaches the discontinuous definition of Eq. (2.69), but the convergence is very slow. Figure 2.16 is very important because it shows that the numerical solution is trustworthy exactly on the surface or sufficiently far from it. Very close to the surface, we can only trust the numerical predictions if the number of panels is enormous. This example is solved by Matlab code `panel_sphere.m`

Example 2.5 has shown that we can use Green's third identity in order to obtain a numerical solution for the potential of the flow around an impermeable sphere. Similar procedures can be used in order to solve for the flow around any closed surface but several improvements can be made. Recall that the discrete form of Green's third identity we used is

$$\begin{aligned} E(\mathbf{x}_0)\phi(\mathbf{x}_0) &= \frac{1}{4\pi}\int_S \mathbf{n}(\mathbf{x}_s)\cdot\left(-\frac{1}{r(\mathbf{x}_s,\mathbf{x}_0)}\nabla\phi(\mathbf{x}_s) + \phi(\mathbf{x}_s)\nabla\left(\frac{1}{r(\mathbf{x}_s,\mathbf{x}_0)}\right)\right)\mathrm{d}S \\ &\approx -\frac{1}{4\pi}\sum_{J=1}^{N}\mathbf{n}(\mathbf{x}_{c_J})\cdot\nabla\phi(\mathbf{x}_{c_J})\frac{1}{r(\mathbf{x}_{c_J},\mathbf{x}_0)}\int_{\Delta s_J}\mathrm{d}S \\ &\quad + \frac{1}{4\pi}\sum_{J=1}^{N}\phi(\mathbf{x}_{c_J})\mathbf{n}(\mathbf{x}_{c_J})\cdot\nabla\left(\frac{1}{r(\mathbf{x}_{c_J},\mathbf{x}_0)}\right)\int_{\Delta s_J}\mathrm{d}S \end{aligned}$$

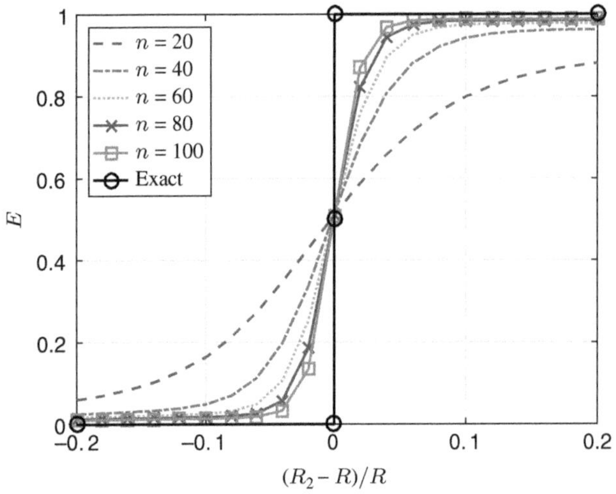

Figure 2.16 Variation of E close to the surface for increasing grid size.

A more accurate approximation can be obtained if we assume that only $\mathbf{n}(\mathbf{x}_s) \cdot \nabla\phi(\mathbf{x}_s)$ and $\phi(\mathbf{x}_s)$ are constant on each panel and equal to their values at the control points. Then, the discrete Green's identity becomes

$$E(\mathbf{x}_0)\phi(\mathbf{x}_0) \approx -\frac{1}{4\pi}\sum_{J=1}^{N}\mathbf{n}(\mathbf{x}_{c_J}) \cdot \nabla\phi(\mathbf{x}_{c_J})\int_{\Delta s_J}\frac{1}{r(\mathbf{x}_s, \mathbf{x}_0)}\mathrm{d}S + \frac{1}{4\pi}\sum_{J=1}^{N}\phi(\mathbf{x}_{c_J})\int_{\Delta s_J}\mathbf{n}(\mathbf{x}_s) \cdot \nabla\left(\frac{1}{r(\mathbf{x}_s, \mathbf{x}_0)}\right)\mathrm{d}S \tag{2.94}$$

Several authors have developed analytical expressions for the source and doublet integrals in this latest expression. One such development is demonstrated in Appendices A.10 and A.11 for flat quadrilateral panels but analytical or quasi-analytical expressions exist even for curved panels. Even higher-order approximations can be obtained if we keep $\mathbf{n}(\mathbf{x}_s) \cdot \nabla\phi(\mathbf{x}_s)$ and $\phi(\mathbf{x}_s)$ inside the integrals but make assumptions about their variation over the panel's surface, e.g. linear or quadratic. In this book, a linear model will be implemented for 2D flows but not for 3D flows, for which the singularity strength will stay constant throughout the panel.

In our formulation of Green's third identity, we stated that we place a sink at a point $\mathbf{x}_0$ and calculate its influence on all points $\mathbf{x}_s$ on the surface S; the sink acts as a sensor that measures the potential at point $\mathbf{x}_0$ by means of the divergence theorem. It is entirely equivalent to say that we place a continuous distribution of sources and doublets at all points $\mathbf{x}_s$ on the surface S and calculate their influence on a general point $\mathbf{x}$, where there is no singularity unless $\mathbf{x}$ lies on the surface. As a consequence, $\mathbf{n}(\mathbf{x}_s) \cdot \nabla\phi(\mathbf{x}_s)$ is the strength of the source distribution and $\phi(\mathbf{x}_s)$ is the strength of the doublet distribution. This latter description of Green's identity is much more common in the literature, but the difference between the two formulations is only conceptual. From this point onwards, we will also adopt the principle of placing a source and doublet distribution on the surface of the body.

Consequently, Green's identity (2.82) is reformulated as

$$E(\mathbf{x})\phi(\mathbf{x}) = -\frac{1}{4\pi}\int_S \mathbf{n}(\mathbf{x}_s)\cdot\nabla_s\phi(\mathbf{x}_s)\frac{1}{r(\mathbf{x},\mathbf{x}_s)}\mathrm{d}S + \frac{1}{4\pi}\int_S \phi(\mathbf{x}_s)\mathbf{n}(\mathbf{x}_s)\cdot\nabla_s\left(\frac{1}{r(\mathbf{x},\mathbf{x}_s)}\right)\mathrm{d}S \tag{2.95}$$

where $\mathbf{x}$ is any point outside, on or inside S and

$$\nabla_s = \left(\frac{\partial}{\partial x_s}, \frac{\partial}{\partial y_s}, \frac{\partial}{\partial z_s}\right) \tag{2.96}$$

$$r(\mathbf{x},\mathbf{x}_s) = \sqrt{(x-x_s)^2+(y-y_s)^2+(z-z_s)^2} \tag{2.97}$$

Usually, the strengths of the source and doublet distributions are denoted by

$$\sigma(\mathbf{x}_s) = \mathbf{n}(\mathbf{x}_s)\cdot\nabla_s\phi(\mathbf{x}_s) \tag{2.98}$$

$$\mu(\mathbf{x}_s) = \phi(\mathbf{x}_s) \tag{2.99}$$

where $\sigma(\mathbf{x}_s)$ is the source strength and $\mu(\mathbf{x}_s)$ the doublet strength. Finally, it should be noted that many authors, such as Epton and Magnus (1990), Katz and Plotkin (2001), Maskew (1982), apply Green's identity (2.95) to both the internal and external flows and then subtract the two. In this case, the source and doublet strengths are defined as

$$\sigma(\mathbf{x}_s) = \mathbf{n}(\mathbf{x}_s)\cdot\nabla_s\phi(\mathbf{x}_s) - \mathbf{n}(\mathbf{x}_s)\cdot\nabla_s\phi_i(\mathbf{x}_s) \tag{2.100}$$

$$\mu(\mathbf{x}_s) = \phi(\mathbf{x}_s) - \phi_i(\mathbf{x}_s) \tag{2.101}$$

where $\phi(\mathbf{x}_s)$ is the potential of the outer flow at point $\mathbf{x}_s$ and $\phi_i(\mathbf{x}_s)$ is the potential of the inner flow at the same point, as seen in Figure 2.17. Nevertheless, the inner potential is usually set to $\phi_i(\mathbf{x}_s) = 0$ so that the source and doublet strength definitions of Eqs. (2.98), (2.99) and (2.100), (2.101) become identical.

The 3D source and doublet panel solutions in this book are based on Eqs. (2.95)–(2.99). In the case of unsteady flow, Eq. (2.95) does not change form, but it becomes dependent on

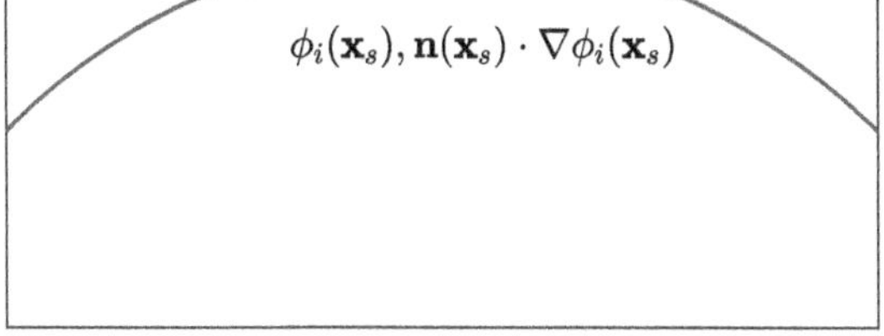

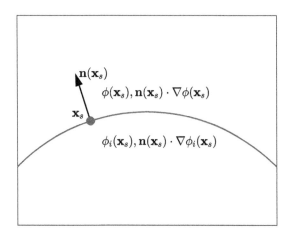

Figure 2.17 Definition of potential and normal velocity either side of the surface of a closed body.

time, such that

$$E(\mathbf{x},t)\phi(\mathbf{x},t) = -\frac{1}{4\pi}\int_{S(t)} \mathbf{n}(\mathbf{x}_s,t)\cdot\nabla_s\phi(\mathbf{x}_s,t)\frac{1}{r(\mathbf{x},\mathbf{x}_s,t)}\mathrm{d}S + \frac{1}{4\pi}\int_{S(t)} \phi(\mathbf{x}_s,t)\mathbf{n}(\mathbf{x}_s,t)\cdot\nabla_s\left(\frac{1}{r(\mathbf{x},\mathbf{x}_s,t)}\right)\mathrm{d}S \tag{2.102}$$

where now the surface $S(t)$ is a function of time and so are the points $\mathbf{x}_s(t)$, the normal vectors $\mathbf{n}(\mathbf{x}_s,t)$ and the distances $r(\mathbf{x}_s,\mathbf{x}_0,t)$. The point $\mathbf{x}$ may be defined as constant in time but, as the surface $S(t)$ moves and deforms it may find itself outside, on or inside $S(t)$ at different time instances so that $E(\mathbf{x},t)$ also depends on t. Now, Eq. (2.102) must be solved at each time instance of interest for the time-varying source and doublet panel strengths $\sigma(\mathbf{x}_s,t)$, $\mu(\mathbf{x}_s,t)$. The perturbation flow velocities at point $\mathbf{x}$ can be obtained by differentiating equation (2.102) with respect to x, y and z. For example, differentiating with respect to x leads to

$$\frac{\partial}{\partial x}\left(E(\mathbf{x},t)\phi(\mathbf{x},t)\right) = -\frac{1}{4\pi}\int_{S(t)} \mathbf{n}(\mathbf{x}_s,t)\cdot\nabla_s\phi(\mathbf{x}_s,t)\frac{\partial}{\partial x}\left(\frac{1}{r(\mathbf{x},\mathbf{x}_s,t)}\right)\mathrm{d}S + \frac{1}{4\pi}\int_{S(t)} \phi(\mathbf{x}_s,t)\frac{\partial}{\partial x}\left(\mathbf{n}(\mathbf{x}_s,t)\cdot\nabla_s\left(\frac{1}{r(\mathbf{x},\mathbf{x}_s,t)}\right)\right)\mathrm{d}S$$

Inside volume V, $E(\mathbf{x},t) = 1$ everywhere and, hence, the perturbation velocities become

$$\begin{aligned}
\phi_x(\mathbf{x},t) &= -\frac{1}{4\pi}\int_{S(t)} \mathbf{n}(\mathbf{x}_s,t)\cdot\nabla_s\phi(\mathbf{x}_s,t)\frac{\partial}{\partial x}\left(\frac{1}{r(\mathbf{x},\mathbf{x}_s,t)}\right)\mathrm{d}S \\
&\quad + \frac{1}{4\pi}\int_{S(t)} \phi(\mathbf{x}_s,t)\frac{\partial}{\partial x}\left(\mathbf{n}(\mathbf{x}_s,t)\cdot\nabla_s\left(\frac{1}{r(\mathbf{x},\mathbf{x}_s,t)}\right)\right)\mathrm{d}S \\
\phi_y(\mathbf{x},t) &= -\frac{1}{4\pi}\int_{S(t)} \mathbf{n}(\mathbf{x}_s,t)\cdot\nabla_s\phi(\mathbf{x}_s,t)\frac{\partial}{\partial y}\left(\frac{1}{r(\mathbf{x},\mathbf{x}_s,t)}\right)\mathrm{d}S \\
&\quad + \frac{1}{4\pi}\int_{S(t)} \phi(\mathbf{x}_s,t)\frac{\partial}{\partial y}\left(\mathbf{n}(\mathbf{x}_s,t)\cdot\nabla_s\left(\frac{1}{r(\mathbf{x},\mathbf{x}_s,t)}\right)\right)\mathrm{d}S \\
\phi_z(\mathbf{x},t) &= -\frac{1}{4\pi}\int_{S(t)} \mathbf{n}(\mathbf{x}_s,t)\cdot\nabla_s\phi(\mathbf{x}_s,t)\frac{\partial}{\partial z}\left(\frac{1}{r(\mathbf{x},\mathbf{x}_s,t)}\right)\mathrm{d}S \\
&\quad + \frac{1}{4\pi}\int_{S(t)} \phi(\mathbf{x}_s,t)\frac{\partial}{\partial z}\left(\mathbf{n}(\mathbf{x}_s,t)\cdot\nabla_s\left(\frac{1}{r(\mathbf{x},\mathbf{x}_s,t)}\right)\right)\mathrm{d}S
\end{aligned} \tag{2.103}$$

On the surface S, $E(\mathbf{x},t)$ is discontinuous so that $\partial\left(E(\mathbf{x},t)\phi(\mathbf{x},t)\right)/\partial x|_{\mathbf{x}_s}$ is not differentiable. The usual solution to this problem is to calculate the derivatives of $\phi(x_s)$ in directions parallel to the surface such that $E(\mathbf{x}_s,t)$ remains equal to $1/2$ and, hence, continuous and differentiable. The practical calculation of the perturbation velocities will be addressed in Section 5.6.

2.3.2 Solutions in Two Dimensions

A large class of interesting aerodynamic problems are defined in two dimensions. This simplification is usually applied to bodies that have very long spans and whose cross-sectional geometry does not change, such as helicopter blades, bridges, towers, electricity cables, and others. The Laplace and Bernoulli equations are still valid in 2D such that Eqs. (2.22) and (2.29) become

$$\frac{\partial^2 \Phi}{\partial x^2} + \frac{\partial^2 \Phi}{\partial y^2} = 0 \tag{2.104}$$

$$\rho \left(\frac{1}{2} \left(u^2 + v^2 \right) + \frac{\partial \Phi}{\partial t} \right) + p = \frac{1}{2} \rho Q_\infty^2 + p_\infty \tag{2.105}$$

where the potential is defined as

$$u = \frac{\partial \Phi}{\partial x}, \; v = \frac{\partial \Phi}{\partial y}$$

and the 3D irrotationality conditions (2.13) reduce to

$$\frac{\partial v}{\partial x} - \frac{\partial u}{\partial y} = 0$$

As there are only two dimensions, it is customary to express points in 2D space using the complex variable $z = x + iy$; similarly, we can define the complex potential $F(z) = \phi + i\psi$, where ϕ is the real potential and ψ is the stream function, defined as

$$u = \frac{\partial \psi}{\partial y}, \; v = -\frac{\partial \psi}{\partial x}$$

Substituting this definition into Eq. (2.104) shows that the stream function automatically satisfies Laplace's equation. The most interesting property of the stream function is the fact that it is constant along a flow streamline. The total differential of ψ is

$$d\psi = \frac{\partial \psi}{\partial x} dx + \frac{\partial \psi}{\partial y} dy = -v dx + u dy$$

On a 2D streamline, Eq. (2.52) reduces to

$$\frac{u}{dx} = \frac{v}{dy} \tag{2.106}$$

so that

$$d\psi = -v dx + u dy = -\frac{u}{dx} dy dx + u dy = -u dy + u dy = 0$$

Consequently, the stream function is a practical means of calculating flow streamlines in 2D. Finally, the complex velocity $V(z)$ is defined as

$$V(z) = u(x, y) - iv(x, y)$$

Laplace's equation has fundamental solutions also in 2D; the free stream for example is given by $\Phi(x, y) = U_\infty x + V_\infty y$. There are also 2D source, sink and doublet solutions, but their forms are different to those of Eqs. (2.49) and (2.55). The 2D point source is given by

$$\phi(\mathbf{x}, \mathbf{x}_0) = \ln \sqrt{(x - x_0)^2 + (y - y_0)^2} = \ln r(\mathbf{x}, \mathbf{x}_0) \tag{2.107}$$

where $\mathbf{x} = (x, y)$, $\mathbf{x}_0 = (x_0, y_0)$ and $r(\mathbf{x}, \mathbf{x}_0) = \sqrt{(x - x_0)^2 + (y - y_0)^2}$. The complex potential induced by the source is given by

$$\phi(z, z_0) = \ln(z - z_0) \tag{2.108}$$

The potential, stream function and velocity distributions induced by the 2D point source are described in more detail in Appendix A.1.

As in the 3D case, the 2D point doublet is the derivative of the 2D point source in any general direction **n**. In this book, we will not make use of the 2D doublet; therefore, we will not discuss it any further. We will use the 2D source and the 2D point vortex, which is introduced in Section 2.7.1. The solution procedure is similar to that for 3D flows; a continuous distribution of singularities is placed on the contour of the 2D body and its strength is chosen such that the impermeability boundary condition is satisfied.

2.4 From Navier–Stokes to Unsteady Compressible Potential Flow

Recall that after applying the inviscid and irrotational flow assumptions we obtained the Euler equations (2.24)–(2.26)

$$\frac{\partial}{\partial t}\left(\frac{\partial \Phi}{\partial x}\right)\mathrm{d}x + \frac{1}{2}\frac{\partial}{\partial x}\left(u^2+v^2+w^2\right)\mathrm{d}x = -\frac{1}{\rho}\frac{\partial p}{\partial x}\mathrm{d}x$$

$$\frac{\partial}{\partial t}\left(\frac{\partial \Phi}{\partial y}\right)\mathrm{d}y + \frac{1}{2}\frac{\partial}{\partial y}\left(u^2+v^2+w^2\right)\mathrm{d}y = -\frac{1}{\rho}\frac{\partial p}{\partial y}\mathrm{d}y$$

$$\frac{\partial}{\partial t}\left(\frac{\partial \Phi}{\partial z}\right)\mathrm{d}z + \frac{1}{2}\frac{\partial}{\partial z}\left(u^2+v^2+w^2\right)\mathrm{d}z = -\frac{1}{\rho}\frac{\partial p}{\partial z}\mathrm{d}z$$

If we do not assume that the flow is incompressible, the expressions above cannot be integrated in space without knowing the form of $\rho = \rho(x, y, z, t)$. We can still exchange the order of the derivatives in the first terms and then add up the three equations to obtain

$$\frac{\partial}{\partial x}\left(\frac{\partial \Phi}{\partial t}\right)\mathrm{d}x + \frac{\partial}{\partial y}\left(\frac{\partial \Phi}{\partial t}\right)\mathrm{d}y + \frac{\partial}{\partial z}\left(\frac{\partial \Phi}{\partial t}\right)\mathrm{d}z + \frac{1}{2}\frac{\partial Q^2}{\partial x}\mathrm{d}x + \frac{1}{2}\frac{\partial Q^2}{\partial y}\mathrm{d}y + \frac{1}{2}\frac{\partial Q^2}{\partial z}\mathrm{d}z$$
$$= -\frac{1}{\rho}\left(\frac{\partial p}{\partial x}\mathrm{d}x + \frac{\partial p}{\partial y}\mathrm{d}y + \frac{\partial p}{\partial z}\mathrm{d}z\right) \tag{2.109}$$

where

$$Q^2 = u^2 + v^2 + w^2 = \left(\frac{\partial \Phi}{\partial x}\right)^2 + \left(\frac{\partial \Phi}{\partial y}\right)^2 + \left(\frac{\partial \Phi}{\partial z}\right)^2 \tag{2.110}$$

Equation (2.109) contains three perfect differentials, such that it can be written in more compact form as

$$\mathrm{d}\left(\frac{\partial \Phi}{\partial t} + \frac{1}{2}Q^2\right) = -\frac{\mathrm{d}p}{\rho} \tag{2.111}$$

2.4.1 The Compressible Bernoulli Equation

We would like to group the left and right-hand sides of Eq. (2.111) into a single differential so that we can integrate it. However, this is not possible at the moment because ρ is not differentiated. In order to achieve this grouping, we must consider the thermodynamics of the flow. For reversible processes, the first and second laws of thermodynamics give

$$T\mathrm{d}S + \frac{\mathrm{d}p}{\rho} = \mathrm{d}h \tag{2.112}$$

where T is the temperature, S is the entropy and h is the enthalpy, defined as

$$h = e + \frac{p}{\rho} \tag{2.113}$$

e being the internal energy of the system. If we now assume that the process is also isentropic, then $\mathrm{d}S = 0$ and Eq. (2.112) becomes

$$\frac{\mathrm{d}p}{\rho} = \mathrm{d}h \tag{2.114}$$

Substituting this latest result into Eq. (2.111) leads to

$$\mathrm{d}\left(\frac{\partial \Phi}{\partial t} + \frac{1}{2}Q^2 + h\right) = 0 \tag{2.115}$$

Consequently, for compressible, inviscid, irrotational and isentropic flow, the Euler equations reduce to

$$\frac{\partial \Phi}{\partial t} + \frac{1}{2}Q^2 + h = \text{constant} \tag{2.116}$$

which is the unsteady compressible Bernoulli equation. The constant can be evaluated far from the body creating the disturbance, where $\partial\Phi/\partial t = 0$, $Q = Q_\infty$, $h = h_\infty$, $p = p_\infty$, $\rho = \rho_\infty$ and $T = T_\infty$. Equation (2.116) becomes

$$\frac{\partial \Phi}{\partial t} + \frac{1}{2}Q^2 + h = \frac{1}{2}Q_\infty^2 + h_\infty \tag{2.117}$$

This latest form of the Bernoulli equation is written in terms of h, but we would like to repose it in terms of ρ because the density is present in the continuity equation (2.1), which we will use later. First, we need to introduce two more thermodynamic concepts, starting with the gas state equation

$$p = \rho R T \tag{2.118}$$

where $R = C_p - C_v$ is the gas constant, C_p the specific heat at constant pressure and C_v the specific heat at constant volume. We will also use the definition of the internal energy

$$e = C_v T \tag{2.119}$$

Substituting Eqs. (2.118) and (2.119) into the enthalpy definition (2.113), we obtain

$$h = \frac{\gamma}{\gamma - 1}\frac{p}{\rho} \tag{2.120}$$

where $\gamma = C_p/C_v$ is the ratio of specific heats. Now, we introduce the speed of sound, a, given by

$$a^2 = \frac{\mathrm{d}p}{\mathrm{d}\rho} \tag{2.121}$$

which under the adiabatic assumption $p/\rho^\gamma = \text{constant}$, becomes

$$a^2 = \gamma\frac{p}{\rho} \tag{2.122}$$

Using the adiabatic process relation, we can state that

$$\frac{p}{\rho^\gamma} = \frac{p_\infty}{\rho_\infty^\gamma}$$

so that

$$p = \frac{p_\infty}{\rho_\infty^\gamma}\rho^\gamma \tag{2.123}$$

Following Balakrishnan (2004), we substitute this latest expression into Eq. (2.120) to obtain

$$h = \frac{\gamma}{\gamma-1}\frac{p_\infty}{\rho_\infty^\gamma}\rho^{\gamma-1} = \frac{\gamma}{\gamma-1}\left(\frac{\rho}{\rho_\infty}\right)^{\gamma-1}\frac{p_\infty}{\rho_\infty}$$

Using Eq. (2.122), the free stream speed of sound is given by $a_\infty^2 = \gamma p_\infty/\rho_\infty$ so that the enthalpy can be written as

$$h = \frac{1}{\gamma-1}\left(\frac{\rho}{\rho_\infty}\right)^{\gamma-1} a_\infty^2 \tag{2.124}$$

Substituting Eq. (2.124) for h and h_∞ into the Bernoulli equation (2.117) and solving for ρ/ρ_∞, we finally obtain

$$\left(\frac{\rho}{\rho_\infty}\right)^{\gamma-1} = 1 + \frac{\gamma-1}{a_\infty^2}\left(\frac{1}{2}Q_\infty^2 - \frac{1}{2}Q^2 - \frac{\partial\Phi}{\partial t}\right) \tag{2.125}$$

or

$$\frac{\rho}{\rho_\infty} = \left(1 + \frac{\gamma-1}{a_\infty^2}\left(\frac{1}{2}Q_\infty^2 - \frac{1}{2}Q^2 - \frac{\partial\Phi}{\partial t}\right)\right)^{1/(\gamma-1)} \tag{2.126}$$

This is another form of the unsteady compressible Bernoulli equation, relating density to potential, since $Q^2 = (\partial\Phi/\partial x)^2 + (\partial\Phi/\partial y)^2 + (\partial\Phi/\partial z)^2$ from Eq. (2.110).

2.4.2 The Full Potential Equation

In order to calculate the potential in the flow, we need to use the continuity equation (2.1)

$$\frac{\partial\rho}{\partial t} + \frac{\partial(\rho u)}{\partial x} + \frac{\partial(\rho v)}{\partial y} + \frac{\partial(\rho w)}{\partial z} = 0 \tag{2.127}$$

Substituting from Eq. (2.17) results in

$$\frac{\partial\rho}{\partial t} + \frac{\partial\rho}{\partial x}\frac{\partial\Phi}{\partial x} + \frac{\partial\rho}{\partial y}\frac{\partial\Phi}{\partial y} + \frac{\partial\rho}{\partial z}\frac{\partial\Phi}{\partial z} + \rho\left(\frac{\partial^2\Phi}{\partial x^2} + \frac{\partial^2\Phi}{\partial y^2} + \frac{\partial^2\Phi}{\partial z^2}\right) = 0$$

Following Balakrishnan (2004) again, we pre-multiply both sides of the continuity equation by $\rho^{\gamma-2}$ such that

$$\begin{aligned}
\rho^{\gamma-2}\frac{\partial\rho}{\partial t} &= -\rho^{\gamma-2}\left(\frac{\partial\rho}{\partial x}\frac{\partial\Phi}{\partial x} + \frac{\partial\rho}{\partial y}\frac{\partial\Phi}{\partial y} + \frac{\partial\rho}{\partial z}\frac{\partial\Phi}{\partial z}\right) - \rho^{\gamma-1}\left(\frac{\partial^2\Phi}{\partial x^2} + \frac{\partial^2\Phi}{\partial y^2} + \frac{\partial^2\Phi}{\partial z^2}\right) \\
&= -\frac{1}{\gamma-1}\left(\frac{\partial\rho^{\gamma-1}}{\partial x}\frac{\partial\Phi}{\partial x} + \frac{\partial\rho^{\gamma-1}}{\partial y}\frac{\partial\Phi}{\partial y} + \frac{\partial\rho^{\gamma-1}}{\partial z}\frac{\partial\Phi}{\partial z}\right) \\
&\quad - \rho^{\gamma-1}\left(\frac{\partial^2\Phi}{\partial x^2} + \frac{\partial^2\Phi}{\partial y^2} + \frac{\partial^2\Phi}{\partial z^2}\right)
\end{aligned} \tag{2.128}$$

Next, we differentiate Eq. (2.125) with respect to time to obtain

$$(\gamma-1)\frac{\rho^{\gamma-2}}{\rho_\infty^{\gamma-1}}\frac{\partial\rho}{\partial t}=\frac{\gamma-1}{a_\infty^2}\left(-\frac{1}{2}\frac{\partial Q^2}{\partial t}-\frac{\partial^2\Phi}{\partial t^2}\right)$$

or

$$\rho^{\gamma-2}\frac{\partial\rho}{\partial t}=-\frac{\rho_\infty^{\gamma-1}}{a_\infty^2}\left(\frac{1}{2}\frac{\partial Q^2}{\partial t}+\frac{\partial^2\Phi}{\partial t^2}\right) \tag{2.129}$$

The left-hand sides of Eqs. (2.128) and (2.129) are identical so that we can equate the right-hand sides

$$\frac{\partial^2\Phi}{\partial t^2}+\frac{1}{2}\frac{\partial Q^2}{\partial t}=\frac{a_\infty^2}{\gamma-1}\left(\frac{\partial(\rho/\rho_\infty)^{\gamma-1}}{\partial x}\frac{\partial\Phi}{\partial x}+\frac{\partial(\rho/\rho_\infty)^{\gamma-1}}{\partial y}\frac{\partial\Phi}{\partial y}+\frac{\partial(\rho/\rho_\infty)^{\gamma-1}}{\partial z}\frac{\partial\Phi}{\partial z}\right)$$
$$+a_\infty^2(\rho/\rho_\infty)^{\gamma-1}\left(\frac{\partial^2\Phi}{\partial x^2}+\frac{\partial^2\Phi}{\partial y^2}+\frac{\partial^2\Phi}{\partial z^2}\right) \tag{2.130}$$

The term $(\rho/\rho_\infty)^{\gamma-1}$ can be substituted directly from Eq. (2.125). Its derivatives are the derivatives of the right-hand side of the same equation such that

$$\frac{\partial(\rho/\rho_\infty)^{\gamma-1}}{\partial x}=-\frac{\gamma-1}{a_\infty^2}\left(\frac{1}{2}\frac{\partial Q^2}{\partial x}+\frac{\partial^2\Phi}{\partial x\partial t}\right)$$

$$\frac{\partial(\rho/\rho_\infty)^{\gamma-1}}{\partial y}=-\frac{\gamma-1}{a_\infty^2}\left(\frac{1}{2}\frac{\partial Q^2}{\partial y}+\frac{\partial^2\Phi}{\partial y\partial t}\right)$$

$$\frac{\partial(\rho/\rho_\infty)^{\gamma-1}}{\partial z}=-\frac{\gamma-1}{a_\infty^2}\left(\frac{1}{2}\frac{\partial Q^2}{\partial z}+\frac{\partial^2\Phi}{\partial z\partial t}\right)$$

We substitute these derivatives along with Eq. (2.125) into expression (2.130) and make use of the notation

$$\Phi_x=\frac{\partial\Phi}{\partial x},\ \Phi_{xt}=\frac{\partial^2\Phi}{\partial x\partial t},\ \Phi_{yz}=\frac{\partial^2\Phi}{\partial y\partial z},\ \text{etc}$$

to obtain

$$\Phi_{tt}+2\Phi_x\Phi_{xt}+2\Phi_y\Phi_{yt}+2\Phi_z\Phi_{zt}$$
$$=a_\infty^2\left(1+\frac{\gamma-1}{a_\infty^2}\left(\frac{1}{2}\left(Q_\infty^2-\Phi_x^2-\Phi_y^2-\Phi_z^2\right)-\Phi_t\right)\right)\left(\Phi_{xx}+\Phi_{yy}+\Phi_{zz}\right)$$
$$-2\left(\Phi_x\Phi_y\Phi_{xy}+\Phi_x\Phi_z\Phi_{xz}+\Phi_y\Phi_z\Phi_{yz}\right)-\Phi_x^2\Phi_{xx}-\Phi_y^2\Phi_{yy}-\Phi_z^2\Phi_{zz} \tag{2.131}$$

Equation (2.131) is known as the full potential equation for inviscid, irrotational and isentropic compressible flow. It is a highly non-linear partial differential equation to be solved for $\Phi(x,y,z,t)$ everywhere in the flowfield. It can be solved numerically by discretising the fluid domain using finite differences, finite elements or finite volumes and time-marching the solution (Bakhle et al. 1993). It can also be time-linearised before it is solved in the frequency domain, thus avoiding the cost of time-marching (Florea et al. 2000). Due to the isentropic assumption, the equation cannot describe flows with shock waves. Nevertheless, the full potential equation is used regularly in the presence of weak shocks. Using the definition of the Mach number,

$$M = \frac{Q}{a} = \frac{\sqrt{u^2 + v^2 + w^2}}{a} \tag{2.132}$$

where u, v and w are the local velocity components and a the local speed of sound, Nixon and Kerlick (1982) state that a shock wave is sufficiently weak to allow the use of the full potential equation if

$$\left(M^2 - 1\right)^{3/2} \ll \sqrt{M^2 - 1}$$

ahead of the shock. The condition $(M^2 - 1)^{3/2} = \sqrt{M^2 - 1}$ is met at $M = 1.41$ so that the generally accepted validity condition is that the Mach number must not exceed 1.3 anywhere in the flow. The equation is usually applied to transonic flows, that is flows for which the Mach number is subsonic far from the body but becomes locally supersonic over the body. Note though that the full potential equation also applies to incompressible flow, for which $a_\infty \to \infty$. In this case, the only important term in Eq. (2.131) is the one multiplied by a_∞ so that we recover Laplace's equation.

2.4.3 The Transonic Small Disturbance Equation

Solutions whereby the fluid domain is discretised are beyond the scope of the present book. We will therefore continue to apply simplifying assumptions to the full potential equation in order to linearise it. This can be achieved by assuming that the thickness of the body is small, as is the amplitude and frequency of its motion. Then, using the approach by Anderson Jr. (1990), the total potential in the flow can be written as

$$\Phi(\mathbf{x}, t) = \phi_\infty(\mathbf{x}) + \phi(\mathbf{x}, t)$$

where $\phi_\infty(\mathbf{x})$ is the potential due to the free stream and $\phi(\mathbf{x}, t)$ is the perturbation potential due to the body's geometry and motion. Further assuming that $\phi_\infty(\mathbf{x}) = U_\infty x$ so that the free stream velocity is parallel to the x axis,

$$\Phi_x = U_\infty + \phi_x,\ \Phi_y = \phi_y,\ \Phi_z = \phi_z,\ \Phi_{xx} = \phi_{xx},\ \text{etc} \tag{2.133}$$

Substituting from Eq. (2.133) in the full potential Eq. (2.131), the left-hand side becomes

$$\begin{aligned}\Phi_{tt} + 2\Phi_x\Phi_{xt} + 2\Phi_y\Phi_{yt} + 2\Phi_z\Phi_{zt} &= \phi_{tt} + 2\phi_x\phi_{xt} + 2U_\infty\phi_{xt} + 2\phi_y\phi_{yt} + 2\phi_z\phi_{zt} \\ &\approx \phi_{tt} + 2U_\infty\phi_{xt}\end{aligned} \tag{2.134}$$

since the magnitude of $U_\infty\phi_{xt}$ is much higher than those of $\phi_x\phi_{xt}$, $\phi_y\phi_{yt}$ and $\phi_z\phi_{zt}$. Similarly, the first bracket in the first term of the right-hand side of Eq. (2.131) yields

$$\begin{aligned}&1 + \frac{\gamma - 1}{a_\infty^2}\left(\frac{1}{2}\left(U_\infty^2 - \Phi_x^2 - \Phi_y^2 - \Phi_z^2\right) - \Phi_t\right) \\ &= 1 + \frac{\gamma - 1}{a_\infty^2}\left(\frac{1}{2}\left(-2U_\infty\phi_x - \phi_x^2 - \phi_y^2 - \phi_z^2\right) - \phi_t\right) \\ &\approx 1 + \frac{\gamma - 1}{a_\infty^2}\left(-U_\infty\phi_x - \phi_t\right)\end{aligned}$$

since $|2U_\infty \phi_x| \gg |\phi_x^2|, |\phi_y^2|, |\phi_z^2|$. In fact, the time derivative in the expression above is usually neglected, such that

$$1 + \frac{\gamma-1}{a_\infty^2}\left(\frac{1}{2}\left(U_\infty^2 - \phi_x^2 - \phi_y^2 - \phi_z^2\right) - \phi_t\right) = 1 - \frac{\gamma-1}{a_\infty^2}U_\infty \phi_x$$

Using the same logic, the last three terms of Eq. (2.131) yield

$$\Phi_x^2\Phi_{xx} + \Phi_y^2\Phi_{yy} + \Phi_z^2\Phi_{zz} \approx U_\infty^2\phi_{xx} + 2U_\infty\phi_x\phi_{xx}$$

while the term $2(\Phi_x\Phi_y\Phi_{xy} + \Phi_x\Phi_z\Phi_{xz} + \Phi_y\Phi_z\Phi_{yz})$ is neglected completely because it contains mixed partial derivatives.

Substituting the simplifications above into Eq. (2.131) yields

$$\phi_{tt} + 2U_\infty\phi_{xt} = a_\infty^2\left(1 - M_\infty^2 - \frac{\gamma+1}{U_\infty}M_\infty^2\phi_x\right)\phi_{xx} + a_\infty^2\phi_{yy} + a_\infty^2\phi_{zz} \tag{2.135}$$

where we have defined the free stream Mach number as $M_\infty = U_\infty/a_\infty$. Equation (2.135) is known as the transonic small disturbance (TSD) equation; it can be written in several different forms (Nixon 1989), depending on the scaling applied to $\phi(x, y, z, t)$. Since the TSD equation is still non-linear, it must be solved numerically using discretisation of the flow domain. As with the full potential equation, the TSD equation is applied mainly to transonic flows.

2.4.4 The Linearised Small Disturbance Equation

We can also apply small perturbation analysis directly to the original continuity and Euler equations (see for example (Ward 1955)). Setting $u = U_\infty + \phi_x$, $v = \phi_y$, $w = \phi_z$, replacing ρ by $\rho_\infty + \rho$, p by $p_\infty + p$ in Eqs. (2.1) and (2.9) and neglecting products of perturbations, we obtain

$$\frac{\partial\rho}{\partial t} + U_\infty\frac{\partial\rho}{\partial x} + \rho_\infty\left(\phi_{xx} + \phi_{yy} + \phi_{zz}\right) = 0$$

$$\phi_{xt} + U_\infty\phi_{xx} + \frac{1}{\rho_\infty}\frac{\partial p}{\partial x} = 0$$

having used the simplification $1/(\rho_\infty + \rho) \approx 1/\rho_\infty$. As ρ_∞ and p_∞ are constants, we can introduce them into any of the derivatives without changing the equations, leading to

$$\frac{\partial\rho}{\partial t} + U_\infty\frac{\partial(\rho - \rho_\infty)}{\partial x} + \rho_\infty\left(\phi_{xx} + \phi_{yy} + \phi_{zz}\right) = 0 \tag{2.136}$$

$$\phi_{xt} + U_\infty\phi_{xx} + \frac{1}{\rho_\infty}\frac{\partial(p - p_\infty)}{\partial x} = 0 \tag{2.137}$$

Following Edward Ehlers et al. (1979), we can expand $\rho - \rho_\infty$ as a Taylor series in terms of the pressure, such that

$$\rho - \rho_\infty = \left.\frac{\mathrm{d}\rho}{\mathrm{d}p}\right|_\infty (p - p_\infty) + \cdots$$

The definition of the free stream speed of sound is $a_\infty^2 = \mathrm{d}p/\mathrm{d}\rho|_\infty$ so that

$$\rho - \rho_\infty \approx \frac{(p - p_\infty)}{a_\infty^2}$$

Substituting this result in Eq. (2.136) leads to

$$\frac{\partial \rho}{\partial t} + \frac{U_\infty}{a_\infty^2}\frac{\partial (p - p_\infty)}{\partial x} + \rho_\infty \left(\phi_{xx} + \phi_{yy} + \phi_{zz}\right) = 0 \tag{2.138}$$

From Eq. (2.137)

$$\frac{\partial (p - p_\infty)}{\partial x} = -\rho_\infty \left(\phi_{xt} + U_\infty \phi_{xx}\right)$$

Substituting this result in Eq. (2.138) leads to

$$\frac{\partial}{\partial t}\left(\frac{\rho}{\rho_\infty}\right) - \frac{U_\infty}{a_\infty^2}\left(\phi_{xt} + U_\infty \phi_{xx}\right) + \phi_{xx} + \phi_{yy} + \phi_{zz} = 0 \tag{2.139}$$

The time derivative of ρ/ρ_∞ can be obtained by differentiating in time Eq. (2.126) after linearising it by means of a Taylor series around ρ_∞. Setting

$$\Delta Q = \frac{\gamma - 1}{a_\infty^2}\left(\frac{1}{2}Q_\infty^2 - \frac{1}{2}Q^2 - \frac{\partial \phi}{\partial t}\right)$$

Equation (2.126) can be expanded as

$$\begin{aligned}\frac{\rho}{\rho_\infty} &= \left.\frac{\rho}{\rho_\infty}\right|_{\Delta Q=0} + \left.\frac{\mathrm{d}}{\mathrm{d}\Delta Q}\left(\frac{\rho}{\rho_\infty}\right)\right|_{\Delta Q=0} + \cdots \\ &\approx 1 + \frac{1}{a_\infty^2}\left(\frac{1}{2}Q_\infty^2 - \frac{1}{2}Q^2 - \frac{\partial \phi}{\partial t}\right)\end{aligned} \tag{2.140}$$

We now set $Q_\infty = U_\infty$ and make the small perturbation assumption, so that $Q^2 = (U_\infty + \phi_x)^2 + \phi_y^2 + \phi_z^2$. Consequently,

$$\frac{1}{2}Q_\infty^2 - \frac{1}{2}Q^2 \approx -U_\infty \phi_x - \frac{\phi_x^2 + \phi_y^2 + \phi_z^2}{2} \tag{2.141}$$

Neglecting the second-order terms and substituting into Eq. (2.140), we finally obtain

$$\frac{\rho}{\rho_\infty} = 1 + \frac{1}{a_\infty^2}\left(-U_\infty \phi_x - \frac{\partial \phi}{\partial t}\right) \tag{2.142}$$

Differentiating this latest result with respect to time and substituting into Eq. (2.139) results in

$$-\frac{1}{a_\infty^2}\phi_{tt} - 2\frac{U_\infty}{a_\infty^2}\phi_{xt} - \frac{U_\infty^2}{a_\infty^2}\phi_{xx} + \phi_{xx} + \phi_{yy} + \phi_{zz}$$

which can be re-arranged in the form

$$\left(1 - M_\infty^2\right)\phi_{xx} + \phi_{yy} + \phi_{zz} - \frac{2M_\infty}{a_\infty}\phi_{xt} - \frac{1}{a_\infty^2}\phi_{tt} = 0 \tag{2.143}$$

and is known as the linearised small disturbance equation. Note that we can obtain the same result by neglecting the nonlinear term in the TSD Eq. (2.135). Again, for incompressible conditions, $a_\infty \to \infty$ and Eq. (2.143) reduces to Laplace's equation. For steady compressible flows, we obtain

$$\left(1 - M_\infty^2\right)\phi_{xx} + \phi_{yy} + \phi_{zz} = 0 \tag{2.144}$$

which is sometimes referred to as the Prandtl–Glauert equation. Both Eqs. (2.143) and (2.144) are linear and have fundamental solutions that can be superimposed, as in the incompressible case.

2.4.5 The Compressible Unsteady Pressure Coefficient

Assuming that we have solved Eq. (2.143) for $\phi(x, y, z, t)$ and, therefore, for $\Phi(x, y, z, t) = \phi_\infty + \phi(x, y, z, t)$, we can then calculate the pressure acting on the surface by integrating Eq. (2.111) from far upstream to any point on the surface, that is from p_∞, ρ_∞, Q_∞ to p, ρ, Q, such that

$$\frac{\partial \phi}{\partial t} + \frac{1}{2}Q^2 - \frac{1}{2}Q_\infty^2 + \int_{p_\infty}^{p} \frac{\mathrm{d}p}{\rho} = 0 \tag{2.145}$$

noting that $\partial\Phi/\partial t = \partial\phi/\partial t$ since ϕ_∞ is constant and that $\partial\phi/\partial t = 0$ very far from the surface. The problem lies in integrating the $\mathrm{d}p/\rho$ term. From Eq. (2.123), the differential $\mathrm{d}p$ can be calculated in terms of ρ as

$$\mathrm{d}p = \gamma \frac{p_\infty}{\rho_\infty^\gamma} \rho^{\gamma-1} \mathrm{d}\rho$$

This means that the integral of $\mathrm{d}p/\rho$ is given by

$$\int_{p_\infty}^{p} \frac{\mathrm{d}p}{\rho} = \gamma \frac{p_\infty}{\rho_\infty^\gamma} \int_{\rho_\infty}^{\rho} \rho^{\gamma-2} \mathrm{d}\rho = \frac{\gamma}{\gamma-1}\frac{p_\infty}{\rho_\infty^\gamma}\rho^{\gamma-1} - \frac{\gamma}{\gamma-1}\frac{p_\infty}{\rho_\infty} \tag{2.146}$$

Recalling that what we want to do is calculate the pressure on the surface, we can use Eq. (2.123) in order to substitute

$$\rho = \left(\frac{p}{p_\infty}\right)^{1/\gamma} \rho_\infty$$

into expression (2.146), such that

$$\int_{p_\infty}^{p} \frac{\mathrm{d}p}{\rho} = \frac{\gamma}{\gamma-1}\frac{p_\infty^{1/\gamma}}{\rho_\infty} p^{(\gamma-1)/\gamma} - \frac{\gamma}{\gamma-1}\frac{p_\infty}{\rho_\infty} \tag{2.147}$$

This latest integral can be substituted into Eq. (2.145) in order to solve for p. However, the $(\gamma-1)/\gamma$ exponent complicates such a solution. Assuming that $p - p_\infty$ is small, we can apply a second-order Taylor expansion to the function $p^{(\gamma-1)/\gamma}$ around p_∞ such that

$$p^{(\gamma-1)/\gamma} \approx p_\infty^{(\gamma-1)/\gamma} + \left.\frac{\mathrm{d}p^{(\gamma-1)/\gamma}}{\mathrm{d}p}\right|_{p_\infty} (p - p_\infty) + \frac{1}{2}\left.\frac{\mathrm{d}^2 p^{(\gamma-1)/\gamma}}{\mathrm{d}p^2}\right|_{p_\infty} (p - p_\infty)^2$$

or

$$p^{(\gamma-1)/\gamma} \approx p_\infty^{(\gamma-1)/\gamma} + \frac{\gamma-1}{\gamma} p_\infty^{-1/\gamma} (p - p_\infty) - \frac{\gamma-1}{2\gamma^2} p_\infty^{-(\gamma+1)/\gamma} (p - p_\infty)^2$$

Substituting into expression (2.147) and carrying out the algebra leads to

$$\int_{p_\infty}^{p} \frac{\mathrm{d}p}{\rho} \approx \frac{p - p_\infty}{\rho_\infty} - \frac{1}{2a_\infty^2}\left(\frac{p - p_\infty}{\rho_\infty}\right)^2 \tag{2.148}$$

recalling that $a_\infty^2 = \gamma p_\infty/\rho_\infty$. Substituting this latest result into Eq. (2.145), we obtain

$$\frac{\partial \phi}{\partial t} + \frac{1}{2}Q^2 - \frac{1}{2}Q_\infty^2 + \frac{p - p_\infty}{\rho_\infty} - \frac{1}{2a_\infty^2}\left(\frac{p - p_\infty}{\rho_\infty}\right)^2 = 0 \tag{2.149}$$

which is a quadratic equation for $(p - p_\infty)/\rho_\infty$. Its solutions are

$$\frac{p - p_\infty}{\rho_\infty} = -a_\infty^2\left(-1 \pm \sqrt{1 + \frac{2}{a_\infty^2}\left(\frac{\partial \phi}{\partial t} + \frac{1}{2}Q^2 - \frac{1}{2}Q_\infty^2\right)}\right)$$

We choose the positive solution and recall that the Maclaurin series of $\sqrt{1+x}$ is given by

$$\sqrt{1+x} = 1 + \frac{x}{2} - \frac{x^2}{8} + \cdots$$

so that

$$\frac{p - p_\infty}{\rho_\infty} \approx -a_\infty^2\left(\frac{1}{a_\infty^2}\left(\frac{\partial \phi}{\partial t} + \frac{1}{2}Q^2 - \frac{1}{2}Q_\infty^2\right) - \frac{4}{8a_\infty^4}\left(\frac{\partial \phi}{\partial t} + \frac{1}{2}Q^2 - \frac{1}{2}Q_\infty^2\right)^2\right) \tag{2.150}$$

For compressible flows, the definition of the pressure coefficient (2.30 is)

$$c_p = \frac{p - p_\infty}{\frac{1}{2}\rho_\infty Q_\infty^2} \tag{2.151}$$

so that, substituting from Eq. (2.150),

$$c_p = -\frac{2}{Q_\infty^2}\left(\frac{\partial \phi}{\partial t} + \frac{1}{2}Q^2 - \frac{1}{2}Q_\infty^2\right) + \frac{1}{Q_\infty^2 a_\infty^2}\left(\frac{\partial \phi}{\partial t} + \frac{1}{2}Q^2 - \frac{1}{2}Q_\infty^2\right)^2$$

After writing $a_\infty = Q_\infty/M_\infty$ and re-arranging we obtain the compressible pressure coefficient as a second-order polynomial in $(1 - Q^2/Q_\infty^2 - 2\phi_t/Q_\infty^2)$, that is

$$c_p = \left(1 - \frac{Q^2}{Q_\infty^2} - \frac{2}{Q_\infty^2}\frac{\partial \phi}{\partial t}\right) + \frac{M_\infty^2}{4}\left(1 - \frac{Q^2}{Q_\infty^2} - \frac{2}{Q_\infty^2}\frac{\partial \phi}{\partial t}\right)^2 \tag{2.152}$$

Substituting $Q^2 = u^2 + v^2 + w^2$, $u = U_\infty + \phi_x$, $v = V_\infty + \phi_y$, $w = W_\infty + \phi_z$, $Q_\infty^2 = U_\infty^2 + V_\infty^2 + W_\infty^2$ we see that Eq. (2.152) is in fact a fourth-order polynomial in ϕ_x, ϕ_y, ϕ_z and ϕ_t. In deriving the transonic and linearised small disturbance equations, we assumed that the free stream only has a component in the x direction, U_∞. We do not need to impose the same assumption to the pressure coefficient, but we can simplify it if we state that the compressibility direction is x, which means that $U_\infty \gg V_\infty$ and $U_\infty \gg W_\infty$. Consequently, V_∞ and W_∞ are essentially perturbations. Neglecting all products of perturbations of order 3 and higher, results in

$$c_p = 1 - \frac{Q^2}{Q_\infty^2} + \frac{M_\infty^2}{Q_\infty^2}\phi_x^2 - \frac{2}{Q_\infty^2}\phi_t + \frac{M_\infty^2}{Q_\infty^4}\phi_t^2 + \frac{2M_\infty^2}{Q_\infty^3}\phi_x\phi_t \tag{2.153}$$

If we also neglect the second-order terms, we can completely linearise the pressure coefficient so that

$$c_p = -2\frac{\phi_x}{Q_\infty} - \frac{2}{Q_\infty^2}\phi_t \tag{2.154}$$

Equations (2.152)–(2.154) are all polynomial approximations of the pressure coefficient that retain different orders of non-linearity. An exact value of c_p can be obtained from the adiabatic relation (2.123) written in the form

$$\frac{p}{p_\infty} = \left(\frac{\rho}{\rho_\infty}\right)^\gamma$$

Substituting for ρ/ρ_∞ from Eq. (2.126) leads to

$$\frac{p}{p_\infty} = \left(1 + \frac{\gamma - 1}{a_\infty^2}\left(\frac{1}{2}Q_\infty^2 - \frac{1}{2}Q^2 - \frac{\partial \Phi}{\partial t}\right)\right)^{\gamma/(\gamma-1)} \tag{2.155}$$

Consequently, the pressure coefficient definition (2.151) becomes

$$c_p = \frac{p_\infty}{\frac{1}{2}\rho_\infty Q_\infty^2}\left(\left(1 + \frac{\gamma - 1}{a_\infty^2}\left(\frac{1}{2}Q_\infty^2 - \frac{1}{2}Q^2 - \frac{\partial \Phi}{\partial t}\right)\right)^{\gamma/(\gamma-1)} - 1\right)$$

This latest expression is usually written in the form (Epton and Magnus 1990)

$$c_p = \frac{2}{\gamma M_\infty^2}\left(\left(1 + \frac{\gamma - 1}{2}M_\infty^2\left(1 - \frac{Q^2}{Q_\infty^2} - \frac{2}{Q_\infty^2}\frac{\partial \Phi}{\partial t}\right)\right)^{\gamma/(\gamma-1)} - 1\right) \tag{2.156}$$

since $a_\infty^2 = \gamma p_\infty/\rho_\infty$ from Eq. (2.122). Equation (2.156) is fully non-linear and exact under the adiabatic and isentropic assumptions. It is straightforward to apply it in the time domain if Q and $\partial\Phi/\partial t$ are known but its use in the frequency domain is problematic, due to its high degree of non-linearity. Consequently, we will be using the polynomial Eqs. (2.152)–(2.154) in this book. Nevertheless, the expression is useful in defining the minimum possible value of the pressure coefficient. While pressure difference can be negative, pressure itself can only be positive or zero. Therefore, the minimum value of p/p_∞ is equal to zero and known as vacuum pressure. The corresponding value of the isentropic pressure coefficient is

$$c_{p_{\min}} = -\frac{2}{\gamma M_\infty^2}$$

Finally, it should be noted that applying a second-order binomial expansion to

$$\left(1 + \frac{\gamma - 1}{2}M_\infty^2\left(1 - \frac{Q^2}{Q_\infty^2} - \frac{2}{Q_\infty^2}\frac{\partial \Phi}{\partial t}\right)\right)^{\gamma/(\gamma-1)}$$

in Eq. (2.156) results in Eq. (2.152).

There is an important difference between linear potential flow in incompressible and compressible conditions. In the former case, Laplace's equation is naturally linear; the basic assumptions of inviscid, incompressible and irrotational flow led directly to a linear equation. In the compressible case, we have had to make the additional assumption that the flow is isentropic, and we performed several linearisations in order to obtain the linearised small perturbation equation. This means that the range of validity of linearised compressible flow is much smaller than that of incompressible flow and difficult to determine exactly. The flow must be subsonic or supersonic but not transonic. Subsonic solutions become invalid if the local Mach number ever reaches or exceeds $M = 1$ anywhere in the flow, a condition known as critical. The flow can become critical at reasonably low Mach numbers and angles of attack, depending on the thickness of the body or the frequency and amplitude of its motion.

2.4.6 Motion in a Compressible, Inviscid, Irrotational Fluid

In Section 2.2.3, we detailed three different descriptions of motion in an incompressible fluid. For compressible flow, the linearisation of the flow equations depends on making the small perturbation assumption, such that there must be a free stream and the motion amplitude and frequency must be even smaller than in the incompressible case. Only the motion schemes of Figures 2.2 and 2.3 can be used. The angles of attack and sideslip are assumed to be so small that, even though U_∞/a_∞ is significant, V_∞/a_∞ and W_∞/a_∞ are negligible. Hence, the direction of compressibility is the x axis.

The far-field boundary condition for compressible flow is identical to that for incompressible flow that is $\phi(\mathbf{x}, t) = 0$ for $r(\mathbf{x}, \mathbf{x}_s) \to 0$. In Section 2.2.3, we imposed an impermeability boundary condition that sets the total flow velocity perpendicular to the surface of a body equal to zero. For compressible flows, there are two options for formulating the impermeability boundary condition:

- For compressible flows around very thin wings the impermeability boundary condition takes the form of Eq. (2.37), but the steady and unsteady impermeability problems are usually solved separately (see for example (Blair 1994; Chen et al. 1993))
- For thicker wings and bodies, an alternative expression is used: the total mass flux normal to the surface is zero (Edward Ehlers et al. 1979; Epton and Magnus 1990).

In the second case, the impermeability boundary condition is expressed in terms of the mass flux, which is the product of the local flow density times the relative velocity between the body and the fluid. Using the motion scheme of Figure 2.2 and Eq. (2.43), the relative flow velocity on the surface is equal to

$$\nabla\phi(\mathbf{x}_s, t) + \mathbf{Q}_\infty - \mathbf{v}_0(t) - \mathbf{\Omega}_0(t) \times (\mathbf{x}_s(t) - \mathbf{x}_f(t))$$

The total mass flux on the surface, J_m, is given by

$$J_m = \rho(\mathbf{x}_s, t)\left(\nabla\phi(\mathbf{x}_s, t) + \mathbf{Q}_\infty - \mathbf{v}_0(t) - \mathbf{\Omega}_0(t) \times (\mathbf{x}_s(t) - \mathbf{x}_f(t))\right)$$

while the free stream mass flux is simply $\rho_\infty \mathbf{Q}_\infty$. The perturbation mass flux, j_m, is the difference between the total and free stream mass fluxes so that

$$j_m = \rho(\mathbf{x}_s, t)\left(\nabla\phi(\mathbf{x}_s, t) + \mathbf{Q}_\infty - \mathbf{v}_0(t) - \mathbf{\Omega}_0(t) \times (\mathbf{x}_s(t) - \mathbf{x}_f(t))\right) - \rho_\infty \mathbf{Q}_\infty$$

We can substitute for $\rho(\mathbf{x}_s, t)$ from Eq. (2.142), leading to

$$j_m = \rho_\infty \left(1 + \frac{1}{a_\infty^2}\left(-U_\infty \phi_x - \frac{\partial \phi}{\partial t}\right)\right)\left(\nabla\phi + \mathbf{Q}_\infty - \mathbf{v}_0(t) - \mathbf{\Omega}_0(t) \times (\mathbf{x}_s(t) - \mathbf{x}_f(t))\right) - \rho_\infty \mathbf{Q}_\infty$$

Expanding this expression and neglecting products of perturbations, while setting $\mathbf{Q}_\infty = (U_\infty, 0, 0)$ and recalling that the motion-induced velocities are small leads to the linearised perturbation mass flux

$$j_m = \rho_\infty \left(\beta^2 \phi_x - \frac{M_\infty}{a_\infty^2}\phi_t, \phi_y, \phi_z\right) - \rho_\infty \left(\mathbf{v}_0(t) + \mathbf{\Omega}_0(t) \times (\mathbf{x}_s(t) - \mathbf{x}_f(t))\right)$$

and to the linearised total mass flux

$$J_m = \rho_\infty \left(\beta^2\phi_x - \frac{M_\infty}{a_\infty^2}\phi_t, \phi_y, \phi_z\right) + \rho_\infty \left(\mathbf{Q}_\infty - \mathbf{v}_0(t) - \boldsymbol{\Omega}_0(t) \times (\mathbf{x}_s(t) - \mathbf{x}_f(t))\right)$$

The impermeability boundary condition then becomes

$$\rho_\infty \left(\left(\beta^2\phi_x - \frac{M_\infty}{a_\infty^2}\phi_t, \phi_y, \phi_z\right) + \mathbf{Q}_\infty - \mathbf{v}_0(t) - \boldsymbol{\Omega}_0(t) \times (\mathbf{x}_s(t) - \mathbf{x}_f(t))\right) \cdot \mathbf{n}(\mathbf{x}_s) = 0 \tag{2.157}$$

on the surface of the body. Nevertheless, for very thin wings and slender bodies at small angles of attack, the x and y components of the unit vector become very small, while the z component is close to 1, that is $n_x \ll n_z, n_y \ll n_z$. For the case of a very thin wing performing pitch and plunge oscillations, the impermeability boundary condition linearises to

$$\phi_z(\mathbf{x}_s, t) + U_\infty \alpha(t) + \dot{\alpha}(t)(x_s - x_f) + \dot{h}(t) = 0 \tag{2.158}$$

which is identical to the incompressible condition of Eq. (2.46).

2.5 Subsonic Linearised Potential Flow Solutions

In subsonic flow, we can write the linearised small disturbance Eq. (2.143) as

$$\beta^2\phi_{xx} + \phi_{yy} + \phi_{zz} - \frac{2M_\infty}{a_\infty}\phi_{xt} - \frac{1}{a_\infty^2}\phi_{tt} = 0 \tag{2.159}$$

where $\beta = \sqrt{1 - M_\infty^2}$ is known as the subsonic compressibility factor. The source function of Eq. (2.49) is a solution of Laplace's equation but not of Eq. (2.159). However, if we define

$$r_\beta(\mathbf{x}, \mathbf{x}_0) = \sqrt{(x - x_0)^2 + \beta^2(y - y_0)^2 + \beta^2(z - z_0)^2} \tag{2.160}$$

then the function

$$\phi(\mathbf{x}) = \frac{1}{\sqrt{(x - x_0)^2 + \beta^2(y - y_0)^2 + \beta^2(z - z_0)^2}} = \frac{1}{r_\beta(\mathbf{x}, \mathbf{x}_0)} \tag{2.161}$$

becomes a solution of both Eq. (2.159) and its steady counterpart

$$\beta^2\phi_{xx} + \phi_{yy} + \phi_{zz} = 0 \tag{2.162}$$

This fact can be verified by substituting expression (2.161) into either of the two equations and showing that their left-hand sides are indeed equal to zero.

In the incompressible case, if the source changes strength at time t, the effects of this change are felt everywhere in the flow immediately. This phenomenon is due to the fact that information travels in a flow at the speed of sound. In the incompressible case, the speed of sound is infinite and therefore perturbations are propagated everywhere in the flow with infinite speed. In the compressible case, the speed of sound is finite and therefore perturbations occurring at time t are not felt everywhere in the flow at the same time.

If the effects of perturbations are felt at different times in different parts of the flowfield, it is logical to guess an unsteady source solution for the linearised small perturbation equation of the form

$$\phi(\mathbf{x}, t) = \frac{F\left(t - \tau(\mathbf{x}, \mathbf{x}_0)\right)}{r_\beta(\mathbf{x}, \mathbf{x}_0)} \tag{2.163}$$

where r_β is given by Eq. (2.160), F is any function of time and $\tau(\mathbf{x}, \mathbf{x}_0)$ is a time delay that depends on the distance between points $\mathbf{x}$ and $\mathbf{x}_0$. We can now calculate the derivatives of $\phi(\mathbf{x}, t)$ and then substitute them into Eq. (2.143). For convenience, we define the delayed time

$$\bar{t} = t - \tau(\mathbf{x}, \mathbf{x}_0)$$

so that $F\left(t - \tau(\mathbf{x}, \mathbf{x}_0)\right) = F(\bar{t})$ is a function of delayed time only and

$$\frac{\partial \bar{t}}{\partial t} = 1, \quad \frac{\partial \bar{t}}{\partial x} = -\frac{\partial \tau}{\partial x}, \quad \frac{\partial \bar{t}}{\partial y} = -\frac{\partial \tau}{\partial y}, \quad \frac{\partial \bar{t}}{\partial z} = -\frac{\partial \tau}{\partial z}, \text{ etc}$$

Then, for example,

$$\phi_x = \frac{\partial}{\partial x}\left(\frac{1}{r_\beta}\right) F(\bar{t}) + \frac{1}{r_\beta}\frac{\partial F(\bar{t})}{\partial x} = \frac{\partial}{\partial x}\left(\frac{1}{r_\beta}\right) F(\bar{t}) + \frac{1}{r_\beta}\frac{\partial F(\bar{t})}{\partial \bar{t}}\frac{\partial \bar{t}}{\partial x}$$

We simplify the notation by denoting $F = F(\bar{t})$ and using F', F'', etc., to denote the derivatives of $F(\bar{t})$ with respect to $\bar{t}$, so that ϕ_x becomes

$$\phi_x = \frac{\partial}{\partial x}\left(\frac{1}{r_\beta}\right) F - \frac{1}{r_\beta} F' \frac{\partial \tau}{\partial x} \tag{2.164}$$

Similarly,

$$\phi_y = \frac{\partial}{\partial y}\left(\frac{1}{r_\beta}\right) F - \frac{1}{r_\beta} F' \frac{\partial \tau}{\partial y}, \quad \phi_z = \frac{\partial}{\partial z}\left(\frac{1}{r_\beta}\right) F - \frac{1}{r_\beta} F' \frac{\partial \tau}{\partial z} \tag{2.165}$$

The second derivative ϕ_{xx} is given by

$$\begin{aligned}\phi_{xx} &= \frac{\partial}{\partial x}\left(\frac{\partial}{\partial x}\left(\frac{1}{r_\beta}\right) F - \frac{1}{r_\beta} F' \frac{\partial \tau}{\partial x}\right) \\ &= \frac{\partial^2}{\partial x^2}\left(\frac{1}{r_\beta}\right) F - 2\frac{\partial}{\partial x}\left(\frac{1}{r_\beta}\right) F' \frac{\partial \tau}{\partial x} + \frac{1}{r_\beta} F'' \left(\frac{\partial \tau}{\partial x}\right)^2 - \frac{1}{r_\beta} F' \frac{\partial^2 \tau}{\partial x^2}\end{aligned} \tag{2.166}$$

Similarly,

$$\phi_{yy} = \frac{\partial^2}{\partial y^2}\left(\frac{1}{r_\beta}\right) F - 2\frac{\partial}{\partial y}\left(\frac{1}{r_\beta}\right) F' \frac{\partial \tau}{\partial y} + \frac{1}{r_\beta} F'' \left(\frac{\partial \tau}{\partial y}\right)^2 - \frac{1}{r_\beta} F' \frac{\partial^2 \tau}{\partial y^2} \tag{2.167}$$

$$\phi_{zz} = \frac{\partial^2}{\partial z^2}\left(\frac{1}{r_\beta}\right) F - 2\frac{\partial}{\partial z}\left(\frac{1}{r_\beta}\right) F' \frac{\partial \tau}{\partial z} + \frac{1}{r_\beta} F'' \left(\frac{\partial \tau}{\partial z}\right)^2 - \frac{1}{r_\beta} F' \frac{\partial^2 \tau}{\partial z^2} \tag{2.168}$$

The mixed derivative ϕ_{xt} is

$$\phi_{xt} = \frac{\partial}{\partial t}\left(\frac{\partial}{\partial x}\left(\frac{1}{r_\beta}\right)F - \frac{1}{r_\beta}F'\frac{\partial \tau}{\partial x}\right) = \frac{\partial}{\partial x}\left(\frac{1}{r_\beta}\right)F' - \frac{1}{r_\beta}F''\frac{\partial \tau}{\partial x} \tag{2.169}$$

since τ and r_β are independent of t. Finally, the second time derivative is given simply by

$$\phi_{tt} = \frac{1}{r_\beta}F'' \tag{2.170}$$

Now, we substitute all the derivatives in Eq. (2.143), repeated here for convenience,

$$\beta^2\phi_{xx} + \phi_{yy} + \phi_{zz} - \frac{2M_\infty}{a_\infty}\phi_{xt} - \frac{1}{a_\infty^2}\phi_{tt} = 0 \tag{2.171}$$

to obtain

$$\begin{aligned}
&F\left(\beta^2\frac{\partial^2}{\partial x^2}\left(\frac{1}{r_\beta}\right) + \frac{\partial^2}{\partial y^2}\left(\frac{1}{r_\beta}\right) + \frac{\partial^2}{\partial z^2}\left(\frac{1}{r_\beta}\right)\right)\\
&+F'\left(-2\beta^2\frac{\partial}{\partial x}\left(\frac{1}{r_\beta}\right)\frac{\partial \tau}{\partial x} - \frac{\beta^2}{r_\beta}\frac{\partial^2 \tau}{\partial x^2} - 2\frac{\partial}{\partial y}\left(\frac{1}{r_\beta}\right)\frac{\partial \tau}{\partial y} - \frac{1}{r_\beta}\frac{\partial^2 \tau}{\partial y^2}\right.\\
&\qquad\left.-2\frac{\partial}{\partial z}\left(\frac{1}{r_\beta}\right)\frac{\partial \tau}{\partial z} - \frac{1}{r_\beta}\frac{\partial^2 \tau}{\partial z^2} - \frac{2M_\infty}{a_\infty}\frac{\partial}{\partial x}\left(\frac{1}{r_\beta}\right)\right)\\
&+F''\left(\frac{\beta^2}{r_\beta}\left(\frac{\partial \tau}{\partial x}\right)^2 + \frac{1}{r_\beta}\left(\frac{\partial \tau}{\partial y}\right)^2 + \frac{1}{r_\beta}\left(\frac{\partial \tau}{\partial z}\right)^2 + \frac{2M_\infty}{a_\infty}\frac{1}{r_\beta}\frac{\partial \tau}{\partial x} - \frac{1}{a_\infty^2 r_\beta}\right) = 0
\end{aligned}$$

One way for this equation to be satisfied is for the coefficients of F, F' and F'' to be all equal to zero. The coefficient of F is in fact the substitution of the steady compressible source of expression (2.161) into the steady compressible flow Eq. (2.144) so that it is indeed equal to zero. Setting the coefficients of F' and F'' to zero yields

$$\begin{aligned}
&-2\left(\beta^2\frac{\partial}{\partial x}\left(\frac{1}{r_\beta}\right)\frac{\partial \tau}{\partial x} - \frac{\partial}{\partial y}\left(\frac{1}{r_\beta}\right)\frac{\partial \tau}{\partial y} - \frac{\partial}{\partial z}\left(\frac{1}{r_\beta}\right)\frac{\partial \tau}{\partial z}\right)\\
&-\frac{1}{r_\beta}\left(\beta^2\frac{\partial^2 \tau}{\partial x^2} + \frac{\partial^2 \tau}{\partial y^2} + \frac{\partial^2 \tau}{\partial z^2}\right) - \frac{2M_\infty}{a_\infty}\frac{\partial}{\partial x}\left(\frac{1}{r_\beta}\right) = 0
\end{aligned} \tag{2.172}$$

and

$$\beta^2\left(\frac{\partial \tau}{\partial x}\right)^2 + \left(\frac{\partial \tau}{\partial y}\right)^2 + \left(\frac{\partial \tau}{\partial z}\right)^2 + \frac{2M_\infty}{a_\infty}\frac{\partial \tau}{\partial x} - \frac{1}{a_\infty^2} = 0 \tag{2.173}$$

Equations (2.172) and (2.173) are two equations with many unknowns, the first and second derivatives of τ in the three directions. We can reduce the number of unknowns by thinking about the nature of the time delay $\tau(x, y, z)$. Consider an acoustic wave generator that travels with constant speed $U = Ma$ in the $-x$ direction, where $M < 1$ and a is the speed of sound. The generator emits constantly acoustic waves that travel with the speed of sound. Figure 2.18 plots the wave pattern after four time instances $t_i = i\Delta t$, for $i = 0, \dots, 3$. The acoustic wave released by the generator at time t_0 and position $x_0 = 0$ has propagated in

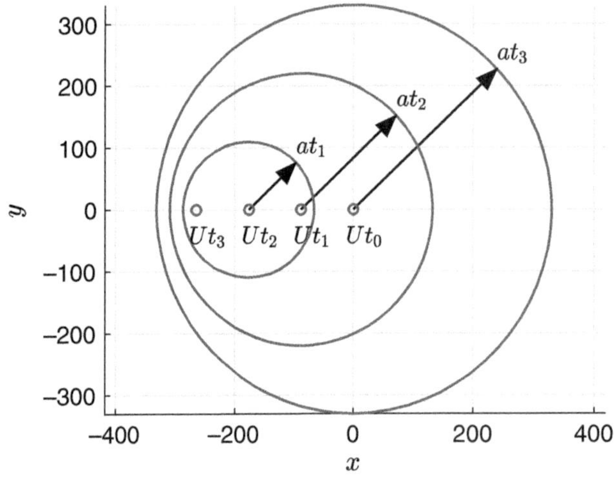

Figure 2.18 Propagation of acoustic disturbances in subsonic flow.

a circle with radius at_3, while the generator itself has move to position $-Ut_3$. The acoustic waves released at times t_1 and t_2 are also shown. Whether a point in space will be affected by a sound wave at each particular time instance depends on the speed of propagation of the sound waves, a, and on the speed of the generator, U. For example, at time t_3, the point $-Ut_1 - at_2$ is affected by the sound wave that was released at time t_1 but not by any other sound wave. No points lying outside the circle with radius at_3 and centred at the origin are affected. Nevertheless, even though the point $Ut_1 + at_2$ lies inside this circle, it is not affected by a sound wave at time t_3 because the generator is moving away from it. Therefore, whether a sound wave will arrive at a particular point in space at the current time instance depends on the radial distance between the point and the location of the source at a previous time instance, as well as the horizontal distance between the point and the location of the source at a different previous time instance.

Returning to the problem of a static source located at x_0, y_0, z_0 in a free stream moving with speed U_∞ in the $+x$ direction, acoustic perturbations are generated constantly by the source, but they are carried downstream by the free stream. There are two possible reasons for these perturbations to arrive at a general point x, y, z with a time delay:

- The time it takes for an acoustic disturbance to cover the distance between the source and the point x, y, z, represented by r_β. Therefore, the resulting time delay will be proportional to r_β.
- The time it takes for the free stream to carry the perturbations downstream. As the horizontal distance between the source and the point is $x - x_0$, the resulting time delay will be proportional to this distance.

From these two time-delay components, we can write that

$$\tau(\mathbf{x}, \mathbf{x}_0) = A_1(x - x_0) + A_2 r_\beta \tag{2.174}$$

Substituting back into Eq. (2.172) and carrying out all the differentiations, we obtain

$$\frac{2(x - x_0)\left(a_\infty \beta^2 A_1 + M_\infty\right)}{a_\infty r_\beta^3} = 0$$

which is always satisfied if

$$A_1 = \frac{-M_\infty}{a_\infty \beta^2}$$

Substituting for $\tau(\mathbf{x}, \mathbf{x}_0)$ and A_1 into Eq. (2.173) and carrying out all the differentiations yields

$$-\frac{A_2^2 a^2 \beta^4 - 1}{a_\infty^2 \beta} = 0$$

which can be solved to give

$$A_2 = \pm \frac{1}{a_\infty \beta^2}$$

If we substitute both of these solutions in Eq. (2.174), we obtain

$$\tau_\pm(\mathbf{x}, \mathbf{x}_0) = \frac{-M_\infty(x - x_0) \pm r_\beta}{a_\infty \beta^2}$$

As $M_\infty < 1$ and r_β is always positive, the time delay $\tau_-(\mathbf{x}, \mathbf{x}_0)$ is always negative, irrespective of the sign of $(x - x_0)$. This would mean that changes at point $\mathbf{x}_0$ are felt at point $\mathbf{x}$ before they even happen. This situation is unphysical so we will discard the negative solution and retain

$$\tau(\mathbf{x}, \mathbf{x}_0) = \frac{-M_\infty(x - x_0) + r_\beta}{a_\infty \beta^2} \tag{2.175}$$

Using this value of the time delay, the subsonic source of Eq. (2.163) becomes

$$\phi_s(\mathbf{x}, t) = \frac{F\left(t + \frac{M_\infty(x - x_0) - r_\beta}{a_\infty \beta^2}\right)}{r_\beta} \tag{2.176}$$

where $\phi_s(\mathbf{x}, t)$ is the source perturbation potential and r_β is given in expression (2.160). Equation (2.176) is our first unsteady fundamental solution of the linearised small disturbance Eq. (2.143) for any differentiable function F, which represents the time-varying strength of the source.

There is a subsonic version of the incompressible doublet function of Eq. (2.55). Its definition is

$$\mathbf{n} \cdot \nabla \left(\frac{F\left(t + \frac{M_\infty(x - x_0) - r_\beta}{a_\infty \beta^2}\right)}{r_\beta} \right) \tag{2.177}$$

where $\mathbf{n}$ is any unit direction vector, but, usually, it is the unit vector normal to the surface of a body immersed in compressible flow. Substituting Eq. (2.177) into the linearised small disturbance Eq. (2.171), we obtain

$$\beta^2 \frac{\partial^2}{\partial x^2}\left(\frac{\partial \phi_s}{\partial \mathbf{n}}\right) + \frac{\partial^2}{\partial y^2}\left(\frac{\partial \phi_s}{\partial \mathbf{n}}\right) + \frac{\partial^2}{\partial z^2}\left(\frac{\partial \phi_s}{\partial \mathbf{n}}\right) - \frac{2M_\infty}{a_\infty}\frac{\partial^2}{\partial x \partial t}\left(\frac{\partial \phi_s}{\partial \mathbf{n}}\right) - \frac{1}{a_\infty^2}\frac{\partial^2}{\partial t^2}\left(\frac{\partial \phi_s}{\partial \mathbf{n}}\right) = 0$$

Changing the order of the differentiations and re-arranging them gives

$$\frac{\partial}{\partial \mathbf{n}}\left(\beta^2 \frac{\partial^2 \phi_s}{\partial x^2} + \frac{\partial^2 \phi_s}{\partial y^2} + \frac{\partial^2 \phi_s}{\partial z^2} - \frac{2M_\infty}{a_\infty}\frac{\partial^2 \phi_s}{\partial x \partial t} - \frac{1}{a_\infty^2}\frac{\partial^2 \phi_s}{\partial t^2}\right) = 0$$

The term inside the brackets is the linearised small disturbance equation applied to the source potential. As the source is a solution of this equation, the term inside the brackets is equal to zero and hence the doublet of expression (2.177) is also a solution of the linearised small disturbance equation.

Recalling that the function F in Eqs. (2.176) and (2.177) can be any differentiable function, the potential induced by the subsonic source and doublet can be defined by analogy to the incompressible equivalents of Eqs. (2.98) and (2.99) as

$$\phi_s(\mathbf{x}, \mathbf{x}_0, t) = \frac{\sigma\left(t + \frac{M_\infty(x-x_0)-r_\beta}{a_\infty \beta^2}\right)}{r_\beta} \tag{2.178}$$

$$\phi_d(\mathbf{x}, \mathbf{x}_0, t) = \mathbf{n} \cdot \nabla \frac{\mu\left(t + \frac{M_\infty(x-x_0)-r_\beta}{a_\infty \beta^2}\right)}{r_\beta} \tag{2.179}$$

where $\sigma(\mathbf{x}_0, \mathbf{x}_0, t)$ and $\mu(\mathbf{x}_0, \mathbf{x}_0, t)$ are the strengths of the source and doublet at point $\mathbf{x}_0$ and time t.

The perturbation velocities induced by the source are obtained by differentiating $\phi_s(\mathbf{x}, \mathbf{x}_0, t)$ in the three directions so that from Eqs. (2.164), (2.165) and (2.175)

$$u_s(\mathbf{x}, \mathbf{x}_0, t) = \frac{\partial \phi_s}{\partial x} = -\sigma \frac{x - x_0}{r_\beta^3} + \frac{\sigma'}{r_\beta}\left(\frac{M_\infty}{a_\infty \beta^2} - \frac{x - x_0}{a_\infty \beta^2 r_\beta}\right)$$

$$v_s(\mathbf{x}, \mathbf{x}_0, t) = \frac{\partial \phi_s}{\partial y} = -\sigma \frac{\beta^2 (y - y_0)}{r_\beta^3} - \frac{\sigma'}{a_\infty r_\beta^2}(y - y_0)$$

$$w_s(\mathbf{x}, \mathbf{x}_0, t) = \frac{\partial \phi_s}{\partial z} = -\sigma \frac{\beta^2 (z - z_0)}{r_\beta^3} - \frac{\sigma'}{a_\infty r_\beta^2}(z - z_0)$$

In order to visualise the compressible source, we will consider two cases, one where the source strength is constant and the free stream airspeed non-zero and one where the source strength is time-varying and the free stream airspeed zero. Figure 2.19 plots a cross-section of the velocity field (for $z = 0$) induced by a source lying at the origin with $\sigma(0, 0, t) = -1$ and immersed in a free stream with two different values of the Mach number. The flowfield of Figure 2.19a for $M_\infty = 0$ is identical to the flow induced by an

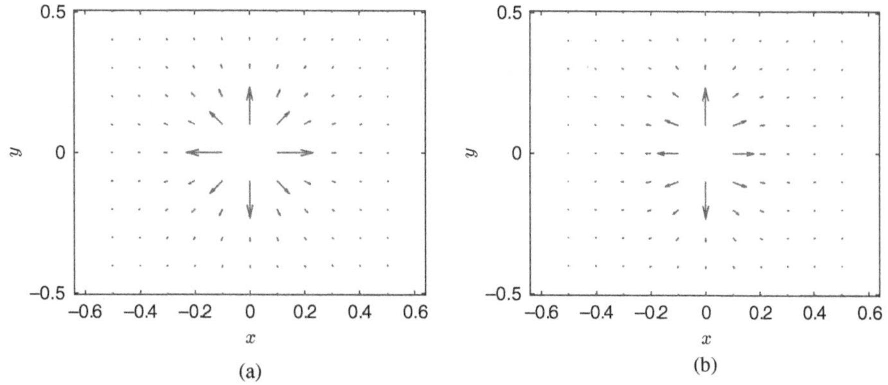

Figure 2.19 2D section of flowfield induced by a compressible point source with constant strength at two Mach numbers. (a) $M_\infty = 0$ and (b) $M_\infty = 0.8$.

incompressible source. Figure 2.19b plots the u_s and v_s velocity vectors for $M_\infty = 0.8$. It can be seen that the flowfield is now squashed in the x direction with respect to the y direction. This phenomenon is due to the form of r_β in Eq. (2.160), which implies a transformation of the flowfield that stretches the y and z axes.

The second illustrative case is a static source lying at the origin with sinusoidally oscillating strength such that

$$\sigma(\mathbf{x}, \mathbf{x}_0, t) = \sin(\omega(t - \tau(\mathbf{x}, \mathbf{x}_0))),\ \sigma'(\mathbf{x}, \mathbf{x}_0, t) = \omega \cos(\omega(t - \tau(\mathbf{x}, \mathbf{x}_0)))$$

for $\omega = 1$. The flow velocity is zero so that $M_\infty = 0$ but the speed of sound is $a_\infty = 330$ 3m/s. Figure 2.20 plots the flowfield induced by this source at two phase angles, $\omega t = 0$ and $\omega t = \pi$. At $\omega t = 0$, the singularity has strength $\sigma(0, 0, 0) = 0$ so that is neither a source nor a sink. A little while earlier however, it was a source because $\sigma(0, 0, t < 0)$ was negative. Therefore, the velocity vectors close to the origin are pointing away from the origin. Very far from the origin, the velocity vectors are pointing inwards because they are affected by the strength of the singularity some time ago, when it was a sink. At $\omega t = \pi$, the singularity has again strength $\sigma(0, 0, \pi/\omega) = 0$, but the near flow is moving towards the origin because slightly earlier $\sigma(0, 0, t)$ was positive, while the far flow is moving away because even earlier $\sigma(0, 0, t)$ was negative. Clearly, in order to observe this phenomenon, we had to plot the flowfield at a great distance from the source, up to 1650 m, and we also scaled the velocity vectors so that the length of the longest vector is always equal to 1. The magnitudes of the flow velocities this far from the source are negligible so that, in practice, we would not feel any measurable reverse flow at this frequency and amplitude of the source strength.

These two test cases illustrate graphically the effect of compressibility, assuming that the perturbations are small and that the flow is subsonic, inviscid and irrotational. The flowfield is scaled with respect to the flow direction and changes occurring in one part of the flow are felt elsewhere with a time delay depending on the speed of sound. In general unsteady

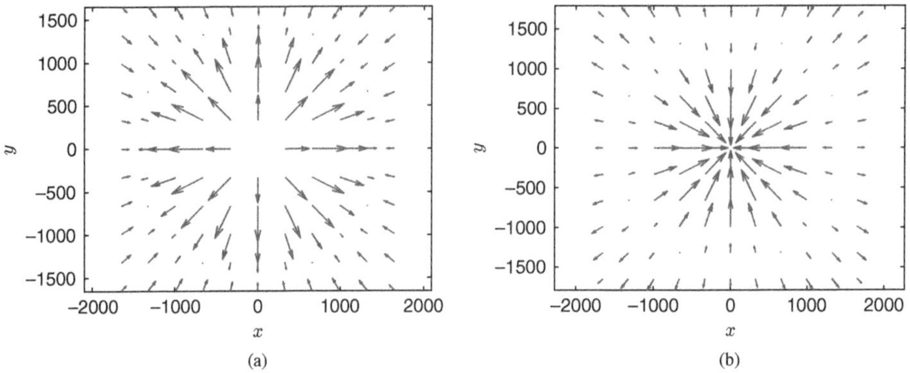

Figure 2.20 2D section of flowfield induced by a compressible point source with sinusoidally oscillating strength in a static fluid at two-phase angles. (a) $\omega t = 0$ and (b) $\omega t = \pi$.

subsonic flows, these two effects occur at the same time. The compressible doublet behaves in the same way, the illustration of the effects of the time delay on the flow induced by a doublet is left as an exercise for the reader.

By analogy to the incompressible case, compressible flowfields can be constructed by superimposing compressible source and doublet distributions on the surface of the body. Morino (1974) derived Green's theorem for compressible subsonic flow over a deforming surface $S(\mathbf{x}_s, t) = 0$. Assuming quasi-fixed aerodynamic modelling of the form of Figure 2.3, Green's identity for compressible subsonic flow over surface $S(\mathbf{x}_s) = 0$ simplifies to

$$\begin{aligned} 4\pi E(\mathbf{x}, t)\phi(\mathbf{x}, t) = &-\int_S \mathbf{n}(\mathbf{x}_s) \cdot \nabla_s \phi(\mathbf{x}_s, t - \tau(\mathbf{x}, \mathbf{x}_s)) \frac{1}{r_\beta(\mathbf{x}, \mathbf{x}_s)} \mathrm{d}S \\ &+ \int_S \phi(\mathbf{x}_s, t - \tau(\mathbf{x}, \mathbf{x}_s)) \mathbf{n}(\mathbf{x}_s) \cdot \nabla_s \left(\frac{1}{r_\beta(\mathbf{x}, \mathbf{x}_s)} \right) \mathrm{d}S \\ &- \frac{\partial}{\partial t} \int_S \frac{\phi(\mathbf{x}_s, t - \tau(\mathbf{x}, \mathbf{x}_s))}{r_\beta(\mathbf{x}, \mathbf{x}_s)} \mathbf{n}(\mathbf{x}_s) \cdot \nabla_s \tau(\mathbf{x}, \mathbf{x}_s) \mathrm{d}S \end{aligned} \tag{2.180}$$

where E is given by Eq. (2.95), while $\tau(\mathbf{x}, \mathbf{x}_s)$ and $r_\beta(\mathbf{x}, \mathbf{x}_s)$ are given by

$$\tau(\mathbf{x}, \mathbf{x}_s) = \frac{-M_\infty (x - x_s) + r_\beta(\mathbf{x}, \mathbf{x}_s)}{a_\infty \beta^2}$$

$$r_\beta(\mathbf{x}, \mathbf{x}_s) = \sqrt{(x - x_s)^2 + \beta^2 (y - y_s)^2 + \beta^2 (z - z_s)^2}$$

Equation (2.180) is the compressible equivalent of Eq. (2.95) for a quasi-fixed surface. The first two terms are source and doublet contributions, respectively, and the source strength is the normal perturbation velocity on the surface while the doublet strength is the potential on the surface. However, not only is there a third term that has no equivalent in the incompressible case, but also Eq. (2.180) expresses the value of the potential at any point in the flow and at every time instance, in terms of the potential and normal velocity on the surface at previous time instances. Therefore, the compressible Green's identity has a clear memory effect that represents the time it takes for information to travel from a point on the surface $\mathbf{x}_s$ to the point in the flow $\mathbf{x}$ where we want to calculate the potential. Memory effects due to wake shedding are still not present in Eq. (2.180) and must be added by including a wake model. Setting up and solving Eq. (2.180) in the time domain is challenging not only due to the fact that the integrals cannot be evaluated analytically for general geometries but also because the time delay τ varies in space. Numerical solutions of the equation in the frequency domain will be presented in Section 6.3. Once the potential on the surface has been calculated, Eq. (2.152) can be used in order to evaluate the pressure distribution.

2.6 Supersonic Linearised Potential Flow Solutions

The linearised small disturbance equation

$$\left(1 - M_\infty^2\right)\phi_{xx} + \phi_{yy} + \phi_{zz} - \frac{2M_\infty}{a_\infty}\phi_{xt} - \frac{1}{a_\infty^2}\phi_{tt} = 0 \tag{2.181}$$

is still valid for supersonic flow but, as $M_\infty > 1$, the ϕ_{xx} term has a negative coefficient. The equation becomes

$$-B^2\phi_{xx} + \phi_{yy} + \phi_{zz} - \frac{2M_\infty}{a_\infty}\phi_{xt} - \frac{1}{a_\infty^2}\phi_{tt} = 0 \tag{2.182}$$

where $B = \sqrt{M_\infty^2 - 1}$ is the supersonic compressibility factor. In the steady case, the linearised small disturbance equation reduces to

$$B^2\phi_{xx} - \phi_{yy} - \phi_{zz} = 0 \tag{2.183}$$

We will first look for solutions to this steady equation before generalising to unsteady flow. Equation (2.183) is a second-order hyperbolic linear partial differential equation. By analogy to the subsonic steady source solution of Eq. (2.161), we can define a supersonic steady source solution of the form

$$\phi_s(\mathbf{x}, \mathbf{x}_0) = \frac{1}{r_B(\mathbf{x}, \mathbf{x}_0)} \tag{2.184}$$

where $r_B(\mathbf{x}, \mathbf{x}_0)$ is the hyperbolic distance between a point $\mathbf{x}$ and a source lying at $\mathbf{x}_0$,

$$r_B(\mathbf{x}, \mathbf{x}_0) = \sqrt{\left(x - x_0\right)^2 - B^2\left(y - y_0\right)^2 - B^2\left(z - z_0\right)^2} \tag{2.185}$$

It can be easily verified that the source function of expression (2.161) is a solution of the steady supersonic linearised small disturbance equation by substitution into Eq. (2.182).

It is crucial to note that the supersonic source only takes real values if

$$(x - x_0)^2 \geq B^2(y - y_0)^2 + B^2(z - z_0)^2 \tag{2.186}$$

As the physical problem of flow around a body is real, we expect real solutions for the potential and velocity of the flow. Therefore, the condition of Eq. (2.186) must be observed,

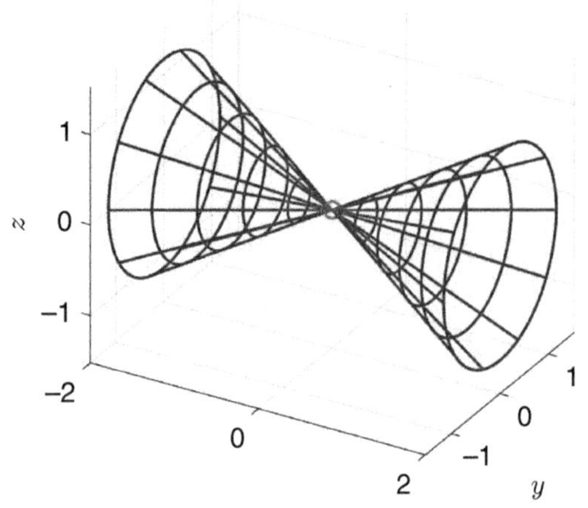

Figure 2.21 Domain of validity of a steady supersonic source for $M_\infty = 2$.

meaning that the source cannot influence all the points in the flow. For each value of x, the domain of influence of the source is delimited by the circle

$$(y - y_0)^2 + (z - z_0)^2 = \frac{(x - x_0)^2}{B^2}$$

whose radius is $(x - x_0)/B$. Figure 2.21 draws the domain of validity of a source lying at $\mathbf{x}_0 = \mathbf{0}$ in a supersonic free stream with $M_\infty = 2$. Clearly, the radius of the circles increases linearly with x so that the domain of influence is a cone whose angle with the x axis is given by

$$\mu = \tan^{-1}\left(\frac{1}{B}\right)$$

known as the Mach angle, not to be confused with the dynamic viscosity, which is also denoted by μ. Using basic trigonometric identities, it can be shown that μ is also given by the more standard expression:

$$\mu = \sin^{-1}\left(\frac{1}{M_\infty}\right) \tag{2.187}$$

It should be stressed that a supersonic source cannot induce potential in both the cones plotted in Figure 2.21. If we consider a supersonic free stream U_∞ in the x direction, then the source can only induce potential in the downstream cone. The upstream cone is the domain of dependence of point $\mathbf{x}_0$, that is the region in space from which an upstream source can induce flow there. The downstream cone is known as the Mach cone.

In order to better understand the existence and role of Mach cones, we consider an acoustic wave generator that travels with constant speed $U = Ma$ in the $-x$ direction, where $M > 1$ and a is the speed of sound. As in the subsonic case, the generator emits constantly acoustic waves that travel with the speed of sound, but it is itself moving faster than that speed. Figure 2.22 plots the wave pattern after four time instances $t_i = i\Delta t$, for $i = 0, \ldots, 3$. The acoustic wave released by the generator at time t_0 and position $x_0 = 0$ has propagated in

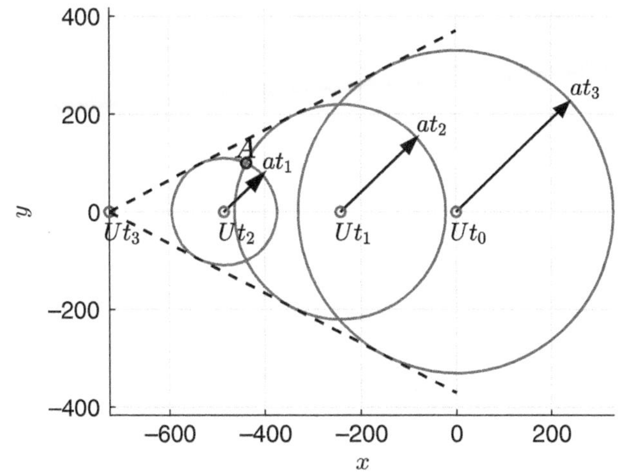

Figure 2.22 Propagation of acoustic disturbances in supersonic flow.

a circle with radius at_3, while the generator itself has moved to position $-Ut_3$. The generator now lies outside the region of space in which even the oldest sound wave has propagated. If we draw the two lines that pass by the generator at $-Ut_3$ and are tangent to all the sound circles, we obtain the Mach cone with angle μ given by Eq. (2.187). Any region of space outside this Mach cone is untouched by the sound waves.

If we now consider a static source in a moving free stream, the inequality (2.186) is necessary but not sufficient to define its domain of influence. If the free stream is supersonic and moving in the positive x direction, it will be carrying away the sound waves created by the source in the same direction. Therefore, the Mach cone is defined by

$$(x - x_0)^2 \geq B^2(y - y_0)^2 + B^2(z - z_0)^2, \text{ and, } x \geq x_0$$

Figure 2.22 also shows that the sound circles created by the source at different time instances intersect, unlike the subsonic case of Figure 2.18. For example, point A is affected by the sound wave generated at time t_1 and travelling forward and by the sound wave generated at time t_2 and travelling backward. This is a very important fact that will be used in the subsequent analysis.

Going back to the potential induced by the supersonic source, the surface of the downstream cone is a discontinuity, since $r_B = 0$ there and the source potential depends on $1/r_B$. On this surface, the potential and velocities induced by the source become infinite. As we move further from the surface and from point $\mathbf{x}_0$, the induced potential and velocities decrease in magnitude very quickly. Figure 2.23 plots a 2D section of the velocity field induced by a supersonic source, showing that the flow is nearly perpendicular to the surface of the Mach cone, although it slopes further downstream as the source location is approached. Away from the cone, the magnitude of the induced velocities is very small in comparison.

The discontinuity of the flow on the surface of the Mach cone can be used to represent oblique shock waves. In reality, oblique shocks will generally occur at angles higher than

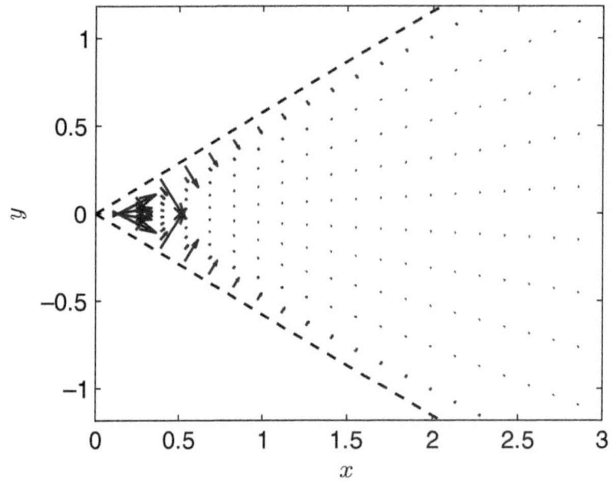

Figure 2.23 The 2D section of the velocity field induced by a steady supersonic source for $M_\infty = 2$.

the Mach angle, unless they are very weak. Hence, linearised supersonic flow can be used to represent supersonic flow either at very low supersonic Mach numbers or around very thin surfaces at very small angles of attack and slightly higher Mach numbers. Mach cones can also be used to represent the effect of expansion waves on the surface but not in the general flowfield (Edward Ehlers et al. 1979).

Given the importance of the Mach angle and Mach cone in supersonic linearised flow, we class surfaces into two categories:

- Subinclined surfaces. Surfaces that are inclined to the flow at an angle lower than the Mach angle
- Superinclined surface. Surface that are inclined to the flow at an angle higher than the Mach angle

Super-inclined surfaces will generally cause strong shocks and are not considered any further in this work, even though there are methods for dealing with them in a potential analysis.

Using the same arguments we applied to the incompressible and subsonic source functions, we can define a supersonic doublet of the form

$$\phi_d(\mathbf{x}, \mathbf{x}_0) = \mathbf{n} \cdot \nabla \left(\frac{1}{r_B} \right) \tag{2.188}$$

which is also a solution of Eq. (2.183). However, as the source function is not continuous, we need to be careful about calculating its derivative $\nabla(1/r_B)$. The supersonic doublet function is valid only inside the two cones of Figure 2.21, just like the source.

In the unsteady supersonic flow case, we can define the source strength by analogy to the subsonic case as

$$\phi(\mathbf{x}, \mathbf{x}_0, t) = \frac{F\left(t - \tau(\mathbf{x}, \mathbf{x}_0)\right)}{r_B(\mathbf{x}, \mathbf{x}_0)} \tag{2.189}$$

We can evaluate $\tau(\mathbf{x}, \mathbf{x}_0)$ using the approach of Section 2.5. We re-write Eqs. (2.172) and (2.173) after replacing β^2 by $-B^2$ to obtain

$$-2\left(-B^2\frac{\partial}{\partial x}\left(\frac{1}{r_B}\right)\frac{\partial \tau}{\partial x} - \frac{\partial}{\partial y}\left(\frac{1}{r_B}\right)\frac{\partial \tau}{\partial y} - \frac{\partial}{\partial z}\left(\frac{1}{r_B}\right)\frac{\partial \tau}{\partial z}\right)$$
$$-\frac{1}{r_B}\left(-B^2\frac{\partial^2 \tau}{\partial x^2} + \frac{\partial^2 \tau}{\partial y^2} + \frac{\partial^2 \tau}{\partial z^2}\right) - \frac{2M_\infty}{a_\infty}\frac{\partial}{\partial x}\left(\frac{1}{r_B}\right) = 0 \tag{2.190}$$

and

$$-B^2\left(\frac{\partial \tau}{\partial x}\right)^2 + \left(\frac{\partial \tau}{\partial y}\right)^2 + \left(\frac{\partial \tau}{\partial z}\right)^2 + \frac{2M_\infty}{a_\infty}\frac{\partial \tau}{\partial x} - \frac{1}{a_\infty^2} = 0 \tag{2.191}$$

As the factors contributing to the time delay are the same in subsonic and supersonic flow, we assume the form of Eq. (2.174) for $\tau(\mathbf{x}, \mathbf{x}_0)$,

$$\tau(\mathbf{x}, \mathbf{x}_0) = A_1(x - x_0) + A_2 r_B \tag{2.192}$$

and substitute it into Eq. (2.191), to obtain

$$\frac{2(x - x_0)\left(-a_\infty B^2 A_1 + M_\infty\right)}{a_\infty r_B^3} = 0$$

Clearly, this equation is always satisfied when

$$A_1 = \frac{M_\infty}{a_\infty B^2}$$

Next, we substitute Eq. (2.192) and the value of A_1 into Eq. (2.190) and obtain

$$\frac{A_2^2 a^2 B^4 - 1}{a_\infty^2 B^2} = 0$$

so that the value of A_2 becomes

$$A_2 = \pm\frac{1}{a_\infty B^2}$$

In the subsonic case, we retained only the positive value of A_2 because the negative value resulted in negative time delays. In the supersonic case, $A_1 > 0$, $M_\infty > 1$ and both values of A_2 can result in positive time delays. Furthermore, Figure 2.22 shows that every point inside the Mach cone is affected by the same source twice at each instance in time, once by a forward travelling wave and once by a backward travelling wave. Therefore, the two solutions for A_2 represent these two waves and we will retain both of them; substituting the values of A_1 and A_2 into expression (2.192), we obtain

$$\tau_{\pm}(\mathbf{x}, \mathbf{x}_0) = \frac{M_\infty(x - x_0) \pm r_B(\mathbf{x}, \mathbf{x}_0)}{a_\infty B^2} \tag{2.193}$$

such that there are two time delays of interest in supersonic flow. Consequently, the potential induced by an unsteady supersonic source is given by

$$\phi(\mathbf{x}, \mathbf{x}_0, t) = \frac{1}{r_B(\mathbf{x}, \mathbf{x}_0)} \left(F\left(t - \tau_+(\mathbf{x}, \mathbf{x}_0)\right) + F\left(t - \tau_-(\mathbf{x}, \mathbf{x}_0)\right)\right) \tag{2.194}$$

and the unsteady supersonic doublet becomes

$$\phi(\mathbf{x}, \mathbf{x}_0, t) = \mathbf{n} \cdot \nabla \left(\frac{1}{r_B(\mathbf{x}, \mathbf{x}_0)} \left(F\left(t - \tau_+(\mathbf{x}, \mathbf{x}_0)\right) + F\left(t - \tau_-(\mathbf{x}, \mathbf{x}_0)\right)\right)\right) \tag{2.195}$$

Morino (1974) shows that Green's theorem for unsteady supersonic flow over a closed, quasi-fixed surface $S(\mathbf{x}_s) = 0$ can be written as

$$\begin{aligned}
4\pi E(\mathbf{x}, t)\phi(\mathbf{x}, t) = & -\int_S \mathbf{n}(\mathbf{x}_s) \cdot \nabla\phi(\mathbf{x}_s, t - \tau_+(\mathbf{x}, \mathbf{x}_s))\frac{1}{r_B(\mathbf{x}, \mathbf{x}_s)} \mathrm{d}S \\
& -\int_S \mathbf{n}(\mathbf{x}_s) \cdot \nabla\phi(\mathbf{x}_s, t - \tau_-(\mathbf{x}, \mathbf{x}_s))\frac{1}{r_B(\mathbf{x}, \mathbf{x}_s)} \mathrm{d}S \\
& +\int_S \phi(\mathbf{x}_s, t - \tau_+(\mathbf{x}, \mathbf{x}_s))\mathbf{n}(\mathbf{x}_s) \cdot \nabla \left(\frac{1}{r_B(\mathbf{x}, \mathbf{x}_s)}\right) \mathrm{d}S \\
& +\int_S \phi(\mathbf{x}_s, t - \tau_-(\mathbf{x}, \mathbf{x}_s))\mathbf{n}(\mathbf{x}_s) \cdot \nabla \left(\frac{1}{r_B(\mathbf{x}, \mathbf{x}_s)}\right) \mathrm{d}S \\
& -\frac{\partial}{\partial t}\int_S \frac{\phi(\mathbf{x}_s, t - \tau_+(\mathbf{x}, \mathbf{x}_s))}{r_B(\mathbf{x} - \mathbf{x}_s)} \mathbf{n}(\mathbf{x}_s) \cdot \nabla\tau_+(\mathbf{x}, \mathbf{x}_s)\mathrm{d}S \\
& -\frac{\partial}{\partial t}\int_S \frac{\phi(\mathbf{x}_s, t - \tau_-(\mathbf{x}, \mathbf{x}_s))}{r_B(\mathbf{x} - \mathbf{x}_s)} \mathbf{n}(\mathbf{x}_s) \cdot \nabla\tau_-(\mathbf{x}, \mathbf{x}_s)\mathrm{d}S
\end{aligned} \tag{2.196}$$

where $E(\mathbf{x}, t)$ is given by Eq. (2.95), while

$$\tau_{\pm}(\mathbf{x}, \mathbf{x}_s) = \frac{M_\infty(x - x_s) \pm r_B(\mathbf{x}, \mathbf{x}_s)}{a_\infty B^2}$$

$$r_B(\mathbf{x}, \mathbf{x}_s) = \sqrt{(x - x_s)^2 - B^2(y - y_s)^2 - B^2(z - z_s)^2}$$

As in the subsonic case, there is a clear memory effect since the potential anywhere in the flow depends on previous values of the potential and normal velocity on the surface. Comparing to the subsonic Green's identity of Eq. (2.180), there are twice as many terms in the supersonic case due to the effect of the forward and backward travelling waves. Equation (2.196) cannot be set up and solved analytically for general geometries; numerical solutions in the frequency domain will be presented in Section 6.4.

2.7 Vorticity and Circulation

Up to this point, all flows have been assumed to be irrotational by enforcing Eq. (2.12). The irrotationality assumption was used to simplify the flow equations such that they become linear and potential solutions of the form (2.95), (2.180), and (2.196) can be derived. Irrotationality is expressed mathematically by Eq. (2.12),

$$\nabla \times \mathbf{u} = \mathbf{0} \tag{2.197}$$

and can be used in regions of the flow where viscous phenomena are not important. Some of the regions where viscous phenomena cannot be ignored are

- a thin flow region in contact with the surface, known as the boundary layer,
- separated flow regions,
- mixing layers between two flows.

Inside these regions, the pattern of the flow is affected by the relative importance between inertial and viscous forces, expressed in terms of the Reynolds number

$$\mathrm{Re} = \frac{\rho U L}{\mu}$$

where ρ, U, L and μ are characteristic values of the flow density, airspeed, length-scale and viscosity. For example, a boundary layer whose Reynolds number is low tends to be laminar, while one with a high *Re* value tends to be turbulent. For inviscid flows, the Reynolds number is infinite. Boundary layers will be discussed in more detail in Chapter 7.

In regions where the flow cannot be treated as irrotational, the vector product of Eq. (2.197) is non-zero and is known as the vorticity

$$\boldsymbol{\omega} = \nabla \times \mathbf{u} \tag{2.198}$$

such that, in cartesian coordinates, the three components of the vorticity vector are given by

$$\omega_x = \frac{\partial w}{\partial y} - \frac{\partial v}{\partial z}, \quad \omega_y = \frac{\partial u}{\partial z} - \frac{\partial w}{\partial x}, \quad \omega_z = \frac{\partial v}{\partial x} - \frac{\partial u}{\partial y} \tag{2.199}$$

The momentum equations can be written in terms of the vorticity, eliminating the pressure. For incompressible flow, continuity is given by Eq. (2.21)

$$\frac{\partial u}{\partial x} + \frac{\partial v}{\partial y} + \frac{\partial w}{\partial z} = 0 \tag{2.200}$$

and the Navier–Stokes equations (2.2)–(2.4) become

$$\frac{\partial u}{\partial t} + u\frac{\partial u}{\partial x} + v\frac{\partial u}{\partial y} + w\frac{\partial u}{\partial z} = -\frac{1}{\rho}\frac{\partial p}{\partial x} + \nu\left(\frac{\partial^2 u}{\partial x^2} + \frac{\partial^2 u}{\partial y^2} + \frac{\partial^2 u}{\partial z^2}\right) \tag{2.201}$$

$$\frac{\partial v}{\partial t} + u\frac{\partial v}{\partial x} + v\frac{\partial v}{\partial y} + w\frac{\partial v}{\partial z} = -\frac{1}{\rho}\frac{\partial p}{\partial y} + \nu\left(\frac{\partial^2 v}{\partial x^2} + \frac{\partial^2 v}{\partial y^2} + \frac{\partial^2 v}{\partial z^2}\right) \tag{2.202}$$

$$\frac{\partial w}{\partial t} + u\frac{\partial w}{\partial x} + v\frac{\partial w}{\partial y} + w\frac{\partial w}{\partial z} = -\frac{1}{\rho}\frac{\partial p}{\partial z} + \nu\left(\frac{\partial^2 w}{\partial x^2} + \frac{\partial^2 w}{\partial y^2} + \frac{\partial^2 w}{\partial z^2}\right) \tag{2.203}$$

where $\nu = \mu/\rho$ is the kinematic viscosity and where we have subtracted Eq. (2.200) multiplied by u, v and w from Eqs. (2.2)–(2.4) in order to obtain Eqs. (2.201)–(2.203), respectively. We differentiate Eq. (2.203) by y, Eq. (2.202) by z and subtract the two to obtain

$$\frac{\partial}{\partial t}\left(\frac{\partial w}{\partial y} - \frac{\partial v}{\partial z}\right) + u\frac{\partial}{\partial x}\left(\frac{\partial w}{\partial y} - \frac{\partial v}{\partial z}\right) + v\frac{\partial}{\partial y}\left(\frac{\partial w}{\partial y} - \frac{\partial v}{\partial z}\right) + w\frac{\partial}{\partial z}\left(\frac{\partial w}{\partial y} - \frac{\partial v}{\partial z}\right)$$

$$+\frac{\partial u}{\partial y}\frac{\partial w}{\partial x}+\frac{\partial v}{\partial y}\frac{\partial w}{\partial y}+\frac{\partial w}{\partial y}\frac{\partial w}{\partial z}-\frac{\partial u}{\partial z}\frac{\partial v}{\partial x}-\frac{\partial v}{\partial z}\frac{\partial v}{\partial y}-\frac{\partial w}{\partial z}\frac{\partial v}{\partial z}$$
$$=\nu\left(\frac{\partial^2}{\partial x^2}\left(\frac{\partial w}{\partial y}-\frac{\partial v}{\partial z}\right)+\frac{\partial^2}{\partial y^2}\left(\frac{\partial w}{\partial y}-\frac{\partial v}{\partial z}\right)+\frac{\partial^2}{\partial z^2}\left(\frac{\partial w}{\partial y}-\frac{\partial v}{\partial z}\right)\right)$$

Note that the pressure term has been eliminated because both equations result in $\partial^2 p/\partial y\partial z$ after the differentiations and they are then subtracted. Substituting for the terms $\partial v/\partial y$ and $\partial w/\partial z$ in the left-hand side from the continuity equation

$$\frac{\partial v}{\partial y}=-\frac{\partial u}{\partial x}-\frac{\partial w}{\partial z},\ \frac{\partial w}{\partial z}=-\frac{\partial u}{\partial x}-\frac{\partial v}{\partial y}$$

and adding and subtracting the term $(\partial u/\partial y)(\partial u/\partial z)$ leads to

$$\frac{\partial}{\partial t}\left(\frac{\partial w}{\partial y}-\frac{\partial v}{\partial z}\right)+u\frac{\partial}{\partial x}\left(\frac{\partial w}{\partial y}-\frac{\partial v}{\partial z}\right)+v\frac{\partial}{\partial y}\left(\frac{\partial w}{\partial y}-\frac{\partial v}{\partial z}\right)+w\frac{\partial}{\partial z}\left(\frac{\partial w}{\partial y}-\frac{\partial v}{\partial z}\right)$$
$$+\left(\frac{\partial v}{\partial z}-\frac{\partial w}{\partial y}\right)\frac{\partial u}{\partial x}+\left(\frac{\partial w}{\partial x}-\frac{\partial u}{\partial z}\right)\frac{\partial v}{\partial y}+\left(\frac{\partial u}{\partial y}-\frac{\partial v}{\partial x}\right)\frac{\partial w}{\partial z}$$
$$=\nu\left(\frac{\partial^2}{\partial x^2}\left(\frac{\partial w}{\partial y}-\frac{\partial v}{\partial z}\right)+\frac{\partial^2}{\partial y^2}\left(\frac{\partial w}{\partial y}-\frac{\partial v}{\partial z}\right)+\frac{\partial^2}{\partial z^2}\left(\frac{\partial w}{\partial y}-\frac{\partial v}{\partial z}\right)\right)$$

Finally, substituting from Eq. (2.199), we obtain

$$\frac{\partial \omega_x}{\partial t}+u\frac{\partial \omega_x}{\partial x}+v\frac{\partial \omega_x}{\partial y}+w\frac{\partial \omega_x}{\partial z}-\omega_x\frac{\partial u}{\partial x}-\omega_y\frac{\partial u}{\partial y}-\omega_z\frac{\partial u}{\partial z}$$
$$=\nu\left(\frac{\partial^2\omega_x}{\partial x^2}+\frac{\partial^2\omega_x}{\partial y^2}+\frac{\partial^2\omega_x}{\partial z^2}\right)$$

which is known as the vorticity transport equation in the x direction. It can be written in more compact vector form as

$$\frac{\partial \omega_x}{\partial t}+\mathbf{u}\cdot\nabla\omega_x=\boldsymbol{\omega}\cdot\nabla u+\nu\nabla^2\omega_x \tag{2.204}$$

Repeating the procedure for the other two directions, we obtain

$$\frac{\partial \omega_y}{\partial t}+\mathbf{u}\cdot\nabla\omega_y=\boldsymbol{\omega}\cdot\nabla v+\nu\nabla^2\omega_y \tag{2.205}$$

$$\frac{\partial \omega_z}{\partial t}+\mathbf{u}\cdot\nabla\omega_z=\boldsymbol{\omega}\cdot\nabla w+\nu\nabla^2\omega_z \tag{2.206}$$

Equations (2.204)–(2.206) are second order nonlinear partial differential equations and they are not easier to solve that than the Navier–Stokes equations. However, they can be used in order to derive important conclusions about the vorticity in a flow. They are often written in vector form

$$\frac{\mathrm{D}\boldsymbol{\omega}}{\mathrm{D}t}=(\boldsymbol{\omega}\cdot\nabla)\mathbf{u}+\nu\nabla^2\boldsymbol{\omega} \tag{2.207}$$

where $\mathrm{D}/\mathrm{D}t$ is the substantial or material or Lagrangian derivative

$$\frac{\mathrm{D}}{\mathrm{D}t}=\frac{\partial}{\partial t}+u\frac{\partial}{\partial x}+v\frac{\partial}{\partial y}+w\frac{\partial}{\partial z}$$

so that

$$\frac{\mathrm{D}\boldsymbol{\omega}}{\mathrm{D}t} = \frac{\partial \boldsymbol{\omega}}{\partial t} + (\mathbf{u} \cdot \nabla)\boldsymbol{\omega}$$

The term $(\boldsymbol{\omega} \cdot \nabla)\mathbf{u}$ in Eq. (2.207) is known as the vorticity stretching term while $\nu\nabla^2\boldsymbol{\omega}$ is known as the vorticity diffusion term.

Consider a 2D flow, for which w does not exist and the only component of vorticity is $\omega \equiv \omega_z$. The single vorticity transport Eq. (2.206) simplifies to

$$\frac{\partial \omega}{\partial t} + u\frac{\partial \omega}{\partial x} + v\frac{\partial \omega}{\partial y} = \nu\left(\frac{\partial^2 \omega}{\partial x^2} + \frac{\partial^2 \omega}{\partial y^2}\right) \tag{2.208}$$

so that the stretching term disappears. If this flow is now assumed to be inviscid, we obtain

$$\frac{\partial \omega}{\partial t} + u\frac{\partial \omega}{\partial x} + v\frac{\partial \omega}{\partial y} = \frac{\mathrm{D}\omega}{\mathrm{D}t} = 0 \tag{2.209}$$

Equation (2.209) states that, in a 2D inviscid and incompressible flow, the vorticity of any fluid element remains constant. It follows that the total vorticity of all the fluid elements in the flow is also constant. The sum of the vorticity of all the elements is known as circulation. The concept of constant circulation can be extended to 3D flows inside bounded regions.

Consider a bounded surface S in 3D space, whose boundary is the contour C, as seen in Figure 2.24. If this surface is a slice of rotational flow, then the value of the vorticity will not be zero on S and, in general, will vary across S and in time. The circulation, Γ, is defined as the sum of the vorticity over S, in a direction normal to S, that is

$$\Gamma = \int_S \boldsymbol{\omega} \cdot \mathbf{n}\mathrm{d}S \tag{2.210}$$

where $\mathbf{n}$ is a unit vector normal to the surface and $\mathrm{d}S$ is an infinitesimal area of S. Substituting for $\boldsymbol{\omega}$ from Eq. (2.198) and applying Stokes' theorem, we obtain

$$\Gamma = \int_S \boldsymbol{\omega} \cdot \mathbf{n}\mathrm{d}S = \int_S (\nabla \times \mathbf{u}) \cdot \mathbf{n}\mathrm{d}S = \oint_C \mathbf{u} \cdot \boldsymbol{\tau}\mathrm{d}s$$

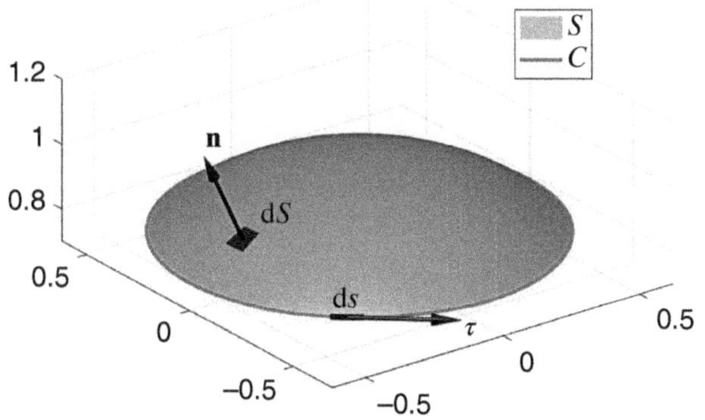

Figure 2.24 A surface S bounded by a contour C for the definition of the circulation.

where $\oint_C$ indicates that the integration is carried out over a complete circuit of the contour C, $\mathrm{d}s = \sqrt{\mathrm{d}x^2 + \mathrm{d}y^2 + \mathrm{d}z^2}$ is an infinitesimal arc length of C and $\boldsymbol{\tau}$ is the unit vector tangent to this contour, given by

$$\boldsymbol{\tau} = \frac{\mathrm{d}\mathbf{x}}{\mathrm{d}s} \tag{2.211}$$

Therefore, circulation is usually defined as the sum of the flow velocity along a closed contour C, in a direction tangent to C, that is

$$\Gamma = \oint_C \mathbf{u} \cdot \boldsymbol{\tau} \mathrm{d}s \tag{2.212}$$

Now consider a case where the fluid elements in surface S translate and rotate with the local flow velocity so that S deforms but remains bounded and contains the same fluid elements. Consequently, C also deforms with the local flow velocity but remains a closed path. The Lagrangian derivative of Γ is given by

$$\frac{\mathrm{D}\Gamma}{\mathrm{D}t} = \frac{\mathrm{D}}{\mathrm{D}t}\oint_C \mathbf{u} \cdot \boldsymbol{\tau} \mathrm{d}s = \oint_C \frac{\mathrm{D}\mathbf{u}}{\mathrm{D}t} \cdot \boldsymbol{\tau} \mathrm{d}s + \oint_C \mathbf{u} \cdot \frac{\mathrm{D}}{\mathrm{D}t}(\boldsymbol{\tau} \mathrm{d}s)$$

Assuming that the flow is inviscid, $\mathrm{D}\mathbf{u}/\mathrm{D}t$ is given by the Euler equations (2.9)–(2.11), that is

$$\frac{\mathrm{D}\mathbf{u}}{\mathrm{D}t} = -\nabla\left(\frac{p}{\rho}\right)$$

Furthermore, it can be shown that

$$\frac{\mathrm{D}}{\mathrm{D}t}(\boldsymbol{\tau} \mathrm{d}s) = (\boldsymbol{\tau} \mathrm{d}s) \cdot \nabla \mathbf{u}$$

since the curve C moves with the local flow velocity (see for example (Paterson 1984)). The Lagrangian derivative of Γ becomes

$$\frac{\mathrm{D}\Gamma}{\mathrm{D}t} = -\oint_C \nabla\left(\frac{p}{\rho}\right) \cdot \boldsymbol{\tau} \mathrm{d}s + \oint_C \mathbf{u} \cdot (\boldsymbol{\tau} \mathrm{d}s) \cdot \nabla \mathbf{u} = \oint_C \nabla\left(-\frac{p}{\rho} + \frac{1}{2}\nabla Q^2\right) \cdot \boldsymbol{\tau} \mathrm{d}s$$

where $Q^2 = \mathbf{u} \cdot \mathbf{u}$. Noting from Eq. (2.211) that $\boldsymbol{\tau} \mathrm{d}s = (\mathrm{d}x, \mathrm{d}y, \mathrm{d}z)$,

$$\frac{\mathrm{D}\Gamma}{\mathrm{D}t} = \oint_C \mathrm{d}\left(-\frac{p}{\rho} + \frac{1}{2}\nabla Q^2\right)$$

The integrand is a perfect differential and the integral is evaluated over a closed path so that its result is zero if p/ρ and $Q^2/2$ are single-valued functions. Therefore,

$$\frac{\mathrm{D}\Gamma}{\mathrm{D}t} = 0 \tag{2.213}$$

Equation (2.213) is known as Kelvin's theorem. Note that, since Γ is an integrated quantity, it does not depend on $\mathbf{x}$, so that Kelvin's theorem can also be written as

$$\frac{\mathrm{d}\Gamma}{\mathrm{d}t} = 0 \tag{2.214}$$

2.7.1 Solutions of the Vorticity Transport Equations

As mentioned earlier, Eq. (2.207) are non-linear partial differential equations so that they have no general solutions. However, if the flow we are modelling has particular properties, the non-linear terms can disappear and the resulting linear equations can be solved analytically. In order to demonstrate such a flow, we write the 2D vorticity transport Eq. (2.208) in polar coordinates, r, θ

$$\frac{\partial \omega}{\partial t} + u_r \frac{\partial \omega}{\partial r} + \frac{u_\theta}{r}\frac{\partial \omega}{\partial \theta} = \nu \left(\frac{1}{r}\frac{\partial}{\partial r}\left(r\frac{\partial \omega}{\partial r} \right) + \frac{1}{r^2}\frac{\partial^2 \omega}{\partial \theta^2} \right) \tag{2.215}$$

where u_r, u_θ are the flow velocities in the r, θ directions, respectively. Consider a flow with the properties that $u_r = 0$ and $u_\theta = u_\theta(r, t)$, $\omega = \omega(r, t)$, such that the velocity in the θ direction and the vorticity are not functions of θ. This flow satisfies automatically the incompressible continuity equation in polar coordinates:

$$\frac{\partial u_r}{\partial r} + \frac{u_r}{r} + \frac{1}{r}\frac{\partial u_\theta}{\partial \theta} = 0$$

Furthermore, the non-linear terms in Eq. (2.215) disappear, such that the vorticity transport equation becomes

$$\frac{\partial \omega}{\partial t} = \nu \left(\frac{1}{r}\frac{\partial \omega}{\partial r} + \frac{\partial^2 \omega}{\partial r^2} \right)$$

which is a linear partial differential equation with known solutions. It can be easily verified that one such solution is

$$\omega(r, t) = \frac{c}{4\pi \nu t} \mathrm{e}^{-r^2/4\nu t} \tag{2.216}$$

where c is a constant. The circulation inside a circle of radius r is obtained by adapting Eq. (2.210) to 2D flows,

$$\Gamma(r, t) = \int_S \omega \mathrm{d}S = \int_0^r \int_0^{2\pi} \omega(r, t) r\mathrm{d}\theta \mathrm{d}r = 2\pi \int_0^r \omega(r, t) r\mathrm{d}r$$

since $\mathrm{d}S = r\mathrm{d}\theta\mathrm{d}r$ and ω does not depend on θ. Substituting from Eq. (2.216) and carrying out the integration yields

$$\Gamma(r, t) = 2\pi \int_0^r \frac{c}{4\pi \nu t} \mathrm{e}^{-r^2/4\nu t} r\mathrm{d}r = c\left(1 - \mathrm{e}^{-r^2/4\nu t}\right) \tag{2.217}$$

At $t = 0$, the circulation for any $r > 0$ is then given by

$$\Gamma(0) = c$$

which means that the constant c in Eq. (2.216) is in fact the initial circulation in the flow, $\Gamma(0)$. The circulation inside the circle r becomes

$$\Gamma(r, t) = \Gamma(0)\left(1 - \mathrm{e}^{-r^2/4\nu t}\right) \tag{2.218}$$

so that it decays to zero as t tends to infinity for any finite value of r. We can then rewrite Eq. (2.216) as

$$\omega(r,t) = \frac{\Gamma(0)}{4\pi\nu t} e^{-r^2/4\nu t} \tag{2.219}$$

All that remains to do is to calculate the velocity u_θ. In 2D, the definition of vorticity (2.199) reduces to

$$\omega = \frac{\partial v}{\partial x} - \frac{\partial u}{\partial y} = \frac{1}{r}\left(\frac{\partial}{\partial r}(ru_\theta) - \frac{\partial u_r}{\partial \phi}\right)$$

In our flow $u_r = 0$, so that we can calculate u_θ from

$$\frac{1}{r}\frac{\partial}{\partial r}(ru_\theta) = \omega \tag{2.220}$$

or after integrating over r,

$$u_\theta(r,t) = \frac{1}{r}\int_0^r r\omega \mathrm{d}r = \frac{\Gamma(0)}{2\pi r}\left(1 - e^{-r^2/4\nu t}\right) \tag{2.221}$$

Equations (2.219) and (2.221) define a flow known as the Lamb–Oseen vortex (Lamb 1932). The fluid rotates around the origin with a speed that depends on the radial distance and on time. Figure 2.25a plots u_θ as a function of r for a vortex with $\Gamma(0) = 1$ in water at four time instances. The quantity $\Gamma(0)$ is known as the initial vortex strength. At $t = 0$, $u_\theta(r, 0)$ decreases monotonically from infinity to zero as r increases. However, at all time instances $t > 0$, $u_\theta(r,t)$ is zero at the origin, increases to a maximum as r reaches a certain value $r_c(t)$ and then decreases towards zero as $r \to \infty$. As time increases, $\max(u_\theta)$ decreases and r_c increases. When $t \to \infty$ both $u_\theta(r)$ and $\omega(r)$ tend to zero for all r. Figure 2.25b plots the instantaneous velocity field of the Lamb–Oseen vortex at $t = 20$ s and at four different radial distances from the origin. The flow is circumferential everywhere, since $u_r = 0$. Furthermore, the velocity vectors are longest at $r = 0.01$ m, since this radial distance lies close to r_c at $t = 20$ s. The section of the vortex between $r = 0$ and r_c, in which u_θ increases, is known

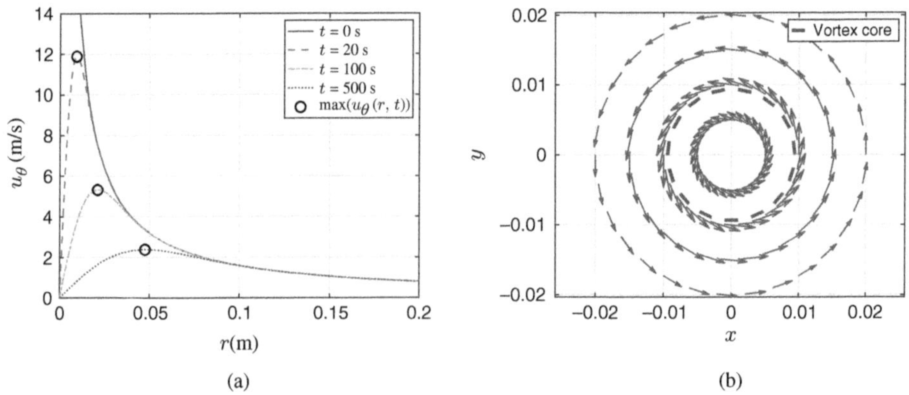

Figure 2.25 Lamb–Oseen vortex. (a) Velocity in θ direction and (b) Instantaneous velocity field, $t = 20$ s.

as the vortex core. Figure 2.25a shows that the radius of the vortex core, r_c, increases with time. The derivative of u_θ with respect to r is

$$\frac{\partial u_\theta}{\partial r} = \frac{\Gamma(0)}{2\pi r^2}\left(e^{-r^2/4\nu t}\left(1+\frac{r^2}{2\nu t}\right)-1\right)$$

The value of r_c is obtained when this derivative is equal to zero, that is when

$$e^{-r_c^2/4\nu t}\left(1+\frac{r_c^2}{2\nu t}\right)-1=0$$

Without loss of generality, we select for r_c the form $r_c(t) = \sqrt{4\kappa\nu t}$, such that the expression above becomes

$$e^{-\kappa}(1+2\kappa)-1=0$$

which is a non-linear equation for κ that can be solved numerically to yield $\kappa = 0$ and $\kappa = 1.256$. The maximum of u_θ occurs for the second solution so that the value of the vortex core radius is

$$r_c(t) = \sqrt{4\kappa\nu t} = 2.242\sqrt{\nu t} \tag{2.222}$$

so that r_c increases indeed monotonically with time.

The Lamb–Oseen vortex demonstrates an aspect of the behaviour of free vortices in viscous flows. The radius of the vortex core increases and the velocities induced by the vortex decrease with time, until the effect of the vortex becomes negligible. For inviscid flows, the kinematic viscosity is zero and therefore Eq. (2.221) simplifies to the inviscid 2D point vortex:

$$u_\theta(r) = \frac{\Gamma}{2\pi r} \tag{2.223}$$

which is independent of time so that the vortex does not dissipate. The vortex strength, Γ, is also a constant. It should be noted that the velocity induced by an inviscid vortex is identical to that induced by the Lamb–Oseen vortex of Eq. (2.221) at $t = 0$. Equation (2.219) is no longer valid, since it is a solution of the viscous vorticity transport equation. The vorticity in the flow can be calculated from Eq. (2.220), after substituting for u_θ from Eq. (2.223), so that

$$\omega = \frac{1}{r}\frac{\partial}{\partial r}\left(\frac{\Gamma}{2\pi}\right) = \frac{0}{r}$$

Therefore, the vorticity of this inviscid flow is zero everywhere, except at $r = 0$ where it is undefined (the same is true for the Lamb–Oseen vortex at $t = 0$). The inviscid 2D point vortex is described in more detail in Section A.2.

Several other vortex models have been developed, based either on solutions of the vorticity transport equations or on empirical observation. An important difficulty with the Lamb–Oseen model is that it induces infinite velocity at the origin for $t = 0$. The same issue affects the inviscid 2D point vortex at all times. Different phenomenological vortex core models have been proposed in order to overcome this difficulty; Vatistas et al. (1991) suggested a generalisation of several of these models, described by

$$u_\theta(r) = \frac{\Gamma(0)}{2\pi}\frac{r}{\left(r_c^{2n}+r^{2n}\right)^{1/n}} \tag{2.224}$$

where n is an exponent. Setting $n = \infty$ gives rise to the Rankine vortex, whereby the core rotates as a rigid body. Setting $n = 1$ yields the Scully vortex, while $n = 2$ is similar to the Lamb-Oseen model. These vortex core models do not give any information on the value of the vortex core radius. If r_c is taken as a constant, the vortices do not dissipate. Anathan and Leishman (2004) combined the Vatistas model with a modified version of the Lamb–Oseen model such that the velocity induced by the vortex is given by Eq. (2.224) with $n = 2$, while the vortex core radius grows in time according to

$$r_c(t) = \sqrt{r_c(0)^2 + 4\kappa\delta\nu t} \tag{2.225}$$

where $r_c(0)$ is the vortex core radius at $t = 0$ and δ is a correction factor accounting for turbulent eddy viscosity. For laminar flows, $\delta = 1$. Figure 2.26 plots the velocities u_θ induced by the Vatistas–Leishman model of Eqs. (2.224) and (2.225), for $r_c(0) = 0.01$ m, $\delta = 1$, and compares them to the predictions of the Lamb–Oseen model in Figure 2.25a, at the same four time instances. The two sets of predictions are nearly identical for $r > 0.05$, but the Vatistas–Leishman values are attenuated for smaller values of r due to the effect of $r_c(0)$, especially at the earlier time instances. Finally, the circulation inside a circle of radius r can be calculated from Eq. (2.212)

$$\Gamma(r,t) = \int_0^{2\pi} u_\theta(r,t) r d\theta = \frac{\Gamma(0) r^2}{\left(r_c(t)^{2n} + r^{2n}\right)^{1/n}} \tag{2.226}$$

As $r_c(t)$ increases monotonically with time, $\Gamma(r,t)$ tends to zero as $t \to \infty$ for any finite value of r. This decrease can be very slow due to the small value of ν in air and water.

Point vortices cannot exist in three-dimensional flows. Vorticity in 3D can be found in vortex filaments, tubes, rings and sheets. As tubes, rings and sheets can be mathematically described in terms of filaments, we will concentrate here on the latter. An inviscid vortex filament is a continuous infinite line in 3D space whose strength, Γ, is constant along its length. Figure 2.27 draws such a filament and a general point P in 3D space. It also draws an infinitesimal segment of the filament, **ds**, where s denotes the coordinate along the filament

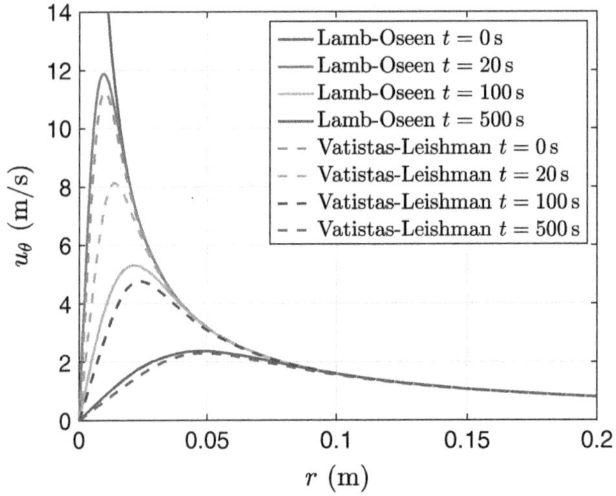

Figure 2.26 Comparison of velocities induced by the Lamb–Oseen and Vatistas-Leishman vortex models.

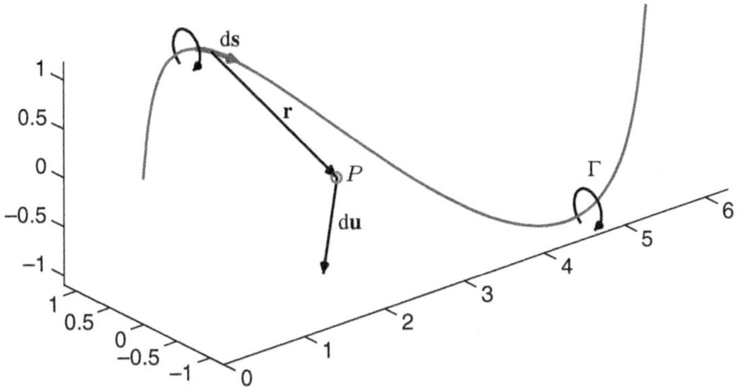

Figure 2.27 Velocity induced by segment of vortex filament on point P.

that induces a velocity $d\mathbf{u}$ at point P. The distance from $d\mathbf{s}$ to P is $\mathbf{r}(s)$. Note that the strength of the vortex filament is Γ around any point s.

The Biot–Savart law states that the velocity $d\mathbf{u}$ induced by segment $d\mathbf{s}$ on point P is given by

$$d\mathbf{u} = \frac{\Gamma}{4\pi} \frac{d\mathbf{s} \times \mathbf{r}(s)}{||\mathbf{r}(s)||^3} \tag{2.227}$$

The function $\mathbf{r}(s)/||\mathbf{r}(s)||^3$ is in fact the vector derivative of the source function of Eq. (2.49)

$$\frac{\mathbf{r}(s)}{||\mathbf{r}(s)||^3} = \nabla \left(\frac{1}{||\mathbf{r}(s)||} \right) \tag{2.228}$$

or a vector made up of doublets in the three directions. Since the doublet is a solution of Laplace's equation, so is the vortex filament.

The total velocity $\mathbf{u}$ induced by the vortex filament at point P is the integral of $d\mathbf{u}$ such that

$$\mathbf{u} = \frac{\Gamma}{4\pi} \int_{-\infty}^{+\infty} \frac{d\mathbf{s} \times \mathbf{r}(s)}{||\mathbf{r}(s)||^3}$$

In order to carry out this integration, we can write $d\mathbf{s}$ as $d\mathbf{s} = \boldsymbol{\tau}(s)ds$, where $\boldsymbol{\tau}(s)$ is the unit vector tangent to the filament at position s. Consequently, the integral becomes

$$\mathbf{u} = \frac{\Gamma}{4\pi} \int_{-\infty}^{+\infty} \frac{\boldsymbol{\tau}(s) \times \mathbf{r}(s)}{||\mathbf{r}(s)||^3} ds \tag{2.229}$$

Vortex filaments obey Helmholtz's theorems:

1. The strength of a vortex filament is constant along its length.
2. A vortex filament cannot end in space. It can either extend to the boundaries of the fluid (or infinity) or form a closed ring.
3. Fluid elements that are initially irrotational remain irrotational if there are not external forces.

Consequently, fluid elements travel with the vortex filament they enclose. As a fluid element will either contain no vorticity or a vortex filament of strength Γ for ever, it follows that the circulation around any fluid element can never change, which means that Helmholtz's third theorem is fully compatible with Kelvin's theorem (Wu 2018). As the vortex filament is a solution to Laplace's equation, it can be used to set up and solve aerodynamic problems. Such a solution will be demonstrated in Section 5.4.

2.7.2 Vorticity-Moment and Kutta–Joukowski Theorems

Wu (1981, 2018) developed a theory that relates the unsteady aerodynamic loads acting on a wing in an incompressible viscous flow to the change in vorticity in that flow. Consider the motion case of Figure 2.2 whereby a rigid solid body with volume V_B immersed in a steady free stream $\mathbf{Q}_\infty$ translates with unsteady velocity $\mathbf{v}_0(t)$ and rotates with rotational velocity $\mathbf{\Omega}_0(t)$ around the centre of rotation $\mathbf{x}_f$. Wu defines the first moment of the vorticity field as

$$\boldsymbol{\alpha} = \int_{V_\infty} (\mathbf{x} - \mathbf{x}_0) \times \boldsymbol{\omega} \mathrm{d}V \tag{2.230}$$

where V_∞ is a very large volume surrounding the flow and $\mathbf{x}_0$ is a static datum. Using Newton's second law, Wu shows that the aerodynamic force, $\mathbf{F}$ acting on the body is given by

$$\mathbf{F} = -\frac{\rho}{2}\frac{\mathrm{d}\boldsymbol{\alpha}}{\mathrm{d}t} + \rho\frac{\mathrm{d}}{\mathrm{d}t}\int_{V_B} \left(\mathbf{v}_0(t) + \mathbf{\Omega}_0(t) \times (\mathbf{x}_0(t) - \mathbf{x}_f)\right) \mathrm{d}V \tag{2.231}$$

where $\mathbf{v}_0(t) + \mathbf{\Omega}_0(t) \times (\mathbf{x}_0(t) - \mathbf{x}_f)$ is the unsteady velocity of all points in the body of volume V_B. Consequently, if $\mathbf{v}_0(t)$ and $\mathbf{\Omega}_0(t)$ are equal to zero, that is if the body does not move with respect to the free stream, the second term in Eq. (2.231) is equal to zero. For 2D flows, Wu shows that the aerodynamic force, $\mathbf{f}$, is given by

$$\mathbf{f} = -\rho\frac{\mathrm{d}\boldsymbol{\alpha}}{\mathrm{d}t} + \rho\frac{\mathrm{d}}{\mathrm{d}t}\int_{S_B} \left(\mathbf{v}_0(t) + \mathbf{\Omega}_0(t) \times (\mathbf{x}_0(t) - \mathbf{x}_f)\right) \mathrm{d}S \tag{2.232}$$

where S_B is the surface area of the body. In developing Eqs. (2.231) and (2.232), inviscid assumptions are only made for the pressure on the boundary surface of V_∞, which lies very far from the body. Therefore, the equations can be used to calculate the aerodynamic forces acting on a body immersed in a viscous flow, as long as the vorticity distribution in this flow is known at all times.

Consider a two-dimensional steady flow, for which $\boldsymbol{\omega} = (0, 0, \partial v/\partial x - \partial u/\partial y)$. For this flow, the first moment of vorticity around a static datum $\mathbf{x}_0$ is given by

$$\boldsymbol{\alpha} = \int_{S_\infty} (\mathbf{x} - \mathbf{x}_0) \times \boldsymbol{\omega} \mathrm{d}S$$

Carrying out the vector product, we obtain

$$a_x = -\int_{S_\infty} (y - y_0)\left(\frac{\partial v}{\partial x} - \frac{\partial u}{\partial y}\right) \mathrm{d}S$$

$$a_y = \int_{S_\infty} (x - x_0)\left(\frac{\partial v}{\partial x} - \frac{\partial u}{\partial y}\right) \mathrm{d}S$$

where $\boldsymbol{\alpha} = (a_x, a_y)$. The total circulation in the flow is given by

$$\Gamma = \int_{S_\infty} \boldsymbol{\omega} \cdot \mathbf{n} \mathrm{d}S = \int_{S_\infty} \left(\frac{\partial v}{\partial x} - \frac{\partial u}{\partial y} \right) \mathrm{d}S$$

since the unit vector normal to the plane S is $\mathbf{n} = (0, 0, 1)$. If we assume that all of the vorticity is concentrated at a single point $(x(t), y(t))$ and travels with it, then the first moment of the vorticity field becomes

$$a_x = -(y(t) - y_0)\Gamma$$
$$a_y = (x(t) - x_0)\Gamma$$

If the point $(x(t), y(t))$ moves with constant velocity $(-U_\infty, -V_\infty)$ and the circulation is constant in time, then, the 2D aerodynamic force of Eq. (2.232) yields

$$f_x = \rho \frac{\mathrm{d}y}{\mathrm{d}t}\Gamma = -\rho V_\infty \Gamma \tag{2.233}$$

$$f_y = -\rho \frac{\mathrm{d}x}{\mathrm{d}t}\Gamma = \rho U_\infty \Gamma \tag{2.234}$$

where $\mathbf{f} = (f_x, f_y)$. Equations (2.233) and (2.234) constitute the Kutta–Joukowski theorem for a steady flow $\mathbf{Q}_\infty = (U_\infty, V_\infty)$ around a static single 2D point vortex with constant strength. The theorem can also be applied to 3D vortex segments.

2.7.3 The Wake and the Kutta Condition

Ludwig Prandtl produced a number of flow visualisation films for educational purposes. One of these films, usually referred to as C1, includes a movie of the flow around an impulsively started wing section immersed in still water (Tollmien et al. 1961). Figure 2.28 reproduces four snapshots from this movie. Initially the wing is at rest, as seen in Figure 2.28a. At a certain time instance, it is set abruptly in motion with constant velocity and the camera follows it. The water contains fine aluminium particles that glow when lit, such that the motion of the fluid can be visualised. Figure 2.28b shows that, immediately after the impulsive start, a clockwise vortex appears just behind the wing's trailing edge. This vortex is known as the starting vortex and is crucial to our understanding of unsteady aerodynamics. As the wing and camera move forward, the starting vortex is left behind. Figures 2.28b to 2.28d show that during the first three seconds after the start of the motion, the flow remains parallel to the wing's surface everywhere, except at the trailing edge. There, both the upper and lower surface flows separate in a direction roughly parallel to the wing's chord line, forming a thin bright flow region known as the wake that extends from the trailing edge to the starting vortex.

As the wing and fluid are both at rest before the start of the motion, the total circulation in the flow must be equal to zero. This statement can be easily verified by substituting $\mathbf{u} = \mathbf{0}$ in Eq. (2.198) to obtain $\boldsymbol{\omega} = \mathbf{0}$ and then substituting this result in Eq. (2.210) to obtain $\Gamma = 0$. According to Kelvin's theorem (2.214), the total circulation in the flow must remain zero at all subsequent time instances, even if the wing and/or fluid start moving. However, the starting vortex clearly has non-zero circulation, which we can denote by $-\Gamma$ because its direction is counter-clockwise. This means that there must be another source of circulation

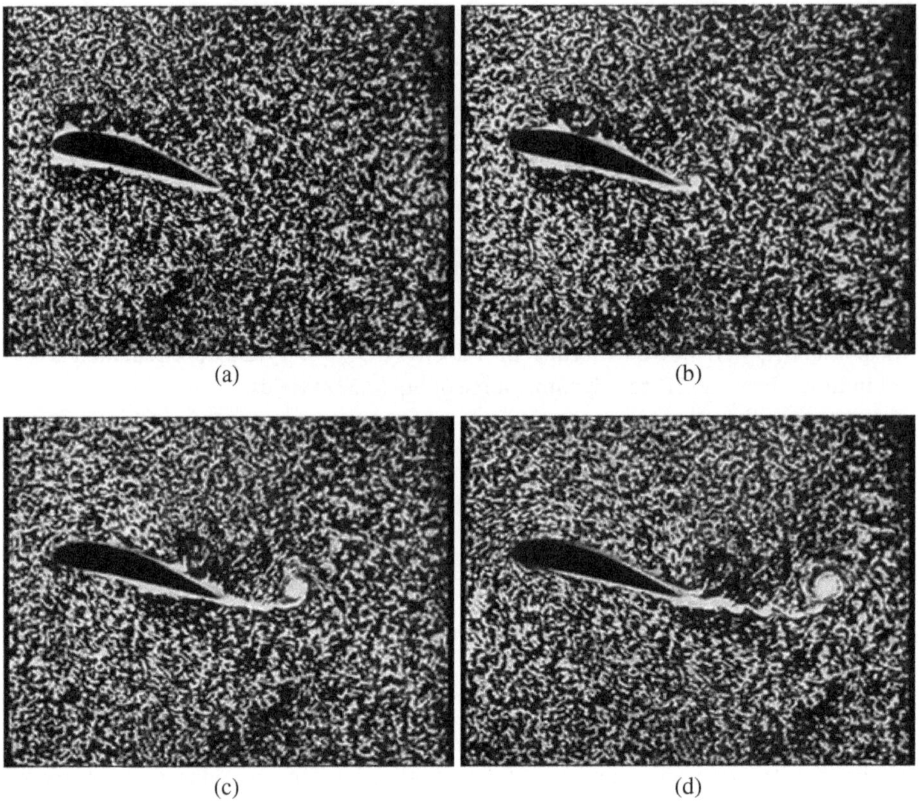

Figure 2.28 Four snapshots of Prandtl's visualisation of the flow around and impulsively started wing section. (a) Snapshot 1, (b) Snapshot 2, (c) Snapshot 3, and (d) Snapshot 4. Source: Prandtl (1936)/Technische Informationsbibliothek (TIB), provided by Technische Informationsbibliothek (TIB).

somewhere in the flow with equal and opposite strength, Γ, such that Kelvin's theorem can be satisfied. Classical aerodynamic theory states that this missing circulation can be found around the wing; as its direction is clockwise, it increases the flow speed on the upper surface and decreases it on the lower surface. There are two consequences:

- The flow behind the wing is pushed downwards, which is why the wake propagates in a direction parallel to the wing's chord and not horizontally.
- If the wing pushes the flow downwards, then by Newton's third law, the flow must be pushing the wing upwards. This force exerted by the flow on the wing in a direction perpendicular to the motion is known as the lift.

The flow around an impulsively started wing is inherently viscous, but many aspects of it can be described using potential theory, as long as we ensure that the wake separates smoothly from the trailing edge and in a direction parallel to the chord. This condition is known as the Kutta condition and is one of the fundamental building blocks of potential flow aerodynamics, both steady and unsteady. It will be discussed in more detail in Chapters 3 and 4, but it will be used throughout this book. It is important to stress the fact that the

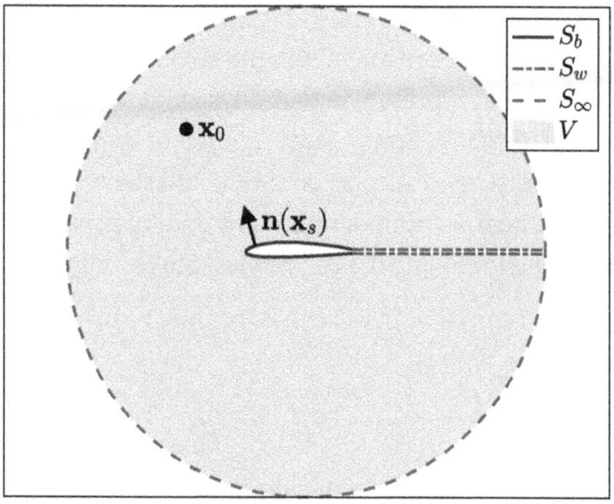

Figure 2.29 Flow domain around a wing and its wake.

imposition of the Kutta condition is only necessary because potential flow is irrotational; rotational flows will naturally separate smoothly from the trailing edge and in the correct direction. Furthermore, as the Kutta condition is an imposition of a rotational phenomenon on a potential flow, it is not uniquely defined. There exist several different forms of the condition that will be discussed in Section 4.7.

Clearly, potential flow methods must model the wake in order to give physical predictions. This means that when applying Green's identities, we need to model not only the body's surface but also the wake's surface. Figure 2.12 must then be completed by including the surface of the wake, as shown in Figure 2.29. Now the control volume V is delimited by the upper and lower surfaces of the wake, S_w, the surface of the body S_b and the far-field surface S_∞. We will discuss potential flow calculations on wings with wakes throughout this book.

The complete video of the flow presented in Figure 2.28 shows that there is a significant change in the form of the flow after around three seconds. The starting vortex has moved outside the field of view of the camera and the motion of the wing is steady, but the flow around it is not; it separates from the upper surface of the wing and becomes highly vortical. The direction of motion of the wing and its geometric shape are such that, after the transient phenomena due to the impulsive start have moved sufficiently far, the flow cannot remain attached to the upper surface. Flow separation is a purely viscous phenomenon and cannot be modelled using potential flow assumptions; such flows will be treated in Chapter 7.

2.8 Concluding Remarks

In this chapter, we have set out the basics of unsteady potential flow around wings and bodies for both incompressible and compressible flows. In each case, we have started from the fundamental equations of fluid flow and their boundary conditions to arrive at an expression for Green's identity that can be used to calculate the potential anywhere in the flow in terms of the potential on the surface of the body. Whenever using this identity, it is important to remember its underlying assumptions: the flow is inviscid and irrotational and all disturbances are small. For the compressible cases, the flow is also assumed to be isentropic. These assumptions are quite restrictive, but, as will be shown later in the book, the resulting analysis can be used to obtain useful solutions that represent actual physical flows.

Potential flow cannot be used to model directly flows with significant separation. For such flow cases, we have introduced the vorticity transport equation and the vorticity-momentum theorem that can be used in conjunction with potential flow results in order to obtain approximate solutions, without resorting to solving the full Navier–Stokes equations all over the flowfield.

References

Anathan, S. and Leishman, J.G. (2004). Role of filament strain in the free-vortex modeling of rotor wakes. *Journal of the American Helicopter Society* 49 (2): 176–191.

Anderson, J.D. Jr. (1985). *Fundamentals of Aerodynamics*. McGraw-Hill International Editions.

Anderson, J.D. Jr. (1990). *Modern Compressible Flow*, 2e. McGraw-Hill Publishing Company.

Bakhle, M.A., Reddy, T.S.R., and Keith, T.G. Jr. (1993). Subsonic/transonic cascade flutter using a full-potential solver. *AIAA Journal* 31 (7): 1347–1349.

Balakrishnan, A.V. (2004). Transonic small disturbance potential equation. *AIAA Journal* 42 (6): 1081–1088.

Blair, M. (1994). A Compilation of the Mathematics Leading to the Doublet-Lattice Method. *Report WL-TR-95-3022*. Air Force Wright Laboratory.

Chen, P.C., Lee, H.W., and Liu, D.D. (1993). Unsteady subsonic aerodynamics for bodies and wings with external stores including wake effect. *Journal of Aircraft* 30 (5): 618–628.

Edward Ehlers, F., Epton, M.A., Johnson, F.T. et al. (1979). A Higher Order Panel Method for Linearized Supersonic Flows. *Contractor Report CR-3062*. NASA.

Epton, M.A. and Magnus, A.E. (1990). PAN AIR- A Computer Program for Predicting Subsonic or Supersonic Linear Potential Flows About Arbitrary Configurations Using a Higher Order Panel Method. *Contractor Report CR-3251*. NASA.

Florea, R., Hall, K.C., and Dowell, E.H. (2000). Eigenmode analysis and reduced-order modeling of unsteady transonic potential flow around airfoils. *Journal of Aircraft* 37 (3): 454–462.

Katz, J. and Plotkin, A. (2001). *Low Speed Aerodynamics*. Cambridge University Press.

Kuethe, A.M. and Chow, C.Y. (1986). *Foundations of Aerodynamics - Basis of Aerodynamic Design*. Wiley.

Lamb, H. (1932). *Hydrodynamics*, 6e. Cambridge: Cambridge University Press.

Maskew, B. (1982). Prediction of subsonic aerodynamic characteristics: a case for low-order panel methods. *Journal of Aircraft* 19 (2): 157–163.

Morino, L. (1974). A General Theory of Unsteady Compressible Potential Aerodynamics. *Contractor Report CR-2464*. NASA.

Nixon, D. (1989). Basic equations for unsteady transonic flow. In: *Unsteady Transonic Aerodynamics, Progress in Astronautics and Aeronautics*, vol. 120 (ed. D. Nixon), 57–73. AIAA.

Nixon, D. and Kerlick, G.D. (1982). Potential equation methods for transonic flow prediction. In: *Transonic Aerodynamics, Progress in Astronautics and Aeronautics*, vol. 81 (ed. D. Nixon), 239–296. AIAA.

Paterson, A.R. (1984). *A First Course in Fluid Dynamics*. Cambridge University Press.

Prandtl, L. (1936). Entstehung von wirbeln bei wasserströmungen - 1. Entstehung von wirbeln und künstliche beeinflussung der wirbelbildung Reichsanstalt für Film und Bild in Wissenschaft und Unterricht (RWU). https://doi.org/10.3203/IWF/C-1.

Tollmien, W., Schlichting, H., Görtler, H., and Riegels, F.W. (1961). Entstehung von wirbeln bei wasserströmungen. In: *Ludwig Prandtl Gesammelte Abhandlungen: zur angewandten Mechanik, Hydro- und Aerodynamik* (ed. F.W. Riegels), 817–818. Berlin, Heidelberg: Springer-Verlag.

Vatistas, G.H., Kozel, V., and Mih, W.C. (1991). A simpler model for concentrated vortices. *Experiments in Fluids* 11 (1): 73–76.

Ward, G.N. (1955). *Linearized Theory of Steady High-Speed Flow*. Cambridge University Press.

Wu, J.C. (1981). Theory for aerodynamic force and moment in viscous flows. *AIAA Journal* 19 (4): 432–441.

Wu, J.C. (2018). *Elements of Vorticity Aerodynamics*. Berlin, Heidelberg: Springer-Verlag.

3

Analytical Incompressible 2D Models

3.1 Introduction

In Chapter 2 we show that for inviscid, incompressible and irrotational flow, the continuity equation reduces to Laplace's equation (2.23)

$$\nabla^2\phi = 0 \tag{3.1}$$

where ϕ is the flow's pertubration potential, while the momentum equations reduce to the unsteady Bernoulli equation (2.27). In this chapter, we will set up and solve these equations for the particular case of a 2D flat plate in pitching and plunging motion. For 2D flow, Laplace's and Bernoulli's equations can be written as

$$\frac{\partial^2\phi}{\partial x^2} + \frac{\partial^2\phi}{\partial y^2} = 0 \tag{3.2}$$

$$\rho\left(\frac{1}{2}\left(u^2 + v^2\right) + \frac{\partial\phi}{\partial t}\right) + p = \text{constant} \tag{3.3}$$

where x and y are the horizontal and vertical directions, while u and v are the horizontal and vertical flow velocity components.

Before starting to analyse unsteady flows, we will present one of the most fundamental theories of 2D steady aerodynamics, namely thin airfoil theory. The reason for this diversion is the fact that this theory makes use of many of the basic modelling choices that we will apply in later parts of this chapter, as well as elsewhere in this book. The rest of the chapter will present different unsteady thin airfoil theories, concentrating on the approaches by Wagner, Theodorsen and Peters.

3.2 Steady Thin Airfoil Theory

Numerous aerodynamics textbooks present thin airfoil theory in detail, such as Anderson Jr. (1985), Glauert (1947) or Kuethe and Chow (1986). As its name suggests, this theory is concerned with the calculation of the aerodynamic loads acting on a thin airfoil in steady motion. The term 'thin airfoil' implies that the aerodynamic modelling methodology ignores completely the thickness of the airfoil. This procedure is demonstrated in figure 3.1, which plots the contour of a NACA 3412 airfoil. NACA four-digit airfoils are a family of

Unsteady Aerodynamics: Potential and Vortex Methods, First Edition. Grigorios Dimitriadis.

Companion website: www.wiley.com/go/dimitriadis/unsteady_aerodynamics

airfoils created by the superposition of a thickness distribution, $f_t(x)$, and a camber line, $f_c(x)$, such that

$$y_{u,l}(x) = \pm f_t(x) + f_c(x) \tag{3.4}$$

where x and y are the horizontal and vertical coordinates of the airfoil's surface and subscripts u and l denote the upper and lower surfaces. The thickness and camber distributions are given by Abbott and Von Doenhoff (1959) as

$$f_t(x) = c\frac{t}{0.2}\left(0.2969\sqrt{\bar{x}} - 0.126\bar{x} - 0.35160\bar{x}^2 + 0.2843\bar{x}^3 - 0.1015\bar{x}^4\right) \tag{3.5}$$

$$f_c(x) = \begin{cases} c\frac{m}{p^2}\left(2p\bar{x} - \bar{x}^2\right) & \text{for } 0 \leq \bar{x} \leq p \\ c\frac{m}{(1-p)^2}\left((1-2p) + 2p\bar{x} - \bar{x}^2\right) & \text{for } p < \bar{x} \leq 1 \end{cases} \tag{3.6}$$

where c is the chord length of the airfoil, that is the horizontal distance between the leading and trailing edges, and $\bar{x} = x/c$. The term t in Eq. (3.5) is the thickness-to-chord ratio of the airfoil, m is the maximum non-dimensional height of the camber line and p is the value of $\bar{x}$ where $f_c(\bar{x}) = m$. The values of m, p and t are denoted by the serial number of the airfoil. For example, the NACA 3412 is a 12% thick airfoil with 3% maximum camber, occurring at 40% of the chord, that is $m = 0.03$, $p = 0.4$ and $t = 0.12$. Its thickness distribution and camber line are plotted in figure 3.1.

Thin airfoil theory assumes that the flow is inviscid, irrotational, incompressible and that it separates from the airfoil's surface at the trailing edge. Furthermore, it ignores completely the thickness distribution, such that it calculates the aerodynamic loads acting on the camber line only. Essentially, the airfoil is completely replaced by its camber line. As discussed in Chapter 2, Laplace's equation is linear and has known fundamental solutions; for 2D flows, such solutions are the point source of Eq. (2.107) and the point vortex of Eq. (2.223). These singularities are presented in much more detail in the appendix, Sections A.1–A.4.

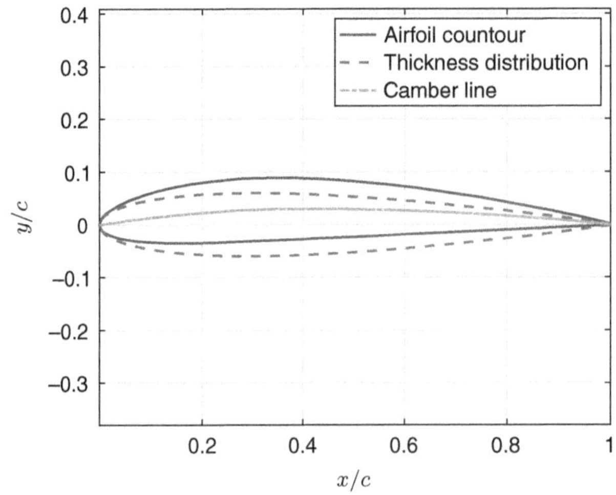

Figure 3.1 Thickness distribution and camber line of a NACA 3412 airfoil.

A 2D flowfield can then be built up from a superposition of such fundamental solutions. The modelling procedure is the following:

- Place fundamental solutions of the 2D Laplace equation in the flowfield in order to represent the airfoil and its steady motion.
- Calculate the strengths of these fundamental solutions such that the impermeability boundary condition is respected and such that the flow separates at the trailing edge.
- Calculate the pressure difference across the camber line and, hence, evaluate the total aerodynamic loads.

The airfoil is modelled as static such that its motion is represented by a constant free stream $\mathbf{Q}_\infty = (U_\infty \cos\alpha,\ U_\infty \sin\alpha)$, where U_∞ is the free stream airspeed and α the angle of attack. This description of the flow conforms to that of Figure 2.3 for a steady free stream. Therefore, the first fundamental solution of Laplace's equation to be inserted into the flow is the free stream $\phi_\infty = U_\infty \cos\alpha x + U_\infty \sin\alpha y$. However, thin airfoil theory assumes that the angle of attack is small, such that $\cos\alpha \approx 1$, $\sin\alpha \approx \alpha$. Consequently, the potential induced by the free stream is given by

$$\phi_\infty \approx U_\infty x + U_\infty \alpha y$$

It remains to insert additional fundamental solutions of Laplace's equation in the flow in order to represent the camber line. For thin airfoil theory, these singularities are a continuous distribution of 2D point vortices. The exact placement of these vortices will be discussed later; for the moment, we remark that the vortices will induce an additional perturbation potential $\phi(x,y)$, such that the total potential anywhere in the flow is given by

$$\Phi(x,y) = \phi_\infty(x,y) + \phi(x,y)$$

and the flow velocity components become

$$u(x,y) = \Phi_x(x,y) = U_\infty + \phi_x(x,y),\ v(x,y) = \Phi_y(x,y) = U_\infty\alpha + \phi_y(x,y)$$

The resulting velocity vector $\mathbf{u}(x,y) = (u(x,y),\ v(x,y))$ is plotted on the camber line in Figure 3.2a. As the camber line must be impermeable, the velocity vector must be tangent

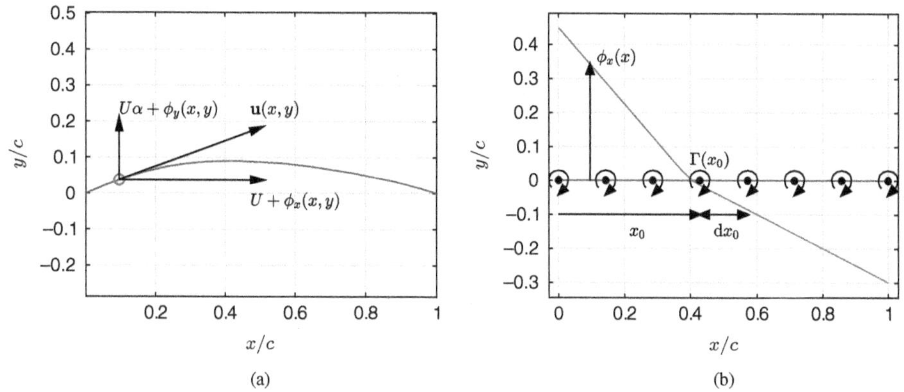

Figure 3.2 Thin airfoil theory representation of an impermeable camber line. (a) Impermeability on the camber line and (b) impermeability on the chord line.

to the camber line, which means that

$$\frac{v(x,y)}{u(x,y)} = \left.\frac{\mathrm{d}f_c}{\mathrm{d}x}\right|_x$$

Substituting for $u(x,y)$, $v(x,y)$ the impermeability boundary condition becomes

$$\frac{U_\infty \alpha + \phi_y(x,y)}{U_\infty + \phi_x(x,y)} = \left.\frac{\mathrm{d}f_c}{\mathrm{d}x}\right|_x$$

Solving for $\phi_y(x,y)$, we obtain

$$\phi_y(x,y) = -U_\infty \alpha + U_\infty \left.\frac{\mathrm{d}f_c}{\mathrm{d}x}\right|_x + \phi_x(x,y) \left.\frac{\mathrm{d}f_c}{\mathrm{d}x}\right|_x$$

but since we assume that the camber is small and that the flow perturbations ϕ_x, ϕ_y are also small, the second term in the right-hand side can be neglected, leading to

$$\phi_y(x) \approx -U_\infty \alpha + U_\infty \left.\frac{\mathrm{d}f_c}{\mathrm{d}x}\right|_x \tag{3.7}$$

It is important to note that we have removed the dependence of $\phi_y(x)$ on y on the left-hand side because the right-hand side of the equation does not depend on y either. Therefore, thanks to the small angle of attack, small camber and small perturbation assumptions, the vertical flow velocity on the camber line is only a function of chordwise position x.

Now, we return to the question of the placement of the vortices. They could be placed on the camber line itself but that would complicate the derivation of the solution. Instead, a continuous distribution of vortices with strength $\Gamma(x)$ is placed on the $y = 0$ line from $x = 0$ to $x = c$, as seen in Figure 3.2b. In order for the solution to represent the camber, this vortex distribution must induce vertical velocities equal to $\phi_y(x)$ in Eq. (3.7) at all points $0 \leq x \leq c$ and $y = 0$. The impermeable camber line has been replaced by the permeable chord line such that the total vertical velocity passing through it is equal to

$$v(x) = U_\infty \alpha + \phi_y(x) = U_\infty \alpha - U_\infty \alpha + U_\infty \left.\frac{\mathrm{d}f_c}{\mathrm{d}x}\right|_x = U_\infty \left.\frac{\mathrm{d}f_c}{\mathrm{d}x}\right|_x$$

which represents the component of the free stream velocity normal to the camber line. For small amounts of camber, the camber line lies close to the chord line so that the imposition of the boundary condition on the latter can be considered approximately correct.

It now remains to calculate the strength of the vortex distribution $\Gamma(x)$ that is necessary for impermeability. In order to do this, we need an expression for the vertical velocity induced by $\Gamma(x)$ that we can substitute in the left-hand side of (3.7). Equation (A.19) shows that the velocity in the y direction induced at point x, y by a point vortex lying on point x_0, y_0 is

$$\mathrm{d}\phi_y(x,y) = -\frac{\Gamma(x_0)}{2\pi} \frac{x - x_0}{(x - x_0)^2 + (y - y_0)^2}$$

where $\mathrm{d}\phi_y(x,y)$ is the contribution of a single vortex belonging to a continuous distribution of vortices. As we are only interested in calculating the velocities on the chord line ($y = 0$) and all the vortices also lie on the chord line,

$$\mathrm{d}\phi_y(x) = -\frac{\Gamma(x_0)}{2\pi} \frac{1}{x - x_0}$$

The total vertical velocity induced by the entire vortex distribution is obtained by summing the contributions of all the vortices lying on $0 \le x_0 \le c$, such that

$$\phi_y(x) = -\int_0^c \frac{\Gamma(x_0)}{2\pi} \frac{1}{x - x_0}$$

In order to evaluate this integral, we can write the vortex strength as $\Gamma(x_0) = \gamma(x_0)\mathrm{d}x_0$, where $\gamma(x_0)$ is circulation per unit length. Then,

$$\phi_y(x) = -\int_0^c \frac{\gamma(x_0)}{2\pi} \frac{1}{x - x_0} \mathrm{d}x_0$$

Finally, substituting from Eq. (3.7), we obtain the fundamental equation of thin airfoil theory

$$\frac{1}{2\pi} \int_0^c \frac{\gamma(x_0)}{x - x_0} \mathrm{d}x_0 = U_\infty \left(\alpha - \left. \frac{\mathrm{d}f_c}{\mathrm{d}x} \right|_x \right) \tag{3.8}$$

which is a single equation with a single unknown, $\gamma(x_0)$, since the slope of the camber line, the chord, the angle of attack and the free stream airspeed are known. Unfortunately, Eq. (3.8) is an integral equation with no general solution. The usual approach for solving it is to assume a series form for $\gamma(x)$, after applying the change of variables

$$x = \frac{c}{2}(1 - \cos\theta), \quad x_0 = \frac{c}{2}\left(1 - \cos\theta_0\right) \tag{3.9}$$

where $\theta = 0$, $\theta_0 = 0$ correspond to the leading edge and $\theta = \pi$, $\theta_0 = \pi$ to the trailing edge. Equation (3.8) becomes

$$-\frac{1}{2\pi} \int_0^\pi \frac{\gamma(\theta_0)\sin\theta_0}{\cos\theta - \cos\theta_0} \mathrm{d}\theta_0 = U_\infty \left(\alpha - \left. \frac{\mathrm{d}f_c}{\mathrm{d}x} \right|_x \right) \tag{3.10}$$

and then the assumed series solution for $\gamma(\theta_0)$ is

$$\gamma(\theta_0) = 2U_\infty \left(A_0 \frac{1 + \cos\theta_0}{\sin\theta_0} + \sum_{n=1}^{\infty} A_n \sin n\theta_0 \right) \tag{3.11}$$

The functions in this series are Glauert expansions and will be discussed in much more detail in Section 3.6.1. An important characteristic of this choice for $\gamma(\theta_0)$ is the fact that the vortex strength becomes zero at the trailing edge. Taking the limit of Eq. (3.11) as $\theta_0 \to \pi$ results in

$$\gamma(\theta_0) = 2U_\infty A_0 \lim_{\theta_0 \to \pi} \frac{1 + \cos\theta_0}{\sin\theta_0} = 2U_\infty A_0 \lim_{\theta_0 \to \pi} \frac{-\sin\theta_0}{\cos\theta_0} = 0$$

having made use of L'Hôpital's rule. Therefore, the vortex strength at the trailing edge is indeed equal to zero, which is a form of the Kutta condition.

In order to arrive at a complete solution, we still need to calculate the values of the coefficients A_0 and A_n. Substituting from Eq. (3.11) into Eq. (3.10) and carrying out the integration using Glauert's integrals (Glauert 1947)

$$\int_0^\pi \frac{\sin n\theta_0 \sin\theta_0}{\cos\theta - \cos\theta_0} \mathrm{d}\theta_0 = \pi \cos n\theta \tag{3.12}$$

$$\int_0^\pi \frac{\cos n\theta_0}{\cos\theta - \cos\theta_0} \mathrm{d}\theta_0 = -\pi \frac{\sin n\theta}{\sin\theta} \tag{3.13}$$

results in

$$A_0 - \sum_{n=1}^{\infty} A_n \cos n\theta = \alpha - \left.\frac{\mathrm{d}f_c}{\mathrm{d}x}\right|_x \tag{3.14}$$

The next step is to write the derivative of the camber line evaluated at x as the symmetric Fourier series

$$\left.\frac{\mathrm{d}f_c}{\mathrm{d}x}\right|_x = B_0 + \sum_{n=1}^{\infty} B_n \cos n\theta$$

where

$$B_0 = \frac{1}{\pi}\int_0^{\pi} \left.\frac{\mathrm{d}f_c}{\mathrm{d}x}\right|_{x_0} \mathrm{d}\theta_0, \quad B_n = \frac{2}{\pi}\int_0^{\pi} \left.\frac{\mathrm{d}f_c}{\mathrm{d}x}\right|_{x_0} \cos n\theta_0 \mathrm{d}\theta_0$$

Substituting back into the impermeability boundary condition leads to

$$A_0 - \sum_{n=1}^{\infty} A_n \cos n\theta = \alpha - B_0 - \sum_{n=1}^{\infty} B_n \cos n\theta$$

so that we can at last solve for the unknown coefficients A_0, A_n, by equating the coefficients of $\cos n\theta$ on the left- and right-hand sides for $n = 0$ to ∞, a process known as harmonic balancing. The result is

$$A_0 = \alpha - \frac{1}{\pi}\int_0^{\pi} \left.\frac{\mathrm{d}f_c}{\mathrm{d}x}\right|_{x_0} \mathrm{d}\theta_0 \tag{3.15}$$

$$A_n = \frac{2}{\pi}\int_0^{\pi} \left.\frac{\mathrm{d}f_c}{\mathrm{d}x}\right|_{x_0} \cos n\theta_0 \mathrm{d}\theta_0 \tag{3.16}$$

Substituting back into Eq. (3.11), we have now fully determined the vortex distribution necessary in order to impose impermeability on the cambered airfoil.

The last step is to calculate the aerodynamic loads acting on the airfoil, but, before doing this, we will discuss the directions of these loads and the usual approximations of these directions. In two-dimensional flows, the total aerodynamic load, $\mathbf{f}$, is defined as the contour integral of the pressure around the airfoil in a direction normal to the surface,

$$\mathbf{f} = -\oint p(s)\mathbf{n}(s)\mathrm{d}s \tag{3.17}$$

where s is arclength distance along the surface of the airfoil, $p(s)$ is the pressure at s and $\mathbf{n}(s)$ is the unit vector normal to the surface at s. Figure 3.3 plots the surface of the airfoil and defines s and $\mathbf{n}(s)$. The arclength distance is measured from the lower trailing edge to the upper trailing edge, passing from the leading edge so that it contours completely the airfoil. The negative sign in Eq. (3.17) is due to this definition of the arclength and the fact that the normal vector is pointing outwards from the airfoil. The force $\mathbf{f}$ is a vector that can be analysed in two components, depending on the chosen frame of reference. There are two main frames of reference used in aerodynamics:

- Body axes. If the body is an airfoil, the body axes are the chordwise axis, that is the line connecting the leading to the trailing edge and an axis normal to the chordwise axis. In this frame of reference, the total aerodynamic load is analysed into the chordwise force, f_c, and the normal force f_n.

- Free stream axes. This reference system consists of the streamwise axis that is parallel to the free stream and an axis normal to it. In this frame of reference, the total aerodynamic load is analysed into the drag, d, that is parallel to the free stream and the lift, l, that is perpendicular to the free stream.

Figure 3.3 plots the total aerodynamic force and its components f_c, f_n and l, d, acting on an airfoil exposed to a free stream Q_∞ at an angle of attack α. The relationship between the two sets of forces is given by

$$\begin{aligned} d &= f_c \cos\alpha + f_n \sin\alpha \\ l &= -f_c \sin\alpha + f_n \cos\alpha \end{aligned} \tag{3.18}$$

so that $\sqrt{d^2 + l^2} = \sqrt{f_c^2 + f_n^2} = \|\mathbf{f}\|$. For attached flows around airfoils, α is small so that $\cos\alpha \approx 1$, $\sin\alpha \approx \alpha$ and f_c is a small force compared to f_n, so that Eq. (3.18) can be approximated by

$$\begin{aligned} d &\approx f_c + f_n\alpha \\ l &\approx f_n \end{aligned} \tag{3.19}$$

This means that the lift and normal force are often considered to be interchangeable for attached flow aerodynamics. In this chapter, we will do the same; the normal force will be referred to as the lift, in line with the usual notation in the literature. Furthermore, the airfoil will always be aligned with the x axis and the free stream will be inclined at an angle α, as shown in Figure 3.3. Consequently, the chordwise and normal forces can be obtained from Eq. (3.17) as

$$\begin{aligned} f_c &= -\oint p(s)n_x(s)\mathrm{d}s \\ f_n &= -\oint p(s)n_y(s)\mathrm{d}s \end{aligned} \tag{3.20}$$

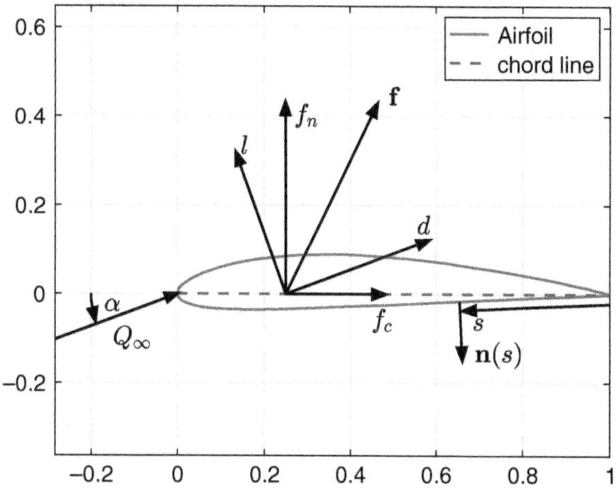

Figure 3.3 Total aerodynamic load and its components.

where $n_x(s)$, $n_y(s)$ are the components of $\mathbf{n}(s)$ in the x and y directions, respectively. Finally, if the airfoil is assumed to be flat and coinciding with the x axis, then $n_x(s) = 0$, $n_y(s) = -1$ on the lower surface and $n_y(s) = 1$ on the upper surface, so that Eq. (3.20) become

$$f_c = 0$$
$$l \approx f_n = \int_0^c \Delta p(x)\mathrm{d}x \tag{3.21}$$

where $\Delta p(x)$ is the pressure difference between the upper and lower surfaces of the airfoil at x. Note that, in two dimensions, $\mathbf{f}, f_c, f_n$, d and l are all forces per unit length.

One of the standard ways to calculate the aerodynamic loads acting on a 2D airfoil is by means of the Kutta–Joukowski theorem of Eq. (2.234). Applying this theorem to one of the vortices in Figure 3.2b yields

$$\mathrm{d}l = \rho U_\infty \Gamma(x) = \rho U_\infty \gamma(x)\mathrm{d}x$$

where $\mathrm{d}l$ is the incremental lift generated by vortex $\Gamma(x)$. The total lift is obtained by integrating $\mathrm{d}l$ over all the vortices from leading to trailing edge, that is

$$l = \rho U_\infty \int_0^c \gamma(x)\mathrm{d}x$$

Comparing this latest result to Eq. (3.21), we see that

$$\Delta p(x) = \rho U_\infty \gamma(x) \tag{3.22}$$

An unsteady version of this equation will be derived in Section 3.6. The total 2D lift is obtained by substituting this pressure difference into Eq. (3.21), so that

$$l = \int_0^c \Delta p(x_0)\mathrm{d}x_0 = \rho U_\infty \frac{c}{2}\int_0^\pi \gamma(\theta_0)\sin\theta_0 \mathrm{d}\theta_0 = \rho U_\infty^2 c\left(\pi A_0 + \frac{\pi}{2}A_1\right)$$

The 2D lift coefficient is defined as

$$c_l = \frac{l}{\frac{1}{2}\rho U_\infty^2 c} = 2\pi A_0 + \pi A_1 \tag{3.23}$$

Substituting from Eqs. (3.15) and (3.16), we obtain the famous result

$$c_l = 2\pi\alpha + 2\int_0^\pi \left.\frac{\mathrm{d}f_c}{\mathrm{d}x}\right|_{x_0} (\cos\theta_0 - 1)\mathrm{d}\theta_0 = a_0\left(\alpha - \alpha_{l=0}\right) \tag{3.24}$$

where $a_0 = 2\pi$ is known as the lift curve slope and

$$\alpha_{l=0} = -\frac{1}{\pi}\int_0^\pi \left.\frac{\mathrm{d}f_c}{\mathrm{d}x}\right|_{x_0} (\cos\theta_0 - 1)\mathrm{d}\theta_0 \tag{3.25}$$

is the zero-lift angle of attack.

The 2D moment of the lift around the leading edge, m_{le}, is calculated from

$$m_{le} = \int_0^c \Delta p(x_0)x_0\mathrm{d}x_0 = -\frac{\pi\rho U_\infty^2 c^2}{4}\left(A_0 + A_1 - \frac{A_2}{2}\right)$$

and is also known as the pitching moment around the leading edge. It is defined such that $m_{le} > 0$ when it has the tendency to increase the angle of attack (nose-up moment). The pitching moment coefficient is defined as

$$c_{m_{le}} = \frac{m_{le}}{\frac{1}{2}\rho U_\infty^2 c^2} = -\frac{\pi}{2}\left(A_0 + A_1 - \frac{A_2}{2}\right) = -\frac{c_l}{4} + \frac{\pi}{4}\left(A_2 - A_1\right)$$

Finally, the pitching moment coefficient around the quarter-chord point, $x = c/4$, can be calculated from

$$c_{m_{c/4}} = c_{m_{le}} + \frac{c_l}{4} = \frac{\pi}{4}\left(A_2 - A_1\right) \tag{3.26}$$

Note that if the wing has no camber, $A_0 = \alpha$ and $A_n = 0$. In this case,

$$c_l = 2\pi\alpha \tag{3.27}$$

$$c_{m_{c/4}} = 0$$

The quarter-chord point is known as the aerodynamic centre because, in the absence of camber, the pitching moment around it is equal to zero. For cambered airfoils, $c_{m_{c/4}} \neq 0$ but is independent of the angle of attack.

Example 3.1 *Calculate the lift coefficient and the pitching moment coefficient around the quarter chord for the NACA four-digit airfoil family, as functions of the angle of attack.*

As we are calculating non-dimensional load coefficients, which are independent of the chord and free stream airspeed, we can set $c = 1$, $U_\infty = 1$ *without loss of generality. Then, the non-dimensional coordinates of the NACA four-digit airfoils given in Eq. (3.6) are in fact dimensional coordinates. The slope of the camber line can be obtained easily by differentiating Eq. (3.6) with respect to* $\bar{x} = x$,

$$\frac{df_c}{dx} = \begin{cases} \frac{m}{p^2}(2p - 2x) & \text{for } 0 \leq x \leq p \\ \frac{m}{(1-p)^2}(2p - 2x) & \text{for } p < x \leq 1 \end{cases}$$

Since we are using thin airfoil theory, we will ignore the thickness distribution of Eq. (3.5). The lift coefficient and the pitching moment coefficient around the quarter chord are given by Eqs. (3.23) and (3.26), respectively, but, in order to evaluate these expressions, we first need to calculate A_0, A_1 *and* A_2. *We start with* A_0, *which is given by Eq. (3.15)*

$$A_0 = \alpha - \frac{1}{\pi}\int_0^\pi \left.\frac{df_c}{dx}\right|_{x_0} d\theta_0$$

We apply the change of variable of Eq. (3.9) with $c = 1$

$$x = \frac{1}{2}(1 - \cos\theta)$$

to the derivative of the camber line to obtain

$$\frac{df_c}{dx} = \begin{cases} \frac{m}{p^2}(2p - (1 - \cos\theta)) & \text{for } 0 \leq \theta \leq \theta_p \\ \frac{m}{(1-p)^2}(2p - (1 - \cos\theta)) & \text{for } \theta_p < \theta \leq \pi \end{cases}$$

where θ_p *is the value of* θ *that corresponds to* $x = p$, *that is*

$$p = \frac{1}{2}\left(1 - \cos\theta_p\right), \quad \text{or,} \quad \theta_p = \cos^{-1}(1 - 2p)$$

Consequently, the expression for A_0 becomes

$$A_0 = \alpha - \frac{1}{\pi}\int_0^{\theta_p} \frac{m}{p^2}(2p + \cos\theta - 1)\,\mathrm{d}\theta_0 - \frac{1}{\pi}\int_{\theta_p}^{\pi} \frac{m}{(1-p)^2}(2p + \cos\theta - 1)\,\mathrm{d}\theta_0$$

which can be integrated easily to give

$$A_0 = \alpha + \frac{m}{\pi(1-p)^2}\left((1-2p)(\pi - \theta_p) + \sin\theta_p\right) + \frac{m}{\pi p^2}\left((1-2p)\theta_p - \sin\theta_p\right)$$

The A_n coefficients are given by Eq. (3.16). Following the same procedure, we obtain

$$A_n = \frac{2}{\pi}\int_0^{\theta_p} \frac{m}{p^2}(2p + \cos\theta - 1)\cos n\theta_0 \mathrm{d}\theta_0 + \frac{2}{\pi}\int_{\theta_p}^{\pi} \frac{m}{(1-p)^2}(2p + \cos\theta - 1)\cos n\theta_0 \mathrm{d}\theta_0$$

Integrating for $n = 1, 2$ yields

$$A_1 = \frac{m}{2\pi p^2}\left(2\theta_p - 4(1-2p)\sin\theta_p + \sin 2\theta_p\right) - \frac{m}{2\pi(1-p)^2}\left(2(\theta_p - \pi) - 4(1-2p)\sin\theta_p + \sin 2\theta_p\right)$$

$$A_2 = -\frac{m(2p-1)}{3\pi p^2(1-p)^2}\left(6\sin\theta_p - 3(1-2p)\sin 2\theta_p - 4\sin^3\theta_p\right) \tag{3.28}$$

We calculate A_0, A_1 and A_2 for various values of α, m and p and then we evaluate the aerodynamic load coefficients from Eqs. (3.23) and (3.26).

Figure 3.4 plots the variation of c_l and $c_{m_{c/4}}$ with α for values of m between 0 and 0.04 and for $p = 0.4$. The lift curves in Figure 3.4a are straight lines with the same slope but with different intercepts. The figure also plots the zero lift angle, calculated using Eq. (3.25). As m increases, the zero lift angle becomes more negative. In fact, the NACA four-digit family is designed such that the value of $\alpha_{l=0}$ in degree is approximately equal to $-100\,m$ if p lies between 0.3 and 0.5. Figure 3.4b demonstrates that $c_{m_{c/4}}$ does not depend on the angle of attack, it depends only on m. As m increases, the pitching moment around the quarter chord becomes increasingly negative, that is nose-down. Subject to the assumptions and simplifications behind the developlusmnent of thin airfoil theory, Figure 3.4 shows that increasing camber decreases the

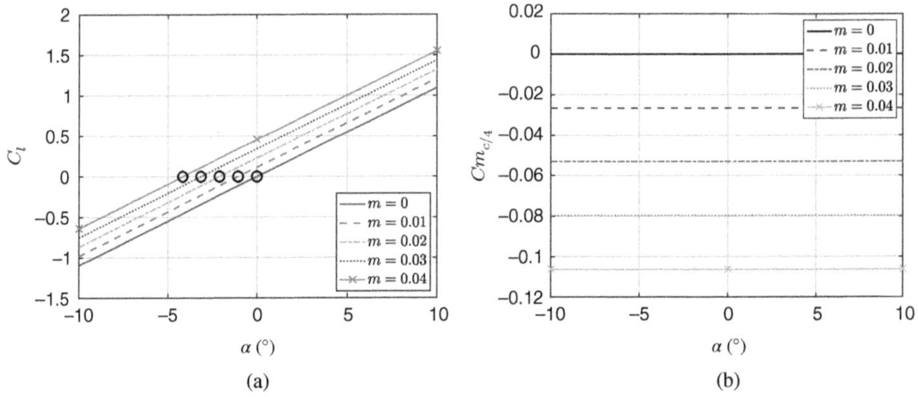

Figure 3.4 Thin airfoil theory lift and moment predictions for different NACA four digit airfoils with $p = 0.4$. (a) Lift coefficient and (b) pitching moment coefficient.

zero lift angle and the pitching moment around the quarter chord but does not affect the slope of the lift curve, which is always equal to 2π. This example is solved by Matlab code `thin_airfoil.m`.

While the entire analysis presented in this section is steady, it has served to introduce a number of important concepts that will be reused in many parts of this book. One such concept is the replacement of the true geometry by an infinitely thin straight line and the modelling of impermeability on the true geometry by means of an imposed vertical flow velocity through this line. This velocity is known as the upwash and, in the case of thin airfoil theory, it can be used to represent the effects of both the angle of attack and the camber of the airfoil on the flow. The remainder of this chapter deals with unsteady thin airfoil theories applied to symmetric airfoils with zero camber; we will still model them as lying on the $y = 0$ line and use the upwash to model their instantaneous pitch angle and other motion-induced velocities. We will return to the use of the upwash to model camber in later chapters.

Several other important concepts have been introduced in this section. The zero trailing edge vorticity formulation of the Kutta condition will be reused several times in this book, although other formulations of the condition will also be applied. The vortex sheet of strength $\gamma(x)$, the calculation of the influence of singular solutions of Laplace's equation on a straight line segment, the Kutta–Joukowski theorem, the integration of the pressure difference in order to calculate the total lift and pitching moment and the aerodynamic centre are all concepts that will be used time and again in this book.

3.3 Fundamentals of Wagner and Theodorsen Theory

We now turn our attention back to unsteady 2D aerodynamic modelling, for which the continuity equation is still (3.2) and Bernoulli's equation is (3.3). The modelling approaches we will use are the ones developed by Wagner (1925) and Theodorsen (1935). The basic assumptions are the following:

- The flow is 2D, inviscid, incompressible and irrotational. The constant free stream airspeed is denoted by U_∞ and is parallel to the x axis.
- The wing is a thin uncambered flat plate with chord c moving in plunge, $h(t)$, and pitch, $\alpha(t)$, around a pitch axis lying at a distance x_f from the leading edge.
- The plate is modelled as quasi-fixed; the motion is represented by the vertical relative velocity (or upwash) $w(t)$ caused by the motion and the free stream. The amplitude of the motion is small so that $|w(t)| \ll U_\infty$ at all times.
- The Kutta condition applies so that the flow separates smoothly at the trailing edge.
- The wake of the plate is a straight line of vorticity that extends downstream from the trailing edge to infinity.
- Kelvin's theorem applies so that the total circulation in the flowfield is constant in time.

The fundamental motion case we will consider is the pitching and plunging flat plate airfoil, which conforms to the motion case presented in Figure 2.3. Figure 3.5 plots the geometry of the flat plate as well as the upwash flow components induced by the free stream and the motion. The plate is centred at its mid-chord point so that its leading edge lies at

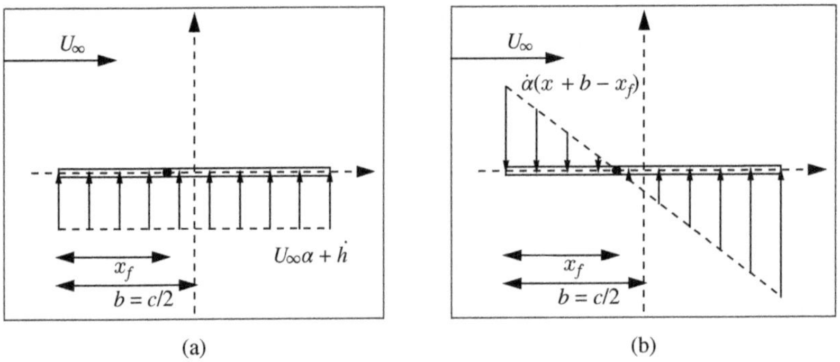

Figure 3.5 Upwash components due to motion of 2D flat plate. (a) Upwash due to free stream and plunging motion and (b) upwash due to pitching motion.

$x = -b$ and its trailing edge at $x = +b$, where $b = c/2$; the free stream is flowing from left to right. The plunge displacement is defined positive downwards so that the wing experiences an upwards relative flow speed equal to $\dot{h}$ when moving in the positive plunge direction, as seen in Figure 3.5a. The pitch angle is taken positive nose up so that the vertical component of the free stream is $U_\infty \sin\alpha \approx U_\infty \alpha > 0$ when $\alpha > 0$. Finally, when $\dot{\alpha} > 0$, the pitching motion induces a relative flow with linearly varying strength that is directed downwards upstream of the pitch axis and upwards downstream, as seen in Figure 3.5b. The total upwash is then given by the sum of all three contributions, such that

$$w(x,t) = U_\infty \alpha(t) + \dot{h}(t) + \dot{\alpha}(t)(x + b - x_f) \tag{3.29}$$

Solving Laplace's equation for this flow consists in superimposing elementary solutions of this equation such that the boundary conditions are satisfied. These are

- Impermeability. The total flow speed normal to the surface of the plate is equal to zero. Adapting Eq. (2.46) to the present case leads to a boundary condition of the form

$$\phi_z(x,t) + w(x,t) = 0 \tag{3.30}$$

  Note that x will always be used to refer to points on the surface of the wing measured from the mid-chord in this chapter.
- Far field. The disturbance caused by the presence and motion of the plate is negligible at a far enough distance.

Furthermore, Kelvin's theorem and the Kutta condition must also be satisfied. Several different choices of elementary solutions exist, but the ones we will use in the present modelling are the free stream, the 2D source and the 2D vortex (see Appendix A). Recall that both the source and the vortex automatically satisfy the far field boundary condition.

There are many possible options for selecting and placing fundamental solutions of Laplace's equation in order to model the flow around a pitching and plunging flat plate. Wagner chose to place vortices in the wake and Theodorsen added sources on the plate. In fact, the choice of the flat plate airfoil was motivated by the fact that it can be transformed into a circle using conformal mapping (see for example Anderson Jr. (1985)). The advantage of this approach is that most of the resulting integrals can be evaluated analytically.

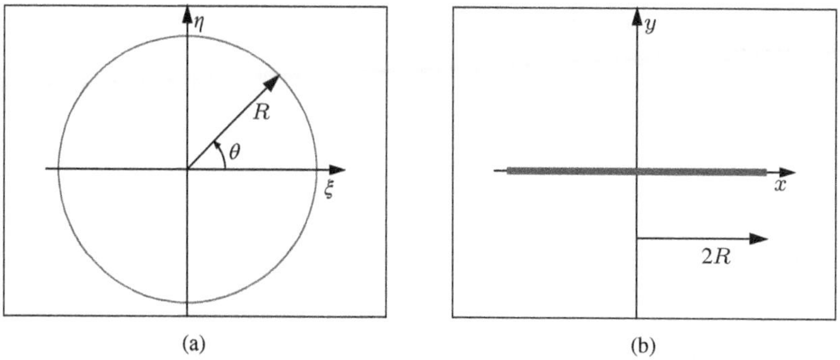

Figure 3.6 Conformal transformation from circle to flat plate. (a) Circle plane and (b) flat plate plane.

Figure 3.6 demonstrates the conformal mapping procedure. It is a transformation of a circle with radius R defined in the ξ, η plane into a flat plate airfoil of half-chord $2R$ defined in the x, y plane. Using the complex variables

$$\zeta = \xi + i\eta$$
$$z = x + iy$$

the operation that will transform the circle $\xi^2 + \eta^2 = R^2$ into a flat plate is given by

$$z = \zeta + \frac{R^2}{\zeta} \tag{3.31}$$

Substituting from the definition of ζ, we obtain that for points lying on the circle,

$$\begin{aligned} z &= \xi + i\eta + \frac{R^2}{(\xi + i\eta)} = \xi + i\eta + \frac{R^2(\xi - i\eta)}{(\xi + i\eta)(\xi - i\eta)} \\ &= \xi + i\eta + \frac{R^2(\xi - i\eta)}{\xi^2 + \eta^2} = \xi + i\eta + \frac{R^2(\xi - i\eta)}{R^2} = 2\xi \end{aligned}$$

Consequently, in the z plane, we have $x = 2\xi$ and $y = 0$. Considering that ξ on the circle ranges from $-R$ to $+R$, then x will range from $-2R$ to $+2R$, while y will always remain equal to zero. In other words, the shape in the z plane defined by this transformation is a flat plate with total length $4R$.

It is much more practical to work with a circle that is mapped onto an airfoil such that the chord length of the latter is equal to the diameter of the former. The conformal transformation that can achieve this is (Scanlan and Rosenbaum 1951)

$$z = \frac{1}{2}\left(\zeta + \frac{b^2}{\zeta}\right) \tag{3.32}$$

where b is the half-chord of the flat plate and the radius of the circle and is used here instead of R for compatibility with the work of Theodorsen (1935)

The inverse of the conformal transformation can be applied to a flat plate airfoil with half-chord b in order to transform it to a circle with radius b. The airfoil is defined by $z = x$, for $-b \leq x \leq b$. Substituting $z = x$ in Eq. (3.32) yields

$$x = \frac{1}{2}\left(\zeta + \frac{b^2}{\zeta}\right), \quad \text{or} \quad \zeta^2 - 2x\zeta + b^2 = 0$$

This quadratic equation has solutions

$$\zeta = x \pm \sqrt{x^2 - b^2} \tag{3.33}$$

Since x lies between $-b$ and b, the argument of the square root is always negative except when it is equal to zero. We can therefore write

$$\zeta = \xi + i\eta = x \pm i\sqrt{b^2 - x^2}$$

so that $\xi = x$ and $\eta = \sqrt{b^2 - x^2}$, which are coordinates that indeed define a circle of radius b.

It is also interesting to investigate what happens when we apply this inverse transformation to the wake behind the flat plate. The wake is defined in the z plane by $z = x$, for x ranging from the trailing edge, $x = b$, to $+\infty$. The inverse transformation will result in the roots of Eq. (3.33) again but, noting that the minimum value of x in the wake is b, it becomes clear that ζ is a real number,

$$\xi = x \pm \sqrt{x^2 - b^2}, \quad \eta = 0$$

If we substitute the first solution back into the conformal transformation of Eq. (3.32), we see that it maps indeed to the airfoil's wake in the z plane. The second solution cannot be used in its entirety; it can be split in three regions:

- $0 \leq \xi \leq b$. This region maps to the wake, as it results in $\infty \geq x \geq b$.
- $-b \leq \xi < 0$. This region maps to $-b \geq x \geq -\infty$, which does not lie on the wake.
- $-\infty < \xi < -b$. This region also maps to $-\infty < x \leq -b$, which does not lie on the wake.

In conclusion, there are two regions of the ζ plane that map onto the wake in the z plane, $\xi \in [b, +\infty]$ and $\xi \in [0, b]$, with $\eta = 0$ in both cases.

The conformal transformation is described graphically in Figure 3.7. Figure 3.7a draws the two regions of the circle plane that map to the flat plate wake in different line styles. Figure 3.7b demonstrates that both the dashed and dash-dot lines from the circle plane have mapped to the same region of the flat plate plane. We will take advantage of this phenomenon in order to place the elementary solutions of Laplace's equations.

In order to model the unsteady flow around the flat plate, we will place the following singularity distributions in the circle plane:

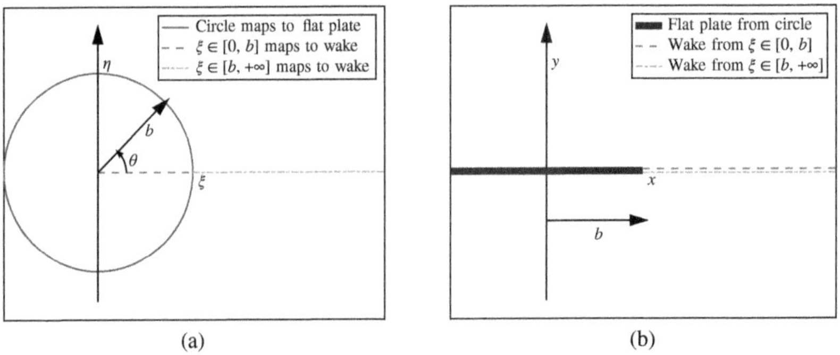

Figure 3.7 Transformation of the wake from the circle plane to the flat plate plane. (a) Circle plane and (b) flat plate plane.

Figure 3.8 Singularity placement in circle plane.

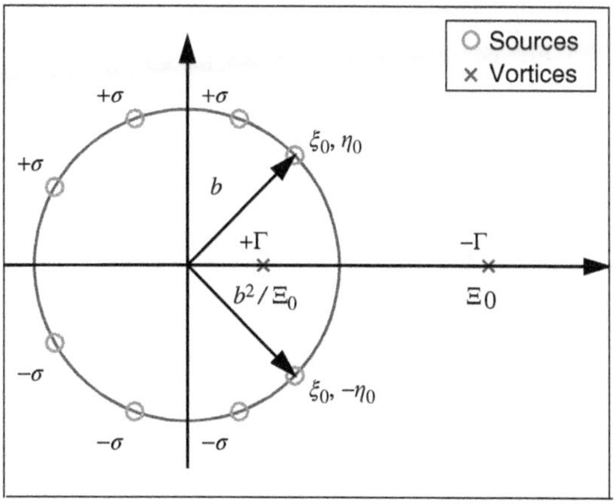

- sources of strength $+\sigma(\xi_0, y_0)$ on every point (ξ_0, y_0) of the upper surface of the circle,
- sources of strength $-\sigma(\xi_0, -y_0)$ on every point $(\xi_0, -y_0)$ of the lower surface of the circle,
- vortices of strength $+\Gamma(b^2/\Xi_0)$ on every point $(b^2/\Xi_0, 0)$ of the line $\xi \in [0, b], \eta = 0$, where $\Xi_0 \in [b, +\infty]$ and
- vortices of strength $-\Gamma(\Xi_0)$ on every point $(\Xi_0, 0)$ of the line $\xi \in [b, +\infty], \eta = 0$.

Note that Wagner placed only the vortices in the wake, while Theodorsen added the sources on the circle. The complete singularity placement scheme in the circle plane can be visualised in Figure 3.8. The source strengths vary with ξ_0, but the strengths of sources placed on opposite sides of the circle and at the same values of ξ_0 are equal and opposite, that is $\sigma(\xi_0, \eta_0) = -\sigma(\xi_0, -\eta_0)$. Similarly, the strengths of the vortices placed at Ξ_0 and b^2/Ξ_0 are also equal and opposite, so that $\Gamma(\Xi_0) = -\Gamma(b^2/\Xi_0)$.

3.3.1 Flow Induced by the Source Distribution

Recall that Laplace's equation is linear and, hence, its solutions can be superimposed. This means that we can study the flow induced by the source distribution on the circle separately from the flow generated by the vortex distribution in the wake. We will explore the flow induced by the source distribution using the concept of the complex potential, as it simplifies the calculations and makes all expressions shorter. The complex potential F is defined as

$$F = \phi + i\psi$$

where ϕ is the potential and ψ the stream function. The complex potential induced at any point on the circle plane $\zeta = \xi + i\eta$ by a source of strength $\sigma(\zeta_0)$ lying on point $\zeta_0 = \xi_0 + i\eta_0$ of the circle is given by

$$F(\zeta, \zeta_0) = \frac{\sigma(\zeta_0)}{2\pi} \ln\left(\zeta - \zeta_0\right)$$

Consequently, the complex potential induced by the source of strength $\sigma(\zeta_0^*)$ lying on $\zeta_0^* = \xi_0 - i\eta_0$, where the $*$ symbol denotes the complex conjugate value, is written as

$$F(\zeta, \zeta_0^*) = -\frac{\sigma(\zeta_0)}{2\pi} \ln\left(\zeta - \zeta_0^*\right)$$

recalling that we have chosen to set $\sigma(\zeta_0^*) = -\sigma(\zeta_0)$. Adding the contributions $F(\zeta, \zeta_0)$ and $F(\zeta, \zeta_0^*)$, the total complex potential induced by the pair of sources lying on ξ_0 is given by

$$F(\zeta, \xi_0) = \frac{\sigma(\xi_0)}{2\pi} \ln\left(\frac{\zeta - \zeta_0}{\zeta - \zeta_0^*}\right) \tag{3.34}$$

The complex velocity is defined as

$$V(\zeta) = \frac{dF}{d\zeta}$$

Using the Cauchy–Riemann relations, the real and imaginary parts of the complex velocity are

$$V(\zeta) = u(\zeta) - iv(\zeta) \tag{3.35}$$

where $u(\zeta)$ and $v(\zeta)$ are the horizontal and vertical velocity components in the ζ plane. Therefore, the complex velocity induced at any point on the circle plane $\zeta = \xi + i\eta$ by the pair of sources lying on ξ_0 is

$$V(\zeta, \xi_0) = \frac{dF(\zeta, \xi_0)}{d\zeta} = \frac{\sigma(\xi_0)}{2\pi}\left(\frac{1}{\zeta - \zeta_0} - \frac{1}{\zeta - \zeta_0^*}\right) \tag{3.36}$$

The actual form of this flow is best demonstrated by means of an example. As the strength of the sources $\sigma(\xi_0)$ is not yet known, we will study the flow for a constant source strength.

Example 3.2 *Draw the potential, stream function and airspeeds induced by a constant source distribution lying on a circle.*

We start by choosing the radius of the circle equal to $b = 1$. *We then place* $2N_0$ *sources on the circle,* N_0 *on the upper surface with strength* $\sigma = 1$ *and* N_0 *on the lower surface with strength* $\sigma = -1$. *Finally, we define a* $N \times N$ *grid of general points* $\xi_{i,j}, \eta_{i,j}$ *on the circle plane ranging from* $-5b$ *to* $+5b$, *where* $i = 1, \ldots, N$ *and* $j = 1, \ldots, N$. *The total complex potential induced at point* $\zeta_{i,j} = \xi_{i,j} + i\eta_{i,j}$ *by all the sources on the circle is given from Eq. (3.34) as*

$$F(\zeta_{i,j}) = \frac{\sigma}{2\pi} \sum_{k=1}^{N_0} \ln\left(\frac{\zeta_{i,j} - \zeta_{0_k}}{\zeta_{i,j} - \zeta_{0_k}^*}\right) \tag{3.37}$$

Note that, for a continuous source distribution, $N_0 \to \infty$ *and therefore* $F(\zeta_{i,j})$ *can only be finite if* σ *is infinitesimal. We will discuss this point later on in this section. For the moment, we will assume that Eq. (3.37) approximates the complex potential induced by a continuous source distribution if* N_0 *is high enough, and we choose* $N_0 = 1001$.

Recalling that $\phi = \Re(F)$ *and* $\psi = \Im(F)$, *Figure 3.9 plots the potential and stream function induced by the sources, as calculated from Eq. (3.37). In the potential plot of Figure 3.9a, the black lines are equipotential lines and the colours (or shades of grey) denote the local value of the potential. There are two distinct regions in the plot, the region outside the circle where the equipotential lines are circular arcs and the region inside the circle where the equipotential lines are straight. In the stream function plot of Figure 3.9b, the black lines denote flow streamlines. The flow inside the circle is directed in a straight line from each source to each corresponding sink while the flow outside the circle follows a circular arc. The existence of the*

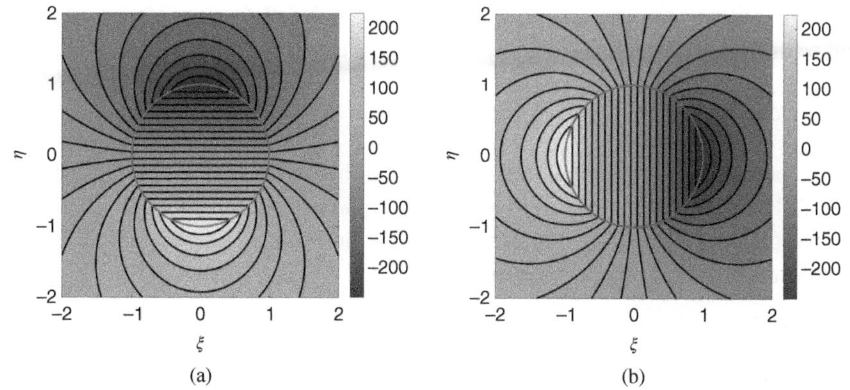

Figure 3.9 Potential (left) and stream function (right) induced by a constant source distribution in the circle plane. (a) $\phi(\xi, \eta)$ and (b) $\psi(\xi, \eta)$.

two distinct flow regions is very important for the conformal transformation procedure and will be discussed in more detail later in this section.

The complex velocity induced at $\zeta_{i,j}$ by the sources can be obtained from Eq. (3.36) as

$$V(\zeta_{i,j}) = \frac{\sigma}{2\pi}\sum_{k=1}^{N_0}\left(\frac{1}{\zeta_{i,j}-\zeta_{0_k}} - \frac{1}{\zeta_{i,j}-\zeta_{0_k}^*}\right) \tag{3.38}$$

This velocity is plotted as a vector field in Figure 3.10, where each vector has elements u, v, calculated from Eq. (3.35), that is $u = \Re(V)$, $v = -\Im(V)$. Clearly, the vector field not only conforms to the streamlines plotted in Figure 3.9b but also demonstrates the direction of the flow. Inside the circle, the flow is directed vertically downwards, while outside the circle, the flow originates at the sources (upper surface) and is directed in a circular path towards the sinks. This example is solved by Matlab code `circle_flow.m`.

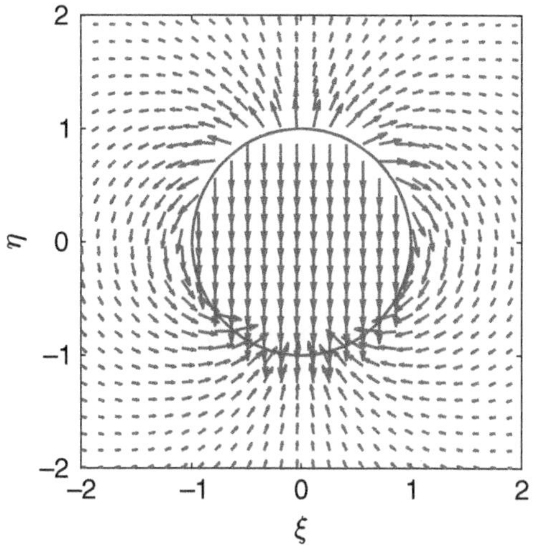

Figure 3.10 Velocity vector field induced by a constant source distribution in the circle plane.

We now want to determine the shape of the flow in the flat plate plane. Recall that the conformal transformation we want to apply is given by Eq. (3.32)

$$z = \frac{1}{2}\left(\zeta + \frac{b^2}{\zeta}\right) \tag{3.39}$$

One of the properties of the conformal transformation is that the complex potential induced by sources, sinks, vortices, etc., in the circle plane is equal to that induced in the flat plate plane at corresponding points (see for example Milne-Thomson (1973)). Then, in the context of a conformal transformation, we have

$$F(z) = F(\zeta) \tag{3.40}$$

that is $\phi(z) = \phi(\zeta)$ and $\psi(z) = \psi(\zeta)$. Furthermore, the complex velocities in the two planes are related by the chain rule

$$\frac{dF}{dz} = \frac{dF}{d\zeta}\frac{d\zeta}{dz}$$

so that after evaluating $dz/d\zeta$ from Eq. (3.39),

$$V(z) = \frac{V(\zeta)}{dz/d\zeta} = 2\frac{V(\zeta)}{1 - b^2/\zeta^2} \tag{3.41}$$

where $V(z)$ is the complex velocity in the flat plate plane and $V(\zeta)$ that in the circle plane. Again, the shape of the flow in the flat plate plane is best demonstrated by means of a constant source example.

Example 3.3 *Draw the potential, stream function and airspeeds induced by a constant source distribution lying on a circle that has been transformed to a flat plate.*

The potential and stream function on the circle plane have already been evaluated in Example 3.2. All we have to do is calculate $z_{i,j}$ for each value of $\zeta_{i,j}$ from Eq. (3.39) and then plot $F(\zeta_{i,j})$ against $z_{i,j}$. However, this treatment would ignore the fact that there are two distinct flow regions in the circle plane. It is interesting to find out what happens to each one of these flows when they are subjected to the conformal transformation. We can do this by identifying all the indices i, j for which $\zeta_{i,j}$ lies inside or outside the unit circle. We define

$$F_{\text{out}_{i,j}} = \begin{cases} F(\zeta_{i,j}) & \text{if } \left\|\zeta_{i,j}\right\| \geq b^2 \\ \text{NaN} & \text{if } \left\|\zeta_{i,j}\right\| < b^2 \end{cases}$$

$$F_{\text{in}_{i,j}} = \begin{cases} F(\zeta_{i,j}) & \text{if } \left\|\zeta_{i,j}\right\| \leq b^2 \\ \text{NaN} & \text{if } \left\|\zeta_{i,j}\right\| > b^2 \end{cases}$$

where $F_{\text{out}_{i,j}}$ is the complex potential at points that lie outside the circle and $F_{\text{in}_{i,j}}$ is the complex potential inside the circle. NaN stands for 'Not a Number' and is a trick to force Matlab to plot nothing for the i, j indices that we are not interested in. We can use the same logic to define $V_{\text{out}_{i,j}}$ and $V_{\text{in}_{i,j}}$. Then we can plot the real and imaginary parts of $F_{\text{out}_{i,j}}$, $F_{\text{in}_{i,j}}$, $V_{\text{out}_{i,j}}$ and $V_{\text{in}_{i,j}}$ against $z_{i,j}$.

Figure 3.11a plots the streamlines and velocity field of the flow outside the circle after transformation to the flat plate plane, while Figure 3.11b represents the transform of the flow inside the circle. The two flows appear to be identical, but there is a significant difference: the outer

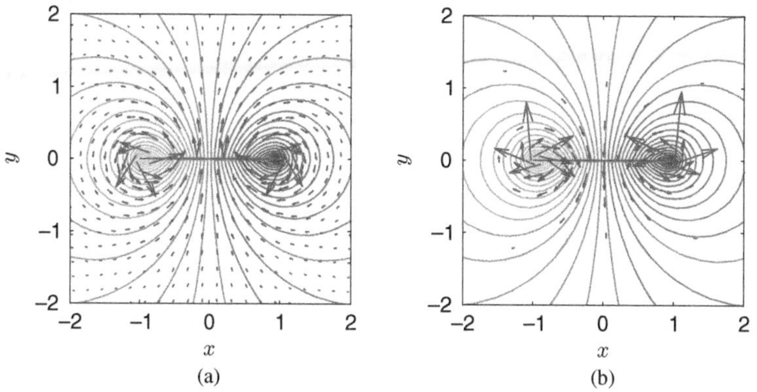

Figure 3.11 Streamlines and velocity vectors of flow originating outside and inside the circle, when transformed to the flat plate plane. (a) Flow from outside the circle and (b) flow from inside the circle.

flow is directed upwards at the mid-chord and downwards at the tips while the inner flow has the opposite direction. The conformal transformation is a nonlinear process so the two flows do not get superimposed and do not cancel each other out in the flat plate plane; they coexist without actually communicating with each other, in the same way that the flows inside and outside the circle do not communicate with each other.

Figure 3.11 also shows that the flat plate is not impermeable; this is normal as the source strength was chosen to be constant and equal to 1, and there is no free stream. We will start discussing how to render the flat plate impermeable in Section 3.3.3. This example is solved by Matlab code `conform.m`.

The most important conclusion that can be drawn from this latest example is the fact that the conformal transformation does note lead to the cancellation of the sources by the equal and opposite sinks. Even though both sets of singularities lie on the same line, the flat plate airfoil, their strengths do not add up to zero. Instead, two co-existing flows are created, one directed upwards through the mid-chord of the plate and one downwards. Bisplinghoff et al. (1996) state that these flows constitute two distinct Riemann surfaces that are not connected to each other; a flow particle lying on one surface can never move onto the other.

3.3.2 Flow Induced by the Vortex Distribution

We start by considering the circle plane, where pairs of vortices lie on points $(b^2/\Xi_0, 0)$ and $(\Xi_0, 0)$ for Ξ_0 between b and $+\infty$, see Figure 3.8. From Appendix A.2, the complex potential induced at any point ζ by a clockwise vortex lying on point ζ_0 is given by

$$F(\zeta, \zeta_0) = \mathrm{i}\frac{\Gamma(\zeta_0)}{2\pi} \ln\left(\zeta - \zeta_0\right)$$

For a pair of vortices of strength Γ (clockwise) and $-\Gamma$ (counter-clockwise) lying on $(b^2/\Xi_0, 0)$ and $(\Xi_0, 0)$ respectively, the total complex potential induced at ζ is given by

$$F(\zeta, \Xi_0) = \mathrm{i}\frac{\Gamma(\Xi_0)}{2\pi} \ln\left(\frac{\zeta - b^2/\Xi_0}{\zeta - \Xi_0}\right) \tag{3.42}$$

The complex velocity induced at ζ by these two vortices is then

$$V(\zeta, \Xi_0) = \frac{dF(\zeta, \Xi_0)}{d\zeta} = i\frac{\Gamma(\Xi_0)}{2\pi}\left(\frac{1}{\zeta - b^2/\Xi_0} - \frac{1}{\zeta - \Xi_0}\right) \tag{3.43}$$

The shape of this flow is demonstrated in the following example for a constant vorticity distribution $\Gamma(\Xi_0) = \Gamma$.

Example 3.4 *Draw the potential, stream function and airspeeds induced by a constant vortex distribution in the circle plane.*

The solution to this example is very similar to that of Example 3.2. We set $b = 1, \Gamma = 1$ and we define $2N_0$ vortices, N_0 in the interval $(0, b)$ and N_0 in the interval $(b, 100b)$. We choose a high value for $N_0 = 1001$ in order to approximate a continuous vortex distribution and calculate the complex potential at points $\zeta_{i,j}$ induced by all vortices from Eq. (3.42), such that

$$F(\zeta_{i,j}) = i\frac{\Gamma}{2\pi}\sum_{k=1}^{N_0} \ln\left(\frac{\zeta_{i,j} - b^2/\Xi_{0_k}}{\zeta_{i,j} - \Xi_{0_k}}\right) \tag{3.44}$$

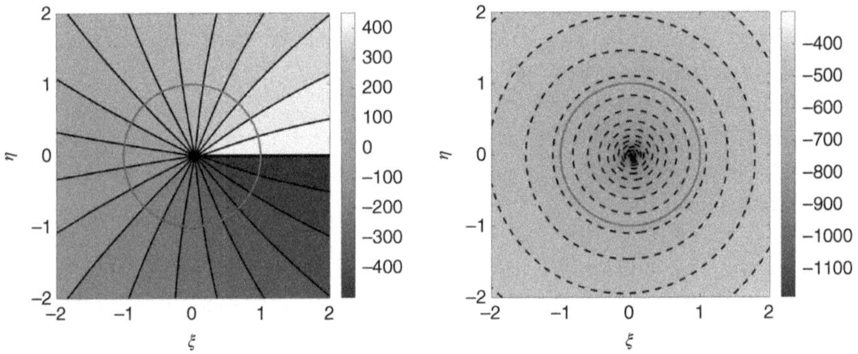

Figure 3.12 Potential (left) and stream function (right) induced by a constant vortex distribution in the circle plane. (a) $\phi(\xi, \eta)$ and (b) $\psi(\xi, \eta)$.

Figure 3.12 plots the potential and stream function induced by the vortices. The potential field is approximately a set of radial lines extending from the origin, while the streamlines (plotted with dashed lines in order to avoid confusion with the circle itself) are approximately a set of coaxial circles around $(0, 0)$. Recall that the maximum value of Ξ_0 was taken as $100b$; if it was infinite and there was an infinite number of vortices, then the streamlines would be perfect circles and the $|z| = b$ circle would be a flow streamline. This is a very important observation because it means that the vortex distribution is so placed such that it does not induce any flow normal to the circle on which the sources are placed.

Figure 3.13 plots the velocity vector field induced by the vortex distribution from

$$V(\zeta_{i,j}) = i\frac{\Gamma}{2\pi}\sum_{k=1}^{N_0}\left(\frac{1}{\zeta_{i,j} - b^2/\Xi_{0_k}} - \frac{1}{\zeta_{i,j} - \Xi_{0_k}}\right) \tag{3.45}$$

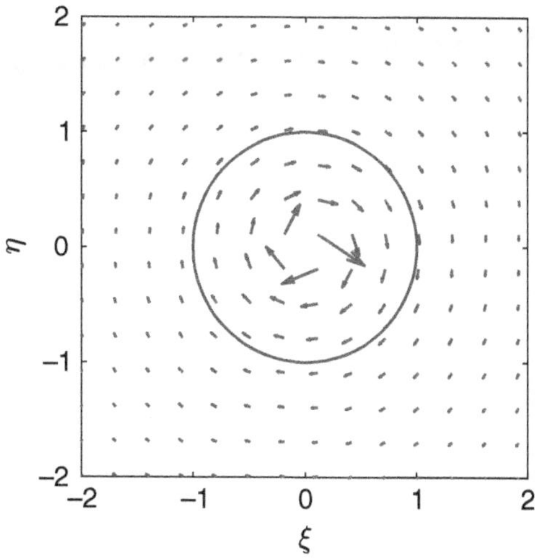

Figure 3.13 Velocity vector field induced by a constant vortex distribution in the circle plane.

As in the case of the flow induced by the source distribution, there are two flow regions, one inside the circle with radius b and one outside. Nevertheless, both the internal and external flow vectors describe circular flow around the origin. This example is solved by Matlab code `circle_flow_vortex.m`.

We will now apply the conformal transformation of Eq. (3.32) to the vortices in the circle plane in order to transform the flow to the flat plate plane. As with the sources, the complex potential is unaltered by the transformation, such that $F(z) = F(\zeta)$. The complex velocity in the flat plate plane is calculated from Eq. (3.41). The resulting flow is demonstrated through the following constant vortex strength example.

Example 3.5 *Draw the potential, stream function and airspeeds induced by a constant vortex strength distribution in the circle plane that has been transformed to the flat plate plane.*

As in the case of Example 3.3, the circle plane flow has regions inside and outside the circle, which must be mapped separately to the flat pate plane. We use exactly the same strategy as in Example 3.3 and then plot the real and imaginary parts of $F_{\text{out}_{i,j}}$, $F_{\text{in}_{i,j}}$, $V_{\text{out}_{i,j}}$ *and* $V_{\text{in}_{i,j}}$ *against* $z_{i,j}$.

Figure 3.14 shows that the flows from both outside and inside the circle transform to flows that are tangent to the flat plate, rotating around it in clockwise and counter-clockwise directions, respectively. Hence, the vortex distribution induces no flow normal to either the circle or the flat plate. This example is solved by Matlab code `conform_vortex.m`.

The discussion of Sections 3.3.1 and 3.3.2 demonstrates that the source distribution induces flow both normal and tangential to the flat plate, while the vortex distribution only induces flow tangential to the plate. Consequently, the impermeability boundary condition, which dictates that there must be no flow normal to the plate, only affects the source distribution. We can take advantage of this fact in order to select the strength of the source distribution such that impermeability is satisfied. This calculation can be carried out independently of the evaluation of the strength of the vortex distribution.

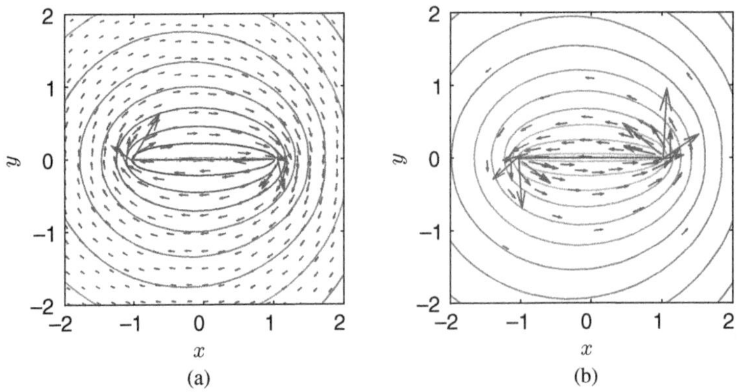

Figure 3.14 Streamlines and velocity vectors of flow originating outside and inside the circle, when transformed to the flat plate plane. (a) Flow from outside the circle and (b) flow from inside the circle.

3.3.3 Imposing the Impermeability Boundary Condition

The impermeability boundary condition must be imposed in the flat plate plane because that is where the upwash velocity is defined. The complex potential induced by the pair of sources lying on ξ_0 on the flat plate's surface is given by Eq. (3.34); it is equal to the potential on the surface of the flat plate, see Eq. (3.40), but we would like to express it in flat plate coordinates z, z_0 and x_0. Solving Eq. (3.39) for ζ yields

$$\zeta = z + \mathrm{i}\sqrt{b^2 - z^2}, \quad \zeta^* = z - \mathrm{i}\sqrt{b^2 - z^2} \tag{3.46}$$

Substituting into Eq. (3.34), the complex potential induced at any point z by the pair of sources lying on $x_0 = \xi_0$ on the flat plate results in

$$F(z, x_0) = \frac{\sigma(x_0)}{2\pi} \ln\left(\frac{z + \mathrm{i}\sqrt{b^2 - z^2} - z_0 - \mathrm{i}\sqrt{b^2 - z_0^2}}{z + \mathrm{i}\sqrt{b^2 - z^2} - z_0 + \mathrm{i}\sqrt{b^2 - z_0^2}}\right) \tag{3.47}$$

The flat plate is described by x in the interval $[-b, b]$ and $y = 0$, so that $z = x$. Therefore, the complex potential on the flat plate's surface becomes

$$F(x, x_0) = \frac{\sigma(x_0)}{2\pi} \ln\left(\frac{(x - x_0) + \mathrm{i}\left(\sqrt{b^2 - x^2} - \sqrt{b^2 - x_0^2}\right)}{(x - x_0) + \mathrm{i}\left(\sqrt{b^2 - x^2} + \sqrt{b^2 - x_0^2}\right)}\right) \tag{3.48}$$

The logarithm of a complex number $x + \mathrm{i}y$ is equal to

$$\ln(x + \mathrm{i}y) = \ln\left(\sqrt{x^2 + y^2}\right) + \mathrm{i}\tan^{-1}\left(\frac{y}{x}\right) \tag{3.49}$$

Recalling that $F = \phi + \mathrm{i}\psi$, the real potential induced by the pair of sources on x_0 is given by

$$\phi(x, x_0) = \frac{\sigma(x_0)}{4\pi} \ln\left(\frac{(x - x_0)^2 + \left(\sqrt{b^2 - x^2} - \sqrt{b^2 - x_0^2}\right)^2}{(x - x_0)^2 + \left(\sqrt{b^2 - x^2} + \sqrt{b^2 - x_0^2}\right)^2}\right) \tag{3.50}$$

and the stream function by

$$\psi(x,x_0) = \frac{\sigma(x_0)}{2\pi}\tan^{-1}\left(\frac{-2\sqrt{b^2-x_0^2}(x-x_0)}{(x-x_0)^2+(b^2-x^2)-(b^2-x_0^2)}\right) \tag{3.51}$$

The complex velocity anywhere in the flat plate plane can be obtained by going back to expression (3.47) for the complex potential and differentiating it with respect to z, such that

$$V(z,x_0) = \frac{\sigma(x_0)}{2\pi}\frac{z^2\sqrt{\frac{b^2-x_0^2}{b^2-z^2}}+\sqrt{b^2-x_0^2}\sqrt{b^2-z^2}}{b^2(z-x_0)} \tag{3.52}$$

On the flat plate's surface, the complex velocity is obtained by substituting $z = x$, as was done for the complex potential. After carrying out this substitution, we obtain

$$V(x,x_0) = \frac{\sigma(x_0)}{2\pi}\frac{x^2\sqrt{\frac{b^2-x_0^2}{b^2-x^2}}+\sqrt{b^2-x_0^2}\sqrt{b^2-x^2}}{b^2(x-x_0)} \tag{3.53}$$

which is a real value. This means that the vertical velocity $v(x,x_0)$ is zero everywhere on the surface of the plate. However, Figure 3.11 shows that there is a significant vertical flow component near the flat plate. This is in fact the key to understanding the flow created by the source distribution; the flat plate is a slit and no flow can cross it. Hence, the vertical velocity exactly on the plate is equal to zero, except at $x = x_0$ where both velocity components are infinite.

In order to estimate the vertical velocity on the flat plate's surface, we calculate $V(z)$ from Eq. (3.52) just off the surface, at $z = x + \mathrm{i}y$, where $y \ll x$. Furthermore, we define $\sigma(x_0) = \bar{\sigma}(x_0)\mathrm{d}x_0$ and calculate the complex velocity induced by all the pairs of sources by taking the integral of Eq. (3.52) for x_0 in the interval $[-b, b]$. We obtain

$$V(x+\mathrm{i}y) = \int_{-b}^{b}\frac{\bar{\sigma}(x_0)}{2\pi}\frac{(x+\mathrm{i}y)^2\sqrt{\frac{b^2-x_0^2}{b^2-(x+\mathrm{i}y)^2}}+\sqrt{b^2-x_0^2}\sqrt{b^2-(x+\mathrm{i}y)^2}}{b^2(x-x_0+\mathrm{i}y)}\mathrm{d}x_0$$

As $y \ll x$, we can simplify this latest expression to

$$V(x+\mathrm{i}y) = \int_{-b}^{b}\frac{\bar{\sigma}(x_0)}{2\pi}\frac{x^2\sqrt{\frac{b^2-x_0^2}{b^2-x^2}}+\sqrt{b^2-x_0^2}\sqrt{b^2-x^2}}{b^2(x-x_0+\mathrm{i}y)}\mathrm{d}x_0$$

noting that, in the case where $x_0 \approx x$, y is of the same order as $x - x_0$. Multiplying both the numerator and the denominator by $(x - x_0 - \mathrm{i}y)$ leads to

$$V(x+\mathrm{i}y) = \int_{-b}^{b}\frac{\bar{\sigma}(x_0)}{2\pi}\frac{x^2\sqrt{\frac{b^2-x_0^2}{b^2-x^2}}+\sqrt{b^2-x_0^2}\sqrt{b^2-x^2}}{b^2((x-x_0)^2+y^2)}(x-x_0-\mathrm{i}y)\mathrm{d}x_0$$

From the definition of the complex velocity $V = u - \mathrm{i}v$, the vertical component $v(x + \mathrm{i}y)$ is given by

$$v(x+\mathrm{i}y) = y\int_{-b}^{b}\frac{\bar{\sigma}(x_0)}{2\pi}\frac{x^2\sqrt{\frac{b^2-x_0^2}{b^2-x^2}}+\sqrt{b^2-x_0^2}\sqrt{b^2-x^2}}{b^2((x-x_0)^2+y^2)}\mathrm{d}x_0$$

The entire integral is multiplied by y so that $v(x + iy)$ is equal to zero as $y \to 0$, except when $x_0 = x$. Therefore, we will apply the procedure by Bisplinghoff et al. (1996), which consists in evaluating the integral using the substitution $x_0 - x = \varepsilon_0$ where $\varepsilon_0 \in [-\epsilon, \epsilon]$ and $\varepsilon \ll 1$; anywhere outside these limits $v(x + iy) = 0$ (see also Appendix A.3). The vertical velocity becomes

$$\lim_{y\to 0} v(x+iy) = \lim_{y,\varepsilon\to 0} y \int_{-\varepsilon}^{\varepsilon} \frac{\bar{\sigma}(x+\epsilon_0)}{2\pi} \frac{x^2\sqrt{\frac{b^2-(x+\epsilon_0)^2}{b^2-x^2}} + \sqrt{b^2-(x+\epsilon_0)^2}\sqrt{b^2-x^2}}{b^2({\varepsilon_0}^2+y^2)} \mathrm{d}\varepsilon_0$$

Assuming that, between these limits, $\sqrt{b^2 - (x + \epsilon_0)^2} \approx \sqrt{b^2 - x^2}$ and, $\sigma(x + \epsilon_0) \approx \sigma(x)$ the integral can be simplified to

$$\lim_{y\to 0} v(x+iy) = \lim_{y,\varepsilon\to 0} y\frac{\bar{\sigma}(x)}{2\pi} \int_{-\varepsilon}^{\varepsilon} \frac{1}{{\epsilon_0}^2+y^2} \mathrm{d}\varepsilon_0$$

After carrying out the integration and substituting the limits, we obtain

$$\lim_{y\to 0} v(x+iy) = \frac{\bar{\sigma}(x)}{\pi} \lim_{y,\varepsilon\to 0} \tan^{-1}\left(\frac{\varepsilon}{y}\right)$$

Now, letting $y \to 0^+$ faster than $\varepsilon \to 0$, the vertical velocity on the flat plate finally becomes

$$v(x+i0^+) = \frac{\bar{\sigma}(x)}{2} \tag{3.54}$$

Furthermore, letting $y \to 0^-$, we obtain

$$v(x+i0^-) = -\frac{\bar{\sigma}(x)}{2} \tag{3.55}$$

This is a very important result because it states that the velocity normal to the flat plate induced by the sources at any point x between $-b$ and b is equal to half the local source strength per unit length $\pm\bar{\sigma}(x)/2$, where the sign depends on whether the velocity is evaluated on the upper or lower surface. Consequently, if we know the normal velocity that will act on the plate, we can easily choose a source distribution in order to impose the impermeability boundary condition.

Impermeability is ensured when the sum of the velocity induced by the sources and the upwash due to the motion, given by Eq. (3.29), is equal to zero. Hence, $v(x, t) + w(x, t) = 0$ and, after substituting from Eq. (3.29), we obtain

$$v(x,t) = -w(x,t) = -\left(U_\infty\alpha(t) + \dot{h}(t) + \dot{\alpha}(t)(x+b-x_f)\right)$$

Consequently, the source strength distribution required to impose impermeability is

$$\sigma(x_0, 0^+, t) = \bar{\sigma}(x_0, 0^+, t)\mathrm{d}x_0 = -2\left(U_\infty\alpha(t) + \dot{h}(t) + \dot{\alpha}(t)(x_0+b-x_f)\right)\mathrm{d}x_0 \tag{3.56}$$

$$\sigma(x_0, 0^-, t) = \bar{\sigma}(x_0, 0^-, t)\mathrm{d}x_0 = 2\left(U_\infty\alpha(t) + \dot{h}(t) + \dot{\alpha}(t)(x_0+b-x_f)\right)\mathrm{d}x_0 \tag{3.57}$$

so that the source strengths on the upper and lower surfaces are equal and opposite. Note that, as discussed in Example 3.2, the value of $\sigma(x_0)$ is infinitesimal such that the total complex potential induced by all the sources remains finite. Furthermore, as the upwash due to the motion changes in time, the source strength is a function of time.

Example 3.6 *Consider a static flat plate at an angle of attack, α, to a constant free stream, Q_∞. Apply a source distribution on the plate and determine the source strength necessary to impose the impermeability boundary condition.*

Let $Q_\infty = 10$ m/s, $\alpha = 20°$, so that the components of the free stream in directions parallel and normal to the plate are $U_\infty = Q_\infty \cos\alpha$ and $V_\infty = Q_\infty \sin\alpha$. In order to impose impermeability, the flow velocity induced by the sources in a direction normal to the plate must be

$$v(x) + V_\infty = 0, \quad \text{or,} \quad v(x) = -Q_\infty \sin\alpha$$

We also set b = 1 m and start the solution as in Example 3.2. We place the sources on the circle at equally spaced values of ξ_0, such that $\xi_{0_k} - \xi_{0_{k-1}} = \Delta\xi$, for all k. We then use Eqs. (3.54) and (3.55) to select

$$\bar{\sigma}(x_0) = \pm 2v(x + i0^{\pm}) = \mp 2Q_\infty \sin\alpha$$

The dimensional source strength on the upper side of the circle is then given by

$$\sigma(x_0) = \bar{\sigma}\Delta\xi = -2Q_\infty \sin\alpha\Delta\xi$$

We proceed to calculate the total complex velocity in the circle plane from Eq. (3.38). Note that, as the flat plate does not move with respect to the free stream, $\sigma(x_0)$ is constant in time. The next step is to apply the conformal transformation, as was done in Example 3.3, and to calculate the complex velocity in the flat plate plane from Eq. (3.41). Finally, we separate the flows from inside and outside the circle, again as in Example 3.3.

Figure 3.15 plots the streamlines and flow vectors around the flat plate lying at an angle of attack of 20° to the free stream. It can be seen that the flow never crosses the plate, wrapping itself around both the leading and trailing edges. Clearly, the source distribution can indeed impose the impermeability boundary condition but not the Kutta condition. The flow wraps itself around the trailing edge and separates on the upper surface, which is not physical. The high value of α used in this example was chosen to demonstrate clearly the wrapping of the flow around the trailing edge. This example is solved by Matlab code `imperm_sigma.m`.

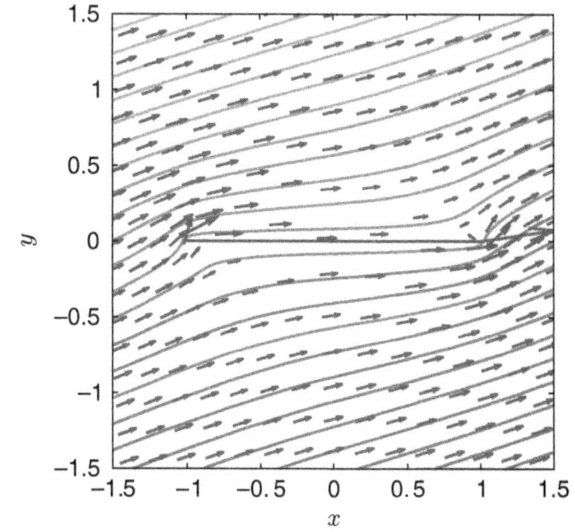

Figure 3.15 Flow around a flat plate at constant angle of attack in a constant airflow.

Figure 3.15 is a classic demonstration of d'Alembert's paradox. The flow is antisymmetric around the origin and, therefore, the aerodynamic loads acting on the plate are equal to zero, in both the horizontal and vertical directions. In section 3.3.4 we will apply Bernoulli's equation in order to calculate these loads, but they will still be equal to zero; the source of the problem lies in the form of the flow around the trailing edge. The Kutta condition must be imposed by selecting an appropriate value for the strength of the vortex distribution in the wake in order to ensure that the flow separates exactly at the trailing edge and in a direction parallel to it. This analysis will be carried out in Section 3.3.5.

3.3.4 Calculating the Loads Due to the Source Distribution

The total aerodynamic loads acting on the flat plate depend on both the sources and the vortices. However, Laplace's equation is linear, and we will also linearise the calculation of the aerodynamic loads; we can therefore evaluate the loads due to the source distribution separately from those due to the vorticity in the wake. In order to do this, we will follow the approach by Theodorsen (1935), which starts with the real potential induced by a pair of sources lying at x_0 on any point on the surface of the flat plate, x; this potential is given by Eq. (3.50). Substituting from the impermeability condition (3.56), we obtain the real potential induced on the upper surface by a pair of sources:

$$\phi(x, x_0, 0^+, t) = -\frac{\left(U_\infty \alpha(t) + \dot{h}(t) + \dot{\alpha}(t)(x_0 + b - x_f)\right)}{2\pi} \times \ln\left(\frac{(x-x_0)^2 + \left(\sqrt{b^2-x^2} - \sqrt{b^2-x_0^2}\right)^2}{(x-x_0)^2 + \left(\sqrt{b^2-x^2} + \sqrt{b^2-x_0^2}\right)^2}\right) \mathrm{d}x_0$$

The total real potential $\phi(x, 0^+, t)$ on the upper surface induced by all the sources is obtained by integrating over x_0 the expression above, from the leading edge to the trailing edge,

$$\phi(x, 0^+, t) = \int_{-b}^{b} \phi(x, x_0, 0^+, t) \tag{3.58}$$

The total potential on the lower surface is simply $\phi(x, 0^-, t) = -\phi(x, 0^+, t)$. Calculating $\phi(x, 0^+, t)$ or $\phi(x, 0^-, t)$ requires the evaluation of the integrals

$$I_1 = \int_{-b}^{b} \ln\left(\frac{(x-x_0)^2 + \left(\sqrt{b^2-x^2} - \sqrt{b^2-x_0^2}\right)^2}{(x-x_0)^2 + \left(\sqrt{b^2-x^2} + \sqrt{b^2-x_0^2}\right)^2}\right) \mathrm{d}x_0$$

$$I_2 = \int_{-b}^{b} (x_0 + b) \ln\left(\frac{(x-x_0)^2 + \left(\sqrt{b^2-x^2} - \sqrt{b^2-x_0^2}\right)^2}{(x-x_0)^2 + \left(\sqrt{b^2-x^2} + \sqrt{b^2-x_0^2}\right)^2}\right) \mathrm{d}x_0$$

which can be re-written in the form:

$$I_1 = b\int_{-1}^{1} \ln\left(\frac{(\bar{x}-\bar{x}_0)^2+(\bar{y}-\bar{y}_0)^2}{(\bar{x}-\bar{x}_0)^2+(\bar{y}+\bar{y}_0)^2}\right) d\bar{x}_0$$

$$I_2 = b^2\int_{-1}^{1} (\bar{x}_0+1)\ln\left(\frac{(\bar{x}-\bar{x}_0)^2+(\bar{y}-\bar{y}_0)^2}{(\bar{x}-\bar{x}_0)^2+(\bar{y}+\bar{y}_0)^2}\right) d\bar{x}_0$$

where $\bar{x} = x/b, \bar{x}_0 = x_0/b, \bar{y} = \sqrt{1-\bar{x}^2}$ and $\bar{y}_0 = \sqrt{1-\bar{x}_0^2}$. Theodorsen (1935) gives the following results for these integrals,

$$I_1 = -2\pi b\sqrt{1-\bar{x}^2} = -2\pi\sqrt{b^2-x^2}$$

$$I_2 = -\pi b^2(\bar{x}+2)\sqrt{1-\bar{x}^2} = -\pi(x+2b)\sqrt{b^2-x^2}$$

Substituting I_1 and I_2 into Eq. (3.58) gives the potential induced by the sources on the upper side of the flat plate

$$\phi(x, 0^+, t) = \left(U_\infty\alpha(t) + \dot{h}(t) - x_f\dot{\alpha}(t)\right)\sqrt{b^2-x^2} + \dot{\alpha}(t)\frac{x+2b}{2}\sqrt{b^2-x^2} \tag{3.59}$$

We recall that there is only a tangential flow velocity component on the surface of the plate, since the normal flow velocity must always be equal to zero in order to enforce impermeability. Therefore, the total velocity on the upper surface is given by

$$U_\infty + u(x, 0^+, t) = U_\infty + \frac{\partial\phi}{\partial x}$$

where U_∞ is the horizontal free stream airspeed and ϕ is given by Eq. (3.59). On the lower surface, the total velocity is simply

$$U_\infty + u(x, 0^-, t) = U_\infty - \frac{\partial\phi}{\partial x}$$

since $\phi(x, 0^-) = -\phi(x, 0^+)$. The unsteady Bernoulli equation (3.3) on the two surfaces gives

$$p(x, 0^+, t) = -\rho\left(\frac{1}{2}\left(U_\infty + \frac{\partial\phi}{\partial x}\right)^2 + \frac{\partial\phi}{\partial t}\right) + \text{constant}$$

$$p(x, 0^-, t) = -\rho\left(\frac{1}{2}\left(U_\infty - \frac{\partial\phi}{\partial x}\right)^2 - \frac{\partial\phi}{\partial t}\right) + \text{constant}$$

where we have assumed that the constants are equal for the two equations since the two streamlines originate at the same location far upstream and should therefore have the same total pressure. The pressure difference between the upper and lower surfaces of the plate at x is then

$$\Delta p(x, t) = p(x, 0^+, t) - p(x, 0^-, t) = -2\rho\left(U_\infty\frac{\partial\phi}{\partial x} + \frac{\partial\phi}{\partial t}\right) \tag{3.60}$$

Finally, the lift is the integral of the pressure difference over the entire chord

$$l_\sigma(t) = \int_{-b}^{b} \Delta p(x, t)\mathrm{d}x = -2\rho U_\infty\, \phi(x)|_{-b}^{b} - 2\rho\int_{-b}^{b}\frac{\partial\phi}{\partial t}\mathrm{d}x \tag{3.61}$$

where we have used the symbol l_σ to denote the fact that this lift value is due to the source distribution only. Substituting from Eq. (3.59) and carrying out the integration yields

$$l_\sigma(t) = -\rho\pi b^2 \left(\ddot{h} - \left(x_f - b\right)\ddot{\alpha} + U_\infty \dot{\alpha}\right)$$

The sign of the lift is negative because, as mentioned in Section 3.3, downwards motion is taken to be positive. Theodorsen defined the non-dimensional distance between the pitching axis and the half-chord as

$$a = \frac{x_f - b}{b} \tag{3.62}$$

so that the lift finally simplifies to

$$l_\sigma(t) = -\rho\pi b^2 \left(\ddot{h} - ba\ddot{\alpha} + U_\infty \dot{\alpha}\right) \tag{3.63}$$

The other load of interest is the pitching moment caused by the lift around the pitching axis, m_{xf}. It is defined as

$$m_{xf_\sigma}(t) = \int_{-b}^{b} \Delta p(x,t)(x - ba)\mathrm{d}x = -2\rho \int_{-b}^{b} \left(U_\infty \frac{\partial \phi}{\partial x} + \frac{\partial \phi}{\partial t}\right)(x - ba)\mathrm{d}x$$

Substituting again from Eq. (3.59) and carrying out the integration results in

$$m_{xf_\sigma}(t) = -\rho\pi b^2 \left(-ba\ddot{h} + b^2\left(\frac{1}{8} + a^2\right)\ddot{\alpha} - U_\infty \dot{h} - U_\infty^2 \alpha\right) \tag{3.64}$$

The lift and moment expressions of Eqs. (3.63) and (3.64) are also known as the non-circulatory loads because they are caused only by the source distribution, which does not create any circulation in the flow. It is interesting to note that the non-circulatory lift is equal to zero when the flat plate is not moving (that is when $\ddot{h} = \ddot{\alpha} = \dot{h} = \dot{\alpha} = 0$), so that Eq. (3.63) is consistent with d'Alembert's paradox, the fact that potential flows do not generate aerodynamic loads in the absence of circulation. In contrast, the lift is non-zero when the airfoil performs pitch and/or plunge oscillations. The term $-\rho\pi b^2\left(\ddot{h} - ba\ddot{\alpha}\right)$ is often referred to as an 'added' (or 'apparent' or 'virtual') mass term, as it is a force due to the acceleration in pitch and in plunge of the mass of a cylinder of air with diameter b surrounding the flat plate. According to Fung (1993), the term $-\rho\pi b^2 U_\infty \dot{\alpha}$ is of centrifugal nature, describing the rotation with speed $\dot{\alpha}$ and translation with speed U_∞ of a body with mass $\rho\pi b^2$.

Added mass terms also appear in the pitching moment equation (3.64), notably the terms $\rho\pi b^2(ba\ddot{h} + b^2(1/8 + a^2)\ddot{\alpha})$. Bisplinghoff et al. (1996) note that the association of the non-circulatory loads with the mass of a cylinder of air with radius b can be dangerous since the moment of inertia of this cylinder around the pitch axis is $\rho\pi b^4(1/4 + a^2)$ and not $\rho\pi b^4(1/8 + a^2)$. The terms $\rho\pi b^2 U_\infty^2 \alpha$ and $\rho\pi b^2 U_\infty \dot{h}$ are equivalent since $U_\infty \alpha + \dot{h}$ is an upwash velocity (see Eq. (3.29)). These terms represent the fact that, even when the angle of attack is constant, the flow is rotated by the presence of the flat plate, such that the stagnation points do not lie on the leading and trailing edges but inboard (see Figure 3.15). If the flat plate applies a moment on the flow, then but Newton's third law, the flow will apply an equal and opposite moment on the flat plate, represented by the terms $\rho\pi b^2 U_\infty^2 \alpha$ and $\rho\pi b^2 U_\infty \dot{h}$; the former is non-zero even when the flat plate is static.

3.3.5 Imposing the Kutta Condition

Recall Eq. (3.42) for the complex potential induced at any point $\zeta = \xi + i\eta$ in the circle plane by a pair of vortices lying at $(b^2/\Xi_0, 0)$ and $(\Xi_0, 0)$,

$$F(\zeta, \Xi_0) = \mathrm{i}\frac{\Gamma(\Xi_0)}{2\pi} \ln\left(\frac{\zeta - b^2/\Xi_0}{\zeta - \Xi_0}\right) \tag{3.65}$$

Applying the conformal transformation of Eq. (3.32) to the location of vortex $(\Xi_0, 0)$, we can determine that this point maps to $(X_0, 0)$ on the flat plate plane, where

$$X_0 = \frac{1}{2}\left(\Xi_0 + \frac{b^2}{\Xi_0}\right) \tag{3.66}$$

and $X_0 \geq b$. The vortex $(b^2/\Xi_0, 0)$ maps to exactly the same point, so that the two distinct vortices $(\Xi_0, 0)$ and $(b^2/\Xi_0, 0)$ in the circle plane are coincident in the flat plate plane. Solving Eq. (3.66) for Ξ_0 yields

$$\Xi_0 = X_0 + \sqrt{X_0^2 - b^2} \tag{3.67}$$

Also, recall from Eq. (3.46) that any point z in the flat plate plane maps to

$$\zeta = z + \mathrm{i}\sqrt{b^2 - z^2} \tag{3.68}$$

in the circle plane.

Substituting from Eq. (3.68) into Eq. (3.65) and recalling that, on the flat plate, $z = x$, the complex potential on the flat plate's surface becomes

$$F(x, \Xi_0) = \mathrm{i}\frac{\Gamma(\Xi_0)}{2\pi} \ln\left(\frac{x + \mathrm{i}\sqrt{b^2 - x^2} - b^2/\Xi_0}{x + \mathrm{i}\sqrt{b^2 - x^2} - \Xi_0}\right) \tag{3.69}$$

Recalling the definition of the complex potential $F = \phi + \mathrm{i}\psi$ and using relation (3.49), we obtain the real potential on the plate's surface

$$\phi(x, \Xi_0) = \frac{\Gamma(\Xi_0)}{2\pi} \tan^{-1} \frac{\sqrt{b^2 - x^2}(\Xi_0 - b^2/\Xi_0)}{2b^2 - x(\Xi_0 + b^2/\Xi_0)} \tag{3.70}$$

From Eq. (3.66), we substitute $\Xi_0 + b^2/\Xi_0 = 2X_0$ in the denominator. Furthermore, using Eq. (3.66), expression (3.67) can be written as

$$\Xi_0 = \frac{1}{2}\left(\Xi_0 + \frac{b^2}{\Xi_0}\right) + \sqrt{X_0^2 - b^2}$$

which can be re-arranged to give

$$\Xi_0 - \frac{b^2}{\Xi_0} = 2\sqrt{X_0^2 - b^2}$$

Substituting this latest result in the numerator of Eq. (3.70) results in the complete expression for the real potential on the flat plate's surface induced by the pair of vortices lying on $(X_0, 0)$

$$\phi(x, X_0) = \frac{\Gamma(X_0)}{2\pi} \tan^{-1} \frac{\sqrt{b^2 - x^2}\sqrt{X_0^2 - b^2}}{b^2 - xX_0} \tag{3.71}$$

The next step is to sum the potential contributions of all the pairs of vortices in the wake, but as the vortex distribution is continuous and extends to infinity, the resulting lift will be infinite unless $\Gamma(X_0)$ is infinitesimal. We will therefore follow the same approach we took for the source strength and define

$$\Gamma(X_0, t) = \bar{\Gamma}(X_0, t)\mathrm{d}X_0 \tag{3.72}$$

where $\bar{\Gamma}$ is circulation strength per unit length. Note that the strength of the vortices has been denoted as a function of both location in the wake and time. Now we can integrate the potential induced by the vortices over the entire wake, such that

$$\phi(x, t) = \frac{1}{2\pi}\int_b^{\infty} \tan^{-1}\frac{\sqrt{b^2 - x^2}\sqrt{X_0^2 - b^2}}{b^2 - xX_0}\bar{\Gamma}(X_0, t)\mathrm{d}X_0 \tag{3.73}$$

The total potential, $\phi_{\text{tot}}(x)$ on the surface of the flat plate is the sum of the potential induced by all the vortices and that induced by all the sources. Adding up the potentials of expressions (3.59) and (3.73) yields for the upper surface

$$\begin{aligned}\phi_{\text{tot}}(x, 0^+, t) = &\left(U_\infty\alpha(t) + \dot{h}(t) - x_f\dot{\alpha}(t)\right)\sqrt{b^2 - x^2} + \dot{\alpha}(t)\frac{x + 2b}{2}\sqrt{b^2 - x^2}\\ &+ \frac{1}{2\pi}\int_b^{\infty} \tan^{-1}\frac{\sqrt{b^2 - x^2}\sqrt{X_0^2 - b^2}}{b^2 - xX_0}\bar{\Gamma}(X_0, t)\mathrm{d}X_0\end{aligned} \tag{3.74}$$

Consequently, the total velocity tangent to the surface induced by all the sources and vortices is given by

$$\begin{aligned}\frac{\partial\phi_{\text{tot}}}{\partial x} = \frac{1}{\sqrt{b^2 - x^2}}\Bigg(&- \left(U_\infty\alpha(t) + \dot{h}(t) - x_f\dot{\alpha}(t)\right)x - \frac{\dot{\alpha}(t)}{2}\left(2x^2 + 2bx - b^2\right)\\ &- \frac{1}{2\pi}\int_b^{\infty}\frac{\sqrt{X_0^2 - b^2}}{x - X_0}\bar{\Gamma}(X_0, t)\mathrm{d}X_0\Bigg)\end{aligned} \tag{3.75}$$

The Kutta condition can be expressed in several different ways. In its most general form, it states that the flow must separate smoothly from the trailing edge. One of the mathematical forms of the condition is that the tangential velocity must be finite at the trailing edge. Going back to Example 3.6 where the Kutta condition is not enforced, the flow wraps itself around the trailing edge and does not separate there. Furthermore, the horizontal flow velocity becomes infinite at the trailing edge. It follows that, if we force the horizontal flow velocity to be finite at the trailing edge, it will not be able to wrap itself around that edge.

Looking at Eq. (3.75), the denominator becomes equal to zero at $x = b$ and therefore, the flow velocity at the trailing edge is in fact infinite if the numerator is finite. The only way to avoid this problem is to ensure that the numerator also goes to zero at the trailing edge, so that the velocity is indeterminate instead of infinite. Hence, the Kutta condition for this particular problem becomes

$$-\frac{1}{2\pi}\int_b^{\infty}\frac{\sqrt{X_0^2 - b^2}}{b - X_0}\bar{\Gamma}(X_0, t)\mathrm{d}X_0 = \left(U_\infty\alpha(t) + \dot{h}(t) - x_f\dot{\alpha}(t)\right)b + \frac{3b^2}{2}\dot{\alpha}(t)$$

or after cleaning up,

$$\frac{1}{2\pi b}\int_b^{\infty}\frac{\sqrt{X_0+b}}{\sqrt{X_0-b}}\bar{\Gamma}(X_0,t)\mathrm{d}X_0 = U_\infty\alpha(t)+\dot{h}(t)+\left(\frac{3b}{2}-x_f\right)\dot{\alpha}(t) \tag{3.76}$$

Equation (3.76) is the central result of Wagner's work on the unsteady flow around a flat plate that is initially a rest and then accelerates impulsively to speed U_∞ (that is for $\dot{h} = \dot{\alpha} = 0$); it was also developed by Theodorsen for a pitching and plunging airfoil. The equation relates the value of the upwash at the three-quarter chord to an integral of the strength and distribution of the vortices in the wake. It can be used in order to evaluate the unknown vortex strength $\bar{\Gamma}(X_0, t)$ necessary to impose the Kutta condition. However, this calculation is not straightforward for two reasons:

- it is an integral equation with no obvious general solution and
- $\bar{\Gamma}(X_0, t)$ is an unknown function of both space and time.

The first issue can be resolved using several analytical and numerical approaches. The second issue requires further reflection on the nature of the wake and its vorticity. We can look at the wake either as vortices of constant strength Γ travelling downstream with unknown velocity $\dot{X}_0(t)$ or as vortices lying on fixed positions X_0 whose strength is an unknown function of time $\Gamma(t)$. Helmholtz's theorems state that a vortex line must travel with the flow and that its strength does not change in time. Therefore, the first of the options mentioned above for the wake description is physical and can certainly be applied to the present problem. Both Wagner and Theodorsen chose to fix the propagation velocity of the wake to

$$\dot{X}_0 = U_\infty$$

for all times and all locations in the wake. This means that a vortex with strength Γ_0 lying at X_0 at time t_0 will move to $X_0 + U_\infty\Delta t$ at time $t + \Delta t$, but its strength will remain unchanged. The intricacies of this wake description can be demonstrated more clearly through the description of Wagner's solution.

3.4 Wagner Theory

Consider a flat plate airfoil at a constant angle α to the horizontal. At time $t = 0$, the horizontal free stream airspeed is increased impulsively from 0 m/s to U_∞, while the vertical free stream airspeed remains zero; the angle of attack remains equal to α at all times and the wing does not undergo any other motion. The flow around an impulsively started thick airfoil was presented graphically in Figure 2.28; here the airfoil is a flat plate and the wake is also modelled as flat. As the plate lies at a non-zero angle of attack to the free stream, it must generate circulation if the Kutta condition is to be satisfied. At each time instance, the change in circulation around the plate must be balanced by the shedding of a wake vortex of equal and opposite strength from the trailing edge, so that Kelvin's theorem is satisfied. Consequently, the continuous wake vortex sheet can be approximated by a set of point vortices shed at distinct times $t = i\Delta t$, for $i = 1, 2, \ldots$. The strength of the vortex shed at the ith time instance is denoted by $\bar{\Gamma}_i$.

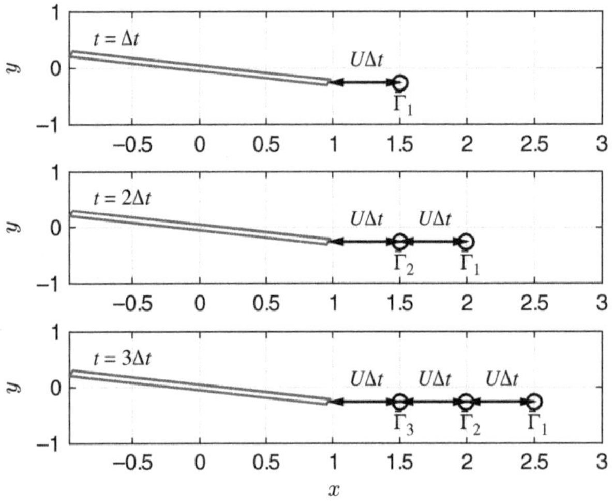

Figure 3.16 Wake history after impulsive start of a flat plate airfoil.

Figure 3.16 demonstrates graphically the wake shedding and propagation scheme. At time $t_1 = \Delta t$ there is a single vortex of strength $\bar{\Gamma}_1$ at location $X_{0_1} = b + U_\infty \Delta t$. At time $t_2 = 2\Delta t$, $\bar{\Gamma}_1$ has moved downstream to position $X_{0_2} = b + 2U_\infty \Delta t$ and a new vortex with strength $\bar{\Gamma}_2$ can be found at position X_{0_1}. One time step later, both $\bar{\Gamma}_1$ and $\bar{\Gamma}_2$ have moved downstream by $U_\infty \Delta t$ and vortex $\bar{\Gamma}_3$ has appeared at X_{0_1}. At time n, there are n vortices, and so on. The strengths of the vortices are still unknown; they can be calculated by means of the Kutta condition of Eq. (3.76).[1] As there is no pitching or plunging motion, the condition can be simplified to

$$\frac{1}{2\pi b} \int_b^\infty \frac{\sqrt{X_0 + b}}{\sqrt{X_0 - b}} \bar{\Gamma}(X_0, t) \mathrm{d}X_0 = U_\infty \alpha \tag{3.77}$$

The wake is described by discrete vortices $\bar{\Gamma}_i$ at discrete time instances t_i so that the integral in the Kutta condition becomes a sum. We write out the condition for the first three time instances t_1 to t_3:

$$\frac{1}{2\pi b} \left(\frac{\sqrt{X_{0_1} + b}}{\sqrt{X_{0_1} - b}} \bar{\Gamma}_1 \right) \Delta X_0 = U_\infty \alpha$$

$$\frac{1}{2\pi b} \left(\frac{\sqrt{X_{0_1} + b}}{\sqrt{X_{0_1} - b}} \bar{\Gamma}_2 + \frac{\sqrt{X_{0_2} + b}}{\sqrt{X_{0_2} - b}} \bar{\Gamma}_1 \right) \Delta X_0 = U_\infty \alpha$$

1 Equation (3.76) was developed here using Theodorsen's approach. The interested reader can find Wagner's derivation in Appendix C.

$$\frac{1}{2\pi b}\left(\frac{\sqrt{X_{0_1}+b}}{\sqrt{X_{0_1}-b}}\bar{\Gamma}_3+\frac{\sqrt{X_{0_2}+b}}{\sqrt{X_{0_2}-b}}\bar{\Gamma}_2+\frac{\sqrt{X_{0_3}+b}}{\sqrt{X_{0_3}-b}}\bar{\Gamma}_1\right)\Delta X_0 = U_\infty \alpha \tag{3.78}$$

where, as mentioned earlier, $X_{0_i} = b + iU_\infty \Delta t$. The above equations are in fact a system of three equations with three unknowns; the first can be solved for $\bar{\Gamma}_1$, the second for $\bar{\Gamma}_2$ etc. In general, at the ith time instance, we can write

$$\frac{\Delta X_0}{2\pi b}\sum_{j=1}^{i}\frac{\sqrt{X_{0_j}+b}}{\sqrt{X_{0_j}-b}}\bar{\Gamma}_{i-j+1} = U_\infty \alpha \tag{3.79}$$

and solve for

$$\bar{\Gamma}_i = \frac{\sqrt{X_{0_1}-b}}{\sqrt{X_{0_1}+b}}\left(\frac{2\pi b U_\infty \alpha}{\Delta X_0} - \sum_{j=2}^{i}\frac{\sqrt{X_{0_j}+b}}{\sqrt{X_{0_j}-b}}\bar{\Gamma}_{i-j+1}\right) \tag{3.80}$$

since the vortex strengths up to $\bar{\Gamma}_{i-1}$ are already known.

Once the vortex strengths have been determined, we can evaluate the circulatory lift acting on the flat plate, that is the lift due to the vortices. The difference in pressure over the flat plate caused by the potential distribution is still given by Eq. (3.60)

$$\Delta p(x,t) = -2\rho\left(U_\infty\frac{\partial\phi}{\partial x}+\frac{\partial\phi}{\partial t}\right)$$

Unlike the source case, the time derivative $\partial\phi/\partial t$ need not be calculated by taking the derivative of the potential induced by the vortices with respect to time. As we have already assumed that the wake is composed of constant strength vortices traveling downstream with velocity $\dot{X}_0 = U_\infty$, we can calculate the time derivative of the vortex-induced potential using the chain rule

$$\frac{\partial\phi}{\partial t} = \frac{\partial\phi}{\partial X_0}\dot{X}_0 = U_\infty\frac{\partial\phi}{\partial X_0}$$

The pressure difference then becomes

$$\Delta p(x,X_0,t) = -2\rho U_\infty\left(\frac{\partial\phi}{\partial x}+\frac{\partial\phi}{\partial X_0}\right) \tag{3.81}$$

After substitution from Eq. (3.71),

$$\Delta p(x,X_0,t) = -\rho U_\infty\frac{\Gamma(X_0,t)}{\pi}\frac{x+X_0}{\sqrt{b^2-x^2}\sqrt{X_0^2-b^2}}$$

The lift caused by a pair of vortices lying at X_0 can then be obtained by integrating the pressure difference over the chord, such that

$$l_\Gamma(X_0,t) = \int_{-b}^{b}\Delta p(x,X_0,t)\mathrm{d}x = -\rho U_\infty\frac{\Gamma(X_0,t)}{\pi\sqrt{X_0^2-b^2}}\int_{-b}^{b}\frac{x+X_0}{\sqrt{b^2-x^2}}\mathrm{d}x$$

where l_Γ denotes the fact that the lift is due to vortices only. The integral in the expression above is the sum of two standard integrals, so that

$$l_\Gamma(X_0,t) = -\rho U_\infty \frac{X_0}{\sqrt{X_0^2-b^2}}\Gamma(X_0,t) \tag{3.82}$$

Now the total vortex-induced lift becomes

$$l_\Gamma(t) = -\rho U_\infty \int_b^\infty \frac{X_0}{\sqrt{X_0^2-b^2}}\bar{\Gamma}(X_0,t)\mathrm{d}X_0 \tag{3.83}$$

Equation (3.63) shows that in the absence of pitching and plunging motion, the lift induced by the sources is equal to zero. This means that Eq. (3.83) gives the total lift acting on the impulsively started plate. If the vortex strengths are only available at discrete times, the integral must be replaced by the sum

$$l_\Gamma(t_i) = -\rho U_\infty \sum_{j=1}^{i} \frac{X_{0_j}}{\sqrt{X_{0_j}^2-b^2}}\bar{\Gamma}_{i-j+1}\Delta X_0 \tag{3.84}$$

The circulatory aerodynamic moment around the pitching axis due to the wake vortices can be calculated from

$$\begin{aligned} m_{xf_\Gamma}(X_0,t) &= \int_{-b}^{b} \Delta p(x,X_0,t)\,(x-ba)\,\mathrm{d}x \\ &= -\rho U_\infty \frac{\Gamma(X_0,t)}{\pi\sqrt{X_0^2-b^2}} \int_{-b}^{b} \frac{\left(x+X_0\right)(x-ba)}{\sqrt{b^2-x^2}}\mathrm{d}x \end{aligned}$$

The integral in this latest expression can be evaluated using Matlab's symbolic toolbox (see Matlab code `derive_Dp_1_m.m`). We obtain

$$m_{xf_\Gamma}(X_0,t) = -\rho U_\infty b \frac{\left(b/2-X_0 a\right)\Gamma(X_0,t)}{\sqrt{X_0^2-b^2}}$$

Integrating for X_0 over the entire wake gives the total circulatory aerodynamic moment around the pitch axis due to the wake vortices,

$$m_{xf_\Gamma}(t) = -\rho U_\infty b \int_b^\infty \frac{\left(b/2-X_0 a\right)\bar{\Gamma}(X_0,t)}{\sqrt{X_0^2-b^2}}\mathrm{d}X_0$$

or after adding and subtracting $X_0/2$ to the bracket in the numerator inside the integral,

$$m_{xf_\Gamma}(t) = -\rho U_\infty b \int_b^\infty \left(\frac{1}{2}\frac{\sqrt{X_0+b}}{\sqrt{X_0-b}} - \left(a+\frac{1}{2}\right)\frac{X_0}{\sqrt{X_0^2-b^2}}\right)\bar{\Gamma}(X_0,t)\mathrm{d}X_0 \tag{3.85}$$

Equation (3.64) shows that the source distribution induces a moment $m_{xf_\sigma} = \rho\pi b^2 U_\infty^2 \alpha$ that must be added to the vortex-induced moment of Eq. (3.85) in order to obtain the total aerodynamic pitching moment.

Example 3.7 *Calculate the wake vortex strengths and lift variation with time for the impulsively started flow around a flat plate at constant angle of attack* α.

We set $b = 0.5$ *m,* $U_\infty = 10$ *m/s and* $\alpha = 5°$. *We also choose* $\Delta t = 0.001$ *seconds, so that* $\Delta X_0 = 0.01$ *m. We calculate* $\bar{\Gamma}_i$ *for* $i = 1, 2, \ldots, n$, *where* $n = 10{,}000$, *using Eq. (3.80). We then plot the vortex strength* $\Gamma_i = \bar{\Gamma}_i \Delta X_0$ *against location in the wake* X_0 *at* $t = n\Delta t = 10$ *seconds, the end of the simulation time. Figure 3.17a shows that the vortex that was released at* $t = \Delta t$ *and now lies at* $X_0 = n\Delta X_0 + b = 100.5$ *m was very strong but subsequent vortices quickly became much weaker.*

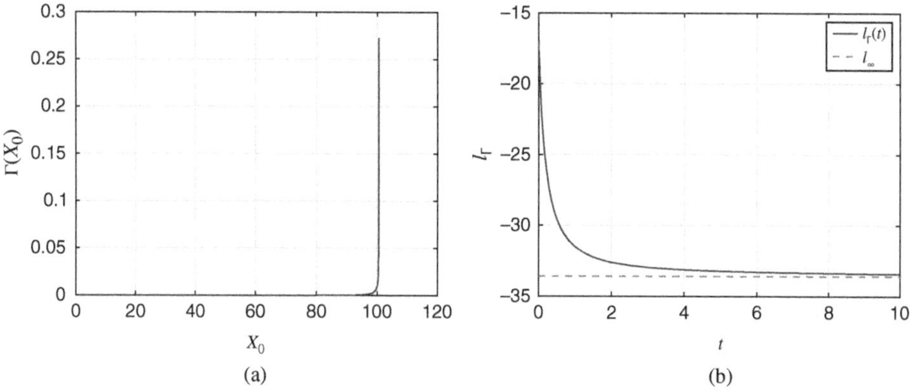

Figure 3.17 Wake vorticity distribution in the wake and lift time response after impulsive start of a flat plate. (a) Wake vorticity and (b) lift response.

We also calculate the lift response to the impulsive start from Eq. (3.84) and plot it against time in Figure 3.17b. At $t = 0$, *the lift has already jumped to around* -16.8 *N/m; subsequently, it keeps decreasing but at decaying rates until it asymptotes towards* -33.6 *N/m. The flow is very unsteady when the impulsive start occurs, since the free stream airspeed jumps from 0 to* $U_\infty = 10$ *m/s instantaneously. However, as there are no further changes either in the airspeed or in the airfoil's motion, the flow becomes gradually steadier, which is why the largest changes in wake vorticity and lift occur at the start of the motion.*

Figure 3.17a shows that if a sufficiently long time has passed after the impulsive start, the only vortices in the wake that have significant strength are the ones lying very far away; the nearby vortices are negligible. The important vortices lie at locations $X_{0_j} \gg b$, *for which*

$$\frac{\sqrt{X_{0_j} + b}}{\sqrt{X_{0_j} - b}} \approx 1$$

Equation (3.79) can then be approximated by

$$\sum_{j=1}^{\infty} \Gamma_j = 2\pi b U_\infty \alpha$$

which means that the total vorticity in the wake must be equal to $2\pi b U_\infty \alpha$ *after a sufficiently long time has passed. The lift equation (3.84) can be treated in the same way; only the far away vortices are significant, which lie at* $X_{0_j} \gg b$, *for which*

$$\frac{X_{0_j}}{\sqrt{X_{0_j}^2 - b^2}} \approx 1$$

This means that after a sufficiently long time has passed from the start of the motion,

$$l_\infty = -\rho U_\infty \sum_{j=1}^{\infty} \Gamma_j = -2\rho U_\infty^2 b\pi\alpha \tag{3.86}$$

where the negative sign is due to the fact that the lift has been defined as positive downwards. Note that the result of Eq. (3.86) is in perfect agreement with the prediction of thin airfoil theory for a symmetric airfoil, Eq. (3.27). Figure 3.17b shows that the unsteady lift response indeed asymptotes towards the l_∞ value as t increases. This example is solved by Matlab code `impulsive_Wagner.m`*.*

In order to convince ourselves that the Kutta condition has indeed been imposed, we can calculate the complete flowfield due to both the sources and the vortices at the end of the motion. In Example 3.6, we calculated the flowfield due to the sources only and saw that impermeability is satisfied, but the flow wraps itself around the trailing edge, so that the Kutta condition is not satisfied (see Figure 3.15). Now we will repeat this example, but we will add the effect of the vortices, as calculated in Example 3.7.

Example 3.8 *Repeat Example 3.6 for an impulsively started flow around a flat plate at a constant angle of attack to a free stream, a long time after the start of the motion, and include the effect of the wake vortices.*

We consider the impulsively started flow around a flat plate in a free stream given by $Q_\infty = 10$ m/s and $\alpha = 20°$ m/s, so that $U_\infty = Q_\infty \cos\alpha$ and $V_\infty = Q_\infty \sin\alpha$. We will calculate and plot the flowfield a long time after the impulsive start. We place $2N_0$ sources on the surface of the circle on points $\zeta_{0_k} = \xi_{0_k} \pm i\eta_{0_k}$ and set up a $N \times N$ grid of general points $\zeta_{i,j} = \xi_{i,j} + i\eta_{i,j}$ on which we will plot the flow, as in Example 3.2.

The source strength needed to impose impermeability is constant and given by

$$\sigma = -2V_\infty \Delta\xi$$

where $\Delta\xi = \xi_{0_k} - \xi_{0_{k-1}}$, as in Example 3.6. We set the time step to $\Delta t = 0.01$ seconds, so that $\Delta X_0 = U_{tot}\Delta t = 0.112$ m. The total number of time instances is set to $n = 1000$, and we calculate the strength of the vortices $\Gamma_k = \bar{\Gamma}_k \Delta X_0$ from Eq. (3.80), for $k = 1, \ldots, n$. Now, we have all the necessary ingredients to calculate and plot the flowfield. Recall that Γ_k are calculated in temporal order, while X_{0_k} are arranged in spatial order. This means that, at time $t = n\Delta t$, Γ_1 lies at X_{0_n} while Γ_n lies at X_{0_1}. The vortex positions in the complex plane are given by Ξ_{0_k} and b^2/Ξ_{0_k}, where from Eq. (3.67)

$$\Xi_{0_k} = X_{0_k} + \sqrt{X_{0_k}^2 - b^2}$$

The total complex potential at a general point $\zeta_{i,j}$ in the complex plane is given from Eqs. (3.37) and (3.44) as

$$F(\zeta_{i,j}) = \frac{\sigma}{2\pi} \sum_{k=1}^{N_0} \ln\left(\frac{\zeta_{i,j} - \zeta_{0_k}}{\zeta_{i,j} - \zeta_{0_k}^*}\right) + \frac{i}{2\pi} \sum_{k=1}^{n} \Gamma_k \ln\left(\frac{\zeta_{i,j} - b^2/\Xi_{0_k}}{\zeta_{i,j} - \Xi_{0_k}}\right) \tag{3.87}$$

where N_0 is the number of sources on the circle. The total complex velocity is obtained from Eq. (3.38) and (3.45) as

$$V(\zeta_{i,j}) = \frac{\sigma}{2\pi}\sum_{k=1}^{N_0}\left(\frac{1}{\zeta_{i,j}-\zeta_{0_k}} - \frac{1}{\zeta_{i,j}-\zeta_{0_k}^*}\right) + \frac{i}{2\pi}\sum_{k=1}^{n}\Gamma_k\left(\frac{1}{\zeta_{i,j}-b^2/\Xi_{0_k}} - \frac{1}{\zeta_{i,j}-\Xi_{0_k}}\right) \tag{3.88}$$

Finally, we carry out the conformal transformation by calculating $z_{i,j}$ from Eq. (3.39) and $V(\zeta_{i,j})$ from Eq. (3.41), while $F(z_{i,j}) = F(\zeta_{i,j})$. We then separate the flows coming from inside and outside the circle, following the approach of Example 3.3.

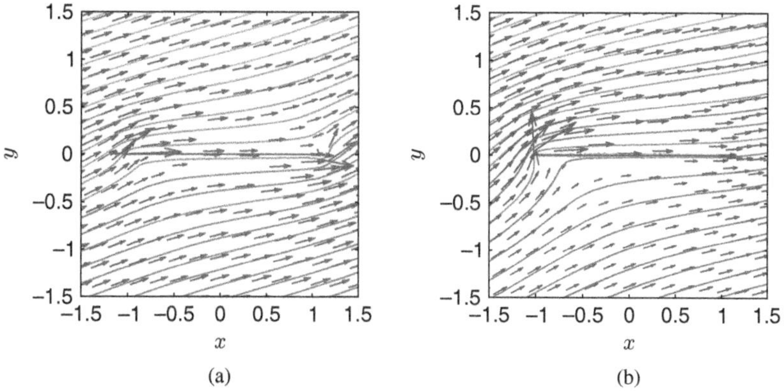

Figure 3.18 Flow around a flat plate in an impulsively started flow. (a) $t = \Delta t$ and (b) $t = n\Delta t$.

Figure 3.18 plots the streamlines and flow vectors around the flat plate at time $t = \Delta t$ and $t = n\Delta t = 9.99$ seconds. Figure 3.18a shows that, at the start of the motion, the flow still wraps itself around the trailing edge, as shown in Figure 3.15. However, after a sufficiently long time has passed, the flow separates smoothly from the trailing edge, in a direction approximately parallel to the chord (see Figure 3.18b). This means that the Kutta condition has indeed been enforced, once the flow has been stabilised. Note however that the front stagnation point lies under the plate and the flow still wraps itself around the leading edge. We will return to this point later in this chapter. The wing is of course impermeable at all time instances. This example is solved by Matlab code `imperm_sigma_Gamma.m`.

The description of an impulsive start given in Figure 3.18 is representative of real flows. When an impulsive start occurs, the flow wraps itself around the trailing edge, creating a strong vortex in a direction opposite to the circulation that starts to build up around the airfoil, as seen in Figure 2.28b. This vortex is known as the starting vortex; it subsequently detaches itself from the trailing edge and travels downstream, see Figure 2.28c,d. Initially, it lies close enough to the wing to have a significant effect on the aerodynamic loads acting on it, hence, the lift is lower than the steady-state lift (see Figure 3.17b). Once it has moved far enough, its effect is negligible and the lift takes its steady-state value.

3.4.1 The Wagner Function

Consider Eq. (3.83) for the total lift at time t. We can divide it by the expression for the Kutta condition, Eq. (3.77), to obtain

$$l_\Gamma(t) = -2\rho U_\infty^2 b\pi\alpha \frac{\int_b^\infty \frac{X_0}{\sqrt{X_0^2-b^2}}\bar{\Gamma}(X_0,t)\mathrm{d}X_0}{\int_b^\infty \frac{\sqrt{X_0+b}}{\sqrt{X_0-b}}\bar{\Gamma}(X_0,t)\mathrm{d}X_0} \tag{3.89}$$

We can define Wagner's function as

$$\Phi(t) = \frac{\int_b^\infty \frac{X_0}{\sqrt{X_0^2-b^2}}\bar{\Gamma}(X_0,t)\mathrm{d}X_0}{\int_b^\infty \frac{\sqrt{X_0+b}}{\sqrt{X_0-b}}\bar{\Gamma}(X_0,t)\mathrm{d}X_0} \tag{3.90}$$

so that the variation of the circulatory lift with time is given by

$$l_\Gamma(t) = -2\rho U_\infty^2 b\pi\alpha\Phi(t) = l_\infty\Phi(t) \tag{3.91}$$

Again, if the wake vorticity is only available at discrete time instances, the definition of the Wagner function becomes

$$\Phi(t_i) = \frac{\sum_{j=1}^{i} \frac{X_{0_j}}{\sqrt{X_{0_j}^2-b^2}}\bar{\Gamma}_{i-j+1}\Delta X_0}{\sum_{j=1}^{i} \frac{\sqrt{X_{0_j}+b}}{\sqrt{X_{0_j}-b}}\bar{\Gamma}_{i-j+1}\Delta X_0} \tag{3.92}$$

Wagner's function describes how the lift acting on an impulsively started plate grows with time, until it reaches the steady state-value, l_∞, when time tends to infinity. The exact expression for this function is given by (see Fung (1993))

$$\Phi(\tau) = 1 - \int_0^\infty \left(\left(K_0(\lambda) - K_1(\lambda)\right)^2 + \pi^2\left(I_0(\lambda) + I_1(\lambda)\right)^2\right)^{-1} \mathrm{e}^{-\lambda\tau}\lambda^{-2}\mathrm{d}\lambda \tag{3.93}$$

where τ is a non-dimensional time defined as

$$\tau = \frac{U_\infty t}{b}$$

I_0, I_1, K_0, K_1 are modified Bessel functions of the first and second kind and λ is an integration variable. Bessel functions can be evaluated using built-in Matlab functions, but the integral cannot be obtained analytically. A much more intuitive estimate of Wagner's function can be obtained from the summation of Eq. (3.92).

Example 3.9 *Calculate Wagner's function for an impulsively started flat plate airfoil.*

We can start to explore the value of Wagner's function by evaluating expression (3.92) at times $t = \Delta t$ and $t \to \infty$. For $t = \Delta t$, we set $i = 1$ into Eq. (3.92) to obtain

$$\Phi(t_1) = \frac{\frac{X_{0_1}}{\sqrt{X_{0_1}^2-b^2}}\bar{\Gamma}_1\Delta X_0}{\frac{\sqrt{X_{0_1}+b}}{\sqrt{X_{0_1}-b}}\bar{\Gamma}_1\Delta X_0} = \frac{\frac{X_{0_1}}{\sqrt{X_{0_1}^2-b^2}}}{\frac{\sqrt{X_{0_1}+b}}{\sqrt{X_{0_1}-b}}} = \frac{X_{0_1}}{X_{0_1}+b}$$

Recall from Example 3.7 that $X_{0_1} = b + \Delta X_0$. *Then, as* $\Delta X_0 \to 0$ *and, hence* $\Delta t \to 0$,

$$\Phi(0) = \frac{1}{2}$$

For the case $t \to \infty$, *we can use the argument of Example 3.7 again, so that all the powerful vortices lie very far away, at locations where* $X_0 \gg b$. *Consequently,*

$$\frac{X_0}{\sqrt{X_0^2 - b^2}} \approx 1, \quad \frac{\sqrt{X_{0_1} + b}}{\sqrt{X_{0_1} - b}} \approx 1$$

The vortices lying close to the wing's trailing edge have negligible strength and can therefore be ignored. As a consequence, Eq. (3.92) becomes

$$\Phi(t_i) = \frac{\sum_{j=1}^{i} \Gamma_i}{\sum_{j=1}^{i} \Gamma_i} = 1$$

for $i \to \infty$. *This latest result is compatible with Eq. (3.86) for the steady-state lift and Eq. (3.91).*

Consequently, Wagner's function takes the value $\Phi(0) = 1/2$ *at the start of the motion and* $\Phi(\infty) = 1$ *after an infinitely long time. The full time response can be obtained from the following algorithm:*

1. *At the ith time instance, use Eq. (3.80) to calculate* $\bar{\Gamma}_i$
2. *Calculate* $l_\Gamma(t)$ *from Eq. (3.84)*
3. *Re-arrange Eq. (3.91) to obtain* $\Phi(t_i)$ *as*

$$\Phi(t_i) = \frac{l_\Gamma(t)}{l_\infty}$$

4. *Increment i and return to step 1*

Figure 3.19 plots Wagner's function against time. The function's value is 0.5 at $t = 0$ *and asymptotes to 1 at* $t = 10$, *as expected. In between, the function grows monotonically and in a quasi-exponential fashion.*

The calculation of Wagner's lift presented here is numerical, based on expressing the integral equations (3.77) and (3.83) as sums. Hence, the calculation of $\Phi(t)$ *is also expressed as a summation and is therefore approximate. Figure 3.20a compares the numerical estimate of Wagner's function plotted in Figure 3.19 to results obtained from the numerical integration of Eq. (3.93). Clearly, the agreement is very good, but neither result is exact. In fact, the Bessel function result is more accurate because it is calculated using trapezoidal integration on a very fine non-linearly spaced grid. The numerical solution can be improved by decreasing* Δt.

Since both the exact expression of Eq. (3.93) and the summation of Eq. (3.92) are not very practical, several authors have proposed approximate analytical expressions for Wagner's function:

- *Garrick (Garrick 1938).*

$$\Phi(\tau) = 1 - \frac{2}{4 + \tau}$$

- *Drela (Izraelevitz et al. 2017).*

$$\Phi(\tau) = 1 - 0.5e^{-0.25\tau}$$

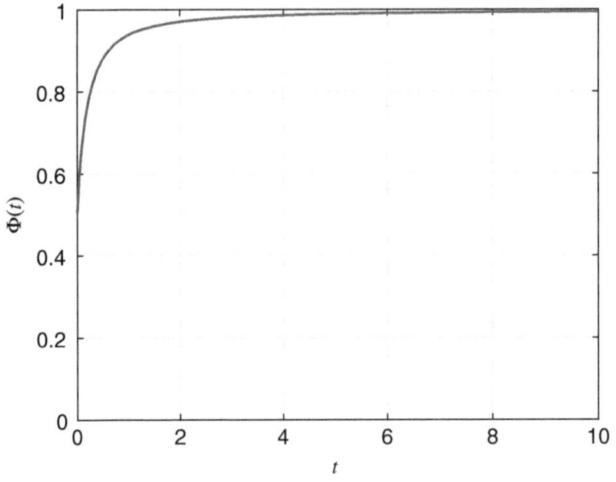

Figure 3.19 Wagner's function.

- *W.P. Jones (Jones 1945).*

$$\Phi(\tau) = 1 - 0.165e^{-0.041\tau} - 0.335e^{-0.32\tau}$$

- *R.T. Jones (Jones 1940).*

$$\Phi(\tau) = 1 - 0.165e^{-0.0455\tau} - 0.335e^{-0.3\tau} \tag{3.94}$$

Figure 3.20b plots these four approximations to Wagner's function, along with the numerical result of Figure 3.19. It can be seen that W. P. Jones' and R. T. Jones' approximations lie close to each other and are more accurate than Garrick's for $\tau \leq 50$. Note that $\tau = 50$ is a significant time duration, as the starting vortex has travelled 25 chord lengths downstream from the trailing edge. Drela's approximation is only accurate for the initial slope and the final asymptote of the true function (Izraelevitz et al. 2017). This example is solved by Matlab code `Wagner_function.m`.

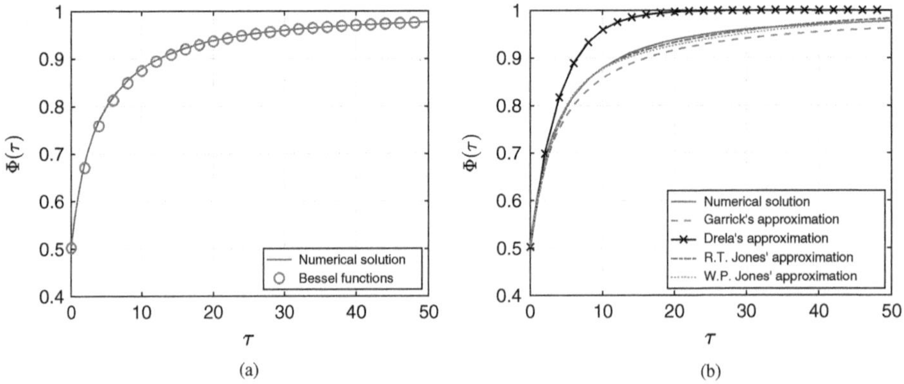

Figure 3.20 Wagner's function estimates. (a) Bessel functions and (b) approximate expressions.

3.4.2 Drag and Thrust

Up to this point, we have only calculated the lift and moment acting on the wing; both these loads depend on the difference in pressure across the airfoil. Strictly speaking, we have not even calculated the lift but the normal force. The lift is defined as the force perpendicular to the free stream, whereas all load estimations to this point have calculated a force acting in a direction perpendicular to the chord. For small angles of attack this is not a problem, as the normal and lift forces are nearly equal. However, there is also a difference in pressure between the front and the rear of the wing, which should give rise to a force acting in the chordwise direction.

Consider Figure 3.18b which plots the flow around a flat plate a long time after an impulsive start. It can be seen that the streamlines wrap themselves around the leading edge and that the flow velocity is very high there and therefore, the pressure is low. We can in fact calculate the pressure coefficient in the entire flowfield around the plate using the steady, 2D version of Eq. (2.31)

$$c_p = 1 - \frac{\left((U_\infty + u)^2 + (V_\infty + v)^2\right)}{Q_\infty^2} \tag{3.95}$$

The resulting pressure coefficient is shown in the contour plot of Figure 3.21. Clearly, there is an area of extremely low pressure around the leading edge, owing to the fact that the flow wraps itself around it and therefore the local velocity takes infinite values. In contrast, as we have imposed the Kutta condition, the trailing edge is a stagnation point and the flow velocity is zero there, so that the pressure coefficient is locally equal to 1. Consequently, the pressure is high at the trailing edge and low at the leading edge, leading to the creation of a thrusting force in the chordwise direction, known as leading edge suction.

It could be argued that the leading edge suction force should be zero for an infinitely thin flat plate, since this force is equal to the pressure difference times the frontal area of the airfoil, which is zero. However, the velocity at the leading edge is infinite. Recall that the

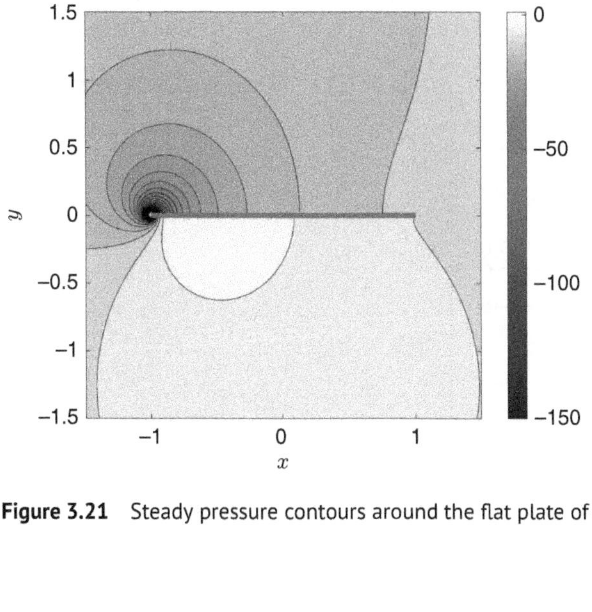

Figure 3.21 Steady pressure contours around the flat plate of Example 3.8.

velocity normal to the plate is $U_\infty + \partial\phi_{tot}/\partial x$, where the derivative of the total potential is given by Eq. (3.74)

$$\frac{\partial\phi_{tot}}{\partial x} = \frac{1}{\sqrt{b^2 - x^2}}\left(-\left(U_\infty\alpha(t) + \dot{h}(t) - x_f\dot{\alpha}(t)\right)x - \frac{\dot{\alpha}(t)}{2}\left(2x^2 + 2bx - b^2\right)\right.$$
$$\left. - \frac{1}{2\pi}\int_b^\infty \frac{\sqrt{X_0^2 - b^2}}{x - X_0}\bar{\Gamma}(X_0, t)\mathrm{d}X_0\right)$$

The denominator in this expression becomes equal to zero at both the leading and trailing edges, where $x = \pm b$. At the trailing edge, we have forced the numerator to be also zero so that the flow velocity remains finite; at the leading edge, the numerator is non-zero and the flow velocity becomes infinite, such that the local pressure, given by Eq. (3.95), tends to minus infinity. Hence, the thickness of the airfoil is zero, but the pressure difference across the front and rear of the plate is infinite; the leading edge suction force is undefined.

Grammel (1917) developed an alternative analysis in order to demonstrate that the leading edge suction force actually has a limiting value as the thickness of the plate tends to zero. The approach is based on the fact that a flat plate can be seen as a very sharp parabola, in which case the flow around it can be modelled using a polynomial function for the complex velocity. The analysis is carried out in the flat plate plane, $z = x + iy$, and the leading edge of the plate is placed at point $z_0 = -b$. Then, the complex potential anywhere in the flow is described by

$$F(z) = 2S\sqrt{z - z_0} \tag{3.96}$$

where S is a constant. Consequently, the complex velocity is

$$V(z) = \frac{S}{\sqrt{z - z_0}} \tag{3.97}$$

Figure 3.22 plots the streamlines and velocity field of the flow defined by Eqs. (3.96) and (3.97). Any of the streamlines can be taken as the contour of the flat plate, but it is

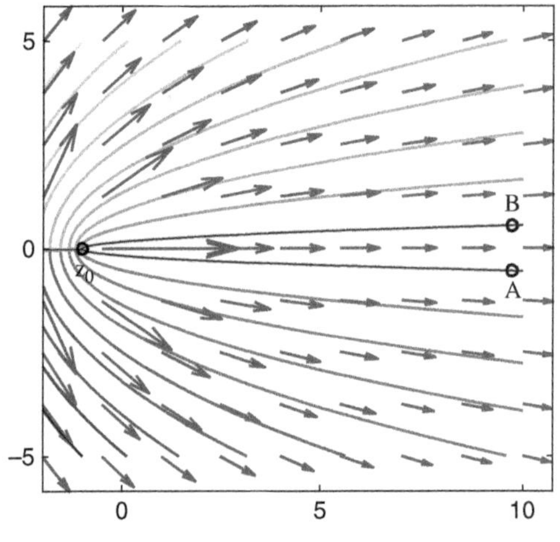

Figure 3.22 Parabolic velocity field.

clear that the sharpest parabola, described by points Az_0B, is the closest to a flat plate. Even sharper streamlines can be calculated, so that points A and B lie closer to each other. Hence, Eq. (3.97) can be taken to represent the flow around a flat plate at zero angle of attack. The velocity is still infinite at the leading edge, but we can use the parabolic distribution in order to determine the forces acting on the flat plate.

The complex aerodynamic force acting on the parabola Az_0B is obtained from the integral of the pressure along its arclength (see Eq. (3.17))

$$\mathcal{F} = -\oint p(z)\mathrm{d}z$$

where $\mathcal{F} = f_{n_s} + \mathrm{i}f_{c_s}$, f_{n_s} and f_{c_s} being the normal and chordwise forces due to leading edge suction. From the steady Bernoulli equation (3.3), the complex pressure acting on the parabola is given by $p(z) = -\rho V(z)^2/2$, having ignored the constant since the integral of the pressure is calculated over a contour. The aerodynamic load becomes

$$\mathcal{F} = \frac{\rho}{2}\oint V(z)^2\mathrm{d}z \tag{3.98}$$

If points A and B lie sufficiently far from z_0, the contour integral can be replaced by an integral between limits A and B, passing from z_0. Substituting for $V(z)$ from expression (3.97), Eq. (3.98) is modified to

$$\mathcal{F} = \frac{\rho}{2}\int_A^B \frac{S^2}{z - z_0}\mathrm{d}z = \frac{\rho S^2}{2}\left(\ln(z_B - z_0) - \ln(z_A - z_0)\right)$$

Using amplitude and phase notation, the points z_A, z_B and z_0 can be written as $z_A = r_A \mathrm{e}^{\mathrm{i}\theta_A}$, $z_B = r_B \mathrm{e}^{\mathrm{i}\theta_B}$ while $z_0 = -b$ is the position of the leading edge. The load becomes

$$\mathcal{F} = \frac{\rho S^2}{2}\left(\ln(r_B \mathrm{e}^{\mathrm{i}\theta_B} + b) - \ln(r_A \mathrm{e}^{\mathrm{i}\theta_A} + b)\right)$$

Writing $\mathrm{e}^{\mathrm{i}\theta_A} = \cos \mathrm{i}\theta_A + \mathrm{i}\sin \mathrm{i}\theta_A$, $\mathrm{e}^{\mathrm{i}\theta_B} = \cos \mathrm{i}\theta_B + \mathrm{i}\sin \mathrm{i}\theta_B$ and making use of Eq. (3.49) results in

$$\begin{aligned}\mathcal{F} = \frac{\rho S^2}{2}\Bigg(&\ln\sqrt{\left(r_B\cos\theta_B + b\right)^2 + r_B^2\sin^2\theta_B} + \mathrm{i}\tan^{-1}\frac{r_B\sin\theta_B}{r_B\cos\theta_B + b}\\ &- \ln\sqrt{\left(r_A\cos\theta_A + b\right)^2 + r_A^2\sin^2\theta_A} - \mathrm{i}\tan^{-1}\frac{r_A\sin\theta_A}{r_A\cos\theta_A + b}\Bigg)\end{aligned}$$

Thanks to the symmetric placement of points A and B, $r_A = r_B$. Furthermore, since A and B lie sufficiently far from z_0, $r_A \gg b$, $r_B \gg b$. The load equation simplifies to

$$\mathcal{F} = \frac{\rho S^2}{2}\left((\ln r_B + \mathrm{i}\theta_B) - (\ln r_A + \mathrm{i}\theta_A)\right) = \frac{\rho S^2}{2}\left(\mathrm{i}\theta_B - \mathrm{i}\theta_A\right)$$

Finally, if the plate has zero thickness, then $\theta_A = 2\pi$ and $\theta_B = 0$. Substituting into the equation for the load, we get

$$\mathcal{F} = -\mathrm{i}\rho\pi S^2$$

so that the chordwise force is a thrust

$$f_{c_s} = -\rho\pi S^2 \tag{3.99}$$

while the normal force is $f_{n_s} = 0$. Note that in the case of a plate with zero thickness whose leading edge lies on $x = -b$, the complex velocity of Eq. (3.97) becomes real and only a horizontal component exists

$$u(x) = \frac{S}{\sqrt{x+b}} \tag{3.100}$$

The remaining question is the value of S; without it we cannot determine the value of the leading edge suction. We can evaluate S by considering the complete flow over a pitching and plunging flat plate, whose horizontal velocity distribution is given by Eq. (3.75). The value of S can be obtained by equating the horizontal airspeed values given by Eqs. (3.75) and (3.100), so that

$$\frac{\partial \phi_{\text{tot}}}{\partial x} = \frac{1}{\sqrt{b^2-x^2}} \left(-\left(U_\infty \alpha(t) + \dot{h}(t) - x_f \dot{\alpha}(t)\right) x - \frac{\dot{\alpha}(t)}{2}\left(2x^2+2bx-b^2\right) - \frac{1}{2\pi}\int_b^\infty \frac{\sqrt{X_0^2-b^2}}{x-X_0} \bar{\Gamma}(X_0,t)\mathrm{d}X_0 \right) = \frac{S}{\sqrt{x+b}}$$

which simplifies to

$$S = \frac{1}{\sqrt{b-x}} \left(-\left(U_\infty \alpha(t) + \dot{h}(t) - x_f \dot{\alpha}(t)\right) x - \frac{\dot{\alpha}(t)}{2}\left(2x^2+2bx-b^2\right) - \frac{1}{2\pi}\int_b^\infty \frac{\sqrt{X_0^2-b^2}}{x-X_0} \bar{\Gamma}(X_0,t)\mathrm{d}X_0 \right)$$

This equation cannot be satisfied everywhere; in order to write it we equated a parabolic velocity profile for the flow around the leading edge of a flat plate at zero angle of attack with a much more complex velocity profile for the flow around a complete pitching and plunging plate. In any case, the left-hand side is a constant while the right-hand side is a function of x. Nevertheless, we can follow Garrick's approach (Garrick 1937) and enforce the equation only at the leading edge, $x = -b$, so that

$$S = \frac{1}{\sqrt{2b}} \left(\left(U_\infty \alpha(t) + \dot{h}(t) - x_f \dot{\alpha}(t)\right) b + b^2 \frac{\dot{\alpha}(t)}{2} + \frac{1}{2\pi}\int_b^\infty \frac{\sqrt{X_0-b}}{\sqrt{X_0+b}} \bar{\Gamma}(X_0,t)\mathrm{d}X_0 \right) \tag{3.101}$$

Example 3.10 *Determine the leading edge suction force for a steady flow around a flat plate at a constant angle of attack α to a free stream U_∞.*

Steady flow occurs a long time after an impulsive start, so that this example is an extension of Example 3.7. The integral in Eq. (3.101) can be replaced by the summation

$$\int_b^\infty \frac{\sqrt{X_0-b}}{\sqrt{X_0+b}} \bar{\Gamma}(X_0,t)\mathrm{d}X_0 \approx \sum_{j=1}^{i} \frac{\sqrt{X_{0_j}-b}}{\sqrt{X_{0_j}+b}} \Gamma_j$$

If $i \to \infty$, then all the powerful vortices will lie at $X_{0_j} \gg b$, while the vortices near the wing will be negligible. Consequently,

$$\frac{\sqrt{X_{0_j} - b}}{\sqrt{X_{0_j} + b}} \approx 1$$

and we already showed in Example 3.7 that

$$\sum_{j=1}^{\infty} \Gamma_j = 2\pi b U_\infty \alpha$$

Equation (3.101) simplifies to

$$S = \sqrt{2b}U_\infty \alpha$$

Then, from Eq. (3.99), the leading edge suction force is equal to

$$f_{c_s} = -2\rho U_\infty^2 b\pi\alpha^2 \tag{3.102}$$

Recalling from Eq. (3.86) that the steady-state lift is given by $l_\infty = -2\rho U_\infty^2 b\pi\alpha$, *it is clear that*

$$f_{c_s} = \alpha l_\infty \tag{3.103}$$

Note that, even though this force appears to be positive, it is actually a thrust since l_∞ *is defined positive downwards.*

We now recall the first of Eq. (3.19), which states that the drag is given in terms of the chordwise and normal aerodynamic forces as

$$d = f_c + f_n\alpha$$

The normal force is obtained from the second of Eq. (3.19), that is $f_n \approx l$. *As the lift is defined positive downwards here,* $f_n \approx -l$. *We substitute* $f_c = f_{c_s}$ *and* $f_n = -l$ *in the expression for the drag to obtain*

$$d = f_{c_s} - \alpha l \tag{3.104}$$

For steady conditions and substituting for f_{c_s} *from Eq. (3.103), the total drag is given by*

$$d = \alpha l_\infty - \alpha l_\infty = 0$$

as expected from d'Alembert's paradox. Further (and more detailed) discussion of this issue can be found in von Kármán and Burgers (1935).

In the unsteady case, the total drag is not equal to zero; from Eqs. (3.99) and (3.104), the total drag is given by

$$d = -\rho\pi S^2 - \alpha l \tag{3.105}$$

and S is obtained from Eq. (3.101). Several simplifications were carried out in order to derive Eq. (3.105). First, the flow is assumed to be parabolic around the leading edge, and compatibility with the rest of the flowfield is enforced only at the leading edge. Second, the suction force is taken to act parallel to the free stream irrespective of the actual angle of attack of the wing. Third, the load equation (3.98) is only applicable to steady flows but, in writing Eq. (3.101), we have introduced unsteadiness. Finally, it should be mentioned that the 2D drag coefficient is defined in the same way as the lift coefficient of Eq. (3.23)

$$c_d = \frac{d}{\frac{1}{2}\rho U_\infty^2 c} \tag{3.106}$$

Equation (3.101) is not very practical because it contains a new integral that we have not yet calculated. Garrick (1937) shows that it can be re-written as

$$S(t) = \sqrt{2b}\left(\left(U_\infty \alpha(t) + \dot{h}(t) + \left(\frac{3b}{2} - x_f\right)\dot{\alpha}(t)\right)\frac{\int_b^\infty \frac{X_0}{\sqrt{X_0^2-b^2}}\bar{\Gamma}(X_0,t)\mathrm{d}X_0}{\int_b^\infty \frac{\sqrt{X_0+b}}{\sqrt{X_0-b}}\bar{\Gamma}(X_0,t)\mathrm{d}X_0} - \frac{b}{2}\dot{\alpha}(t)\right) \tag{3.107}$$

which, from the definition of Wagner's function (Eq. (3.90)), becomes

$$S(t) = \frac{\sqrt{2b}}{2}\left(2\left(U_\infty \alpha(t) + \dot{h}(t) + \left(\frac{3b}{2} - x_f\right)\dot{\alpha}(t)\right)\Phi(t) - b\dot{\alpha}(t)\right) \tag{3.108}$$

For the impulsively started flow around a flat plate, it is now easy to see that $S(t)$ jumps to the value

$$S(0) = \frac{\sqrt{2b}}{2}U_\infty \alpha$$

at time $t = 0$, so that the total drag jumps to

$$d(0) = -\rho\pi\frac{b}{2}U_\infty^2\alpha^2 + \rho U_\infty^2 b\pi\alpha^2 = \frac{1}{2}\rho U_\infty^2 b\pi\alpha^2$$

This means that an impulsive start causes an initial drag which, according to Example 3.10 should eventually drop to zero.

Example 3.11 *Calculate the unsteady drag acting on an impulsively started flat plate at constant angle of attack α*

We repeat the calculations of Example 3.7, but this time we also evaluate S from Eq. (3.101). As argued already in Example 3.10, the integral in this equation can be written as a sum so that

$$S(t_i) = \frac{1}{\sqrt{2b}}\left(U_\infty \alpha b + \frac{1}{2\pi}\sum_{j=1}^{i}\frac{\sqrt{X_{0_j} - b}}{\sqrt{X_{0_j} + b}}\Gamma_j\right) \tag{3.109}$$

where t_i is the ith time instance. We have already seen how to calculate the values of Γ_i from Eq. (3.80) in Example 3.7; these can be inserted directly into Eq. (3.109) to calculate $S(t_i)$. Then, the total drag is given by

$$d(t_i) = -\rho\pi S(t_i)^2 - \alpha l_\Gamma(t_i)$$

and $l_\Gamma(t_i)$ is obtained from Eq. (3.84). We also calculate Wagner's function, $\Phi(t_i)$, as shown in Example 3.9 and, hence, $S_G(t_i)$ from Eq. (3.108). This evaluation gives rise to a second estimate for the drag,

$$d_G(t_i) = -\rho\pi S_G(t_i)^2 - \alpha l_\Gamma(t_i)$$

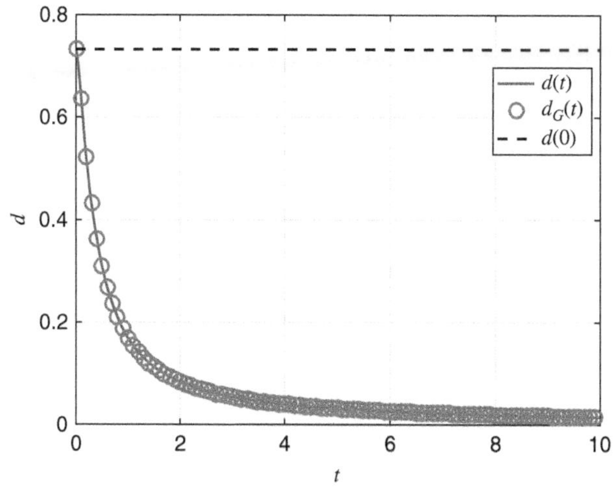

Figure 3.23 Drag variation with time acting on an impulsively started flat plate.

Figure 3.23 plots the two estimates for the drag against time. As argued earlier, the drag jumps to the value $d(0) = 1/2\rho U_\infty^2 b\pi\alpha^2$ at time $t = 0$. It then decays monotonously with time, asymptoting towards $d(\infty) = 0$. This example is solved by Matlab code `impulsive_Wagner_drag.m`.

3.4.3 General Motion

Wagner's solution concerns one particular type of unsteady flow, the impulsive start of a flat plate airfoil at angle of attack α. In this section, we will show that Wagner's theory can be used to represent any pitching and/or plunging motion of a 2D airfoil. First, we return to the Kutta condition expression for a pitching and plunging motion, Eq. (3.76)

$$\frac{1}{2\pi b}\int_b^\infty \frac{\sqrt{X_0+b}}{\sqrt{X_0-b}}\bar{\Gamma}(X_0,t)\mathrm{d}X_0 = U_\infty\alpha(t) + \dot{h}(t) + \left(\frac{3b}{2} - x_f\right)\dot{\alpha}(t) \tag{3.110}$$

If we divide the lift equation (3.83) by the Kutta condition, we obtain

$$\begin{aligned} l_\Gamma(t) &= -2\rho U_\infty b\pi\left(U_\infty\alpha(t) + \dot{h}(t) + \left(\frac{3b}{2} - x_f\right)\dot{\alpha}(t)\right)\frac{\int_b^\infty \frac{X_0}{\sqrt{X_0^2-b^2}}\bar{\Gamma}(X_0,t)\mathrm{d}X_0}{\int_b^\infty \frac{\sqrt{X_0+b}}{\sqrt{X_0-b}}\bar{\Gamma}(X_0,t)\mathrm{d}X_0} \\ &= -\,2\rho U_\infty b\pi w_{3/4}\Phi(t) \end{aligned} \tag{3.111}$$

where $w_{3/4}$ is the upwash evaluated at the 3/4 chord location

$$w_{3/4}(t) = U_\infty\alpha(t) + \dot{h}(t) + \left(\frac{3c}{4} - x_f\right)\dot{\alpha}(t) \tag{3.112}$$

Equation (3.111) is only valid for a flow whereby the upwash is zero for $t < 0$ and jumps to a constant value $w_{3/4}$ for $t \geq 0$; this is still an impulsive motion, not a general motion. However, we can model the latter as a superposition of many small impulsive jumps taking place at different time instances. Figure 3.24 demonstrates the principle by plotting

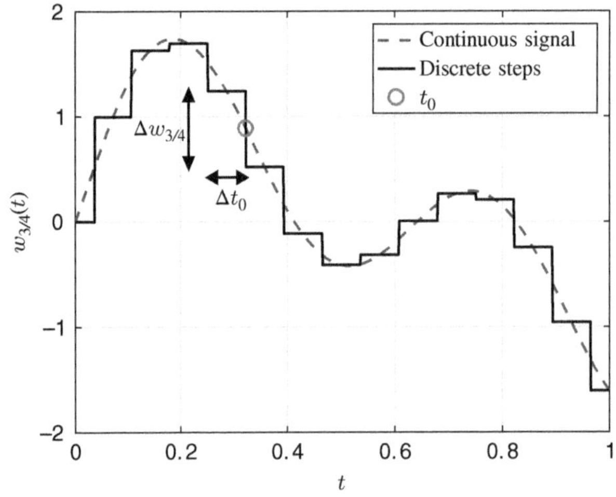

Figure 3.24 Modelling a continuous time signal by discrete steps.

a smooth continuous time signal and its representation as 14 discrete steps occurring at regular intervals Δt_0. The increments in upwash are denoted by $\Delta w_{3/4}$, and they have different magnitudes while the time increments are always constant. Clearly, as Δt_0 decreases, the two signals become increasingly similar, while if $\Delta t_0 \to 0$, they become identical.

Consider a small change in upwash, $\Delta w_{3/4}(t_0)$, occurring at time t_0. The triangle defined by $\Delta w_{3/4}$ and Δt_0 at point t_0 in Figure 3.24 has been deliberately constructed such that

$$\frac{\Delta w_{3/4}}{\Delta t_0} = \left.\frac{\mathrm{d}w_{3/4}}{\mathrm{d}t}\right|_{t_0}$$

From Eq. (3.111), the increase in lift Δl around the airfoil due to $\Delta w_{3/4}$ at any time $t \geq t_0$ is

$$\Delta l = -2\rho U_\infty b\pi\Phi(t-t_0)\Delta w_{3/4}(t_0) = 2\rho U_\infty b\pi\Phi(t-t_0)\dot{w}_{3/4}(t_0)\Delta t_0$$

where $\dot{w}_{3/4}(t_0) = \mathrm{d}w_{3/4}/\mathrm{d}t|_{t_0}$. Then, as $\Delta t_0 \to 0$ and the number of discrete steps tends to infinity,

$$l_\Gamma(t) = -2\rho U_\infty b\pi \int_{-\infty}^{t} \Phi(t-t_0)\dot{w}_{3/4}(t_0)\mathrm{d}t_0 \tag{3.113}$$

which states that the lift at time t is due to all the infinitesimal jumps in $w_{3/4}$ that happened since the start of the motion. Assuming that $w_{3/4}(t) = 0$ at all times $t < 0$ and that the upwash jumps to the value $w_{3/4}(0)$ at $t = 0$, Eq. (3.113) becomes (Fung 1993)

$$l_\Gamma(t) = -2\rho U_\infty b\pi \left(w_{3/4}(0)\Phi(t) + \int_0^t \Phi(t-t_0)\dot{w}_{3/4}(t_0)\mathrm{d}t_0 \right) \tag{3.114}$$

Applying integration by parts, we obtain

$$l_\Gamma(t) = -2\rho U_\infty b\pi \left(w_{3/4}(0)\Phi(t) + w_{3/4}(t_0)\Phi(t-t_0)\Big|_0^t - \int_0^t \dot{\Phi}(t-t_0)w_{3/4}(t_0)\mathrm{d}t_0 \right)$$
$$= -2\rho U_\infty b\pi \left(w_{3/4}(t)\Phi(0) - \int_0^t \dot{\Phi}(t-t_0)w_{3/4}(t_0)\mathrm{d}t_0 \right)$$

Finally, substituting from Eq. (3.112) for the upwash, the circulatory lift predicted by Wagner theory due to a general pitching and plunging motion of a 2D flat plate becomes

$$l_\Gamma(t) = -2\rho U_\infty b\pi \left(\left(U_\infty \alpha(t) + \dot{h}(t) + \left(\frac{3c}{4} - x_f\right)\dot{\alpha}(t)\right)\Phi(0) - \int_0^t \dot{\Phi}(t-t_0)\left(U_\infty \alpha(t_0) + \dot{h}(t_0) + \left(\frac{3c}{4} - x_f\right)\dot{\alpha}(t_0)\right) \mathrm{d}t_0 \right) \quad (3.115)$$

The circulatory moment around the pitching axis is given by Eq. (3.85)

$$m_{xf_\Gamma}(t) = -\rho U_\infty b \int_b^\infty \left(\frac{1}{2}\frac{\sqrt{X_0+b}}{\sqrt{X_0-b}} - \left(a+\frac{1}{2}\right)\frac{X_0}{\sqrt{X_0^2-b^2}} \right) \bar{\Gamma}(X_0,t)\mathrm{d}X_0$$

Dividing by the Kutta condition of expression (3.76), we obtain

$$m_{xf_\Gamma}(t) = -2\rho U_\infty b^2 \pi w_{3/4}\left(\frac{1}{2} - \left(a+\frac{1}{2}\right)\Phi(t)\right)$$

Substituting from Eq. (3.111), the pitching moment becomes

$$m_{xf_\Gamma}(t) = -\rho U_\infty b^2 \pi \left(U_\infty \alpha(t) + \dot{h}(t) + \left(\frac{3c}{4} - x_f\right)\dot{\alpha}(t)\right) - b\left(a+\frac{1}{2}\right) l_\Gamma(t) \quad (3.116)$$

where $l_\Gamma(t)$ is evaluated from Eq. (3.111) for an impulsive start or from Eq. (3.115) for a general pitching and plunging motion. Note that $b(a+1/2)$ is the distance between the pitching axis and the quarter-chord, the latter being the aerodynamic centre according to thin airfoil theory. In other words, the steady aerodynamic centre still has a role to play in unsteady aerodynamics.

3.4.4 Total Loads

The total aerodynamic lift and pitching moment acting on the pitch-plunge wing can be obtained by summing the contributions from the sources and the vortices. The total lift $l(t)$ is obtained from Eqs. (3.63) and (3.115) as

$$\begin{aligned} l(t) = l_\sigma(t) + l_\Gamma(t) &= -\rho\pi b^2 \left(\ddot{h} - ba\ddot{\alpha} + U_\infty\dot{\alpha}\right) \\ &- 2\rho U_\infty b\pi \left(\left(U_\infty \alpha(t) + \dot{h}(t) + \left(\frac{3c}{4} - x_f\right)\dot{\alpha}(t)\right)\Phi(0) \right. \\ &\left. - \int_0^t \dot{\Phi}(t-t_0)\left(U_\infty \alpha(t_0) + \dot{h}(t_0) + \left(\frac{3c}{4} - x_f\right)\dot{\alpha}(t_0)\right)\mathrm{d}t_0 \right) \end{aligned} \quad (3.117)$$

Similarly, the total moment is obtained from Eqs. (3.64) and (3.116) as

$$\begin{aligned} m_{xf}(t) = m_{xf_\sigma}(t) + m_{xf_\Gamma}(t) &= -\rho\pi b^2 \left(-ba\ddot{h} + b^2\left(\frac{1}{8}+a^2\right)\ddot{\alpha} - U_\infty\dot{h} - U_\infty^2\alpha\right) \\ &- \rho U_\infty b^2\pi \left(U_\infty\alpha(t) + \dot{h}(t) + \left(\frac{3c}{4} - x_f\right)\dot{\alpha}(t)\right) - b\left(a+\frac{1}{2}\right) l_\Gamma(t) \end{aligned} \quad (3.118)$$

Note that the non-circulatory terms proportional to α and $\dot{h}$ drop out so that

$$\begin{aligned} m_{xf}(t) = &-\rho\pi b^2 \left(-ba\ddot{h} + b^2\left(\frac{1}{8}+a^2\right)\ddot{\alpha}\right) - \rho U_\infty b^2\pi\left(\frac{3c}{4} - x_f\right)\dot{\alpha}(t) \\ &+ 2\rho U_\infty \pi b^2\left(a+\frac{1}{2}\right)\left(\left(U_\infty\alpha(t) + \dot{h}(t) + \left(\frac{3c}{4} - x_f\right)\dot{\alpha}(t)\right)\Phi(0) \right. \\ &\left. - \int_0^t \dot{\Phi}(t-t_0)\left(U_\infty\alpha(t_0) + \dot{h}(t_0) + \left(\frac{3c}{4} - x_f\right)\dot{\alpha}(t_0)\right)\mathrm{d}t_0 \right) \end{aligned} \quad (3.119)$$

From Eq. (3.62), it can be seen that

$$\frac{3c}{4} - x_f = b\left(\frac{1}{2} - a\right)$$

so that Eqs. (3.117) and (3.119) can also be written as

$$l(t) = -\rho\pi b^2\left(\ddot{h} - ba\ddot{\alpha} + U_\infty\dot{\alpha}\right) - 2\rho U_\infty b\pi\left(\left(U_\infty\alpha(t) + \dot{h}(t) + b\left(\frac{1}{2} - a\right)\dot{\alpha}(t)\right)\Phi(0)\right.$$
$$\left. - \int_0^t \dot{\Phi}(t - t_0)\left(U_\infty\alpha(t_0) + \dot{h}(t_0) + b\left(\frac{1}{2} - a\right)\dot{\alpha}(t_0)\right)\mathrm{d}t_0\right) \tag{3.120}$$

and

$$m_{xf}(t) = -\rho\pi b^2\left(-ba\ddot{h} + b^2\left(\frac{1}{8} + a^2\right)\ddot{\alpha}\right) - \rho U_\infty b^3\pi\left(\frac{1}{2} - a\right)\dot{\alpha}(t)$$
$$+ 2\rho U_\infty\pi b^2\left(a + \frac{1}{2}\right)\left(\left(U_\infty\alpha(t) + \dot{h}(t) + b\left(\frac{1}{2} - a\right)\dot{\alpha}(t)\right)\Phi(0)\right.$$
$$\left. - \int_0^t \dot{\Phi}(t - t_0)\left(U_\infty\alpha(t_0) + \dot{h}(t_0) + b\left(\frac{1}{2} - a\right)\dot{\alpha}(t_0)\right)\mathrm{d}t_0\right) \tag{3.121}$$

which is a form compatible with Theodorsen's lift and moment expressions (Theodorsen 1935).

The drag for an impulsively started motion can be obtained by combining Eqs. (3.105), (3.108) and (3.112)

$$d(t) = -\rho\pi\frac{b}{2}\left(2w_{3/4}(t)\Phi(t) - b\dot{\alpha}(t)\right)^2 - \alpha(t)l(t)$$

Using the arguments applied to the circulatory lift, the drag for a general motion becomes

$$d(t) = -\rho\pi\frac{b}{2}\left(2\left[\left(U_\infty\alpha(t) + \dot{h}(t) + b\left(\frac{1}{2} - a\right)\dot{\alpha}(t)\right)\Phi(0)\right.\right.$$
$$\left.\left. - \int_0^t \dot{\Phi}(t - t_0)\left(U_\infty\alpha(t_0) + \dot{h}(t_0) + b\left(\frac{1}{2} - a\right)\dot{\alpha}(t_0)\right)\mathrm{d}t_0\right] - b\dot{\alpha}(t)\right)^2$$
$$- \alpha(t)l(t) \tag{3.122}$$

where $l(t)$ is evaluated from Eq. (3.120). Note that the drag contains non-linear terms in α, $\dot{\alpha}$ and $\dot{h}$, while the lift and moment expressions are all linear. The expression for the drag can be written in much more compact form by substituting from Eq. (3.115)

$$d(t) = -\rho\pi\frac{b}{2}\left(\frac{l_\Gamma(t)}{\rho U_\infty b\pi} + b\dot{\alpha}(t)\right)^2 - \alpha(t)l(t) \tag{3.123}$$

Equations (3.120)–(3.123) give the total lift, pitching moment and drag acting on the 2D pitch-plunge wing at any moment in time for any prescribed motion $h(t)$ and $\alpha(t)$. However, they are integro-differential expressions and therefore, it is difficult to use them to evaluate directly $l(t)$, $m_{xf}(t)$ and $d(t)$. Nevertheless, a simple change of variable proposed by Lee et al. (1999) can be used to write the expressions in purely differential form, taking advantage of the exponential approximation for Wagner's function,

$$\Phi(t) = 1 - \Psi_1 \mathrm{e}^{-\varepsilon_1 U_\infty t/b} - \Psi_2 \mathrm{e}^{-\varepsilon U_\infty t/b}$$

where for R. T. Jones' expression, $\Psi_1 = 0.165$, $\Psi_2 = 0.335$, $\varepsilon_1 = 0.0455$ and $\varepsilon_2 = 0.3$ (see Example 3.9). Following the approach by Lee et al. (1999), we define four new variables

$$\begin{aligned} w_1(t) &= \int_0^t e^{-\varepsilon_1 U_\infty (t-t_0)/b} h(t_0) \mathrm{d}t_0 \\ w_2(t) &= \int_0^t e^{-\varepsilon_2 U_\infty (t-t_0)/b} h(t_0) \mathrm{d}t_0 \\ w_3(t) &= \int_0^t e^{-\varepsilon_1 U_\infty (t-t_0)/b} \alpha(t_0) \mathrm{d}t_0 \\ w_4(t) &= \int_0^t e^{-\varepsilon_2 U_\infty (t-t_0)/b} \alpha(t_0) \mathrm{d}t_0 \end{aligned} \tag{3.124}$$

known as the aerodynamic states. Now, we look at the integral

$$-\int_0^t \dot{\Phi}(t-t_0)\left(U_\infty \alpha(t_0) + \dot{h}(t_0) + b\left(\frac{1}{2}-a\right)\dot{\alpha}(t_0)\right)\mathrm{d}t_0 = -\int_0^t \dot{\Phi}(t-t_0) w_{3/4}(t) \mathrm{d}t_0$$

that is present in Eqs. (3.120) and (3.121). Substituting expressions (3.124) and applying integration by parts, we obtain

$$\begin{aligned} -\int_0^t \dot{\Phi}(t-t_0) w_{3/4}(t)\mathrm{d}t_0 = &-\left(h(0) + b\left(\frac{1}{2}-a\right)\alpha(0)\right)\dot{\Phi}(t) \\ &+ \left(h(t) + b\left(\frac{1}{2}-a\right)\alpha(t)\right)\dot{\Phi}(0) \\ &- \Psi_1\left(\frac{\varepsilon_1 U_\infty}{b}\right)^2 w_1(t) - \Psi_2\left(\frac{\varepsilon_2 U_\infty}{b}\right)^2 w_2(t) \\ &+ \Psi_1 \frac{\varepsilon_1 U_\infty^2}{b}\left(1 - \varepsilon_1\left(\frac{1}{2}-a\right)\right) w_3(t) \\ &+ \Psi_2 \frac{\varepsilon_2 U_\infty^2}{b}\left(1 - \varepsilon_2\left(\frac{1}{2}-a\right)\right) w_4(t) \end{aligned} \tag{3.125}$$

Now, we can replace the integral in Eqs. (3.120) and (3.124) by expression (3.125) so that

$$\begin{pmatrix} l(t) \\ m_{xf}(t) \end{pmatrix} = \rho \boldsymbol{B}\ddot{\mathbf{y}} + \rho U_\infty \boldsymbol{D}\dot{\mathbf{y}} + \rho U_\infty^2 \boldsymbol{F}\mathbf{y} + \rho U_\infty^3 \boldsymbol{W}\mathbf{w} + \rho U_\infty \mathbf{g}\dot{\Phi}(t) \tag{3.126}$$

where $\mathbf{y} = (h, \alpha)^T$, $\mathbf{w} = (w_1, w_2, w_3, w_4)^T$, $\boldsymbol{B}$ is the aerodynamic mass matrix

$$\boldsymbol{B} = b^2 \begin{pmatrix} -\pi & \pi ab \\ \pi ab & -\pi b^2(1/8 + a^2) \end{pmatrix}$$

$\boldsymbol{D}$ is the aerodynamic damping matrix $\boldsymbol{D} = \boldsymbol{D}_1 + \boldsymbol{D}_2$ where

$$\boldsymbol{D}_1 = b^2 \begin{pmatrix} 0 & -\pi \\ 0 & -\pi(1/2 - a)b \end{pmatrix}, \quad \boldsymbol{D}_2 = \begin{pmatrix} -\pi b & -\pi b^2(1/2-a) \\ \pi b^2(a+1/2) & \pi b^3(a+1/2)(1/2-a) \end{pmatrix}$$

$\boldsymbol{F}$ is the aerodynamic stiffness matrix $\boldsymbol{F} = \boldsymbol{F}_1 + 2\left(\Psi_1\varepsilon_1/b + \Psi_2\varepsilon_2/b\right)\boldsymbol{D}_2$, where

$$\boldsymbol{F}_1 = \begin{pmatrix} 0 & -\pi b \\ 0 & \pi b^2(a+1/2) \end{pmatrix}$$

$\boldsymbol{W}$ is the aerodynamic state influence matrix, where

$$\boldsymbol{W} = \begin{pmatrix} -2\pi b \boldsymbol{W}_0 \\ 2\pi b^2(a+1/2)\boldsymbol{W}_0 \end{pmatrix}, \quad \boldsymbol{W}_0 = \begin{pmatrix} -\Psi_1(\varepsilon_1/b)^2 \\ -\Psi_2(\varepsilon_2/b)^2 \\ \Psi_1\varepsilon_1(1-\varepsilon_1(1/2-a))/b \\ \Psi_2\varepsilon_2(1-\varepsilon_2(1/2-a))/b \end{pmatrix}^T$$

and $\mathbf{g}$ is the initial condition excitation vector

$$\mathbf{g} = b\left(h(0) + b\left(\frac{1}{2} - a\right)\alpha(0)\right)\begin{pmatrix} 2\pi \\ -2\pi b(a+1/2) \end{pmatrix} \tag{3.127}$$

Equation (3.126) gives the total lift and moment acting on the 2D flat plate wing in terms of the motion $h(t)$, $\alpha(t)$ and the aerodynamic states $w_1(t), \ldots, w_4(t)$. The latter are still unknown quantities, but they can be calculated by applying Leibniz's rule for integrals with variable limits, according to which (see for example (Dimitriadis 2017; Stephenson 1973))

$$\frac{\mathrm{d}}{\mathrm{d}t}\int_{a(t)}^{b(t)} f(t_0, t)\mathrm{d}t_0 = \frac{\mathrm{d}b(t)}{\mathrm{d}t}f(b(t), t) - \frac{\mathrm{d}a(t)}{\mathrm{d}t}f(a(t), t) + \int_{a(t)}^{b(t)} \frac{\partial f(t_0, t)}{\partial t}\mathrm{d}t_0 \tag{3.128}$$

where $a(t)$ and $b(t)$ are continuous and differentiable functions of t and $f(t_0, t)$ is a continuous and differentiable function of both t_0 and t. Applying this rule to the equation for $w_1(t)$, we set $a(t) = 0$, $b(t) = t$ and $f(t_0, t) = e^{-\varepsilon_1 U_\infty (t-t_0)/b} h(t_0)$, so that

$$\dot{w}_1(t) = h(t) - \frac{\varepsilon_1 U_\infty}{b}\int_0^t e^{-\varepsilon_1 U_\infty (t-t_0)/b} h(t_0)\mathrm{d}t_0 = h(t) - \frac{\varepsilon_1 U_\infty}{b} w_1(t)$$

Using the same procedure, four equations can be written, one for each aerodynamic state,

$$\begin{aligned} \dot{w}_1(t) &= h(t) - \frac{\varepsilon_1 U_\infty}{b} w_1(t) \\ \dot{w}_2(t) &= h(t) - \frac{\varepsilon_2 U_\infty}{b} w_2(t) \\ \dot{w}_3(t) &= \alpha(t) - \frac{\varepsilon_1 U_\infty}{b} w_3(t) \\ \dot{w}_4(t) &= \alpha(t) - \frac{\varepsilon_2 U_\infty}{b} w_4(t) \end{aligned} \tag{3.129}$$

or, in matrix form,

$$\dot{\mathbf{w}} - \boldsymbol{W}_1\mathbf{y} - U_\infty \boldsymbol{W}_2\mathbf{w} = \mathbf{0} \tag{3.130}$$

where

$$\boldsymbol{W}_1 = \begin{pmatrix} 1 & 0 \\ 1 & 0 \\ 0 & 1 \\ 0 & 1 \end{pmatrix}, \quad \boldsymbol{W}_2 = \begin{pmatrix} -\varepsilon_1/b & 0 & 0 & 0 \\ 0 & -\varepsilon_2/b & 0 & 0 \\ 0 & 0 & -\varepsilon_1/b & 0 \\ 0 & 0 & 0 & -\varepsilon_2/b \end{pmatrix} \tag{3.131}$$

Assuming that the motion starts at $t = 0$, the initial conditions for Eq. (3.130) are $\mathbf{w}(0) = \mathbf{0}$. Eqs. (3.126) and (3.130) constitute a set of six linear algebraic equations with six unknowns, $l(t)$, $m_{xf}(t)$ and $w_1(t)$ to $w_4(t)$.

Example 3.12 *Calculate the lift and moment acting on a 2D flat plate wing undergoing sinusoidal pitching and plunging motion.*

We choose a wing with $c = 0.25$ *m, so that* $b = 0.125$ *m, and we set* $x_f = 0.1$ *m. The air density is* $\rho = 1.225$ kg/m^3 *and the values of the Wagner function approximation parameters are* $\Psi_1 = 0.165$, $\Psi_2 = 0.335$, $\varepsilon_1 = 0.0455$, $\varepsilon_2 = 0.3$. *Now, we select sinusoidal pitching and plunging motion such that*

$$h(t) = h_0 + h_1 \sin\left(\omega_0 t + \phi_h\right), \quad \alpha(t) = \alpha_0 + \alpha_1 \sin\left(\omega_0 t + \phi_\alpha\right) \tag{3.132}$$

where h_0, α_0 *are the mean values in plunge and pitch,* h_1, α_1 *are the oscillation amplitudes in plunge and pitch,* ω_0 *is the frequency and* ϕ_h, ϕ_α *are phase angles. We set* $h_0 = \alpha_0 = 0$ *so that both the plunge and pitch motions are symmetric around zero. The first step is to solve Eq. (3.129) in order to obtain the values of* $\mathbf{w}(t)$ *for all times. The equation is a linear non-homogeneous Ordinary Differential Equation (ODE) and has an analytical solution. It can be written in the form*

$$\dot{\mathbf{w}} - U_\infty \boldsymbol{W}_2 \mathbf{w} = \boldsymbol{W}_1 \mathbf{y}(t)$$

and pre-multiplied by $\mathrm{e}^{-U_\infty \boldsymbol{W}_2 t}$, *such that*

$$\frac{\mathrm{d}\left(\mathrm{e}^{-U_\infty \boldsymbol{W}_2 t}\mathbf{w}\right)}{\mathrm{d}t} = \mathrm{e}^{-U_\infty \boldsymbol{W}_2 t}\boldsymbol{W}_1 \mathbf{y}(t)$$

Recalling that the motion is sinusoidal, we let $\mathbf{y}(t) = (h(t), \alpha(t))^T$. *We can now integrate both sides of the equation from time 0 to time t to obtain*

$$\mathbf{w}(t) = \mathrm{e}^{U_\infty \boldsymbol{W}_2 t}\mathbf{w}(0) + \int_0^t \mathrm{e}^{U_\infty \boldsymbol{W}_2 (t-t_0)}\boldsymbol{W}_1 \mathbf{y}(t_0)\mathrm{d}t_0$$

Choosing initial conditions $\mathbf{w}(0) = \mathbf{0}$ *and using the properties of the matrix exponential (see (Dimitriadis 2017)), we can write*

$$\mathbf{w}(t) = \int_0^t \boldsymbol{V}\mathrm{e}^{U_\infty \boldsymbol{L}(t-t_0)}\boldsymbol{V}^{-1}\boldsymbol{W}_1 \mathbf{y}(t_0)\mathrm{d}t_0 \tag{3.133}$$

where $\boldsymbol{V}$ *and* $\boldsymbol{L}$ *are the eigenvector and eigenvalue matrices of* $\boldsymbol{W}_2$, *respectively, such that* $\boldsymbol{W}_2 = \boldsymbol{V}\boldsymbol{L}\boldsymbol{V}^{-1}$. *Equation (3.131) shows that* $\boldsymbol{W}_2$ *is a diagonal matrix, such that* $\boldsymbol{V} = \boldsymbol{I}$ *and*

$$U_\infty \boldsymbol{L} = \begin{pmatrix} \lambda_1 & 0 & 0 & 0 \\ 0 & \lambda_2 & 0 & 0 \\ 0 & 0 & \lambda_3 & 0 \\ 0 & 0 & 0 & \lambda_4 \end{pmatrix}$$

where

$$\lambda_1 = -\varepsilon_1 U_\infty / b, \quad \lambda_2 = -\varepsilon_2 U_\infty / b, \quad \lambda_3 = -\varepsilon_1 U_\infty / b, \quad \lambda_4 = -\varepsilon_2 U_\infty / b$$

Substituting from Eq. (3.132) into Eq. (3.133), $\mathbf{w}(t)$ *is obtained as*

$$\mathbf{w}(t) = \begin{pmatrix} h_0 \int_0^t \mathrm{e}^{\lambda_1(t-t_0)}\mathrm{d}t_0 + h_1 \int_0^t \mathrm{e}^{\lambda_1(t-t_0)} \sin\left(\omega_0 t_0 + \phi_h\right) \mathrm{d}t_0 \\ h_0 \int_0^t \mathrm{e}^{\lambda_2(t-t_0)}\mathrm{d}t_0 + h_1 \int_0^t \mathrm{e}^{\lambda_2(t-t_0)} \sin\left(\omega_0 t_0 + \phi_h\right) \mathrm{d}t_0 \\ \alpha_0 \int_0^t \mathrm{e}^{\lambda_3(t-t_0)}\mathrm{d}t_0 + \alpha_1 \int_0^t \mathrm{e}^{\lambda_3(t-t_0)} \sin\left(\omega_0 t_0 + \phi_\alpha\right) \mathrm{d}t_0 \\ \alpha_0 \int_0^t \mathrm{e}^{\lambda_4(t-t_0)}\mathrm{d}t_0 + \alpha_1 \int_0^t \mathrm{e}^{\lambda_4(t-t_0)} \sin\left(\omega_0 t_0 + \phi_\alpha\right) \mathrm{d}t_0 \end{pmatrix}$$

The integrals can be evaluated analytically from

$$\int_0^t \mathrm{e}^{\lambda_i(t-t_0)}\mathrm{d}t_0 = \frac{1}{\lambda_i}\left(\mathrm{e}^{\lambda_i t} - 1\right)$$

$$\int_0^t e^{\lambda_i(t-t_0)} \sin\left(\omega_0 t_0 + \phi_{h,\alpha}\right) dt_0 = \frac{e^{\lambda_i t}(\omega_0 \cos\phi_{h,\alpha} + \lambda_i \sin\phi_{h,\alpha})}{\lambda_i^2 + \omega_0^2} - \frac{\omega_0 \cos\left(\omega_0 t + \phi_{h,\alpha}\right) + \lambda_i \sin\left(\omega_0 t + \phi_{h,\alpha}\right)}{\lambda_i^2 + \omega_0^2}$$

Once we have calculated $\mathbf{w}(t)$*, we can substitute it, along with* $\mathbf{y}(t)$*, into Eq. (3.126) in order to calculate the lift and moment acting on the wing. Recall that we have set* $h_0 = \alpha_0 = 0$ *so that the initial value of* $\mathbf{y}(t) = \mathbf{0}$ *and, therefore,* $\mathbf{g} = \mathbf{0}$*. Consequently, the initial condition excitation term has no effect on the solution.*

The loads in Eq. (3.126) can be split into non-circulatory and circulatory as follows:

$$\begin{pmatrix} l_\sigma(t) \\ m_{xf_\sigma}(t) \end{pmatrix} = \rho \mathbf{B}\ddot{\mathbf{y}} + \rho U_\infty \mathbf{D}_1 \dot{\mathbf{y}} \tag{3.134}$$

$$\begin{pmatrix} l_\Gamma(t) \\ m_{xf_\Gamma}(t) \end{pmatrix} = \rho U_\infty \mathbf{D}_2 \dot{\mathbf{y}} + \rho U_\infty^2 \mathbf{F}\mathbf{y} + \rho U_\infty^3 \mathbf{W}\mathbf{w} + \rho U_\infty \mathbf{g}\dot{\Phi}(t) \tag{3.135}$$

The drag depends on both the circulatory and non-circulatory lift and can be calculated directly from Eq. (3.123) once the lift and moment have been evaluated.

Figure 3.25 plots the aerodynamic states and loads acting on the wing for $h_0 = \alpha_0 = 0$*,* $h_1 = 0.1c$*,* $\alpha_1 = 2°$*,* $\phi_h = \phi_\alpha = 0°$ *and for an oscillation frequency of 2 Hz, so that* $\omega_0 = 4\pi$ *rad/s. The aerodynamic state responses show a clear transient component before settling down to a steady state. A transient component is also visible on the aerodynamic load responses; it is most pronounced in the drag signal. Note that the mean values of the lift and pitching moment are zero; in contrast, the mean drag is slightly negative, which means that the oscillation is producing a net thrust. Also note that the frequency of the drag response is equal to twice the frequency of oscillation due to the square term in Eq. (3.123).*

It is interesting to examine how non-circulatory and circulatory loads behave as the frequency of oscillation is changed. To this end, we repeat this example for 50 different frequencies ranging from $\omega_0 = 0.16$ *to 480 rad/s. In order to evaluate the variation of the aerodynamic loads with frequency more objectively, we define the reduced frequency*

$$k = \frac{\omega_0 b}{U_\infty} \tag{3.136}$$

The frequency values mentioned above correspond to k *values between 0.001 and 3. We run each simulation for five cycles of the pitch and plunge oscillation, setting the time step such that there are 30 time instances per cycle,* $\Delta t = 2\pi/30\omega_i$ *for the ith frequency value. We then calculate the circulatory and non-circulatory aerodynamic loads from Eqs. (3.134) and (3.135). Finally, we evaluate the amplitudes in lift and pitching moment as the maxima of these two loads over the last cycle of oscillation.*

Figure 3.26 plots the variation of the non-circulatory, circulatory and total loads with reduced frequency. The circulatory loads for both the lift and moment are much higher than the non-circulatory loads for $k < 1$*. In contrast, the non-circulatory loads become significantly higher for* $k > 1.5$*, and this difference increases quadratically with reduced frequency. Equation (3.134) shows that the non-circulatory loads depend on velocity and acceleration; this means that for a sinusoidal motion, they are proportional to* k^2*. The circulatory loads on the other hand only depend on displacement and velocity so that they are proportional to* k^0

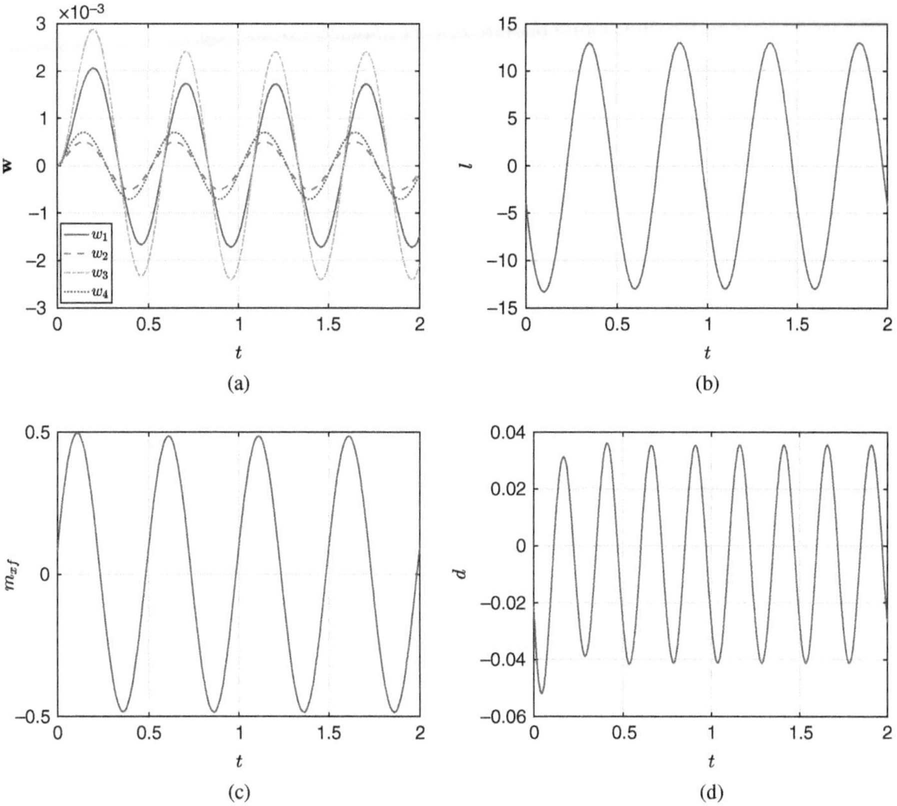

Figure 3.25 Aerodynamic states and loads of sinusoidally oscillating 2D pitch-plunge wing. (a) $\mathbf{w}(t)$, (b) $l(t)$, (c) $m_{xf}(t)$ and (d) $d(t)$.

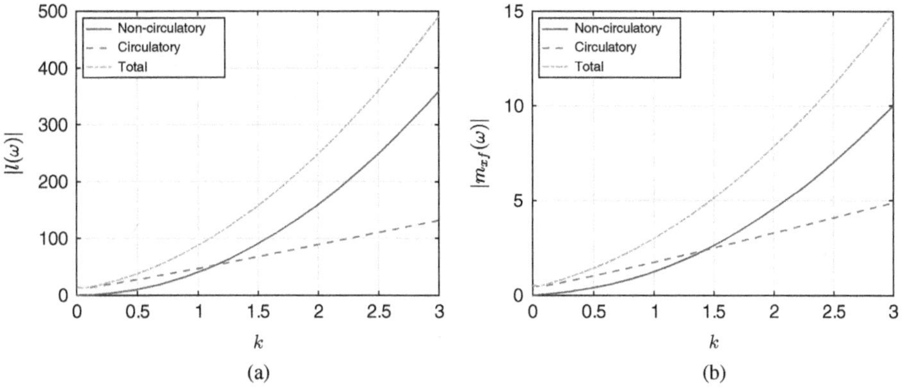

Figure 3.26 Amplitude variation of circulatory and non-circulatory loads with reduced frequency. (a) Lift amplitude and (b) pitching moment amplitude.

and k^1. The non-circulatory loads vary with the square of the frequency and hence become very important at high k values while the circulatory loads depend linearly on the frequency and will therefore be lower at high k.

From a physical point of view, the flow remains attached at all times on the surface of the wing at low reduced frequencies. At higher values of k, dynamic stall occurs, which means that the flow separates and reattaches periodically from the surface of the airfoil. This phenomenon will be treated in Chapter 7. Finally, it should be mentioned that the Kutta condition is not necessarily valid at high values of the reduced frequency. Satyanarayana and Davis (1978) show experimentally that the Kutta condition is not valid for $k > 0.6$. This means that the theories presented in this chapter become increasingly inaccurate as this value of the reduced frequency is exceeded. This example is solved by Matlab code `pitchplunge_Wagner.m`.

Equations (3.126) and (3.129) form a state-space system whereby the inputs are the pitch and plunge time histories, $\mathbf{y}(t)$, the system states are the aerodynamic states $\mathbf{w}(t)$ and the outputs are the lift $l(t)$ and pitching moment $m_{xf}(t)$. In Example 3.12, we solved this state-space system analytically but numerical solutions can also be applied, such as Euler or Runge–Kutta methods.

3.4.5 Quasi-Steady Aerodynamics

We return to Eqs. (3.120) and (3.121) that give the time responses of the lift and pitching moment acting on the pitch-plunge wing. These expressions can be significantly simplified if it is assumed that $\Phi = 1$ at all times. Then,

$$l(t) = -\rho\pi b^2(\ddot{h} - ba\ddot{\alpha} + U_\infty\dot{\alpha}) - 2\rho U_\infty b\pi\left(U_\infty\alpha(t) + \dot{h}(t) + b\left(\frac{1}{2} - a\right)\dot{\alpha}(t)\right) \quad (3.137)$$

$$\begin{aligned} m_{xf}(t) = &-\rho\pi b^2\left(-ba\ddot{h} + b^2\left(\frac{1}{8} + a^2\right)\ddot{\alpha}\right) - \rho U_\infty b^3\pi\left(\frac{1}{2} - a\right)\dot{\alpha}(t) \\ &+2\rho U_\infty\pi b^2\left(a + \frac{1}{2}\right)\left(U_\infty\alpha(t) + \dot{h}(t) + b\left(\frac{1}{2} - a\right)\dot{\alpha}(t)\right) \end{aligned} \quad (3.138)$$

According to the discussion of Section 3.4.1, setting $\Phi = 1$ means that the instantaneous circulatory lift is always equal to the steady-state circulatory lift. This observation constitutes the basis of the quasi-steady aerodynamic assumption, which is the simplest possible form of unsteady aerodynamic modelling. Quasi-steadiness can be used in situations where the reduced frequency is very low but can significantly overestimate the circulatory loads for higher k values.

Example 3.13 *Calculate the difference between the quasi-steady and unsteady circulatory loads for the problem of Example 3.12*

Using the notation of Eq. (3.126), the quasi-steady lift and pitching moment are given by

$$\begin{pmatrix} l_{qs}(t) \\ m_{xf_{qs}}(t) \end{pmatrix} = \rho\mathbf{B}\ddot{\mathbf{y}} + \rho U_\infty\left(\mathbf{D}_1 + 2\mathbf{D}_2\right)\dot{\mathbf{y}} + 2\rho U_\infty^2\mathbf{F}_1\mathbf{y} \quad (3.139)$$

where the subscript qs denotes quasi-steady values. Separating out the non-circulatory loads, which are still given by Eq. (3.134), we obtain

$$\begin{pmatrix} l_{\Gamma,qs}(t) \\ m_{xf_{\Gamma,qs}}(t) \end{pmatrix} = 2\rho U_\infty\mathbf{D}_2\dot{\mathbf{y}} + 2\rho U_\infty^2\mathbf{F}_1\mathbf{y} \quad (3.140)$$

We repeat Example 3.12, but this time we also calculate the quasi-steady circulatory loads from Eq. (3.140) and their resulting amplitudes for all values of the reduced frequency.

Figure 3.27 plots the ratio of the amplitude of the unsteady circulatory lift to that of the quasi-steady circulatory lift, as the reduced frequency is varied. This ratio is equal to one at $k = 0$, which means that the unsteady and quasi-steady theories give the same result at steady conditions. However, as the frequency is increased, the ratio drops significantly and asymptotes towards 1/2 at very high k. Consequently, the quasi-steady lift overestimates the unsteady lift by a factor of up to 2, depending on the frequency. We will see the curve of Figure 3.27 again in Section 3.5 when we discuss Theodorsen's function. This example is solved by Matlab code `pitchplunge_Wagner.m`.

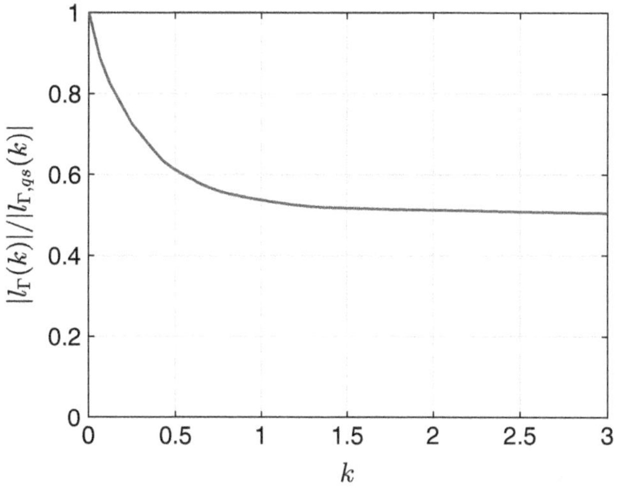

Figure 3.27 Ratio of unsteady to quasi-steady load amplitudes.

3.5 Theodorsen Theory

We return to Eq. (3.111) for the circulatory lift

$$l_\Gamma(t) = -2\rho U_\infty b\pi \left(U_\infty \alpha(t) + \dot{h}(t) + b\left(\frac{1}{2} - a\right)\dot{\alpha}(t)\right) \frac{\int_b^\infty \frac{X_0}{\sqrt{X_0^2-b^2}}\bar{\Gamma}(X_0, t)\mathrm{d}X_0}{\int_b^\infty \frac{\sqrt{X_0+b}}{\sqrt{X_0-b}}\bar{\Gamma}(X_0, t)\mathrm{d}X_0} \quad (3.141)$$

In Section 3.4.3, we used this equation to evaluate the circulatory lift acting on a pitching and plunging flat plate. We used the concept of superposition of an infinite number of infinitesimal impulsive changes in upwash in order to describe a general motion in terms of the Wagner function. We then presented an example for a sinusoidally pitching and plunging plate. In this section, we will consider the circulatory lift response to sinusoidal motion directly. First we note that ratio of integrals in equation 3.141 is not a function of time if $\bar{\Gamma}(X_0, t)$ is a sunsoidal function of time with zero mean and frequency ω_0, since the integrations are carried out over space only and the time variations in the numerator and denominator cancel out. Therefore, when we take the Fourier Transform of Eq. (3.141) we obtain

$$l_\Gamma(\omega) = -2\rho U_\infty b\pi \left(\mathrm{i}\omega h(\omega) + \left(U_\infty + b\left(\frac{1}{2} - a\right)\mathrm{i}\omega \right) \alpha(\omega) \right) \frac{\int_b^\infty \frac{X_0}{\sqrt{X_0^2-b^2}} \bar{\Gamma}(X_0,\omega_0)\mathrm{d}X_0}{\int_b^\infty \frac{\sqrt{X_0+b}}{\sqrt{X_0-b}} \bar{\Gamma}(X_0,\omega_0)\mathrm{d}X_0} \tag{3.142}$$

We now choose $h(t)$ and $\alpha(t)$ to be sinusoidal functions of the form of Eq. (3.132)

$$h(t) = h_0 + h_1 \sin\left(\omega_0 t + \phi_h\right),\ \alpha(t) = \alpha_0 + \alpha_1 \sin\left(\omega_0 t + \phi_\alpha\right)$$

Recall that h_0, α_0 are mean values, h_1, α_1 are oscillation amplitudes, ω_0 is the frequency and ϕ_h, ϕ_α are phase angles. The Fourier Transforms of these signals are

$$h(\omega) = h_0\delta(\omega) + \frac{h_1}{2\mathrm{i}}\mathrm{e}^{\mathrm{i}\phi_h}\delta(\omega-\omega_0) - \frac{h_1}{2\mathrm{i}}\mathrm{e}^{-\mathrm{i}\phi_h}\delta(\omega+\omega_0) \tag{3.143}$$

$$\alpha(\omega) = \alpha_0\delta(\omega) + \frac{\alpha_1}{2\mathrm{i}}\mathrm{e}^{\mathrm{i}\phi_\alpha}\delta(\omega-\omega_0) - \frac{\alpha_1}{2\mathrm{i}}\mathrm{e}^{-\mathrm{i}\phi_\alpha}\delta(\omega+\omega_0) \tag{3.144}$$

where $\delta(\omega - \omega_0)$ is the Kronecker delta function, which is defined as

$$\delta(\omega-\omega_0) = \begin{cases} 1 & \text{if}\omega = \omega_0 \\ 0 & \text{if}\omega \neq \omega_0 \end{cases} \tag{3.145}$$

and has the sifting property

$$\int_{-\infty}^{\infty} f(\omega)\delta(\omega-\omega_0) = f(\omega_0) \tag{3.146}$$

Therefore, functions $h(\omega)$ and $\alpha(\omega)$ take non-zero values only at $\omega = 0$, $\omega = \omega_0$ and $\omega = -\omega_0$, which is the definition of a sinusoidal function in the frequency domain. Note that $h(-\omega_0)$ and $\alpha(-\omega_0)$ are the complex conjugates of $h(\omega_0)$ and $\alpha(\omega_0)$, respectively. The vortex strength function $\bar{\Gamma}(X_0, \omega)$ in Eq. (3.142) is the solution of the Fourier transform of the Kutta condition (3.76)

$$\frac{1}{2\pi b}\int_b^\infty \frac{\sqrt{X_0+b}}{\sqrt{X_0-b}}\bar{\Gamma}(X_0,\omega)\mathrm{d}X_0 = \mathrm{i}\omega h(\omega) + \left(U_\infty + b\left(\frac{1}{2}-a\right)\mathrm{i}\omega \right)\alpha(\omega) \tag{3.147}$$

As the left-hand side is a linear superposition of $\bar{\Gamma}(X_0, \omega)$ for all X_0 and the right-hand side is a linear relationship in $h(\omega)$ and $\alpha(\omega)$, we expect $\bar{\Gamma}(X_0, \omega)$ to be sinusoidal too.

We now make the same assumption that we made for Wagner theory: once a wake vortex is shed from the trailing edge, it travels downstream at the free stream airspeed and its strength is unchanged. In other words, the wake oscillates both in time and in space, as seen in Figure 3.28, which plots the distribution of vortex strength in the wake of a flat plate with $b = 1$ m that oscillates with frequency $f_0 = 2$ Hz in a free stream $U_\infty = 10$ m/s. At the trailing edge, the vortex strength oscillates in time with mean $\bar{\Gamma}_0(b) = 1$, amplitude $\bar{\Gamma}_1(b) = 1$, frequency $\omega_0 = 2\pi f_0$ and phase $\phi_\Gamma(b) = 0$, so that

$$\bar{\Gamma}(b,t) = \bar{\Gamma}_0(b) + \bar{\Gamma}_1(b)\sin\omega_0 t \tag{3.148}$$

At time $t = 0$, a vortex of strength 1 is shed at $X_0 = 1$, denoted by a circle. Since all vortices travel with speed U_∞, the same circle can be found at position $\Delta X_0 = X_0 - b$ at a later time $\Delta t = (X_0 - b)/U_\infty$. All the circles in the figure denote vortices shed from the trailing

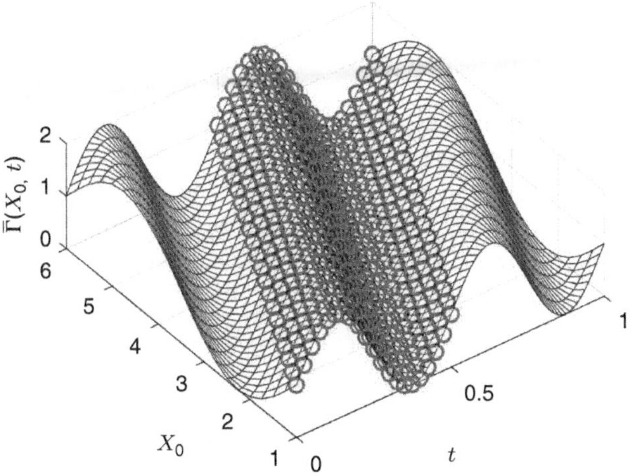

Figure 3.28 Wake oscillation in time and space.

edge over a single cycle with period $T = 1/f_0 = 0.5$ seconds and found downstream at later times. The meshed surface plots $\bar{\Gamma}(X_0, t)$ for times between $t = 0$ and $t = 2T$ and at positions between $X_0 = 1$ and $X_0 = 1 + \lambda$, where $\lambda = U_\infty T$ is the wavelength of the spatial oscillation. Looking at $\bar{\Gamma}(X_0, 0)$, we see that it describes a complete cycle over a distance λ. This oscillation is described by

$$\bar{\Gamma}(X0, 0) = \bar{\Gamma}_0(b) + \bar{\Gamma}_1(b) \sin \frac{-2\pi(X_0 - b)}{U_\infty T} = \bar{\Gamma}_0(b) + \bar{\Gamma}_1(b) \sin \frac{-\omega_0(X_0 - b)}{U_\infty} \tag{3.149}$$

Combining Eqs. (3.148) and (3.149), the oscillation of the wake strength at all positions X_0 and times t is given by

$$\bar{\Gamma}(b, t) = \bar{\Gamma}_0(b) + \bar{\Gamma}_1(b) \sin \left(\omega_0 t - \frac{\omega_0(X_0 - b)}{U_\infty} \right) \tag{3.150}$$

Taking the Fourier transform of this sinusoidal function, we obtain

$$\bar{\Gamma}(X_0, \omega) = \bar{\Gamma}_0(b)\delta(\omega) + \frac{\bar{\Gamma}_1(b)}{2\mathrm{i}} \mathrm{e}^{\mathrm{i}(\phi_{\bar{\Gamma}}(b) - \omega_0 X_0/U_\infty)} \delta(\omega - \omega_0)$$
$$- \frac{\bar{\Gamma}_1(b)}{2\mathrm{i}} \mathrm{e}^{-\mathrm{i}(\phi_{\bar{\Gamma}}(b) - \omega_0 X_0/U_\infty)} \delta(\omega + \omega_0) \tag{3.151}$$

where we have replaced $\omega_0 b/U_\infty$ by a more general phase $\phi_{\bar{\Gamma}}(b)$ to also allow for an initial phase shift of the time oscillation at the trailing edge.

As we have assumed that both the motion and the vortex strength response are sinusoidal, we will proceed to assume that the circulatory lift response is also a sine wave, such that

$$l_\Gamma(\omega) = l_{\Gamma_0}\delta(\omega) + \frac{l_{\Gamma_1}}{2\mathrm{i}} \mathrm{e}^{\mathrm{i}\phi_{l_\Gamma}} \delta(\omega - \omega_0) - \frac{l_{\Gamma_1}}{2\mathrm{i}} \mathrm{e}^{-\mathrm{i}\phi_{l_\Gamma}} \delta(\omega + \omega_0) \tag{3.152}$$

where l_{Γ_0} is the mean value, l_{Γ_1} is the amplitude and ϕ_{l_Γ} is the phase of the circulatory lift oscillation. Equation (3.152) can be written in more compact form as

$$l_\Gamma(\omega) = l_{\Gamma_0}\delta(\omega) + l_\Gamma(\omega_0)\delta(\omega - \omega_0) + l_\Gamma^*(\omega_0)\delta(\omega + \omega_0) \tag{3.153}$$

where

$$l_\Gamma(\omega_0) = \frac{l_{\Gamma_1}}{2\mathrm{i}}\mathrm{e}^{\mathrm{i}\phi_{l_\Gamma}}, \quad l_\Gamma^*(\omega_0) = -\frac{l_{\Gamma_1}}{2\mathrm{i}}\mathrm{e}^{-\mathrm{i}\phi_{l_\Gamma}}$$

and the superscript * denotes the complex conjugate. In order to fully determine Eq. (3.152), we only need l_{Γ_0} and $l_\Gamma(\omega_0)$, since $l_\Gamma^*(\omega_0)$ can be obtained by conjugating $l_\Gamma(\omega_0)$. Therefore, Eq. (3.142) need only be solved at two frequencies, $\omega = 0$ and $\omega = \omega_0$.

We start by setting $\omega = 0$ in Eq. (3.142) so that, after substituting for $h(0)$, $\alpha(0)$, $\bar{\Gamma}(X_0, 0)$ and $l_\Gamma(0)$ from Eqs. (3.143), (3.144), (3.151) and (3.152), respectively, we obtain

$$l_{\Gamma_0} = -2\rho U_\infty b\pi U_\infty \alpha_0 \frac{\int_b^\infty \frac{X_0}{\sqrt{X_0^2-b^2}}\mathrm{d}X_0}{\int_b^\infty \frac{\sqrt{X_0+b}}{\sqrt{X_0-b}}\mathrm{d}X_0} \tag{3.154}$$

since $\bar{\Gamma}_0(b)$ is constant with X_0 and can be taken outside the integrals and cancelled out. The integrals can be calculated analytically,

$$\int_b^\infty \frac{X_0}{\sqrt{X_0^2 - b^2}}\mathrm{d}X_0 = \sqrt{X_0^2 - b^2}\Big|_b^\infty$$

$$\int_b^\infty \frac{\sqrt{X_0 + b}}{\sqrt{X_0 - b}}\mathrm{d}X_0 = \int_b^\infty \frac{X_0 + b}{\sqrt{X_0^2 - b^2}}\mathrm{d}X_0 = \left(b\ln\left(X_0 + \sqrt{X_0^2 - b^2}\right) + \sqrt{X_0^2 - b^2}\right)\Big|_b^\infty$$

Both integrals become infinite at the upper limit, but their ratio can be calculated by the limiting procedure:

$$\frac{\int_b^\infty \frac{X_0}{\sqrt{X_0^2-b^2}}\mathrm{d}X_0}{\int_b^\infty \frac{\sqrt{X_0+b^2}}{\sqrt{X_0-b}}\mathrm{d}X_0} = \lim_{X_0\to\infty} \frac{\sqrt{X_0^2 - b^2}}{b\ln\left(X_0 + \sqrt{X_0^2 - b^2}\right) + \sqrt{X_0^2 - b} - b\ln b} = 1$$

Substituting this latest result in Eq. (3.154) yields

$$l_{\Gamma_0} = -2\rho U_\infty^2 b\pi\alpha_0 \tag{3.155}$$

which is the familiar steady lift result of Eq. (3.86). Therefore, if a wing is oscillating around a non-zero mean pitch angle α_0, the resulting mean lift is equal to the lift that would have acted on a steady wing at the same angle of attack.

We now set $\omega = \omega_0$ in Eq. (3.142) and substitute for $h(\omega_0)$, $\alpha(\omega_0)$ and $\bar{\Gamma}(X_0, \omega_0)$ from Eqs. (3.143), (3.144) and (3.151) to obtain

$$l_\Gamma(\omega_0) = -2\rho U_\infty b\pi\left(\mathrm{i}\omega_0\frac{h_1}{2\mathrm{i}}\mathrm{e}^{\mathrm{i}\phi_h} + \left(U_\infty + b\left(\frac{1}{2} - a\right)\mathrm{i}\omega_0\right)\frac{\alpha_1}{2\mathrm{i}}\mathrm{e}^{\mathrm{i}\phi_\alpha}\right) \times \frac{\int_b^\infty \frac{X_0}{\sqrt{X_0^2-b^2}}\mathrm{e}^{-\mathrm{i}\omega_0 X_0/U_\infty}\mathrm{d}X_0}{\int_b^\infty \frac{\sqrt{X_0+b}}{\sqrt{X_0-b}}\mathrm{e}^{-\mathrm{i}\omega_0 X_0/U_\infty}\mathrm{d}X_0} \tag{3.156}$$

where, again, $\bar{\Gamma}_1(b)\mathrm{e}^{\phi_\Gamma}/2\mathrm{i}$ is constant with X_0 and can be taken outside the integrals, thus cancelling out. If we can calculate the value of the ratio of integrals in Eq. (3.156), then we can substitute $l_\Gamma(\omega_0)$ and l_{Γ_0} from Eq. (3.155) back into Eq. (3.153) and calculate the circulatory lift.

3.5.1 Theodorsen's Function

The ratio of integrals in Eq. (3.156) is known as Theodorsen's function, defined as

$$C(\omega) = \frac{\int_b^\infty \frac{X_0}{\sqrt{X_0^2-b^2}} \mathrm{e}^{-\mathrm{i}\omega X_0/U_\infty} \mathrm{d}X_0}{\int_b^\infty \frac{\sqrt{X_0+b}}{\sqrt{X_0-b}} \mathrm{e}^{-\mathrm{i}\omega X_0/U_\infty} \mathrm{d}X_0} \tag{3.157}$$

for any general frequency ω. If we define the non-dimensional distance $\bar{X}_0 = X_0/b$, then,

$$C(\omega) = C(k) = \frac{\int_1^\infty \frac{\bar{X}_0}{\sqrt{\bar{X}_0^2-1^2}} \mathrm{e}^{-\mathrm{i}k\bar{X}_0} \mathrm{d}\bar{X}_0}{\int_1^\infty \frac{\sqrt{\bar{X}_0+1}}{\sqrt{\bar{X}_0-1}} \mathrm{e}^{-\mathrm{i}k\bar{X}_0} \mathrm{d}\bar{X}_0} \tag{3.158}$$

where $k = \omega b/U_\infty$ is the reduced frequency of Eq. (3.136). When Theodorsen's function is written in terms of the reduced frequency, it becomes a function of a single variable, k. Nevertheless, $C(\omega) = C(k)$ and the two notations will be used interchangeably in this book.

According to Theodorsen, the integrals can be evaluated in terms of Bessel functions, such that

$$C(k) = \frac{-J_1(k) + \mathrm{i}Y_1(k)}{-(J_1(k) + Y_0(k)) + \mathrm{i}(Y_1(k) - J_0(k))} \tag{3.159}$$

where J_i and Y_i are Bessel functions of the first and second kind, respectively, and k is a general reduced frequency. Substituting Eq. (3.157) into the circulatory lift equation (3.156), we obtain

$$\begin{aligned} l_\Gamma(\omega_0) &= -2\rho U_\infty b\pi \left(\mathrm{i}\omega_0 \frac{h_1}{2\mathrm{i}} \mathrm{e}^{\mathrm{i}\phi_h} + \left(U_\infty + b\left(\frac{1}{2} - a\right) \mathrm{i}\omega_0 \right) \frac{\alpha_1}{2\mathrm{i}} \mathrm{e}^{\mathrm{i}\phi_\alpha} \right) C(\omega_0) \\ &= -2\rho U_\infty b\pi w_{3/4}(\omega_0) C(\omega_0) \end{aligned} \tag{3.160}$$

where

$$w_{3/4}(\omega_0) = \mathrm{i}\omega_0 \frac{h_1}{2\mathrm{i}} \mathrm{e}^{\mathrm{i}\phi_h} + \left(U_\infty + b\left(\frac{1}{2} - a\right) \mathrm{i}\omega_0 \right) \frac{\alpha_1}{2\mathrm{i}} \mathrm{e}^{\mathrm{i}\phi_\alpha} \tag{3.161}$$

is the Fourier transform of the downwash at the 3/4 chord, evaluated at $\omega = \omega_0$.

It is interesting to compare Eq. (3.160) to the circulatory lift predicted by Wagner theory. Equation (3.113) states that

$$l_\Gamma(t) = -2\rho U_\infty b\pi \int_{-\infty}^t \Phi(t - t_0) \dot{w}_{3/4}(t_0) \mathrm{d}t_0 \tag{3.162}$$

In order to compare Eq. (3.160) with Eq. (3.162), we need to apply the inverse Fourier Transform to $l_\Gamma(\omega) = l_\Gamma(\omega_0)\delta(\omega - \omega_0)$. As the latter has only one non-zero frequency component, we take advantage of the sifting property of the Kronecker delta function (3.146) to obtain

$$l_\Gamma(t) = -2\rho U_\infty b\pi w_{3/4}(\omega_0) \mathrm{e}^{\mathrm{i}\omega_0 t} C(\omega_0) \tag{3.163}$$

where $w_{3/4}(\omega_0)\mathrm{e}^{\mathrm{i}\omega_0 t}$ is the complex exponential time-domain form of $w_{3/4}(t)$. Substituting from Eq. (3.162) yields

$$w_{3/4}(\omega_0) \mathrm{e}^{\mathrm{i}\omega_0 t} C(\omega_0) = \int_{-\infty}^t \Phi(t - t_0) \dot{w}_{3/4}(t_0) \mathrm{d}t_0 \tag{3.164}$$

If $w_{3/4}(t) = w_{3/4}(\omega_0)e^{i\omega_0 t}$, then

$$\dot{w}_{3/4}(t_0) = i\omega_0 w_{3/4}(\omega_0)e^{i\omega_0 t_0}$$

Substituting back into Eq. (3.164) leads to

$$w_{3/4}(\omega_0)e^{i\omega_0 t}C(\omega_0) = i\omega w_{3/4}(\omega_0)\int_{-\infty}^{t} \Phi(t - t_0)e^{i\omega_0 t_0}\,dt_0$$

If we now introduce a non-dimensional time $\tau = U_\infty t/b$ and divide both sides by $w_{3/4}(\omega_0)e^{i\omega_0 t}$, the relationship between Theodorsen's and Wagner's functions becomes

$$C(k) = ik\int_{-\infty}^{\tau} \Phi(\tau - \tau_0)e^{-ik(\tau-\tau_0)}\,d\tau_0 \tag{3.165}$$

where k represents any general reduced frequency. In other words, Wagner's and Theodorsen's functions form essentially a Fourier Transform pair. Fung (1993) points out the intricacy of calculating this integral as $\tau_0 \to -\infty$. This issue is addressed in the next example.

Example 3.14 *Calculate Theodorsen's function from Eq. (3.165) using R. T. Jones' and W. P. Jones' approximations for Wagner's function.*

In Example 3.9, we presented R. T. Jones' and W. P. Jones' approximations for Wagner's functions, which are both of the form

$$\Phi(\tau) = 1 - \Psi_1 e^{-\varepsilon_1 \tau} - \Psi_2 e^{-\varepsilon_2 \tau} \tag{3.166}$$

where Ψ_1, Ψ_2, ε_1 and ε_2 are constants that take different values depending on which approximation is used. For this form of Wagner's function, calculating the indefinite integral in Eq. (3.165) is straightforward, leading to

$$\begin{aligned} I &= \int \left(1 - \Psi_1 e^{-\varepsilon_1(\tau-\tau_0)} - \Psi_2 e^{-\varepsilon_2(\tau-\tau_0)}\right) e^{-ik(\tau-\tau_0)}\,d\tau_0 \\ &= e^{-ik(\tau-\tau_0)}\left(\frac{1}{ik} - \frac{\Psi_1 e^{-\varepsilon_1(\tau-\tau_0)}}{\varepsilon_1 + ik} - \frac{\Psi_2 e^{-\varepsilon_2(\tau-\tau_0)}}{\varepsilon_2 + ik}\right) \end{aligned} \tag{3.167}$$

Applying the $\tau_0 = \tau$ limit leads to

$$I(\tau) = \frac{1}{ik} - \frac{\Psi_1}{\varepsilon_1 + ik} - \frac{\Psi_2}{\varepsilon_2 + ik} \tag{3.168}$$

When $\tau_0 \ll 0$, $\tau - \tau_0 \gg 0$ and the $e^{-\varepsilon_{1,2}(\tau-\tau_0)}$ terms vanish so that $I(-\infty)$ can be written as

$$I(-\infty) = \frac{1}{ik}\lim_{\tau_0 \to -\infty} e^{ik(\tau-\tau_0)}$$

Unfortunately, this limit does not exist, since sinusoidal functions are bounded. In order to resolve this issue, we split the definite integral in the form

$$\begin{aligned} I &= \int_{\tau_1}^{\tau} \left(1 - \Psi_1 e^{-\varepsilon_1(\tau-\tau_0)} - \Psi_2 e^{-\varepsilon_2(\tau-\tau_0)}\right) e^{-ik(\tau-\tau_0)}\,d\tau_0 \\ &+ \int_{-\infty}^{\tau_1} \left(1 - \Psi_1 e^{-\varepsilon_1(\tau-\tau_0)} - \Psi_2 e^{-\varepsilon_2(\tau-\tau_0)}\right) e^{-ik(\tau-\tau_0)}\,d\tau_0 \end{aligned} \tag{3.169}$$

where $\tau_1 \ll 0$ *is a finite time so negative that*

$$\left(1 - \Psi_1 e^{-\varepsilon_1(\tau-\tau_0)} - \Psi_2 e^{-\varepsilon_2(\tau-\tau_0)}\right) e^{-ik(\tau-\tau_0)} \approx e^{-ik(\tau-\tau_0)}$$

and

$$I \approx \frac{1}{ik} e^{ik(\tau-\tau_0)}$$

for $-\infty < \tau_0 \leq \tau_1$. *Furthermore,* τ_1 *is selected such that* $e^{ik(\tau-\tau_1)} = 0$, *so that time instance* τ_1 *is the start of a sinusoidal cycle. Then, using the result of Eq. (3.168) and* $I(\tau_1) = 0$ *for the first integral in expression (3.169), we obtain*

$$I = \frac{1}{ik} - \frac{\Psi_1}{\varepsilon_1 + ik} - \frac{\Psi_2}{\varepsilon_2 + ik} + \int_{-\infty}^{\tau_1} e^{-ik(\tau-\tau_0)} \, d\tau_0$$

Finally, in order to overcome the problem of integrating an unbounded function, we state that the remaining integral is evaluated over an infinite number of complete cycles. The integral of a harmonic function over any number of complete cycles is equal to zero, so that

$$I = \frac{1}{ik} - \frac{\Psi_1}{\varepsilon_1 + ik} - \frac{\Psi_2}{\varepsilon_2 + ik}$$

Substituting back into Eq. (3.165), we obtain

$$C(k) = 1 - \frac{\Psi_1}{1 - i\frac{\varepsilon_1}{k}} - \frac{\Psi_2}{1 - i\frac{\varepsilon_2}{k}} \tag{3.170}$$

Figure 3.29 plots the variation of the magnitude and phase of Theodorsen's function with reduced frequency. The exact result of Eq. (3.159) is plotted along with the result of Eq. (3.170) with R. T. Jones' and W. P. Jones' values for $\Psi_{1,2}$ *and* $\varepsilon_{1,2}$. *The agreement is very good for the magnitude but appears to be less satisfactory for the phase. Nevertheless, the maximum error in phase angle is only around 1.3°; consequently, the approximations (particularly R. T. Jones') are used quite often in practice. Brunton and Rowley (2013) give a much more complete overview of approximations to Theodorsen's function. Figure 3.29 is plotted by Matlab code* `Theodorsen_Function.m`.

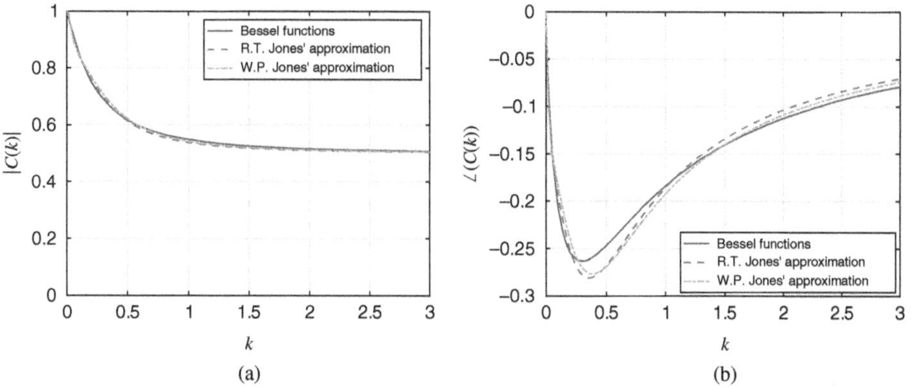

Figure 3.29 Variation of magnitude and phase of Theodorsen's function with reduced frequency. (a) $\Re(C(k))$ and (b) $\Im(C(k))$.

Figure 3.29a should be compared to Figure 3.27, which plots the ratio of the unsteady to the steady circulatory lift of a flat plate undergoing sinusoidal oscillations. In fact, both figures represent Theodorsen's function. From Eq. (3.137), the circulatory quasi-steady lift is given by

$$l_{\Gamma_{qs}} = -2\rho U_\infty b\pi w_{3/4}(t)$$

Comparing to Eq. (3.163), it is clear that

$$\frac{l_\Gamma}{l_{\Gamma_{qs}}} = C(k)$$

In other words, Theodorsen's function is a filter that attenuates and phase-shifts the quasi-steady circulatory lift. The maximum attenuation occurs for $k \to \infty$, where the amplitude of the quasi-steady lift is divided by two; the maximum phase shift occurs for $k \approx 0.3$. As $C(0) = 1$, the steady result at $k = 0$ is unaltered.

3.5.2 Total Loads for Sinusoidal Motion

Recall that the circulatory lift is given by Eq. (3.153) as

$$l_\Gamma(\omega) = l_{\Gamma_0}\delta(\omega) + l_\Gamma(\omega_0)\delta(\omega - \omega_0) + l_\Gamma^*(\omega_0)\delta(\omega + \omega_0) \tag{3.171}$$

where from Eqs. (3.155) and (3.160),

$$l_{\Gamma_0} = -2\rho U_\infty^2 b\pi\alpha_0 \tag{3.172}$$

$$l_\Gamma(\omega_0) = -2\rho U_\infty b\pi\left(\mathrm{i}\omega_0\frac{h_1}{2\mathrm{i}}\mathrm{e}^{\mathrm{i}\phi_h} + \left(U_\infty + b\left(\frac{1}{2} - a\right)\mathrm{i}\omega_0\right)\frac{\alpha_1}{2\mathrm{i}}\mathrm{e}^{\mathrm{i}\phi_\alpha}\right)C(\omega_0) \tag{3.173}$$

and $l_\Gamma^*(\omega_0)$ is the complex conjugate of $l_\Gamma(\omega_0)$.

The total lift for sinusoidal motion is obtained by adding the circulatory lift $l_\Gamma(\omega)$ to the Fourier Transform of the non-circulatory lift of Eq. (3.63), which is

$$l_\sigma(\omega) = -\rho\pi b^2\left(-\omega^2 h(\omega) + \omega^2 ba\alpha(\omega) + \mathrm{i}\omega U_\infty\alpha(\omega)\right)$$

Substituting for $h(\omega)$ and $\alpha(\omega)$ from Eqs. ((3.143)) and (3.144), we obtain

$$l_\sigma(\omega) = l_\sigma(\omega_0)\delta(\omega - \omega_0) + l_\sigma^*(\omega_0)\delta(\omega + \omega_0) \tag{3.174}$$

where

$$l_\sigma(\omega_0) = -\rho\pi b^2\left(-\omega_0^2\frac{h_1}{2\mathrm{i}}\mathrm{e}^{\mathrm{i}\phi_h} + \left(\omega_0^2 ba + \mathrm{i}\omega_0 U_\infty\right)\frac{\alpha_1}{2\mathrm{i}}\mathrm{e}^{\mathrm{i}\phi_\alpha}\right)$$

Note that $l_\sigma(\omega)$ has no zero frequency term. The total lift is then given by

$$l(\omega) = l_\Gamma(\omega) + l_\sigma(\omega_0) = l_0\delta(\omega) + l(\omega_0)\delta(\omega - \omega_0) + l^*(\omega_0)\delta(\omega + \omega_0) \tag{3.175}$$

where

$$l_0 = -2\rho U_\infty^2 b\pi\alpha_0 \tag{3.176}$$

$$l(\omega_0) = -\rho\pi b^2\left(-\omega_0^2\frac{h_1}{2\mathrm{i}}\mathrm{e}^{\mathrm{i}\phi_h} + \left(\omega_0^2 ba + \mathrm{i}\omega_0 U_\infty\right)\frac{\alpha_1}{2\mathrm{i}}\mathrm{e}^{\mathrm{i}\phi_\alpha}\right)$$

$$- 2\rho U_\infty b\pi \left(\mathrm{i}\omega_0 \frac{h_1}{2\mathrm{i}} \mathrm{e}^{\mathrm{i}\phi_h} + \left(U_\infty + b\left(\frac{1}{2} - a\right) \mathrm{i}\omega_0 \right) \frac{\alpha_1}{2\mathrm{i}} \mathrm{e}^{\mathrm{i}\phi_\alpha} \right) C(\omega_0) \tag{3.177}$$

The total pitching moment of Eq. (3.118) can be simplified to

$$m_{xf}(t) = -\rho\pi b^2 \left(-ba\ddot{h} + b^2 \left(\frac{1}{8} + a^2 \right) \ddot{\alpha} + b\left(\frac{1}{2} - a \right) U_\infty \dot{\alpha} \right) - b\left(a + \frac{1}{2} \right) l_\Gamma(t) \tag{3.178}$$

Taking the Fourier Transform of this expression, we obtain

$$m_{xf}(\omega) = -\rho\pi b^2 \left(ba\omega^2 h(\omega) - \left(b^2 \left(\frac{1}{8} + a^2 \right) \omega^2 - b\left(\frac{1}{2} - a \right) U_\infty \mathrm{i}\omega \right) \alpha(\omega) \right) - b\left(a + \frac{1}{2} \right) l_\Gamma(\omega) \tag{3.179}$$

Substituting for $h(\omega)$, $\alpha(\omega)$ and $l_\Gamma(\omega)$ from Eqs. (3.143), (3.144) and (3.171)–(3.173), we obtain

$$m_{xf}(\omega) = m_{xf_0}\delta(\omega) + m_{xf}(\omega_0)\delta(\omega - \omega_0) + m_{xf}^*(\omega_0)\delta(\omega + \omega_0) \tag{3.180}$$

where

$$m_{xf_0} = -2\rho U_\infty^2 \pi b^2 \left(a + \frac{1}{2} \right) \alpha_0 \tag{3.181}$$

$$\begin{aligned} m_{xf}(\omega_0) = &-\rho\pi b^2 \left(ba\omega_0^2 \frac{h_1}{2\mathrm{i}} \mathrm{e}^{\mathrm{i}\phi_h} - \left(b^2 \left(\frac{1}{8} + a^2 \right) \omega_0^2 - b\left(\frac{1}{2} - a \right) U_\infty \mathrm{i}\omega_0 \right) \frac{\alpha_1}{2\mathrm{i}} \mathrm{e}^{\mathrm{i}\phi_\alpha} \right) \\ &+ 2\rho U_\infty \pi b^2 \left(a + \frac{1}{2} \right) \\ &\times \left(\mathrm{i}\omega_0 \frac{h_1}{2\mathrm{i}} \mathrm{e}^{\mathrm{i}\phi_h} + \left(U_\infty + b\left(\frac{1}{2} - a \right) \mathrm{i}\omega_0 \right) \frac{\alpha_1}{2\mathrm{i}} \mathrm{e}^{\mathrm{i}\phi_\alpha} \right) C(\omega_0) \end{aligned} \tag{3.182}$$

The lift and moment can be written in more compact form as

$$\begin{pmatrix} l(\omega_0) \\ m_{xf}(\omega_0) \end{pmatrix} = \frac{1}{2\mathrm{i}} \boldsymbol{Q}(\omega_0) \begin{pmatrix} h_1 \mathrm{e}^{\mathrm{i}\phi_h} \\ \alpha_1 \mathrm{e}^{\mathrm{i}\phi_\alpha} \end{pmatrix} \tag{3.183}$$

where

$$\boldsymbol{Q}(\omega_0) = -\rho \boldsymbol{B}\omega_0^2 + \rho U_\infty \left(\boldsymbol{D}_1 + 2\boldsymbol{D}_2 C(\omega_0) \right) \mathrm{i}\omega + 2\rho U_\infty^2 \boldsymbol{F}_1 C(\omega_0)$$

and matrices $\boldsymbol{B}$, $\boldsymbol{D}_1$, $\boldsymbol{D}_2$ and $\boldsymbol{F}_1$ are defined after Eq. (3.126).

The drag is obtained from Eq. (3.105)

$$d(t) = -\rho\pi S(t)^2 - \alpha(t) l(t) \tag{3.184}$$

where $S(t)$ is given by Eq. (3.107)

$$S(t) = \frac{\sqrt{2b}}{2} \left(2\left(U_\infty \alpha(t) + \dot{h}(t) + b\left(\frac{1}{2} - a \right) \dot{\alpha}(t) \right) \frac{\int_b^\infty \frac{X_0}{\sqrt{X_0^2 - b^2}} \bar{\Gamma}(X_0, t) \mathrm{d}X_0}{\int_b^\infty \frac{\sqrt{X_0 + b}}{\sqrt{X_0 - b}} \bar{\Gamma}(X_0, t) \mathrm{d}X_0} - b\dot{\alpha}(t) \right)$$

Taking the Fourier transform of this expression and recognising that the ratio of integrals becomes Theodorsen's function leads to

$$S(\omega) = \frac{\sqrt{2b}}{2} \left(2\left(\mathrm{i}\omega h(\omega) + \left(U_\infty + b\left(\frac{1}{2} - a \right) \mathrm{i}\omega \right) \alpha(\omega) \right) C(\omega_0) - b\mathrm{i}\omega\alpha(\omega) \right)$$

Substituting for $h(\omega)$, $\alpha(\omega)$ from Eqs. (3.143) and (3.144), we obtain

$$S(\omega) = S_0\delta(\omega) + S(\omega_0)\delta(\omega - \omega_0) + S^*(\omega_0)\delta(\omega + \omega_0) \tag{3.185}$$

where

$$S_0 = \sqrt{2b}U_\infty\alpha_0 \tag{3.186}$$

$$\begin{aligned} S(\omega_0) = \frac{\sqrt{2b}}{2}\left(2\left(\mathrm{i}\omega_0\frac{h_1}{2\mathrm{i}}\mathrm{e}^{\mathrm{i}\phi_h} + \left(U_\infty + b\left(\frac{1}{2} - a\right)\mathrm{i}\omega_0\right)\frac{\alpha_1}{2\mathrm{i}}\mathrm{e}^{\mathrm{i}\phi_\alpha}\right)C(\omega_0)\right. \\ \left. -b\mathrm{i}\omega_0\frac{\alpha_1}{2\mathrm{i}}\mathrm{e}^{\mathrm{i}\phi_\alpha}\right) \end{aligned} \tag{3.187}$$

The drag of Eq. (3.184) must be Fourier transformed, but it is a non-linear expression. Multiplication in the time domain corresponds to convolution in the frequency domain, such that

$$d(\omega) = -\rho\pi S(\omega) * S(\omega) - \alpha(\omega) * l(\omega) \tag{3.188}$$

where the operator symbol $*$ denotes convolution. The convolution of two functions $f(\omega)$ and $g(\omega)$ is defined as

$$f(\omega) * g(\omega) = \int_{-\infty}^{\infty} f(\Omega)g(\omega - \Omega)\mathrm{d}\Omega = \int_{-\infty}^{\infty} f(\omega - \Omega)g(\Omega)\mathrm{d}\Omega$$

where Ω is a variable of integration. The convolution integral is simplified for the case where the two functions $f(\omega)$ and $g(\omega)$ are sinusoidal with the same frequency, thanks to the sifting property of the Kronecker delta (3.146). If we write these functions as

$$\begin{aligned} f(\omega) &= f_0\delta(\omega) + f_1\delta(\omega - \omega_0) + f_1^*\delta(\omega + \omega_0) \\ g(\omega) &= g_0\delta(\omega) + g_1\delta(\omega - \omega_0) + g_1^*\delta(\omega + \omega_0) \end{aligned}$$

then the first of the convolution integrals becomes

$$\begin{aligned} f(\omega) * g(\omega) = \int_{-\infty}^{\infty} & \big(f_0\delta(\Omega)g_0\delta(\omega - \Omega) + f_0\delta(\Omega)g_1\delta(\omega - \omega_0 - \Omega) \\ & + f_0\delta(\Omega)g_1^*\delta(\omega + \omega_0 - \Omega) + f_1\delta(\Omega - \omega_0)g_0\delta(\omega - \Omega) \\ & + f_1\delta(\Omega - \omega_0)g_1\delta(\omega - \omega_0 - \Omega) + f_1\delta(\Omega - \omega_0)g_1^*\delta(\omega + \omega_0 - \Omega) \\ & + f_1^*\delta(\Omega + \omega_0)g_0\delta(\omega - \Omega) + f_1^*\delta(\Omega + \omega_0)g_1\delta(\omega - \omega_0 - \Omega) \\ & + f_1^*\delta(\Omega + \omega_0)g_1^*\delta(\omega + \omega_0 - \Omega)\big)\,\mathrm{d}\Omega \end{aligned}$$

Every term in the integrand is equal to zero except at the frequencies Ω at which the first Kronecker delta function in each term is equal to one. Consequently, the integral simplifies to

$$\begin{aligned} f(\omega) * g(\omega) &= f_0g_0\delta(\omega) + f_0g_1\delta(\omega - \omega_0) + f_0g_1^*\delta(\omega + \omega_0) \\ &\quad + f_1g_0\delta(\omega - \omega_0) + f_1g_1\delta(\omega - 2\omega_0) + f_1g_1^*\delta(\omega) \\ &\quad + f_1^*g_0\delta(\omega + \omega_0) + f_1^*g_1\delta(\omega) + f_1^*g_1^*\delta(\omega + 2\omega_0) \\ &= f_1^*g_1^*\delta(\omega + 2\omega_0) + \left(f_0g_1^* + f_1^*g_0\right)\delta(\omega + \omega_0) \\ &\quad + \left(f_0g_0 + f_1g_1^* + f_1^*g_1\right)\delta(\omega) + \left(f_0g_1 + f_1g_0\right)\delta(\omega - \omega_0) \\ &\quad + f_1g_1\delta(\omega - 2\omega_0) \end{aligned} \tag{3.189}$$

so that the convoluted signal has five frequency components, $\omega = -2\omega_0$, $\omega = -\omega_0$, $\omega = 0$, $\omega = \omega_0$ and $\omega = 2\omega_0$. Note that the frequency components at $-2\omega_0$ and $-\omega_0$ are the complex conjugates of those at $2\omega_0$ and ω_0, respectively.

In practice, we can calculate the convolution of signals with few frequency components using the convolution of vectors. We write function $S(\omega)$ as

$$S(\omega) = S_0\delta(\omega) + S(\omega_0)\delta(\omega - \omega_0) + S^*(\omega_0)\delta(\omega + \omega_0)$$

which can be denoted by means of the following vector:

$$S(\omega) = \left(S^*(\omega_0),\ S_0,\ S(\omega_0)\right)$$

where the first element corresponds to $\omega = -\omega_0$, the second to $\omega = 0$ and the third to $\omega = \omega_0$. Consequently, the convolution $S(\omega) * S(\omega)$ becomes a convolution of vectors

$$\begin{aligned} S(\omega) &= \left(S^*(\omega_0),\ S_0,\ S(\omega_0)\right) * \left(S^*(\omega_0),\ S_0,\ S(\omega_0)\right) \\ &= \left(S^*(\omega_0)^2,\ 2S^*(\omega_0)S_0,\ S_0^2 + 2S^*(\omega_0)S(\omega_0),\ 2S_0S(\omega_0),\ S(\omega_0)^2\right) \end{aligned}$$

The resulting vector has five elements, corresponding to $-2\omega_0$, $-\omega_0$, 0, ω_0 and $2\omega_0$. Note that the convolution of vectors is equivalent to the multiplication of polynomials. Similarly, the convolution $\alpha(\omega) * l(\omega)$ becomes

$$\begin{aligned} \alpha(\omega) * l(\omega) &= \left(\alpha^*(\omega_0),\ \alpha_0,\ \alpha(\omega_0)\right) * \left(l^*(\omega_0),\ l_0,\ l(\omega_0)\right) \\ &= \begin{pmatrix} \alpha^*(\omega_0)l^*(\omega_0) \\ \alpha^*(\omega_0)l_0 + \alpha_0 l^*(\omega_0) \\ \alpha_0 l_0 + \alpha^*(\omega_0)l(\omega) + \alpha(\omega_0)l^*(\omega_0) \\ \alpha_0 l(\omega_0) + \alpha(\omega_0)l_0 \\ \alpha(\omega_0)l(\omega_0) \end{pmatrix} \end{aligned}$$

where $\alpha(\omega_0) = \alpha_1 e^{i\phi_\alpha}/2i$, $\alpha^*(\omega_0) = -\alpha_1 e^{-i\phi_\alpha}/2i$ from Eq. (3.144). Substituting these two results in Eq. (3.188) leads to

$$\begin{aligned} \begin{pmatrix} d(-2\omega_0) \\ d(-\omega_0) \\ d_0 \\ d(\omega_0) \\ d(2\omega_0) \end{pmatrix} &= -\rho\pi \begin{pmatrix} S^*(\omega_0) \\ S_0 \\ S(\omega_0) \end{pmatrix} * \begin{pmatrix} S^*(\omega_0) \\ S_0 \\ S(\omega_0) \end{pmatrix} - \begin{pmatrix} \alpha^*(\omega_0) \\ \alpha_0 \\ \alpha(\omega_0) \end{pmatrix} * \begin{pmatrix} l^*(\omega_0) \\ l_0 \\ l(\omega_0) \end{pmatrix} \\ &= - \begin{pmatrix} \rho\pi S^*(\omega_0)^2 + \alpha^*(\omega_0)l^*(\omega_0) \\ 2\rho\pi S^*(\omega_0)S_0 + \alpha^*(\omega_0)l_0 + \alpha_0 l^*(\omega_0) \\ \rho\pi\left(S_0^2 + 2S^*(\omega_0)S(\omega_0)\right) + \alpha_0 l_0 + \alpha^*(\omega_0)l(\omega_0) + \alpha(\omega_0)l^*(\omega_0) \\ 2\rho\pi S_0 S(\omega_0) + \alpha_0 l(\omega_0) + \alpha(\omega_0)l_0 \\ \rho\pi S(\omega_0)^2 + \alpha(\omega_0)l(\omega_0) \end{pmatrix} \end{aligned} \quad (3.190)$$

where $d(0)$, $d(\omega_0)$, $d(2\omega_0)$ and their complex conjugates are the five frequency components of the drag force.

Theodorsen theory yields frequency responses, that is the frequency components of the lift, pitching moment and drag responses to sinusoidal motion. In order to calculate the corresponding time responses, we must apply the inverse Fourier transform to these frequency components. The following example demonstrates how to do this.

Example 3.15 *Calculate the aerodynamic load responses of a NACA 0012 airfoil undergoing sinusoidal pitching oscillations around the quarter chord and compare to experimental results.*

McAlister et al. (1982) published a series of experimental measurements of the pressure and aerodynamic load responses of several oscillating airfoils. The vast majority of test cases featured very high instantaneous pitch angles and, therefore, unsteady flow separation. The test case we will model here was a pitching oscillation of a NACA 0012 airfoil of chord $c = 0.61$ *m around its quarter chord with mean* $\alpha_0 = 2.64°$, *amplitude* $\alpha_1 = 10.16°$ *and reduced frequency* $k_0 = 0.099$. *The Reynolds number was* 3.9×10^6 *and the free stream Mach number was 0.3; some compressibility effects are expected at this Mach number, but they are not very strong; therefore, we can use an incompressible method to approximate this test case. Furthermore, the NACA 0012 may be symmetric, but it is also 12% thick, while the flat plate modelled by Theodorsen's and Wagner's theories is infinitely thin. By comparing the predictions to the experimental data, we can observe the effect of the simplifying assumptions behind the methods presented in this chapter.*

We start by setting up the problem. The half-chord is $b = c/2$ *and the pitch axis lies at* $x_f = 0.25c$, *so that* $a = -1/2$. *The oscillation frequency was* $f_0 = 5.35$ *Hz, so that* $\omega_0 = 2\pi f_0$, *and we can estimate the flow speed from*

$$U_\infty = \omega_0 b / k_0$$

The experimental results are given in the form of non-dimensional coefficients so that choosing the right chord length and flow speed is not essential. We set the pitch mean and amplitude to the values given above and the plunge amplitude h_1 *to zero. We also set the pitch phase to* $\phi_\alpha = 0°$ *for compatibility with the experimental time responses.*

Next, we calculate Theodorsen's function from R. T. Jones' approximation of Eq. (3.170), using $\Psi_1 = 0.165$, $\Psi_2 = 0.334$, $\varepsilon_1 = 0.0455$ *and* $\varepsilon_2 = 0.3$. *Then we evaluate*

- *the downwash at the 3/4 chord point,* $w_{3/4}(\omega_0)$ *from Eq. (3.161),*
- *the two-frequency components of the lift,* l_0 *and* $l(\omega_0)$ *from Eqs. (3.176) and (3.177),*
- *the two-frequency components of the pitching moment,* m_{xf_0} *and* $m_{xf}(\omega_0)$ *from Eqs. (3.181) and (3.182) and*
- *the two-frequency components of the leading edge suction function,* S_0 *and* $S(\omega_0)$ *from Eqs. (3.186) and (3.187).*

Finally, we calculate the five components of the drag from Eq. (3.190). Instead of coding explicitly the result of the convolutions, we use Matlab's built-in `conv.m` *function.*

At this stage, the aerodynamic calculation process is complete, but the aerodynamic loads come in the form of frequency components. In order to compare them to the experimental results, we need to apply the inverse Fourier Transform to obtain time responses. The pitch angle, lift and pitching moment variations with time are given by

$$\alpha(t) = \alpha_0 + \alpha(\omega_0)e^{i\omega t} + \alpha(\omega_0)^* e^{-i\omega t}$$
$$l(t) = l_0 + l(\omega_0)e^{i\omega t} + l(\omega_0)^* e^{-i\omega t}$$
$$m_{xf}(t) = m_{xf_0} + m_{xf}(\omega_0)e^{i\omega t} + m_{xf}(\omega_0)^* e^{-i\omega t}$$

for any value of t. The drag has five frequency components so that its inverse Fourier transform is given by

$$d(t) = d_0 + d(\omega_0)e^{i\omega t} + d(\omega_0)^* e^{-i\omega t} + d(2\omega_0)e^{i\omega t} + d(2\omega_0)^* e^{-i\omega t}$$

As all the non-zero frequency terms come in complex conjugate pairs, the time responses are real.

Figure 3.30 plots the three aerodynamic load responses obtained from Theodorsen theory over a single cycle of oscillation against the phase ωt in degrees and compares them to the experimental measurements. It should be noted that the experimental loads were calculated by integrating the measured pressure distributions, so that skin friction is not included in the measured drag values. It can be seen that the amplitude of the lift response predicted by Theodorsen's method is lower than the experimental measurement due to the fact that the former neglects compressibility effects. As a consequence, the predicted drag amplitude is underestimated, especially during the second half of the upstroke ($0° < \omega t < 100°$). Furthermore, the predicted pitching moment signal is around 45° out of phase with the experimental measurements and its amplitude is lower. Nevertheless, given that both compressibility and airfoil

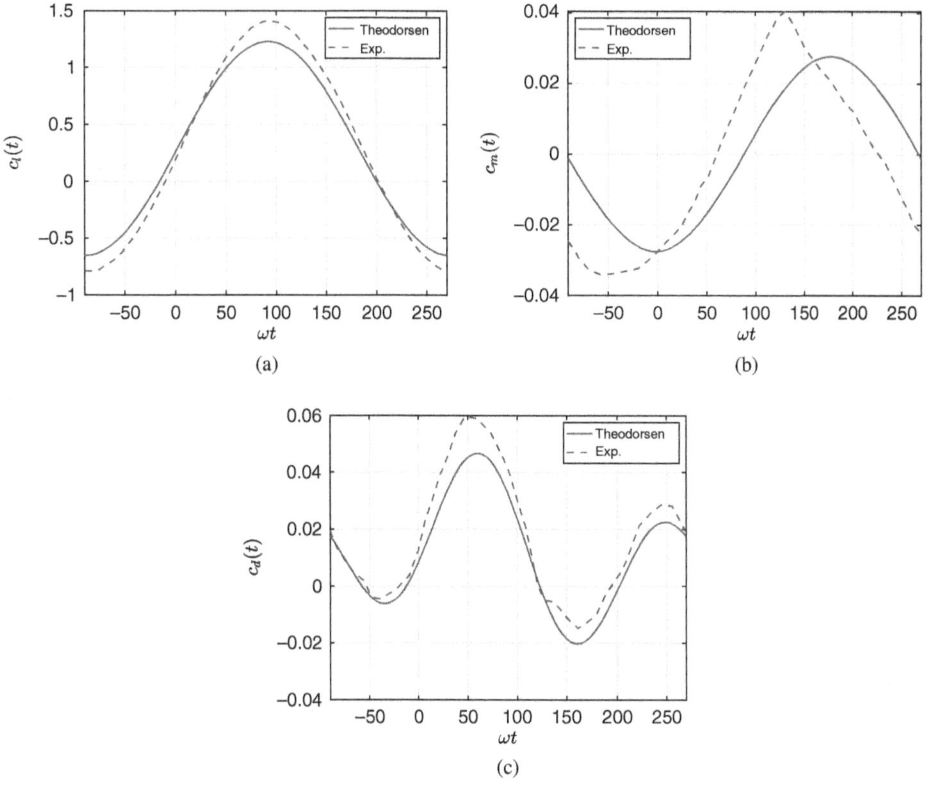

Figure 3.30 Aerodynamic load response predictions for sinusoidally pitching 2D NACA 0012 airfoil and comparison to experimental results by McAlister et al. (1982). (a) $c_l(t)$, (b) $c_{m_{xf}}(t)$ and (c) $c_d(t)$. Source: Adapted from McAlister et al. (1982).

thickness were ignored, the agreement between the predicted and measured load responses is not unacceptable.

Some more information about the accuracy of the modelling can be obtained from Figure 3.31 which plots the same load responses against the instantaneous pitch angle. The figures also plot the steady and unsteady experimental measurements. Figure 3.31a shows that both the slope and the thickness of the lift ellipse have been underestimated by Theodorsen theory. This underestimation is mainly due to the fact that the thickness of the NACA 0012 airfoil has been ignored. The force due to the pressure at any point on a thick airfoil points in the direction normal to the surface, which can be very different to the direction normal to the chord, particularly around the leading edge. In contrast, Theodorsen theory only calculates force due to pressure difference across the upper and lower surface, and this force is always normal to the chord.

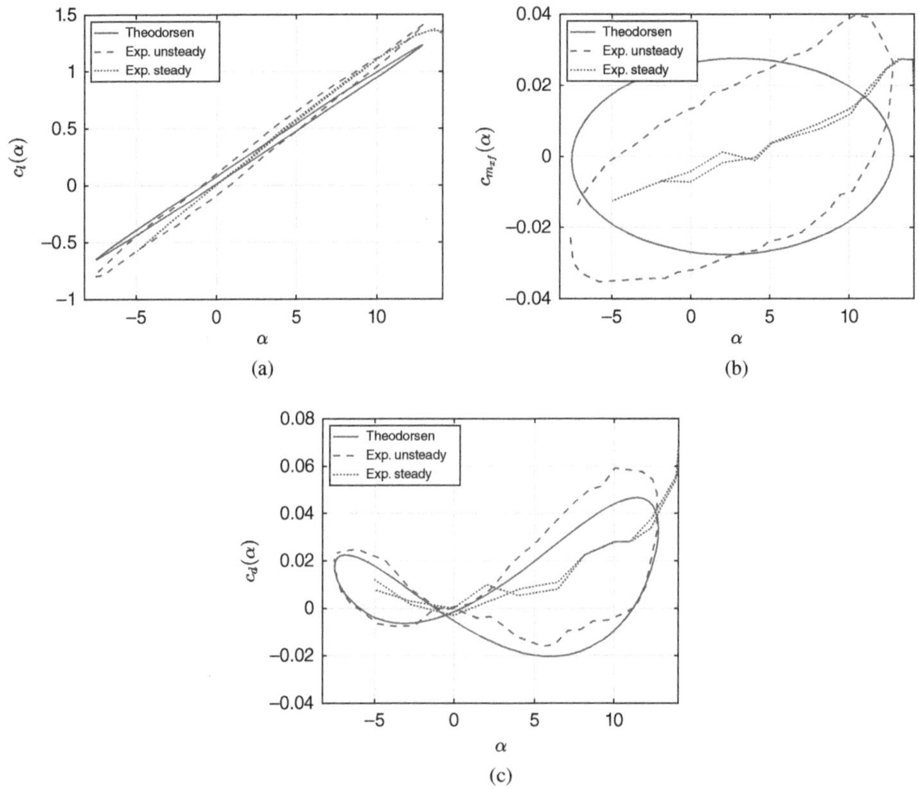

Figure 3.31 Predictions of aerodynamic load variation with instantaneous pitch angle for sinusoidally pitching NACA 0012 airfoil. Experimental results by McAlister et al. (1982). (a) $c_l(\alpha)$, (b) $c_{m_{xf}}(\alpha)$ and (c) $c_d(\alpha)$. Source: Adapted from McAlister et al. (1982), credit: NASA.

Figure 3.31b shows that the ellipse predicted by Theodorsen theory has its major axis aligned with the α axis, while the major axis of the experimental ellipse is inclined to this axis. This discrepancy is due to the lack of thickness and compressibility modelling. Both these phenomena have an effect on the position of the steady aerodynamic centre, such that the *steady experimental pitching moment curve has a positive slope. In the Theodorsen model, the steady aerodynamic centre lies exactly on the quarter chord, resulting in a horizontal pitching moment ellipse. Finally, Figure 3.31c shows that the drag prediction is quite accurate for low values of α. At the top of the upstroke, the drag is under-predicted; this phenomenon is not only due to the under-prediction of the lift but also due to moderate flow separation. It should be noted that the experimental drag and pitching moment curves are slightly distorted for α values between 8° and 12°, suggesting that some flow separation must be occurring.*

We will finish the discussion by stating that Wagner's time domain model yields identical results to Theodorsen's frequency domain approach. The reader is invited to repeat this example using the implementation of Wagner's model in Example 3.12 and to verify this statement. The present example is solved by Matlab code `pitchplunge_Theodorsen.m`.

In this example, we used convolution in the frequency domain in order to calculate the non-linear drag and then we applied the inverse Fourier transform in order to compare the resulting drag time response to the experimental measurements. We could have also applied the inverse Fourier transform to $\alpha(\omega)$, $l(\omega)$ and $S(\omega)$ and then used the time-domain drag equation (3.184) to calculate the drag by multiplication instead of convolution. The latter approach is better suited to signals with rich frequency content, such as those obtained from aeroelastic responses of complete wing or aircraft models, see for example Reyes et al. (2019). For simple harmonic response to simple harmonic excitation, the present convolution technique is more computationally efficient.

3.5.3 General Motion

Wagner's model was developed for impulsively started flows but can be made to calculate the unsteady aerodynamic loads acting on a pitching and plunging flat plate for any general motion using the convolution integral of Eq. (3.113). Theodorsen's model was developed for sinusoidal motion, but it can also be used to calculate the aerodynamic loads due to general motion, described as the superposition of many sinusoids at different frequencies.

As an example, we will use the aeroelastic response of a 2D flat plate wing with pitch and plunge degrees of freedom. This is the only aeroelastic test case that will be presented in this book. The model is known as the typical aeroelastic section and is shown in Figure 3.32. The flat plate of mass m and chord c has uniform thickness and density distributions, such that its centre of gravity lies at the half-chord point, $c/2$. The wing's motion in the h and α directions is restrained by an extension spring of stiffness K_h and a rotational spring of stiffness K_α, respectively. Both springs are attached to the wing at the pitch axis location, x_f. The equations of motion of the system are given by (see for example (Fung 1993))

$$\begin{pmatrix} m & S \\ S & I_\alpha \end{pmatrix} \begin{pmatrix} \ddot{h}(t) \\ \ddot{\alpha}(t) \end{pmatrix} + \begin{pmatrix} K_h & 0 \\ 0 & K_\alpha \end{pmatrix} \begin{pmatrix} h(t) \\ \alpha(t) \end{pmatrix} = \begin{pmatrix} l(t) \\ m_{xf}(t) \end{pmatrix} \tag{3.191}$$

where

$$S = m\left(\frac{c}{2} - x_f\right), \quad I_\alpha = \frac{m}{3}\left(c^2 - 3cx_f + 3x_f^2\right)$$

are the static imbalance and moment of inertia around the pitch axis, respectively. The aerodynamic lift and moment around the pitch axis are denoted by $l(t)$ and $m_{xf}(t)$, as usual.

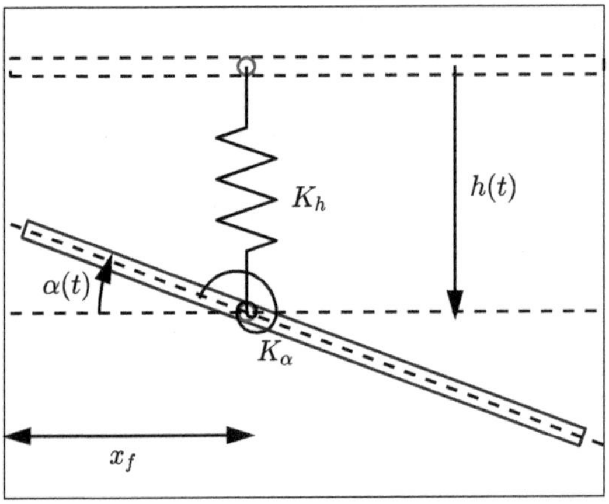

Figure 3.32 Typical aeroelastic section with pitch and plunge degrees of freedom.

If we impose initial conditions, $\dot{h}(0) = \dot{h}_0$, $\dot{\alpha}(0) = \dot{\alpha}_0$, $h(0) = h_0$ and $\alpha(0) = \alpha_0$ and then release the system, it will vibrate freely. The objective is to determine the time responses $h(t)$ and $\alpha(t)$ for all times of interest. We can do this in the time domain using Wagner's model for the aerodynamic loads, but we can also do it in the frequency domain. We use the one-sided Fourier transform of a general function $f(t)$, that is

$$f(\omega) = \int_0^\infty f(t)e^{-i\omega t}\,dt$$

since there is no motion for times $t < 0$. Then, the Fourier transform of $\dot{f}(t)$ can be obtained using integration by parts as

$$\dot{f}(\omega) = \int_0^\infty \dot{f}(t)e^{-i\omega t}\,dt = \left(f(t)e^{-i\omega t}\right)_0^\infty + i\omega\int_0^\infty f(t)e^{-i\omega t}\,dt = -f(0) + i\omega f(\omega) \quad (3.192)$$

Similarly, the Fourier transform of $\ddot{f}(t)$ is given by

$$\ddot{f}(\omega) = \int_0^\infty \ddot{f}(t)e^{-i\omega t}\,dt = -\dot{f}(0) - i\omega f(0) - \omega^2 f(\omega) \quad (3.193)$$

Applying this version of the Fourier transform to Eq. (3.191), we obtain

$$\boldsymbol{A}\left(-\dot{\mathbf{q}}_0 - i\omega\mathbf{q}_0 - \omega^2\mathbf{q}(\omega)\right) + \boldsymbol{E}\mathbf{q}(\omega) = \begin{pmatrix} l(\omega) \\ m_{xf}(\omega) \end{pmatrix} \quad (3.194)$$

where

$$\boldsymbol{A} = \begin{pmatrix} m & S \\ S & I_\alpha \end{pmatrix}, \quad \boldsymbol{E} = \begin{pmatrix} K_h & 0 \\ 0 & K_\alpha \end{pmatrix}, \quad \mathbf{q}(\omega) = \begin{pmatrix} h(\omega) \\ \alpha(\omega) \end{pmatrix}$$

The Fourier transform of the non-circulatory lift, $l_\sigma(\omega)$, can be obtained in the same way from Eq. (3.63), such that

$$\begin{aligned} l_\sigma(\omega) = &-\rho\pi b^2\left(-\omega^2 h(\omega) + ba\omega^2\alpha(\omega) + U_\infty i\omega\alpha(\omega)\right) \\ &+ \rho\pi b^2\left(\dot{h}_0 + i\omega h_0 - ba(\dot{\alpha}_0 + i\omega\alpha_0) + U_\infty\alpha_0\right) \end{aligned} \quad (3.195)$$

The Fourier transform of the downwash at the 3/4 chord is obtained from Eq. (3.112)

$$w_{3/4}(\omega) = U_\infty\alpha(\omega) + i\omega h(\omega) + b\left(\frac{1}{2} - a\right)i\omega\alpha(\omega) - h_0 - b\left(\frac{1}{2} - a\right)\alpha_0$$

The circulatory lift can be obtained by generalising Eq. (3.160) for all frequencies ω,

$$\begin{aligned} l_\Gamma(\omega) &= -2\rho U_\infty b\pi w_{3/4}(\omega)C(\omega) \\ &= -2\rho U_\infty b\pi \left(U_\infty \alpha(\omega) + i\omega h(\omega) + b\left(\frac{1}{2} - a\right) i\omega\alpha(\omega)\right) C(\omega) \\ &+ 2\rho U_\infty b\pi \left(h_0 + b\left(\frac{1}{2} - a\right)\alpha_0\right) C(\omega) \end{aligned} \tag{3.196}$$

The frequency response of the pitching moment is obtained by taking the Fourier transform of Eq. (3.178), such that

$$\begin{aligned} m_{xf}(\omega) &= -\rho\pi b^2 \left(ba\left(\dot{h}_0 + i\omega h_0\right) - b^2\left(\frac{1}{8} + a^2\right) ba\left(\dot{\alpha}_0 + i\omega\alpha_0\right) - b\left(\frac{1}{2} - a\right) U_\infty \alpha_0\right) \\ &- \rho\pi b^2 \left(ba\omega^2 h(\omega) - b^2\left(\frac{1}{8} + a^2\right)\omega^2\alpha(\omega) + b\left(\frac{1}{2} - a\right) U_\infty i\omega\alpha(\omega)\right) \\ &- b\left(a + \frac{1}{2}\right) l_\Gamma(\omega) \end{aligned} \tag{3.197}$$

We can now substitute Eqs. (3.195)–(3.197) in the aeroelastic equations of motion (3.194) and group the terms containing $\mathbf{q}(\omega)$ to obtain

$$\left(-\omega^2 \boldsymbol{A} + \boldsymbol{E} - \rho U_\infty^2 \boldsymbol{Q}(\omega)\right)\mathbf{q}(\omega) = (\boldsymbol{A} + \rho\boldsymbol{B})\left(\dot{\mathbf{q}}_0 + i\omega\mathbf{q}_0\right) - \rho U_\infty \left(\mathbf{D}_1 + 2\mathbf{D}_2 C(\omega)\right)\mathbf{q}_0 \tag{3.198}$$

where matrices $\boldsymbol{B}$, $\boldsymbol{D}_1$, $\boldsymbol{D}_2$ are given just after Eq. (3.126) and $\boldsymbol{Q}(\omega)$ is defined just after Eq. (3.183). Equation (3.198) can be solved at each frequency of interest in order to obtain $\mathbf{q}(\omega)$, the frequency response of the system.

Example 3.16 *Calculate the pitch and plunge response time histories of a typical aeroelastic section.*

The aeroelastic model looks identical to that of Figure 3.32. The wing is a flat plate of chord $c = 0.25$ m, thickness $d = 0.02$ m and is made of aluminium. We calculate its mass per unit area from $m = \rho_a cd$, where $\rho_a = 2700$ kg/m^3 is the density of aluminium. The pitch axis is selected to lie at $x_f = 0.4c$ and the static imbalance S and moment of inertia, I_α, of the plate are calculated from the expressions just after Eq. (3.191). Finally, we select the spring stiffnesses such that $K_h = m(4\pi)^2$, $K_\alpha = I_\alpha(16\pi)^2$, so that the uncoupled wind-off natural frequencies in plunge and pitch are 2 and 8 Hz, respectively.

We calculate the matrices A, E from the expressions just after Eq. (3.194) and $\mathbf{B}$, $\mathbf{D}_1$, $\mathbf{D}_2$ from the expressions just after (3.126). Next, we need to define the frequencies of interest; one way to do this is by defining the times at which we would like to calculate the time response of the system. We select time instances $t_i = i\Delta t$ for $i = 0, \dots, n-1$. For Eqs. (3.192) and (3.193) to be valid, $t_n = (n-1)\Delta t$ must be high enough that the upper limit of integration becomes negligible. Consequently, we choose $\Delta t = 0.01$ seconds and $n = 8192$, so that $t_n = 81.91$ seconds, which is an adequate value for this system.

Using the principles of the discrete time Fourier transform, the frequencies corresponding to the time instances t_i are given by

$$\omega_i = 2\pi i \frac{1}{n\Delta t}$$

where $i = 0, \dots, n/2$. For each one of these frequencies, we calculate $C(k_i)$ and then solve Eq. (3.198) for $\mathbf{q}(\omega_i)$. At this stage, the solution procedure is complete, but the response of

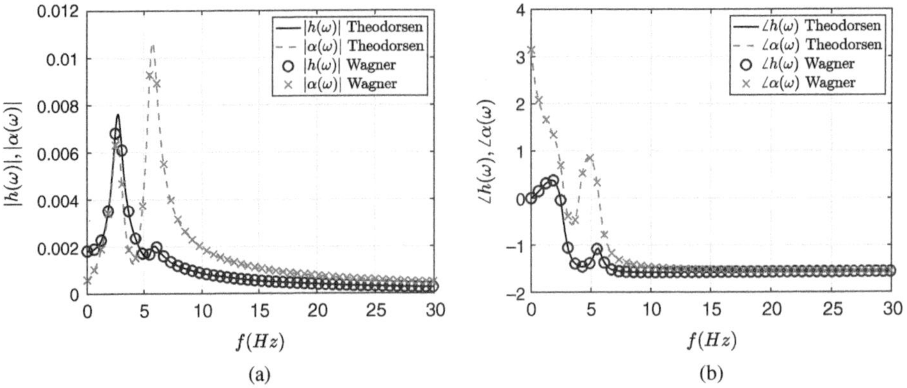

Figure 3.33 Magnitude and phase of frequency response of typical aeroelastic section at $U_\infty = 45$ m/s. (a) Magnitude and (b) phase.

the system has been calculated in the frequency domain. Figure 3.33 plots the magnitude and phase of $h(\omega)$ and $\alpha(\omega)$ against frequency $f = \omega/2\pi$ for $U_\infty = 45$ m/s and initial conditions $\dot{h}_0 = \dot{\alpha}_0 = 0$, $h_0 = 0.2c$, $\alpha_0 = 5°$. We have also calculated the time responses using Wagner function aerodynamics, by substituting Eq. (3.126) into the time domain equations of motion (3.191), solving them in the time domain and then applying the discrete Fourier transform to the resulting $h(t)$ and $\alpha(t)$. It can be seen that the Theodorsen and Wagner solutions are nearly identical. The magnitude plots of Figure 3.33a have two peaks each, corresponding to the two modes of vibration of the system at this airspeed. The frequencies at which the peaks occur are the natural frequencies of the aeroelastic system at this airspeed.

In order to obtain the time response from the Theodorsen model, we need to apply the discrete time inverse Fourier transform to $\mathbf{q}(\omega_i)$. We have only calculated the frequency response for positive frequencies, but we do not need to also calculate it for negative frequencies; we impose that $\mathbf{q}(-\omega_i) = \mathbf{q}^*(\omega_i)$, where the superscript * denotes the complex conjugate. In order to use Matlab's built-in inverse Fourier transform function `ifft.m`, we need to create the vector

$$\bar{\mathbf{q}}(\omega) = \left(\mathbf{q}(\omega_0)\ \mathbf{q}(\omega_1)\ \dots\ \mathbf{q}(\omega_{n-2})\ \Re(\mathbf{q}(\omega_{n-1}))\ \mathbf{q}^*(\omega_{n-2})\ \dots\ \mathbf{q}^*(\omega_1)\right)$$

whose total length is n. We take the real part of $\mathbf{q}(\omega_{n-1})$ to ensure that the output of the `ifft.m` function will have zero imaginary part. The resulting vector $\bar{\mathbf{q}}(t_i)$ will be of the form

$$\bar{\mathbf{q}}(t_i) = \Delta t \begin{pmatrix} h(t_i) \\ \alpha(t_i) \end{pmatrix} \quad \text{for } i = 1, \dots, n-1, \quad \text{while } \bar{\mathbf{q}}(t_1) = \frac{\Delta t}{2} \begin{pmatrix} h(t_1) \\ \alpha(t_1) \end{pmatrix}$$

Figure 3.34a plots the time responses $h(t)$ and $\alpha(t)$ at $U_\infty = 45$ m/s, as calculated using both the Theodorsen and Wagner models. It can be seen that both signals are sinusoidally decaying functions; despite the fact that Theodorsen theory was developed for periodic sinusoidal motion, it can clearly be used to predict the aerodynamic loads for non-periodic motion. Again, the predictions of the two models are nearly identical, reinforcing the earlier statement that they constitute a Fourier transform pair. Figure 3.34b plots the system's time response at $U_\infty = 50.46$ m/s. At this higher airspeed, the signals change in nature; the steady-state response is sinusoidal. This change is due to the well-known phenomenon of flutter, whereby the pitch and plunge motions combine to extract energy from the free stream and to cause self-excited vibrations. The interested reader can find much more information on flutter and aeroelasticity in general in several textbooks, such as Bisplinghoff et al. (1996), Dowell (2004), Fung

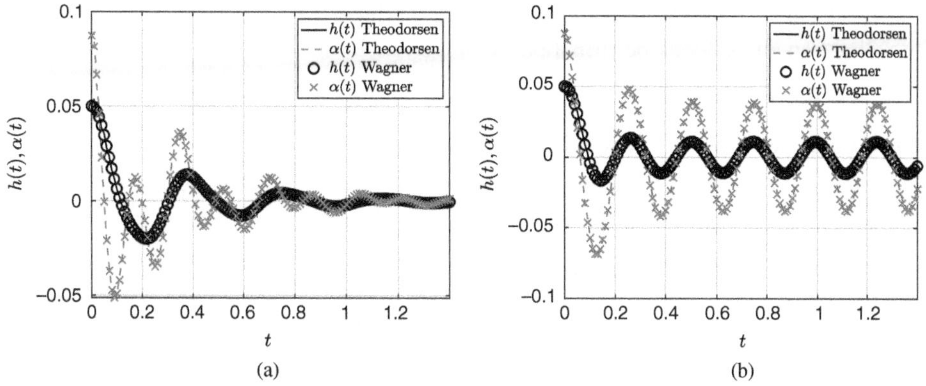

Figure 3.34 Time response of typical aeroelastic section at two airspeeds. (a) $U_\infty = 45$ m/s and (b) $U_\infty = 50.46$ m/s.

(1993), and Wright and Cooper (2015). Note that, at the flutter speed, the agreement between the Theodorsen and Wagner responses is not as good as at lower airspeeds. This is due the fact that the steady state response does not decay to nearly zero and therefore, both the Fourier and inverse Fourier transforms suffer from truncation error. This example is solved by Matlab code `dof2_Theodorsen.m`.

3.6 Finite State Theory

Up to now, the 2D flat plate has always been modelled by sources on the surface and vortices in the wake. Now, we will use a different modelling approach, whereby only vortices are used. Several authors have followed this route, but the description here is based on Peters (2008). The airfoil is represented as a vortex sheet with strength $\gamma_b(x,t)$, known as the bound vorticity, extending from $x = -b$ to $x = b$. There is an additional wake vortex sheet, with strength $\gamma_w(x,t)$, extending from $x = b$ to $x = \infty$. The strength of a vortex sheet (bound or wake) is given by

$$\gamma(x,t) = u_u(x,t) - u_l(x,t) \tag{3.199}$$

where $u_u(x,t)$ and $u_l(x,t)$ are the tangent flow velocities on the upper and lower surfaces of the sheet, respectively (see for example (Kuethe and Chow 1986)). Furthermore, the pressure difference across the vortex sheet is simply

$$\Delta p(x,t) = p_l(x,t) - p_u(x,t)$$

The 2D incompressible unsteady Euler equations are given by

$$\begin{aligned} \frac{\partial u}{\partial x} + \frac{\partial v}{\partial y} &= 0 \\ \frac{\partial u}{\partial t} + u\frac{\partial u}{\partial x} + v\frac{\partial u}{\partial y} &= -\frac{1}{\rho}\frac{\partial p}{\partial x} \\ \frac{\partial v}{\partial t} + u\frac{\partial v}{\partial x} + v\frac{\partial v}{\partial y} &= -\frac{1}{\rho}\frac{\partial p}{\partial y} \end{aligned} \tag{3.200}$$

They can be linearised by writing u and v as sums of the constant free stream U_∞ acting in the x direction and velocity perturbations u' and v', such that

$$u = U_\infty + u', \quad v = v'$$

Substituting back into the Euler equations, we obtain

$$\begin{aligned}\frac{\partial(U_\infty + u')}{\partial x} + \frac{\partial v'}{\partial y} &= 0\\ \frac{\partial(U_\infty + u')}{\partial t} + (U_\infty + u')\frac{\partial(U_\infty + u')}{\partial x} + v'\frac{\partial(U_\infty + u')}{\partial y} &= -\frac{1}{\rho}\frac{\partial p}{\partial x}\\ \frac{\partial v'}{\partial t} + (U_\infty + u')\frac{\partial v'}{\partial x} + v'\frac{\partial v'}{\partial y} &= -\frac{1}{\rho}\frac{\partial p}{\partial y}\end{aligned}$$

Neglecting all products of perturbations, the linearised Euler equations become

$$\begin{aligned}&\frac{\partial u'}{\partial x} + \frac{\partial v'}{\partial y} = 0\\ &\frac{\partial u'}{\partial t} + U_\infty\frac{\partial u'}{\partial x} = -\frac{1}{\rho}\frac{\partial p}{\partial x}\\ &\frac{\partial v'}{\partial t} + U_\infty\frac{\partial v'}{\partial x} = -\frac{1}{\rho}\frac{\partial p}{\partial y}\end{aligned} \tag{3.201}$$

Writing the x-momentum equation for the upper and lower surfaces leads to

$$\begin{aligned}&\frac{\partial u'_u}{\partial t} + U_\infty\frac{\partial u'_u}{\partial x} = -\frac{1}{\rho}\frac{\partial p_u}{\partial x}\\ &\frac{\partial u'_l}{\partial t} + U_\infty\frac{\partial u'_l}{\partial x} = -\frac{1}{\rho}\frac{\partial p_l}{\partial x}\end{aligned}$$

Subtracting the lower from the upper equation[2] and substituting from Eq. (3.199) gives

$$\frac{\partial \gamma}{\partial t} + U_\infty\frac{\partial \gamma}{\partial x} = \frac{1}{\rho}\frac{\partial \Delta p}{\partial x} \tag{3.202}$$

The pressure difference accross the surface is obtained by integrating this equation using the integration variable x_0 that take values from $-\infty$ to x, resulting in

$$\Delta p(x, t) = \rho\frac{\partial}{\partial t}\int_{-\infty}^{x} \gamma(x_0, t)\,\mathrm{d}x_0 + \rho U_\infty \gamma(x, t) \tag{3.203}$$

which is an unsteady version of the Kutta–Joukowski theorem; its steady version was presented in Eq. (3.22). The airfoil lies from $x = -b$ to $x = +b$; the pressure difference across it is then

$$\Delta p_b(x, t) = \rho\frac{\partial}{\partial t}\int_{-b}^{x} \gamma_b(x_0, t)\,\mathrm{d}x_0 + \rho U_\infty \gamma_b(x, t) \tag{3.204}$$

For $x > b$, Eq. (3.203) gives the pressure difference across the wake as

$$\Delta p_w(x, t) = \rho\dot{\Gamma}(t) + \rho\frac{\partial}{\partial t}\int_{b}^{x} \gamma_w(x_0, t)\,\mathrm{d}x_0 + \rho U_\infty \gamma_w(x, t) \tag{3.205}$$

---

2 Note that, up to this point we have always defined the pressure difference as $\Delta p = p_u - p_l$. Here, we invert the definition to $\Delta p = p_l - p_u$ in order to be consistent with the developlusmnents in Peters et al. (1995), Peters (2008) and Johnson (1980). We will therefore also invert the calculation of the lift and moment in order to obtain results consistent with the rest of this chapter.

where

$$\Gamma(t) = \int_{-b}^{b} \gamma_b(x_0, t)\,dx_0 \tag{3.206}$$

is the total bound circulation. Furthermore, the wake is force-free, which means that $\Delta p_w(x,t) = 0$. Equation (3.205) becomes

$$\dot{\Gamma}(t) + \frac{\partial}{\partial t}\int_{b}^{x} \gamma_w(x_0, t)\,dx_0 + U_\infty \gamma_w(x,t) = 0 \tag{3.207}$$

Note also that, since $\Delta p_w = 0$, Eq. (3.202) can be written as

$$\frac{\partial \gamma_w}{\partial t} + U_\infty \frac{\partial \gamma_w}{\partial x} = 0 \tag{3.208}$$

across the wake. This is a first-order linear wave equation with general solution $\gamma_w(x,t) = f(x - U_\infty t)$ for any differentiable function f. In order to determine f, we apply Eq. (3.207) evaluated at $x = b$ as the boundary condition (Peters 2008), such that $\gamma_w(b,t) = -\dot{\Gamma}(t)/U_\infty$.[3] Then, the wake vorticity is given by

$$\gamma_w(x,t) = -\frac{\dot{\Gamma}(t - (x-b)/U_\infty)}{U_\infty} \tag{3.209}$$

Finally, for the bound vorticity, Eq. (3.202) yields

$$\frac{\partial \gamma_b}{\partial t} + U_\infty \frac{\partial \gamma_b}{\partial x} = \frac{1}{\rho}\frac{\partial \Delta p_b}{\partial x} \tag{3.210}$$

From Eq. (A.19), the vertical velocity induced on any point x on the airfoil by a strip of length dx_0 of the vortex sheet lying at x_0 is given by

$$\upsilon(x,t) = -\frac{1}{2\pi}\int_{-b}^{\infty} \frac{\gamma(x_0,t)}{x - x_0}dx_0 \tag{3.211}$$

where $\gamma(x_0)dx_0$ denotes the strength of a strip on either the bound or the wake vortex sheets. This integral can be calculated in two parts, over the airfoil and in the wake, such that

$$\upsilon(x,t) = -\nu(x,t) - \lambda(x,t) \tag{3.212}$$

where

$$\nu(x,t) = \frac{1}{2\pi}\int_{-b}^{b} \frac{\gamma_b(x_0,t)}{x - x_0}dx_0 \tag{3.213}$$

$$\lambda(x,t) = \frac{1}{2\pi}\int_{b}^{\infty} \frac{\gamma_w(x_0,t)}{x - x_0}dx_0 \tag{3.214}$$

Example 3.17 *Write a partial differential equation for the downwash induced by the wake vorticity, λ, in terms of the rate of change of the total bound vorticity $\dot{\Gamma}$.*

A partial differential equation for the downwash induced by the wake vorticity was presented by Peters (2008). Consider the time derivative $\partial\lambda/\partial t$: from definition (3.214),

$$\frac{\partial \lambda}{\partial t} = \frac{1}{2\pi}\int_{b}^{\infty} \frac{1}{x - x_0}\frac{\partial \gamma_w(x_0)}{\partial t}dx_0 \tag{3.215}$$

---

3 This expression for $\gamma_w(b,t)$ can be obtained by setting $x = b$ in Eq. (3.207) and recognising that the integral of $\gamma_w(x)$ from b to b is equal to zero. Alternatively, we can solve Eq. (3.208) for $\partial\gamma_w/\partial t = -U_\infty \partial\gamma_w/\partial x$ and then substitute into Eq. (3.207) and evaluate the integral to obtain $\gamma_w(b,t) = -\dot{\Gamma}(t)/U_\infty$.

Evaluating Eq. (3.208) at x_0 and solving for $\partial\gamma_w(x_0)/\partial t$, we obtain

$$\frac{\partial\gamma_w(x_0)}{\partial t} = -U_\infty \frac{\partial\gamma_w(x_0)}{\partial x_0}$$

Substituting into Eq. (3.215) yields

$$\frac{\partial\lambda}{\partial t} = -\frac{U_\infty}{2\pi}\int_b^\infty \frac{1}{x-x_0}\frac{\partial\gamma_w(x_0)}{\partial x_0}\mathrm{d}x_0$$

Using integration by parts, we can re-write this as

$$\frac{\partial\lambda}{\partial t} = -\frac{U_\infty}{2\pi}\left(\left.\frac{\gamma_w(x_0)}{x-x_0}\right|_b^\infty - \int_b^\infty \frac{\gamma_w(x_0)}{(x-x_0)^2}\mathrm{d}x_0\right)$$

Assuming that $\gamma_w(\infty)$ is finite,

$$\frac{\partial\lambda}{\partial t} = -\frac{U_\infty}{2\pi}\left(-\frac{\gamma_w(b)}{x-b} - \int_b^\infty \frac{\gamma_w(x_0)}{(x-x_0)^2}\mathrm{d}x_0\right) \tag{3.216}$$

The integral in this latest expression can be written as

$$-\int_b^\infty \frac{\gamma_w(x_0)}{(x-x_0)^2}\mathrm{d}x_0 = \frac{\partial}{\partial x}\int_b^\infty \frac{\gamma_w(x_0)}{x-x_0}\mathrm{d}x_0$$

so that Eq. (3.216) becomes

$$\frac{\partial\lambda}{\partial t} = -\frac{U_\infty}{2\pi}\left(-\frac{\gamma_w(b)}{x-b} + \frac{\partial}{\partial x}\int_b^\infty \frac{\gamma_w(x_0)}{x-x_0}\mathrm{d}x_0\right)$$

Substituting from Eqs. (3.209) and (3.214), we obtain

$$\frac{\partial\lambda}{\partial t} = -\frac{1}{2\pi}\frac{\dot{\Gamma}(t)}{x-b} - U_\infty\frac{\partial\lambda}{\partial x}$$

Finally, after re-arranging, we obtain

$$\frac{\partial\lambda}{\partial t} + U_\infty\frac{\partial\lambda}{\partial x} = \frac{1}{2\pi}\frac{\dot{\Gamma}(t)}{b-x} \tag{3.217}$$

This equation will be used in order to derive the finite state model by Peters.

The determination of the bound and wake vorticity is carried out by imposing the impermeability boundary condition, such that the total flow normal to the surface is equal to zero over the airfoil. Then, the lift is calculated by integrating Eq. (3.203) over the chord. The total upwash, $w(x,t)$ is given by Eq. (3.29),

$$w(x,t) = U_\infty\alpha(t) + \dot{h}(t) + \dot{\alpha}(t)(x + b - x_f) \tag{3.218}$$

The impermeability boundary condition becomes $v(x,t) + w(x,t) = 0$, where $v(x,t)$ is given by Eq. (3.211). Substituting for $v(x,t)$ from Eq. (3.212) and using definition (3.213), we obtain

$$\frac{1}{2\pi}\int_{-b}^{b} \frac{\gamma_b(x_0,t)}{x-x_0}\mathrm{d}x_0 = w(x,t) - \lambda(x,t) \tag{3.219}$$

This is one equation with a single unknown, γ_b, since γ_w (and, hence, lambda) is related to γ_b from Eq. (3.209). As it features an integral of the unknown vorticity distribution γ_b, it is not simple to solve. Furthermore, the solution for $\gamma_b(x,t)$ must also satisfy the Kutta condition that the flow must separate smoothly at the trailing edge. For the vorticity-based model used in this section, the Kutta condition can be written simply as

$$\gamma_b(b) = 0 \tag{3.220}$$

Dowell (2004) shows that, for sinusoidal motion, Theodorsen's lift and moment expressions can be derived from Eqs. (3.219) and (3.220). Gülçat (2016) obtains the same result using Laplace transforms. Here, we will use the approach described by Johnson (1980) and extended by Peters et al. (1995) in order to derive a time-domain finite state theory. Note that Eq. (3.219) is an unsteady version of the steady thin airfoil theory.

3.6.1 Glauert Expansions

In order to solve Eq. (3.219) for the bound vorticity, Peters et al. (1995) proposed to use a functional representation of the pressure distribution around the airfoil and then express other flow quantities, such as the bound vorticity, in terms of the same functional basis. This representation is written in elliptic coordinates, η, β, where $\eta \geq 0$ and $0 \leq \beta \leq 2\pi$. We can transform from elliptic to Cartesian coordinates using

$$x = b\cosh\eta\cos\beta \tag{3.221}$$

$$y = b\sinh\eta\sin\beta \tag{3.222}$$

where b is the airfoil's half-chord. It follows that $\eta =$ constant defines ellipses while $\beta =$ constant defines hyperbolae. Figure 3.35 plots the ellipses and hyperbolae for $b = 2$, as well as the airfoil, which is defined by $\eta = 0$ and extends from $-b$ to $+b$.

Example 3.18 *Write the 2D Laplace equation in elliptic coordinates.*

Laplace's equation in 2D is given in Cartesian coordinates by

$$\frac{\partial^2\phi}{\partial x^2} + \frac{\partial^2\phi}{\partial y^2} = 0$$

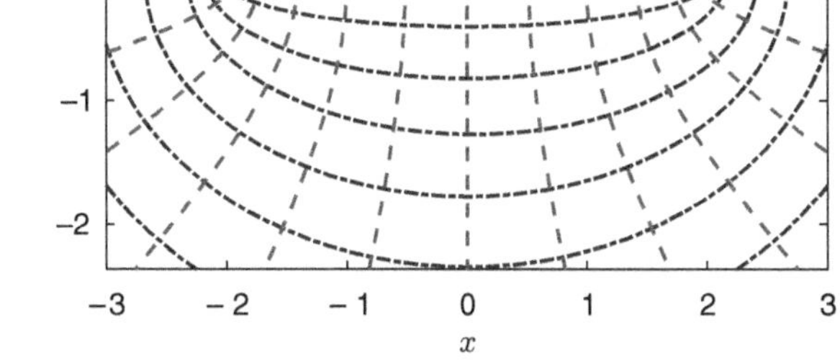

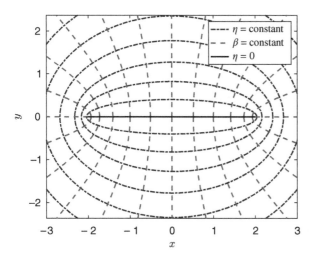

Figure 3.35 Elliptic coordinate system.

In order to transform it to elliptic coordinates, we need to calculate from Eqs. (3.221) and (3.222)

$$\frac{\partial x}{\partial \eta} = b \sinh \eta \cos \beta, \quad \frac{\partial x}{\partial \beta} = -b \cosh \eta \sin \beta \tag{3.223}$$

$$\frac{\partial y}{\partial \eta} = b \cosh \eta \sin \beta, \quad \frac{\partial y}{\partial \beta} = b \sinh \eta \cos \beta \tag{3.224}$$

Using the chain rule,

$$\frac{\partial \phi}{\partial \eta} = \frac{\partial \phi}{\partial x}\frac{\partial x}{\partial \eta} + \frac{\partial \phi}{\partial y}\frac{\partial y}{\partial \eta}$$

and, hence,

$$\begin{aligned}\frac{\partial^2 \phi}{\partial \eta^2} &= \frac{\partial}{\partial x}\left(\frac{\partial \phi}{\partial x}\frac{\partial x}{\partial \eta} + \frac{\partial \phi}{\partial y}\frac{\partial y}{\partial \eta}\right)\frac{\partial x}{\partial \eta} + \frac{\partial}{\partial y}\left(\frac{\partial \phi}{\partial x}\frac{\partial x}{\partial \eta} + \frac{\partial \phi}{\partial y}\frac{\partial y}{\partial \eta}\right)\frac{\partial y}{\partial \eta} \\ &= \frac{\partial^2 \phi}{\partial x^2}\left(\frac{\partial x}{\partial \eta}\right)^2 + 2\frac{\partial^2 \phi}{\partial x \partial y}\frac{\partial x}{\partial \eta}\frac{\partial y}{\partial \eta} + \frac{\partial^2 \phi}{\partial y^2}\left(\frac{\partial y}{\partial \eta}\right)^2\end{aligned} \tag{3.225}$$

since $\partial x/\partial \eta$ and $\partial x/\partial \beta$ are functions of η and β only and, therefore, are independent of x and y. Similarly, the second derivative of ϕ with respect to β is given by

$$\begin{aligned}\frac{\partial^2 \phi}{\partial \beta^2} &= \frac{\partial}{\partial x}\left(\frac{\partial \phi}{\partial x}\frac{\partial x}{\partial \beta} + \frac{\partial \phi}{\partial y}\frac{\partial y}{\partial \beta}\right)\frac{\partial x}{\partial \beta} + \frac{\partial}{\partial y}\left(\frac{\partial \phi}{\partial x}\frac{\partial x}{\partial \beta} + \frac{\partial \phi}{\partial y}\frac{\partial y}{\partial \beta}\right)\frac{\partial y}{\partial \beta} \\ &= \frac{\partial^2 \phi}{\partial x^2}\left(\frac{\partial x}{\partial \beta}\right)^2 + 2\frac{\partial^2 \phi}{\partial x \partial y}\frac{\partial x}{\partial \beta}\frac{\partial y}{\partial \beta} + \frac{\partial^2 \phi}{\partial y^2}\left(\frac{\partial y}{\partial \beta}\right)^2\end{aligned} \tag{3.226}$$

Summing Eqs. (3.225) and (3.226), we obtain

$$\begin{aligned}\frac{\partial^2 \phi}{\partial \eta^2} + \frac{\partial^2 \phi}{\partial \beta^2} = {}& \frac{\partial^2 \phi}{\partial x^2}\left[\left(\frac{\partial x}{\partial \eta}\right)^2 + \left(\frac{\partial x}{\partial \beta}\right)^2\right] + 2\frac{\partial^2 \phi}{\partial x \partial y}\left(\frac{\partial x}{\partial \eta}\frac{\partial y}{\partial \eta} + \frac{\partial x}{\partial \beta}\frac{\partial y}{\partial \beta}\right) \\ &+ \frac{\partial^2 \phi}{\partial y^2}\left[\left(\frac{\partial y}{\partial \eta}\right)^2 + \left(\frac{\partial y}{\partial \beta}\right)^2\right]\end{aligned} \tag{3.227}$$

From Eqs. (3.223) and (3.224),

$$\begin{aligned}\left[\left(\frac{\partial x}{\partial \eta}\right)^2 + \left(\frac{\partial x}{\partial \beta}\right)^2\right] &= b^2\left(\sinh^2\eta + \sin^2\beta\right) \\ \left[\left(\frac{\partial y}{\partial \eta}\right)^2 + \left(\frac{\partial y}{\partial \beta}\right)^2\right] &= b^2\left(\sinh^2\eta + \sin^2\beta\right) \\ \left(\frac{\partial x}{\partial \eta}\frac{\partial y}{\partial \eta} + \frac{\partial x}{\partial \beta}\frac{\partial y}{\partial \beta}\right) &= 0\end{aligned}$$

where we have used the basic identities $\sin^2\beta + \cos^2\beta = 1$ and $\cosh^2\eta - \sinh^2\eta = 1$. Substituting back into Eq. (3.227), we obtain

$$\frac{1}{b^2\left(\sinh^2\eta + \sin^2\beta\right)}\left(\frac{\partial^2 \phi}{\partial \eta^2} + \frac{\partial^2 \phi}{\partial \beta^2}\right) = 0$$

Assuming that b and η are finite, the left-hand side can only be equal to zero if

$$\frac{\partial^2\phi}{\partial\eta^2}+\frac{\partial^2\phi}{\partial\beta^2}=0 \tag{3.228}$$

which is Laplace's equation in elliptic coordinates.

Solutions for Laplace's equation in elliptic coordinates can be obtained using the method of separation of variables. We set $\phi(\eta,\beta)=f(\eta)g(\beta)$, where f and g are functions of single variables. Substituting back into Eq. (3.228), we obtain

$$\frac{d^2f}{d\eta^2}g(\beta)+f(\eta)\frac{d^2g}{d\beta^2}=0$$

which can be re-arranged such that

$$\frac{1}{f(\eta)}\frac{d^2f}{d\eta^2}=-\frac{1}{g(\beta)}\frac{d^2g}{d\beta^2}$$

As the left-hand side depends only on η and the right-hand side only on β, keeping in mind that η and β are independent variables, there are two possibilities:

$$\frac{d^2f}{d\eta^2}=\frac{d^2g}{d\beta^2}=0 \tag{3.229}$$

$$\frac{1}{f(\eta)}\frac{d^2f}{d\eta^2}=-\frac{1}{g(\beta)}\frac{d^2g}{d\beta^2}=n^2 \tag{3.230}$$

where n is a constant. Equation (3.229) gives

$$f(\eta)=1,\quad f(\eta)=\eta,\quad g(\beta)=1,\quad g(\beta)=\beta \tag{3.231}$$

Equation (3.230) leads to

$$\frac{d^2f}{d\eta^2}-n^2f=0 \quad \text{and} \quad \frac{d^2g}{d\beta^2}+n^2g=0$$

These are two second-order ODEs whose eigenvalues are $\pm n$ for the η equation and $\pm in$ for the β equation. Hence, the general solutions for $f(\eta)$ and $g(\beta)$ are

$$f(\eta)=e^{\pm n\eta},\quad g(\beta)=\sin n\beta,\quad g(\beta)=\cos n\beta \tag{3.232}$$

Assembling expressions (3.231) and (3.232) with the assumption $\phi(\eta,\beta)=f(\eta)g(\beta)$, we obtain the following solutions for Laplace's equation:

$$\phi(\eta,\beta)=1,\ \eta,\ \beta,\ \eta\beta,\ e^{\pm n\eta}\sin n\beta,\ e^{\pm n\eta}\cos n\beta \tag{3.233}$$

Recall that we are looking for functions that can represent the pressure around the airfoil due to the bound vorticity $\gamma_b(x)$, in order to solve Eq. (3.219). We can select the most appropriate of the functions in expression (3.233) by examining their form when plotted in Cartesian coordinates.

Example 3.19 *Plot the solutions of the elliptical Laplace equation in cartesian coordinates and select the most appropriate for representing the pressure distribution around a 2D airfoil.*

We first set out the criteria for a function that can represent the pressure around a 2D airfoil due to a vorticity distribution on its surface:

- *The pressure due to the vorticity must go to zero far from the airfoil (far field boundary condition).*

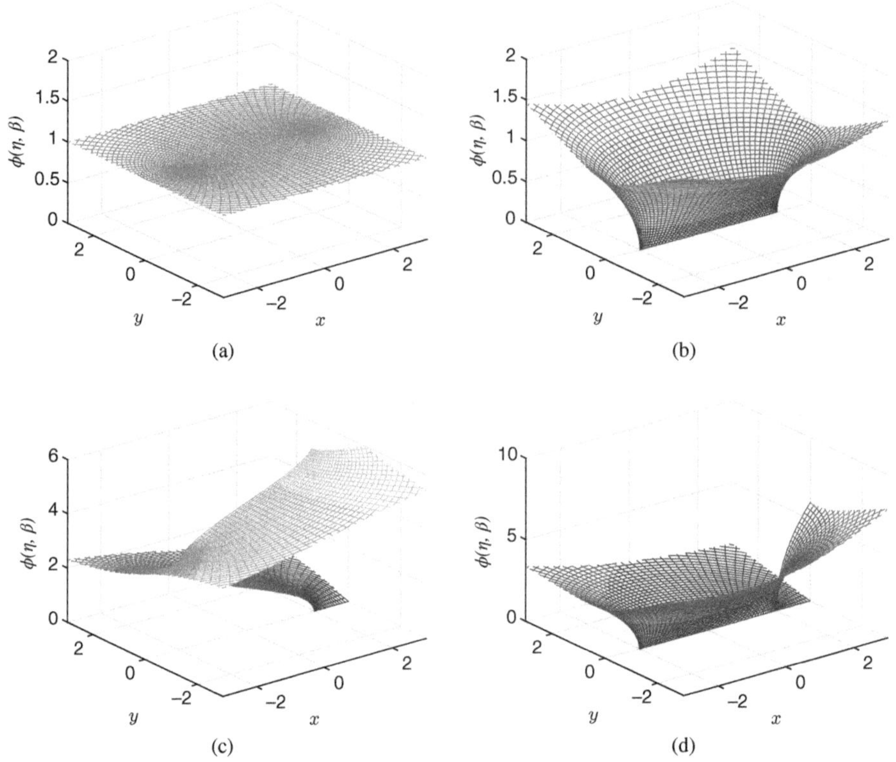

Figure 3.36 Polynomial functions of Eq. (3.231). (a) $\phi(\eta, \beta) = 1$, (b) $\phi(\eta, \beta) = \eta$, (c) $\phi(\eta, \beta) = \beta$ and (d) $\phi(\eta, \beta) = \eta\beta$.

- *There must be a pressure jump (discontinuity) between the upper and lower surfaces. This discontinuity must only exist along the chord, the pressure distribution must be continuous everywhere else.*

Keeping these criteria in mind, we go on to plot the functions of expressions (3.231). We select $b = 2$ and set up a vector η_i going from 0 to $2b$ for $i = 0, \ldots, 300$. We also set up a vector β_i going from 0 to 2π, again for $i = 0, \ldots, 300$. The corresponding cartesian coordinates are given from Eqs. (3.221) and (3.222) as $x_i = b\cosh\eta_i \cos\beta_i$ and $y_i = b\sinh\eta_i \sin\beta_i$. We first evaluate the functions $\phi(\eta, \beta) = 1, \eta, \beta, \eta\beta$, plotted against x_i and y_i, as seen in Figure 3.36. Recall that the airfoil lies on $y = 0$, $-b \leq x \leq b$.

Function $\phi(\eta, \beta) = 1$ of Figure 3.36a does not satisfy any of the criteria set out above, as it is constant for all x and y. Function $\phi(\eta, \beta) = \eta$ of Figure 3.36b is not discontinuous across the airfoil and its value increases with η. Function $\phi(\eta, \beta) = \beta$ of Figure 3.36c is discontinuous across the airfoil, but it is also discontinuous in the wake of the airfoil, for $x > b$. Finally, function $\phi(\eta, \beta) = \eta\beta$ (Figure 3.36d) is only discontinuous across the wake, not across the chord of the airfoil. Hence, none of the functions plotted in Eq. (3.231) are suitable for representing the pressure distribution around a 2D airfoil.

Next, we plot functions $\phi(\eta, \beta) = e^{\pm n\eta} \sin n\beta$ and $\phi(\eta, \beta) = e^{\pm n\eta} \cos n\beta$ for $n = 1$ in Figure 3.37. Function $\phi(\eta, \beta) = e^{-\eta} \sin\beta$ of Figure 3.37a satisfies both criteria, as it is discontinuous across the airfoil only and goes to zero for high values of x and y. Function $\phi(\eta, \beta) = e^{\eta} \sin\beta$ (Figure 3.37b) is also discontinuous across the airfoil but does not satisfy

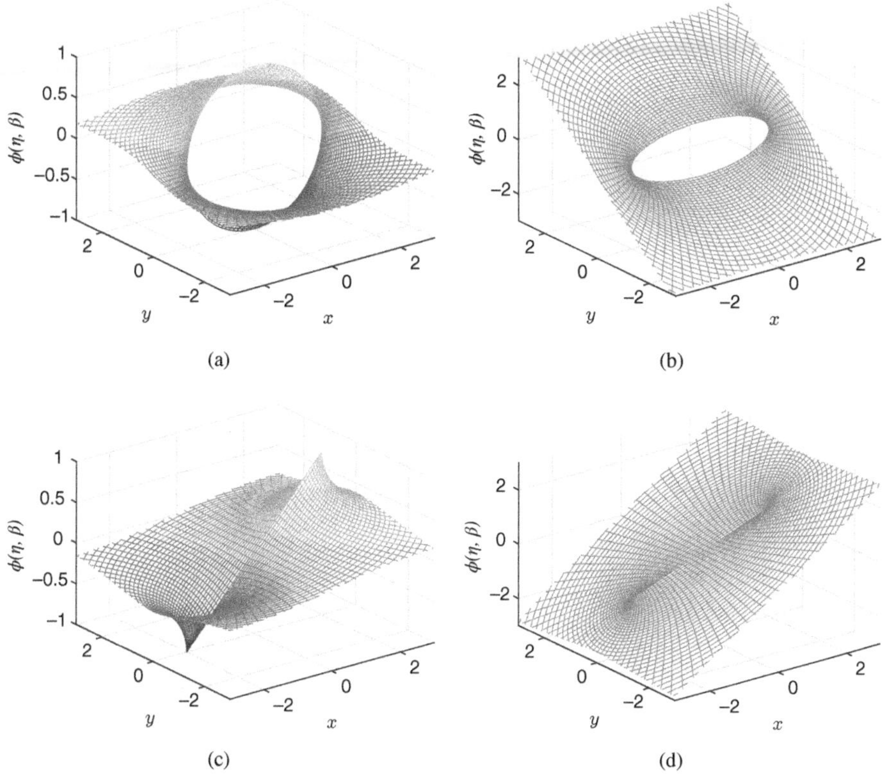

Figure 3.37 Damped sinusoidal functions of Eq. (3.231). (a) $\phi(\eta, \beta) = e^{-\eta} \sin \beta$, (b) $\phi(\eta, \beta) = e^{\eta} \sin \beta$, (c) $\phi(\eta, \beta) = e^{-\eta} \cos \beta$ and (d) $\phi(\eta, \beta) = e^{\eta} \cos \beta$.

the far field boundary condition. Function $\phi(\eta, \beta) = e^{-\eta} \cos \beta$ (Figure 3.37c) satisfies the far field condition but is not discontinuous. Finally, function $\phi(\eta, \beta) = e^{\eta} \cos \beta$ (Figure 3.37d) does not satisfy any of the two criteria.

In conclusion, of all the solutions of Laplace's equation given in expression (3.231), only function $\phi(\eta, \beta) = e^{-\eta} \sin \beta$ is suitable for expressing the pressure distribution around a 2D airfoil. We now look at the same function for values of $n = 1, \dots, 4$, as plotted in Figure 3.38. It can be seen that the function satisfies the pressure distribution criteria for all these values of n. We can therefore extrapolate that it will also satisfy them for any general n. Hence,

$$\phi_n(\eta, \beta) = e^{-n\eta} \sin n\beta \tag{3.234}$$

are all suitable functions for representing the pressure distribution around a 2D airfoil. Note that for $n = 1, 3$, the function $\phi(\eta, \beta)$ is symmetric around $y = 0$, while for $n = 2, 4$, it is anti-symmetric.

If ϕ_n for $n = 1, \dots$ are all solutions of Laplace's equation and valid functions for representing the pressure distribution around a 2D airfoil, it stands to reason that linear combinations of ϕ_n are also such valid functions. Peters et al. (1995) state that the sum of all symmetric terms is a convergent series,

$$\phi_s(\eta, \beta) = \sum_{n=1,3,5}^{\infty} 2\phi_n = \frac{\cosh \eta \sin \beta}{\sinh^2 \eta + \sin^2 \beta} \tag{3.235}$$

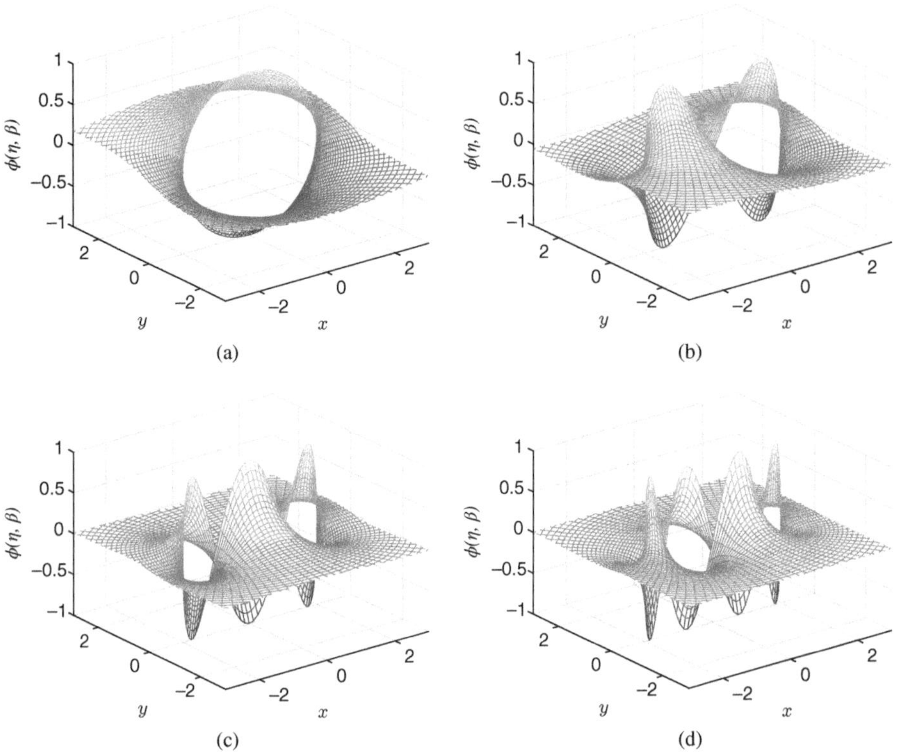

Figure 3.38 Functions $\phi_n(\eta, \beta) = e^{-n\eta} \sin n\beta$ for $n = 1, \ldots, 4$. (a) $n = 1$, (b) $n = 2$, (c) $n = 3$ and (d) $n = 4$.

while the sum of all anti-symmetric terms is also convergent and given by

$$\phi_a(\eta, \beta) = \sum_{n=2,4,6}^{\infty} 2\phi_n = \frac{\sin\beta\cos\beta}{\sinh^2\eta + \sin^2\beta} \tag{3.236}$$

Functions $\phi_s(\eta, \beta)$ and $\phi_a(\eta, \beta)$ are plotted in Figure 3.39. It can be seen that, as expected, they satisfy the criteria given at the beginning of this example. Note that both functions are singular at the leading and trailing edge of the airfoil. These singularities are not problematic; they also feature in the Wagner/Theodorsen solution approach.

As a final thought, it should be recalled that the form of Laplace's equation is identical in Cartesian and in elliptic coordinates. Consequently, functions of the form $\phi(x, y) = e^{\pm nx} \sin ny$ or $\phi(x, y) = e^{\pm ny} \sin nx$ are also solutions of Laplace's equation. However, these functions are not discontinuous across the airfoil because both $\sin ny$ and $e^{\pm ny}$ are continuous around $y = 0$. In contrast, consider function $\phi(\eta, \beta) = e^{-\eta} \sin\beta$; on the surface of the airfoil, it becomes $\phi(0, \beta) = \sin\beta$ so that plotting $\phi(0, \beta)$ against $x = b\cos\beta$ results in a circle. In other words, the function takes different values on the upper and lower surfaces, positive for $0 < \beta < \pi$ and negative for $\pi < \beta < 2\pi$.[4] Therefore, it is the use of elliptic coordinates that renders these

4 Note that the function $\phi(\eta, \beta) = e^{-\eta} \cos\beta$ is not discontinuous across the airfoil because, by the same argument, for $\eta = 0$ we would be plotting $\phi(0, \beta) = \cos\beta$ against $x = b\cos\beta$, which is a straight line.

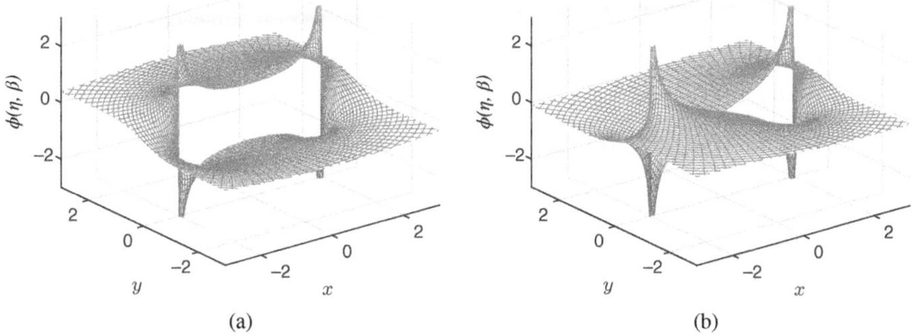

Figure 3.39 Sums of all symmetric and antisymmetric functions $\phi_n(\eta,\beta)$. (a) $\phi_s(\eta,\beta)$ and (b) $\phi_a(\eta,\beta)$.

functions viable for solving the flow around a 2D airfoil (Peters et al. 1995). This example is solved by Matlab code `elliptical.m`.

Example 3.19 shows that functions of the form given in Eqs. (3.234)–(3.236), known as Glauert expansions, can be used to express the pressure distribution around a 2D airfoil, for any n. Note that these are only spatial functions; the unsteady pressure distribution must also be expressed in terms of functions of time. This can be done by expressing the pressure as a sum of functions ϕ_n, ϕ_s and ϕ_a, multiplied by time-varying coefficients. Peters (2008) chose the following form for the pressure:

$$p(x,y,t) = 2\left(\tau_s(t)\phi_s(\eta,\beta) - \tau_a(t)\phi_a(\eta,\beta) + \sum_{n=1}^{N}\tau_n(t)\phi_n(\eta,\beta)\right)$$

where $\tau_s(t)$, $\tau_a(t)$ and $\tau_n(t)$ are functions of time. Recalling that the airfoil's surface is defined by $\eta = 0$, the pressure on the upper and lower surfaces of the airfoil is given by

$$p_u(x,y,t) = 2\left(\tau_s(t)\phi_s(0,\beta) - \tau_a(t)\phi_a(0,\beta) + \sum_{n=1}^{N}\tau_n(t)\phi_n(0,\beta)\right), \quad \text{for } \pi \geq \beta \geq 0$$

$$p_l(x,y,t) = 2\left(\tau_s(t)\phi_s(0,\beta) + \tau_a(t)\phi_a(0,\beta) + \sum_{n=1}^{N}\tau_n(t)\phi_n(0,\beta)\right), \quad \text{for } \pi \leq \beta \leq 2\pi$$

Figure 3.38 indicates that, for $n = 1, \dots, 4$, $\phi_n(0,\beta)$ on the upper surface is equal to $-\phi_n(0,\beta)$ on the lower surface. This observation can be generalised using the form of the function in Eq. (3.234); since $\eta = 0$, we see that $\phi_n(0,\beta) = \sin n\beta$. One of the properties of the sine function is that $\sin n\beta = -\sin n(2\pi - \beta)$. Hence, only the upper surface values are necessary to describe the pressure difference across the surface of the airfoil. Consequently, after setting $\eta = 0$,

$$\Delta p_b = 2\left(\frac{\tau_s(t)}{\sin\beta} - \frac{\tau_a(t)\cos\beta}{\sin\beta} + \sum_{n=1}^{N}\tau_n(t)\sin n\beta\right) \tag{3.237}$$

for $\pi \geq \beta \geq 0$. Peters et al. (1995) chose the same functional form to express the bound vorticity,

$$\gamma_b = 2\left(\frac{\gamma_s(t)}{\sin\beta} - \frac{\gamma_a(t)\cos\beta}{\sin\beta} + \sum_{n=1}^{N}\gamma_n(t)\sin n\beta\right) \tag{3.238}$$

The reader should note the similarity between this latest expression and Eq. (3.11) for steady thin airfoil theory.

Now, it is possible to start evaluating some of the flow quantities in Eqs. (3.204)–(3.210). For example, the total bound vorticity of Eq. (3.206) is given by

$$\Gamma = \int_{-b}^{b} 2\left(\frac{\gamma_s(t)}{\sin\beta} - \frac{\gamma_a(t)\cos\beta}{\sin\beta} + \sum_{n=1}^{N}\gamma_n(t)\sin n\beta\right)\mathrm{d}x$$

On the surface of the airfoil $\eta = 0$, so that Eq. (3.221) becomes

$$x = b\cos\beta$$

Differentiating this expression, we obtain $\mathrm{d}x = -b\sin\beta\mathrm{d}\beta$. Furthermore, a general point on the airfoil $(x, 0)$ corresponds to $\eta = 0, \beta = \cos^{-1}(x/b)$, the leading edge $(-b, 0)$ occurs at $\eta = 0, \beta = \pi$ and the trailing edge $(b, 0)$ at $\eta = 0, \beta = 0$. The integral for the total bound vorticity becomes

$$\Gamma = -2b\int_{\pi}^{0}\left(\frac{\gamma_s(t)}{\sin\beta} - \frac{\gamma_a(t)\cos\beta}{\sin\beta} + \sum_{n=1}^{N}\gamma_n(t)\sin n\beta\right)\sin\beta\mathrm{d}\beta \tag{3.239}$$

We will use the standard integral

$$\int \sin n\beta\sin\beta\mathrm{d}\beta = \begin{cases} \frac{\beta}{2} - \frac{\sin 2\beta}{4} & \text{if } n = 1 \\ \frac{\sin\beta(n-1)}{2(n-1)} - \frac{\sin\beta(n+1)}{2(n+1)} & \text{if } n > 1 \end{cases}$$

Substituting into Eq. (3.239) for the total bound circulation and carrying out the other integrations, we obtain

$$\Gamma = 2\pi b\left(\gamma_s + \frac{\gamma_1}{2}\right) \tag{3.240}$$

The pressure difference across the surface is related to the bound vorticity by Eq. (3.204). After applying the same change of variable from x to β, Eq. (3.204) becomes

$$\Delta p_b(\beta) = -\rho b\frac{\partial}{\partial t}\int_{\pi}^{\beta}\gamma_b\sin\beta\mathrm{d}\beta + \rho U_\infty\gamma_b(\beta) \tag{3.241}$$

Substituting from Eqs. (3.237) and (3.238), we obtain

$$\frac{\tau_s}{\sin\beta} - \frac{\tau_a\cos\beta}{\sin\beta} + \sum_{n=1}^{N}\tau_n\sin n\beta = -\rho b\int_{\pi}^{\beta}\left(\frac{\dot\gamma_s}{\sin\beta} - \frac{\dot\gamma_a\cos\beta}{\sin\beta} + \sum_{n=1}^{N}\dot\gamma_n\sin n\beta\right)\sin\beta\mathrm{d}\beta$$
$$+\rho U_\infty\left(\frac{\gamma_s(t)}{\sin\beta} - \frac{\gamma_a(t)\cos\beta}{\sin\beta} + \sum_{n=1}^{N}\gamma_n(t)\sin n\beta\right)$$

or after carrying out the integration,

$$\frac{\tau_s}{\sin\beta} - \frac{\tau_a\cos\beta}{\sin\beta} + \sum_{n=1}^{N}\tau_n\sin n\beta = -\rho b\dot\gamma_s(\beta - \pi) + \rho b\dot\gamma_a\sin\beta - \rho b\dot\gamma_1\left(\frac{\beta - \pi}{2} - \frac{\sin 2\beta}{4}\right)$$

$$- \rho \sum_{n=2}^{N} \dot{\gamma}_n \left(\frac{\sin \beta(n-1)}{2(n-1)} - \frac{\sin \beta(n+1)}{2(n+1)} \right)$$

$$+ \rho U_\infty \left(\frac{\gamma_s}{\sin \beta} - \frac{\gamma_a \cos \beta}{\sin \beta} + \sum_{n=1}^{N} \gamma_n \sin n\beta \right) \quad (3.242)$$

The objective is now to eliminate β from this latest equation in order to obtain expressions that relate the τ_s, τ_a and τ_n coefficients to γ_s, γ_a and γ_n. This can be done by applying a Galerkin procedure, which involves multiplying both sides of the equation by $\sin \beta$, $\sin 2\beta$, $\sin 3\beta$, etc., and then integrating between 0 and 2π. We start by multiplying Eq. (3.242) by $\sin \beta$ and integrating, which results in

$$2\pi\tau_s + \pi\tau_1 = 2\pi\rho b\dot{\gamma}_s + \pi\rho b\dot{\gamma}_a + \pi\rho b\dot{\gamma}_1 - \frac{\pi\rho b}{2}\dot{\gamma}_2 + 2\pi\rho U_\infty\gamma_s + \pi\rho U_\infty\gamma_1$$

After substituting from Eq. (3.240) and re-arranging, we obtain

$$\rho b \left(\dot{\gamma}_a - \frac{1}{2}\dot{\gamma}_2 \right) + \rho U_\infty \gamma_1 = 2\left(\tau_s - \rho U_\infty \gamma_s\right) + \tau_1 - \frac{\rho}{\pi}\dot{\Gamma} \quad (3.243)$$

Now, we multiply Eq. (3.242) by $\sin 2\beta$ and integrate, so that

$$-2\pi\tau_a + \pi\tau_2 = \pi\rho b\dot{\gamma}_s + \frac{3\pi\rho b}{4}\dot{\gamma}_1 - \frac{\pi\rho b}{4}\dot{\gamma}_3 - 2\pi\rho U_\infty\gamma_a + \pi\rho U_\infty\gamma_2$$

Writing $3\pi\rho b\dot{\gamma}_1/4$ as $\pi\rho b\dot{\gamma}_1/2 + \pi\rho b\dot{\gamma}_1/4$, substituting from Eq. (3.240) and re-arranging, we obtain

$$\frac{\rho b}{4}\left(\dot{\gamma}_1 - \dot{\gamma}_3\right) + \rho U_\infty \gamma_2 = -2\left(\tau_a - \rho U_\infty \gamma_a\right) + \tau_2 - \frac{\rho}{2\pi}\dot{\Gamma} \quad (3.244)$$

Generalising this process, we multiply Eq. (3.242) by $\sin n\beta$ and integrate between 0 and 2π to obtain

$$\frac{\rho b}{2n}\left(\dot{\gamma}_{n-1} - \dot{\gamma}_{n+1}\right) + \rho U_\infty \gamma_N = 2\left(\tau_s - \rho U_\infty \gamma_s\right) + \tau_n - \frac{\rho}{n\pi}\dot{\Gamma} \text{ if } n \text{ is even} \quad (3.245)$$

$$\frac{\rho b}{2n}\left(\dot{\gamma}_{n-1} - \dot{\gamma}_{n+1}\right) + \rho U_\infty \gamma_N = -2\left(\tau_a - \rho U_\infty \gamma_a\right) + \tau_n - \frac{\rho}{n\pi}\dot{\Gamma} \text{ if } n \text{ is odd} \quad (3.246)$$

In order to simplify these expressions, we can go back to Eq. (3.242) and evaluate its limit as β tends to 0. The terms with $\sin \beta$ in the denominator will dominate while all the other terms will be either zero or negligible. Hence, the equation will become

$$\frac{\tau_s}{\sin \beta} - \frac{\tau_a \cos \beta}{\sin \beta} = \rho U_\infty \left(\frac{\gamma_s}{\sin \beta} - \frac{\gamma_a \cos \beta}{\sin \beta} \right)$$

or

$$\left(\tau_s - \rho U_\infty \gamma_s\right) - \left(\tau_a - \rho U_\infty \gamma_a\right) = 0$$

One way to satisfy this equation, which also results in a simplification of Eqs. (3.243)–(3.246), is to set $\tau_s = \rho U_\infty \gamma_s$ and $\tau_a = \rho U_\infty \gamma_a$. Then, the full set of expressions relating the τ_s, τ_a and τ_n coefficients to γ_s, γ_a and γ_n become

$$\tau_s = \rho U_\infty \gamma_s \quad (3.247)$$

$$\tau_a = \rho U_\infty \gamma_a \quad (3.248)$$

$$\rho b\left(\dot{\gamma}_a - \frac{1}{2}\dot{\gamma}_2\right) + \rho U_\infty \gamma_1 = \tau_1 - \frac{\rho}{\pi}\dot{\Gamma} \tag{3.249}$$

$$\frac{\rho b}{2n}\left(\dot{\gamma}_{n-1} - \dot{\gamma}_{n+1}\right) + \rho U_\infty \gamma_n = \tau_n - \frac{\rho}{n\pi}\dot{\Gamma}, \quad \text{for } n \geq 2 \tag{3.250}$$

3.6.2 Solution of the Impermeability Equation

We are now in a position to start solving the impermeability problem of Eq. (3.219)

$$\frac{1}{2\pi}\int_{-b}^{b} \frac{\gamma_b(x_0, t)}{x - x_0} \mathrm{d}x_0 = w(x, t) - \lambda(x, t) \tag{3.251}$$

subject to the Kutta condition of Eq. (3.220), $\gamma_b(b) = 0$. The bound vorticity has already been expressed using the functional form of Eq. (3.238). For the downwash $w(x, t)$ and wake vorticity $\lambda(x, t)$, Johnson (1980) proposed to form

$$\lambda = \sum_{n=0}^{N} \lambda_n \cos n\beta \tag{3.252}$$

$$w = \sum_{n=0}^{N} w_n \cos n\beta \tag{3.253}$$

Substituting all these functional forms into Eq. (3.251) and setting $x = b\cos\beta$, $\mathrm{d}x = -b \sin\beta \mathrm{d}\beta$, we obtain

$$-\frac{1}{2\pi}\int_{\pi}^{0} \frac{2\left(\gamma_s - \gamma_a \cos\beta_0 + \sum_{n=1}^{N} \gamma_n \sin n\beta_0 \sin\beta_0\right)}{\cos\beta - \cos\beta_0} \mathrm{d}\beta_0 = \sum_{n=0}^{N}(w_n - \lambda_n)\cos n\beta$$

The integral on the left-hand side can be evaluated using Glauert's integrals of Eqs. (3.12) and (3.13) so that the impermeability condition becomes

$$\gamma_a + \sum_{n=1}^{N} \gamma_n \cos n\beta = \sum_{n=0}^{N}(w_n - \lambda_n)\cos n\beta \tag{3.254}$$

Integrating this equation between $\beta = 0$ and $\beta = 2\pi$ yields $\gamma_a = w_0 - \lambda_0$. Multiplying both sides of Eq. (3.254) by $\cos n\beta$ and then integrating between 0 and 2π yields $\gamma_n = w_n - \lambda_n$. Hence, the integral equation (3.251) reduces to the set of algebraic equations:

$$\gamma_a = w_0 - \lambda_0 \tag{3.255}$$

$$\gamma_n = w_n - \lambda_n, \quad \text{for } n \geq 1 \tag{3.256}$$

Substituting these latest expressions into Eqs. (3.248)–(3.250) yields

$$\tau_a = \rho U_\infty (w_0 - \lambda_0) \tag{3.257}$$

$$\tau_1 = \left[\rho b\left(\dot{w}_0 - \frac{1}{2}\dot{w}_2\right) + \rho U_\infty w_1\right] - \left[\rho b\left(\dot{\lambda}_0 - \frac{1}{2}\dot{\lambda}_2\right) + \rho U_\infty \lambda_1\right] + \frac{\rho}{\pi}\dot{\Gamma} \tag{3.258}$$

$$\tau_n = \left[\frac{\rho b}{2n}\left(\dot{w}_{n-1} - \dot{w}_{n+1}\right) + \rho U_\infty w_n\right] - \left[\frac{\rho b}{2n}\left(\dot{\lambda}_{n-1} - \dot{\lambda}_{n+1}\right) + \rho U_\infty \lambda_n\right] + \frac{\rho}{n\pi}\dot{\Gamma} \tag{3.259}$$

Recalling that τ_a and τ_n are the coefficients of Eq. (3.237), we now have expressions for evaluating the pressure difference across the airfoil, $\Delta p_b(x,t)$, which we can integrate over the chord to obtain the lift and moment. However, there are still two problems:

- we know w_n because it is the effect of the airfoil's motion, but we do not have values for λ_0 and λ_n, the vertical velocity induced by the wake vorticity,
- the time derivative of the total bound circulation $\dot{\Gamma}$ is related to γ_s and γ_1 by Eq. (3.240), but these are still unknown.

Some of these issues can be resolved by looking at Eq. (3.217)

$$\frac{\partial \lambda}{\partial t} + U_\infty \frac{\partial \lambda}{\partial x} = \frac{1}{2\pi} \frac{\dot{\Gamma}}{b - x} \tag{3.260}$$

Substituting from Eq. (3.252) into Eq. (3.260), using the chain rule

$$\frac{\partial \lambda}{\partial x} = \frac{\partial \lambda}{\partial \beta}\frac{\partial \beta}{\partial x} = \frac{\partial \lambda}{\partial \beta}\frac{1}{\frac{\partial x}{\partial \beta}} = -\frac{\partial \lambda}{\partial \beta}\frac{1}{b \sin\beta}$$

and recalling that $x = b\cos\beta$, we obtain

$$\sum_{n=0}^{N} \dot{\lambda}_n \cos n\beta + \frac{nU_\infty}{b}\sum_{n=0}^{N} \lambda_n \frac{\sin n\beta}{\sin\beta} = \frac{1}{2\pi}\frac{\dot{\Gamma}}{b(1-\cos\beta)}$$

Multiplying both sides of this equation by $(1 - \cos\beta)$

$$(1-\cos\beta)\sum_{n=0}^{N} \dot{\lambda}_n \cos n\beta + (1-\cos\beta)\frac{nU_\infty}{b}\sum_{n=0}^{N} \lambda_n \frac{\sin n\beta}{\sin\beta} = \frac{1}{2\pi b}\dot{\Gamma}$$

We will now perform a Galerkin procedure, multiplying this equation by $\cos k\beta$ for $k = 0, 1, 2, \ldots$ and integrating both sides between $\beta = 0$ and $\beta = 2\pi$. The results of this procedure for $k = 0, 1, 2, 3$ are as follows:

$$2\pi\dot{\lambda}_0 - \pi\dot{\lambda}_1 + \frac{U_\infty}{b}\left(2\pi\lambda_1 - 4\pi\lambda_2 + 6\pi\lambda_3 - 8\pi\lambda_4 + \cdots\right) = \frac{1}{b}\dot{\Gamma} \tag{3.261}$$

$$-\pi\dot{\lambda}_0 + \pi\dot{\lambda}_1 - \frac{\pi}{2}\dot{\lambda}_2 + \frac{U_\infty}{b}\left(-\pi\lambda_1 + 4\pi\lambda_2 - 6\pi\lambda_3 + 8\pi\lambda_4 + \cdots\right) = 0 \tag{3.262}$$

$$-\frac{\pi}{2}\dot{\lambda}_1 + \pi\dot{\lambda}_2 - \frac{\pi}{2}\dot{\lambda}_3 + \frac{U_\infty}{b}\left(-2\pi\lambda_2 + 6\pi\lambda_3 - 8\pi\lambda_4 + 10\lambda_5 + \cdots\right) = 0 \tag{3.263}$$

$$-\frac{\pi}{2}\dot{\lambda}_2 + \pi\dot{\lambda}_3 - \frac{\pi}{2}\dot{\lambda}_4 + \frac{U_\infty}{b}\left(-3\pi\lambda_3 + 8\pi\lambda_4 - 10\lambda_5 + 12\lambda_6 + \cdots\right) = 0 \tag{3.264}$$

Adding up Eqs. (3.261) and (3.262), we obtain

$$\pi\dot{\lambda}_0 - \frac{\pi}{2}\dot{\lambda}_2 + \pi\frac{U_\infty}{b}\lambda_1 = \frac{1}{b}\dot{\Gamma} \tag{3.265}$$

Adding up Eqs. (3.262), (3.263) and (3.265) yields

$$\frac{\pi}{2}\dot{\lambda}_1 - \frac{\pi}{2}\dot{\lambda}_3 + 2\pi\frac{U_\infty}{b}\lambda_2 = \frac{1}{b}\dot{\Gamma} \tag{3.266}$$

Similarly, adding up Eqs. (3.263), (3.264) and (3.266) results in

$$\frac{\pi}{2}\dot{\lambda}_2 - \frac{\pi}{2}\dot{\lambda}_4 + 3\pi\frac{U_\infty}{b}\lambda_3 = \frac{1}{b}\dot{\Gamma} \tag{3.267}$$

Continuing this process, it becomes evident that Eq. (3.260) can be represented by

$$\dot{\lambda}_0 - \frac{1}{2}\dot{\lambda}_2 + \frac{U_\infty}{b}\lambda_1 = \frac{\dot{\Gamma}}{b\pi} \text{ for } n = 1 \tag{3.268}$$

$$\frac{1}{2n}\dot{\lambda}_{n-1} - \frac{1}{2n}\dot{\lambda}_{n+1} + \frac{U_\infty}{b}\lambda_n = \frac{\dot{\Gamma}}{nb\pi} \text{ for } n \geq 2 \tag{3.269}$$

This is a set of $N-1$ first-order ODEs with $N+1$ unknowns, $\lambda_0, \lambda_1, \ldots, \lambda_N$. It can be derived with the help of Matlab code `equation_lambda_derive.m`.

3.6.3 Completing the Equations

Equations (3.268) and (3.269) are incomplete. If $\dot{\lambda}_{n-1}$ and $\dot{\lambda}_n$ are known, then we can solve for $\dot{\lambda}_{n+1}$, but this solution scheme cannot be started because there is no explicit equation for λ_0. Therefore, the equations must be completed in order to solve for the impermeability condition. In order to do this, we can return to the functional form of expression (3.252):

$$\lambda = \sum_{n=0}^{N} \lambda_n \cos n\beta$$

and invert it using a Galerkin procedure. Integrating both sides between $\beta = 0$ and $\beta = \pi$, we obtain

$$\int_0^\pi \lambda \, d\beta = \pi\lambda_0$$

Furthermore, multiplying both sides by $\cos n\beta$ and integrating between $\beta = 0$ and $\beta = \pi$ results in

$$\int_0^\pi \lambda \cos n\beta \, d\beta = \frac{\pi}{2}\lambda_n$$

Re-arranging the last two equations, we obtain

$$\lambda_0 = \frac{1}{\pi}\int_0^\pi \lambda \, d\beta$$

$$\lambda_n = \frac{2}{\pi}\int_0^\pi \lambda \cos n\beta \, d\beta$$

Following the procedure described in Johnson (1980), we can substitute for the definition of λ (Eq. (3.214)), such that

$$\lambda_0 = \frac{1}{\pi}\int_0^\pi \left[\frac{1}{2\pi}\int_b^\infty \frac{\gamma_w(x_0, t)}{x - x_0} dx_0\right] d\beta$$

$$\lambda_n = \frac{2}{\pi}\int_0^\pi \left[\frac{1}{2\pi}\int_b^\infty \frac{\gamma_w(x_0, t)}{x - x_0} dx_0\right] \cos n\beta \, d\beta$$

Recall that x_0 is the x-coordinate of the wake, while x is the x-coordinate of the flat plate, both y-coordinates being equal to zero. Furthermore, from the definition of the elliptical coordinates (3.221), $x = b\cos\beta$. We substitute this definition and change the order of the integrals to obtain

$$\lambda_0 = -\frac{1}{2\pi}\int_b^\infty \gamma_w(x_0, t)\left[\frac{1}{\pi}\int_0^\pi \frac{1}{x_0 - b\cos\beta} d\beta\right] dx_0$$

$$\lambda_n = -\frac{1}{\pi}\int_b^\infty \gamma_w(x_0,t)\left[\frac{1}{\pi}\int_0^\pi \frac{\cos n\beta}{x_0 - b\cos\beta}\mathrm{d}\beta\right]\mathrm{d}x_0$$

Johnson (1980) gives the following result:

$$\lambda_0 = -\frac{1}{2\pi}\int_b^\infty \gamma_w(x_0,t)\left[\frac{1}{\sqrt{x_0^2 - b^2}}\right]\mathrm{d}x_0$$

$$\lambda_n = -\frac{1}{\pi}\int_b^\infty \gamma_w(x_0,t)\left[\frac{\left(x_0 - \sqrt{x_0^2 - b^2}\right)^n}{b^n\sqrt{x_0^2 - b^2}}\right]\mathrm{d}x_0$$

Now, we turn our attention to the x_0 coordinate. As it lies along the wake at $y = 0$, in elliptical coordinates it is defined by $\beta = 0$ and $\eta = 0, \dots, \infty$. Then, from the definition of Eq. (3.221), $x_0 = b\cosh\eta$ and $\mathrm{d}x_0 = b\sinh\eta\,\mathrm{d}\eta$. Hence, λ_0 and λ_n become

$$\lambda_0 = -\frac{1}{2\pi}\int_0^\infty \gamma_w(\eta,t)\left[\frac{1}{\sqrt{b^2\cosh^2\eta - b^2}}\right] b\sinh\eta\,\mathrm{d}\eta$$

$$\lambda_n = -\frac{1}{\pi}\int_0^\infty \gamma_w(\eta,t)\left[\frac{\left(b\cosh\eta - \sqrt{b^2\cosh^2\eta - b^2}\right)^n}{b^n\sqrt{b^2\cosh^2\eta - b^2}}\right] b\sinh\eta\,\mathrm{d}\eta$$

Using the hyperbolic function identities $\cosh^2\eta - 1 = \sinh^2\eta$ and $\cosh\eta - \sin\eta = \mathrm{e}^{-\eta}$, we obtain

$$\lambda_0 = -\frac{1}{2\pi}\int_0^\infty \gamma_w(\eta,t)\mathrm{d}\eta \tag{3.270}$$

$$\lambda_n = -\frac{1}{\pi}\int_0^\infty \gamma_w(\eta,t)\mathrm{e}^{-n\eta}\,\mathrm{d}\eta \tag{3.271}$$

At this stage, Peters et al. (1995) introduce the approximate series

$$1 \approx \sum_{n=1}^{N} b_n \mathrm{e}^{-n\eta} \tag{3.272}$$

for $0 < \mathrm{e}^{-n\eta} \leq 1$, where b_n are coefficients to be evaluated such that the equation above is satisfied. They then substitute it into Eq. (3.270), such that

$$\lambda_0 \approx -\frac{1}{2\pi}\int_0^\infty \gamma_w(\eta,t)\sum_{n=1}^{N} b_n\mathrm{e}^{-n\eta}\,\mathrm{d}\eta$$

Finally, substituting from Eq. (3.271) yields

$$\lambda_0 \approx \frac{1}{2}\sum_{n=1}^{N} b_n\lambda_n \tag{3.273}$$

It remains to evaluate the correct values of the b_n coefficients. Peters (2008) suggests two different sets of values for these coefficients,

$$b_{n_1} = (-1)^{n-1}\frac{N!}{n!(N-n)!} \tag{3.274}$$

$$b_{n_2} = \begin{cases} (-1)^{n-1} \frac{(N+n-1)!}{(N-n-1)!} \frac{1}{(n!)^2} & \text{for } 1 \le n \le N-1 \\ (-1)^{N+1} & \text{for } n = N \end{cases} \tag{3.275}$$

Equation (3.273) then completes the set of equations for λ.

Example 3.20 *Investigate the series approximation of Eq. (3.272)*

We define first $X = \mathrm{e}^{-\eta}$, so that Eq. (3.272) becomes

$$1 \approx \sum_{n=1}^{N} b_n X^n \tag{3.276}$$

where X takes values between 0 and 1, since η ranges from 0 to ∞. Note that the value $X = 0$ corresponds to $\eta = \infty$, that is the far wake, while $X = 1$ corresponds to $\eta = 0$, that is the trailing edge. We expect that the sum in Eq. (3.276) will be approximately equal to 1 for all values of X in the interval $[0, 1]$. In order to verify this assertion, we can evaluate the two sets of b_n coefficient values from Eqs. (3.274) and (3.275) and then plot the sum of Eq. (3.276) against X. The result of this calculation is given in Figure 3.40, where the sums are plotted, one for b_{n_1} and one for b_{n_2}, for $N = 2, 8, 16, 24$.

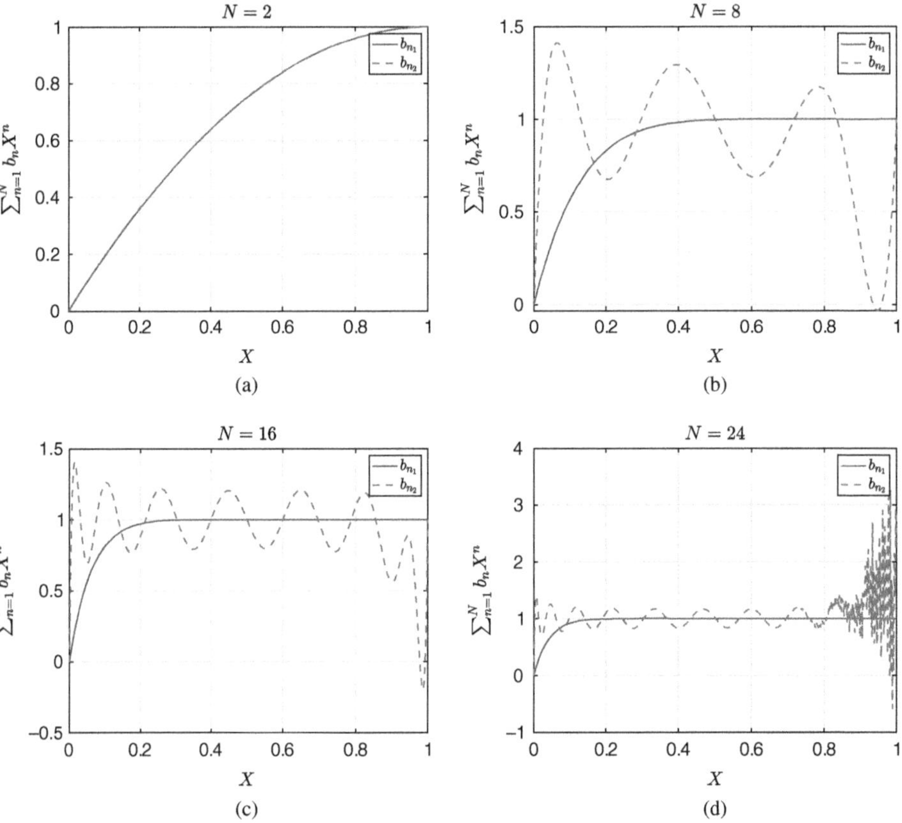

Figure 3.40 Accuracy of Eq. (3.276) for four values of N. (a) $N = 2$, (b) $N = 8$, (c) $N = 16$ and (d) $N = 24$.

For $N = 2$, b_{n_1} and b_{n_2} give exactly the same result, which is totally unsatisfactory; the sum is only equal to 1 at the trailing edge, $X = 1$. As the number of terms in the series increases, the b_{n_1} result starts to asymptote towards 1 at smaller values of X, reaching 0.99 at $X = 0.175$ for $N = 24$, which corresponds to $\eta = 1.743$ or $x = 2.95b$. This means that, for $N = 24$, the approximation of Eq. (3.276) is correct up to a distance of 1.5 chords behind the airfoil, if we use the b_{n_1} coefficients.

The b_{n_2} coefficients result in sums that are also equal to 0 at $X = 0$ but grow much faster, oscillating around unity. Unfortunately, they are calculated from factorials, which are prone to numerical instability at high values of N. Figure 3.40d shows that, for $N = 24$, the value of the sum oscillates between 0.8 and 1.2 up to $X = 0.6$ but then a numerical instability starts to develop, resulting in divergent behaviour near the trailing edge. It should be noted that the b_{n_1} coefficients also lead to instability for sufficiently high N (e.g. $N > 50$). The conclusion is that the approximation of Eq. (3.276) and, hence, (3.273) cannot be satisfactory throughout the wake. This example is solved by Matlab function `testbn.m`.

3.6.4 Kutta Condition and Aerodynamic Loads

It is now possible to derive expressions for the unsteady lift and moment by means of finite state theory. Recall that the impermeability equation has been reduced to the ordinary differential equations (3.257)–(3.259). If we substitute into them Eqs. (3.270) and (3.271), we obtain

$$\tau_a = \rho U_\infty (w_0 - \lambda_0) \tag{3.277}$$

$$\tau_1 = \rho b \left(\dot{w}_0 - \frac{1}{2}\dot{w}_2 \right) + \rho U_\infty w_1 \tag{3.278}$$

$$\tau_n = \frac{\rho b}{2n} \left(\dot{w}_{n-1} - \dot{w}_{n+1} \right) + \rho U_\infty w_n \tag{3.279}$$

The lift is given by the integral of the pressure difference

$$l = -\int_{-b}^{b} \Delta p_b \, \mathrm{d}x \tag{3.280}$$

where the minus sign denotes that the pressure difference definition has been inverted (see footnote just before Eq. (3.202)) and Δp_b is given by the functional form of Eq. (3.237)

$$\Delta p_b = 2 \left(\frac{\tau_s}{\sin\beta} - \frac{\tau_a \cos\beta}{\sin\beta} + \sum_{n=1}^{N} \tau_n \sin n\beta \right) \tag{3.281}$$

First, we need to impose the Kutta condition, such that the pressure difference at the trailing edge is equal to zero. At the trailing edge, $\beta = 0$ and

$$\Delta p_b = 2 \left(\frac{\tau_s - \tau_a}{\sin 0} \right)$$

The only way to satisfy the Kutta condition is to choose $\tau_a = \tau_s$, so that the pressure difference is undefined but finite at the trailing edge. Then, substituting Eq. (3.281) and $x = b\cos\beta$ into Eq. (3.280), we obtain

$$l = b \int_{\pi}^{0} 2 \left(\frac{\tau_a}{\sin\beta} - \frac{\tau_a \cos\beta}{\sin\beta} + \sum_{n=1}^{N} \tau_n \sin n\beta \right) \sin\beta \mathrm{d}\beta$$

Applying the same integration procedure used to derive Eq. (3.240) leads to

$$l = -2\pi b\left(\tau_a + \frac{\tau_1}{2}\right) \tag{3.282}$$

Finally, after substituting from expressions (3.277) and (3.278), we obtain

$$l = -2\pi b\left(\rho U_\infty(w_0 - \lambda_0) + \frac{\rho b\left(\dot{w}_0 - \frac{1}{2}\dot{w}_2\right) + \rho U_\infty w_1}{2}\right)$$

or after re-arranging

$$l = -2\pi\rho b\left(U_\infty\left(w_0 + \frac{w_1}{2} - \lambda_0\right) + \frac{1}{2}\left(\dot{w}_0 - \frac{1}{2}\dot{w}_2\right)\right) \tag{3.283}$$

This result is remarkable because it states that the lift depends only on the downwash due to the motion and the first term of the downwash due to the wake, λ_0. Recall that, for a pitching and plunging airfoil, the downwash is given by Eq. (3.29) or, equivalently,

$$w(x,t) = U_\infty\alpha(t) + \dot{h}(t) + \dot{\alpha}(t)(x - ab) \tag{3.284}$$

where $a = (x_f - b)/b$, see the definition of expression (3.62). Finite state theory assumes the expansion of Eq. (3.253) for w, which can be expressed as (Johnson 1980)

$$w = w_0 + \left(\frac{x}{b}\right)w_1 + \left(\frac{2x^2}{b^2} - 1\right)w_2 + \cdots$$

using the definition $x = b\cos\beta$ and the cosine multiple angle formulae. Comparing this latest expression to Eq. (3.284) we see that

$$w_0(t) = U_\infty\alpha(t) + \dot{h}(t) - ab\dot{\alpha}(t), \quad w_1(t) = b\dot{\alpha}(t), \quad w_2(t) = w_3(t) = \cdots = 0 \tag{3.285}$$

Substituting back into Eq. (3.283), we obtain

$$l = -\rho\pi b^2\left(\ddot{h} - ba\ddot{\alpha} + U_\infty\dot{\alpha}\right) - 2\pi\rho U_\infty b\left(U_\infty\alpha(t) + \dot{h}(t) + b\left(\frac{1}{2} - a\right)\dot{\alpha}(t) - \lambda_0(t)\right) \tag{3.286}$$

which is of the same form as Wagner's lift (Eq. (3.120)) and Theodorsen's lift (Eq. (3.177)). Setting $\lambda_0(t) = 0$ results in the quasi-steady lift of Eq. (3.137). Clearly, the first bracket in Eq. (3.286) is the non-circulatory lift, while the second term is the circulatory lift, which depends on the downwash at the 3/4 chord point and on λ_0.

Recall that λ_0 is approximated by Eq. (3.273) where b_n are obtained from Eq. (3.274) or (3.275), while λ_n are calculated from Eqs. (3.268) and (3.269). Furthermore, $\dot{\Gamma}$ can be expressed from Eqs. (3.240) and (3.256) as

$$\dot{\Gamma} = 2\pi b\left(\dot{\gamma}_s + \frac{\dot{w}_1 - \dot{\lambda}_1}{2}\right) \tag{3.287}$$

In order to enforce the Kutta condition, we have chosen $\tau_s = \tau_a$ so that, from Eqs. (3.247) and (3.248), $\gamma_s = \gamma_a$. Hence, Eq. (3.287) becomes

$$\dot{\Gamma} = 2\pi b\left(\dot{w}_0 + \frac{1}{2}\dot{w}_1 - \left(\dot{\lambda}_0 + \frac{1}{2}\dot{\lambda}_1\right)\right)$$

after substituting from Eq. (3.255). Finally, using expression (3.285), $\dot{\Gamma}$ is given by

$$\dot{\Gamma} = 2\pi b\left(\left(U_{\infty}\dot{\alpha}(t) + \ddot{h}(t) + b\left(\frac{1}{2} - a\right)\ddot{\alpha}(t)\right) - \left(\dot{\lambda}_0 + \frac{1}{2}\dot{\lambda}_1\right)\right) \tag{3.288}$$

Assembling Eqs. (3.268), (3.269), (3.273) and (3.288), we have

$$\frac{3}{2}\sum_{n=1}^{N} b_n\dot{\lambda}_n + \dot{\lambda}_1 - \frac{1}{2}\dot{\lambda}_2 + \frac{U_{\infty}}{b}\lambda_1 = 2\left(U_{\infty}\dot{\alpha}(t) + \ddot{h}(t) + b\left(\frac{1}{2} - a\right)\ddot{\alpha}(t)\right)$$

$$\frac{1}{n}\sum_{n=1}^{N} b_n\dot{\lambda}_n + \frac{1}{n}\dot{\lambda}_1 + \frac{1}{2n}\dot{\lambda}_{n-1} - \frac{1}{2n}\dot{\lambda}_{n+1} + \frac{U_{\infty}}{b}\lambda_n$$
$$= \frac{2}{n}\left(U_{\infty}\dot{\alpha}(t) + \ddot{h}(t) + b\left(\frac{1}{2} - a\right)\ddot{\alpha}(t)\right)$$

noting that the last equation will be missing a term, since λ_{N+1} does not exist. Peters et al. (1995) and Hodges and Pierce (2002) propose to write these equations in matrix form as

$$\boldsymbol{A}\dot{\boldsymbol{\lambda}} + \frac{U_{\infty}}{b}\boldsymbol{\lambda} = \mathbf{c}\left(U_{\infty}\dot{\alpha}(t) + \ddot{h}(t) + b\left(\frac{1}{2} - a\right)\ddot{\alpha}(t)\right) \tag{3.289}$$

where $\boldsymbol{A} = \boldsymbol{D} + \mathbf{d}\mathbf{b}^T + \mathbf{c}\mathbf{d}^T + \frac{1}{2}\mathbf{c}\mathbf{b}^T$, $\boldsymbol{D}$ is a $N \times N$ matrix with elements

$$D_{n,m} = \left\{\begin{array}{cl} 1/2n & \text{if } n = m+1 \\ -1/2n & \text{if } n = m-1 \\ 0 & \text{if } n \neq m \pm 1 \end{array}\right\}$$

for $n = 1, \dots, N$, $m = 1, \dots, N$, $\mathbf{d}$ is a $N \times 1$ vector with elements

$$d_n = \left\{\begin{array}{cl} 1/2 & \text{if } n = 1 \\ 0 & \text{if } n \neq 1 \end{array}\right\}$$

$\mathbf{b}$ is a $N \times 1$ vector with elements b_n given by Eq. (3.274) or (3.275), $\mathbf{c}$ is a $N \times 1$ vector with elements $c_n = 2/n$ and, finally, $\boldsymbol{\lambda}$ is a $N \times 1$ vector with elements λ_n. Using the same notation, the lift equation (3.286) becomes

$$l = -\rho\pi b^2\left(\ddot{h} - ba\ddot{\alpha} + U_{\infty}\dot{\alpha}\right) - 2\pi\rho U_{\infty} b\left(U_{\infty}\alpha(t) + \dot{h}(t) + b\left(\frac{1}{2} - a\right)\dot{\alpha}(t) - \frac{1}{2}\mathbf{b}^T\boldsymbol{\lambda}\right) \tag{3.290}$$

Example 3.21 *Calculate the lift on an impulsively started flat plate airfoil using the finite state approximation.*

Here, we will repeat Example 3.7 using finite state theory. The finite state equations (3.289) are continuous so we need to express the time variation of the downwash due to the motion using a continuous function. An impulsive start denotes that the angle of attack of the airfoil is constant and equal to α, the plunge motion is always equal to zero, $h = 0$, while the airspeed jumps from zero to U_{∞} impulsively. Therefore, we can write the downwash around the three-quarter chord as

$$w_{3/4}(t) = U_{\infty}u(t)\alpha$$

where $u(t)$ is the Heaviside step function, defined as

$$u(t) = \left\{\begin{array}{ll} 1 & \text{if } t > 0 \\ 0 & \text{if } t \leq 0 \end{array}\right.$$

The derivative of the Heaviside function is the Dirac delta function, $\delta(t)$, so that $\dot{w}_{3/4}(t) = U_\infty\delta(t)\alpha$. Hence, Eq. (3.289) becomes

$$\boldsymbol{A}\dot{\boldsymbol{\lambda}} + \frac{U_\infty}{b}\boldsymbol{\lambda} = \mathbf{c}U_\infty\delta(t)\alpha$$

and can be re-written as

$$\dot{\boldsymbol{\lambda}} + \bar{\boldsymbol{A}}\boldsymbol{\lambda} = \bar{\mathbf{c}}U_\infty\delta(t)\alpha$$

where $\bar{\mathbf{A}} = (U_\infty/b)\boldsymbol{A}^{-1}$ and $\bar{\mathbf{c}} = \boldsymbol{A}^{-1}\mathbf{c}$. This latest equation can be solved analytically by pre-multiplying both sides by the matrix exponential $e^{\bar{\mathbf{A}}t}$ such that

$$e^{\bar{\mathbf{A}}t}\dot{\boldsymbol{\lambda}} + e^{\bar{\mathbf{A}}t}\bar{\boldsymbol{A}}\boldsymbol{\lambda} = e^{\bar{\mathbf{A}}t}\bar{\mathbf{c}}U_\infty\delta(t)\alpha$$

Thanks to the properties of the matrix exponential, $e^{\bar{\mathbf{A}}t}\bar{\boldsymbol{A}} = \bar{\mathbf{A}}e^{\bar{\mathbf{A}}t}$ and, hence, the left-hand side is a total derivative, leading to

$$\frac{\mathrm{d}}{\mathrm{d}t}\left(e^{\bar{\mathbf{A}}t}\boldsymbol{\lambda}\right) = e^{\bar{\mathbf{A}}t}\bar{\mathbf{c}}U_\infty\delta(t)\alpha$$

We can integrate both sides of this latest expression between times 0 and t to obtain

$$e^{\bar{\mathbf{A}}t}\boldsymbol{\lambda}(t) - \boldsymbol{\lambda}(0) = U_\infty\alpha\int_0^t e^{\bar{\mathbf{A}}t_0}\bar{\mathbf{c}}\delta(t_0)\mathrm{d}t_0$$

where t_0 is an integration variable. We can make the assumption that

$$\int_0^t e^{\bar{\mathbf{A}}t_0}\bar{\mathbf{c}}\delta(t_0)\mathrm{d}t_0 = \int_{-\infty}^{\infty} e^{\bar{\mathbf{A}}t_0}\bar{\mathbf{c}}\delta(t_0)\mathrm{d}t_0$$

because the only motion occurred at time $t = 0$. Combining this assumption with the properties of the Dirac delta function, we obtain

$$\int_0^t e^{\bar{\mathbf{A}}t_0}\bar{\mathbf{c}}\delta(t_0)\mathrm{d}t_0 = e^{\bar{\mathbf{A}}\times 0}\bar{\mathbf{c}} = \bar{\mathbf{c}}$$

Finally, setting the initial conditions to $\boldsymbol{\lambda}(0) = 0$, the solution for $\boldsymbol{\lambda}(t)$ becomes

$$\boldsymbol{\lambda}(t) = U_\infty\alpha e^{-\bar{\mathbf{A}}t}\bar{\mathbf{c}}$$

The property of the matrix exponential $e^{-\bar{\mathbf{A}}t} = \mathbf{V}e^{\mathbf{M}t}\boldsymbol{V}^{-1}$, where $\boldsymbol{V}$ is the matrix of eigenvectors of $-\bar{\mathbf{A}}$ and $\boldsymbol{M}$ is its eigenvalue matrix, can be used to write a more practical form of the solution for $\boldsymbol{\lambda}(t)$, that is

$$\boldsymbol{\lambda}(t) = U_\infty\alpha\sum_{n=1}^{N}\mathbf{v}_n e^{\mu_n t}\bar{\bar{c}}_n$$

where $\mathbf{v}_n$ is the nth column of $\boldsymbol{V}$, μ_n is the nth diagonal element of $\boldsymbol{M}$ and $\bar{\bar{c}}_n$ is the nth element of vector $\bar{\bar{\mathbf{c}}} = \boldsymbol{V}^{-1}\bar{\mathbf{c}}$.

Using Eq. (3.290), the lift acting on the impulsively started plate becomes

$$l = -2\pi\rho U_\infty b\left(U_\infty\alpha - \frac{1}{2}\mathbf{b}^T\boldsymbol{\lambda}\right) = -2\rho U_\infty^2\pi b\alpha\left(1 - \frac{1}{2}\mathbf{b}^T\sum_{n=1}^{N}\mathbf{v}_n e^{\mu_n t}\bar{\bar{c}}_n\right) \tag{3.291}$$

while the steady-state lift is given by $l_\infty = -2\rho U_\infty^2 \pi b\alpha$ *(Eq. (3.86)). Comparing to the definition of the Wagner function, Eq. (3.91), it becomes clear that*

$$\phi_{fs} = 1 - \frac{1}{2}\mathbf{b}^T \sum_{n=1}^{N} \mathbf{v}_n e^{\mu_n t} \bar{\bar{c}}_n \tag{3.292}$$

is the finite state approximation of Wagner's function.

We calculate the impulsive lift for $U_\infty = 10$ *m/s,* $\alpha = 5°$ *and* $b = 0.5$ *m, using* $N = 8$ *and the* b_n *values from Eq. (3.275), for times between 0 and 10 seconds. Figure 3.41a plots the lift time response evaluated from Eq. (3.291), along with the result obtained by means of Wagner theory in Example 3.7. It can be seen that the two lift estimates are very similar. Furthermore, Figure 3.41b plots the finite state estimate for Wagner's function from Eq. (3.292), along with the result from Wagner theory (numerical evaluation of Eq. (3.93)). Again, the finite state estimate lies quite close to the Wagner result, particularly for non-dimensional times* $\tau < 50$*. For* $\tau > 50$*, the finite state estimate of the Wagner function approaches 1 faster than the Bessel function result. However,* $\tau = 50$ *means that the airfoil has travelled 25 chord-lengths since the start of the motion. For a general unsteady motion, any newer vortices will be much more important than the starting vortex that lies 25 chords downstream.*

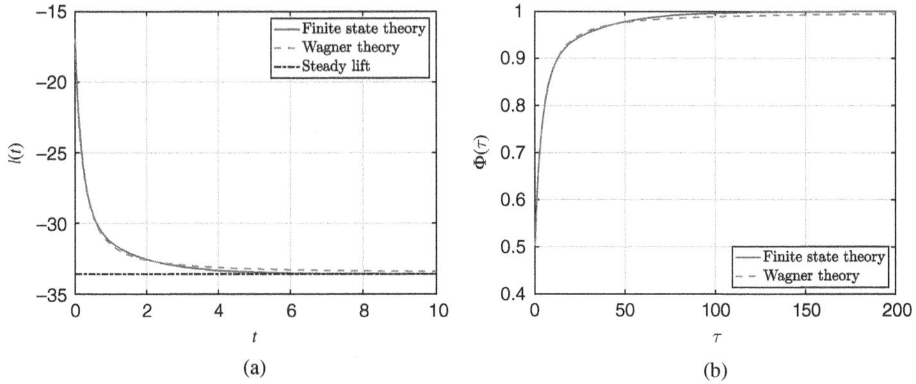

Figure 3.41 Lift acting on an impulsively started airfoil and Wagner function for $N = 8$. (a) Impulsive lift and (b) Wagner function.

It should be kept in mind that finite state theory is an approximation of Wagner theory and its results depend on the choice of the number of states, N, and on the values of b_n*. Increasing N does not always result in better aerodynamic estimates. In fact, for better results, the number of states must be even and must not exceed 12. If* $N > 12$*, the solution for* $\lambda(t)$ *becomes unstable, due to the instability inherent in the calculation of* b_n *discussed in Example 3.20. This example is solved by Matlab code* `impulsive_finitestate.m`*.*

The pitching moment around the pitching axis is given by

$$m_{xf} = -\int_{-b}^{b} \Delta p_b (x - ba) \mathrm{d}x$$

where, again, the minus sign is due to the inversion of the definition of the pressure difference. Substituting from Eq. (3.281) with $\tau_a = \tau_s$ and $x = b\cos\beta$, we obtain

$$m_{xf} = b^2 \int_{\pi}^{0} 2\left(\frac{\tau_a}{\sin\beta} - \frac{\tau_a\cos\beta}{\sin\beta} + \sum_{n=1}^{N}\tau_n \sin n\beta\right)(\cos\beta - a)\sin\beta\, d\beta$$

Carrying out the integration yields

$$m_{xf} = \pi b^2 \left(\tau_a - \frac{\tau_2}{2} + 2a\left(\tau_a + \frac{\tau_1}{2}\right)\right)$$

Substituting from the lift expression (3.282)

$$m_{xf} = \pi b^2 \left(\tau_a - \frac{\tau_2}{2}\right) - abl$$

Furthermore, substituting from expressions (3.277) and (3.279)

$$m_{xf} = \pi b^2 \left(\rho U_\infty (w_0 - \lambda_0) - \frac{\frac{\rho b}{4}\left(\dot{w}_1 - \dot{w}_3\right) + \rho U_\infty w_2}{2}\right) - abl$$

Finally, Eqs. (3.285) and (3.286) can be substituted into the expression above to obtain

$$\begin{aligned} m_{xf} &= -\rho\pi b^2 \left(-ba\ddot{h} + b^2\left(\frac{1}{8} + a^2\right)\ddot{\alpha}\right) - \rho U_\infty b^3 \pi \left(\frac{1}{2} - a\right)\dot{\alpha}(t) \\ &\quad + 2\rho U_\infty \pi b^2 \left(a + \frac{1}{2}\right)\left(U_\infty \alpha(t) + \dot{h}(t) + b\left(\frac{1}{2} - a\right)\dot{\alpha}(t) - \lambda_0\right) \end{aligned} \tag{3.293}$$

which is of the same form as the pitching moment obtained from Wagner theory (Eq. (3.121)). Again, setting $\lambda_0 = 0$ results in the quasi-steady moment equation (3.138). Recalling that $\lambda_0 = \frac{1}{2}\mathbf{b}^T\lambda$ and assembling Eqs. (3.289), (3.290) and (3.293), we obtain the complete lift and moment equations for the finite state approximation as

$$\begin{aligned} l &= -\rho\pi b^2 \left(\ddot{h} - ba\ddot{\alpha} + U_\infty \dot{\alpha}\right) \\ &\quad - 2\pi\rho U_\infty b\left(U_\infty \alpha(t) + \dot{h}(t) + b\left(\frac{1}{2} - a\right)\dot{\alpha}(t) - \frac{1}{2}\mathbf{b}^T\lambda\right) \end{aligned} \tag{3.294}$$

$$\begin{aligned} m_{xf} &= -\rho\pi b^2 \left(-ba\ddot{h} + b^2\left(\frac{1}{8} + a^2\right)\ddot{\alpha}\right) - \rho U_\infty b^3 \pi \left(\frac{1}{2} - a\right)\dot{\alpha}(t) \\ &\quad + 2\rho U_\infty \pi b^2 \left(a + \frac{1}{2}\right)\left(U_\infty \alpha(t) + \dot{h}(t) + b\left(\frac{1}{2} - a\right)\dot{\alpha}(t) - \frac{1}{2}\mathbf{b}^T\lambda\right) \end{aligned} \tag{3.295}$$

$$\boldsymbol{A}\dot{\lambda} + \frac{U_\infty}{b}\lambda = \mathbf{c}\left(U_\infty \dot{\alpha}(t) + \ddot{h}(t) + b\left(\frac{1}{2} - a\right)\ddot{\alpha}(t)\right) \tag{3.296}$$

The drag can also be calculated using the finite state approximation. Equation (3.123) gives the drag in terms of the circulatory lift $l_\Gamma(t)$, as obtained from Wagner theory. The finite state circulatory lift is given by

$$l_\Gamma(t) = -2\pi\rho U_\infty b\left(U_\infty \alpha(t) + \dot{h}(t) + b\left(\frac{1}{2} - a\right)\dot{\alpha}(t) - \frac{1}{2}\mathbf{b}^T\lambda\right)$$

so that the drag becomes

$$d(t) = -\rho\pi\frac{b}{2}\left(-2\left(U_\infty \alpha(t) + \dot{h}(t) + b\left(\frac{1}{2} - a\right)\dot{\alpha}(t) - \frac{1}{2}\mathbf{b}^T\lambda\right) + b\dot{\alpha}(t)\right)^2 - \alpha(t)l(t) \tag{3.297}$$

where $l(t)$ is obtained from Eq. (3.294).

Example 3.22 *Calculate the lift and moment acting on a 2D flat plate wing undergoing sinusoidal pitching and plunging motion using finite state theory and compare to the Wagner, Theodorsen and quasi-steady results.*

We repeat Examples 3.12 and 3.13 using finite state theory. The first task is to solve Eq. (3.296) for sinusoidal motion. We write

$$\dot{w}_{3/4}(t) = U_\infty \dot{\alpha}(t) + \ddot{h}(t) + b\left(\frac{1}{2} - a\right)\ddot{\alpha}(t)$$

and then follow the approach of Example 3.21 to write Eq. (3.296) as

$$e^{\bar{A}t}\boldsymbol{\lambda}(t) - \boldsymbol{\lambda}(0) = \int_0^t e^{\bar{A}t_0}\bar{\mathbf{c}}\dot{w}_{3/4}(t_0)dt_0$$

Choosing $\boldsymbol{\lambda}(0) = \mathbf{0}$ *and pre-multiplying the entire equation by* $e^{-\bar{A}t}$ *results in*

$$\boldsymbol{\lambda}(t) = \int_0^t e^{-\bar{A}(t-t_0)}\bar{\mathbf{c}}\dot{w}_{3/4}(t_0)dt_0$$

or after applying the eigendecomposition $e^{-\bar{A}(t-t_0)} = \boldsymbol{V}e^{-\boldsymbol{M}(t-t_0)}\boldsymbol{V}^{-1}$,

$$\boldsymbol{\lambda}(t) = \int_0^t \sum_{n=1}^{N}\mathbf{v}_n e^{\mu_n(t-t_0)}\bar{\bar{c}}_n \dot{w}_{3/4}(t_0)dt_0$$

where $\mathbf{v}_n$ *is the nth column of* $\boldsymbol{V}$, μ_n *is the nth diagonal element of* $\boldsymbol{M}$ *and* $\bar{\bar{c}}_n$ *is the nth element of vector* $\bar{\bar{\mathbf{c}}} = \boldsymbol{V}^{-1}\bar{\mathbf{c}}$.

We write sinusoidal expressions for $h(t)$ *and* $\alpha(t)$ *(see Eq. (3.132))*

$$h(t) = h_0 \sin\left(\omega t + \phi_h\right), \quad \alpha(t) = \alpha_0 \sin\left(\omega t + \phi_\alpha\right) \tag{3.298}$$

such that

$$\dot{w}_{3/4}(t) = U_\infty \omega\alpha_0 \cos\left(\omega t + \phi_\alpha\right) - \omega^2 h_0 \sin(\omega t + \phi_h) - \omega^2 b\left(\frac{1}{2} - a\right)\alpha_0 \sin\left(\omega t + \phi_\alpha\right)$$

and

$$\begin{aligned}\boldsymbol{\lambda}(t) = U_\infty\omega\alpha_0 &\int_0^t \sum_{n=1}^{N}\bar{\bar{c}}_n\mathbf{v}_n e^{\mu_n(t-t_0)}\cos\left(\omega t_0 + \phi_\alpha\right)dt_0 \\ &- \omega^2 h_0 \int_0^t \sum_{n=1}^{N}\bar{\bar{c}}_n\mathbf{v}_n e^{\mu_n(t-t_0)}\sin\left(\omega t_0 + \phi_h\right)dt_0 \\ &- \omega^2 b\left(\frac{1}{2} - a\right)\alpha_0 \int_0^t \sum_{n=1}^{N}\bar{\bar{c}}_n\mathbf{v}_n e^{\mu_n(t-t_0)}\sin\left(\omega t_0 + \phi_\alpha\right)dt_0\end{aligned}$$

The three integrals in this latest expression can be evaluated easily from

$$\begin{aligned}\int_0^t e^{\mu_n(t-t_0)}\sin\left(\omega t_0 + \phi_{h,\alpha}\right)dt_0 = &\frac{e^{\mu_n t}(\omega\cos\phi_{h,\alpha} + \mu_n\sin\phi_{h,\alpha})}{\mu_n^2 + \omega^2} \\ &- \frac{\omega\cos\left(\omega t + \phi_{h,\alpha}\right) + \mu_n\sin\left(\omega t + \phi_{h,\alpha}\right)}{\mu_n^2 + \omega^2}\end{aligned}$$

$$\int_0^t e^{\mu_n(t-t_0)}\cos\left(\omega t_0+\phi_{h,\alpha}\right)dt_0 = \frac{e^{\mu_n t}(\mu_n\cos\phi_{h,\alpha}-\omega\sin\phi_{h,\alpha})}{\mu_n^2+\omega^2} - \frac{\mu_n\cos\left(\omega t+\phi_{h,\alpha}\right)-\omega\sin\left(\omega t+\phi_{h,\alpha}\right)}{\mu_n^2+\omega^2}$$

Finally, $\lambda(t)$ is substituted into Eqs. (3.294), (3.295) and (3.297) to obtain the lift, pitching moment and drag, respectively.

Figure 3.42 plots the finite state solution for $N = 8$ and the b_n values from Eq. (3.275), along with the Wagner theory and quasi-steady predictions. The steady-state responses of the three aerodynamic loads obtained from finite state theory are nearly identical with the Wagner results, but there is a discrepancy between the two theories at the start of the motion. This is due to the fact that, at $t = 0$, the finite state approximation reduces to the quasi-steady result, since we have chosen $\boldsymbol{\lambda}(0) = \mathbf{0}$ and, hence, $\lambda_0(0) = 0$. The initial values of the Wagner aerodynamic loads are not equal to the quasi-steady estimates, hence, the discrepancy mentioned above.

We repeat the calculation for 50 reduced frequency values between $k = 0.001$ and $k = 3$. For each frequency, we calculate the ratio between the circulatory finite state and quasi-steady lift

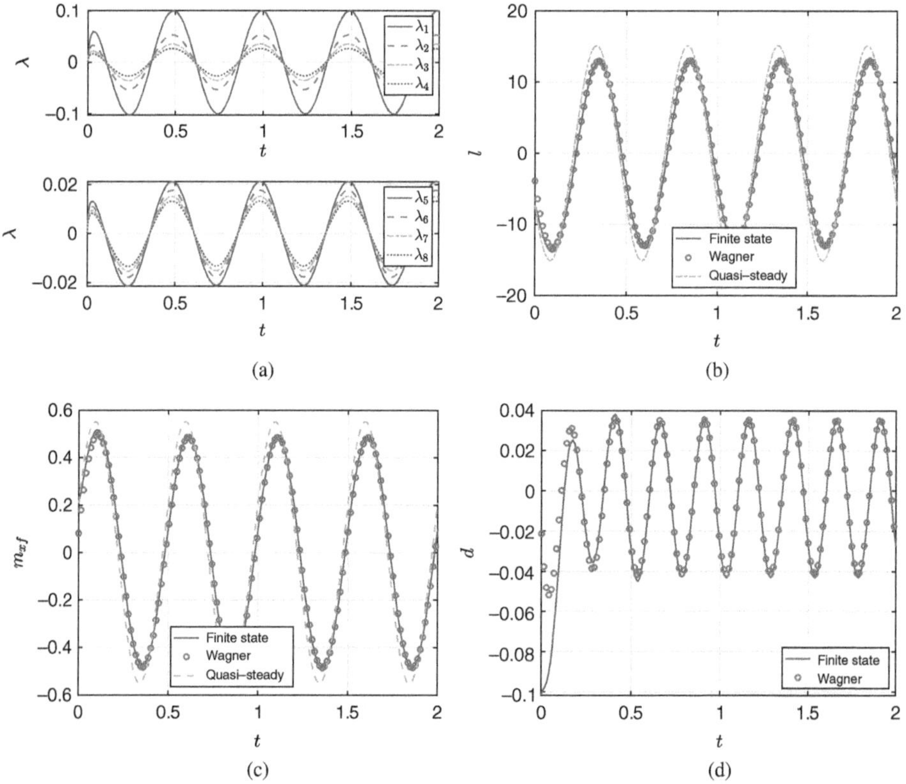

Figure 3.42 Aerodynamic states and loads of sinusoidally oscillating 2D pitch-plunge wing. (a) $\lambda(t)$, (b) $l(t)$, (c) $m_{xf}(t)$ and (d) $d(t)$.

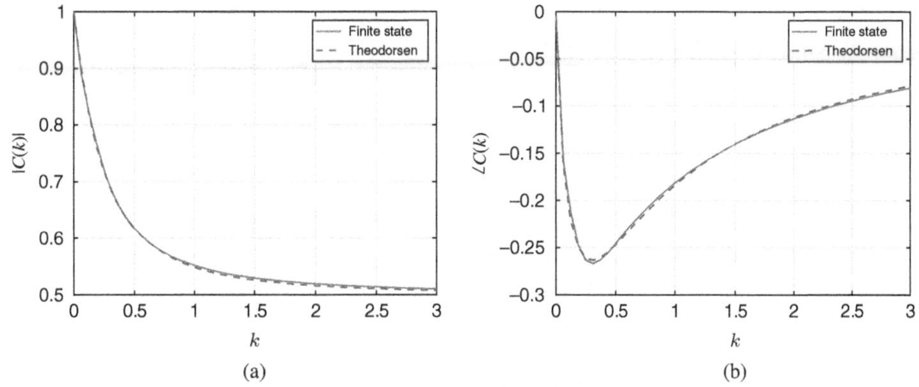

Figure 3.43 Theodorsen's function estimate from finite state theory. (a) Amplitude and (b) phase.

amplitudes and plot it against k in Figure 3.43a. It can be seen that this amplitude ratio approximates quite well the amplitude of Theodorsen's function. According to Peters et al. (1995), Theodorsen's function can be estimated from finite state theory using

$$C(k) = \frac{1 + \frac{1}{2}\bar{\lambda}_1}{1 + \bar{\lambda}_a + \frac{1}{2}\bar{\lambda}_1}$$

where

$$\bar{\lambda} = \left((\boldsymbol{D} + \mathbf{d}\mathbf{b}^T)ik + \boldsymbol{I}\right)^{-1}\mathbf{c}ik, \quad \bar{\lambda}_a = \frac{1}{2}\mathbf{b}^T\bar{\lambda}, \quad \frac{1}{2}\bar{\lambda}_1 = \mathbf{d}^T\bar{\lambda}$$

Figure 3.43 shows that both the amplitude and phase angle of Theodorsen's function are estimated accurately by finite state theory in the chosen frequency range. This example is solved by Matlab code `pitchplunge_Finitestate.m`.

3.7 Concluding Remarks

In this chapter, we have presented four analytical models for predicting the unsteady aerodynamic loads acting on 2D flat plates undergoing two types of motion: impulsive start and pitching and plunging. Other types of motion can be easily implemented, such as chordwise deformation or surging (unsteady streamwise motion). The four analytical models are Wagner theory, Theodorsen theory, quasi-steady aerodynamics and finite state theory. The fundamental assumptions behind the models must always be kept in mind before any application can be attempted; all of them assume that the motion is of small amplitude and frequency, the airfoil is a thin flat plate, the flow is inviscid, incompressible and remains attached to the airfoil at all times, except at the trailing edge, where it separates smoothly. For quasi-steady aerodynamic theory to be applicable, the reduced frequency must be very low. Note that the quasi-steady model presented here includes non-circulatory loads, but other forms of this model do not include these loads since their magnitude is very small at low reduced frequencies, see for example Berci (2021).

Analytical models can represent flat plates or even cambered flat plates, but they cannot model the effect of the thickness distribution of an airfoil. In order to achieve the latter, we must resort to numerical panel methods, which will be introduced in Chapter 4.

3.8 Exercises

1 Calculate the unsteady aerodynamic loads acting on an impulsively started flat plate lying at a zero angle of attack to the free stream but with a control surface that has been deflected to a constant angle β. Use the load equations given in Theodorsen (1935).

2 Repeat Example 3.15 using Wagner's model of Example 3.12. How similar are the predictions of Theodorsen's and Wagner's theory?

3 Repeat Example 3.16 using Wagner theory. Try airspeeds higher than $U_\infty = 50.47$; how similar are the time and frequency responses predicted by Wagner theory and Theodorsen theory at such airspeeds?

References

Abbott, I.H. and Von Doenhoff, A.E. (1959). *Theory of Wing Sections: Including a Summary of Airfoil Data*. New York: Dover Publications, Inc.

Anderson, J.D. Jr. (1985). *Fundamentals of Aerodynamics*. McGraw-Hill International Editions.

Berci, M. (2021). On aerodynamic models for flutter analysis: a systematic overview and comparative assessment. *Applied Mechanics* 2: 516–541.

Bisplinghoff, R.L., Ashley, H., and Halfman, R.L. (1996). *Aeroelasticity*. New York: Dover Publications, Inc.

Brunton, S.L. and Rowley, C.W. (2013). Empirical state-space representations for Theodorsen's lift model. *Journal of Fluids and Structures* 38: 174–186.

Dimitriadis, G. (2017). *Introduction to Nonlinear Aeroelasticity*. Chichester, West Sussex: Wiley.

Dowell, E.H. (ed.) (2004). *A Modern Course in Aeroelasticity*, 4e. Kluwer Academic Publishers.

Fung, Y.C. (1993). *An Introduction to the Theory of Aeroelasticity*. Dover Publications, Inc.

Garrick, I.E. (1937). Propulsion of a Flapping and Oscillating Airfoil. *Technical Report TR-567*. NACA.

Garrick, I.E. (1938). On Some Reciprocal Relations in the Theory of Nonstationary Flows. *Technical Report TR-629*. NACA.

Glauert, H. (1947). *The Elements of Aerofoil and Airscrew Theory*, 2e. New York: Cambridge University Press.

Grammel, R. (1917). *Die hydrodynamischen Grundlagen des Fluges*. Braunschweig: F. Vieweg & sohn.

Gülçat, U. (2016). *Fundamentals of Modern Unsteady Aerodynamics*, 2e. Springer.

Hodges, D.H. and Pierce, G.A. (2002). *Introduction to Structural Dynamics and Aeroelasticity*. Cambridge: Cambridge University Press.

Izraelevitz, J., Zhu, Q., and Triantafyllou, M.S. (2017). State-space adaptation of unsteady lifting line theory: twisting/flapping wings of finite span. *AIAA Journal* 55 (4): 1279–1294.

Johnson, W. (1980). *Helicopter Theory*. New York: Dover Publications, Inc.

Jones, R.T. (1940). The Unsteady Lift of a Wing of Finite Aspect Ratio. *Technical Report TR-681*. NACA.

Jones, W.P. (1945). Aerodynamic Forces on Wings in Non-Uniform Motion. *Technical Report R&M 2217*. Aeronautical Research Council.

Kuethe, A.M. and Chow, C.Y. (1986). *Foundations of Aerodynamics - Basis of Aerodynamic Design*. Wiley.

Lee, B.H.K., Price, S.J., and Wong, Y.S. (1999). Nonlinear aeroelastic analysis of airfoils: bifurcation and chaos. *Progress in Aerospace Sciences* 35 (3): 205–334.

McAlister, K.W., Pucci, S.L., McCroskey, W.J., and Carr L.W. (1982). An Experimental Study of Dynamic Stall on Advanced Airfoil Sections, Volume 2. Pressure and Force Data. Technical Memorandum TM 84245. NASA.

Milne-Thomson, L.M. (1973). *Theoretical Aerodynamics*, 4e. New York: Dover Publications, Inc.

Peters, D.A. (2008). Two-dimensional incompressible unsteady airfoil theory - an overview. *Journal of Fluids and Structures* 24: 295–312.

Peters, D.A., Karunamoorthy, S., and Cao, W.M. (1995). Finite state induced flow models part I: two-dimensional thin airfoil. *Journal of Aircraft* 32 (2): 313–322.

Reyes, M., Climent, H., Karpel, M. et al. (2019). Examples on increased-order aeroservoelastic modeling. *CEAS Aeronautical Journal* 10: 1071–1087.

Satyanarayana, B. and Davis, S. (1978). Experimental studies of unsteady trailing-edge conditions. *AIAA Journal* 16 (2): 125–129.

Scanlan, R.H. and Rosenbaum, R. (1951). *Aircraft Vibration and Flutter*. New York: The MacMillan Company.

Stephenson, G. (1973). *Mathematical Methods for Science Students*, 2e. Longman Scientific & Technical.

Theodorsen, T. (1935). General Theory of Aerodynamic Instability and the Mechanism of Flutter. *Technical Report TR-496*. NACA.

von Kármán, T. and Burgers, J.M. (1935). General aerodynamic theory - perfect fluids. In: *Aerodynamic Theory*, vol. II (ed. W.F. Durand). Berlin: Julius Springer. 52–53.

Wagner, H. (1925). Über die entstehung des dynamischen auftriebes von tragflügeln. *Zeitschrift für Angewandte Mathematik und Mechanik* 5 (1): 17–35.

Wright, J.R. and Cooper, J.E. (2015). *Introduction to Aircraft Aeroelasticity and Loads*, 2e. Chichester: Wiley.

4

Numerical Incompressible 2D Models

4.1 Introduction

The analytical unsteady aerodynamic models presented in Chapter 3 are reliable and fast to calculate, but they have their limitations. Thickness is ignored, camber is represented by imposing an upwash on the chord line, all angles are small and the wake is modelled as flat and propagating at the free stream airspeed. These modelling choices are dictated by the need to obtain analytical results for the integrals involved in the calculation of the influence of the singularities on the surface of the airfoil. Numerical methods do not require analytical integral evaluations, at least not for the complete airfoil. As a consequence, many of these assumptions can be relaxed and the resulting models can reflect more of the physics of the unsteady aerodynamic problem. In this chapter, we will present numerical modelling methods of increasing complexity for 2D incompressible unsteady aerodynamics. We will therefore be able to model the effects of camber and thickness more accurately and to allow the wake to deform and propagate at the local flow velocity. In fact, the numerical methods that are presented in this chapter do not even require small motion amplitudes, but they can only represent flows that remain attached to the airfoil's surface at all times and only separate at the trailing edge.

4.2 Lumped Vortex Method

The lumped vortex method is essentially a discrete unsteady thin airfoil theory. Recall that the finite state theory of Section 3.6 represents the flow using continuous bound and wake vorticity distributions, leading to the integral equation (3.219), which is then solved analytically by expressing these vorticity distributions as finite series. The numerical alternative is to place a finite number of vortices both on the airfoil's surface and in the wake and thus replace the integral impermeability equation by a sum. Both approaches ignore the thickness of the airfoil, but the lumped vortex method imposes impermeability near the camber line while thin airfoil theory (see Section 3.2) imposes it on the chord line, modelling the camber by means of the upwash it generates.

Unsteady Aerodynamics: Potential and Vortex Methods, First Edition. Grigorios Dimitriadis.

Companion website: www.wiley.com/go/dimitriadis/unsteady_aerodynamics

Figure 4.1 represents the lumped vortex modelling scheme on a cambered airfoil immersed in a free stream of speed Q_∞ at an angle of attack α. Only the camber of the airfoil is modelled, and the thickness is assumed to be infinitesimal. There are 10 bound vortices on the camber line and 4 vortices in the wake; the strengths of all 14 vortices must be determined in order to fully define the flowfield and calculate the aerodynamic loads. We will discuss the placement of these vortices shortly but, first, the equations for calculating the vortex strengths will be presented. The n bound vortices are denoted by Γ_{b_j} and lie on points x_{b_j}, y_{b_j} for $j = 1, \dots, n$, while the k wake vortices are denoted by Γ_{w_j} and lie on points x_{w_j}, y_{w_j} for $j = 1, \dots, k$. Appendix A.2 gives the complex velocity $V = u - iv$ induced by a single point vortex of strength Γ lying on point $z_0 = x_0 + iy_0$ at a general point $z = x + iy$ as

$$V(z) = i\frac{\Gamma}{2\pi}\frac{1}{z - z_0}$$

Then, the horizontal and vertical velocities are given by

$$\begin{aligned} u &= \frac{\Gamma}{2\pi}\frac{y - y_0}{(x - x_0)^2 + (y - y_0)^2} \\ v &= -\frac{\Gamma}{2\pi}\frac{x - x_0}{(x - x_0)^2 + (y - y_0)^2} \end{aligned} \tag{4.1}$$

The total velocity induced by all the bound vortices and wake vortices on a general point x, y is then given by

$$\begin{aligned} u(x,y) &= \sum_{j=1}^{n}\frac{\Gamma_{b_j}}{2\pi}\frac{y - y_{b_j}}{(x - x_{b_j})^2 + (y - y_{b_i})^2} + \sum_{j=1}^{k}\frac{\Gamma_{w_j}}{2\pi}\frac{y - y_{w_j}}{(x - x_{w_j})^2 + (y - y_{w_j})^2} \\ v(x,y) &= -\sum_{j=1}^{n}\frac{\Gamma_{b_j}}{2\pi}\frac{x - x_{b_j}}{(x - x_{b_j})^2 + (y - y_{b_j})^2} - \sum_{j=1}^{k}\frac{\Gamma_{w_j}}{2\pi}\frac{x - x_{w_j}}{(x - x_{w_j})^2 + (y - y_{w_j})^2} \end{aligned} \tag{4.2}$$

The impermeability boundary condition is not imposed on the airfoil's surface but on its camber line. Hence, if point $\mathbf{x}_c = (x_c, y_c)$ lies on the camber line, the strengths of the vortices Γ_{b_j} and Γ_{w_j} must be such that the total velocity normal to the camber line is equal to zero at all times. Using the motion scheme of Figure 2.2, the impermeability boundary condition of Eq. (2.43) becomes

$$\left(\nabla\phi(\mathbf{x}_c, t) + \mathbf{Q}_\infty - \mathbf{v}_0(t) - \boldsymbol{\Omega}_0(t) \times (\mathbf{x}_c(t) - \mathbf{x}_f(t))\right) \cdot \mathbf{n}(\mathbf{x}_c, t) = 0 \tag{4.3}$$

where $\mathbf{n}(\mathbf{x}_c, t)$ is a unit vector normal to the camber line at point $\mathbf{x}_c$, $\nabla\phi(\mathbf{x}_c, t)$ is the perturbation flow velocity at $\mathbf{x}_c$ and

$$\mathbf{u}_m(t) = \mathbf{Q}_\infty - \mathbf{v}_0(t) - \boldsymbol{\Omega}_0(t) \times (\mathbf{x}_c(t) - \mathbf{x}_f(t)) \tag{4.4}$$

is the relative velocity between the wing and the flow due to both the free stream and the body's motion. If Eq. (4.3) is evaluated at n points on the camber line, it will become a system of n linear algebraic equations involving the n bound vortices and the k wake vortices. Three issues must be highlighted:

- As discussed in Section 3.4 and elsewhere in Chapter 3, the wake vorticity depends on the bound vorticity at previous time instances.

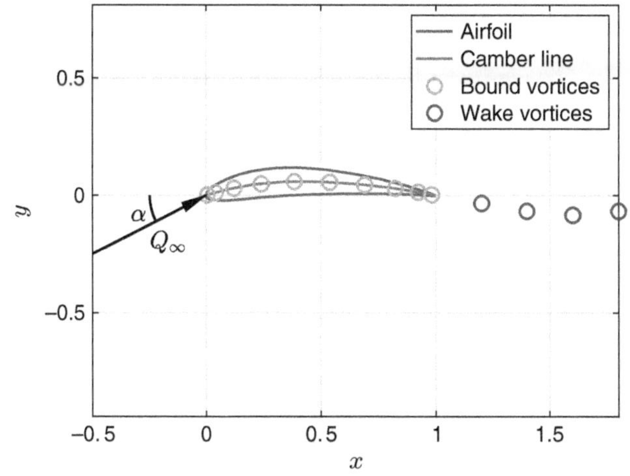

Figure 4.1 Lumped vortex representation of unsteady flow around 2D airfoil.

- The positions of the bound and wake vortices must be selected carefully.
- The points on the wing where the impermeability boundary condition is evaluated must also be selected carefully. These points are known as the control or collocation points.

The first issue is linked to Kelvin's theorem and will be addressed later. The last two issues are linked to the imposition of the Kutta condition. In order to address them, we will consider the simplest possible case, a static uncambered flat plate modelled by a single bound vortex.

Example 4.1 *Solve the lifting flow on a static flat plate airfoil at angle of attack α to a free stream Q_∞ using a single bound vortex.*

We will use a single bound vortex and no wake vortices. Furthermore, we will arbitrarily place the bound vortex on the quarter chord of the airfoil and the collocation point on the 3/4 chord point. Figure 4.2 draws the flat plate, aligned with the x axis, the free stream at an angle α, the vortex and the collocation point.

As the flow is steady, we can represent it without a wake model. Furthermore, there is a single bound vortex Γ_b lying on point $x_b = c/4$, $y_v = 0$ and the collocation point lies on $x_c = 3c/4$, $y_c = 0$. If we evaluate Eq. (4.2) at x_c, y_c, they become

$$
\begin{aligned}
u(x_c, y_c) &= 0 \\
v(x_c, y_c) &= -\frac{\Gamma_b}{2\pi} \frac{1}{x_c - x_b} = -\frac{\Gamma_b}{\pi c}
\end{aligned}
\tag{4.5}
$$

since $x_c - x_b = c/2$. The unit vector normal to the surface is $\mathbf{n} = (0, 1)$, the free stream velocity is $\mathbf{Q}_\infty = (Q_\infty \cos\alpha, Q_\infty \sin\alpha)$, the constant perturbation velocity is $\nabla\phi(\mathbf{x}_c) = (u(x_c, y_c), v(x_c, y_c))$, and there is no translation or rotation of the airfoil, so that Eq. (4.3) becomes

$$Q_\infty \sin\alpha - \frac{\Gamma_b}{\pi c} = 0$$

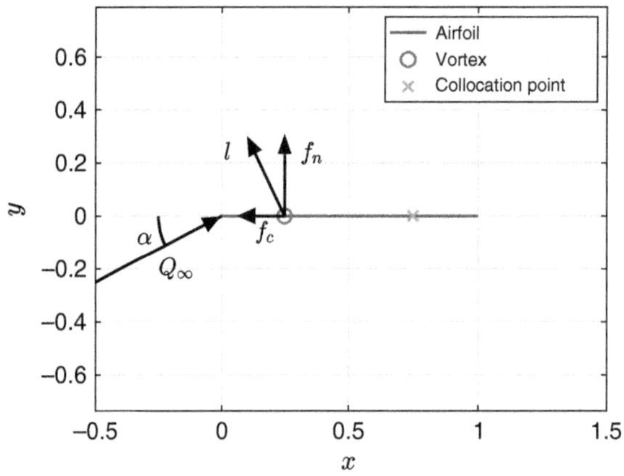

Figure 4.2 Single lumped vortex representation of steady flow around 2D flat plate.

from which we obtain

$$\Gamma_b = \pi c Q_\infty \sin\alpha$$

The normal force acting on the airfoil is obtained from the Kutta–Joukowski theorem of Eq. (2.234), applied to the single vortex, Γ_b*,*

$$f_n = \rho Q_\infty \cos\alpha \Gamma_b = \rho Q_\infty^2 c\pi \sin\alpha\cos\alpha$$

If α *is so small that* $\cos\alpha \approx 1$, $Q_\infty \approx U_\infty$ *and* $\sin\alpha \approx \alpha$*, this normal force becomes identical to the result of Eq. (3.86) for the lift acting on a 2D flat plate airfoil, if the lift is defined as positive downwards. In this chapter, we will define the lift as positive upwards, as this is the normal convention used when discussing panel methods.*

Therefore, by modelling the flat plate as a single lumped vortex placed on the quarter-chord and by enforcing the impermeability condition at the 3/4-chord, we have obtained the exact thin airfoil theory solution for the steady lift acting on a flat plate. However, this does not mean that the flowfield has been modelled accurately. Figure 4.3 plots the velocity vectors around the

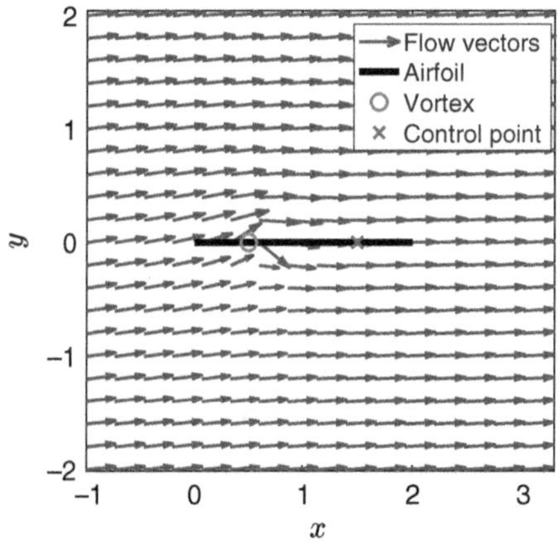

Figure 4.3 Flow vectors around 2D flat plate modelled by a single lumped vortex.

flat plate, calculated for $Q_\infty = 10$ *m/s and* $\alpha = 5°$, $c = 2$ *m. It can be seen that the flat plate is not impermeable; the only point on the surface where impermeability has been enforced is the collocation point. In essence, selecting the appropriate placement for the vortex and the collocation point has allowed a simplistic model to give the exact answer for the lift. This type of approach is quasi-empirical, as the model parameters have been chosen based on the prior knowledge of the answer.*

It should be noted that the value of Γ_b *depends on the distance between the vortex and the collocation point; the position of the vortex itself is irrelevant, as long as it lies at* $c/2$ *from the collocation point. In other words, the vortex could have been placed at the leading edge and the collocation point at the half-chord point. The vortex is placed at the quarter-chord because, according to thin airfoil theory, the lift acts on that point for a flat plate airfoil. The Kutta–Joukowski lift acts on the vortex itself so, as we have replaced the airfoil by a single vortex, the latter must be placed at the quarter chord in order to simulate properly the pitching moment around the leading edge. The single lumped vortex model is usually employed as a far-field approach in order to simulate the effect of an airfoil lying far away from the flow of interest. This example is solved by Matlab code* `steady_lv_single.m`.

Note that the lift direction in Figure 4.2 is perpendicular to the free stream; we only calculate the normal force, which we assumed is approximately equal to the lift. If we use both of the Kutta–Joukowski equations (2.233) and (2.234)

$$\begin{aligned} f_x &= -\rho V_\infty \Gamma_b \\ f_y &= \rho U_\infty \Gamma_b \end{aligned} \tag{4.6}$$

we obtain the lift from

$$l = \sqrt{f_x^2 + f_y^2} = \rho \sqrt{V_\infty^2 + U_\infty^2}\,\Gamma_b = \rho Q_\infty \Gamma_b \tag{4.7}$$

In Chapter 3, we systematically approximated the lift by the aerodynamic force normal to the plate, since the airfoil was aligned with the x axis and the pitch angle was always small. In the present chapter, we will often incline the flat plate to its true angle, so we will be calculating the exact lift vector.

Example 4.1 showed that a single lumped vortex results in an airfoil that is permeable. Furthermore, this vortex cannot model cambered airfoils and is calibrated for steady flows. More physical solutions can be obtained by placing several vortices on the surface of the camber line. Figure 4.4 draws the camber line of a 2D airfoil and its discretisation into four linear panels. By analogy to the single-vortex solution of Example 4.1, we place a vortex on the quarter-chord of each panel. As there are now four vortices, there need to be four boundary conditions in order to solve for the vortex strengths. Consequently, impermeability is enforced on four collocation points, placed on the quarter chord of each panel.

It should be noted that the linear panels in Figure 4.4 are notional; they do not exist in the flowfield. The only singularities present in the flow are the four vortices. The panels serve only to define the positions of the vortices and of the collocation points. As the flow is steady, there is no need for a wake model. Figure 4.4 also draws four unit vectors at each of the collocation points, normal to the panels. As the number of panels is small, the unit vectors are not normal to the true camber line. Increasing the number of panels reduces the

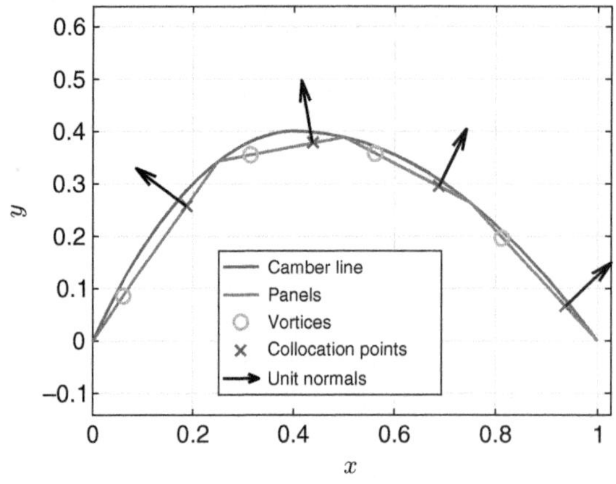

Figure 4.4 Discretisation of camber line by lumped vortex panels.

distances between the true camber line, the vortices and the collocation points and reduces the angle between the vectors normal to the linear panels and normal to the camber line. If the number of panels becomes infinite, the camber line will become truly impermeable and its geometry exact.

The vortex and control point placement scheme described above means that, for steady flows, the Kutta condition can be enforced without the need for a wake model. Consequently, only bound vortices are necessary; the velocities they induce at the ith collocation point can be calculated from Eq. (4.2), such that

$$u(x_{c_i}, y_{c_i}) = \sum_{j=1}^{n} \frac{\Gamma_{b_j}}{2\pi} \frac{y_{c_i} - y_{b_j}}{(x_{c_i} - x_{b_j})^2 + (y_{c_i} - y_{b_j})^2} \tag{4.8}$$

$$v(x_{c_i}, y_{c_i}) = -\sum_{j=1}^{n} \frac{\Gamma_{b_j}}{2\pi} \frac{x_{c_i} - x_{b_j}}{(x_{c_i} - x_{b_j})^2 + (y_{c_i} - y_{b_j})^2} \tag{4.9}$$

Writing the unit normal vector of the ith panel as $\mathbf{n}_i = (n_{x_i}, n_{y_i})$ and substituting Eqs. (4.8) and (4.9) into the boundary condition of Eq. (4.3) yields

$$\sum_{j=1}^{n} \frac{\Gamma_{b_j}}{2\pi} \frac{n_{x_i}(y_{c_i} - y_{b_j}) - n_{y_i}(x_{c_i} - x_{b_j})}{(x_{c_i} - x_{b_j})^2 + (y_{c_i} - y_{b_j})^2} = -Q_\infty \cos \alpha n_{x_i} - Q_\infty \sin \alpha n_{y_i} \tag{4.10}$$

for $i = 1, \dots, n$. This latest expression is a system of n linear equations to be solved for the n unknown vortex strengths Γ_{b_i}. Once the latter have been determined, the incremental lift acting on the ith panel is given by

$$\Delta l_i = \rho Q_\infty \Gamma_{b_i} \tag{4.11}$$

recalling that the lift is defined to be normal to the free stream. The total lift is simply

$$l = \sum_{i=1}^{n} \Delta l_i = \rho Q_\infty \sum_{i=1}^{n} \Gamma_{b_i} \tag{4.12}$$

From Eq. (4.6), the incremental chordwise and normal forces acting on the ith panel are given by

$$\Delta f_{x_i} = -\rho Q_\infty \sin \alpha \Gamma_{b_i}, \quad \Delta f_{y_i} = \rho Q_\infty \cos \alpha \Gamma_{b_i}$$

The incremental pitching moment around the quarter chord acting on the ith panel is given by

$$\Delta m_{c/4_i} = \Delta \mathbf{f}_i \times \mathbf{r}_i \tag{4.13}$$

where $\Delta \mathbf{f}_i = (\Delta f_{x_i}, \Delta f_{y_i})$ is the vectorial incremental aerodynamic load acting on the panel and $\mathbf{r}_i(t_k) = (x_{b_i} - c/4, y_{b_i})$ is the vectorial distance between the ith bound vortex and the quarter chord. Working out the vector product leads to

$$\Delta m_{c/4_i} = \Delta f_{x_i} y_{b_i} - \Delta f_{y_i}(x_{b_i} - c/4)$$

Finally, noting that $\Delta f_{x_i} = -\Delta l_j \sin\alpha$, $\Delta f_{y_i} = \Delta l_j \cos\alpha$ and summing over all panels yields

$$m_{c/4} = -\sum_{j=1}^{n} \Delta l_j y_{b_j} \sin\alpha - \sum_{j=1}^{n} \Delta l_j (x_{b_j} - c/4)\cos\alpha \tag{4.14}$$

Example 4.2 *Model the steady flow around a cambered NACA four-digit airfoil inclined at an angle α to a free stream Q_∞ using n lumped vortices.*

Recall that the lumped vortex approach only models camber lines, not thickness distributions. We will therefore select the camber line as the y coordinate, such that $y(x) = f_c(x)$. Substituting into Eq. (3.6) for the camber line of NACA four-digit airfoils, we obtain

$$\bar{y}(\bar{x}) = \begin{cases} \frac{m}{p^2}\left(2p\bar{x} - \bar{x}^2\right) & \text{for } 0 \le \bar{x} \le p \\ \frac{m}{(1-p)^2}\left((1-2p) + 2p\bar{x} - \bar{x}^2\right) & \text{for } p < \bar{x} \le 1 \end{cases} \tag{4.15}$$

where $\bar{x} = x/c$ is the non-dimensional x coordinate, $\bar{y} = y/c$ is the non-dimensional height of the camber line, m is the maximum camber and p is the chordwise location where the maximum height occurs.

Before applying the lumped vortex approach, we model the flow around the wing using the steady thin airfoil theory of Section 3.2. We calculate the lift and pitching moment around the quarter chord predicted by the thin airfoil theory using the approach of Example 3.1. The next step is to discretise the camber line of the airfoil into n panels. The lengths of the panels are selected to be small near the leading and trailing edges and large near the mid-chord. In order to do this, we select the non-dimensional x-coordinates of the panel vertices from

$$\bar{x}_i = \frac{1}{2}\left(1 - \cos\theta_i\right) \tag{4.16}$$

where θ_i are $n+1$ equally spaced angles lying between 0 and π. The ordinates of the panel vertices are then obtained from Eq. (4.15). Dimensional coordinates are obtained simply from $x = \bar{x}c$ and $y = \bar{y}c$.

The bound vortices and collocation points lie on the quarter chord and 3/4 chord points of each panel, respectively, so their positions are given by

$$x_{b_i} = x_i + (x_{i+1} - x_i)/4, \quad y_{b_i} = y_i + (y_{i+1} - y_i)/4 \tag{4.17}$$

$$x_{c_i} = x_i + 3(x_{i+1} - x_i)/4, \quad y_{c_i} = y_i + 3(y_{i+1} - y_i)/4 \tag{4.18}$$

for i = 1, …, n. Note that there are n + 1 panel end-points, but n panels, bound vortices and collocation points. Unit vectors normal to each panel are obtained from

$$n_{x_i} = -\frac{y_{i+1} - y_i}{\sqrt{(x_{i+1} - x_i)^2 + (y_{i+1} - y_i)^2}}, \quad n_{y_i} = \frac{x_{i+1} - x_i}{\sqrt{(x_{i+1} - x_i)^2 + (y_{i+1} - y_i)^2}} \tag{4.19}$$

for i = 1, …, n. The velocities induced by the bound vortices are written in the form of aerodynamic influence coefficient matrices, $\mathbf{A}_u$ and $\mathbf{A}_v$, which are n × n matrices with elements

$$A_{u_{i,j}} = \frac{1}{2\pi} \frac{y_{c_i} - y_{b_j}}{(x_{c_i} - x_{b_j})^2 + (y_{c_i} - y_{b_j})^2} \tag{4.20}$$

$$A_{v_{i,j}} = -\frac{1}{2\pi} \frac{x_{c_i} - x_{b_j}}{(x_{c_i} - x_{b_j})^2 + (y_{c_i} - y_{b_j})^2} \tag{4.21}$$

for i = 1, …, n, j = 1, …, n, so that Eq. (4.10) becomes

$$\mathbf{A}_n \boldsymbol{\Gamma}_b = \boldsymbol{b} \tag{4.22}$$

where $\mathbf{A}_n$ is the n × n aerodynamic influence coefficient matrix for the direction normal to the camber line, whose elements are given by

$$A_{n_{i,j}} = A_{u_{i,j}} n_{x_i} + A_{v_{i,j}} n_{y_i} \tag{4.23}$$

*while $\boldsymbol{\Gamma}_b = (\Gamma_{b_1}, \ldots, \Gamma_{b_n})^T$ and **b** is a n × 1 vector whose elements are given by*

$$b_i = -Q_\infty \cos \alpha n_{x_i} - Q_\infty \sin \alpha n_{y_i} \tag{4.24}$$

Note that the aerodynamic influence coefficient matrices are purely geometric, since they are not functions of the vortex strengths Γ_{b_j}. Equation (4.22) is identical to Eq. (4.10) but is written in matrix form. In this way, the Γ_{b_j} unknowns are separated from the expressions for the velocities they induce and can be solved for. Once their values are known, the velocities induced on each panel can be obtained by multiplying the appropriate aerodynamic influence coefficient matrix by vector $\boldsymbol{\Gamma}_b$. Hence, the velocities induced on the panels in the x-direction $\boldsymbol{u} = (u(x_{c_1}, y_{c_1}), \ldots, u(x_{c_n}, y_{c_n}))^T$ are given by

$$\boldsymbol{u} = \mathbf{A}_u \boldsymbol{\Gamma}_b$$

while the velocities induced in the y-direction $\boldsymbol{v} = (v(x_{c_1}, y_{c_1}), \ldots, v(x_{c_n}, y_{c_n}))^T$ are given by

$$\boldsymbol{v} = \mathbf{A}_v \boldsymbol{\Gamma}_b$$

These last two expression are matrix versions of Eqs. (4.8) and (4.9).

*The advantage of using aerodynamic influence coefficient matrices is that they are constant with angle of attack, free stream airspeed, airfoil motion, etc. In Eq. (4.22), only vector **b** depends on the angle of attack; $\mathbf{A}_n$ is only a function of airfoil geometry and number of panels. As long as the airfoil's geometry does not change, we can keep using the same value of $\mathbf{A}_n$.*

As an example, we select the NACA 24XX airfoil family, for which m = 0.02 and p = 0.4, with chord c = 0.3 m. Note that the XX in the airfoil name denotes the thickness; as the lumped vortex method does not model thickness, it is irrelevant here. We select the number of panels as $n = 150$ *and apply the complete lumped vortex methodology for a free stream airspeed* $Q_\infty = 10$ *m/s, angle of attack* $\alpha = 10°$ *and* $\rho = 1.225$ kg/m^2.

Figure 4.5a plots the lift vectors calculated using the lumped vortex method, Δl_i*, in a direction perpendicular to the free stream. It can be seen that the lift vectors are longest near the first 8% of the chord and drop to zero at the trailing edge. This situation is consistent with the Kutta condition, which states that the local vorticity is zero at the trailing edge; hence, the local lift should also be zero. Consequently, the lumped vortex method satisfies implicitly the Kutta condition, thanks to the placement of the lumped vortices and control points. Figure 4.5a, which plots the flow vectors around the airfoil, demonstrates that the camber line looks impermeable and that the flow leaves the trailing edge in a direction parallel to the chord line, that is the x axis.*

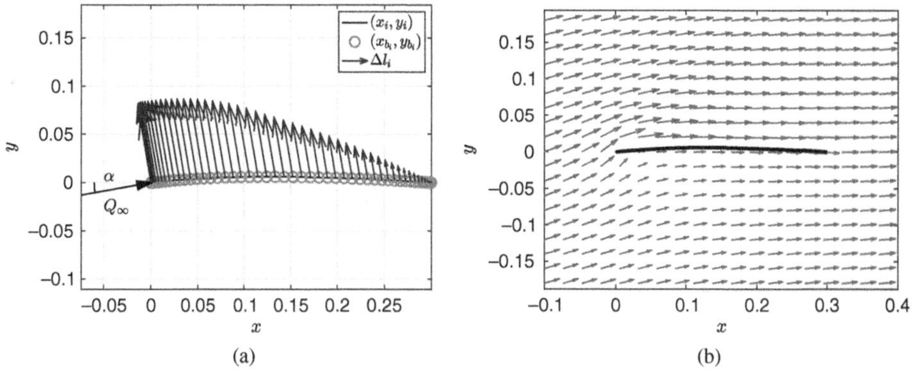

Figure 4.5 Solution of flow around static NACA 24XX airfoil using the lumped vortex method. (a) Lift vectors and (b) flowfield.

We now use both the lumped vortex method and thin airfoil theory to calculate the lift and pitching moment coefficients for a range of angles of attack between $-10°$ *and* $10°$*. For each angle, we re-calculate vector* **b** *and then solve Eq. (4.22) to obtain the vortex strengths. Finally, we compute the lift and pitching moment around the quarter chord from Eqs. (4.11) to (4.14). The thin airfoil theory result is given by Eqs. (3.23) and (3.26).*

Figure 4.6 compares the variations of lift coefficient and pitching moment coefficient with angle of attack calculated from the two methods. It also plots experimental data measured in the wind tunnel for the NACA 2408 airfoil at a Reynolds number of 9×10^6 *(Loftin Jr. and Cohen 1948). It can be seen that the lift curves predicted by the two methods are nearly coincident and in good agreement with the experimental data for angles of attack between* $-10°$ *and* $5°$*. The pitching moment curve predicted by the lumped vortex method lies very close to the thin airfoil theory result between* $-10°$ *and* $0°$*, but it then starts to move away from it. Both methods predict lower pitching moment coefficient values than the experimental measurements and fail to model the increase for angles higher than* $5°$ *that is due to the thickening of the boundary layer.*

The lift curve slope predicted by thin airfoil theory is constant and equal to 2π*, as discussed in Section 3.2. In contrast, the lift curve slope estimated by thin airfoil theory varies with angle*

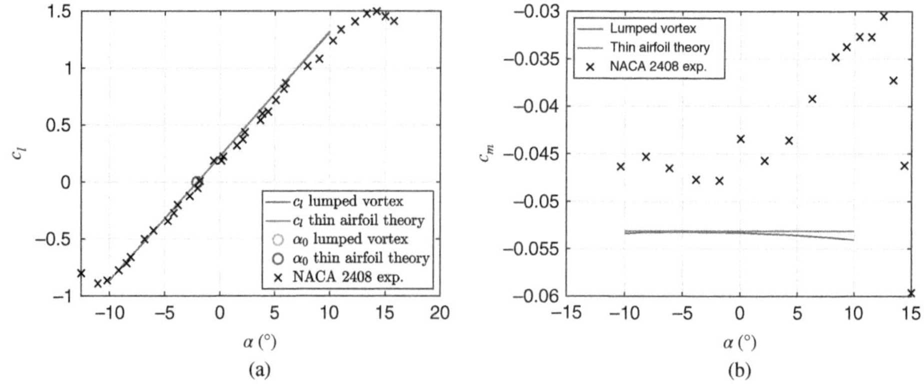

Figure 4.6 Lift and pitching moment coefficient variation with angle of attack for NACA 24XX airfoil. Experimental data by Loftin Jr. and Cohen (1948). (a) $c_l(\alpha)$ and (b) $c_{m_{c/4}}(\alpha)$. Source: Loftin Jr. and Cohen (1948), credit: NACA.

of attack, increasing from 6.2 to 6.3 as α increases from $-10°$ to $\alpha_{l=0}$ and then decreasing again to 6.15 as the angle of attack increases to $10°$. The non-linearity of the lift curve slope does not disappear as the number of panels increases; the lumped vortex method is inherently non-linear.

Recall from Eq. (3.25) that the zero-lift angle, $\alpha_{l=0}$, is always negative for airfoils with positive camber. It can be estimated by interpolating the c_l vs α variation using a cubic spline. The lumped vortex method predicts $\alpha_{l=0} = -2.06°$, while thin airfoil theory gives $\alpha_{l=0} = -2.08°$. The difference is minor and decreases as the number of panels increases. For example, for $n = 500$, the lumped vortex estimate becomes $\alpha_{l=0} = -2.07°$.

As mentioned earlier, both approaches predict lower $c_{m_{c/4}}$ values than the ones measured in the wind tunnel. This is due to the fact that these techniques model the airfoil as infinitely thin so that the aerodynamic centre lies at the quarter-chord. For thick airfoils, the aerodynamic centre lies near but not exactly at the quarter-chord. This example is solved by Matlab code `steady_lv.m`.

The mechanics of the lumped vortex method and thin airfoil theory differ in several ways:

- The lumped vortex method becomes increasingly accurate as the number of panels increases while the thin airfoil result is analytical, based on a continuous distribution of vorticity along the airfoil.
- The lumped vortex results depends on the sine and cosine of the angle of attack, while thin airfoil theory assumes that the angle of attack is small enough to write $\sin\alpha \approx \alpha$, $\cos\alpha \approx 1$. This means that the lift vs α curve predicted by the lumped vortex method is slightly non-linear at higher angles of attack. Furthermore, the pitching moment predicted by the lumped vortex method varies slightly with angle of attack.
- The lift predicted by the lumped vortex method is perpendicular to the free stream, while thin airfoil theory lift is perpendicular to the chord of the airfoil. Again, this difference is negligible for small angles of attack.

- Thin airfoil theory imposes the upwash necessary for impermeability on the chord line, while the lumped vortex method imposes impermeability on control points that lie very close to the camber line. This difference does not appear to have a measurable effect on the predictions of the two approaches as long as the number of lumped vortex panels is high enough.

4.2.1 Unsteady Flows

We now extend the lumped vortex methodology to unsteady attached flows around 2D airfoils with pitch and plunge degrees of freedom. In this case, there must be a vorticity distribution in the wake, which we will model using lumped vortices, just like the bound vorticity. In order to do this, we must determine the strengths and positions of the wake vortices Γ_{w_j}. Both the Kutta condition and Kelvin's theorem must be satisfied at each time instance.

We will start the discussion of unsteady flows by revisiting the impulsively started flat plate of Section 3.4. Recall that the flat plate is static for $t < 0$ and is impulsively set to motion with velocity Q_∞ and angle of attack α at time $t = 0$. Consequently, the total circulation in the entire flowfield, $\Gamma(t)$ at times $t < 0$ is equal to zero. By Kelvin's theorem (Eq. (2.214)), $\mathrm{d}\Gamma/\mathrm{d}t = 0$, so that

$$\Gamma(t) = 0 \tag{4.25}$$

at all time instances $t \geq 0$. At the first time instance, we could impose the impermeability boundary condition without placing any vortices in the wake so that the bound vorticity is given by Eq. (4.10)

$$\sum_{j=1}^{n} \frac{\Gamma_{b_j}(0)}{2\pi} \frac{n_{x_i}(y_{c_i} - y_{b_j}) - n_{y_i}(x_{c_i} - x_{b_j})}{(x_{c_i} - x_{b_j})^2 + (y_{c_i} - y_{b_j})^2} = -Q_\infty \cos \alpha n_{x_i} - Q_\infty \sin \alpha n_{y_i} \tag{4.26}$$

noting that the strengths Γ_{b_j} are now functions of time. Solving this system of equations, we will obtain $\Gamma_{b_j}(0)$ and a total circulation

$$\Gamma(0) = \sum_{j=1}^{n} \Gamma_{b_j}(0)$$

that will be non-zero and equal to the steady-state total bound circulation (see Example 4.3). Therefore, we will violate Kelvin's theorem and the result will be unphysical. The solution to this problem is to place one single wake vortex in the flowfield, directly behind the trailing edge. The position of this wake vortex is (x_{w_1}, y_{w_1}) and its strength Γ_{w_1}. Adding to Eq. (4.26) the velocities induced by this wake vortex in a direction normal to the panels, we obtain

$$\sum_{j=1}^{n} \frac{\Gamma_{b_j}(0)}{2\pi} \frac{n_{x_i}(y_{c_i} - y_{b_j}) - n_{y_i}(x_{c_i} - x_{b_j})}{(x_{c_i} - x_{b_j})^2 + (y_{c_i} - y_{b_j})^2} + \frac{\Gamma_{w_1}}{2\pi} \frac{n_{x_i}(y_{c_i} - y_{w_1}) - n_{y_i}(x_{c_i} - x_{w_1})}{(x_{c_i} - x_{w_1})^2 + (y_{c_i} - y_{w_1})^2}$$

$$= -Q_\infty \cos \alpha n_{x_i} - Q_\infty \sin \alpha n_{y_i} \tag{4.27}$$

This is a system of n equations with $n+1$ unknowns, $\Gamma_{b_j}(0)$ for $j = 1, \dots, n$ and Γ_{w_1}. We can complete the system by writing Kelvin's theorem of Eq. (4.25) for the total circulation in the flow at the first time instance

$$\sum_{j=1}^{n} \Gamma_{b_j}(0) + \Gamma_{w_1} = 0 \tag{4.28}$$

so that we allow the airfoil to develop non-zero total bound circulation (and, hence, non-zero lift), but we use the wake vortex to ensure that Kelvin's theorem is satisfied in the entire flowfield. The problem is that we have not yet defined the position of the first wake vortex. Katz and Plotkin (2001) recommend to shed the first vortex behind the trailing edge, such that

$$x_{w_1} = c + \Delta x_w Q_\infty \cos\alpha\Delta t, \quad y_{w_1} = \Delta x_w Q_\infty \sin\alpha\Delta t \tag{4.29}$$

where Δx_w takes values between 0.2 and 0.3. This means that the first vortex is placed behind the leading edge at a distance equal to 20–30% of the distance travelled by the airfoil during a single time step, Δt. Note that we are implicitly assuming that the Kutta condition is still satisfied automatically thanks to the placement of the bound vortices and collocation points. The situation at the first time instance is drawn in Figure 4.7a, which plots the airfoil with four bound vortices and the first wake vortex, for $\alpha = 0°$ and $\Delta x_w = 0.3$.

We are now in a position to solve Eq. (4.27) for the unknown vortex strengths $\Gamma_{b_j}(0)$ and Γ_{w_1}, thus completely defining the flowfield at the first time instance $t_1 = 0$. Next, we continue to evaluate the flow at subsequent time instances $t_k = (k-1)\Delta t$, for $k \geq 2$. At instance $k = 2$, vortex Γ_{w_1} will move to a new position, (x_{w_2}, y_{w_2}). There are two options for the motion of the wake vortices:

- Prescribed wake. The wake vortices move with a prescribed velocity, usually the free stream velocity.
- Free wake. The wake vortices move with the local flow velocity, which must be calculated.

The prescribed wake is the simplest and computationally cheapest approach, so we will adopt it for the moment. Each wake vortex will then move such that its position at time

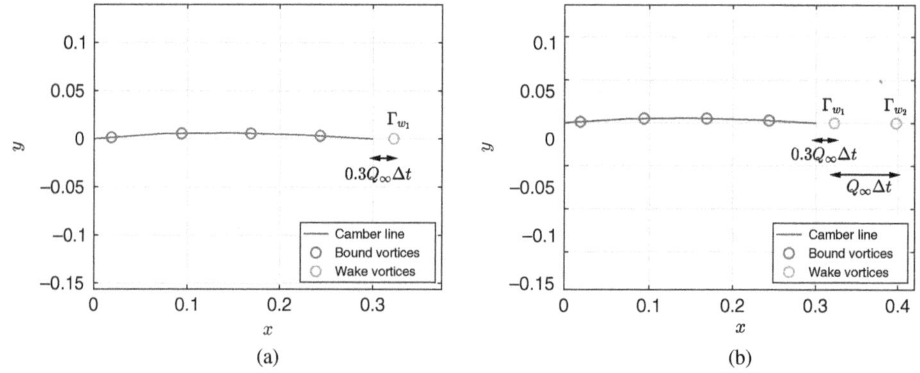

Figure 4.7 Shedding procedure for wake lumped vortices. (a) $t = 0$ and (b) $t = \Delta t$.

instance k is

$$x_{w_j}(t_k) = x_{w_{j-1}}(t_{k-1}) + Q_\infty \cos\alpha\Delta t, \quad y_{w_j}(t_k) = y_{w_{j-1}}(t_{k-1}) + Q_\infty \sin\alpha\Delta t \tag{4.30}$$

for $j = 2, \ldots, k$; position $x_{w_1}(t_k), y_{w_1}(t_k)$ is given by Eq. (4.29) at all time instances.

The strength of the vortices will not change as they propagate downstream because we assumed that the flow is inviscid and therefore free vortices do not dissipate. For an impulsively started cambered plate with angle of attack $\alpha = 0°$, the first two time instances are depicted in Figure 4.7. At $t = 0$, there is only one wake vortex at a distance $0.3Q_\infty\Delta t$ behind the trailing edge. At $t = \Delta t$, the first wake vortex has moved to $1.3Q_\infty\Delta t$ and a new vortex has appeared at $0.3Q_\infty\Delta t$. Note that the name of the original vortex has changed; Γ_{w_1} at $t = 0$ is called Γ_{w_2} at $t = \Delta t$, but its strength has not changed. The vortex nearest to the trailing edge is always called Γ_{w_1}. In other words, $\Gamma_{w_j}(t)$ are functions of time, but they are related to each other by

$$\Gamma_{w_j}(t_k) = \Gamma_{w_{j-1}}(t_{k-1}) \tag{4.31}$$

for $j = 2, \ldots, k$.

For the general case of the kth time instance, Eq. (4.27) is no longer representative of the flow because there are k wake vortices that induce velocities normal to the panels. The impermeability boundary condition becomes

$$\sum_{j=1}^{n} \frac{\Gamma_{b_j}(t_k)}{2\pi} \frac{n_{x_i}(y_{c_i} - y_{b_j}) - n_{y_i}(x_{c_i} - x_{b_j})}{(x_{c_i} - x_{b_j})^2 + (y_{c_i} - y_{b_j})^2}$$
$$+ \sum_{j=1}^{k} \frac{\Gamma_{w_j}(t_k)}{2\pi} \frac{n_{x_i}(y_{c_i} - y_{w_j}) - n_{y_i}(x_{c_i} - x_{w_j})}{(x_{c_i} - x_{w_j})^2 + (y_{c_i} - y_{w_j})^2} = -Q_\infty \cos\alpha n_{x_i} - Q_\infty \sin\alpha n_{y_i} \tag{4.32}$$

Of the k vortex strengths, only $\Gamma_{w_1}(t_k)$ is unknown; all the others were determined at earlier time instances and can be moved to the right-hand side, keeping only the terms involving the unknown $\Gamma_{b_j}(t_k)$ and $\Gamma_{w_1}(t_k)$ on the left-hand side, such that

$$\sum_{j=1}^{n} \frac{\Gamma_{b_j}(t_k)}{2\pi} \frac{n_{x_i}(y_{c_i} - y_{b_j}) - n_{y_i}(x_{c_i} - x_{b_j})}{(x_{c_i} - x_{b_j})^2 + (y_{c_i} - y_{b_j})^2} + \frac{\Gamma_{w_1}(t_k)}{2\pi} \frac{n_{x_i}(y_{c_i} - y_{w_1}) - n_{y_i}(x_{c_i} - x_{w_1})}{(x_{c_i} - x_{w_1})^2 + (y_{c_i} - y_{w_1})^2}$$
$$= -Q_\infty \cos\alpha n_{x_i} - Q_\infty \sin\alpha n_{y_i} - \sum_{j=2}^{k} \frac{\Gamma_{w_j}(t_k)}{2\pi} \frac{n_{x_i}(y_{c_i} - y_{w_j}) - n_{y_i}(x_{c_i} - x_{w_j})}{(x_{c_i} - x_{w_j})^2 + (y_{c_i} - y_{w_j})^2} \tag{4.33}$$

In order to derive the most useful expression for Kelvin's theorem, we write Eq. (4.25) out for time instance t_k as

$$\sum_{j=1}^{n} \Gamma_{b_j}(t_k) + \sum_{j=1}^{k} \Gamma_{w_j}(t_k) = 0, \quad \text{or,} \quad \sum_{j=1}^{n} \Gamma_{b_j}(t_k) + \Gamma_{w_1}(t_k) = -\sum_{j=2}^{k} \Gamma_{w_j}(t_k) \tag{4.34}$$

At time instance t_{k-1}, Eq. (4.25) was

$$\sum_{j=1}^{n} \Gamma_{b_j}(t_{k-1}) + \sum_{j=1}^{k-1} \Gamma_{w_j}(t_{k-1}) = 0$$

and, from the wake propagation expression (4.31),

$$\sum_{j=2}^{k}\Gamma_{w_j}(t_k) = \sum_{j=1}^{k-1}\Gamma_{w_j}(t_{k-1})$$

Hence, at time instance t_k, Kelvin's theorem is expressed as

$$\sum_{j=1}^{n}\Gamma_{b_j}(t_k) + \Gamma_{w_1}(t_k) = \sum_{j=1}^{n}\Gamma_{b_j}(t_{k-1}) \tag{4.35}$$

which states that the change in bound vorticity between time instances k and $k-1$ must be counteracted by the shedding of a new vortex in the wake. Equations (4.32) and (4.35) can now be solved simultaneously for the unknown $\Gamma_{b_j}(t_k)$ and $\Gamma_{w_1}(t_k)$.

In matrix form, the horizontal and vertical velocities induced by the wake vortices on the control points at the kth time instance are given by the $n \times 1$ vectors

$$\boldsymbol{u}_w(t_k) = \boldsymbol{B}_u(t_k)\boldsymbol{\Gamma}_w(t_k) \tag{4.36}$$

$$\boldsymbol{v}_w(t_k) = \boldsymbol{B}_v(t_k)\boldsymbol{\Gamma}_w(t_k) \tag{4.37}$$

where the elements of the $k \times 1$ vector $\boldsymbol{\Gamma}_w(t_k)$ are $\Gamma_{w_j}(t_k)$. Replacing x_{b_j} by x_{w_j} and y_{b_j} by y_{w_j} in Eqs. (4.8) and (4.9), the aerodynamic influence of the wake on the collocation points is given by the $n \times k$ matrices $\boldsymbol{B}_u(t_k)$ and $\boldsymbol{B}_v(t_k)$, whose elements are calculated from

$$B_{u_{i,j}}(t_k) = \frac{1}{2\pi}\frac{y_{c_i} - y_{w_j}(t_k)}{(x_{c_i} - x_{w_j(t_k)})^2 + (y_{c_i} - y_{w_j}(t_k))^2} \tag{4.38}$$

$$B_{v_{i,j}}(t_k) = -\frac{1}{2\pi}\frac{x_{c_i} - x_{w_j}(t_k)}{(x_{c_i} - x_{w_j}(t_k))^2 + (y_{c_i} - y_{w_j}(t_k))^2} \tag{4.39}$$

for $i = 1, \dots, n$ and $j = 1, \dots, k$. The impermeability equation (4.32) is then written as

$$\boldsymbol{A}_n\boldsymbol{\Gamma}_b(t_k) + \boldsymbol{B}_n(t_k)\boldsymbol{\Gamma}_w(t_k) = \boldsymbol{b} \tag{4.40}$$

where matrix $\boldsymbol{A}_n$ is defined in Eqs. (4.20), (4.21) and (4.23), vector $\boldsymbol{b}$ in Eq. (4.24) and $\boldsymbol{B}_n$ is a $n \times n$ matrix whose elements are

$$B_{n_{i,j}}(t_k) = B_{u_{i,j}}(t_k)n_{x_i} + B_{v_{i,j}}(t_k)n_{y_i} \tag{4.41}$$

for $i = 1, \dots, n$, $j = 1, \dots, n$. Then, Eqs. (4.32) and (4.35) can be written as the single matrix system

$$\begin{pmatrix} \boldsymbol{A}_n & \boldsymbol{b}_{n_1} \\ \boldsymbol{I}_{1\times n} & 1 \end{pmatrix}\begin{pmatrix} \boldsymbol{\Gamma}_b(t_k) \\ \Gamma_{w_1}(t_k) \end{pmatrix} = \begin{pmatrix} \boldsymbol{b} - \boldsymbol{B}_n^*(t_k)\boldsymbol{\Gamma}_w^*(t_k) \\ \sum_{j=1}^{n}\Gamma_{b_j}(t_{k-1}) \end{pmatrix} \tag{4.42}$$

where $\boldsymbol{I}_{1\times n}$ is a $1 \times n$ vector whose elements are all equal to 1, $\boldsymbol{b}_{n_1}$ is the first column of matrix $\boldsymbol{B}_n(t_k)$ and $\boldsymbol{B}_n^*(t_k)$ is the $n \times (k-1)$ matrix containing columns 2 to k of matrix $\boldsymbol{B}_n^*$, such that

$$\boldsymbol{B}_n(t_k) = \begin{pmatrix} \boldsymbol{b}_{n_1} & \boldsymbol{B}_n^*(t_k) \end{pmatrix}$$

Note that $\boldsymbol{b}_{n_1}$ is not a function of time because the position of the first wake vortex is constant. Similarly, $\boldsymbol{\Gamma}^*_w(t_k)$ is the $(k-1)\times 1$ vector containing elements 2 to k of vector $\boldsymbol{\Gamma}_w(t_k)$, such that

$$\boldsymbol{\Gamma}_w(t_k) = \begin{pmatrix} \Gamma_{w_1}(t_k) \\ \boldsymbol{\Gamma}^*_w(t_k) \end{pmatrix}$$

The lumped vortex with prescribed wake algorithm can then be summarised as follows:

1. At time instance k propagate the positions of the wake vortices using Eq. (4.30) and their strengths using Eq. (4.31).
2. Shed a new vortex $\Gamma_{w_1}(t_k)$ at position (x_{w_1}, y_{w_1}), given by Eq. (4.29).
3. Solve Eq. (4.42) for the strengths of the bound vortices $\boldsymbol{\Gamma}_b(t_k)$ and that of the latest shed vortex $\Gamma_{w_1}(t_k)$.
4. Increment time index k and go back to step 1. Stop when k has reached the desired value.

At the end of the procedure described above, the flowfield is fully defined at all time instances of interest. Now we need to calculate the aerodynamic forces acting on the airfoil. We will start with a simple case, a flat plate airfoil aligned with the x axis. We have already developed an expression for the pressure difference across such an airfoil modelled using continuous vortex sheets (one for the airfoil and one for the wake) in Section 3.6. Equation (3.204) states that

$$\Delta p_b(x,t) = \rho \frac{\partial}{\partial t}\int_0^x \gamma_b(x_0,t)\,\mathrm{d}x_0 + \rho Q_\infty \gamma_b(x) \tag{4.43}$$

after changing the integral limits to reflect the fact that $x = 0$ at the leading edge. In the case of the lumped vortex method, the airfoil and wake are not modelled as continuous vortex sheets but as lumped vortices. Nevertheless, there is an analogy between the two models; the total bound vorticity is given by Eq. (3.206)

$$\Gamma_b(\mathrm{t}) = \int_0^c \gamma_b(x_0,t)\,\mathrm{d}x_0 \tag{4.44}$$

for the continuous vortex sheet and by

$$\Gamma_b(t_k) = \sum_{j=1}^{n} \Gamma_{b_j}(t_k)$$

for the lumped vortices. As the two models represent the same physical phenomenon, the two values of the bound vorticity must be equal, that is

$$\int_0^c \gamma_b(x_0,t_k)\,\mathrm{d}x_0 = \sum_{j=1}^{n} \Gamma_{b_j}(t_k) \tag{4.45}$$

If we now model the continuous bound vortex sheet as a series of n vortex panels with constant vorticity γ_{b_j} for $j = 1, \ldots, n$, the integral on the left-hand side becomes a sum, such that

$$\sum_{j=1}^{n} \gamma_{b_j} s_j = \sum_{j=1}^{n} \Gamma_{b_j}$$

where s_j is the length of each panel. Furthermore, Eq. (4.45) can be written for shorter chordwise distances, such that

$$\int_0^{s_1} \gamma_b \,\mathrm{d}x = \Gamma_{b_1}, \quad \int_0^{s_2} \gamma_b \,\mathrm{d}x = \Gamma_{b_1} + \Gamma_{b_2}$$

etc. Again, assuming that γ_b is constant on each panel leads to

$$\gamma_{b_j} = \frac{\Gamma_{b_j}}{s_j}$$

for $j = 1, \dots, n$. After substituting these results in Eq. (4.43), we obtain

$$\Delta p_b(x_i, t_k) = \rho \frac{\partial}{\partial t} \sum_{j=1}^{i} \Gamma_{b_j}(t_k) + \rho Q_\infty \frac{\Gamma_{b_i}(t_k)}{s_i} = \rho \sum_{j=1}^{i} \dot{\Gamma}_{b_j}(t_k) + \rho Q_\infty \frac{\Gamma_{b_i}(t_k)}{s_i} \tag{4.46}$$

for $i = 1, \dots, n$. Finally, the incremental lift acting on each panel is given by

$$\Delta l_i(t_k) = \Delta p_b(x_i, t_k) s_i = \rho \left(Q_\infty \Gamma_{b_i}(t_k) + \sum_{j=1}^{i} \dot{\Gamma}_{b_j}(t_k) s_i \right) \tag{4.47}$$

The first term in this result, $\rho Q_\infty \Gamma_{b_i}(t_k)$, is the Kutta–Joukowski lift acting on the ith panel, while the second term is the fluid acceleration term related to the unsteady term in the unsteady Bernoulli equation (Katz and Plotkin 2001). As discussed earlier, in order to derive Eq. (4.43) and, hence, Eq. (4.47), the free stream, airfoil and wake are all assumed to be aligned with the x axis, so that the two terms in Eq. (4.47) act in the same direction, perpendicular to the flat plate.

In the case of the lumped vortex method, as presented in this section, the flat plate is aligned with the x axis, but the free stream and wake are not. Furthermore, the free stream is not the only external flow component seen by airfoil; the wake vortices induce additional velocity components u_{w_i}, v_{w_i} on the collocation points, given by Eqs. (4.36) and (4.37). The total velocity seen by the airfoil is then $Q_\infty \cos\alpha + u_{w_i}(t_k)$ in the x direction and $Q_\infty \sin\alpha + v_{w_i}(t_k)$ in the y direction. Using the principle of Eq. (4.6), we can define chordwise and normal incremental aerodynamic forces acting on each panel such that

$$\Delta f_{x_i}(t_k) = -\rho \left(Q_\infty \sin\alpha + v_{w_i}(t_k) \right) \Gamma_{b_i}(t_k)$$

$$\Delta f_{y_i}(t_k) = \rho \left(\left(Q_\infty \cos\alpha + u_{w_i}(t_k) \right) \Gamma_{b_i}(t_k) + \sum_{j=1}^{i} \dot{\Gamma}_{b_j}(t_k) s_i \right) \tag{4.48}$$

where the equation for the normal force also includes the unsteady Bernoulli term discussed earlier. The incremental lift and drag forces are then given by the components of $\Delta f_{x_i}(t_k)$ and $\Delta f_{y_i}(t_k)$ perpendicular and parallel to the free stream, that is

$$\Delta l_i(t_k) = \Delta f_{y_i}(t_k) \cos\alpha - \Delta f_{x_i}(t_k) \sin\alpha$$

$$\Delta d_i(t_k) = \Delta f_{y_i}(t_k) \sin\alpha + \Delta f_{x_i}(t_k) \cos\alpha \tag{4.49}$$

Substituting from Eq. (4.48), we obtain

$$\Delta l_i(t_k) = \rho\left(Q_\infty\Gamma_{b_i}(t_k) + \left(u_{w_i}(t_k)\cos\alpha + v_{w_i}(t_k)\sin\alpha\right)\Gamma_{b_i}(t_k) + \sum_{j=1}^{i}\dot{\Gamma}_{b_j}(t_k)s_i\cos\alpha\right)$$

$$\Delta d_i(t_k) = \rho\left(\left(u_{w_i}(t_k)\sin\alpha - v_{w_i}(t_k)\cos\alpha\right)\Gamma_{b_i}(t_k) + \sum_{j=1}^{i}\dot{\Gamma}_{b_j}(t_k)s_i\sin\alpha\right)$$

Assuming that, $u_{w_i} \ll Q_\infty$, $v_{w_i} \ll Q_\infty$, $\sin\alpha \approx \alpha$ and $\cos\alpha \approx 1$ and neglecting products of small quantities these latest expressions become

$$\Delta l_i(t_k) = \rho\left(\left(Q_\infty + u_{w_i}(t_k)\right)\Gamma_{b_i}(t_k) + \sum_{j=1}^{i}\dot{\Gamma}_{b_j}(t_k)s_i\right)$$

$$\Delta d_i(t_k) = \rho\left(-v_{w_i}(t_k)\Gamma_{b_i}(t_k) + \sum_{j=1}^{i}\dot{\Gamma}_{b_j}(t_k)s_j\alpha\right) \tag{4.50}$$

Note that the first of Eq. (4.50) is similar to expression (4.47) but includes the term $u_{w_i}(t_k)$ since the wake is no longer aligned with the x axis and the horizontal velocities it induces on the control points are not equal to zero. The total lift and drag at each time instance are then given by

$$d(t_k) = \sum_{i=1}^{n}\Delta d_i(t_k), \quad l(t_k) = \sum_{i=1}^{n}\Delta l_i(t_k)$$

The incremental pitching moment around the quarter chord acting on the ith panel is given by the unsteady version of Eq. (4.13)

$$\Delta m_{c/4_i}(t_k) = \Delta\mathbf{f}_i(t_k) \times \mathbf{r}_i \tag{4.51}$$

so that

$$m_{c/4}(t_k) = \sum_{i=1}^{n}\Delta f_{x_i}(t_k)y_{b_i} - \Delta f_{y_i}(t_k)(x_{b_i} - c/4) \tag{4.52}$$

Finally, it should be noted that the $\dot{\Gamma}_{b_j}(t_k)$ term in Eqs. (4.46)–(4.48) and (4.50) can only be calculated numerically, since the values of $\Gamma_{b_j}(t_k)$ are only available at discrete time instances. As the entire lumped vortex scheme is first-order accurate in the time domain, $\dot{\Gamma}_{b_j}(t_k)$ can be evaluated by means of a first-order finite difference scheme, that is

$$\dot{\Gamma}_{b_j}(t_k) = \frac{\Gamma_{b_j}(t_k) - \Gamma_{b_j}(t_{k-1})}{\Delta t} \tag{4.53}$$

Example 4.3 *Calculate the aerodynamic loads acting on an impulsively started flat plate using the lumped vortex method with a prescribed wake.*

Here, we are repeating Examples 3.7, 3.9 and 3.11 using the lumped vortex approach and comparing with results from Wagner's and Garrick's theories. The flat plate has a half-chord of $b = 0.5$ *m and an angle of attack* $\alpha = 5°$. *After the impulsive start, the free stream airspeed is* $Q_\infty = 10$ *m/s.*

In the examples of Chapter 3, the airfoil and wake were modelled as lying on the x axis; the effect of the angle of attack was represented by an upwash equal to $Q_\infty \sin\alpha$ *acting on the plate. Here, the airfoil still lies on the x axis but, as the free stream is coming from an angle* α, *the prescribed wake will be propagating in the direction of the free stream, not the x axis.*

The flat plate is discretised into $n = 50$ *non-linearly spaced panels with coordinates* x_i, y_i *for* $i = 1, \dots, n+1$, *where* $x_i = c\bar{x}_i$ *is obtained from Eq. (4.16) and* $y_i = 0$. *The coordinates of the bound vortices and collocation points are calculated from Eqs. (4.17) and (4.18), respectively. We also calculate the unit normal vectors to each panel at the collocation points from Eq. (4.19), vector* $\mathbf{b}$ *from Eq. (4.24) and the aerodynamic influence coefficient matrices* $\mathbf{A}_u$, $\mathbf{A}_v$, $\mathbf{A}_n$ *from Eqs. (4.20), (4.21) and (4.23).*

At the first time instance, $k = 1$, *there is only one wake vortex in the flow with strength* $\Gamma_{w_1}(t_1)$, *lying at* x_{w_1}, y_{w_1}. *Equation (4.42) reduces to*

$$\begin{pmatrix} \mathbf{A}_n & \mathbf{b}_{n_1} \\ \mathbf{I}_{1\times n} & 1 \end{pmatrix} \begin{pmatrix} \mathbf{\Gamma}_b(t_1) \\ \Gamma_{w_1}(t_1) \end{pmatrix} = \begin{pmatrix} \mathbf{b} \\ 0 \end{pmatrix} \tag{4.54}$$

which can be solved readily for $\mathbf{\Gamma}_b(t_1)$ *and* $\Gamma_{w_1}(t_1)$.

At a general time instance, k, we first propagate the strength of the wake vortices using Eq. (4.31) and then we calculate the new positions of wake vortices $2, \dots, k$ *from Eq. (4.30). Finally, we shed vortex* $\Gamma_{w_1}(t_k)$ *at point* x_{w_1}, y_{w_1}, *evaluated from Eq. (4.29) with* $\Delta x_w = 0.25$. *We calculate matrices* $\mathbf{B}_u(t_k)$, $\mathbf{B}_v(t_k)$ *and* $\mathbf{B}_n(t_k)$ *from Eqs. (4.38), (4.39) and (4.41), respectively, and then we solve Eq. (4.42) for* $\mathbf{\Gamma}_b(t_k)$ *and* $\Gamma_{w_1}(t_k)$.

The time step Δt *must be chosen such that the distance between the wake vortices is compatible with the distance between the bound vortices. A good choice suggested by Simpson and Palacios (2013) is*

$$\Delta t = \frac{c}{nQ_\infty} \tag{4.55}$$

The first time instance corresponds to $t_1 = 0$; *we continue the simulation until the airfoil has moved by 10 chord lengths, such that the simulation end time is* $t_f = 10c/Q_\infty$. *Once the simulation has finished we can calculate the time derivatives of the bound vortex strengths from expression (4.53). For* $k = 1$, *we use* $\dot{\Gamma}_{b_j}(0) = 0$ *as there is no obvious way to calculate the derivative of the circulation at* $t = 0$. *The final step is to evaluate the aerodynamic loads from Eqs. (4.50) and (4.52).*

Figure 4.8a plots the wake shape after the end of the simulation. It can be seen that the airfoil is aligned with the x axis while the wake is inclined to it by an angle α, *so that it is parallel to the free stream. The shape of the wake is a straight line due to the prescribed wake modelling choice. In Section 4.2.2, we will calculate the free wake for the same problem.*

The steady lift, l_∞, *can be calculated directly by solving Eq. (4.22) for the steady values of* Γ_{b_j} *and then evaluating expression (4.11). The ratio of the unsteady lift to the steady lift is the Wagner function, as discussed in Section 3.4.1. Figure 4.8b plots the estimate of the Wagner function obtained from the lumped vortex method against non-dimensional time* $\tau = Q_\infty t/b$ *and compares it to the values calculated from Eq. (3.93) (see Example 3.9). It can be seen that the lumped vortex estimate is in good agreement with Wagner theory, except for* $\tau = 0$, *where the ratio of unsteady to steady lift takes the value 0.04 instead of 0.5. This low value of the*

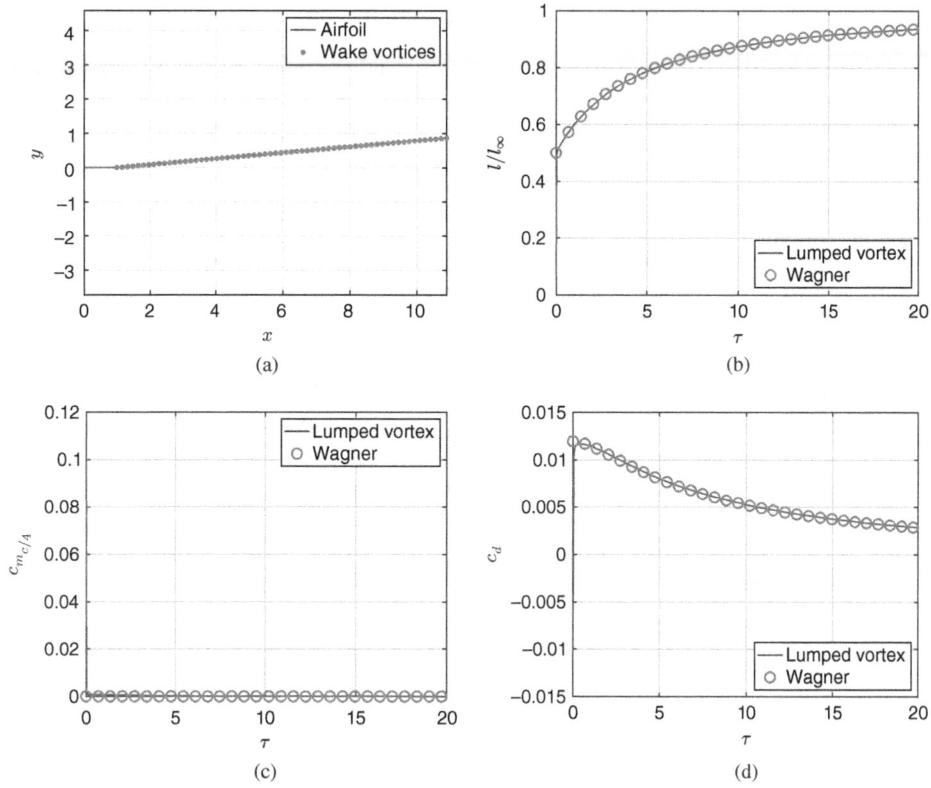

Figure 4.8 Prescribed wake shape and aerodynamic loads for an impulsively started flat plate using the lumped vortex method. (a) $x_w(t), y_w(t)$, (b) $\Phi(t)$, (c) $c_{m_{c/4}}(t)$ and (d) $c_d(t)$.

estimated Wagner function is caused by the use of $\dot{\Gamma}_{b_j}(0) = 0$ in the calculation of the time derivative of the bound circulation at $t = 0$. Nevertheless, for $\tau > 0$ the lumped vortex prediction is very accurate.

Wagner theory predicts zero pitching moment around the quarter chord at all times for the impulsively started airfoil. This result can be obtained from Eq. (3.126) for the Wagner lift and pitching moment, after setting $\dot{\alpha} = \dot{h} = \ddot{\alpha} = \ddot{h} = 0$ and substituting $a = -1/2$, which is the value of a for $x_f = c/4$. Figure 4.8c plots the pitching moment coefficient estimate obtained from the lumped vortex method, along with the Wagner value. Clearly, the initial error in lift also causes an initial error in pitching moment. Nevertheless, the lumped vortex estimate for the pitching moment becomes nearly zero for $\tau > 0$.

Figure 4.8d plots the drag coefficient variation with τ, as estimated by the lumped vortex method, and compares it to the Wagner theory value. The latter is obtained from Eq. (3.104), where S is calculated from expression (3.108) after substituting $\dot{\alpha} = \dot{h} = 0$ and l is obtained from Eq. (3.91). The lumped vortex drag value is a thrust at the first time instance but becomes quite accurate for $\tau > 0$.

As with all numerical methods, the accuracy of the results depends on the chosen numerical parameters. The estimates plotted in Figure 4.8 were obtained for $n = 50$ and for non-linearly *spaced panels. The reader should experiment with the simulation parameter values to investigate their effect on the accuracy of the solution. Nevertheless, increasing the value of n or changing panel spacing will not improve the lift estimate at $\tau = 0$. This example is solved by Matlab code* `impulsive_lv.m`.

4.2.2 Free Wakes

In Section 4.2.1, we modelled the impulsively started flat plate problem using the lumped vortex approach with a prescribed wake, travelling at the free stream airspeed. Here, we will introduce the concept of the free wake and we will apply it to several different problems. We will begin the discussion with the impulsively started airfoil and then extend it to pitching and plunging wings. Real wakes (both steady and unsteady) are force-free, which means that there can be no pressure difference across the wake. If there were a non-zero pressure difference, then the wake would move until the pressure difference dropped to zero. Consequently, wakes are flow streamlines and are always parallel to the local flow velocity, as defined in Eq. (2.51). In two dimensions, the streamline definition simplifies to

$$\frac{\mathrm{d}x}{\mathrm{d}s}u - \frac{\mathrm{d}y}{\mathrm{d}s}v = 0$$

Assume that we know the positions of the wake vortices at time instance $k-1$, and that we would like to calculate their positions at time instance k. We define $\Delta x = x_w(t_k) - x_w(t_{k-1})$, $\Delta y = y_w(t_k) - y_w(t_{k-1})$, so that $\Delta s = \sqrt{\Delta x^2 + \Delta y^2}$. Then, the streamline definition becomes approximately

$$\frac{\Delta x}{\Delta s}v(t_k) - \frac{\Delta y}{\Delta s}u(t_k) = 0$$

One way of satisfying this equation is to set $\Delta x = u(t_k)\Delta t$ and $\Delta y = v(t_k)\Delta t$ so that we obtain

$$\frac{u(t_k)}{\sqrt{u(t_k)^2 + v(t_k)^2}}v(t_k) - \frac{v(t_k)}{\sqrt{u(t_k)^2 + v(t_k)^2}}u(t_k) = 0$$

Therefore, propagating the wake vortices at the local flow speed over the time interval Δt ensures that the wake remains a flow streamline. This free wake propagation scheme is approximate because Δx and Δy are calculated using previous and current values of x_w and y_w, while the flow velocities are only evaluated at the current time instance.

Consider an airfoil at an angle of attack α to a free stream Q_∞. In Section 4.2.1, we placed the airfoil parallel to the x axis and rotated the free stream. Now that we will be propagating the wake at its local flow velocity, it is preferable to rotate the airfoil to an angle α to the x axis and to place the free stream parallel to the x axis, such that $\mathbf{Q}_\infty = (U_\infty, 0)$. This practice is particularly relevant to the case where the airfoil is pitching continuously, that is $\alpha(t)$ is a function of time; it is more realistic to keep rotating the airfoil itself rather than the free stream. Using the free wake propagation scheme defined above, the wake vortex propagation equation becomes

$$\begin{aligned} x_{w_i}(t_k) &= x_{w_{i-1}}(t_{k-1}) + \left(U_\infty + u_{wb_i}(t_k) + u_{ww_i}(t_k)\right)\Delta t \\ y_{w_i}(t_k) &= y_{w_{i-1}}(t_{k-1}) + \left(v_{wb_i}(t_k) + v_{ww_i}(t_k)\right)\Delta t \end{aligned} \tag{4.56}$$

for $i = 1, \dots, k$, where u_{wb_i}, v_{wb_i} are the horizontal and vertical velocities induced by the bound vortices on the ith wake vortex,

$$u_{wb_i}(t_k) = \sum_{j=1}^{n} \frac{\Gamma_{b_j}(t_k)}{2\pi} \frac{y_{w_i}(t_k) - y_{b_j}}{(x_{w_i}(t_k) - x_{b_j})^2 + (y_{w_i}(t_k) - y_{b_j})^2}$$

$$v_{wb_i}(t_k) = -\sum_{j=1}^{n} \frac{\Gamma_{b_j}(t_k)}{2\pi} \frac{x_{w_i}(t_k) - x_{b_j}}{(x_{w_i}(t_k) - x_{b_j})^2 + (y_{w_i}(t_k) - y_{b_j})^2} \tag{4.57}$$

while u_{ww_i}, v_{ww_i} are the horizontal and vertical velocities induced by the wake vortices on the ith wake vortex,

$$u_{ww_i}(t_k) = \sum_{j=1, j\neq i}^{k} \frac{\Gamma_{w_j}(t_k)}{2\pi} \frac{y_{w_i}(t_k) - y_{w_j}(t_k)}{(x_{w_i}(t_k) - x_{w_j}(t_k))^2 + (y_{w_i}(t_k) - y_{w_j}(t_k))^2}$$

$$v_{ww_i}(t_k) = -\sum_{j=1, j\neq i}^{k} \frac{\Gamma_{w_j}(t_k)}{2\pi} \frac{x_{w_i}(t_k) - x_{w_j}(t_k)}{(x_{w_i}(t_k) - x_{w_j}(t_k))^2 + (y_{w_i}(t_k) - y_{w_j}(t_k))^2} \tag{4.58}$$

recalling that the velocity induced by a vortex on itself is equal to zero. The cost of the computation of a free wake is proportional to k^2, so it can become quite significant when many vortices have been shed in the wake. The number of wake vortices depends on the number of panels n and on the desired total simulation time. One way of reducing the cost of the simulation is to 'forget' vortices that lie too far from the wing's trailing edge. This means that vortices that lie, say, further than 10 chords downstream of the trailing edge are deleted. However, in order to continue to satisfy Kelvin's theorem, the strength of the deleted vortices must be added to the right-hand side of Eq. (4.35).

Example 4.4 *Repeat Example 4.3 using a free wake.*

The wing parameters are identical to those of Example 4.3, but, as soon as we calculate the flat plate's panel coordinates x_i, y_i, for $i = 1, \dots, n+1$, we perform a rotation of the plate around the quarter chord by an angle α. We denote by x_{0_i} and y_{0_i} the coordinates of the panel vertices when the flat plate still lies parallel to the x axis. Then the rotation is expressed mathematically as

$$\begin{pmatrix} x_i \\ y_i \end{pmatrix} = \begin{pmatrix} \cos\alpha & \sin\alpha \\ -\sin\alpha & \cos\alpha \end{pmatrix} \begin{pmatrix} x_{0_i} - c/4 \\ y_{0_i} \end{pmatrix} + \begin{pmatrix} c/4 \\ 0 \end{pmatrix} \tag{4.59}$$

where x_i, y_i are the coordinates of the panel vertices after the rotation. We substitute these rotated coordinates into Eqs. (4.17)– (4.21) in order to calculate the positions of the bound vortices, the positions of the collocation points, the unit normal vectors and the aerodynamic influence coefficient matrices.

The position of the first wake vortex is given by

$$x_{w_1} = x_{n+1} + \Delta x_w U_\infty \Delta t, \quad y_{w_1} = y_{n+1} \tag{4.60}$$

where x_{n+1}, y_{n+1} is the position of the trailing edge, so that the flow is exactly equivalent to the non-rotated airfoil version of Example 4.3, and $\Delta x_w = 0.25$. The rest of the wake is propagated using Eq. (4.56). Equation (4.42) is used to calculate the bound and wake vortex strengths at *all time instances, but* $\boldsymbol{b}$ *is calculated from*

$$\boldsymbol{b} = -U_\infty \boldsymbol{n}_x$$

since the free stream is parallel to the x axis. The steady solution is obtained from Eq. (4.22), again with the value of $\mathbf{b}$ *given above.*

The rest of the solution process is identical to the one used in Example 4.3, including the simulation parameter values (time step, final time, etc.), except that, at the end of each time instance, the velocities induced on the wake are calculated from Eq. (4.58), and the wake is propagated using Eq. (4.56). The aerodynamic loads are calculated from Eq. (4.50) or from the slightly less linearised version

$$\Delta l_i(t_k) = \rho\left(\left(U_\infty + u_{w_i}(t_k)\cos\alpha\right)\Gamma_{b_i}(t_k) + \sum_{j=1}^{i}\dot{\Gamma}_{b_j}(t_k)s_i\cos\alpha\right)$$

$$\Delta d_i(t_k) = \rho\left(-v_{w_i}(t_k)\cos\alpha\Gamma_{b_i}(t_k) + \sum_{j=1}^{i}\dot{\Gamma}_{b_j}(t_k)s_i\sin\alpha\right) \tag{4.61}$$

The pitching moment around the quarter chord is calculated from Eq. (4.51), but now the directions x and y are aligned with the free stream and not with the airfoil. This means that the force in the x direction is the drag and that in the y direction is the lift. Therefore, $\Delta\mathbf{f}_i(t_k) = (\Delta d_i(t_k), \Delta l_i(t_k))$. *Carrying out the vector product in Eq. (4.51) leads to*

$$m_{c/4}(t_k) = \sum_{i=1}^{n}\Delta d_i(t_k)y_{b_i} - \Delta l_i(t_k)(x_{b_i} - c/4) \tag{4.62}$$

Figure 4.9 plots the results of the simulation. It can be seen that the values of the aerodynamic loads have not changed visibly as a result of the free wake calculation; they are still very similar to the Wagner theory predictions. In contrast, Figure 4.9a shows that the airfoil is now inclined at an angle α *to the free stream and that the wake is propagating backwards and downwards. At the end of the wake, the lumped vortices form a big vortical structure, which is the starting vortex that has travelled 10 chord lengths downstream. Note the similarity between this vortical structure and the one in the snapshots of Figure 2.28. Recall that, at the start of the motion, the single wake vortex was very strong, since its strength was equal to the sum of the strengths of all the bound vortices due to Kelvin's theorem. All the vortices shed after the first one are weaker, as their strengths are equal to the change in bound vorticity between two time instances. Hence, the first vortex creates a strong velocity field, around which all the neighbouring vortices rotate.*

As discussed in Chapter 3, the starting vortex has a strong effect on the aerodynamic loads, but, as it travels downstream, its effect is reduced and the loads tend asymptotically towards their steady-state values. We can observe the starting vortex in more detail if we decrease the time step. We carry out a new simulation with $n = 500$ *until the airfoil has moved by a single chord length. The complete wake shape of Figure 4.10a is now more detailed, as there are more vortices in it. However, if we zoom around the starting vortex, as shown in Figure 4.10b, we can see that the wake has become irregular, with groups of vortices coming together and rotating around each other, especially at distances higher than* $x = 1.8$ *m. This phenomenon has a physical equivalent in the Kelvin–Helmholtz instability occurring at the interface between two*

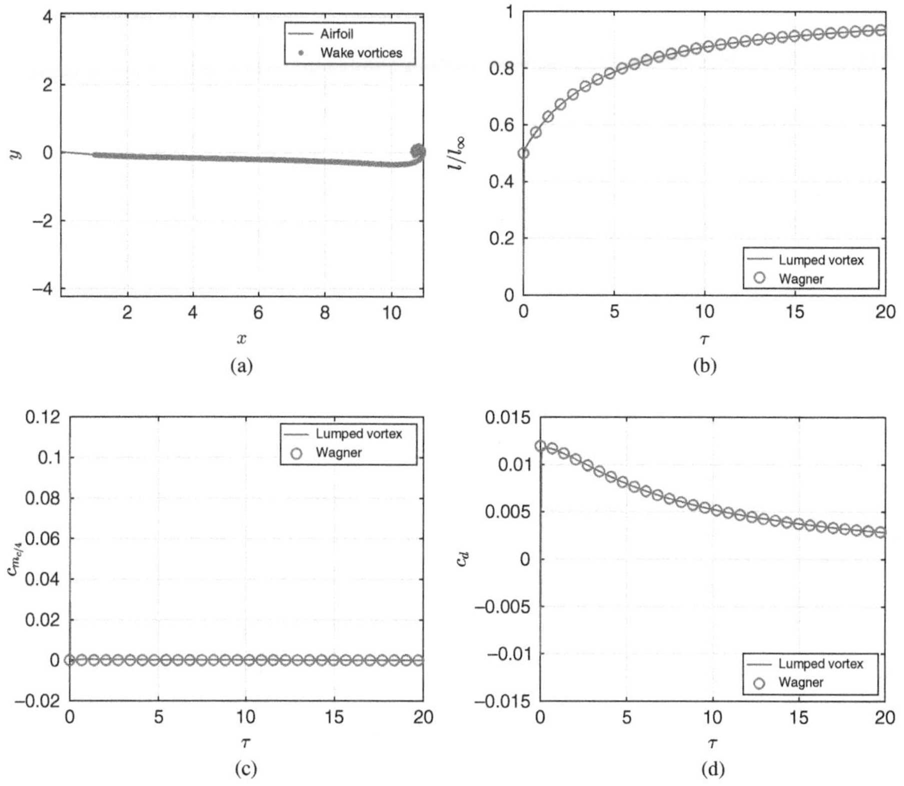

Figure 4.9 Free wake shape and aerodynamic loads for an impulsively started flat plate using the lumped vortex method. (a) Wake shape, (b) $\Phi(t)$, (c) $c_{m_{c/4}}(t)$ and (d) $c_d(t)$.

different fluid flows and can also be observed in Figures 2.28c,d. Nevertheless, in the present case, the instability is numerical because the velocity induced by a vortex at a particular point tends to infinity as the point approaches the vortex, as shown by Eq. (A.11)

$$V(z) = \mathrm{i}\frac{\Gamma}{2\pi}\frac{1}{z} \tag{4.63}$$

where $z = r\mathrm{e}^{\mathrm{i}\theta}$ is the complex distance between the point and the vortex, with magnitude r and phase θ. The induced velocity $V(z)$ is tangential to the circle with radius r and its magnitude is given by Eq. (2.223) as

$$u_\theta(r) = \frac{\Gamma}{2\pi r} \tag{4.64}$$

Therefore, if we shed many closely spaced vortices in the wake, the velocities they induce on each other will be high and, given enough time, they will form groups and start to rotate around each other.

In order to avoid the numerical instability seen in Figure 4.10, the Lamb–Oseen or Vatistas models presented in Section 2.7.1 can be used to calculate the magnitude of the velocity

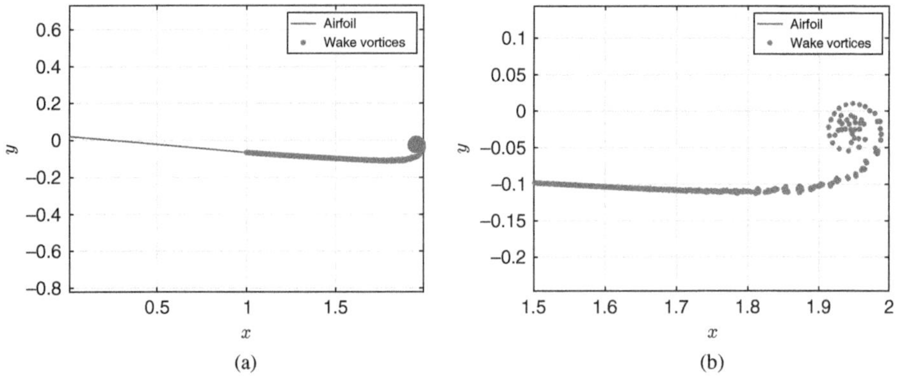

Figure 4.10 Detailed wake shape after impulsive start for $n = 500$. (a) Detailed wake shape and (b) zoom around starting vortex.

induced by a vortex instead of Eq. (4.64). For instance, the Vatistas model of Eq. (2.224) can be written as

$$u_\theta(r) = \frac{\Gamma}{2\pi r_c} \frac{\bar{r}}{\left(1 + \bar{r}^{2n}\right)^{1/n}} \tag{4.65}$$

where $n = 4$ is an exponent, $\bar{r} = r/r_c$ and r_c is the radius of the vortex core. The choice of r_c depends on the physics of the particular problem being investigated, but it can also be used as a free numerical parameter whose value is chosen such as to regularise the behaviour of the free vortices in a discrete vortex simulation.

The Vatistas model can be implemented in the lumped vortex calculation by denoting $r_{i,j}(t_k) = \sqrt{(x_{w_i}(t_k) - x_{w_j}(t_k))^2 + (y_{w_i}(t_k) - y_{w_j}(t_k))^2}$ in Eq. (4.58) so that they become

$$u_{ww_i}(t_k) = \sum_{j=1, j\neq i}^{k} \frac{u_{\theta_{i,j}}(t_k)}{r_{i,j}(t_k)} \left(y_{w_i}(t_k) - y_{w_j}(t_k)\right)$$

$$v_{ww_i}(t_k) = -\sum_{j=1, j\neq i}^{k} \frac{u_{\theta_{i,j}}(t_k)}{r_{i,j}(t_k)} \left(x_{w_i}(t_k) - x_{w_j}(t_k)\right) \tag{4.66}$$

where $u_{\theta_{i,j}}(t_k)$ is the magnitude of the velocity induced by vortex j on vortex i at time t_k and is obtained from Eq. (4.65)

$$u_{\theta_{i,j}}(t_k) = \frac{\Gamma_{w_j}(t_k)}{2\pi r_c} \frac{\bar{r}_{i,j}(t_k)}{\sqrt{1 + \bar{r}_{i,j}(t_k)^4}}$$

while $\bar{r}_{i,j}(t_k) = r_{i,j}(t_k)/r_c$. Repeating the calculation giving rise to Figure 4.10 using the Vatisats et al. model leads to the wake shape plotted in Figure 4.11. For r_c, we have chosen the value $r_c = c/50$. It can be seen that the resulting wake shape is much clearer and more ordered than the one obtained using the standard wake model. In addition to smoothing and ordering the wake, the Vatistas et al. model also smooths the aerodynamic load responses at the early stages of the motion, as seen in Figure 4.12, which plots the Wagner function, pitching moment and drag responses obtained from the Vatistas et al. lumped vortex model and compares

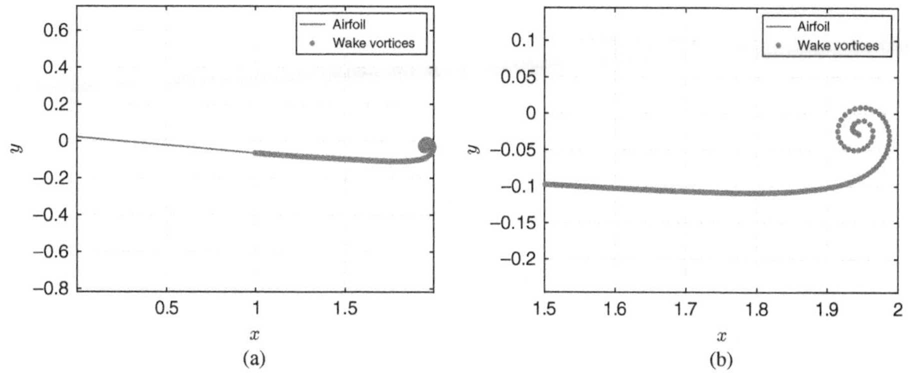

Figure 4.11 Detailed wake shape after impulsive start using Vatistas et al. vortex model for $n = 500$. (a) Detailed wake shape and (b) zoom around starting vortex.

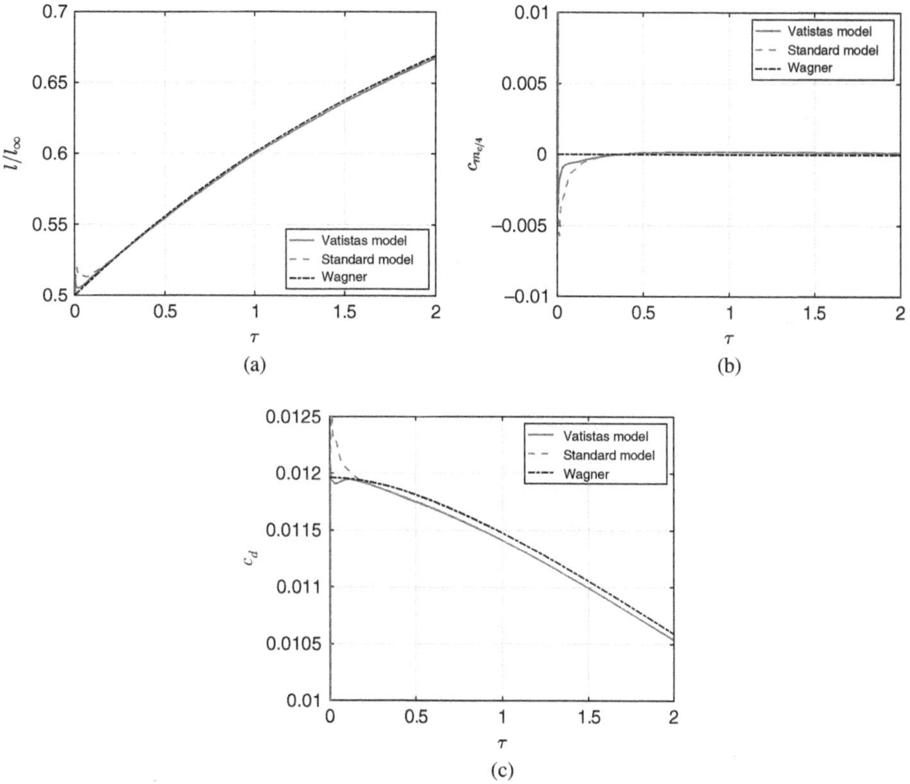

Figure 4.12 Aerodynamic load dependence on vortex model for an impulsively started flat plate using the lumped vortex method. (a) $\Phi(t)$, (b) $c_{m_{c/4}}(t)$ and (c) $c_d(t)$.

them to those calculated using the standard model and Wagner theory. Unfortunately, the Vatistas et al. model increases the computational cost of the lumped vortex method and should only be used when necessary, for example when the spacing between the vortices is small enough to cause severe numerical instabilities. This example is solved by Matlab code `impulsive_lv_free.m`.

This latest example has shown that while the free wake approach reveals more of the physics present in the impulsively started flow by allowing the visualisation of the starting vortex, it does not affect significantly the estimation of the aerodynamic loads. The lift, moment and drag predictions of Figures 4.8 and 4.9 are nearly identical. As the free wake calculation is computationally expensive, it should only be used in cases where the wake shape must be visualised or where it has an important effect on the loads. A 3D example where the wake must be calculated accurately in order to obtain good load estimates is the hovering rotor.

4.3 Gust Encounters

Up to this point, we have been treating the free stream as constant, representing the relative velocity between a wing or aircraft travelling at constant speed and the still air around it. The earth's atmosphere is not still; it is characterised by 3D turbulent flow with many time and length scales. When the intensity of atmospheric turbulence is low, it is usually ignored and replaced by a steady free stream. However, many meteorological and geological phenomena can cause severe turbulent intensity that does not allow the use of the steady free stream approximation.

Atmospheric turbulence is continuous in time and space, but features local events that induce very high velocity fluctuations for a given duration, known as gusts. The time constants of these events can be high compared to the passing time of an aircraft, so that atmospheric turbulence is often treated as frozen. This means that atmospheric turbulence is represented as spatial fluctuations of the free stream velocity in the three directions that are constant in time. Fluctuations in the direction of motion of the aircraft are known as longitudinal turbulence while fluctuations in the other two directions are known as transverse or lateral turbulence.

The simplest model of a transverse turbulent gust is the sharp-edged gust, plotted in Figure 4.13a. A flat plate airfoil is advancing in still air with constant velocity U_∞, approaching a region where the vertical velocity V_g suddenly becomes non-zero and constant. Figure 4.13b plots a smooth transverse gust, known as the 1-cosine gust, whose vertical velocity variation is given by

$$V_g(x) = \frac{v_g}{2}\left(1 - \cos(2\pi x/l_g)\right) \tag{4.67}$$

where v_g is the amplitude of V_g and l_g is the wavelength of the gust. In both cases, the initial lift acting on the airfoil is zero. As the airfoil enters the gust, parts of it see a positive upwash and start to produce lift. Assuming that this lift does not affect the motion of the airfoil, Küssner (1936) developed an analytical solution for the sharp-edged gust encounter. Here, we will model the same problem using the lumped vortex method.

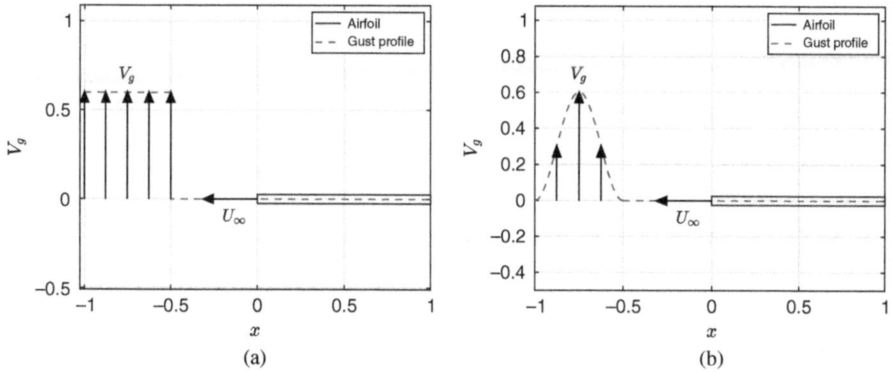

Figure 4.13 Two simple transverse gust models. (a) Sharp-edged gust and (b) 1-cosine gust.

Consider a flat plate airfoil at zero angle of attack, advancing with airspeed $-U_\infty$ towards a sharp-edge gust with vertical velocity V_g. This situation is entirely equivalent to a static airfoil in a free stream with velocity U_∞ that blows towards the airfoil a transverse gust. The gust advances with the free stream airspeed so that the relative velocity between the airfoil and the flow is

$$u_{m_i}(t_k) = U_\infty \tag{4.68}$$

$$v_{m_i}(t_k) = \begin{cases} V_g & \text{if } x_{c_i} \leq U_\infty t_k \\ 0 & \text{if } x_{c_i} > U_\infty t_k \end{cases} \tag{4.69}$$

assuming that the leading edge of the airfoil is in contact with the sharp edge of the gust at time $t = 0$. In other words, as time goes by, more and more of the airfoil lies inside the gust and is affected by the velocity V_g. When $U_\infty t = c$ all of the airfoil lies inside the gust. If $U_\infty t > c$, then parts of the wake also lie inside the gust and are affected by the vertical velocity V_g. Therefore, Eq. (4.56) must be modified to

$$y_{w_i}(t_k) = \begin{cases} y_{w_{i-1}}(t_{k-1}) + \left(V_g + v_{wb_i}(t_k) + v_{ww_i}(t_k)\right)\Delta t & \text{if } x_{w_i}(t_{k-1}) \leq U_\infty t_{k-1} \\ y_{w_{i-1}}(t_{k-1}) + \left(v_{wb_i}(t_k) + v_{ww_i}(t_k)\right)\Delta t & \text{if } x_{w_i}(t_{k-1}) > U_\infty t_{k-1} \end{cases} \tag{4.70}$$

Example 4.5 *Use the lumped vortex method to solve the sharp-edged gust encounter problem and compare to the predictions of Küssner theory.*

We start by setting up the impulsive response Example 4.4. It is very important to choose a linear panel spacing, so that an equal number of control points enter the gust at each time instance. Therefore, the non-dimensional coordinates of the panel vertices are given by

$$\bar{x}_i = \frac{i-1}{n}$$

for $i = 1, \ldots, n+1$. *As the airfoil is not cambered,* $\bar{y}_i = 0$. *Furthermore, as the angle of attack is zero we do not rotate the airfoil. At each time instance, we calculate* $x_g(t_k) = U_\infty t_k$, *where* $x_g(t_k)$ *is the position of the sharp edge of the gust. We evaluate the relative velocities of*

Eqs. (4.68) and (4.69) and then apply the impermeability boundary condition (4.42) with a single modification:

$$\begin{pmatrix} \mathbf{A}_n & \mathbf{b}_{n_1} \\ \mathbf{I}_{1\times n} & 1 \end{pmatrix} \begin{pmatrix} \mathbf{\Gamma}_b(t_k) \\ \Gamma_{w_1}(t_k) \end{pmatrix} = \begin{pmatrix} \mathbf{b}(t_k) - \mathbf{B}_n^*(t_k)\mathbf{\Gamma}_w^*(t_k) \\ \sum_{j=1}^n \Gamma_{b_j}(t_{k-1}) \end{pmatrix} \tag{4.71}$$

where $\mathbf{b}(t_k)$ is given by

$$\mathbf{b}(t_k) = -(\boldsymbol{u}_m(t_k) \circ \boldsymbol{n}_x + \boldsymbol{v}_m(t_k) \circ \boldsymbol{n}_y)$$

where $\boldsymbol{u}_m(t_k)$ and $\boldsymbol{v}_m(t_k)$ are the $n \times 1$ vectors $\boldsymbol{u}_m(t_k) = (u_{m_1}(t_k), u_{m_2}(t_k), \ldots, u_{m_n}(t_k))^T$ and $\boldsymbol{v}_m(t_k) = (v_{m_1}(t_k), v_{m_2}(t_k), \ldots, v_{m_n}(t_k))^T$. The notation $\circ$ denotes the Hadamard product, that is element-by-element multiplication of vectors or matrices of the same size.

The rest of the procedure is identical to that of Example 4.4, except that we propagate the vertical positions of the wake vortices using Eq. (4.70). The definition of the lift and drag is complicated in the present case because these loads are defined with respect to the free stream and the latter changes from $(U_\infty, 0)$ to (U_∞, V_g) as the gust passes over the airfoil. It is therefore simpler to calculate the normal and chordwise forces from Eq. (4.48)

$$\Delta f_{x_i}(t_k) = -\rho\left(v_{m_i}(t_k) + v_{w_i}(t_k)\right)\Gamma_{b_i}(t_k)$$
$$\Delta f_{y_i}(t_k) = \rho\left(\left(u_{m_i}(t_k) + u_{w_i}(t_k)\right)\Gamma_{b_i}(t_k) + \sum_{j=1}^{i}\dot{\Gamma}_{b_j}(t_k)s_i\right) \tag{4.72}$$

where we have replaced the free stream by the relative velocities $u_{m_i}(t_k)$ and $v_{m_i}(t_k)$ of Eqs. (4.68) and (4.69). The pitching moment around the quarter-chord is given by Eq. (4.52) which, simplifies to

$$\Delta m_{c/4_i}(t_k) = -\sum_{i=1}^{n}\Delta f_{y_i}(t_k)(x_{b_i} - c/4)$$

Küssner theory states that the lift response to a sharp-edged gust encounter is given by

$$l(t) = 2\pi\rho U_\infty b V_g \psi(\tau)$$

where $\psi(\tau)$ is the Küssner function and $\tau = U_\infty t/b$ is non-dimensional time. Bisplinghoff et al. (1996) give the following approximation for $\psi(\tau)$

$$\psi(\tau) = \frac{\tau^2 + \tau}{\tau^2 + 2.82\tau + 0.80}$$

If $V_g \ll U_\infty$, then the lift and normal force are approximately equal.

We carry out the lumped vortex calculations with $n = 100$, $U_\infty = 10$ m/s, $V_g = 1$ m/s and $t_f = 10c/U_\infty$. Figure 4.14a plots the wake shape at the end of the motion, along with $x_g(t_f)$, the current position of the sharp-edge front of the gust. Up to the gust front, the wake shape looks like that of Figure 4.9a for an impulsively started airfoil. However, the gust front has split the wake so that there are two counter-rotating vortices on either side of the former. This is due to the fact that upstream of the wavefront the wake is affected by the V_g velocity, whereas

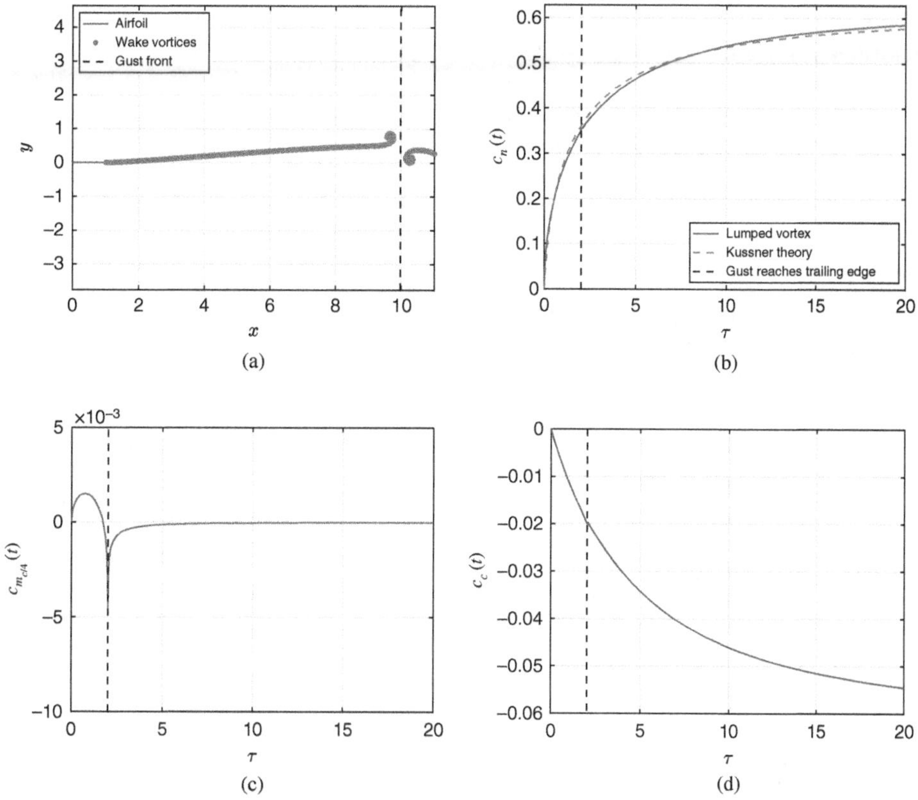

Figure 4.14 Free wake shape and aerodynamic loads for a sharp-edge gust encounter using the lumped vortex method. (a) Wake shape, (b) $c_l(t)$, (c) $c_{m_{c/4}}(t)$ and (d) $c_d(t)$.

downstream it is not. It must also be stated that the strongest wake vortex is shed at the instant the gust front reaches the trailing edge.

Figure 4.14b plots the time response of the normal force coefficient and compares it to the lift prediction of Küssner theory. It can be seen that the two estimates are in very good agreement. The lift is equal to zero at $t = 0$ since only the leading edge touches the gust front at that instant. As more and more of the airfoil is affected by the gust, the lift increases smoothly until the gust reaches the trailing edge. At that instance, given by $\tau = 2$ for all values of c and U_∞, the lumped vortex lift prediction features a small discontinuity. The rest of the response is a slightly delayed impulsive start lift response.

Figure 4.14c shows that the pitching moment around the quarter chord reaches a maximum nose-down value at the moment the gust arrives at the trailing edge and then increases very quickly towards zero. The gust front causes a strong local normal force while it lies on the surface. As the front moves towards the trailing edge, this normal force moves with it, causing a strong nose-down moment around the quarter chord. As soon as the gust front clears the *trailing edge, the local normal force disappears and the pitching moment becomes quickly zero. It must be stressed that the amplitude of the pitching moment response decays with increasing n. Finally, Figure 4.14d shows that the chordwise force becomes increasingly negative as more and more of the airfoil enters the gust. When the gust front clears the trailing edge, the chordwise force starts to level out and to approach asymptotically its steady-state value, $-2\rho Q_\infty^2 b\pi\alpha^2$ (see Eq. (3.102)), where $\alpha = \tan^{-1}(V_g/U_\infty)$ and $Q_\infty^2 = U_\infty^2 + V_g^2$. This example is solved by Matlab code* `gust_1v_free.m`.

The 1-cosine gust of Eq. (4.67) can be modelled in the same manner, as long as the velocity $v_{m_i}(t_k)$ is adapted accordingly. At each time instance, the control points that lie between $x_g(t_k) - l_g/2$ and $x_g(t_k) + l_g/2$ must be identified where, $x_g(t_k) = U_\infty(t_k) - l_g/2$ such that the gust front lies at the leading edge at $t = 0$. Then, the vertical relative velocity is given by

$$v_{m_i}(t_k) = \begin{cases} \frac{v_g}{2}\left(1 - \cos\left(2\pi\frac{x_{c_i}+l_g/2-x_g(t_k)}{l_g}\right)\right) & \text{if } x_g(t_k) - \frac{l_g}{2} \leq x_{c_i} \leq x_g(t_k) + \frac{l_g}{2} \\ 0 & \text{otherwise} \end{cases}$$

The same vertical velocity must be applied to the parts of the wake that lie inside the gust, after replacing x_{c_i} by $x_{w_i}(t_k)$. The rest of the procedure is identical to that of Example 4.5.

4.3.1 Pitching and Plunging Wings

We will now apply the lumped vortex methodology to the pitching and plunging wing problem that we have already addressed in Chapter 3. The motion of the airfoil is described by

$$h(t) = h_1 \sin\left(\omega t + \phi_h\right), \quad \alpha(t) = \alpha_0 + \alpha_1 \sin\left(\omega t + \phi_\alpha\right) \tag{4.73}$$

where α_0 is the mean pitch angle, h_1, α_1 are the amplitudes of the plunging and pitching motions, while ϕ_h and ϕ_α are their phases. There are two main differences between the impulsive start of Section 4.2 and the pitching and plunging problem:

- The airfoil is pitching and plunging constantly and so it must be rotated and translated at each time instance. It follows that the coordinates of the bound vortices and of the control points, the direction of the normal vectors and the position of the first wake vortex will be functions of time.
- The pitching and plunging motion creates an additional relative velocity between the wing and the flow.

First, we will discuss the motion of the airfoil. Physically, the coordinates of the panel vertices, $x_i(t_k), y_i(t_k)$, are now functions of time, but we can follow the quasi-fixed modelling approach of Figure 2.3 and keep the airfoil static, in which case the motion is only represented by the upwash it induces. The other option is to adopt the motion scheme of Figure 2.2 and move the coordinates of the panel vertices in time while keeping the free stream fixed and horizontal, $\mathbf{Q}_\infty = (U_\infty, 0)$. We denote by x_{0_i}, y_{0_i} the panel coordinates aligned with the x axis and, at each time instance t_k, we rotate them by $\alpha(t_k)$ around the pitch axis, x_f, in a manner similar to Eq. (4.59). We also translate them by $-h(t_k)$, the negative sign being there because h is defined positive downwards. Hence,

$$\begin{pmatrix} x_i(t_k) \\ y_i(t_k) \end{pmatrix} = \begin{pmatrix} \cos\alpha(t_k) & \sin\alpha(t_k) \\ -\sin\alpha(t_k) & \cos\alpha(t_k) \end{pmatrix} \begin{pmatrix} x_{0_i} - xf \\ y_{0_i} \end{pmatrix} + \begin{pmatrix} x_f \\ -h(t_k) \end{pmatrix} \tag{4.74}$$

Next, we re-calculate the coordinates of the bound vortices and collocation points as well as the elements of the normal vectors using Eqs. (4.17) to (4.19). As $x_{b_i}(t_k), y_{b_i}(t_k), x_{c_i}(t_k), y_{c_i}(t_k)$ and $n_{x_i}(t_k), n_{y_i}(t_k)$ are now functions of time, the aerodynamic influence coefficient matrices of Eqs. (4.20) and (4.21) also vary in time. However, the normal influence coefficient matrix of equation (4.23) is constant because it only depends on the relative distances between the bound vortices and collocation points; the airfoil is assumed to be rigid so that these distances remain constant in time, despite the pitching and plunging motion. Hence, we only calculate $A_{n_{i,j}}$ once, before the start of the time iterations. Finally, at each time instance, the position of the first wake vortex is calculated with respect to the current coordinates of the trailing edge,

$$\begin{aligned} x_{w_1}(t_k) &= x_{n+1}(t_k) + \Delta x_w U_\infty \Delta t \\ y_{w_1}(t_k) &= y_{n+1}(t_k) \end{aligned} \tag{4.75}$$

since the free stream only has an x component.

Now, we turn our attention to calculating the total flow velocity seen by the airfoil due to its motion. The origins of this flow are, as usual, the free stream, the plunge velocity and the pitch velocity, as seen in Figure 4.15. Note that the pitch-induced flow is not assumed to be aligned with the plunge direction as was the case in Chapter 3 (see Figure 3.5). Plunging and pitching are a translation and a rotation, respectively, so that the relative velocity between the flow and the wing's control points is given by Eq. (4.4)

$$\mathbf{u}_m(t) = \mathbf{Q}_\infty - \mathbf{v}_0(t) - \mathbf{\Omega}_0(t) \times (\mathbf{x}_c(t) - \mathbf{x}_f(t)) \tag{4.76}$$

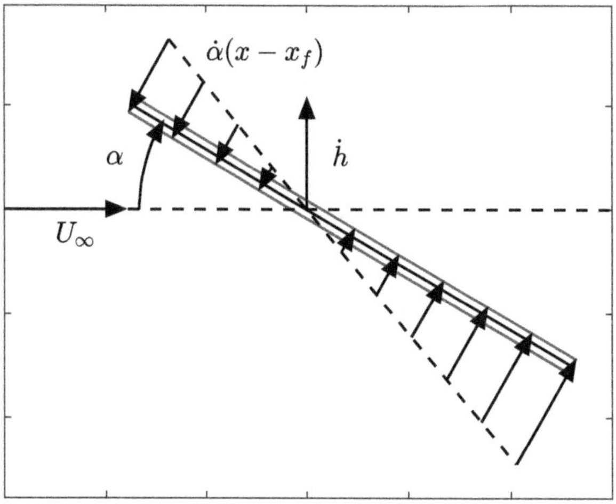

Figure 4.15 Motion-induced flow velocities on flat plate at an angle of attack.

where $\mathbf{v}_0(t)$ is the translation velocity, $\mathbf{\Omega}_0(t)$ is the rotation velocity and $\mathbf{x}_f(t)$ is the axis of rotation. In the 2D pitching and plunging case, the rotation is carried out around the z axis, $Q_\infty = (U_\infty, 0, 0)$, $\mathbf{v}_0(t) = (0, -\dot{h}(t), 0)$, $\mathbf{\Omega}_0(t) = (0, 0, \dot{\alpha}(t))$ and $\mathbf{x}_f(t) = (x_f, -h(t), 0)$. Evaluating the vector product leads to

$$u_{m_i}(t) = U_\infty - \dot{\alpha}(t)(y_{c_i}(t) + h(t)) \tag{4.77}$$

$$v_{m_i}(t) = \dot{h}(t) + \dot{\alpha}(t)(x_{c_i}(t) - x_f) \tag{4.78}$$

where $u_{m_i}(t)$ and $v_{m_i}(t)$ are the horizontal and components, respectively, of $\mathbf{u}_m(t)$.

For pitching and plunging motion, Eq. (4.42) is modified to

$$\begin{pmatrix} \boldsymbol{A}_n & \boldsymbol{b}_{n_1}(t_k) \\ \boldsymbol{I}_{1\times n} & 1 \end{pmatrix} \begin{pmatrix} \boldsymbol{\Gamma}_b(t_k) \\ \Gamma_{w_1}(t_k) \end{pmatrix} = \begin{pmatrix} \boldsymbol{b}(t_k) - \boldsymbol{B}_n^*(t_k)\boldsymbol{\Gamma}_w^*(t_k) \\ \sum_{j=1}^{n} \Gamma_{b_j}(t_{k-1}) \end{pmatrix} \tag{4.79}$$

There are two differences between Eqs. (4.79) and (4.42). Firstly, the distances between the first wake vortex and the control points are no longer constant so that $\boldsymbol{b}_{n_1}(t_k)$ is now a function of time. Secondly, the vector $\boldsymbol{b}(t_k)$ is given by

$$\boldsymbol{b}(t_k) = -\boldsymbol{u}_m(t_k) \circ \boldsymbol{n}_x(t_k) - \boldsymbol{v}_m(t_k) \circ \boldsymbol{n}_y(t_k) \tag{4.80}$$

Note that the components of the normal vector are functions of time since the airfoil is rotating.

Finally, we discuss the modelling of the wake, which can be prescribed or free. A prescribed wake is propagated using the free stream airspeed only, that is

$$x_{w_j}(t_k) = x_{w_{j-1}}(t_{k-1}) + U_\infty \Delta t, \quad y_{w_j}(t_k) = y_{w_{j-1}}(t_{k-1}) \tag{4.81}$$

Nevertheless, the wake shape will depend on the choice we make for the wing motion simulation. If we choose quasi-fixed modelling, the trailing edge will not move and, as all wake vortices are shed from the trailing edge, the wake will be a straight line. This is the flat wake modelled by Theodorsen and Wagner theory and plotted in Figure 4.16a. If, on the other hand, we choose to move the wing, its trailing edge will also move so that the wake vortices will be shed at different positions. Consequently, the wake will have a sinusoidal shape if the wing is moving sinusoidally, as seen in Figure 4.16b. In this work, when using the term prescribed wake, we will refer to wakes of the form seen in Figure 4.16b.

The other choice is to simulate a free wake, propagating it using Eq. (4.56). The wake shape again depends on whether the wing is static or moving. In the present work, we will always choose to move the wing when modelling the wake as free. Figure 4.16c plots the free wake shape behind a sinusoidally pitching and plunging airfoil. The shape is not sinusoidal because the wake vortices interact to create larger vortical structures. We will discuss this type of wake in more detail in the next example, when we compare it to experimental flow visualisations.

The equations for the incremental lift and drag acting on each panel are adapted from expressions (4.61) such that

$$\Delta l_i(t_k) = \rho\left(\left(u_{m_i}(t_k) + u_{w_i}(t_k)\cos\alpha(t_k)\right)\Gamma_{b_i}(t_k) + \sum_{j=1}^{i}\dot{\Gamma}_{b_j}(t_k)s_i\cos\alpha(t_k)\right)$$

$$\Delta d_i(t_k) = \rho\left(-\left(v_{m_i}(t_k) + v_{w_i}(t_k)\cos\alpha(t_k)\right)\Gamma_{b_i}(t_k) + \sum_{j=1}^{i}\dot{\Gamma}_{b_j}(t_k)s_i\sin\alpha(t_k)\right) \tag{4.82}$$

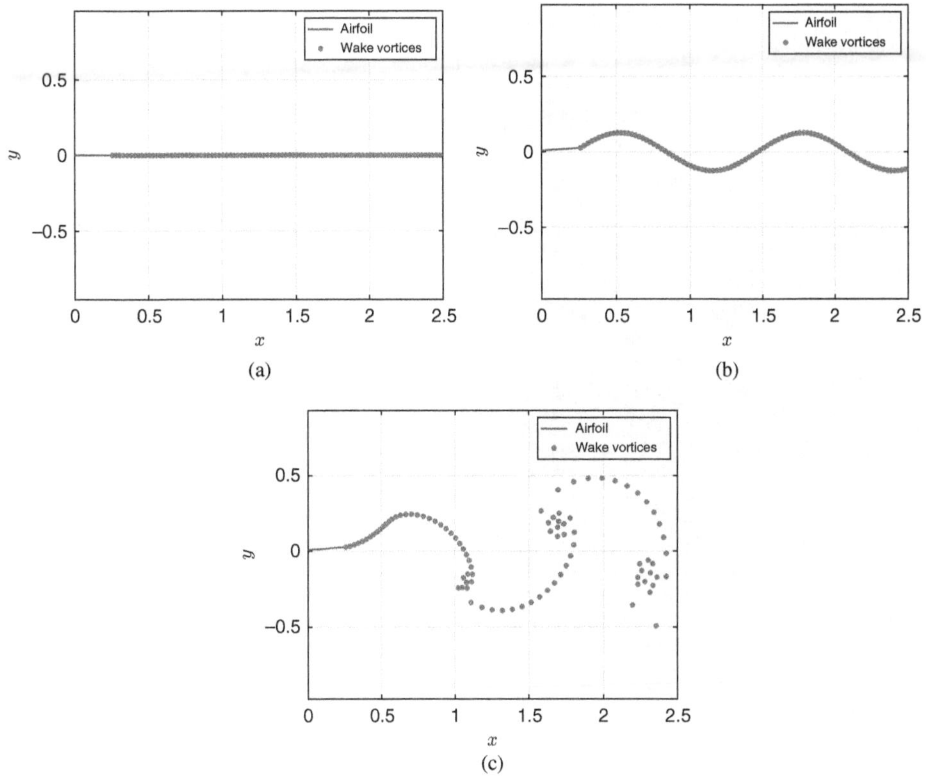

Figure 4.16 Types of wake modelling for a sinusoidally pitching and plunging airfoil. (a) Prescribed wake without wing motion (flat wake), (b) prescribed wake with wing motion and (c) free wake with wing motion.

Finally, generalising Eq. (4.62) the aerodynamic pitching moment around the pitch axis, $(x_f(t_k), y_f(t_k))$, is given by

$$m_{x_f}(t_k) = \sum_{i=1}^{n} \Delta d_i(t_k)(y_{b_i}(t_k) - y_f(t_k)) - \sum_{i=1}^{n} \Delta l_i(t_k)(x_{b_i}(t_k) - x_f(t_k)) \tag{4.83}$$

Example 4.6 *Calculate the free wake shape behind a plunging airfoil*

Over the years, there have been numerous experimental investigations of unsteady aerodynamic phenomena, mainly within the context of dynamic stall research. However, unsteady flow visualisations behind oscillating airfoils are far less numerous. One such investigation by Freymuth (1988) presents flow snapshots around NACA 0015 wings performing pitch or plunge oscillations in a wind tunnel, visualised using fumes from a strip of liquid titanium tetrachloride.

We will address the plunging oscillations first. The wing has a chord of $c = 0.15$ *m, lies at a constant angle of attack of* $5°$ *to a free stream of* $U_\infty = 0.61$ *m/s and plunges with amplitude* $h_1 = 0.2c = 0.03$ *m and frequency 3.5 Hz. The airspeed is very low in order to allow good quality flow visualisations, leading to a high reduced frequency value of* $k = 2.7$*. Figure 4.17a plots the 10 snapshots of the flow that were taken during the experiments, the time between*

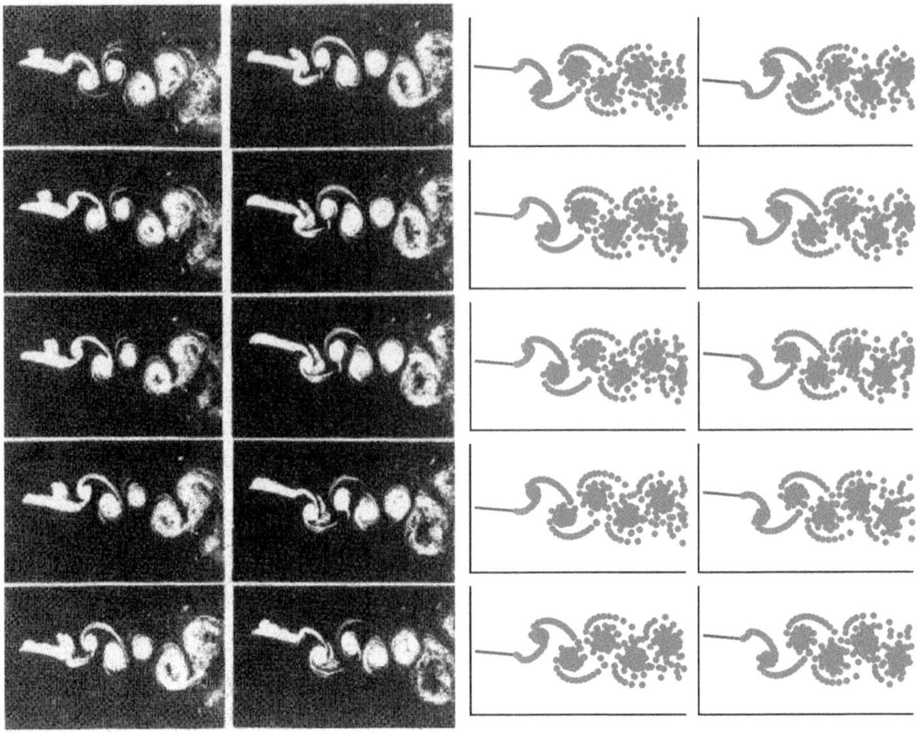

Figure 4.17 Snapshots of the wake behind a plunging airfoil. (a) Experiment (Freymuth 1988). Source: Reprinted by permission of the American Institute of Aeronautics and Astronautics, Inc. (b) Lumped vortex.

consecutive shots being 1/32 seconds. The airfoil is descending during the first four snapshots of the first column (top to bottom), ascending during the last frame of the first column and the first four frames of the second (again top to bottom) and descending in the last shot. A series of vortices can be seen in the wake, rotating in a clockwise direction under the airfoil and in a counter-clockwise direction above it. Note that the wake is becoming wider and less coherent as it travels downstream.

We simulate this flow using $n = 150$ *non-linearly spaced panels on a flat plate (the NACA 0015 airfoil is symmetrical) and a time step of* $\Delta t = c/U_\infty n$*, in order to obtain a good density of wake vortices. In fact,* Δt *is adjusted slightly so that the experimental time step of 1/32 becomes an integer multiple of* Δt*, in order to allow easier comparison between the simulated and experimental time frames. The panel vertices are defined parallel to the x axis and rotated to 5°; they are then translated by* $-h(t) = -h_0 \sin\left(\omega t + \phi_h\right)$*. The circulation in the flow at times* $t < 0$ *is set to zero so that the airfoil is set in motion impulsively at* $t = 0$*. Ten cycles of the motion are simulated, but only the last cycle is used to plot flow snapshots in order to allow for the transient effects of the impulsive start to die out before visualising the wake. The wake is treated as free, using the standard vortex model and the first wake vortex position parameter is chosen as* $\Delta x_w = 0.25$*. The motion-induced upwash is calculated from Eqs. (4.77) and (4.78)* *with $\dot{\alpha}(t_k) = 0$ and the aerodynamic loads from expressions (4.82) and (4.83), noting that the pitching moment is calculated around the quarter chord.*

The wake is plotted at 10 time instances that are equally spaced over the last cycle of oscillation. The phase of the plunging motion is set to $\phi_h = -45°$, so that the simulated snapshots correspond approximately to the experimental frames of Figure 4.17a. The axes of the plot windows are also selected in order to resemble those of the experimental results; in the absence of a grid on the photos, the comparison between the simulated and experimental wakes can only be qualitative. Figure 4.17b plots the resulting simulated snapshots; it can be seen that they are very similar to their experimental counterparts, but there is an important difference. The experimental results show that a clockwise vortex is formed close to the leading edge as the airfoil descends; it then travels downstream until it merges with a clockwise wake vortex. This phenomenon is known as the Leading Edge Vortex (LEV) and is related to the viscous dynamic stall phenomenon (see for example (Carr et al. 1977)). As the lumped vortex calculation performed here is inviscid, it cannot model the LEV.

Figure 4.18 plots a longer snapshot of the wake shape at the end of the motion and the time responses of the aerodynamic loads, compared to those obtained from Wagner theory. Note that the agreement between the two methods is good but not perfect. This phenomenon is due to the

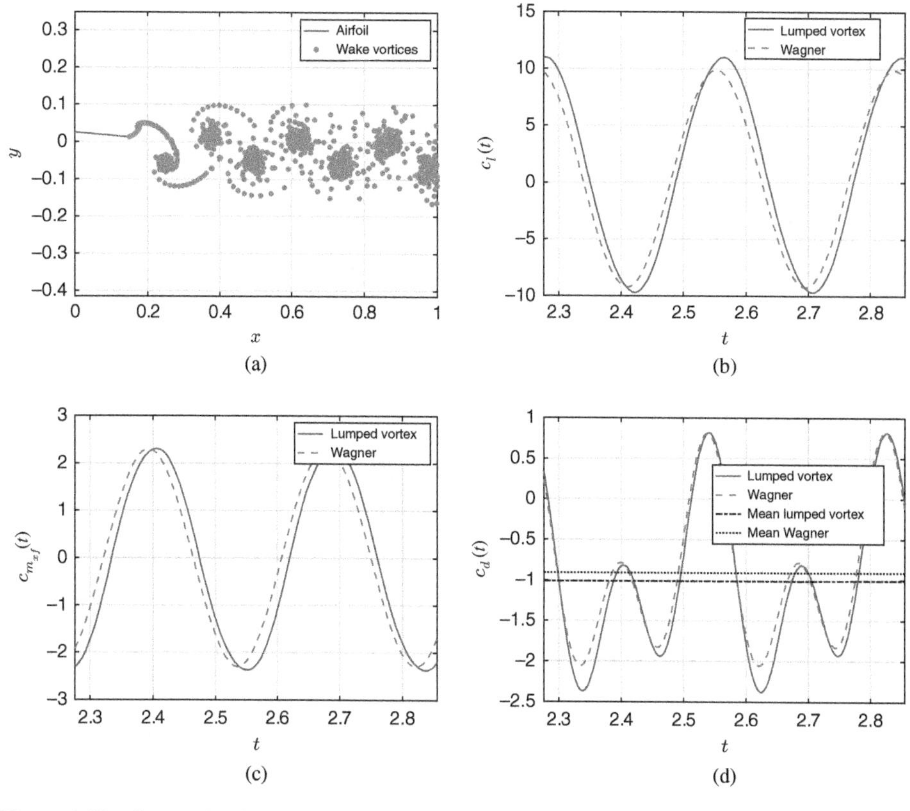

Figure 4.18 Free wake shape and aerodynamic loads for a plunging NACA 0015 airfoil using the lumped vortex method. (a) $x_w(t), y_w(t)$, (b) $c_l(t)$, (c) $c_{m_{x_f}}(t)$ and (d) $c_d(t)$.

fact that the lumped vortex method models the motion of the flat plate, which is no longer aligned with the x axis at all times. As the oscillation amplitude is quite high, irrespective of whether the wake is free or prescribed, the wake vortices can lie significantly above or below the airfoil. This means that the velocities induced by the wake on the airfoil will be different to those predicted by flat wake theories. Hence, the aerodynamic load estimates will also be different to some extent.

As the lumped vortex approach is non-linear and takes into account more of the physics of the motion than linearised analytical methods, it is more reliable for such highly non-linear flows. Nevertheless, neither of the two techniques model viscous effects and therefore none of the load responses presented in Figure 4.18 are reliable. The experimental flow visualisation demonstrates the presence of a LEV, which is not modelled by inviscid methods and which has a significant effect on the aerodynamic loads. Dynamic stall will be discussed in more detail later in this book. This example is solved by Matlab code `plunge_lv_free.m`.

It is interesting to note that the drag responses predicted by both the lumped vortex and Wagner/Garrick method in Figure 4.18d have negative mean value, that is the airfoil is generating a mean thrust. Figure 4.17 demonstrates the physical mechanism for thrust generation; the vortices below the airfoil rotate in a clockwise direction, while those over it rotate in a counter-clockwise direction. The consequence is that:

- The vortices below the airfoil induce velocities pointing in the downstream direction on the wake above them.
- The vortices above the wake induce velocities pointing in the downstream direction on the wake below them.

The net effect is that the wake behind the airfoil is pushed backwards at most time instances.

The velocity distribution behind the airfoil is demonstrated graphically in Figure 4.19, where we have replotted the snapshots of Figure 4.17b, but we have also superimposed the velocity vectors induced by the bound and wake vorticity at four positions in the wake at a distance of zero, one, two and three chords from the trailing edge. Note that the free stream has not been included in this calculation. It can be seen that the velocity vectors directly behind the airfoil point downstream on almost all of the snapshots. This phenomenon is known as the wake's jet effect (Jones et al. 2001). The motion of the airfoil is clearly pushing backward the air behind it; by Newton's third law of motion, the air behind the airfoil is pushing the airfoil forward. Two cases can be distinguished:

- Thrust is produced when the velocity vectors on the upper side of the trailing edge point downstream; this phenomenon occurs through most of the cycle.
- Drag is produced when the velocity vectors on the upper side of the trailing edge point upstream. This phenomenon occurs at the start of the downstroke, while there is no strong counter-clockwise vortex behind the upper side of the trailing edge.

The vorticity immediately behind the upper side of the airfoil is important because of the positive constant pitch attack. Figure 4.18b shows that the mean lift is slightly positive, again due to the positive pitch angle. If the pitch was negative, the mean lift would be slightly negative, but the airfoil would still be producing thrust because the wake would be a mirrored version of Figure 4.17. This calculation is left as an exercise for the reader.

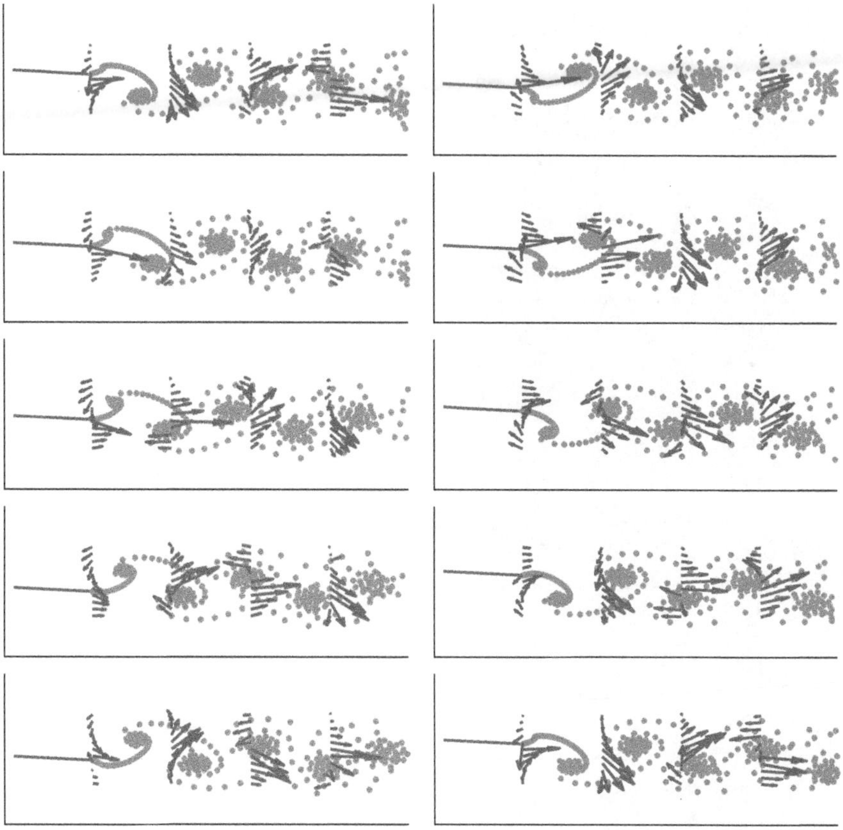

Figure 4.19 Wake jet effect.

The ability of a sinusoidally plunging airfoil to produce thrust is sometimes referred to as the Knoller–Betz effect (Jones et al. 1998).

The present physical explanation of thrust production is based on momentum considerations while the development of Garrick's drag/thrust model in Section 3.4.2 is based on the leading edge suction phenomenon, that is on pressure considerations. The two approaches are compatible, since pressure and momentum are related through the Euler equations (3.200).

Example 4.7 *Calculate the wake shape behind a pitching airfoil*

We revisit the work presented in Freymuth (1988) in order to simulate the second test case, a NACA 0015 airfoil with chord $c = 0.356$ *m, immersed in a free stream of* $U_\infty = 0.61$ *m/s and pitching around its quarter chord at a frequency of 1.6 Hz, such that the instantaneous pitch angle is given by*

$$\alpha(t) = \alpha_0 + \alpha_1 \sin(\omega t + \phi_\alpha)$$

where $\alpha_0 = 5°$ *is the constant mean pitch angle,* $\alpha_1 = 20°$ *is the pitch amplitude and* ϕ_α *is the pitch phase. Figure 4.20a plots 12 snapshots separated by a time increment of 1/16 seconds;*

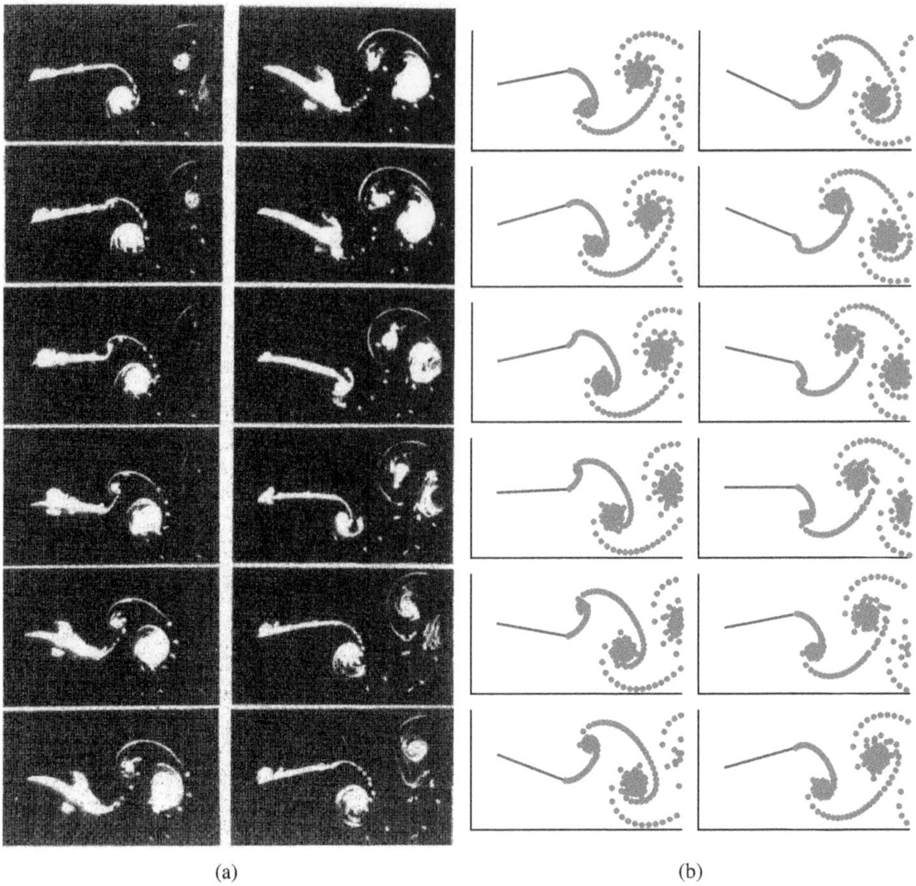

Figure 4.20 Snapshots of the wake behind a pitching airfoil. (a) Experiment (Freymuth 1988). Source: Reprinted by permission of the American Institute of Aeronautics and Astronautics, Inc. (b) Lumped vortex.

the airfoil is pitching up in the frames of the first column and down in the second column. The last two frames of the second column occur at the same phase angles as the first two frames of the first columns, so they are repeated frames. During pitch-up, a counter-clockwise vortex is shed from the trailing edge and a LEV grows near the leading edge and starts to travel downstream. There is significant flow separation over the upper surface of the airfoil during the latter part of the pitch-up. The LEV is shed into the wake in the third frame of the second column and merges with the clockwise vortex that is formed at the trailing edge as the wing starts to pitch down. A new LEV appears to form as the old one is shed into the wake.

The lumped vortex simulation procedure is identical to the one used in Example 4.6, except that $\dot{\alpha}$ is not equal to zero; only four cycles of oscillation are simulated, as a smaller section of the flow is to be visualised than in the case of the plunge oscillations. The phase of the pitch oscillation is set to $\phi_\alpha = 230°$ in order to match the experimental flow visualisations as closely as possible. Figure 4.20b plots 12 snapshots taken during the fourth cycle, corresponding *to the 12 frames of Figure 4.20a. The comparison between the numerical and experimental wake shapes is very good. As we have used the standard vortex model, the lumped vortices lying between the large counter-rotating vortical structure tend to break up into smaller groups of vortices. The same phenomenon was observed experimentally by Freymuth (1988), who attributed it to the Helmholtz instability, but there were two different fluids, air and fumes, involved in the process. In the simulations, there is no viscous mixing and the phenomenon is caused by the numerical instability inherent in the standard vortex model.*

Figure 4.21 plots a longer snapshot of the final wake shape, along with the lift, pitching moment and drag time responses over the last two cycles. The differences between the lumped vortex and Wagner theory load predictions are more important than in the pure plunging case. Nevertheless, neither of the results plotted in Figure 4.21 are reliable, as they both ignore the viscous effects observed in the flow visualisations of Figure 4.20a. It is still interesting to note that, despite the lack of viscous modelling, the lumped vortex method can predict trailing edge vortex shedding patterns that are visually correct. This fact was used by Katz (1981) and others in order to extend the lumped vortex approach and other potential methods to the modelling of separated flows; such methodologies will be discussed in Chapter 7.

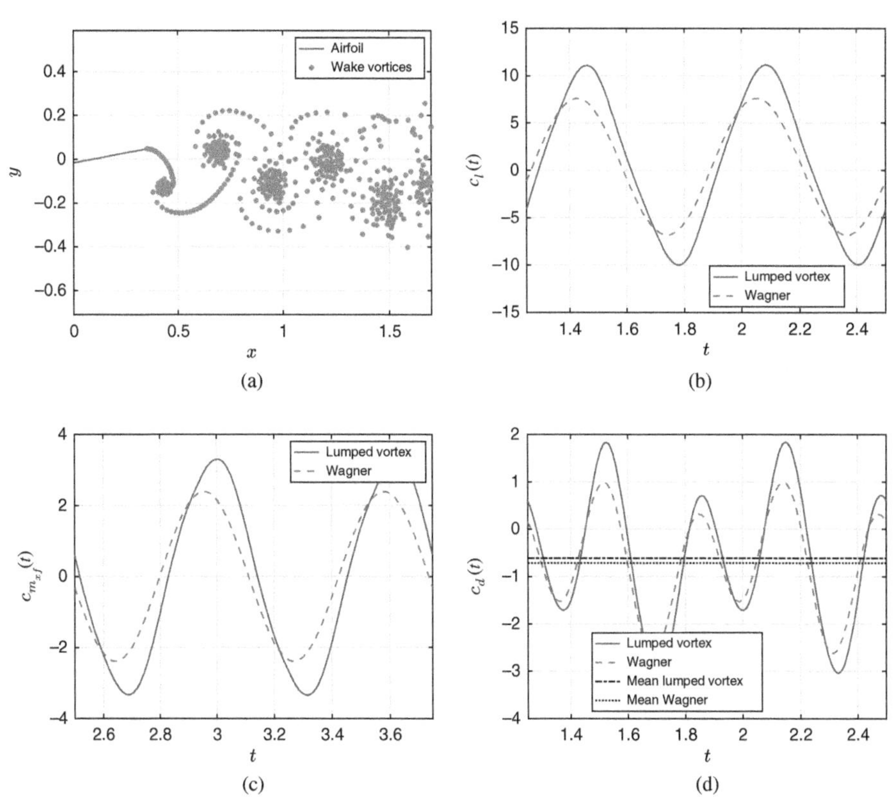

Figure 4.21 Free wake shape and aerodynamic loads for a pitching NACA 0015 airfoil using the lumped vortex method. (a) $x_w(t), y_w(t)$, (b) $c_l(t)$, (c) $c_{m_{c/4}}(t)$ and (d) $c_d(t)$.

The wake shapes in Figure 4.20 feature clockwise vortices on the lower side and counter-clockwise vortices on the upper side, exactly as in the case of the plunge oscillations. This means that, again, we expect that the airfoil will be generating a mean thrust due to the jet effect. This is indeed the case, as seen in Figure 4.18d, which shows that the mean drag is clearly negative. Drag is generated during the middle sections of both pitchup and pitchdown; over the rest of the cycle, thrust is produced. The reader is invited to calculate and plot velocity vectors in the wake over the snapshots of Figure 4.20b, as was done in Example 4.6, in order to investigate the mechanisms of both drag and thrust production.

The mean of the lift response plotted in Figure 4.21b is positive due to the positive mean pitch angle $\alpha_0 = 5°$. As was discussed for the plunge oscillations, changing the sign of the mean pitch angle will also change the sign of the mean lift, but the airfoil will still produce mean thrust. Again, this calculation is left as an exercise for the reader. This example is solved by Matlab code `pitch_lv_free.m`.

It must be stressed that, while plunging oscillations always produce a net thrust, pitching oscillations produce thrust conditionally, depending on the position of the pitch axis and the reduced frequency of the motion. Figure 4.22 plots the mean drag produced by a sinusoidally pitching flat plate at reduced frequency values between $k = 0$ and $k = 2$ and for pitch axis locations between the leading edge and the trailing edge. For low values of k, the wing produces drag for all x_f. When the pitching axis lies at the leading edge, thrust is produced for $k > 0.63$. Higher values of k are required if $x_f > 0$.

The vortices shed from the trailing edge during pitching oscillations induce velocities in the downstream direction in parts of the wake, just as in the case of plunging oscillations (see Figure 4.19). However, at low frequencies, a pitching airfoil spends most of its time on the wrong side of the wake vortices, where they induce velocities in the upstream direction. This situation is exemplified in Figure 4.23a, where the counter-clockwise wake vortices induce a jet effect underneath the airfoil. It is only when the frequency is high enough for the wake to roll, as shown in Figure 4.23b, that the airfoil starts to spend a lot of time on the appropriate part of the flowfield and can therefore profit from the jet effect.

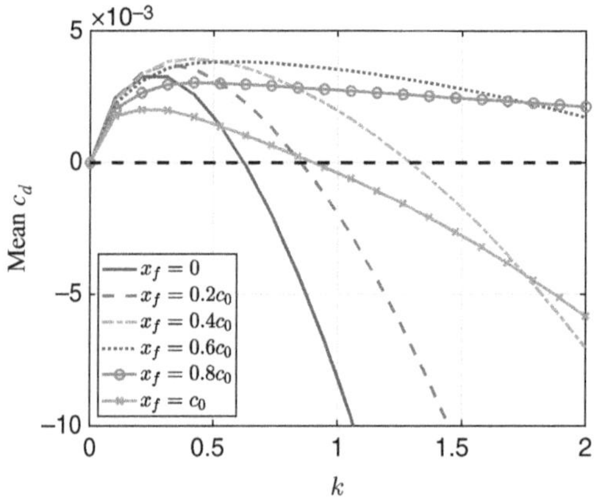

Figure 4.22 Mean drag/thrust produced by sinusoidally pitching wing.

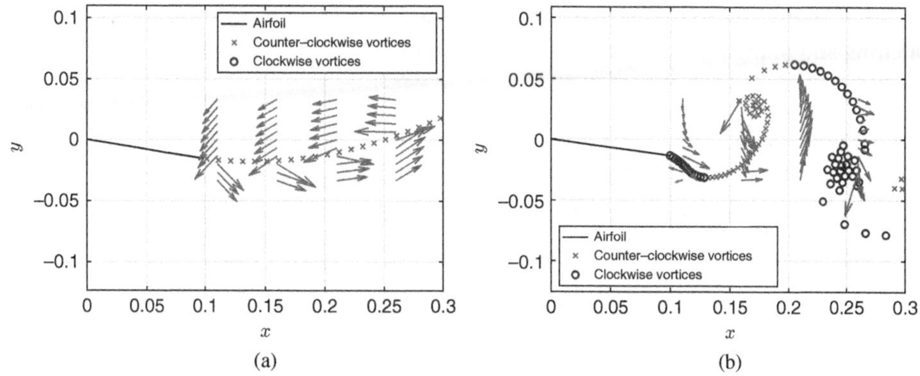

Figure 4.23 Wake behind sinusoidally pitching airfoil at two values of the reduced frequency. (a) $k = 0.4$ and (b) $k = 2.0$.

4.4 Frequency Domain Formulation of the Lumped Vortex Method

Up to this point, the unsteady lumped vortex problem has been solved using a time marching procedure. This means that in order to evaluate the aerodynamic loads at time t_k, the flow must first be solved at all time instances t_{k-i} for $i = k, \dots, 1$. In contrast, the analytical solutions of Chapter 3 are not time marching; the aerodynamic loads can be directly calculated at any time instance. Numerical models can also be formulated in such a manner, after linearising them and transforming them to the frequency domain.

First, we revert to the quasi-fixed formulation used in Chapter 3, whereby the airfoil remains static and its motion is represented only by the relative motion between the airfoil and the fluid. In the case of a pitching and plunging airfoil, this relative motion is given by Eqs. (4.77) and (4.78), but since the airfoil is no longer pitching, x_{c_i} and y_{c_i} are constants, the position of the pitch axis is $(x_f, 0)$, and we must rotate the free stream. Consequently,

$$\begin{aligned} u_{m_i}(t) &= Q_\infty \cos\alpha(t) - \dot{\alpha}(t) y_{c_i} \\ v_{m_i}(t) &= Q_\infty \sin\alpha(t) + \dot{h}(t) + \dot{\alpha}(t)(x_{c_i} - x_f) \end{aligned} \tag{4.84}$$

We would like to apply the continuous-time Fourier transform to these expressions, but the cosine and sine terms complicate this application. Therefore, we decide to linearise Eq. (4.84) to obtain

$$\begin{aligned} u_{m_i}(t) &\approx Q_\infty - \dot{\alpha}(t) y_{c_i} \\ v_{m_i}(t) &\approx Q_\infty \alpha(t) + \dot{h}(t) + \dot{\alpha}(t)(x_{c_i} - x_f) \end{aligned}$$

We can now readily apply the Fourier transform, resulting in

$$\begin{aligned} u_{m_i}(\omega) &= Q_\infty \delta(\omega) - \mathrm{i}\omega\alpha(\omega) y_{c_i} \\ v_{m_i}(\omega) &= Q_\infty \alpha(\omega) + \mathrm{i}\omega h(\omega) + \mathrm{i}\omega\alpha(\omega)(x_{c_i} - x_f) \end{aligned} \tag{4.85}$$

where $\delta(\omega)$ is the Kronecker delta function.

The impermeability boundary condition for the unsteady lumped vortex modelling of a pitching and plunging flat plate is the first of Eq. (4.79). It can be re-written in continuous time as

$$\boldsymbol{A}_n\boldsymbol{\Gamma}_b(t) + \boldsymbol{B}_n\boldsymbol{\Gamma}_w(t) = \boldsymbol{b}(t) \tag{4.86}$$

recalling that $\boldsymbol{b}_{n_1}\Gamma_{w_1}(t) + \boldsymbol{B}^*_n\boldsymbol{\Gamma}^*_w(t) = \boldsymbol{B}_n\boldsymbol{\Gamma}_w(t)$. Note that $\boldsymbol{B}_n$ is now treated as constant thanks to the quasi-fixed modelling. We apply the Fourier transform to Eq. (4.86) to obtain

$$\boldsymbol{A}_n\boldsymbol{\Gamma}_b(\omega) + \boldsymbol{B}_n\boldsymbol{\Gamma}_w(\omega) = \boldsymbol{b}(\omega) \tag{4.87}$$

where $\boldsymbol{b}(\omega)$ is given by the Fourier transform of Eq. (4.80)

$$\boldsymbol{b}(\omega) = -\boldsymbol{u}_m(\omega) \circ \boldsymbol{n}_x - \boldsymbol{v}_m(\omega) \circ \boldsymbol{n}_y \tag{4.88}$$

Note that, again, the components of the normal vectors are constant since the geometry is modelled as quasi-fixed.

The wake is also linearised, such that it is flat and travels with the free stream airspeed. The first wake vortex is always placed at position $x_{w_1} = \Delta x_w Q_\infty \Delta t$, $y_{w_1} = 0$, and all subsequent vortices at $x_{w_j} = \left(\Delta x_w + j - 1\right) Q_\infty \Delta t$, $y_{w_j} = 0$, for $j = 2, \dots, k$. The wake strength is propagated using Eq. (4.31), so that, in continuous time

$$\Gamma_{w_j}(t) = \Gamma_{w_{j-1}}(t - \Delta t) \tag{4.89}$$

Writing out this equation for all $j = 1, \dots, k$ leads to

$$\begin{aligned}
\Gamma_{w_1}(t) &= \Gamma_{w_1}(t)\\
\Gamma_{w_2}(t) &= \Gamma_{w_1}(t - \Delta t)\\
\Gamma_{w_3}(t) &= \Gamma_{w_2}(t - \Delta t) = \Gamma_{w_1}(t - 2\Delta t)\\
&\vdots\\
\Gamma_{w_k}(t) &= \Gamma_{w_1}(t - (k-1)\Delta t)
\end{aligned}$$

Applying the Fourier transform to these equations results in

$$\boldsymbol{\Gamma}_w(\omega) = \begin{pmatrix} 1 \\ \mathrm{e}^{-\mathrm{i}\omega\Delta t} \\ \mathrm{e}^{-\mathrm{i}\omega 2\Delta t} \\ \vdots \\ \mathrm{e}^{-\mathrm{i}\omega(k-1)\Delta t} \end{pmatrix} \Gamma_{w_1}(\omega) = \boldsymbol{P}_e(\omega)\Gamma_{w_1}(\omega) \tag{4.90}$$

where $\boldsymbol{P}_e(\omega) = (1, \mathrm{e}^{-\mathrm{i}\omega\Delta t}, \mathrm{e}^{-\mathrm{i}\omega 2\Delta t}, \dots \mathrm{e}^{-\mathrm{i}\omega(k-1)\Delta t})^T$. Substituting for $\boldsymbol{\Gamma}_w(\omega)$ from this latest expression into the impermeability boundary condition (4.87) gives

$$\boldsymbol{A}_n\boldsymbol{\Gamma}_b(\omega) + \boldsymbol{B}_n\boldsymbol{P}_e(\omega)\Gamma_{w_1}(\omega) = \boldsymbol{b}(\omega) \tag{4.91}$$

Kelvin's theorem is given by Eq. (4.35); in continuous time

$$\sum_{j=1}^{n}\Gamma_{b_j}(t) + \Gamma_{w_1}(t) = \sum_{j=1}^{n}\Gamma_{b_j}(t - \Delta t)$$

Applying the Fourier Transform results in

$$\sum_{j=1}^{n}\Gamma_{b_j}(\omega) + \Gamma_{w_1}(\omega) = \mathrm{e}^{-\mathrm{i}\omega\Delta t}\sum_{j=1}^{n}\Gamma_{b_j}(\omega) \tag{4.92}$$

We can now assemble the frequency domain versions of the impermeability boundary condition (4.91) and Kelvin's theorem (4.92) to obtain the complete system

$$\begin{pmatrix} \boldsymbol{A}_n & \boldsymbol{B}_n\boldsymbol{P}_e(\omega) \\ \boldsymbol{I}_{1\times n} - \mathrm{e}^{-\mathrm{i}\omega\Delta t} & 1 \end{pmatrix} \begin{pmatrix} \boldsymbol{\Gamma}_b(\omega) \\ \Gamma_{w_1}(\omega) \end{pmatrix} = \begin{pmatrix} \boldsymbol{b}(\omega) \\ 0 \end{pmatrix} \tag{4.93}$$

This is a system of $n+1$ equations with $n+1$ unknowns, $\boldsymbol{\Gamma}_b(\omega)$ and $\Gamma_{w_1}(\omega)$, that can be solved readily.

The horizontal and vertical velocities induced by the wake vortices on the control points are given by Eq. (4.37), which after transformation to the frequency domain and substitution from Eq. (4.90) become

$$\boldsymbol{u}_w(\omega) = \boldsymbol{B}_u\boldsymbol{P}_e(\omega)\Gamma_{w_1}(\omega) \tag{4.94}$$

$$\boldsymbol{v}_w(\omega) = \boldsymbol{B}_v\boldsymbol{P}_e(\omega)\Gamma_{w_1}(\omega) \tag{4.95}$$

The incremental lift acting on each bound vortex is obtained by taking the Fourier Transform of the first of Eq. (4.82). In continuous time, the linearised version of the lift equation becomes

$$\Delta l_i(t) = \rho\left(Q_\infty\Gamma_{b_i}(t) + \sum_{j=1}^{i}\dot{\Gamma}_{b_j}(t)s_i\right)$$

Taking the Fourier transform leads to

$$\Delta l_i(\omega) = \rho\left(Q_\infty\Gamma_{b_i}(\omega) + \mathrm{i}\omega\sum_{j=1}^{i}\Gamma_{b_j}(\omega)s_i\right) \tag{4.96}$$

The incremental drag on the ith panel is obtained from the second of Eq. (4.82), namely

$$\Delta d_i(t) = \rho\left(-\left(v_{m_i}(t) + v_{w_i}(t)\cos\alpha(t)\right)\Gamma_{b_i}(t) + \sum_{j=1}^{i}\dot{\Gamma}_{b_j}(t)s_i\sin\alpha(t)\right)$$

where $v_{m_i}(t) = \dot{h}(t) + \dot{\alpha}(t)(x_{c_i} - x_f)$ is the vertical velocity induced only by the motion and not the free stream. We cannot linearise this expression since the drag is a second-order phenomenon. Approximating $\sin\alpha(t_k) \approx \alpha(t_k)$, $\cos\alpha(t_k) \approx 1$ and applying the Fourier transform leads to

$$\Delta d_i(\omega) = \rho\left(-\left(v_{m_i}(\omega) + v_{w_i}(\omega)\right) * \Gamma_{b_i}(\omega) + \mathrm{i}\omega s_i\sum_{j=1}^{i}\Gamma_{b_j}(\omega) * \alpha(\omega)\right) \tag{4.97}$$

where the $*$ operator denotes convolution and $v_{m_i}(\omega) = \mathrm{i}\omega h(\omega) + \mathrm{i}\omega\alpha(\omega)(x_{c_i} - x_f)$. Finally, the pitching moment around the pitch axis expression (4.62) becomes

$$m_{c/4}(\omega) = \sum_{i=1}^{n}\Delta d_i(\omega)y_{b_i} - \sum_{i=1}^{n}\Delta l_i(\omega)(x_{b_i} - c/4) \tag{4.98}$$

Example 4.8 *Calculate the aerodynamic loads acting on a sinusoidally pitching and plunging flat plate using the frequency domain version of the lumped vortex method. Compare the resulting aerodynamic load predictions to those of Theodorsen theory.*

We consider a flat plate of chord c plunging and pitching around a pitch axis lying on the plate at a distance x_f behind the leading edge. The plunge and pitch motions are given by

$$h(t) = h_0 + h_1 \sin(\omega_0 t + \phi_h), \quad \alpha(t) = \alpha_0 + \alpha_1 \sin(\omega_0 t + \phi_\alpha)$$

where ω_0 is the frequency of oscillation. Taking the Fourier transform of these expressions (see Section 3.5), we obtain

$$h(\omega) = h_0\delta(\omega) + \frac{h_1}{2\mathrm{i}}\mathrm{e}^{\mathrm{i}\phi_h}\delta(\omega - \omega_0) - \frac{h_1}{2\mathrm{i}}\mathrm{e}^{-\mathrm{i}\phi_h}\delta(\omega + \omega_0) \tag{4.99}$$

$$\alpha(\omega) = \alpha_0\delta(\omega) + \frac{\alpha_1}{2\mathrm{i}}\mathrm{e}^{\mathrm{i}\phi_\alpha}\delta(\omega - \omega_0) - \frac{\alpha_1}{2\mathrm{i}}\mathrm{e}^{-\mathrm{i}\phi_\alpha}\delta(\omega + \omega_0) \tag{4.100}$$

First, we discretise the airfoil into n panels as we did in Example 4.3 and then we calculate the influence coefficient matrix $\mathbf{A}_n$. The wake is set up in the same way as Theodorsen's wake; it is flat, it travels with the free stream and its strength oscillates sinusoidally in time and space. There are two considerations to take into account when setting up the wake:

- *The wake must be long enough to represent the history of the motion but*
- *The wake must contain an integer number of oscillations of the wake strength; otherwise, the aerodynamic loads will contain numerical oscillations due to the imposition of Kelvin's theorem.*

We can satisfy both these conditions by setting a desired wake length, $n_c c$, where n_c is the number of chord lengths and then adjusting it such that it becomes an integer multiple of the spatial wake oscillation period, which is the wavelength of the wake. In Section 3.5, it is shown that the wake's wavelength is

$$\lambda = \frac{2\pi Q_\infty}{\omega_0} = \frac{c\pi}{k_0}$$

where $k_0 = \omega_0 b/Q_\infty$ is the reduced frequency of the oscillation. Consequently, the number of integer periods in $n_c c$ is $n_p = \lfloor n_c c/\lambda \rceil = \lfloor n_c k_0/pi \rceil$, where $\lfloor \rceil$ denotes rounding to the nearest integer. Therefore, the total wake length will be $n_p\lambda$. The final consideration is the number of vortices to place in this wake. In the time domain version of the lumped vortex method, we selected a time step of $\Delta t = c/nQ_\infty$ so that the spacing of the wake vortices is equal to the mean spacing of the bound vortices. The period of the oscillations is $T = 2\pi/\omega_0$ so that the integer number of time instances per cycle is

$$p_{pc} = \left\lfloor \frac{T}{\Delta t} \right\rceil = \left\lfloor \frac{2\pi n Q_\infty}{c\omega_0} \right\rceil = \left\lfloor \frac{\pi n}{k_0} \right\rceil$$

leading to a spatial distance between vortices of

$$\Delta x = \frac{\lambda}{p_{pc}}$$

and a time step of $\Delta t = \Delta x/Q_\infty$. Finally, the x-coordinates of the wake vortices are

$$x_{w_i} = (i-1)\Delta x + \Delta x_w \Delta x + c$$

for $i = 1, \ldots, p_{pc}n_p$, where $\Delta x_w \Delta x$ is the usual distance between the trailing edge and the first wake vortex. The y-coordinates of the wake are all set to $y_i = 0$.

Once we have defined the wake, we can calculate the normal velocity wake influence matrix $\mathbf{B}_n$ from Eq. (4.41) and create matrix $\mathbf{P}_e(\omega_0)$. We substitute $h(\omega)$ and $\alpha(\omega)$ from Eqs. (4.99) and (4.100) into Eq. (4.93) to obtain two sets of equations, one for $\omega = 0$ and one for $\omega = \omega_0$. The former is the steady case, which features no motion and no wake, so that it can be simplified to

$$\boldsymbol{A}_n \boldsymbol{\Gamma}_b(0) = -Q_\infty \boldsymbol{n}_x - Q_\infty \alpha_0 \boldsymbol{n}_y$$

and solved readily for $\boldsymbol{\Gamma}_b(0)$. Then, for $\omega = \omega_0$ Eq. (4.93) become

$$\begin{pmatrix} \boldsymbol{A}_n & \boldsymbol{B}_n \boldsymbol{P}_e(\omega_0) \\ \boldsymbol{I}_{1\times n} - \mathrm{e}^{-\mathrm{i}\omega_0 \Delta t} & 1 \end{pmatrix} \begin{pmatrix} \boldsymbol{\Gamma}_b(\omega_0) \\ \Gamma_{w_1}(\omega_0) \end{pmatrix} = \begin{pmatrix} \mathbf{b}(\omega_0) \\ 0 \end{pmatrix} \tag{4.101}$$

where $\mathbf{b}(\omega_0)$ is obtained by substituting $\omega = \omega_0$ in Eqs. (4.85) and (4.88). Equation (4.101) can now be solved for $\boldsymbol{\Gamma}_b(\omega_0)$ and $\Gamma_{w_1}(\omega_0)$. We then substitute $\omega = \omega_0$ in Eqs. (4.94) and (4.95) to calculate the horizontal and vertical velocities induced by the wake on the control points.

The final step is to calculate the aerodynamic loads. The incremental lift on each panel is given by Eq. (4.96) after substituting $\omega = 0$ and $\omega = \omega_0$:

$$\Delta l_i(0) = \rho Q_\infty \Gamma_{b_i}(0)$$

$$\Delta l_i(\omega_0) = \rho \left(Q_\infty \Gamma_{b_i}(\omega_0) + \mathrm{i}\omega_0 \sum_{j=1}^{i} \Gamma_{b_j}(\omega_0) s_i \right)$$

The drag is calculated from Eq. (4.97) using the approach of Section 3.5.2. On the ith panel, we write

$$\begin{aligned} \Gamma_{b_i}(\omega) &= \left(\Gamma^*_{b_i}(\omega),\ \Gamma_{b_i}(0),\ \Gamma_{b_i}(\omega_0)\right) \\ \dot{\Gamma}_{b_i}(\omega) &= \left((\mathrm{i}\omega_0 \Gamma_{b_i}(\omega_0))^*,\ 0,\ \mathrm{i}\omega_0 \Gamma_{b_i}(\omega_0)\right) \\ v_{m_i}(\omega) + v_{w_i}(\omega) &= \left(v^*_{m_i}(\omega_0) + v^*_{w_i}(\omega),\ v_{m_i}(0) + v_{w_i}(0),\ v_{m_i}(\omega_0) + v_{w_i}(\omega)\right) \\ \alpha(\omega) &= \left(\alpha^*(\omega_0),\ \alpha(0),\ \alpha(\omega_0)\right) \end{aligned} \tag{4.102}$$

Then, the convolutions in Eq. (4.97) are calculated by convolving vectors, as was done in Eq. (3.190). The resulting incremental drag will have three frequency components, $\Delta d_i(0)$, $\Delta d_i(\omega_0)$ and $\Delta d_i(2\omega_0)$. Finally, the pitching moment coefficient is calculated from Eq. (4.98) after substituting $\Delta l_i(0)$, $\Delta l_i(\omega_0)$, $\Delta d_i(0)$, $\Delta d_i(\omega_0)$ and $\Delta d_i(2\omega_0)$, such that

$$m_{c/4}(0) = \sum_{i=1}^{n} \Delta l_i(0)(x_{b_i} - c/4) - \sum_{i=1}^{n} \Delta d_i(0) y_{b_i}$$

$$m_{c/4}(\omega_0) = \sum_{i=1}^{n} \Delta l_i(\omega_0)(x_{b_i} - c/4) - \sum_{i=1}^{n} \Delta d_i(\omega_0) y_{b_i}$$

$$m_{c/4}(\omega_0) = -\sum_{i=1}^{n} \Delta d_i(2\omega_0) y_{b_i}$$

We apply the frequency domain lumped vortex method to the wing of Example 4.12 for reduced frequency values between $k = 0.2$ and $k = 2.2$. We select the number of panels as $n = 100$, and we choose the length of the wake as $n_c = 20$ chords. We also calculate the lift drag

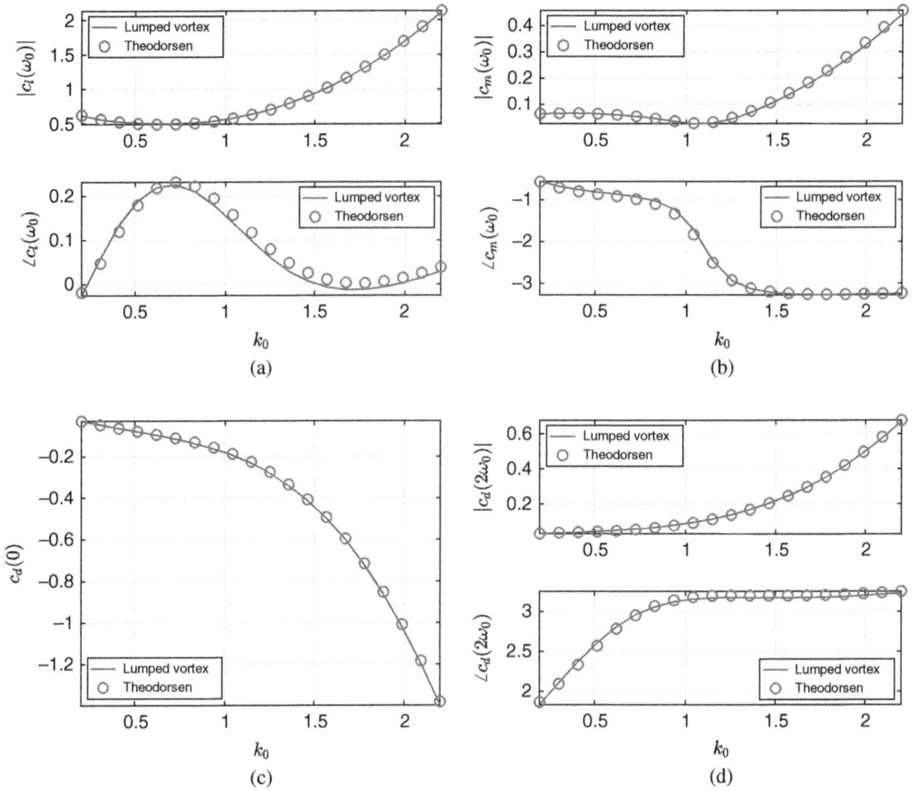

Figure 4.24 Amplitude and phase of lift, drag and pitching moment coefficients as functions of reduced frequency. (a) $c_l(\omega_0)$, (b) $c_{m_{xf}}(\omega_0)$, (c) $c_d(0)$ and (d) $c_d(\omega_0)$.

and moment using Theodorsen theory, as in Example 3.15. Figure 4.24 plots the amplitudes and phases of the lift, moment and drag coefficients as functions of reduced frequency. As $\alpha_0 = 0$, the loads $c_l(0)$ and $c_{m_{xf}}(0)$ are also equal to zero and are not shown. In fact, $c_{m_{xf}}(2\omega_0)$ is also zero because the plate is flat and $y_{b_i} = 0$ for all i. Clearly, the load predictions obtained from the lumped vortex method and Theodorsen theory are very similar, which is a result of the quasi-fixed motion and flat wake modelling. The assumptions behind the frequency domain lumped vortex method and Theodorsen theory are identical, except for the formulation of the Kutta condition for the former, which remains implicit. As a consequence, it is logical that the resulting load predictions will be in agreement. This example is solved by Matlab code `pitchplunge_lv_freq.m`.

The frequency domain lumped vortex approach is much faster than its time-domain counterpart for sinusoidal motion cases, but it yields nearly the same load predictions as Theodorsen theory; hence, its utility may be thought of as questionable. However, the same Fourier transformation procedure can be applied to panel methods that can model airfoil thickness, resulting in fast frequency domain approaches that represent the true geometry of airfoils. Furthermore, as will be shown in Chapter 5, the same treatment can be applied to 3D panel methods, for which no general analytical solution exists.

4.5 Source and Vortex Panel Method

Up to this point, all the aerodynamic models we presented in Chapters 3 and 4 have modelled the airfoil as a flat or cambered line with negligible thickness. Here, we will model the exact shape of airfoils for the first time, using the source and vortex panel method (SVPM). The complete airfoil geometry will be represented by a series of straight line segments known as singularity panels, following the approach by Basu and Hancock (1978). A 1D singularity panel is a straight line segment with a continuous distribution of singularities, such as the vortex line used to model the airfoil in Section 3.6.

Figure 4.25 draws the contour of an airfoil and its discretisation into n non-linearly spaced panels with coordinates x_i, y_i, for $i = 1, \dots, n+1$. Half of the panels lie on the upper side and the other half on the lower side. They are numbered in a clockwise direction, from the lower to the upper trailing edge. Collocation points x_{c_i}, y_{c_i} for $i = 1, \dots, n$ are placed on the midpoint of each panel, with coordinates

$$x_{c_i} = x_i + (x_{i+1} - x_i)/2, \quad y_{c_i} = y_i + (y_{i+1} - y_i)/2 \tag{4.103}$$

Unit vectors tangent, $\boldsymbol{\tau}_i = (\tau_{x_i}, \tau_{y_i})$, and normal, $\mathbf{n}_i = (n_{x_i}, n_{y_i})$, to the ith panel are drawn at the collocation points. The components of these vectors are given by

$$\tau_{x_i} = \frac{x_{i+1} - x_i}{s_i}, \quad \tau_{y_i} = \frac{y_{i+1} - y_i}{s_i} \tag{4.104}$$

$$n_{x_i} = -\tau_{y_i}, \quad n_{y_i} = \tau_{x_i} \tag{4.105}$$

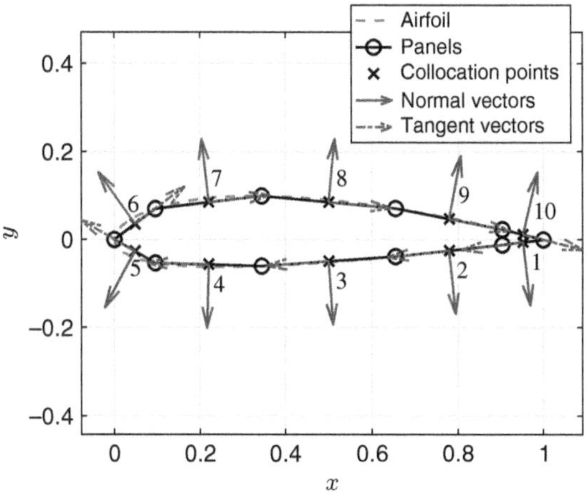

Figure 4.25 Panel discretisation of a 2D airfoil.

where

$$s_i = \sqrt{(x_{i+1} - x_i)^2 + (y_{i+1} - y_i)^2} \tag{4.106}$$

are the lengths of the panels.

We will first consider a steady flow with a free stream Q_∞ inclined at an angle α to the airfoil. We will place source and vortex distributions on the panels, such that each panel has different but constant source strength along its length, σ_i, while the vortex strength per unit length, γ, is equal for all the panels. The objective is to calculate the strengths of the source and vortex panels by imposing the impermeability boundary condition at the panel control points and the Kutta condition at the trailing edge. In order to do this, first we need a way to calculate the velocities induced by all the panels at the control points. Note that, unlike Theodorsen theory, both the source and vortex panels contribute to the imposition of the impermeability and Kutta conditions.

Appendix A.3 gives expressions for the potential and velocities induced at a general point x, y by a source panel of strength σ_0 lying on the x axis and extending from $(x_1, 0)$ to $(x_2, 0)$. The potential is given by

$$\begin{aligned}\phi(x,y) = \frac{\sigma_0}{4\pi}\Bigg(&2(x_1 - x_2) + y\left(\tan^{-1}\left(\frac{x-x_1}{y}\right) - \tan^{-1}\left(\frac{x-x_2}{y}\right)\right)\\ &+ (x-x_1)\ln\left((x-x_1)^2 + y^2\right) - (x-x_2)\ln\left((x-x_2)^2+y^2\right)\Bigg)\end{aligned} \tag{4.107}$$

and the horizontal and vertical velocities by

$$u(x,y) = -\frac{\sigma_0}{4\pi}\ln\left(\frac{(x-x_2)^2+y^2}{(x-x_1)^2+y^2}\right) \tag{4.108}$$

$$v(x,y) = \frac{\sigma_0}{2\pi}\left(\tan^{-1}\left(\frac{x-x_1}{y}\right) - \tan^{-1}\left(\frac{x-x_2}{y}\right)\right) \tag{4.109}$$

Furthermore, Appendix A.4 gives the potential and velocities induced by a vortex panel of strength γ_0, again lying between $(x_1, 0)$ and $(x_2, 0)$. The potential is given by

$$\begin{aligned}\phi(x,y) = \frac{\gamma_0}{2\pi}\Bigg(&(x-x_2)\tan^{-1}\left(\frac{y}{x-x_2}\right) - (x-x_1)\tan^{-1}\left(\frac{y}{x-x_1}\right)\\ &+\frac{y}{2}\ln\left(\frac{(x-x_2)^2+y^2}{(x-x_1)^2+y^2}\right)\Bigg)\end{aligned} \tag{4.110}$$

and the horizontal and vertical velocities by

$$u(x,y) = \frac{\gamma_0}{2\pi}\left(\tan^{-1}\left(\frac{x-x_1}{y}\right) - \tan^{-1}\left(\frac{x-x_2}{y}\right)\right) \tag{4.111}$$

$$v(x,y) = \frac{\gamma_0}{4\pi}\ln\left(\frac{(x-x_2)^2+y^2}{(x-x_1)^2+y^2}\right) \tag{4.112}$$

None of the panels of Figure 4.25 lie on the x axis; in order to calculate the potential and velocities induced by them at a general point (x, y), we need to perform the calculations of Eqs. (4.107) to (4.112) in the reference frame of the panel. This means that we place the origin of the axis system at the midpoint of the jth panel and we replace x and y by the tangent and normal distances from (x, y) to this new origin, as shown in Figure 4.26. The vectorial distance from point (x, y) to the midpoint of panel j is $\mathbf{r} = ((x - x_{c_j}), (y - y_{c_j}))$,

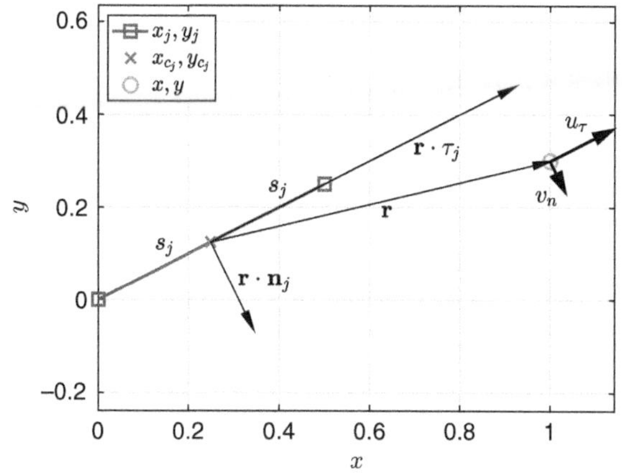

Figure 4.26 Velocities induced by source panel j on point x, y.

so that the tangent and normal distances between the panel's midpoint and point (x, y) are $x_\tau = \mathbf{r} \cdot \boldsymbol{\tau}_j$ and $y_n = \mathbf{r} \cdot \mathbf{n}_j$, respectively. As the origin of the axis system has been placed at the centre of the panel, point x_1 becomes $-s_j/2$ and point x_2 becomes $s_j/2$, while $x_1 - x_2 = s_j$.

Equations (4.107)–(4.109) then become

$$\phi_\sigma(x,y) = \frac{\sigma_j}{4\pi}\left(-2s_j + y_n\left(\tan^{-1}\left(\frac{x_\tau + s_j/2}{y_n}\right) - \tan^{-1}\left(\frac{x_\tau - s_j/2}{y_n}\right)\right)\right.$$
$$\left. + (x_\tau + s_j/2)\ln\left((x_\tau + s_j/2)^2 + y_n^2\right) - (x_\tau - s_j/2)\ln\left((x_\tau - s_j/2)^2 + y_n^2\right)\right) \tag{4.113}$$

$$u_{\sigma_\tau}(x,y) = -\frac{\sigma_j}{4\pi}\ln\left(\frac{(x_\tau - s_j/2)^2 + y_n^2}{(x_\tau + s_j/2)^2 + y_n^2}\right) \tag{4.114}$$

$$v_{\sigma_n}(x,y) = \frac{\sigma_j}{2\pi}\left(\tan^{-1}\left(\frac{x_\tau + s_j/2}{y_n}\right) - \tan^{-1}\left(\frac{x_\tau - s_j/2}{y_n}\right)\right) \tag{4.115}$$

where s_j is the length of the panel and u_{σ_τ}, v_{σ_n} are source-induced flow velocities tangent and normal to panel j. Similarly, Eqs. (4.110) and (4.112) become

$$\phi_\gamma(x,y) = \frac{\gamma_j}{2\pi}\left((x_\tau - s_j/2)\tan^{-1}\left(\frac{y_n}{x_\tau - s_j/2}\right) - (x_\tau + s_j/2)\tan^{-1}\left(\frac{y_n}{x_\tau + s_j/2}\right)\right.$$
$$\left. + \frac{y_n}{2}\ln\left(\frac{(x_\tau - s_j/2)^2 + y_n^2}{(x_\tau + s_j/2)^2 + y_n^2}\right)\right) \tag{4.116}$$

$$u_{\gamma_\tau}(x,y) = \frac{\gamma_j}{2\pi}\left(\tan^{-1}\left(\frac{x_\tau + s_j/2}{y_n}\right) - \tan^{-1}\left(\frac{x_\tau - s_j/2}{x_\tau}\right)\right) \tag{4.117}$$

$$v_{\gamma_n}(x,y) = \frac{\gamma_j}{4\pi}\ln\left(\frac{(x_\tau - s_j/2)^2 + y_n^2}{(x_\tau + s_j/2)^2 + y_n^2}\right) \tag{4.118}$$

where u_{γ_τ}, v_{γ_n} are vortex-induced flow velocities tangent and normal to panel j. If panel j features both a source and a vortex distribution, the velocities induced by the panel in the original x and y axes are obtained from

$$u(x,y) = u_{\sigma_\tau}(x,y)\tau_{x_j} + v_{\sigma_n}(x,y)n_{x_j} + u_{\gamma_\tau}(x,y)\tau_{x_j} + v_{\gamma_n}(x,y)n_{x_j} \tag{4.119}$$

$$v(x,y) = u_{\sigma_\tau}(x,y)\tau_{y_j} + v_{\sigma_n}(x,y)n_{y_j} + u_{\gamma_\tau}(x,y)\tau_{y_j} + v_{\gamma_n}(x,y)n_{y_j} \tag{4.120}$$

Most of the time, we will be calculating the velocities induced by panel j, known as the influencing panel, on the control points of panel i, known as the influenced panel. This means that x and y must be replaced by x_{c_i} and y_{c_i} in Eqs. (4.113)–(4.118), while r, x_τ, y_n must be replaced by $\mathbf{r}_{i,j} = ((x_{c_i} - x_{c_j}), (y_{c_i} - y_{c_j})), x_{\tau_{i,j}} = \mathbf{r}_{i,j} \cdot \boldsymbol{\tau}_j$ and $y_{n_{i,j}} = \mathbf{r}_{i,j} \cdot \mathbf{n}_j$, respectively. The velocities induced by all n source and vortex panels on all n control points can be written in compact matrix form as

$$\boldsymbol{u} = \boldsymbol{A}_{u_\sigma}\boldsymbol{\sigma} + \boldsymbol{A}_{u_\gamma}\boldsymbol{\gamma} \tag{4.121}$$

$$\boldsymbol{v} = \boldsymbol{A}_{v_\sigma}\boldsymbol{\sigma} + \boldsymbol{A}_{v_\gamma}\boldsymbol{\gamma} \tag{4.122}$$

In these expressions, $\mathbf{u}$ is the $n \times 1$ vector with elements $u_i = u(x_{c_i}, y_{c_i})$, $\mathbf{v}$ is the $n \times 1$ vector with elements $v_i = v(x_{c_i}, y_{c_i})$, $\boldsymbol{\sigma}$ is the $n \times 1$ vector with elements σ_i and, since the vortex strength is equal on all the panels, $\boldsymbol{\gamma} = \boldsymbol{I}_{n\times 1}\gamma$ is the $n \times 1$ vector with elements γ. Using Eqs. (4.114)–(4.120) the elements of the $n \times n$ source influence coefficient matrices $\boldsymbol{A}_{u_\sigma}$, $\boldsymbol{A}_{v_\sigma}$ are given by

$$\begin{aligned} A_{u_{\sigma_{i,j}}} &= -\frac{1}{4\pi}\ln\left(\frac{(x_{\tau_{i,j}} - s_j/2)^2 + y_{n_{i,j}}^2}{(x_{\tau_{i,j}} + s_j/2)^2 + y_{n_{i,j}}^2}\right)\tau_{x_j} \\ &+ \frac{1}{2\pi}\left(\tan^{-1}\left(\frac{x_{\tau_{i,j}} + s_j/2}{y_{n_{i,j}}}\right) - \tan^{-1}\left(\frac{x_{\tau_{i,j}} - s_j/2}{y_{n_{i,j}}}\right)\right)n_{x_j} \end{aligned} \tag{4.123}$$

$$\begin{aligned} A_{v_{\sigma_{i,j}}} &= -\frac{1}{4\pi}\ln\left(\frac{(x_{\tau_{i,j}} - s_j/2)^2 + y_{n_{i,j}}^2}{(x_{\tau_{i,j}} + s_j/2)^2 + y_{n_{i,j}}^2}\right)\tau_{y_j} \\ &+ \frac{1}{2\pi}\left(\tan^{-1}\left(\frac{x_{\tau_{i,j}} + s_j/2}{y_{n_{i,j}}}\right) - \tan^{-1}\left(\frac{x_{\tau_{i,j}} - s_j/2}{y_{n_{i,j}}}\right)\right)n_{y_j} \end{aligned} \tag{4.124}$$

where $\mathbf{r}_{i,j} = ((x_{c_i} - x_{c_j}), (y_{c_i} - y_{c_j})), x_{\tau_{i,j}} = \mathbf{r}_{i,j} \cdot \boldsymbol{\tau}_j$ and $y_{n_{i,j}} = \mathbf{r}_{i,j} \cdot \mathbf{n}_j$. The elements of the $n \times n$ vortex influence coefficient matrices $\boldsymbol{A}_{u_\gamma}, \boldsymbol{A}_{v_\gamma}$ can be obtained in the same manner. However, looking at the pairs of Eqs. (4.114), (4.118) and (4.115), (4.117), we can see that $u_{\gamma_n}(x,y)/\gamma_j = v_{\sigma_\tau}(x,y)/\sigma_j$ and $v_{\gamma_n}(x,y)/\gamma_j = -u_{\sigma_\tau}(x,y)/\sigma_j$. Furthermore, Eq. (4.105) states that $n_{x_i} = -\tau_{y_i}$ and $n_{y_i} = \tau_{x_i}$. Assembling these pieces of information, we obtain

$$\boldsymbol{A}_{u_\gamma} = \boldsymbol{A}_{v_\sigma} \tag{4.125}$$

$$\boldsymbol{A}_{v_\gamma} = -\boldsymbol{A}_{u_\sigma} \tag{4.126}$$

The influences of the ith panel on the ith control point are known as self-influences. Appendix A.3 shows that the normal influence of a source panel on itself is $1/2$, while the tangential influence is 0. Conversely, Appendix A.4 shows that the tangential influence of a vortex panel on itself is $1/2$ and the normal influence is 0. Consequently,

$$A_{u_{\sigma_{i,i}}} = 1/2 n_{x_i}, \quad A_{v_{\sigma_{i,i}}} = 1/2 n_{y_i}, \quad A_{u_{\gamma_{i,i}}} = 1/2 n_{y_i}, \quad A_{v_{\gamma_{i,i}}} = -1/2 n_{x_i}$$

We now substitute Eqs. (4.121) and (4.122) into the impermeability boundary condition (4.3). For steady flow with a free stream $\mathbf{Q}_\infty = (Q_\infty \cos\alpha,\ Q_\infty \sin\alpha)$, we obtain

$$\boldsymbol{A}_{n_\sigma}\boldsymbol{\sigma} + \boldsymbol{A}_{n_\gamma}\boldsymbol{I}_{n\times1}\gamma = \boldsymbol{b} \tag{4.127}$$

where $\boldsymbol{A}_{n_\sigma}, \boldsymbol{A}_{n_\gamma}$ are the $n \times n$ normal influence coefficient matrices with elements

$$A_{n_{\sigma_{i,j}}} = A_{u_{\sigma_{i,j}}} n_{x_i} + A_{v_{\sigma_{i,j}}} n_{y_i} \tag{4.128}$$

$$A_{n_{\gamma_{i,j}}} = A_{u_{\gamma_{i,j}}} n_{x_i} + A_{v_{\gamma_{i,j}}} n_{y_i} \tag{4.129}$$

The $n \times 1$ vector $\boldsymbol{b} = -Q_\infty \cos\alpha \boldsymbol{n}_x - Q_\infty \sin\alpha \boldsymbol{n}_y$, as in Eq. (4.24).

Equation (4.127) is a set of n equations for $n+1$ unknowns, the source strengths σ_j for $j = 1, \dots, n$ and γ. The set can be completed by means of the Kutta condition. In the present case, we can express this condition by stating that the pressure on the upper trailing edge panel must be equal to that on the lower trailing edge panel, $c_{p_n} = c_{p_1}$. The steady 2D version of Bernoulli's equation (2.31) is written for the ith panel as

$$c_{p_i} = 1 - \frac{(Q_\infty \cos\alpha + u_i)^2 + (Q_\infty \sin\alpha + v_i)^2}{Q_\infty^2} \tag{4.130}$$

noting that u_i, v_i represent perturbation velocities here while u, v represent total velocities in Eq. (2.31). The pressure coefficient can also be written in the frame of reference of panel i, such that

$$c_{p_i} = 1 - \frac{(\mathbf{Q}_\infty \cdot \boldsymbol{\tau}_i + u_{\tau_i})^2 + (\mathbf{Q}_\infty \cdot \mathbf{n}_i + v_{n_i})^2}{Q_\infty^2}$$

where $u_{\tau_i} = u_i \tau_{x_i} + v_i \tau_{y_i}$ is the perturbation flow velocity tangent to the panel and $v_{n_i} = u_i n_{x_i} + v_i n_{y_i}$ is the perturbation flow velocity normal to the panel. The impermeability boundary condition dictates that $\mathbf{Q}_\infty \cdot \mathbf{n}_i + v_{n_i} = 0$, so that the pressure coefficient simplifies to

$$c_{p_i} = 1 - \frac{(\mathbf{Q}_\infty \cdot \boldsymbol{\tau}_i + u_{\tau_i})^2}{Q_\infty^2} \tag{4.131}$$

As the Kutta condition is $c_{p_n} = c_{p_1}$, it follows that the total tangent velocities on the two trailing edge panels must also be equal, or

$$\mathbf{Q}_\infty \cdot \boldsymbol{\tau}_n + u_{\tau_n} = -(\mathbf{Q}_\infty \cdot \boldsymbol{\tau}_1 + u_{\tau_1}) \tag{4.132}$$

noting that the tangent vector is pointing upstream for panel 1 and downstream for panel n; hence, the opposite signs in expression (4.132). The tangential perturbation velocities on all the panels are obtained from

$$\boldsymbol{u}_\tau = \boldsymbol{A}_{\tau_\sigma}\boldsymbol{\sigma} + \boldsymbol{A}_{\tau_\gamma}\boldsymbol{I}_{n\times1}\gamma \tag{4.133}$$

where $\boldsymbol{A}_{\tau_\sigma}, \boldsymbol{A}_{\tau_\gamma}$ are the $n \times n$ tangential influence coefficient matrices with elements

$$A_{\tau_{\sigma_{i,j}}} = A_{u_{\sigma_{i,j}}} \tau_{x_i} + A_{v_{\sigma_{i,j}}} \tau_{y_i} \tag{4.134}$$

$$A_{\tau_{\gamma_{i,j}}} = A_{u_{\gamma_{i,j}}} \tau_{x_i} + A_{v_{\gamma_{i,j}}} \tau_{y_i} \tag{4.135}$$

Equation (4.132) can therefore be written in terms of the source and vortex strengths as

$$(\boldsymbol{a}_{\tau_{\sigma 1}} + \boldsymbol{a}_{\tau_{\sigma n}})\boldsymbol{\sigma} + (a_{\tau_{\gamma 1}} + a_{\tau_{\gamma n}})\gamma = -\mathbf{Q}_\infty \cdot \boldsymbol{\tau}_1 - \mathbf{Q}_\infty \cdot \boldsymbol{\tau}_n \tag{4.136}$$

where $\boldsymbol{a}_{\tau_{\sigma 1}}, \boldsymbol{a}_{\tau_{\sigma n}}$ are the first and nth rows of matrix $\boldsymbol{A}_{\tau_\sigma}$, respectively, and

$$a_{\tau_{\gamma 1}} = \sum_{j=1}^{n} A_{\tau_{\gamma 1,j}}, \quad a_{\tau_{\gamma n}} = \sum_{j=1}^{n} A_{\tau_{\gamma n,j}}$$

Equations (4.127) and (4.136) can now be solved simultaneously for the unknown singularity strengths. The two equations can be assembled into the single system:

$$\begin{pmatrix} \boldsymbol{A}_{n_\sigma} & \boldsymbol{A}_{n_\gamma}\boldsymbol{I}_{n\times 1} \\ \boldsymbol{a}_{\tau_{\sigma 1}} + \boldsymbol{a}_{\tau_{\sigma n}} & a_{\tau_{\gamma 1}} + a_{\tau_{\gamma n}} \end{pmatrix} \begin{pmatrix} \boldsymbol{\sigma} \\ \gamma \end{pmatrix} = \begin{pmatrix} \boldsymbol{b} \\ -\mathbf{Q}_\infty \cdot \boldsymbol{\tau}_1 - \mathbf{Q}_\infty \cdot \boldsymbol{\tau}_n \end{pmatrix} \tag{4.137}$$

Equation (4.137) can be solved readily for the unknown $\boldsymbol{\sigma}$ and γ. Once the source and vortex strengths on the panels have been evaluated, the tangential velocities on the control points are given by Eq. (4.133) and the pressure coefficient on each panel can be calculated from Eq. (4.131).

As the wing is aligned with the x axis, the chordwise and normal aerodynamic loads acting on it are given by Eq. (3.20)

$$f_c = -\oint p(s)n_x(s)\mathrm{d}s$$

$$f_n = -\oint p(s)n_y(s)\mathrm{d}s$$

For the present discrete calculation, the pressure is constant over each panel so that the integral becomes a sum. Dividing throughout by $1/2\rho U_\infty^2 c$, we obtain

$$c_c = -\frac{1}{c}\sum_{i=1}^{n} c_{p_i} s_i n_{x_i} \tag{4.138}$$

$$c_n = -\frac{1}{c}\sum_{i=1}^{n} c_{p_i} s_i n_{y_i} \tag{4.139}$$

where c_c is the chordwise force coefficient and c_n is the normal force coefficient. Finally, the lift and drag coefficients are obtained from Eq. (3.18), such that

$$c_d = c_c \cos\alpha + c_n \sin\alpha$$

$$c_l = -c_c \sin\alpha + c_n \cos\alpha \tag{4.140}$$

Another estimate for the lift can be obtained from the Kutta–Joukowski theorem $l = \rho Q_\infty \Gamma$, where $\Gamma = \gamma \sum_{i=1}^{n} s_i$ is the total circulation around the airfoil. Dividing by $1/2\rho Q_\infty^2 c$ leads to

$$c_l = \frac{2\gamma}{Q_\infty c}\sum_{i=1}^{n} s_i \tag{4.141}$$

Finally, the pitching moment coefficient around the quarter chorded is given by Eq. (4.13), where $\mathbf{r}_i(t_k) = (x_{c_i} - c/4, y_{c_i})$. The non-dimensional form of this equation after summation is

$$c_{m_{c/4}} = -\frac{1}{c^2}\sum_{i=1}^{n} c_{p_i} s_i n_{x_i} y_{c_i} + \frac{1}{c^2}\sum_{i=1}^{n} c_{p_i} s_i n_{y_i}(x_{c_i} - c/4) \tag{4.142}$$

Example 4.9 *Calculate the steady pressure distribution around a NACA 2412 airfoil and the aerodynamic loads acting on it at several angles of attack and compare to experimental data.*

Seetharam et al. (1977) report on results from a series of wind tunnel experiments on a NACA 2412 airfoil at a Mach number of 0.13 and Reynolds number of 2.2×10^6. *Both pressure distributions and integrated aerodynamic load results are given for a range of angles of attack from* $-3.9°$ *to* $16.4°$. *The airfoil has a chord of* $c = 0.61$ *m and was fitted with 53 pressure taps to measure the pressure distribution on both sides. The lift and pitching moment were measured using an aerodynamic balance. The objective of this example is to compare both the pressure distributions and aerodynamic loads calculated using the SVPM to the experimental measurements.*

We start by calculating the coordinates of the airfoil using the NACA four-digit family Eqs. (3.4)–(3.6). The trailing edge of NACA four-digit airfoils is thick so that the upper and lower trailing edge points do not coincide. Recall that the thickness distribution of these airfoils is given by Eq. (3.5) so that the non-dimensional vertical coordinates are given by

$$\bar{y}_{\pm}(\bar{x}) = \pm\frac{t}{0.2}\left(0.2969\sqrt{\bar{x}} - 0.126\bar{x} - 0.35160\bar{x}^2 + 0.2843\bar{x}^3 - 0.1015\bar{x}^4\right) \tag{4.143}$$

where $\bar{y}_+(\bar{x})$ *denotes the upper surface,* $\bar{y}_-(\bar{x})$ *the lower and* $\bar{x} = x/c$. *Substituting* $\bar{x} = 1$ *gives* $\bar{y}_{\pm}(1) = \pm 0.105t$, *so that the distance between the upper and lower trailing edge points is* $0.210t$. *This geometry can cause convergence issues for panel methods; the predicted pressure distribution around the airfoil is acceptable, but the integrated aerodynamic loads converge slowly with increasing number of panels, if at all. One solution to this problem is to add a panel connecting the upper and lower trailing edge points and to place a source distribution on it in order to impose impermeability.*

A simpler solution is to alter the geometry of the trailing edge in order to force the two trailing edge points to coincide. This can be achieved by modifying slightly the leading coefficient of Eq. (4.143), such that

$$\bar{y}_{\pm}(\bar{x}) = \pm\frac{t}{0.2}\left(0.2969\sqrt{\bar{x}} - 0.126\bar{x} - 0.3516\bar{x}^2 + 0.2843\bar{x}^3 - 0.1036\bar{x}^4\right) \tag{4.144}$$

Now, $\bar{y}_{\pm}(1) = 0$ *and the trailing edge thickness is zero. The overall shape of the airfoil is not affected because* $\bar{x} \leq 1$ *and therefore* $\bar{x}^4$ *is only significant very close to the trailing edge. Figure 4.27a plots the trailing edge of the NACA 2412 airfoil, for which* $m = 0.02$, $p = 0.4$ *and* $t = 0.12$, *computed using the original geometry (labelled Thick trailing edge) and the modified Eq. (4.144) (labelled Closed trailing edge). Clearly, the objective of closing off the trailing edge has been achieved. Furthermore, Figure 4.27b shows that the change in the overall geometry is imperceptible. The* $\bar{x}$ *coordinates are non-linearly spaced, such that*

$$\bar{x}_i = \frac{1}{2}\left(1 - \cos\theta_i\right) \tag{4.145}$$

for $i = 1, \dots, n+1$, *where* θ_i *are* $n+1$ *linearly spaced angles between* $-\pi$ *and* π. *This calculation ensures that* $\bar{x}_1 = \bar{x}_{n+1} = 1, \bar{x}_{n/2+1} = 0$ *and that the panels are more closely spaced around the leading and trailing edges. The first* $n/2$ *panels lie on the lower side of the airfoil, while the last* $n/2$ *lie on the upper side, such that the panel arrangement is as shown in Figure 4.25. Then, dimensional coordinates are calculated from*

$$x = \bar{x}c, \quad y = \bar{y}c \tag{4.146}$$

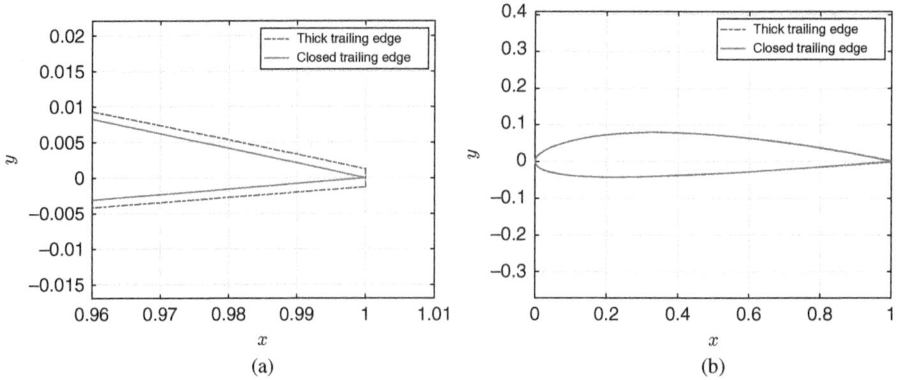

Figure 4.27 Original and modified geometries of the NACA 2412 airfoil. (a) Trailing edge region and (b) complete airfoil.

The next step is to calculate the control points, tangent and normal vectors and panel lengths from Eqs. (4.103)–(4.106). Then, we calculate the aerodynamic influence coefficient matrices $\mathbf{A}_{u_\sigma}$, $\mathbf{A}_{v_\sigma}$, $\mathbf{A}_{u_\gamma}$, $\mathbf{A}_{v_\gamma}$ *from expressions (4.123)–(4.126),* $\mathbf{A}_{n_\sigma}$, $\mathbf{A}_{n_\gamma}$ *from Eqs. (4.128) and (4.129) and* $\mathbf{A}_{\tau_\sigma}$, $\mathbf{A}_{\tau_\gamma}$ *from Eqs. (4.134) and (4.135). Next, we set up and solve the combined impermeability/Kutta condition system of Eq. (4.137) for* $\boldsymbol{\sigma}$ *and* γ*. Finally, we evaluate the tangential velocities from expression (4.133), the pressure coefficients from (4.131) and the aerodynamic loads from Eqs. (4.138)–(4.142).*

We repeat the SVPM calculations for all the experimental angles of attack using $n = 200$ *for the number of panels. Figure 4.28 plots the simulated and experimental pressure distributions around the airfoil at four steady angles of attack. The pressures on the upper and lower surfaces of the airfoil are denoted by different symbols and line styles. The point on the airfoil where the lowest pressure occurs is known as the suction peak. In Figure 4.28a, the angle of attack is negative and the pressure peak occurs on the lower side; for all positive angles of attack, it occurs on the upper side. The agreement between the simulated and experimental pressures is quite good, but there are two important differences:*

- *The value of the pressure coefficient at the trailing edge is not the same. Equation (4.131) shows that* $c_p = 1$ *at points where the total flow velocity is equal to zero, that is at stagnation points. Theoretically, there are two such points, one near the leading edge and one at the trailing edge. The SVPM results follow this theory, but only if the size of the panels near the trailing edge is very small.*[1] *Conversely, the experimental pressure lies close to zero at the trailing edge. This disagreement shows that, in real flows, the trailing edge is not really a stagnation point; the flow velocity there is unsteady due to the merging of the boundary layers from the upper and lower surfaces, even when the airfoil is not moving and the free stream is constant.*

1 Recall that the SVPM calculates the pressure on the control points of the upper and lower trailing edge panels. These points only lie exactly at the trailing edge if the number of panels is infinite. Hence, the pressure at the trailing edge stagnation point is not calculated by the method.

- *The suction peak is overestimated by the SVPM. This phenomenon is again due to viscous effects. In the experimental case, the growing boundary layer displaces the flow and causes slightly lower pressures than the inviscid predictions. At near-stall and post-stall angles of attack, this difference is much more important.*

Figure 4.29 plots the lift and pitching moment coefficients evaluated using the SVPM at all angles of attack and compares them to the data presented in Seetharam et al. (1977). The measurements denoted by Exp. grit concern experiments whereby the boundary layer was tripped near the leading edge on both surfaces by a strip of carborundum grit. The clean data were obtained without this strip. Figure 4.29a shows that the lift-curve slope predicted by the SVPM is slightly higher than the experimental values; this phenomenon is due to the overestimation of the suction peak discussed earlier. The lift coefficient estimates obtained using Eqs. (4.140) (labelled SVPM Pressure) and (4.141) (labelled SVPM Kutta–Joukowski) are nearly identical. The SVPM pitching moment predictions of Figure 4.29b are lower than the experimental measurements and their slope is negative. Unlike the stipulation of thin airfoil theory, the experimental results show that the pitching moment around the quarter chord is not exactly constant with angle of attack, even at attached flow conditions. The reader should also note that the

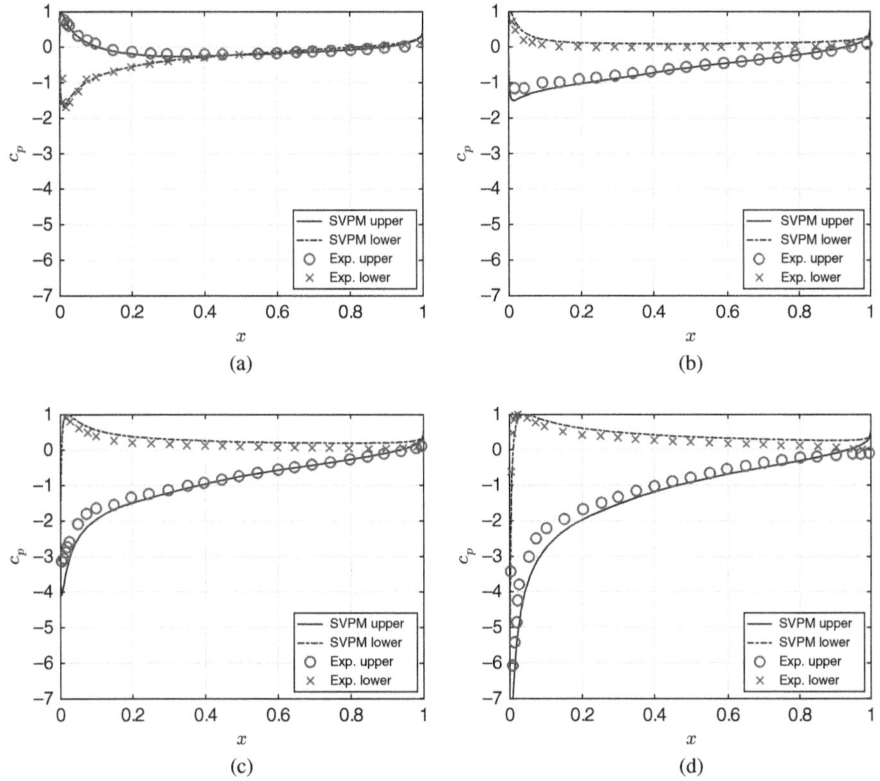

Figure 4.28 Pressure distribution around static NACA2412 airfoil calculated using the source and vortex panel method. Experimental data by Seetharam et al. (1977). (a) $\alpha = -3.9°$, (b) $\alpha = 4.3°$, (c) $\alpha = 8.3°$ and (d) $\alpha = 12.4°$. Source: Seetharam et al. (1977), credit: NASA.

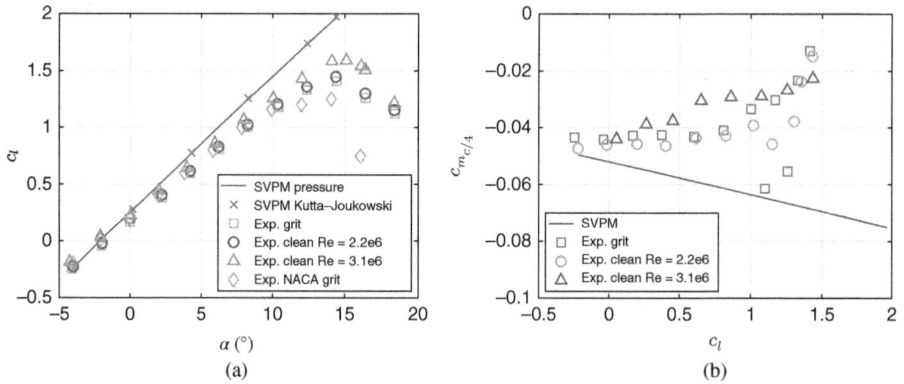

Figure 4.29 Aerodynamic loads acting on static NACA2412 airfoil using the source and vortex panel method. Experimental data by Seetharam et al. (1977). (a) c_l vs α and (b) $c_{m_{c/4}}$ vs c_l. Source: Seetharam et al. (1977), credit: NASA.

repeatability of wind tunnel experiments on airfoils at high angles of attack leaves a lot to be desired.

All numerical solutions must be checked for convergence. Figure 4.30 plots the SVPM predictions for the lift and pitching moment coefficients at $\alpha = 5°$ for increasing number of panels n. It can be seen that both the lift and the moment converge nicely. The circles denote the results for $n = 200$ presented earlier; they are sufficiently converged since $c_l(200) = 0.996c_l(1500)$ and $c_m(200) = 0.982c_m(1500)$. The drag coefficient also converges, although its value never becomes equal to zero. Even for $n = 1500$, $c_d = -3.6 \times 10^{-5}$. This example is solved by Matlab code `steady_SVPM.m`.

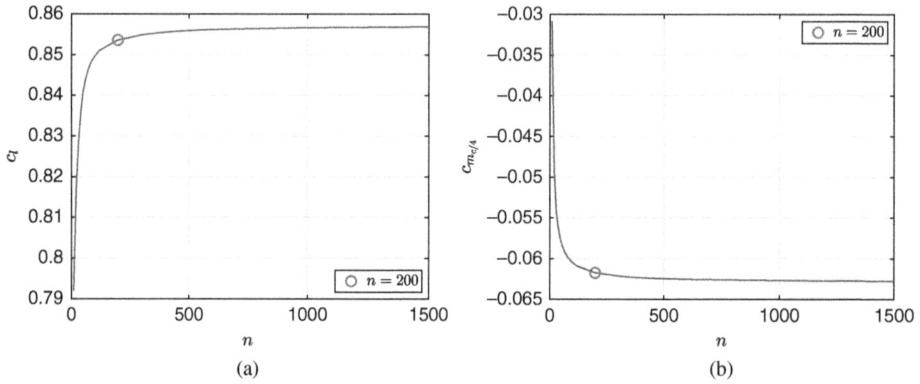

Figure 4.30 Convergence of SVPM predictions for lift and pitching moment with increasing number of panels. (a) c_l vs n and (b) $c_{m_{c/4}}$ vs n.

The SVPM has formed the basis of a popular 2D unsteady aerodynamic modelling approach developed by Basu and Hancock (1978) and used by several other researchers, notably Jones and Platzer (1996). We will present this method shortly, but first, we will deal with the question of the evaluation of the perturbation potential on the surface of the wing.

The solution of Example 4.9 calculates the velocity distribution around the airfoil and, hence, the pressure distribution and the aerodynamic loads. Even though it is a potential flow solution, the potential itself is never calculated explicitly. Only the derivatives of the potential are necessary, since the impermeability boundary condition is imposed in terms of the velocity normal to the surface. However, for the unsteady version of the SVPM, we will need to calculate the potential and its time derivative. We will present this calculation here for the steady case by revisiting Example 4.9.

Example 4.10 *Calculate the perturbation potential on the surface of the airfoil of Example 4.9.*

The perturbation potential, ϕ, is defined as the potential induced by the singularities in the flow only, excluding the effect of the free stream. We start by solving for the strengths of the source and vortex panels in exactly the same way as we did in Example 4.9 for $\alpha = 5°$ and $n = 200$. Once these strengths are available, we could attempt to calculate the potential from Eq. (4.107) for the source panels and Eq. (4.110) for the vortex panels. However, the inverse tangent terms in expression (4.110) introduce a discontinuity around $\theta = \pm\pi$. The cumulative effect of the discontinuities induced by each of the panels leads to numerically unstable values for the potential.

Consider a flow streamline $s(x, y)$ and the velocity tangent to it, $u_\tau(s) = \partial\phi/\partial s$. The perturbation potential at any point on this streamline can be obtained by integrating the tangential velocity from any point very far away (where the perturbation potential is negligible) all the way to the point of interest. The surface of the airfoil is a flow streamline; hence, we can calculate the potential on it by integrating the tangential velocity induced by the source and vortex panels on the surface (Basu and Hancock 1978; Cebeci et al. 2005). The tangent velocities on all the panels are obtained from Eq. (4.133). From the definition of the potential, the tangent perturbation velocity is given by

$$u_\tau = \frac{\partial\phi}{\partial s}$$

where differentiation with respect to the arclength s denotes differentiation with respect to a coordinate parallel to the surface. Hence, the perturbation potential on the surface can be calculated by integrating numerically the equation above, along the flow direction. We assume that at the leading edge, the perturbation potential is equal to ϕ_{le} and then integrate from the leading edge to the trailing edge on the upper and lower surfaces separately. Consequently,

$$\phi(x_i, y_i) = \begin{cases} \phi_{le} - \sum_{j=i}^{n/2} u_{\tau_j} s_j & \text{if } 1 \le i \le n/2 \\ \phi_{le} & \text{if } i = n/2+1 \\ \phi_{le} + \sum_{j=n/2+1}^{i-1} u_{\tau_j} s_j & \text{if } n/2+2 \le i \le n+1 \end{cases} \tag{4.147}$$

for $i = 1, \ldots, n+1$. Note that we have chosen the leading edge as the start of two streamlines along the surface, which is not strictly speaking correct; the start of the streamlines should have been the stagnation point. However, the stagnation point is not one of the panel vertices and, for small angles of attack, it is expected to lie very close to the leading edge.

The numerical integration of Eq. (4.147) gives the values of the potential at the panel vertices, if ϕ_{le} is known. The latter can also be obtained by numerically integrating the velocities induced by the source and vortex panels on the airfoil, along a flow streamline starting far upstream and *stopping at the leading edge. Strictly speaking, we should use the stagnation streamline, which stops at the stagnation point, but we have already approximated the stagnation point by the leading edge. Furthermore, computing the stagnation streamline is computationally expensive. Instead, we choose to perform the numerical integration on a straight line extending from a point far upstream in the direction of the free stream to the leading edge.*

Figure 4.31 plots the airfoil along with a set of non-linearly spaced points extending up to 10 chord-lengths upstream of the airfoil's leading edge in the direction $\pi + \alpha$. These x_{u_i}, y_{u_i} points for $i = 1, \dots, m + 1$ define m panels; they are spaced such that the panels are shortest near the leading edge. The panels have lengths s_{u_i} such that s_{u_1} is the longest and s_{u_m} is the shortest, equal to $s(n/2)$. We define control points $x_{c_{u_i}}, y_{c_{u_i}}$ for $i = 1, \dots, m$ at the midpoint of each of the upstream panels and calculate the velocities induced there by the source and vortex panels on the airfoil in the same way we did for the airfoil panel control points. Finally, we calculate the tangential velocities on these upstream panels, $u_{u_{\tau_i}}$, and the leading edge perturbation potential from

$$\phi_{le} = \sum_{i=1}^{n_u} u_{u_{\tau_i}} s_{u_i}$$

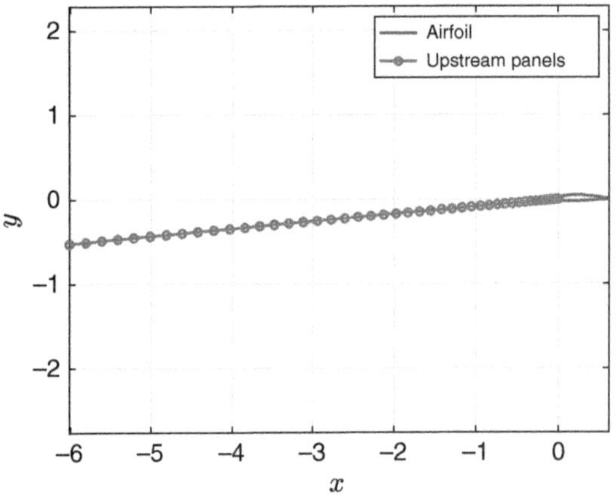

Figure 4.31 Points upstream of the trailing edge, along which the potential is calculated.

Once Eq. (4.147) has been evaluated, the perturbation potential on the control points of the airfoil panels can be calculated from $\phi(x_{c_i}, y_{c_i}) = \big(\phi(x_i, y_i) + \phi(x_{i+1}, y_{i+1})\big)/2$. Figure 4.32a plots the perturbation potential on the airfoil's surface. It can be seen that there is a potential jump across the trailing edge, the potential being negative on the lower surface and positive on the upper surface. This potential jump is due to the vortex panels, as seen in Figure 4.32b which plots separately the perturbation potential contributions of the source and vortex panels. It can be seen that there is no discontinuity in the potential induced by the source panels, and the only discontinuity is caused by the vortex panels.

The definition of circulation is given by Eq. (2.212) as

$$\Gamma = \oint_C \mathbf{u} \cdot \boldsymbol{\tau}\, ds$$

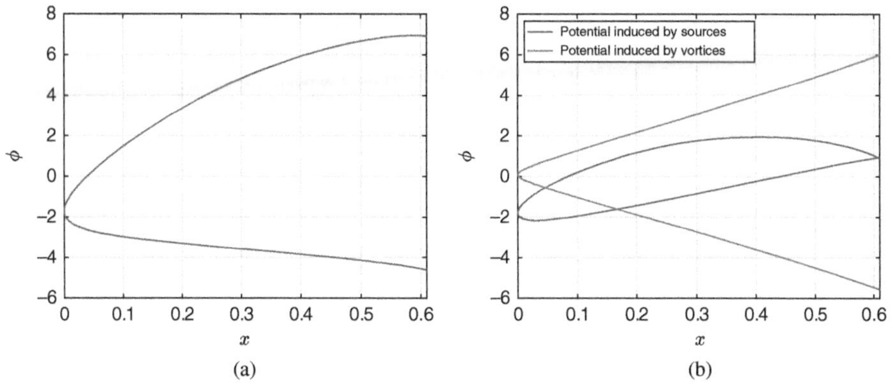

Figure 4.32 Calculation of perturbation potential on airfoil's surface. (a) Total potential and (b) potential contributions from source and vortex panels.

which is the line integral along a closed path C of the vector product of the flow velocity $\mathbf{u} = (u\ v\ w)$ *times the tangential length vector* $\boldsymbol{\tau}ds$. *If the closed path is the airfoil's surface, the only flow velocity is the tangential velocity and, hence, the velocity and the tangential length vector are co-linear. For n panels, the circulation becomes*

$$\Gamma = \sum_{i=1}^{n} u_{\tau_j} s_j$$

From Eq. (4.147), it can be easily seen that

$$\phi(x_{n+1}, y_{n+1}) - \phi(x_1, y_1) = \sum_{i=1}^{n} u_{\tau_j} s_j$$

Consequently,

$$\Delta\phi_{te} = \phi(x_{n+1}, y_{n+1}) - \phi(x_1, y_1) = \Gamma = \gamma \sum_{i=1}^{n} s_i \tag{4.148}$$

so that the jump in perturbation potential across the trailing edge is equal to the total circulation around the airfoil. This example is solved by Matlab code `steady_SVPM_potential.m`.

4.5.1 Impulsively Started Flow

Basu and Hancock's 1978 unsteady SVPM is similar to the steady approach but the strengths of the panels are functions of time. The wake is modelled by point vortices, as in the case of the lumped vortex approach, but the shedding of the first vortex is handled differently. Recall that in Section 4.2, we stated that the lumped vortex technique sheds the first vortex of the wake at an arbitrary position, lying between 20% and 30% of the distance travelled by the trailing edge over one time increment Δt. The unsteady SVPM employs a more sophisticated strategy that is demonstrated graphically in Figure 4.33.

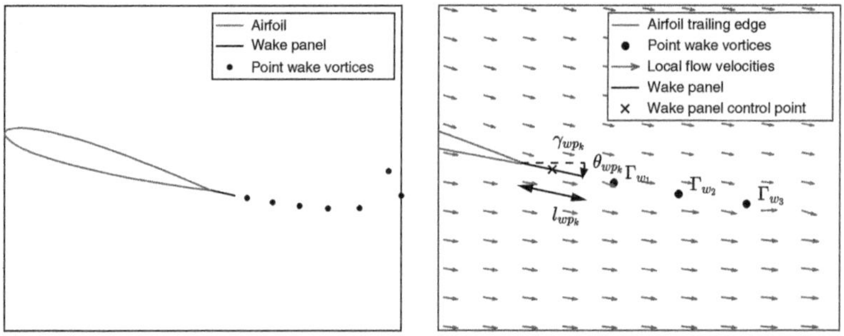

(a) (b)

Figure 4.33 The unsteady source and vortex panel scheme of Basu and Hancock (1978). (a) Airfoil and wake and (b) trailing edge. Source: Adapted from Basu and Hancock (1978).

The wake is composed of a number of point vortices and a single vortex panel attached to the trailing edge and aligned with the instantaneous local flow velocity. This means that the length and orientation of the panel change with time. The local flow velocity is calculated at the control point of the panel, $x_{c_{wp}}(t_k), y_{c_{wp}}(t_k)$, and its horizontal and vertical components are $u_{wp}(t_k)$ and $v_{wp}(t_k)$, including contributions from all the singularities in the flow and the free stream. The length of the panel is denoted by $l_{wp_k} = l_{wp}(t_k)$ and is calculated as the distance travelled by the local flow over one time interval Δt. Hence, the length and orientation angle, $\theta_{wp}(t_k)$, are given by

$$l_{wp}(t_k) = \sqrt{u_{wp}(t_k)^2 + v_{wp}(t_k)^2}\Delta t \tag{4.149}$$

$$\theta_{wp}(t_k) = \tan^{-1}\left(\frac{v_{wp}(t_k)}{u_{wp}(t_k)}\right) \tag{4.150}$$

The vortex strength on the wake panel $\gamma_{wp_k} = \gamma_{wp}(t_k)$ is constant along the panel but changes in time. The rest of the wake is made up of point vortices that have strengths equal to the total circulation of the wake panel at previous time instances, that is $\Gamma_{w_1}(t_k) = l_{wp_{k-1}}\gamma_{wp_{k-1}}$, $\Gamma_{w_2}(t_k) = l_{wp_{k-2}}\gamma_{wp_{k-2}}$, etc.

At the kth time instance, the wake is composed of the wake panel and $k-1$ point vortices. The free stream is taken to be parallel to the x axis, such that $\mathbf{Q}_\infty = (U_\infty, 0)$. The total horizontal and vertical velocities at the control point of the wake panel are given by

$$u_{wp}(t_k) = \boldsymbol{A}_{u_{\sigma,wp}}(t_k)\boldsymbol{\sigma}(t_k) + \boldsymbol{A}_{u_{\gamma,wp}}(t_k)\gamma + \boldsymbol{B}_{u_{wp}}(t_k)\boldsymbol{\Gamma}_w(t_k) + U_\infty \tag{4.151}$$

$$v_{wp}(t_k) = \boldsymbol{A}_{v_{\sigma,wp}}(t_k)\boldsymbol{\sigma} + \boldsymbol{A}_{v_{\gamma,wp}}(t_k)\gamma + \boldsymbol{B}_{v_{wp}}(t_k)\boldsymbol{\Gamma}_w(t_k) \tag{4.152}$$

where the $1 \times n$ influence coefficient vectors $\boldsymbol{A}_{u_{\sigma,wp}}$, $\boldsymbol{A}_{v_{\sigma,wp}}$, $\boldsymbol{A}_{u_{\gamma,wp}}$ and $\boldsymbol{A}_{v_{\gamma,wp}}$ are calculated from Eqs. (4.123)–(4.126) with $\mathbf{r}_{wpj} = ((x_{c_{wp}} - x_{c_j}), (y_{c_{wp}} - y_{c_j}))$, $x_{\tau_{wp,j}} = \mathbf{r}_{wp,j} \cdot \boldsymbol{\tau}_j$ and $y_{n_{wp,j}} = \mathbf{r}_{wp,j} \cdot \mathbf{n}_j$. The velocities induced on the wake panel by the point wake vortices are calculated from Eqs. (4.36) and (4.37) after replacing x_{c_i} and y_{c_i} by $x_{c_{wp}}$ and $y_{c_{wp}}$, respectively, so that vectors $\boldsymbol{B}_{u_{wp}}(t_k)$ and $\boldsymbol{B}_{v_{wp}}(t_k)$ have dimensions $1 \times (k-1)$. Note that the influence of the wake panel on itself has been neglected in Eqs. (4.151) and (4.152).

At each time instance, the unknowns are the source and vortex strengths of the airfoil panels, $\sigma_j(t_k)$ for $j = 1, \dots, n$ and $\gamma(t_k)$, as well as the strength, length and orientation angle of the wake vortex panel, $\gamma_{wp}(t_k)$, l_{wp_k} and θ_{wp_k}. The conditions we have in order to calculate these unknowns are:

- Impermeability on the n airfoil panels
- The Kutta condition
- Kelvin's theorem
- Equations (4.149) and (4.150)

These conditions constitute a total of $n+4$ equations with $n+4$ unknowns, but they cannot be solved directly, as the instantaneous local flow velocities on the wake panel, $u_{wp_k} = u_{wp}(t_k)$ and $v_{wp_k} = v_{wp}(t_k)$, depend on the unknown strengths $\sigma_j(t_k)$, $\gamma(t_k)$ and $\gamma_{wp}(t_k)$. Hence, the equations are non-linear and must be solved iteratively.

First of all we must write out all the equations explicitly. We will be rotating the airfoil to its actual pitch angle at each time instance, as we did for the time-domain lumped vortex method. If the airfoil coordinates at $\alpha = 0$ are given by x_{0_i} and y_{0_i} (from Eqs. (3.5) and (3.6) for a NACA four-digit series airfoil for example), then for any non-zero angle of attack, the rotated airfoil coordinates are obtained from Eq. (4.74). The normal influence coefficients of the bound source and vortex panels, $\boldsymbol{A}_{n_\sigma}$ and $\boldsymbol{A}_{n_\gamma}$ are obtained from Eqs. (4.128) and (4.129). As the normal influence coefficients are calculated from the relative distances between the bound panels and control points, they only need to be calculated once, they do not change value if the airfoil is rotated.

The unit vectors tangent and normal to the wake panels are denoted by $\mathbf{n}_{wp}(t_k) = (n_{x_{wp}}(t_k), n_{y_{wp}}(t_k))$ and $\boldsymbol{\tau}_{wp}(t_k) = (\tau_{x_{wp}}(t_k), \tau_{y_{wp}}(t_k))$. The horizontal and vertical velocity influence coefficients of the wake panel on the control points of the bound panels are denoted by $\boldsymbol{A}_{u_{wp}}(t_k)$ and $\boldsymbol{A}_{v_{wp}}(t_k)$. They are $n \times 1$ vectors and their elements are calculated from Eqs. (4.125) and (4.126), after substituting $\mathbf{r}_i(t_k) = ((x_{c_i}(t_k) - x_{c_{wp}}(t_k)), (y_{c_i}(t_k) - y_{c_{wp}}(t_k)))$, $x_{\tau_i}(t_k) = \mathbf{r}_i(t_k) \cdot \boldsymbol{\tau}_{wp}(t_k)$, $y_{n_i}(t_k) = \mathbf{r}_i(t_k) \cdot \mathbf{n}_{wp}(t_k)$ and $s_j(t_k) = l_{wp}(t_k)$. The normal and tangential influence coefficients of the wake panel on the control points of the bound panels are denoted by $\boldsymbol{A}_{n_{wp}}(t_k)$ and $\boldsymbol{A}_{\tau_{wp}}(t_k)$ and are obtained from Eqs. (4.129) and (4.135). We also calculate the $n \times (k-1)$ horizontal, $\boldsymbol{B}_u(t_k)$, and vertical, $\boldsymbol{B}_v(t_k)$, influence coefficient matrices of the point wake vortices on the control points of the wing panels from Eqs. (4.38) and (4.39), respectively. The normal influence coefficient matrix of the point vortices on the control points of the airfoil, $\boldsymbol{B}_n(t_k)$, is then given by Eq. (4.41), while the elements of the tangential influence coefficient matrix, $\boldsymbol{B}_\tau(t_k)$, are obtained from

$$B_{\tau_{i,j}}(t_k) = B_{u_{i,j}}(t_k)\tau_{x_i} + B_{v_{i,j}}(t_k)\tau_{y_i} \tag{4.153}$$

For an impulsively started airfoil with constant angle of attack α and free stream airspeed U_∞, the impermeability boundary condition is obtained by combining Eqs. (4.127) and (4.40) and adding the influence of the wake panel, such that

$$\boldsymbol{A}_{n_\sigma}\boldsymbol{\sigma}(t_k) + \boldsymbol{A}_{n_\gamma}\boldsymbol{I}_{n\times 1}\gamma(t_k) + \boldsymbol{A}_{n_{wp}}(t_k)\gamma_{wp}(t_k) + \boldsymbol{B}_n(t_k)\boldsymbol{\Gamma}_w(t_k) = \boldsymbol{b} \tag{4.154}$$

Kelvin's theorem is expressed in the same way as for the lumped vortex method, that is the change in bound vorticity between time instances k and $k-1$ must be counteracted by the vortex strength of the wake panel. Equation (4.35) is adapted for the unsteady SVPM as follows:

$$\sum_{j=1}^{n} s_j\gamma(t_k) + l_{wp_k}\gamma_{wp}(t_k) = \sum_{j=1}^{n} s_j\gamma(t_{k-1}) \tag{4.155}$$

The pressure coefficient on each panel is obtained from the 2D version of Eq. (2.31)

$$c_p = 1 - \frac{(u^2+v^2)}{Q_\infty^2} - \frac{2}{Q_\infty^2}\frac{\partial \Phi}{\partial t}$$

On the surface of the ith panel, $u^2 + v^2$ is the square of the total tangential velocity, $\mathbf{Q}_\infty \cdot \boldsymbol{\tau}_n + u_{\tau_i}$, so that

$$c_{p_i} = 1 - \frac{(\mathbf{Q}_\infty \cdot \boldsymbol{\tau}_i + u_{\tau_i})^2}{Q_\infty^2} - \frac{2}{Q_\infty^2}\frac{\partial \phi_i}{\partial t} \tag{4.156}$$

At the kth time instance, the tangential perturbation velocities on all the panels are obtained by adding the effects of the wake panel and point wake vortices to Eq. (4.133)

$$\boldsymbol{u}_\tau(t_k) = \boldsymbol{A}_{\tau_\sigma}(t_k)\sigma(t_k) + \boldsymbol{A}_{\tau_\gamma}(t_k)\boldsymbol{I}_{n\times 1}\gamma(t_k) + \boldsymbol{A}_{\tau_{wp}}(t_k)\gamma_{wp}(t_k) + \boldsymbol{B}_\tau(t_k)\boldsymbol{\Gamma}_w(t_k) \tag{4.157}$$

The time derivative of the perturbation potential can be calculated using a backward finite difference scheme

$$\left.\frac{\partial \phi_i}{\partial t}\right|_{t_k} = \frac{\phi_i(t_k) - \phi_i(t_{k-1})}{\Delta t}$$

The perturbation potential on the panel vertices can be calculated numerically from the unsteady version of Eq. (4.147)

$$\phi(x_i, y_i, t_k) = \begin{cases} \phi_{le}(t_k) - \sum_{j=i}^{n/2} u_{\tau_j}(t_k)s_j & \text{if } 1 \le i \le n/2 \\ \phi_{le}(t_k) & \text{if } i = n/2+1 \\ \phi_{le}(t_k) + \sum_{j=n/2+1}^{i-1} u_{\tau_j}(t_k)s_j & \text{if } n/2+2 \le i \le n+1 \end{cases} \tag{4.158}$$

The perturbation potential on the control points is then simply

$$\phi(x_{c_i}, y_{c_i}, t_k) = \frac{1}{2}\left(\phi(x_{i+1}, y_{i+1}, t_k) + \phi(x_i, y_i, t_k)\right) \tag{4.159}$$

Hence, the pressure coefficient on the ith panel becomes

$$c_{p_i}(t_k) = 1 - \frac{(\mathbf{Q}_\infty \cdot \boldsymbol{\tau}_i + u_{\tau_i})^2}{Q_\infty^2} - \frac{2}{Q_\infty^2}\frac{\phi(x_{c_i}, y_{c_i}, t_k) - \phi(x_{c_i}, y_{c_i}, t_{k-1})}{\Delta t} \tag{4.160}$$

The Kutta condition is identical to the one used for the steady case, that is the pressure must be equal on the two trailing edge panels at all time instances, $c_{p_1}(t_k) = c_{p_n}(t_k)$ or, from Eq. (4.160),

$$(\mathbf{Q}_\infty \cdot \boldsymbol{\tau}_1 + u_{\tau_1}(t_k))^2 + 2\frac{\phi(x_1, y_1, t_k) - \phi(x_1, y_1, t_{k-1})}{\Delta t}$$

$$= (\mathbf{Q}_\infty \cdot \boldsymbol{\tau}_n + u_{\tau_n}(t_k))^2 + 2\frac{\phi(x_{n+1}, y_{n+1}, t_k) - \phi(x_{n+1}, y_{n+1}, t_{k-1})}{\Delta t}$$

Recall that the jump in potential across the trailing edge is equal to the total circulation on the airfoil (see Eq. ((4.148)), so that

$$\phi(x_{n+1}, y_{n+1}, t_k) - \phi(x_1, y_1, t_k) = \Gamma(t_k) = \gamma(t_k)\sum_{j=1}^{n} s_j$$

$$\phi(x_{n+1}, y_{n+1}, t_{k-1}) - \phi(x_1, y_1, t_{k-1}) = \Gamma(t_{k-1}) = \gamma(t_{k-1})\sum_{j=1}^{n} s_j$$

Therefore, we can re-write the Kutta condition as

$$(\mathbf{Q}_\infty \cdot \boldsymbol{\tau}_1 + u_{\tau_1}(t_k))^2 = (\mathbf{Q}_\infty \cdot \boldsymbol{\tau}_n + u_{\tau_n}(t_k))^2 + 2\sum_{j=1}^{n} s_j \frac{\gamma(t_k) - \gamma(t_{k-1})}{\Delta t} \tag{4.161}$$

Note that in developing this latest expression we have calculated the tangent velocities at the control points of the trailing edge panels and the potential difference at the vertices of the trailing edge panels. This small inconsistency does not have an effect on the accuracy of the resulting aerodynamic loads.

Equations (4.154), (4.155), (4.161), (4.149) and (4.150) make up the system of $n+4$ non-linear algebraic equations to be solved for the unknown values of $\boldsymbol{\sigma}(t_k)$, $\gamma(t_k)$, $\gamma_{wp}(t_k)$, l_{wp_k} and θ_{wp_k}. If the values of l_{wp_k} and θ_{wp_k} are known, Eqs. (4.154) and (4.155) are linear and can be solved for $\boldsymbol{\sigma}(t_k)$ and $\gamma_{wp}(t_k)$ in terms of $\gamma(t_k)$. First, we solve Eq. (4.155) for $\gamma_{wp}(t_k)$ such that

$$\gamma_{wp}(t_k) = -\frac{S}{l_{wp_k}}\gamma(t_k) + \frac{\Gamma(t_{k-1})}{l_{wp_k}} \tag{4.162}$$

where $S = \sum_{j=1}^{n} s_j$, then we substitute into Eq. (4.154) and solve for $\boldsymbol{\sigma}(t_k)$

$$\boldsymbol{\sigma}(t_k) = \mathbf{b}_\gamma(t_k)\gamma(t_k) + \mathbf{b}_0(t_k) \tag{4.163}$$

where

$$\mathbf{b}_\gamma(t_k) = \boldsymbol{A}_{n_\sigma}^{-1}\left(-\boldsymbol{A}_{n_\gamma}\boldsymbol{I}_{n\times 1} + \frac{S}{l_{wp_k}}\boldsymbol{A}_{n_{wp}}(t_k)\right)$$

$$\mathbf{b}_0(t_k) = \boldsymbol{A}_{n_\sigma}^{-1}\left(-\frac{\Gamma(t_{k-1})}{l_{wp_k}}\boldsymbol{A}_{n_{wp}}(t_k) - \boldsymbol{B}_n(t_k)\boldsymbol{\Gamma}_w(t_k) + \mathbf{b}\right)$$

Now, we can substitute the expressions for $\boldsymbol{\sigma}(t_k)$ and $\gamma_{wp}(t_k)$ in equation (4.157) to obtain the tangential perturbation velocity as a function of $\gamma(t_k)$ only,

$$\boldsymbol{u}_\tau(t_k) = \mathbf{c}_\gamma(t_k)\gamma(t_k) + \mathbf{c}_0(t_k) \tag{4.164}$$

where

$$\mathbf{c}_\gamma(t_k) = \boldsymbol{A}_{\tau_\sigma}(t_k)\boldsymbol{b}_\gamma(t_k) + \boldsymbol{A}_{\tau_\gamma}(t_k)\boldsymbol{I}_{n\times 1} - \frac{S}{l_{wp_k}}\boldsymbol{A}_{\tau_{wp}}(t_k)$$

$$\mathbf{c}_0(t_k) = \boldsymbol{A}_{\tau_\sigma}(t_k)\boldsymbol{b}_0(t_k) + \frac{\Gamma(t_{k-1})}{l_{wp_k}}\boldsymbol{A}_{\tau_{wp}}(t_k) + \boldsymbol{B}_\tau(t_k)\boldsymbol{\Gamma}_w(t_k)$$

Substituting $u_{\tau_1}(t_k)$ and $u_{\tau_n}(t_k)$ into the Kutta condition of Eq. (4.161) leads to

$$\left(U_\infty \tau_{x_n} + c_{\gamma_1}\gamma(t_k) + c_{0_1}\right)^2 = \left(U_\infty \tau_{x_1} + c_{\gamma_n}\gamma(t_k) + c_{0_n}\right)^2 + 2S\frac{\gamma(t_k) - \gamma(t_{k-1})}{\Delta t}$$

Rearranging, we obtain a quadratic equation in $\gamma(t_k)$

$$\left(c_{\gamma_n}^2 - c_{\gamma_1}^2\right)\gamma^2(t_k) + 2\left(c_{\gamma_n}(c_{0_n} + U_\infty \tau_{x_n}) - c_{\gamma_1}(c_{0_1} + U_\infty \tau_{x_1}) + \frac{S}{\Delta t}\right)\gamma(t_k) + (c_{0_n} + U_\infty \tau_{x_n})^2 - (c_{0_1} + U_\infty \tau_{x_1})^2 - \frac{2\Gamma(t_{k-1})}{\Delta t} = 0 \tag{4.165}$$

The iterative solution procedure is the following:

1. Guess values for l_{wp_k} and θ_{wp_k}.
2. Solve Eq. (4.165) for $\gamma(t_k)$ and then calculate $\sigma(t_k)$ from Eq. (4.163) and $\gamma_{wp}(t_k)$ from Eq. (4.162).
3. Calculate the local flow velocities on the control points of the wake panel, u_{wp} and v_{wp}, from Eqs. (4.151) and (4.152).
4. Calculate new values for l_{wp_k} and θ_{wp_k} from Eqs. (4.149) and (4.150).
5. If the differences between the new and old wake panel length and angle are small, the scheme has converged. Otherwise, go back to step 2.

Once the algorithm has converged, we can calculate the aerodynamic loads, propagate the wake and then go to the next time instance. We will illustrate the practical implementation of this procedure in the following example.

Example 4.11 *Calculate the impulsively started flow around a NACA 0012 airfoil with chord $c = 0.2$ m at angle $\alpha = 5°$ to a horizontal free stream with airspeed $U_\infty = 10$ m/s.*

We select to model the airfoil with $n = 200$ panels and choose a time step of $\Delta t = 16c/nU_\infty$. Since we are now calculating the strength of the vortices shed in the wake from the vorticity in the wake panel, we do not need to select a time step that equates the distance between wake and bound vortices, as we did for the lumped vortex method. The first steps are to calculate the airfoil shape from Eqs. (3.6) and (4.144) and to rotate the airfoil around its quarter-chord using Eq. (4.59). The next step is to evaluate the coordinates of the control points, x_{c_i}, y_{c_i}, from Eq. (4.103), the normal vectors n_{x_i}, n_{y_i}, from Eq. (4.105), the tangential vectors τ_{x_i}, τ_{y_i}, from Eq. (4.104) and the lengths, s_i, of each panel from Eq. (4.106). Then, we calculate all the steady influence coefficient matrices, as we did in Example 4.9

The steady solution is obtained by solving the flow without wake panel or wake vortices using Eq. (4.137), after substituting $\alpha = 0$ since the free stream is parallel to the x axis. Then, the steady-state tangential velocity on the control points is given by Eq. (4.133), again with $\alpha = 0$. At steady-state conditions, the pressure coefficient, $c_{p_{\infty_i}}$, is given by Eq. (4.131) and the lift and drag coefficients can be calculated from the integral of the pressure distribution around the airfoil, such that

$$c_{l_\infty} = -\frac{1}{c}\sum_{i=1}^{n} c_{p_{\infty_i}} s_i n_{y_i} \tag{4.166}$$

$$c_{d_\infty} = -\frac{1}{c}\sum_{i=1}^{n} c_{p_{\infty_i}} s_i n_{x_i} \tag{4.167}$$

since the free stream is aligned with the x axis.

The unsteady solution begins at $t_0 = 0$. The length and orientation angle of the wake panel are not known, but we can guess values of $l_{wp_0} = U_\infty \Delta t$ for the length and $\theta_{wp_0} = 0$ for the *orientation angle. We place the leading edge of the wake panel on either of the two trailing edge points, since they coincide. Hence, the coordinates of the vertices of the wake panel are given by*

$$x_{wp_{1,2}} = x_1 + (0,\ l_{wp_k} \cos\theta_{wp_k}), \quad y_{wp_{1,2}} = y_1 + (0,\ l_{wp_k} \sin\theta_{wp_k})$$

Figure 4.34 plots the trailing edge of the airfoil and the wake panel at its initially guessed position. The collocation point lies midway between the vertices

$$x_{c_{wp}} = \frac{x_{wp_1} + x_{wp_2}}{2}, \quad y_{c_{wp}} = \frac{y_{wp_1} + y_{wp_2}}{2}$$

and the tangent and normal unit vectors to the wake panel are calculated from

$$\boldsymbol{\tau}_{wp}(t_k) = (\cos\theta_{wp_k}, \sin\theta_{wp_k})^T, \quad \mathbf{n}_{wp}(t_k) = (-\sin\theta_{wp_k}, \cos\theta_{wp_k})^T$$

We can now calculate all the influence coefficient matrices and start iterating for l_{wp_k} and θ_{wp_k}, as detailed in the algorithm above. However, at t = 0 the value of $\gamma(t_{k-1})$ in Eq. (4.165) does not exist. We can set it to zero but then the change in circulation across the trailing edge at the first time instance will be equal to $S\gamma(0)/\Delta t$ and its numerical value will be enormous due to the division by Δt. Here we choose to ignore the change in circulation (and potential) across the trailing edge at the first time instance, so that the Kutta condition of Eq. (4.165) becomes

$$\left(c_{\gamma_n}^2 - c_{\gamma_1}^2\right)\gamma^2(0) + 2\left(c_{\gamma_n}(c_{0_n} + U_\infty \tau_{x_n}) - c_{\gamma_1}(c_{0_1} + U_\infty \tau_{x_1})\right)\gamma(0) + (c_{0_n} + U_\infty \tau_{x_n})^2 - (c_{0_1} + U_\infty \tau_{x_1})^2 = 0 \tag{4.168}$$

This expression (and Eq. (4.165) at subsequent time steps) is a quadratic equation of the form $a_2\gamma(t_k)^2 + a_1\gamma(t_k) + a_0 = 0$ that has two solutions. The solution we retain is

$$\gamma(t_k) = \frac{-a_1 + \sqrt{a_1^2 - 4a_0a_2}}{2a_2} \tag{4.169}$$

while the other solution leads to a non-physical flowfield, as demonstrated in Figure 4.35 for a steady flow case.

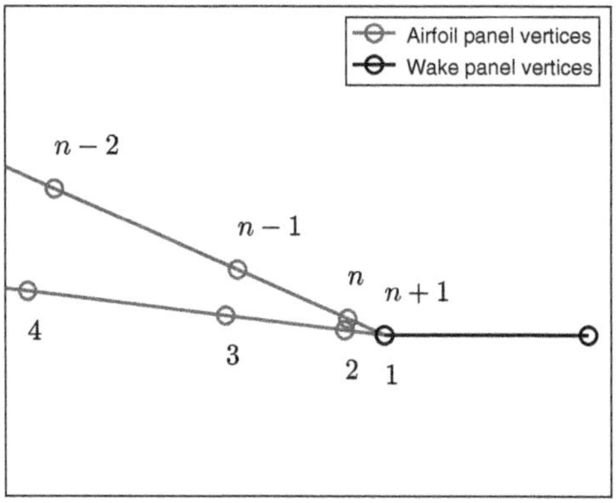

Figure 4.34 Wake panel placement at the trailing edge of NACA four-digit airfoils.

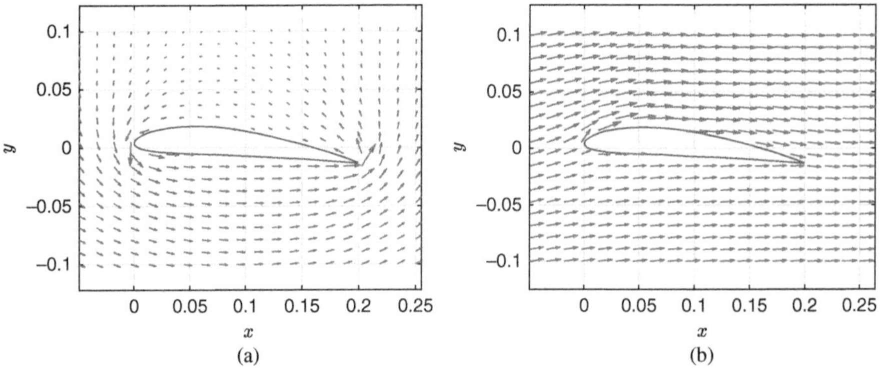

Figure 4.35 Effect of using the two different roots of Eq. (4.165) on the steady flow around an airfoil. (a) Discarded root for γ and (b) retained root for γ.

If we denote the length and orientation angle of the wake panel after the mth iteration by $l_{wp_{k,m}}$ *and* $\theta_{wp_{k,m}}$*, then the convergence criterion can be defined as*

$$\sqrt{\left(l_{wp_{k,m}} - l_{wp_{k,m-1}}\right)^2 + \left(\theta_{wp_{k,m}} - \theta_{wp_{k,m-1}}\right)^2} < \varepsilon$$

where ε *is a small real number, e.g.* $\varepsilon = 10^{-4}$*. Once convergence has been achieved, we recalculate the tangential flow velocities on the airfoil panels from Eq. (4.157) and the pressure coefficient distribution around the airfoil from Eq. (4.160), setting the time derivative of the potential to zero for the first time instance, that is*

$$c_{p_i}(0) = 1 - \left(\frac{\boldsymbol{Q}_\infty \cdot \boldsymbol{\tau}_i + u_{t_i}(0)}{Q_\infty}\right)^2$$

The lift and drag coefficients are evaluated from

$$c_l(t_k) = -\frac{1}{c}\sum_{i=1}^{n} c_{p_i}(t_k) s_i n_{y_i} \tag{4.170}$$

$$c_d(t_k) = -\frac{1}{c}\sum_{i=1}^{n} c_{p_i}(t_k) s_i n_{x_i} \tag{4.171}$$

At a general time instance, k, first, we shed the vorticity of the wake panel to a point vortex with strength $\Gamma_{w_1}(t_k) = \gamma_{wp}(t_{k-1}) l_{wp_{k-1}}$ *lying at*

$$x_{w_1}(t_k) = x_{c_{wp}}(t_{k-1}) + u_{c_{wp}}(t_{k-1})\Delta t, \quad y_{w_1}(t_k) = y_{c_{wp}}(t_{k-1}) + v_{c_{wp}}(t_{k-1})\Delta t$$

and then we propagate the wake point vortices

$$\begin{aligned} x_{w_i}(t_k) &= x_{w_{i-1}}(t_{k-1}) + \left(U_\infty + u_{w\sigma_i}(t_k) + u_{w\gamma_i}(t_k) + u_{w\gamma_{wp_i}}(t_k) + u_{ww_i}(t_k)\right)\Delta t \\ y_{w_i}(t_k) &= y_{w_{i-1}}(t_{k-1}) + \left(v_{w\sigma_i}(t_k) + v_{w\gamma_i}(t_k) + v_{w\gamma_{wp_i}}(t_k) + v_{ww_i}(t_k)\right)\Delta t \end{aligned} \tag{4.172}$$

where $u_{w\sigma_i}(t_k)$, $v_{w\sigma_i}(t_k)$ *are the velocities induced on the point wake vortices by the source panels,* $v_{w\gamma_i}(t_k)$, $v_{w\gamma_i}(t_k)$ *those induced by the wake panels,* $u_{w\gamma_{wp_i}}(t_k)$, $v_{w\gamma_{wp_i}}(t_k)$ *are induced by the wake panel and* $u_{ww_i}(t_k)$, $v_{ww_i}(t_k)$ *by the point vortices themselves. The latter can be calculated using either the classical vortex model of Eq. (4.58) or the Vatistas model of expressions (4.66). We then iterate for* l_{wp_k} *and* θ_{wp_k}, *choosing as initial guesses the previously converged values,* $l_{wp_{k-1}}$ *and* $\theta_{wp_{k-1}}$.

Figure 4.36 plots the resulting lift and drag responses against non-dimensional time $\tau = tU_\infty/b$, *along with the Wagner function predictions of Chapter 3. The ratio of the unsteady to steady lift is equal to 0.10 at time* $t = 0$ *(due to the fact that changes in potential are ignored at this time instance) and jumps to 0.44 at* $t = \Delta t$; *it then increases to join the Wagner solution asymptotically. Furthermore, the drag response predicted by the Basu and Hancock method is overestimated with respect to the Garrick solution. Recall that the Basu and Hancock solution concerns a 12% thick airfoil while the Wagner solution ignores all thickness and models the airfoil as a flat plate. Figure 4.37a plots the lift response of four different airfoils with progressively smaller thickness, from the NACA 0018 to the NACA 0003. Clearly, as the thickness of the airfoil decreases, the lift response predicted by the*

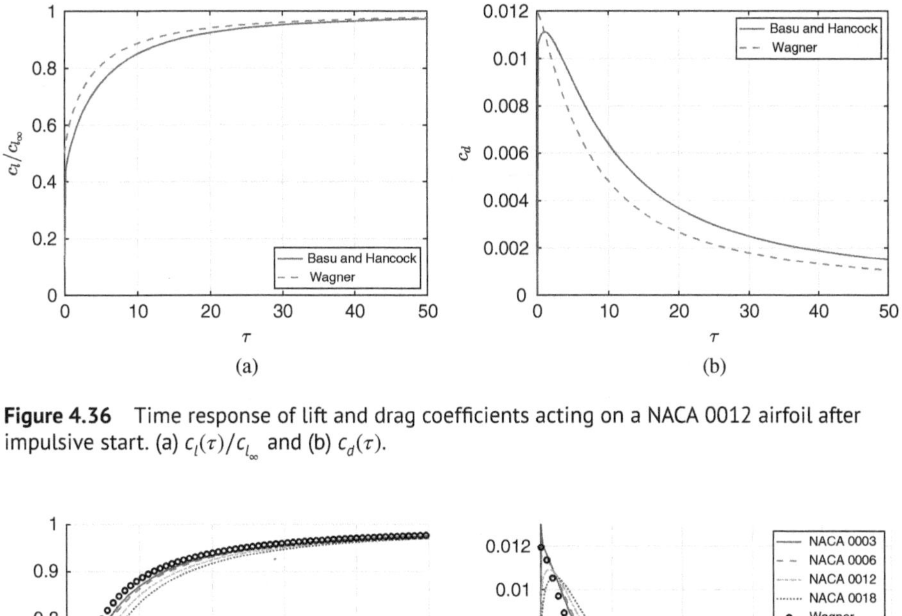

Figure 4.36 Time response of lift and drag coefficients acting on a NACA 0012 airfoil after impulsive start. (a) $c_l(\tau)/c_{l_\infty}$ and (b) $c_d(\tau)$.

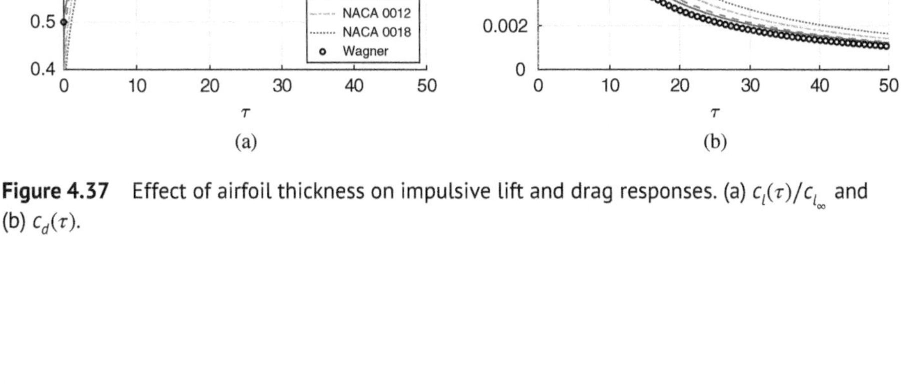

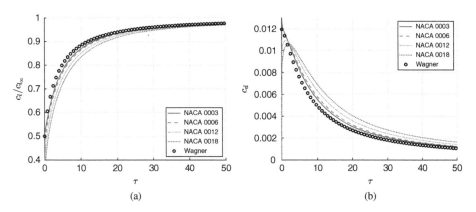

Figure 4.37 Effect of airfoil thickness on impulsive lift and drag responses. (a) $c_l(\tau)/c_{l_\infty}$ and (b) $c_d(\tau)$.

Basu and Hancock technique approaches the Wagner solution. In other words, increasing the thickness delays lift production during the early stages of the motion. This observation demonstrates that modelling the thickness can result in additional physical insights that cannot be obtained from flat plate approaches. Figure 4.37b plots the drag response of the same four airfoils and compares them to the Garrick/Wagner prediction. It can be seen that the Basu and Hancock drag estimates also approach the flat plate exact solution as the thickness decreases. Note that the SVPM converges slower on thin airfoils so all the results in Figure 4.37 were obtained from $n = 400$. This example is solved by Matlab code `impulsive_SVPM.m`.

4.5.2 Thrust and Propulsive Efficiency

We will now apply the Basu and Hancock model to a pitching and plunging wing in order to further study thrust generation. The methodology is similar to the one presented in Section 4.5.1, but some adaptations are necessary. The wing now has the usual plunge, $h(t)$, and pitch, $\alpha(t)$, degrees of freedom, the latter around a pitch axis lying at $(x_f, 0)$. If the coordinates of the ith point on the airfoil at rest are denoted by x_{0_i} and y_{0_i}, then, at time t, the new position of this point is calculated from Eq. (4.74). The control point positions and the unit vectors normal and tangent to the panels are recalculated at each time instance from Eqs. (4.103)–(4.105), so that they are functions of time.

Recall that, for a pitching and plunging airfoil, the relative velocity between the flow and the airfoil is given by Eqs. (4.77) and (4.78), repeated here for convenience

$$u_{m_i}(t) = U_\infty - \dot{\alpha}(t)(y_{c_i}(t) + h(t))$$

$$v_{m_i}(t) = \dot{h}(t) + \dot{\alpha}(t)(x_{c_i}(t) - x_f)$$

This relative velocity contributes to the normal and tangential flow velocities on the airfoil, so that the term $\boldsymbol{b}$ in Eq. (4.154) is now unsteady and given by Eq. (4.80):

$$\mathbf{b}(t_k) = -\boldsymbol{u}_m(t_k) \circ \boldsymbol{n}_x(t_k) - \mathbf{v}_m(t_k) \circ \boldsymbol{n}_y(t_k) \tag{4.173}$$

Furthermore, the pressure coefficient of expression (4.160) must be modified to include the tangential component of the relative motion, so that

$$c_{p_i}(t_k) = 1 - \frac{(u_{m_i}\tau_{x_i} + v_{m_i}\tau_{y_i} + u_{\tau_i})^2}{Q_\infty^2} - \frac{2}{Q_\infty^2}\frac{\phi(x_{c_i}, y_{c_i}, t_k) - \phi(x_{c_i}, y_{c_i}, t_{k-1})}{\Delta t} \tag{4.174}$$

Consequently, the Kutta condition of Eq. (4.161) is modified to

$$(u_{m_1}\tau_{x_1} + v_{m_1}\tau_{y_1} + u_{\tau_1}(t_k))^2 = (u_{m_n}\tau_{x_n} + v_{m_n}\tau_{y_n} + u_{\tau_n}(t_k))^2 + 2\sum_{j=1}^{n} s_j \frac{\gamma(t_k) - \gamma(t_{k-1})}{\Delta t} \tag{4.175}$$

Substituting for $u_{\tau_1}(t_k)$ and $u_{\tau_n}(t_k)$ from expression (4.164) and re-arranging, we obtain the quadratic equation:

$$a_2(t_k)\gamma^2(t_k) + a_1(t_k)\gamma(t_k) + a_0(t_k) = 0 \tag{4.176}$$

where

$$a_2(t_k) = \left(c_{\gamma_n}^2 - c_{\gamma_1}^2\right)$$

$$a_1(t_k) = 2\left(c_{\gamma_n}(c_{0_n} + u_{m_n}\tau_{x_n} + v_{m_n}\tau_{y_n}) - c_{\gamma_1}(c_{0_1} + u_{m_1}\tau_{x_1} + v_{m_1}\tau_{y_1}) + \frac{S}{\Delta t}\right)$$

$$a_0(t_k) = (c_{0_n} + u_{m_n}\tau_{x_n} + v_{m_n}\tau_{y_n})^2 - (c_{0_1} + u_{m_1}\tau_{x_1} + v_{m_1}\tau_{y_1})^2 - \frac{2\Gamma(t_{k-1})}{\Delta t}$$

The rest of the methodology is identical to the one presented in Section 4.5.1, but we will carry out the iterations in a different way in order to improve the numerical performance of the scheme. Recall that we are iterating for l_{wp} and θ_{wp} at each time instance t_k, such that Eqs. (4.149) and (4.150) are satisfied. The iteration scheme we used in the previous section was quite basic; we guessed values for l_{wp} and θ_{wp}, calculated u_{wp} and v_{wp}, obtained new values for l_{wp} and θ_{wp} from Eqs. (4.149) and (4.150) and continued until convergence was achieved. This scheme is simple but inefficient because it is not directed; we have no guarantee that the new values for l_{wp} and θ_{wp} are closer to the correct solution than the original values.

A Newton–Raphson scheme can be defined by looking at Eqs. (4.149) and (4.150) and setting up the function

$$\mathbf{f}(l_{wp}, \theta_{wp}) = \left(l_{wp} - \sqrt{u_{wp}^2 + v_{wp}^2},\ \theta_{wp} - \tan^{-1}\left(\frac{v_{wp}}{u_{wp}}\right)\right)^T = \mathbf{0}$$

Since we are first guessing the values of l_{wp}, θ_{wp} and then using them to evaluate u_{wp} and v_{wp}, at the end of the mth iteration, $\mathbf{f}(l_{wp_m}, \theta_{wp_m}) \neq 0$. We aim to estimate corrections Δl_{wp} and $\Delta\theta_{wp}$ such that

$$\mathbf{f}(l_{wp_m} + \Delta l_{wp}, \theta_{wp_m} + \Delta\theta_{wp}) = \mathbf{0}$$

Expanding this latest expression as a first-order Taylor series around l_{wp_m}, θ_{wp_m}, we obtain

$$\mathbf{f}(l_{wp_m} + \Delta l_{wp}, \theta_{wp_m} + \Delta\theta_{wp}) \approx \mathbf{f}(l_{wp_m}, \theta_{wp_m}) + \left.\frac{\partial \mathbf{f}}{\partial l_{wp}}\right|_m \Delta l_{wp} + \left.\frac{\partial \mathbf{f}}{\partial \theta_{wp}}\right|_m \Delta\theta_{wp} = \mathbf{0}$$

where $\partial\mathbf{f}/\partial l_{wp}\big|_m$ and $\partial\mathbf{f}/\partial\theta_{wp}\big|_m$ are the derivatives of $\mathbf{f}$ evaluated using l_{wp_m}, θ_{wp_m}. Consequently, the corrections Δl_{wp} and $\Delta\theta_{wp}$ can be estimated from

$$\begin{pmatrix} \Delta l_{wp} \\ \Delta\theta_{wp} \end{pmatrix} = -\left(\left.\frac{\partial \mathbf{f}}{\partial l_{wp}}\right|_m \quad \left.\frac{\partial \mathbf{f}}{\partial \theta_{wp}}\right|_m \right)^{-1} \mathbf{f}(l_{wp_m}, \theta_{wp_m}) \tag{4.177}$$

The $m + 1$th estimates of the wake panel length and orientation angle are given by

$$l_{wp_{m+1}} = l_{wp_m} + \Delta l_{wp}, \quad \theta_{wp_{m+1}} = \theta_{wp_m} + \Delta\theta_{wp} \tag{4.178}$$

The only problem is that the derivatives $\left.\frac{\partial \mathbf{f}}{\partial l_{wp}}\right|_m$ and $\left.\frac{\partial \mathbf{f}}{\partial \theta_{wp}}\right|_m$ are difficult to calculate analytically. The solution is to calculate them numerically at each iteration from

$$\left.\frac{\partial \mathbf{f}}{\partial l_{wp}}\right|_m = \frac{\mathbf{f}(l_{wp_m} + \varepsilon, \theta_{wp_m}) - \mathbf{f}\left(l_{wp_m}, \theta_{wp_m}\right)}{\varepsilon}, \quad \left.\frac{\partial \mathbf{f}}{\partial l_{wp}}\right|_m = \frac{\mathbf{f}(l_{wp_m}, \theta_{wp_m} + \varepsilon) - \mathbf{f}\left(l_{wp_m}, \theta_{wp_m}\right)}{\varepsilon}$$

where ε is a small positive number, e.g. $\varepsilon = 10^{-8}$.

The advantage of the Newton–Raphson method is that it is a directed search, which means that the new estimate is obtained from the direction vectors defined by the derivatives. Convergence is faster and smoother, as long as the original guesses for l_{wp} and θ_{wp} do not lie too far from the correct solution. We will demonstrate the application of the technique in the next example.

Example 4.12 *Calculate the thrust and propulsive efficiency of a pitching and plunging NACA 0012 airfoil.*

Anderson et al. (1998) carried out a series of water tunnel experiments on a NACA 0012 wing oscillating in pitch and plunge. The motion was sinusoidal in plunge and pitch around the 1/3 chord point, while the chord of the wing was $c = 0.1$ *m. We will concentrate on a series of tests carried out at a Reynolds number of* $Re = 40{,}000$*, with plunge amplitude* $h_1 = 0.25c$*, pitch amplitude:*

$$\alpha_1 = 15^\circ - \tan^{-1}\left(\frac{2kh_1}{c}\right) \tag{4.179}$$

and reduced frequencies ranging from $k = 0.3$ *to* $k = 2.1$*. The phases of the motions were chosen such that the pitch led the plunge by 90°. The constraint on the pitch amplitude ensures that the amplitude of the effective angle of attack, defined as*

$$\alpha_{\text{eff}}(t) = \alpha(t) + \tan^{-1}\left(\frac{\dot{h}(t)}{U_\infty}\right)$$

is around 15° throughout the reduced frequency range.

Anderson et al. (1998) present measurements of the mean thrust, mean power coefficient and propulsive efficiency for all the tested frequencies. The mean thrust coefficient is defined as

$$C_T = -\frac{1}{T}\int_0^T c_d(t)\mathrm{d}t \tag{4.180}$$

so that it is the mean of $-c_d(t)$ *over a complete cycle with period T. The mean power coefficient required to keep the motion going is given by*

$$C_P = -\frac{1}{TU_\infty}\int_0^T \left(-c_l(t)\dot{h}(t) + c_m(t)\dot{\alpha}(t)c\right)\mathrm{d}t \tag{4.181}$$

assuming that the lift is positive upwards and h positive downwards. Finally, the propulsive efficiency is defined as

$$\eta = \frac{C_T}{C_P}$$

and expresses the ratio of the propulsive power generated by the thrust to the power required to keep the pitching and plunging motion going.

The pitching moment around the pitch axis is calculated from the unsteady version of Eq. (4.142)

$$c_m(t_k) = \frac{1}{c^2}\sum_{i=1}^{n} c_{p_i}(t_k)s_i\left(n_{y_i}(t_k)\left(x_{c_i}(t_k) - x_f\right) - n_{x_i}(t_k)\left(y_{c_i}(t_k) + h(t_k)\right)\right) \tag{4.182}$$

since the pitch axis lies at $x_f = c/3$, $y_f = -h(t_k)$ *as the airfoil is plunging. The lift and drag coefficients* $c_l(t)$ *and* $c_m(t)$ *are obtained from Eqs. (4.170) and (4.171) but with time-varying values of the normal vector components, that is*

$$c_l(t_k) = -\frac{1}{c}\sum_{i=1}^{n} c_{p_i}(t_k)s_i n_{y_i}(t_k) \tag{4.183}$$

$$c_d(t_k) = -\frac{1}{c}\sum_{i=1}^{n} c_{p_i}(t_k)s_i n_{x_i}(t_k) \tag{4.184}$$

Using standard values for the density and viscosity of water, the Reynolds number gives the flow speed as $Q_\infty = 0.4$ *m/s. The plunge and pitch displacements are*

$$h(t) = h_1 \sin(\omega t + \phi_h), \quad \alpha(t) = \alpha_1 \sin(\omega t + \phi_\alpha) \tag{4.185}$$

with the pitch phase $\phi_\alpha = \phi_h + 90°$. *The plunge phase* ϕ_h *is set to zero without loss of generality.*

The reduced frequency is a useful non-dimensional measure of the frequency, but it does not include any information on the amplitude of the motion. For pitching and plunging airfoils, the Strouhal number is defined as

$$Str = \frac{A_{te}\omega}{2\pi Q_\infty} \tag{4.186}$$

where A_{te} *is the peak-to-peak amplitude of oscillation of the trailing edge of the airfoil. For a given airfoil, depending on the amplitude, the Strouhal number can be high even if the frequency is low. The amplitude* A_{te} *is calculated from*

$$A_{te} = \max(y_1(t)) - \min(y_1(t)) = \max(y_{n+1}(t)) - \min(y_{n+1}(t))$$

for t ranging between 0 and T, recalling that points 1 and $n+1$ *are the two trailing edge indices and that* $y_i(t)$ *are panel vertices, so that* $y_1(t)$ *and* $y_{n+1}(t)$ *denote the instantaneous heights of the lower and upper trailing edge points, respectively.*

We simulate the pitching and plunging wing with a number of panels $n = 200$ *and a time step* $\Delta t = 2\pi/40\omega$, *so that the number of time instances per cycle is* $p_{pc} = 40$. *A total of eight cycles are simulated for each value of the reduced frequency. Note that, unlike the frequency domain approach of Example 4.8, we do not have to ensure that the wake includes an integer number of wavelengths. Eleven simulations are carried out for k values between 0.2 and 2.2.*

Figure 4.38 plots the mean thrust coefficient, mean power coefficient and propulsive efficiency, as calculated from Wagner (linear) theory, the lumped vortex method of Section 4.2 and the Basu and Hancock approach of the present section. The numerical results are compared to the experimental data obtained by Anderson et al. (1998). The Basu and Hancock predictions for the thrust coefficient (Figure 4.38a) lie very close to the experimental measurements in the Srouhal range $0.15 < Str < 0.5$; *outside this range, all the models overestimate the thrust, although the Basu and Hancock approach is the most accurate. According to Young and Lai (2007), at low Strouhal numbers viscous drag forces such as friction drag and pressure drag due to separation at the trailing edge constitute a significant component of the total force in the x direction and, hence, decrease the thrust and propulsive efficiency. At intermediate*

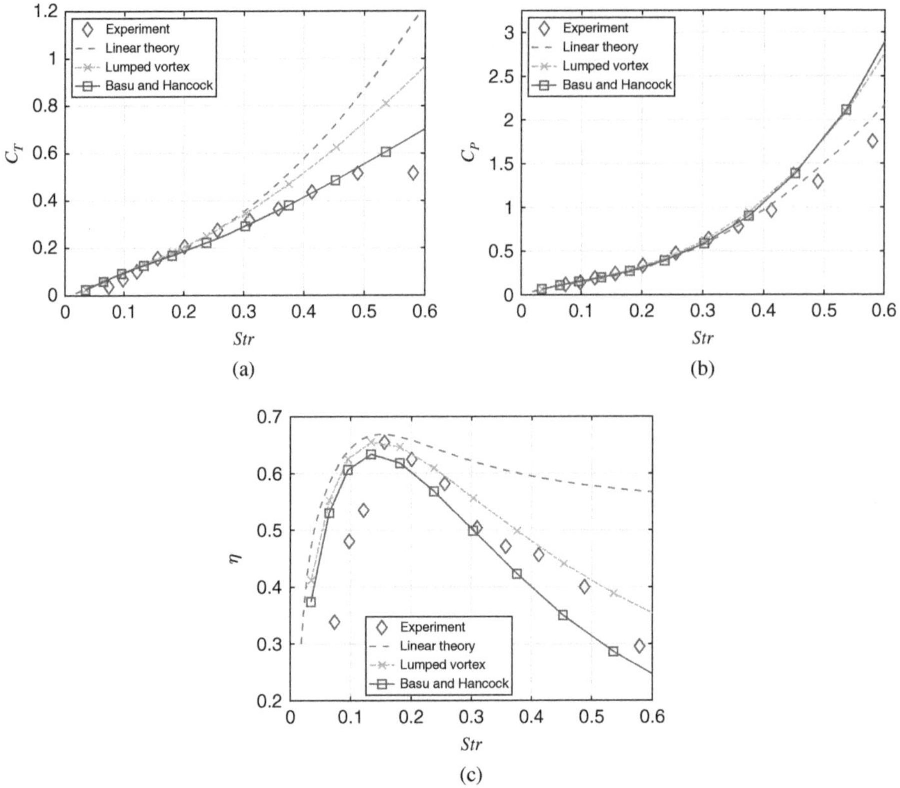

Figure 4.38 Mean thrust coefficient, mean power coefficient and propulsive efficiency as functions of Strouhal number. Experimental data from Anderson et al. (1998). (a) Mean thrust coefficient, (b) mean power coefficient and (c) propulsive efficiency. Source: Data from Anderson et al. (1998).

Strouhal numbers, skin friction becomes negligible compared to the inviscid propulsive force, so that the agreement between the Basu and Hancock thrust predictions and the experimental measurements improves. At high Str values, massive flow separation occurs over the airfoil and the mean thrust is reduced significantly.

The power coefficient (see Figure 4.38b) is predicted accurately by all the methods for Str < 0.4. *Again, outside this range all the predictions are overestimated although, in this case, the linear theory is the most representative of the experimental data. As mentioned earlier, such high values of the Strouhal number result in flows that feature massive unsteady flow separation, a phenomenon known as dynamic stall. It is not clear exactly why the linear theory predictions for the power show better agreement with the experimental results than the numerical estimates. It may be just a result of the particular kinematics used for the present experimental test case. Different test cases reported in Anderson et al. (1998) show that non-linear predictions lie closer to the experimental data than linear estimates at high Str.*

The propulsive efficiency is the ratio of the thrust and power coefficients so that the results of Figure 4.38c reflect those of Figure 4.38a,b. The most accurate predictions for η *are obtained by the Basu and Hancock method in the range* $0.15 < Str < 0.4$. *For* $Str < 0.15$, *the thrust is overestimated and so is the propulsive efficiency. For Str* ≥ 0.4, *the power is overestimated and, consequently, the propulsive efficiency is underestimated.*

Figure 4.38c shows that, for the present kinematics, there is an optimal Strouhal number of around Str $= 0.15$ *that leads to the best propulsive efficiency of* $\eta = 0.65$. *Other kinematics do not always lead to an efficiency peak. Furthermore, Young and Lai (2007) argue that the efficiency peak is caused by the pitch angle constraint of Eq. (4.179) that was imposed during the course of the experiments by Anderson et al. (1998). A different constraint, or the complete absence of a constraint could have suppressed the efficiency peak. This example is solved by Matlab code* `pitchplunge_SVPM.m`.

4.6 Theodorsen's Function and Wake Shape

Theodorsen's function was developed using the assumption that the wake is flat. Numerical modelling with a free wake is a tool that can be used in order to investigate the effect of this assumption on the aerodynamic loads. In Example 3.13, we calculated the amplitude of Theodorsen's function from the ratio of the circulatory unsteady lift to the circulatory quasi-steady lift at different reduced frequencies. We will do the same here using lift responses obtained from Basu and Hancock's method. It is not easy to obtain the circulatory lift directly from this numerical method, but we can estimate it from the difference between the total lift and the non-circulatory lift.

Theodorsen theory and Basu and Hancock's approach share the same basic ingredients: the non-circulatory lift is calculated by imposing impermeability using a source distribution while the circulatory lift is calculated by imposing the Kutta condition using a vortex distribution. This means that the non-circulatory lift can be readily calculated by ignoring all the vortices, both on the surface and in the wake. The impermeability boundary condition of Eq. (4.154) simplifies to

$$\boldsymbol{A}_{n_\sigma}\boldsymbol{\sigma}(t_k) = \boldsymbol{b}(t_k)$$

where $\boldsymbol{b}(t_k)$ is given by Eq. (4.173), $\boldsymbol{u}_m(t_k)$ and $\boldsymbol{v}_m(t_k)$ being calculated from Eqs. (4.77) and (4.78), respectively. Solving for $\boldsymbol{\sigma}(t_k)$ we can then calculate the tangential velocities on the surface from

$$\boldsymbol{u}_\tau(t_k) = \boldsymbol{A}_{\tau_\sigma}(t_k)\boldsymbol{\sigma}(t_k) + \boldsymbol{u}_m(t_k) \circ \boldsymbol{\tau}_x(t_k) + \boldsymbol{v}_m(t_k) \circ \boldsymbol{\tau}_y(t_k)$$

We calculate the perturbation potential from Eqs. (4.158) and (4.159) and the pressure coefficient distribution on the surface from Eq. (4.174). Finally, the non-circulatory lift coefficient, $l_\sigma(t_k)$, at each time instance is obtained from integrating the pressure distribution, as shown in Eq. (4.166). Since we have ignored the wake and we do not impose the Kutta condition, the solution procedure is linear. The circulatory lift is estimated from the complete unsteady Basu and Hancock lift value after subtracting the non-circulatory lift, that is

$$l_\Gamma(t_k) = l(t_k) - l_\sigma(t_k)$$

We repeat Example 4.12 for $k = 1.5$ and plot the time responses of the circulatory and non-circulatory lift over a complete cycle, comparing them to Theodorsen's predictions. The non-circulatory lift coefficients of Figure 4.39a are in very good agreement, but the

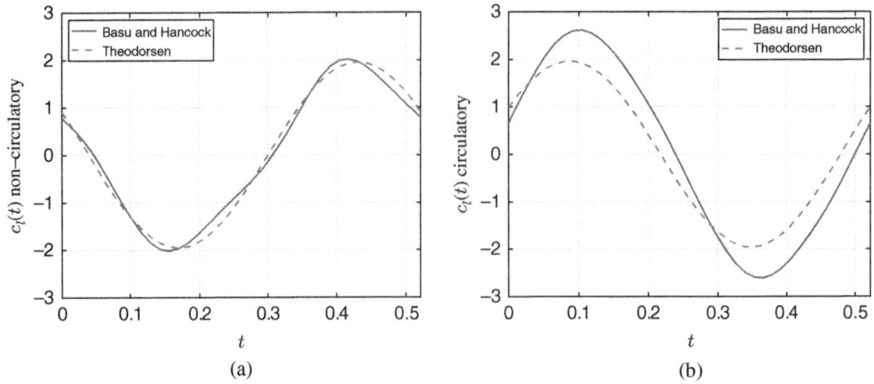

Figure 4.39 Circulatory and non-circulatory lift acting on a pitching and plunging airfoil oscillating at $k = 1.5$. (a) Non-circulatory lift and (b) circulatory lift.

circulatory lift values of Figure 4.39b are not; the amplitude of the Basu and Hancock circulatory lift is significantly higher than the value predicted by Theodorsen theory.

In order to examine the reason for the difference in circulatory lift between the source and vortex panel results and Theodorsen theory's predictions, we carry out a number of simulations with reduced frequency values in the range $k = 0.1$ to $k = 2.2$ and plot the ratio of the resulting lift amplitude to the amplitude of the quasi-steady circulatory lift, given by

$$l_{\Gamma,qs}(t) = -2\rho U_\infty b\pi \left(U_\infty \alpha(t) + \dot{h}(t) + b\left(\frac{1}{2} - a\right)\dot{\alpha}(t)\right)$$

Essentially, we are reconstructing Figure 3.27 using the Basu and Hancock approach. We select three different plunge amplitudes, $h_0 = 0.25c$, $0.125c$ and $0.0625c$, and the pitch amplitude is given by expression (4.179). Figure 4.40 plots the results, along with the magnitude of Theodorsen's function $|C(k)|$. It can be seen that the Basu and Hancock result for the case with the lowest oscillation amplitude is a good approximation of Theodorsen's

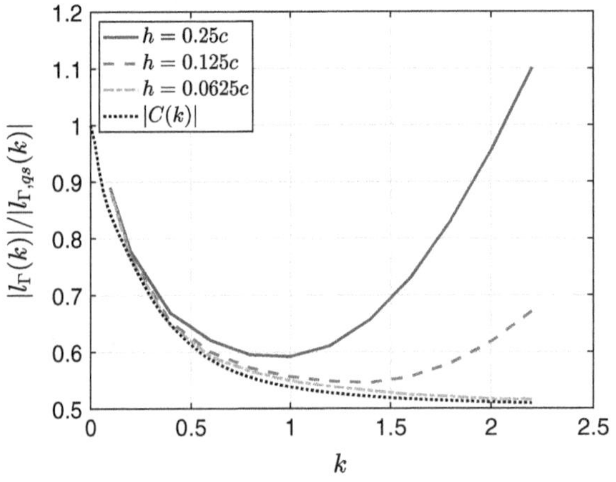

Figure 4.40 Ratio of unsteady to quasi-steady circulatory lift amplitude for sinusoidally pitching and plunging airfoil.

function. However, as the oscillation amplitude increases, the numerical method predicts higher and higher circulatory lift amplitude ratios for $k > 0.5$.

Figure 4.40 plots a generalised Theodorsen's function, whose magnitude depends not only on the reduced frequency but also on the oscillation amplitude. It shows that the classic Theodorsen function is only accurate for low oscillation amplitudes since it significantly underestimates the magnitude of the circulatory lift for higher values of the oscillation amplitude. Mathematically, there are two differences between the three cases plotted in Figure 4.40:

- The wake shape whose vertical width increases significantly with oscillation amplitude and therefore cannot be approximated as flat for all combinations of reduced frequency and oscillation amplitude.
- The non-linearity induced by the rotation of the airfoil at each time instance. For the amplitudes considered here, the effect of this non-linearity is small.

The reader is invited to run these simulations and observe the differences in wake shape using Matlab code `pitchplunge_SVPM.m`. The general conclusion from Figure 4.40 is that a generalised Theodorsen function does not asymptote towards 0.5 as k tends to infinity; its magnitude decreases to a minimum and then increases again, with the potential to exceed unity. Nevertheless, it should be kept in mind that high reduced frequencies and/or oscillation amplitudes can lead to flow separation, which is not modelled by the Basu and Hancock approach.

4.7 Steady and Unsteady Kutta Conditions

Up to this point, we have been enforcing the Kutta condition in different ways but without dwelling on the nature of this condition or, even, on its applicability to unsteady flows. The Kutta condition is a means of enforcing the physically observed phenomenon whereby, for steady attached flows, the boundary layers growing on the two sides of an airfoil separate at the trailing edge and merge behind it. Inviscid calculations cannot represent this phenomenon and, therefore, approximate it in one or more of the following ways:

1. The flow speed at the trailing edge is finite
2. The vorticity is zero at the trailing edge
3. The pressure on the upper side of the trailing edge is equal to that on the lower side. Alternatively, the pressure jump across the trailing edge is equal to zero.
4. The flow separates from the trailing edge in a direction parallel to the bisector of the trailing edge angle

Condition 1 was used in Section 3.3.5 for Wagner/Theodorsen theory, while condition 2 was applied in Section 3.6.4 for finite state theory. The lumped vortex method of Section 4.2 makes use of an implicit Kutta condition, whereby the placement of the bound vortices and collocation points is selected such that the resulting lift takes the value predicted by thin airfoil theory. Finally, the SVPM of Section 4.5 uses condition 3.

In the present section, we will explore the steady and unsteady Kutta condition in more detail by means of another panel technique, the linearly varying vortex panel method

(VPM). With this approach, we can enforce the Kutta condition using either condition 2 or 3. A steady version of the VPM is detailed in Kuethe and Chow (1986) and an unsteady version was developed by Kim and Mook (1986). Here, we will use a different unsteady version in order to demonstrate in more detail the unsteady form of the Kutta condition.

The basis of the method is identical to the one presented in Figure 4.25: the geometry is discretised into n straight panels, with control points lying on the midpoint of each panel and unit normal and tangential vectors defined on the control points. However, only vortex panels are used and their strength varies linearly along each panel, as shown in Figure 4.41. The vortex strength is defined on the panel vertices, so that on panel 1 the strength varies linearly from γ_1 to γ_2, on panel 2 it varies from γ_2 to γ_3, etc. until panel n where the strength varies from γ_n to γ_{n+1}. This means that, if the trailing edge is closed, there are two values of the vortex strength there, γ_1 and γ_{n+1}.

Consider a vortex panel with linearly varying strength that lies on the x axis whose vertices have coordinates $(x_1, 0)$ and $(x_2, 0)$ and strengths γ_1 and γ_2. The flow velocities induced by this panel on any point in the flow (x, y) are derived in Appendix A.4 and given by

$$u(x,y) = u_1(x,y)\gamma_1 + u_2(x,y)\gamma_2$$
$$v(x,y) = v_1(x,y)\gamma_1 + v_2(x,y)\gamma_2$$

where $u_1(x,y)$, $u_2(x,y)$, $v_1(x,y)$ and $v_2(x,y)$ are calculated from Eq. (A.39). For a general panel that is not parallel to the x axis, we can calculate the induced velocities using the approach of Section 4.5. Considering the jth panel with length is s_j, control point (x_{c_j}, y_{c_j}), normal unit vector $\boldsymbol{n}_j$, tangent unit vector $\boldsymbol{\tau}_j$ and vortex strengths at its vertices γ_j and γ_{j+1}, the velocities it induces on a general point (x, y), are

$$u_\tau(x,y) = u_{\tau_1}(x,y)\gamma_j + u_{\tau_2}(x,y)\gamma_{j+1} \tag{4.187}$$

$$v_n(x,y) = v_{n_1}(x,y)\gamma_j + v_{n_2}(x,y)\gamma_{j+1} \tag{4.188}$$

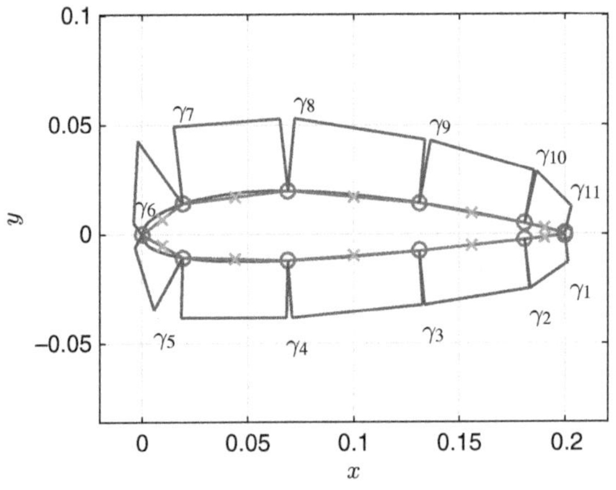

Figure 4.41 Discretisation of a 2D airfoil by vortex panels with linearly varying strength.

where $u_\tau(x,y)$ and $v_n(x,y)$ are parallel to $\boldsymbol{\tau}_j$ and $\boldsymbol{n}_j$, respectively. The velocity components $u_{\tau_1}(x,y)$ to $v_{n_2}(x,y)$ are given by

$$u_{\tau_1}(x,y) = \frac{1}{2\pi(-s_j)}\left\{\left(\tan^{-1}\left(\frac{x_\tau + s_j/2}{y_n}\right) - \tan^{-1}\left(\frac{x_\tau - s_j/2}{y_n}\right)\right)(x_\tau - s_j/2) + \frac{y_n}{2}\ln\left(\frac{(x_\tau - s_j/2)^2 + y_n^2}{(x_\tau + s_j/2)^2 + y_n^2}\right)\right\} \tag{4.189}$$

$$u_{\tau_2}(x,y) = -\frac{1}{2\pi(-s_j)}\left\{\left(\tan^{-1}\left(\frac{x_\tau + s_j/2}{y_n}\right) - \tan^{-1}\left(\frac{x_\tau - s_j/2}{y_n}\right)\right)(x_\tau + s_j/2) + \frac{y_n}{2}\ln\left(\frac{(x_\tau - s_j/2)^2 + y_n^2}{(x_\tau + s_j/2)^2 + y_n^2}\right)\right\} \tag{4.190}$$

$$v_{n_1}(x,y) = \frac{1}{2\pi(-s_j)}\left\{s_j - y_n\left(\tan^{-1}\left(\frac{x_\tau + s_j/2}{y_n}\right) - \tan^{-1}\left(\frac{x_\tau - s_j/2}{y_n}\right)\right) + \frac{(x_\tau - s_j/2)}{2}\ln\left(\frac{(x_\tau - s_j/2)^2 + y_n^2}{(x_\tau + s_j/2)^2 + y_n^2}\right)\right\} \tag{4.191}$$

$$v_{n_2}(x,y) = -\frac{1}{2\pi(-s_j)}\left\{s_j - y_n\left(\tan^{-1}\left(\frac{x_\tau + s_j/2}{y_n}\right) - \tan^{-1}\left(\frac{x_\tau - s_j/2}{y_n}\right)\right) + \frac{(x_\tau + s_j/2)}{2}\ln\left(\frac{(x_\tau - s_j/2)^2 + y_n^2}{(x_\tau + s_j/2)^2 + y_n^2}\right)\right\} \tag{4.192}$$

where $x_\tau = \mathbf{r}\cdot\boldsymbol{\tau}_j$, $y_n = \mathbf{r}\cdot\mathbf{n}_j$, $\mathbf{r} = ((x - x_{c_j}), (y - y_{c_j}))$,

The horizontal and vertical velocities induced by the jth panel on the control point of the ith panel are obtained by projecting the tangential and normal velocities of Eqs. (4.187) and (4.188) onto the x and y axes, so that

$$u(x_{c_i}, y_{c_i}) = A_{u_{1_{i,j}}}\gamma_j + A_{u_{2_{i,j}}}\gamma_{j+1} \tag{4.193}$$

$$v(x_{c_i}, y_{c_i}) = A_{v_{1_{i,j}}}\gamma_j + A_{v_{2_{i,j}}}\gamma_{j+1} \tag{4.194}$$

where

$$\begin{aligned}
A_{u_{1_{i,j}}} &= u_{\tau_1}(x_{c_i}, y_{c_i})\tau_{x_j} + v_{n_1}(x_{c_i}, y_{c_i})n_{x_j}\\
A_{u_{2_{i,j}}} &= u_{\tau_2}(x_{c_i}, y_{c_i})\tau_{x_j} + v_{n_2}(x_{c_i}, y_{c_i})n_{x_j}\\
A_{v_{1_{i,j}}} &= u_{\tau_1}(x_{c_i}, y_{c_i})\tau_{y_j} + v_{n_1}(x_{c_i}, y_{c_i})n_{y_j}\\
A_{v_{2_{i,j}}} &= u_{\tau_2}(x_{c_i}, y_{c_i})\tau_{y_j} + v_{n_2}(x_{c_i}, y_{c_i})n_{y_j}
\end{aligned} \tag{4.195}$$

and $u_{\tau_1}(x_{c_i}, y_{c_i})$ to $v_{n_2}(x_{c_i}, y_{c_i})$ are obtained from Eqs. (4.189)–(4.192) after substituting $x = x_{c_i}$ and $y = y_{c_i}$. As in the case of the SVPM, the influence of a panel on its own control point must be calculated as a special case. Using the discussion in Sections A.3 and A.4 it can be shown that, for $i = j$,

$$\ln\left(\frac{(x_\tau - s_j/2)^2 + y_n^2}{(x_\tau + s_j/2)^2 + y_n^2}\right) = 0$$

$$\tan^{-1}\left(\frac{x_\tau + s_j/2}{y_n}\right) - \tan^{-1}\left(\frac{x_\tau - s_j/2}{y_n}\right) = 0$$

Equations (4.193) and (4.194) can be written in matrix form as

$$\boldsymbol{u} = \boldsymbol{A}_u\boldsymbol{\gamma}, \quad \boldsymbol{v} = \boldsymbol{A}_v\boldsymbol{\gamma}$$

where $\boldsymbol{u}$, $\boldsymbol{v}$ are the $n \times 1$ vectors with elements $u(x_{c_i}, y_{c_i})$, $v(x_{c_i}, y_{c_i})$, respectively, $\boldsymbol{\gamma}$ is the $(n+1) \times 1$ vector with elements γ_j and

$$\boldsymbol{A}_u = \left(\boldsymbol{A}_{u_1} \ \boldsymbol{0}_{n\times 1}\right) + \left(\boldsymbol{0}_{n\times 1} \ \boldsymbol{A}_{u_2}\right)$$
$$\boldsymbol{A}_v = \left(\boldsymbol{A}_{v_1} \ \boldsymbol{0}_{n\times 1}\right) + \left(\boldsymbol{0}_{n\times 1} \ \boldsymbol{A}_{v_2}\right)$$

so that $\boldsymbol{A}_u$ and $\boldsymbol{A}_v$ have dimensions $n \times (n+1)$ and $\boldsymbol{0}_{n,1}$ denotes a $n \times 1$ vector full of zeros.

For steady flow in a free stream $\mathbf{Q}_\infty$ with angle α, the impermeability boundary condition is written as

$$\boldsymbol{A}_n\boldsymbol{\gamma} = \boldsymbol{b} \tag{4.196}$$

where $\boldsymbol{b} = -Q_\infty \cos\alpha \mathbf{n}_x - Q_\infty \sin\alpha \mathbf{n}_y$ and the elements of the $n \times (n+1)$ matrix $\boldsymbol{A}_n$ are given by

$$A_{n_{i,j}} = A_{u_{i,j}} n_{x_i} + A_{v_{i,j}} n_{y_i} \tag{4.197}$$

for $i = 1, \dots, n, j = 1, \dots, n+1$. Equation (4.196) constitutes a set of n equations with $n+1$ unknowns, the vortex strengths γ_j for $j = 1, \dots, n+1$. The equations are completed by the Kutta condition for the steady case only. Using the condition that the pressure must be equal at the control points of the upper and lower sections, $c_{p_n} = c_{p_1}$, we obtain Eq. (4.132)

$$\mathbf{Q}_\infty \cdot \boldsymbol{\tau}_n + u_{\tau_n} = -(\mathbf{Q}_\infty \cdot \boldsymbol{\tau}_1 + u_{\tau_1}) \tag{4.198}$$

where u_{τ_i} is the tangential velocity on the ith panel. The tangential perturbation velocities on all the panels are given by

$$\boldsymbol{u}_\tau = \boldsymbol{A}_\tau\boldsymbol{\gamma} \tag{4.199}$$

where the elements of the $n \times (n+1)$ matrix $\boldsymbol{A}_\tau$ are given by

$$A_{\tau_{i,j}} = A_{u_{i,j}} \tau_{x_i} + A_{v_{i,j}} \tau_{y_i} \tag{4.200}$$

Substituting into Eq. (4.198) we obtain

$$\left(\boldsymbol{a}_{\tau_1} + \boldsymbol{a}_{\tau_n}\right)\boldsymbol{\gamma} = -\mathbf{Q}_\infty \cdot \boldsymbol{\tau}_1 - \mathbf{Q}_\infty \cdot \boldsymbol{\tau}_n \tag{4.201}$$

where $\boldsymbol{a}_{\tau_1}$ and $\boldsymbol{a}_{\tau_n}$ are the first and last rows of matrix $\boldsymbol{A}_\tau$, respectively.

Alternatively, we can apply the condition that the vorticity at the trailing edge must be equal to zero. Then, the Kutta condition becomes simply

$$\gamma_1 + \gamma_{n+1} = 0 \tag{4.202}$$

We will demonstrate the usage of the VPM with the Kutta conditions of Eqs. (4.201) and (4.202) in the following example.

Example 4.13 *Repeat Example 4.9 using the VPM with linearly varying strength*

The determination and discretisation of the geometry of the airfoil is carried out as in Example 4.9. We calculate the panel vertices, panel lengths, control points and normal and tangential unit vectors as usual. Then we evaluate the horizontal and vertical influence coefficients of Eq. (4.195), the normal influence coefficient matrix $\mathbf{A}_n$ *from expression (4.197) and the tangential normal influence coefficient matrix* $\mathbf{A}_\tau$ *from expression (4.200).*

Starting with the equal pressure at the trailing edge Kutta condition, we assemble Eqs. (4.196) and (4.201) to obtain

$$\begin{pmatrix} \mathbf{A}_n \\ \boldsymbol{a}_{\tau_1} + \boldsymbol{a}_{\tau_n} \end{pmatrix} \gamma = \begin{pmatrix} \boldsymbol{b} \\ -\mathbf{Q}_\infty \cdot \boldsymbol{\tau}_1 - \mathbf{Q}_\infty \cdot \boldsymbol{\tau}_n \end{pmatrix} \tag{4.203}$$

If, on the other hand, we make use of the zero vorticity Kutta condition of Eq. (4.202), we obtain the alternative system

$$\begin{pmatrix} \mathbf{A}_n \\ 10\ldots01 \end{pmatrix} \gamma = \begin{pmatrix} \boldsymbol{b} \\ 0 \end{pmatrix} \tag{4.204}$$

Equations (4.203) and (4.204) are both sets of $n+1$ *equations with* $n+1$ *unknowns, the elements of* $\boldsymbol{\gamma}$*. Once the latter have been evaluated, we can calculate the tangential perturbation velocities on the panels from Eq. (4.199), the pressure distribution at the control points from Eq. (4.131) and the lift and drag coefficients from Eq. (4.138)–(4.140). Alternatively, we can evaluate the lift from the Kutta–Joukowski theorem,*

$$l = \rho Q_\infty \Gamma$$

where the total circulation around the airfoil, Γ*, is calculated using trapezoidal integration of the vorticity distribution at the panel vertices, such that*

$$\Gamma = \frac{1}{2}\sum_{i=1}^{n} \left(\gamma_i + \gamma_{i+1}\right) s_i \tag{4.205}$$

Figure 4.42 plots the pressure distributions for $\alpha = 4.3^\circ$ *and* $n = 400$ *obtained from the two systems of Eqs. (4.203) and (4.204) and compares them to the experimental measurements presented in Seetharam et al. (1977). Clearly, the two simulated pressure distributions are nearly identical. However, Table 4.1 shows that the equal trailing edge pressure Kutta condition results in non-zero vorticity at the trailing edge, while the zero vorticity Kutta condition results in non-zero pressure difference at the collocation points of the trailing edge panels. These differences do not disappear as the number of panels is increased. Nevertheless, the lift and drag coefficients are nearly identical. The table also gives the values of* $\Delta\theta_{\mathrm{TE}}$*, the difference between the flow angle at the trailing edge and the bisector of the trailing edge angle. With both Kutta conditions* $\Delta\theta_{\mathrm{TE}}$ *is small. Figure 4.43 plots the flow around the trailing edge for the equal pressure Kutta condition, showing that it is parallel to the trailing edge angle bisector. Note that the flow velocities are calculated at a point* $c/5000$ *behind the trailing edge in order to avoid singularities.*

Poling and Telionis (1986) state that satisfying one version of the Kutta condition should automatically satisfy all the others but, as this example has shown, this is not the case in the

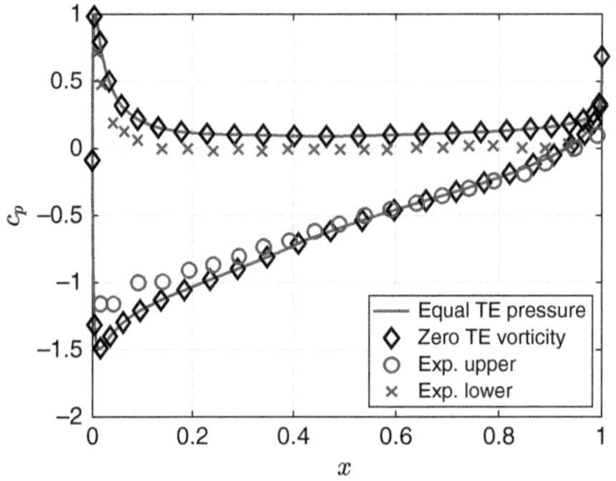

Figure 4.42 Pressure distribution around a NACA 2412 airfoil at $\alpha = 4.3°$ by the vortex panel method using two different Kutta conditions. Experimental data in Seetharam et al. (1977). Source: Seetharam et al. (1977), credit: NASA.

Table 4.1 Aerodynamic results for NACA 2412 airfoil at $\alpha = 5°$ by the vortex panel method using two different Kutta conditions.

Kutta condition	$c_{p_1} - c_{p_n}$	$c(\gamma_1 + \gamma_n)/\Gamma$	c_l	c_d	$\Delta\theta_{\mathrm{TE}}$
Equal TE pressure	0.00	0.03	0.77	0.001	−0.097°
Zero TE vorticity	−0.01	0	0.77	0.001	−0.004°

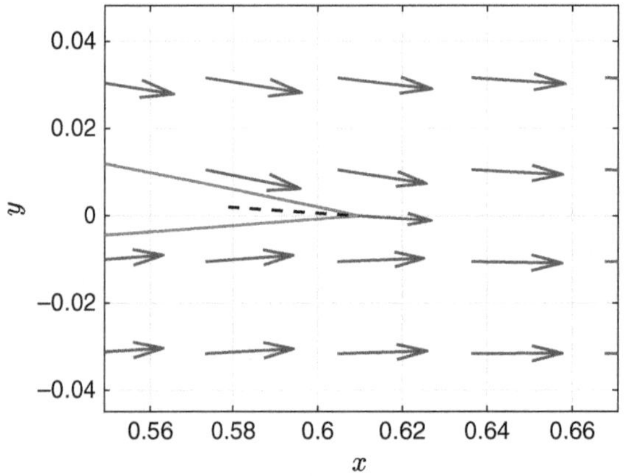

Figure 4.43 Flow around the trailing edge of a NACA 2412 airfoil at $\alpha = 5°$ by the vortex panel method using the equal trailing edge pressure Kutta condition

context of a numerical panel solution; satisfying one form of the Kutta condition does not satisfy the others exactly.

Finally, it should be mentioned that the linearly varying vortex panel approach displays better convergence behaviour than the constant SVPM. Figure 4.44 plots the lift and pitching moment coefficients for the same test case whose convergence was presented in Figure 4.30. The VPM converges as quickly as the SVPM for the lift but significantly quicker for the pitching moment. This statement is true for both formulations of the Kutta condition. This example is solved by Matlab code `steady_VPM.m`.

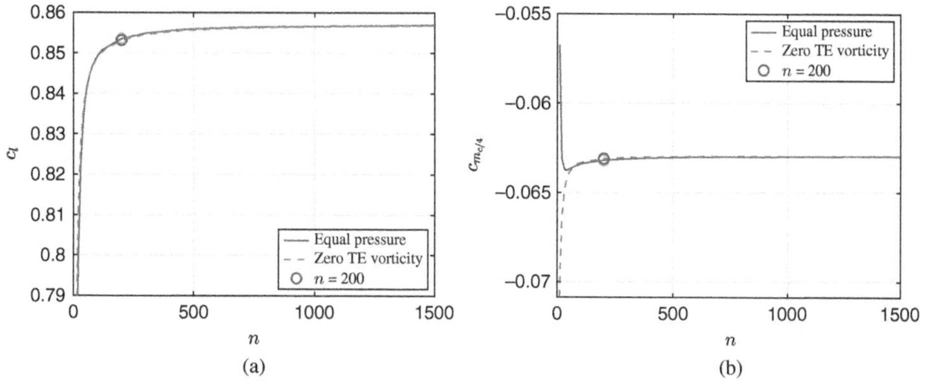

Figure 4.44 Convergence of VPM predictions for steady lift and pitching moment with increasing number of panels. (a) c_l vs n and (b) $c_{m_{c/4}}$ vs n.

4.7.1 The Unsteady Kutta Condition

Poling and Telionis (1986) presented experimental data demonstrating that, under unsteady conditions, the flow at the trailing edge is no longer parallel to the trailing edge bisector. They used these results in order to provide physical evidence for the validity of an alternative, unsteady Kutta condition, which they attribute to Giesing and Maskell. This condition states that the trailing edge flow velocity during unsteady motion is parallel to either the upper or lower trailing edge surface, depending on the sign of the instantaneous shed vorticity. As noted by Poling and Telionis (1986), this unsteady Kutta condition has the weakness of not converging asymptotically to the steady-state case as the motion decays. Even for a negligible amount of unsteadiness, the flow at the trailing edge will be aligned with either the upper or lower trailing edge surfaces and not with the trailing edge bisector.

An alternative to the Giesing and Maskell unsteady Kutta condition is the condition already implemented in the Basu and Hancock method (Basu and Hancock 1978). In this case, the trailing edge flow direction is not prescribed; the orientation of the wake panel is calculated such that the latter is aligned with the trailing edge streamline. It is possible to apply this procedure using both the zero trailing edge vorticity and the equal trailing edge pressure versions of the Kutta condition. In this section, we will use the unsteady VPM in order to demonstrate the similarities and differences between these two Kutta condition formulations.

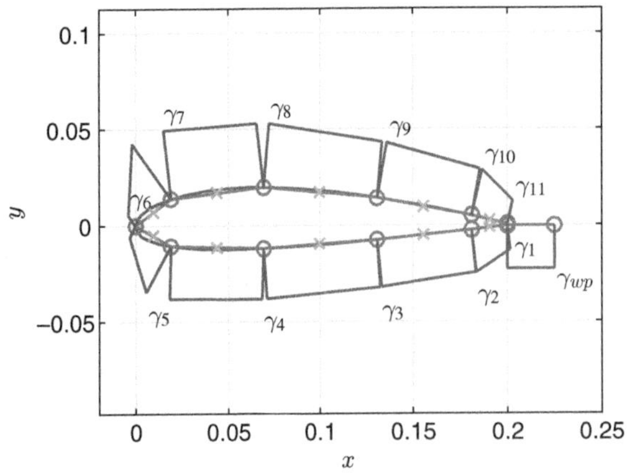

Figure 4.45 Discretisation of a 2D airfoil by vortex panels with linearly varying strength and a wake panel.

The unsteady version of the linear VPM presented here is based on Basu and Hancock's addition of a single wake vortex panel, whose total strength is shed into the wake in the form of a point vortex at subsequent time instances. Figure 4.45 shows the discretisation of the steady approach seen in Figure 4.41 but with an additional vortex panel with constant strength, γ_{wp}, placed at the trailing edge and pointing downstream. The trailing edge has zero thickness so that nodes 1 and $n+1$ (node 11 in the case of Figure 4.45) coincide.

At the kth time instance, the vortex panel has length $l_{wp}(t_k)$ and orientation $\theta_{wp}(t_k)$ so that its vertices are given by

$$x_{wp_{1,2}}(t_k) = x_1(t_k) + [0 \; l_{wp}(t_k)\cos\theta_{wp}(t_k)]$$

$$y_{wp_{1,2}}(t_k) = y_1(t_k) + [0 \; l_{wp}(t_k)\sin\theta_{wp}(t_k)]$$

The conditions for the length and angle of the wake panel are still given by Eqs. (4.149) and (4.150), respectively. The horizontal and vertical flow velocities on the control point of the wake panel are given by

$$u_{wp}(t_k) = \boldsymbol{A}_{u_{\gamma,wp}}(t_k)\boldsymbol{\gamma}(t_k) + \boldsymbol{B}_{u_{wp}}(t_k)\boldsymbol{\Gamma}_w(t_k) + U_\infty \tag{4.206}$$

$$v_{wp}(t_k) = \boldsymbol{A}_{v_{\gamma,wp}}(t_k)\boldsymbol{\gamma}(t_k) + \boldsymbol{B}_{v_{wp}}(t_k)\boldsymbol{\Gamma}_w(t_k) \tag{4.207}$$

where $\boldsymbol{A}_{u_{wp}}(t_k)$, $\boldsymbol{A}_{v_{wp}}(t_k)$ are the $1 \times (n+1)$ influence coefficient vectors of the airfoil panels on the control point of the wake panel and $\boldsymbol{B}_{u_{wp}}(t_k)$, $\boldsymbol{B}_{v_{wp}}(t_k)$ are the $1 \times (k-1)$ influence coefficient vectors of the point wake vortices on the control point of the wake panel.

Combining Eqs. (4.196) and (4.154), the impermeability boundary condition for a pitching and plunging airfoil becomes

$$\boldsymbol{A}_n\boldsymbol{\gamma}(t_k) + \boldsymbol{A}_{n_{wp}}(t_k)\gamma_{wp}(t_k) + \boldsymbol{B}_n(t_k)\boldsymbol{\Gamma}_w(t_k) = \boldsymbol{b}(t_k) \tag{4.208}$$

where $\boldsymbol{b}(t_k)$ is obtained from Eq. (4.173) and the relative velocities between the wing and the flow are given by Eqs. (4.77) and (4.78). Kelvin's theorem is expressed as

$$\Gamma(t_k) + l_{wp}(t_k)\gamma_{wp}(t_k) = \Gamma(t_{k-1}) \tag{4.209}$$

where the total bound circulation $\Gamma(t_k)$ is obtained from Eq. (4.205) and can be written as

$$\Gamma(t_k) = \bar{\mathbf{s}}\boldsymbol{\gamma}(t_k) \tag{4.210}$$

Here, $\bar{\mathbf{s}} = \frac{1}{2}((\mathbf{s}, 0) + (0, \mathbf{s}))$ and $\mathbf{s}$ is the $1 \times n$ vector with elements s_j.

The Kutta condition can be enforced in two ways:

- Setting the total bound vorticity at the trailing edge equal to zero at all times. There are now three vortex panels meeting at the trailing edge: the lower surface panel with trailing edge strength γ_1, the upper surface panel with trailing edge strength γ_{n+1} and the wake panel with strength γ_{wp} everywhere. Consequently, the Kutta condition becomes

$$\gamma_1(t_k) + \gamma_{n+1}(t_k) + \gamma_{wp}(t_k) = 0 \tag{4.211}$$

- Equating the pressures at the control points of the upper and lower trailing edge panels at all times as was done for the Basu and Hancock method, that is

$$c_{p_1}(t_k) = c_{p_n}(t_k) \tag{4.212}$$

We will implement both versions of the Kutta condition in order to demonstrate the fact that they are not necessarily equivalent in the unsteady case. We will start with the zero trailing edge vorticity condition; for general motions, Eqs. (4.208), (4.209) and (4.211) can be written together as

$$\begin{pmatrix} \mathbf{A}_n & \mathbf{A}_{n_{wp}}(t_k) \\ 10\dots01 & 1 \\ \bar{\mathbf{s}} & l_{wp}(t_k) \end{pmatrix} \begin{pmatrix} \boldsymbol{\gamma}(t_k) \\ \gamma_{wp}(t_k) \end{pmatrix} = \begin{pmatrix} \mathbf{b}(t_k) - \mathbf{B}_n(t_k)\boldsymbol{\Gamma}_w(t_k) \\ 0 \\ \Gamma(t_k - 1) \end{pmatrix} \tag{4.213}$$

Equations (4.213) constitute a set of $n+2$ linear algebraic equations with $n+4$ unknowns, $\gamma_1, \dots, \gamma_{n+1}, \gamma_{wp}, l_{wp}(t_k)$ and $\theta_{wp}(t_k)$. They are completed by Eqs. (4.149) and (4.150) and solved iteratively, as in Example 4.12.

Alternatively, the equal pressure at the trailing edge Kutta condition is enforced using Eq. (4.212). We must first calculate the tangent velocities on the airfoil, which are obtained by adding the contributions of the wake panel and point wake vortices to Eq. (4.199), such that

$$\mathbf{u}_\tau = \mathbf{A}_\tau\boldsymbol{\gamma}(t_k) + \mathbf{A}_{\tau_{wp}}(t_k)\gamma_{wp}(t_k) + \mathbf{B}_\tau(t_k)\boldsymbol{\Gamma}_w(t_k) \tag{4.214}$$

The value of the pressure coefficient at the ith panel is given by Eq. (4.160) and, using the arguments applied to the Basu and Hancock approach concerning the potential jump across the trailing edge, the Kutta condition is similar to Eq. (4.175)

$$(u_{m_1}\tau_{x_1} + v_{m_1}\tau_{y_1} + u_{\tau_1}(t_k))^2 = (u_{m_1}\tau_{x_1} + v_{m_1}\tau_{y_1} + u_{\tau_n}(t_k))^2 + \frac{2}{\Delta t}\bar{\mathbf{s}}\left(\boldsymbol{\gamma}(t_k) - \boldsymbol{\gamma}(t_{k-1})\right) \tag{4.215}$$

In order to enforce this latest expression we can solve the impermeability condition and Kelvin's theorem for $\gamma_1, \dots, \gamma_n$ and γ_{wp} in terms of γ_{n+1} and then substitute into (4.215) to obtain a scalar quadratic equation for γ_{n+1}. Kelvin's theorem is the last line of Eq. (4.213)

$$\bar{\mathbf{s}}\boldsymbol{\gamma}(t_k) + l_{wp}(t_k)\gamma_{wp}(t_k) = \Gamma(t_k - 1)$$

This latest expression can be solved for $\gamma_{wp}(t_k)$ such that

$$\gamma_{wp}(t_k) = \frac{\Gamma(t_k - 1)}{l_{wp}(t_k)} - \frac{1}{l_{wp}(t_k)}\bar{\mathbf{s}}\boldsymbol{\gamma}(t_k) \tag{4.216}$$

The impermeability condition is the first line of Eq. (4.213)

$$\boldsymbol{A}_n\boldsymbol{\gamma}(t_k) + \boldsymbol{A}_{n_{wp}}(t_k)\gamma_{wp}(t_k) = \boldsymbol{b}(t_k) - \boldsymbol{B}_n(t_k)\boldsymbol{\Gamma}_w(t_k)$$

Substituting for $\gamma_{wp}(t_k)$ from Eq. (4.216) we obtain

$$\left(\boldsymbol{A}_n - \frac{1}{l_{wp}(t_k)}\boldsymbol{A}_{n_{wp}}(t_k)\bar{\mathbf{s}}\right)\boldsymbol{\gamma}(t_k) = \boldsymbol{b}(t_k) - \boldsymbol{B}_n(t_k)\boldsymbol{\Gamma}_w(t_k) - \boldsymbol{A}_{n_{wp}}(t_k)\frac{\Gamma(t_k - 1)}{l_{wp}(t_k)}$$

This is a set of n equations with $n + 1$ unknowns, the $n + 1$ elements of $\boldsymbol{\gamma}(t_k)$. We therefore choose to solve it for the values of $\gamma_1(t_k), \ldots, \gamma_n(t_k)$ in terms of $\gamma_{n+1}(t_k)$. First, we write the equation in the form

$$\overline{\boldsymbol{A}}_n(t_k)\boldsymbol{\gamma}(t_k) = \overline{\boldsymbol{a}}_{0_n}(t_k) \tag{4.217}$$

where

$$\overline{\boldsymbol{A}}_n(t_k) = \boldsymbol{A}_n - \frac{1}{l_{wp}(t_k)}\boldsymbol{A}_{n_{wp}}(t_k)\bar{\mathbf{s}}$$

$$\overline{\boldsymbol{a}}_{0_n}(t_k) = \boldsymbol{b}(t_k) - \boldsymbol{B}_n(t_k)\boldsymbol{\Gamma}_w(t_k) - \boldsymbol{A}_{n_{wp}}(t_k)\frac{\Gamma(t_k - 1)}{l_{wp}(t_k)}$$

Then, we rewrite Eq. (4.217) as =

$$\overline{\boldsymbol{A}}_n^*(t_k)\boldsymbol{\gamma}^*(t_k) + \overline{\boldsymbol{a}}_n^*(t_k)\gamma_{n+1}(t_k) = \overline{\boldsymbol{a}}_{0_n}(t_k) \tag{4.218}$$

where $\overline{\boldsymbol{A}}_n^*(t_k)$ is the $n \times n$ matrix of the first n columns of $\overline{\boldsymbol{A}}_n(t_k)$, $\boldsymbol{\gamma}^* = (\gamma_1, \ldots, \gamma_n)^T$ and $\overline{\boldsymbol{a}}_n^*(t_k)$ is the last column of $\overline{\boldsymbol{A}}_n(t_k)$. We can now solve Eq. (4.218) for $\boldsymbol{\gamma}^*(t_k)$ to obtain

$$\boldsymbol{\gamma}^*(t_k) = \boldsymbol{b}_\gamma(t_k)\gamma_{n+1}(t_k) + \boldsymbol{b}_0(t_k) \tag{4.219}$$

where

$$\boldsymbol{b}_\gamma(t_k) = -\overline{\boldsymbol{A}}_n^*(t_k)^{-1}\overline{\boldsymbol{a}}_n^*(t_k)$$

$$\boldsymbol{b}_0(t_k) = \overline{\boldsymbol{A}}_n^*(t_k)^{-1}\overline{\boldsymbol{a}}_{0_n}(t_k)$$

Next, we substitute Eq. (4.216) into the equation for the tangential perturbation velocities (4.214), such that

$$\boldsymbol{u}_\tau(t_k) = \left(\boldsymbol{A}_\tau - \frac{1}{l_{wp}(t_k)}\boldsymbol{A}_{\tau_{wp}}\bar{\mathbf{s}}\right)\boldsymbol{\gamma} + \boldsymbol{A}_{\tau_{wp}}\frac{\Gamma(t_k - 1)}{l_{wp}(t_k)} + \boldsymbol{B}_\tau(t_k)\boldsymbol{\Gamma}_w(t_k)$$

which is of the form

$$\boldsymbol{u}_\tau(t_k) = \overline{\boldsymbol{A}}_\tau(t_k)\boldsymbol{\gamma}(t_k) + \overline{\boldsymbol{a}}_{0_\tau}(t_k)$$

where

$$\overline{\boldsymbol{A}}_\tau(t_k) = \boldsymbol{A}_\tau - \frac{1}{l_{wp}(t_k)}\boldsymbol{A}_{\tau_{wp}}\bar{\boldsymbol{s}}$$

$$\overline{\boldsymbol{a}}_{0_\tau}(t_k) = \boldsymbol{A}_{\tau_{wp}}\frac{\Gamma(t_k-1)}{l_{wp}(t_k)} + \boldsymbol{B}_\tau(t_k)\boldsymbol{\Gamma}_w(t_k)$$

Again, we split the matrix $\overline{\boldsymbol{A}}_\tau(t_k)$ such that

$$\boldsymbol{u}_\tau(t_k) = \overline{\boldsymbol{A}}^*_\tau(t_k)\boldsymbol{\gamma}^*(t_k) + \overline{\boldsymbol{a}}^*_\tau(t_k)\gamma_{n+1}(t_k) + \overline{\boldsymbol{a}}_{0_\tau}(t_k) \tag{4.220}$$

where $\overline{\boldsymbol{A}}^*_\tau$ is the matrix made up of the first n columns of $\overline{\boldsymbol{A}}_\tau$ while $\overline{\boldsymbol{a}}^*_\tau$ is the last column of $\overline{\boldsymbol{A}}_\tau$. We now substitute Eq. (4.219) into Eq. (4.220) to obtain

$$\boldsymbol{u}_\tau(t_k) = \boldsymbol{c}_\gamma(t_k)\gamma_{n+1}(t_k) + \boldsymbol{c}_0(t_k) \tag{4.221}$$

where

$$\boldsymbol{c}_\gamma(t_k) = \overline{\boldsymbol{A}}^*_\tau(t_k)\boldsymbol{b}_\gamma(t_k) + \overline{\boldsymbol{a}}^*_\tau(t_k), \quad \boldsymbol{c}_0(t_k) = \overline{\boldsymbol{A}}^*_\tau(t_k)\boldsymbol{b}_0(t_k) + \overline{\boldsymbol{a}}_{0_\tau}(t_k)$$

The total bound circulation at time t_k is given by Eq. (4.210) as

$$\Gamma(t_k) = \bar{\boldsymbol{s}}\boldsymbol{\gamma}(t_k) = \bar{\boldsymbol{s}}^*\boldsymbol{\gamma}^*(t_k) + \frac{s_n}{2}\gamma_{n+1}(t_k)$$

where $\bar{\boldsymbol{s}}^* = \frac{1}{2}\left(\boldsymbol{s} + [0\ s_1\ s_2\ \dots\ s_{n-1}]\right)$ is the vector containing the first n elements of $\boldsymbol{s}$ and $s_n/2$ is the last element of $\boldsymbol{s}$. Substituting from Eq. (4.219) we obtain

$$\Gamma(t_k) = d_\gamma(t_k)\gamma_{n+1}(t_k) + d_0(t_k) \tag{4.222}$$

where

$$d_\gamma(t_k) = \bar{\boldsymbol{s}}^*\boldsymbol{b}_\gamma(t_k) + \frac{s_n}{2}, \quad d_0(t_k) = \bar{\boldsymbol{s}}^*\boldsymbol{b}_0(t_k)$$

Finally, substituting Eqs. (4.221) and (4.222) into the Kutta condition of expression (4.215) and re-arranging, we obtain the quadratic equation in $\gamma_{n+1}(t_k)$

$$a_2(t_k)\gamma^2_{n+1}(t_k) + a_1(t_k)\gamma_{n+1}(t_k) + a_0(t_k) = 0 \tag{4.223}$$

where

$$a_2(t_k) = \left(c^2_{\gamma_n} - c^2_{\gamma_1}\right)$$

$$a_1(t_k) = 2\left(c_{\gamma_n}(c_{0_n} + u_{m_n}\tau_{x_n} + v_{m_n}\tau_{y_n}) - c_{\gamma_1}(c_{0_1} + u_{m_1}\tau_{x_1} + v_{m_1}\tau_{y_1}) + \frac{d_\gamma(t_k)}{\Delta t}\right)$$

$$a_0(t_k) = (c_{0_n} + u_{m_n}\tau_{x_n} + v_{m_n}\tau_{y_n})^2 - (c_{0_1} + u_{m_1}\tau_{x_1} + v_{m_1}\tau_{y_1})^2 + \frac{2(d_0 - \Gamma(t_{k-1}))}{\Delta t}$$

This quadratic equation is solved using the procedure of Eq. (4.169). Once γ_{n+1} is evaluated, it can be substituted into Eqs. (4.218) and (4.216) in order to calculate $\gamma_1, \dots, \gamma_n$ and γ_{wp}.

Whether we use the Kutta condition of Eq. (4.211) or that of Eq. (4.212), the solution procedure is similar to that of the Basu and Hancock approach but there is a significant advantage to using the zero trailing edge vorticity Kutta condition: the complete system of Eq. (4.213) is linear.

The complete unsteady linear vortex panel algorithm is as follows:

1. At time t_k we propagate the point wake vortices and shed the total strength of the wake panel at the previous time step, $l_{wp}(t_{k-1})\gamma_{wp}(t_{k-1})$, as a point vortex lying at position $(x_{c_{wp}}(t_{k-1}) + u_{wp}(t_{k-1})\Delta t, y_{c_{wp}}(t_{k-1}) + v_{wp}(t_{k-1})\Delta t)$.
2. We guess values for $l_{wp}(t_k)$ and $\theta_{wp}(t_k)$, usually from the converged values at t_{k-1}.
3. We calculate all the influence coefficient matrices and vectors related to the wake panel.
4. We solve simultaneously the impermeability, Kutta and Kelvin equations for $\gamma(t_k)$ and $\gamma_{wp}(t_k)$. If the zero vorticity at the trailing edge condition is selected, then we solve Eq. (4.213). If the equal trailing edge pressure condition is selected, we solve Eq. (4.223) for $\gamma_{n+1}(t_k)$ and then substitute back into (4.219) to evaluate the complete $\gamma(t_k)$ vector and into (4.216) to obtain $\gamma_{wp}(t_k)$.
5. We calculate the total horizontal and vertical velocities at the control point of the wake panel, $u_{wp}(t_k)$ and $v_{wp}(t_k)$, from Eqs. (4.206) and (4.207).
6. From Eqs. (4.149) and (4.150) we calculate new values for $l_{wp}(t_k)$ and $\theta_{wp}(t_k)$ respectively and go back to step 3.
7. We stop iterating once the values of $l_{wp}(t_k)$ and $\theta_{wp}(t_k)$ have converged.
8. We calculate the aerodynamic loads at time t_k, increment k and go back to step 2, using the converged values for $l_{wp}(t_{k-1})$ and $\theta_{wp}(t_{k-1})$ as first guesses.

The effect of the two different formulations of the Kutta condition will be explored in the next example.

Example 4.14 *Explore the effect of using different formulations of the unsteady Kutta condition by applying them to Example 4.12.*

This example is solved in exactly the same way as Example 4.12 except that we now use the VPM with either the zero trailing edge vorticity or the equal trailing edge pressure Kutta conditions. Hence, we obtain two different solutions from the VPM that we compare to the Basu and Hancock, lumped vortex and linear theory predictions as well as to the experimental results by Anderson et al. (1998). Figure 4.46 plots the mean thrust and mean power coefficients, as well as the propulsive efficiency, as functions of Strouhal number. The results obtained by the zero trailing edge vorticity Kutta condition are labelled as K1 and those obtained from the equal trailing edge pressure condition as K2.

Figure 4.46a shows that the C_T estimates obtained from the equal pressure condition (K2) are nearly identical to the predictions of the Basu and Hancock approach. The estimates obtained from the zero vorticity at the trailing edge condition (K1) are similar to all the other data up to Str $= 0.3$, but they are overestimated for Str > 0.3. Note that, in this case, Str $= 0.3$ corresponds to a reduced frequency of around k $= 1.4$.

The power coefficient results of Figure 4.46b show that the K2 and Basu and Hancock predictions are very similar. In contrast, the K1 data are overestimated for Str > 0.3 and k > 1.4. Finally, the propulsive efficiency plot of Figure 4.46c shows that the K1, K2 and Basu and Hancock yield predictions have the same form. However, the K1 predictions are lower than those of Basu and Hancock because of the overestimation of C_P, while the K2 estimates lie between the other two predictions.

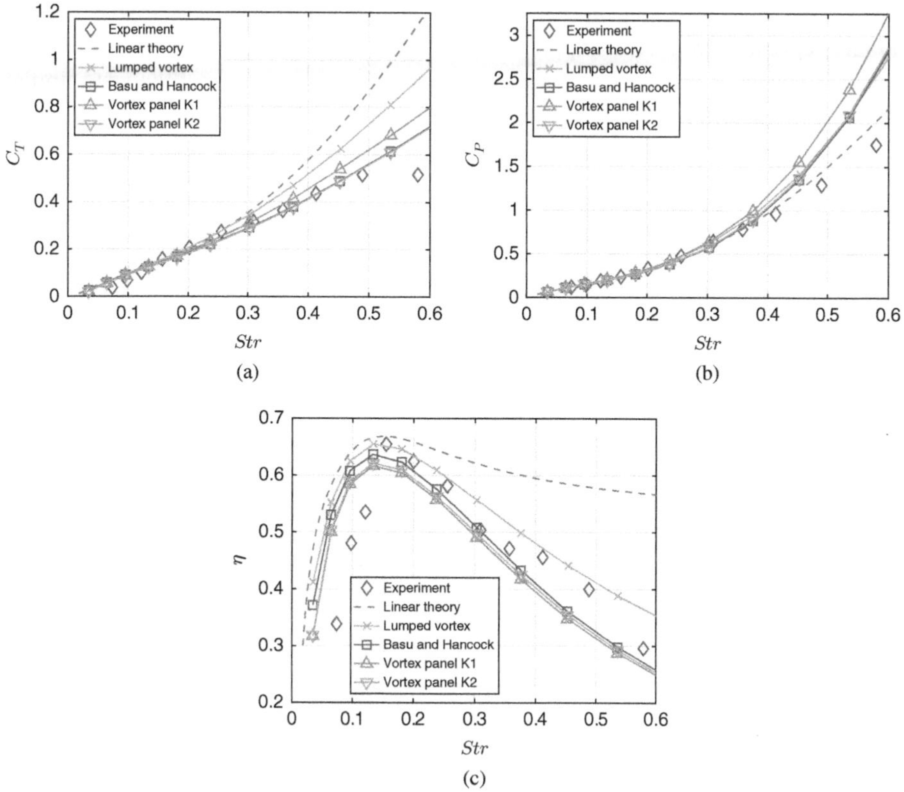

Figure 4.46 Mean thrust coefficient, mean power coefficient and propulsive efficiency as functions of Strouhal number for two different Kutta conditions. Experimental data from Anderson et al. (1998). (a) Mean thrust coefficient, (b) mean power coefficient and (c) propulsive efficiency. Source: Data from Anderson et al. (1998).

The conclusion that can be drawn from Figure 4.46 is that the two Kutta conditions give similar results for Str < 0.3 *and k* < 1.4. *At higher Strouhal and reduced frequency values the equal trailing edge pressure condition predicts* C_T *and* C_P *values that are closer to the experimental results (up to Str* $= 0.54$ *for* C_T *and Str* $= 0.4$ *for* C_P*). The K2 and Basu and Hancock methods make use of the same Kutta condition so that their predictions are very similar. Hence, while the two formulations of the Kutta condition are equivalent at steady conditions, they are not equivalent if the airfoil is pitching and plunging at high Strouhal numbers. The most physically correct formulation appears to be the equal pressure at the trailing edge, as the resulting mean estimates are closer to the experimental data. This example is solved by Matlab code* `pitchplunge_VPM.m`.

The example has shown that, at *Str* ≥ 0.3 and $k \geq 1.4$, the two Kutta conditions start to give different predictions. It should be stressed though that applying a Kutta condition at such high Strouhal numbers and reduced frequencies is not necessarily physical. Anderson et al. (1998) show that vortices start to appear near the leading edge on the suction side at *Str* $= 0.2$ and that significant flow separation is evident at *Str* > 0.3. Hence, the attached flow assumptions underlying the vortex panel and Basu and Hancock approaches are no longer valid, and we cannot expect these methods to match experimental results. For $Str < 0.3$ on the other hand, the two versions of the unsteady Kutta condition yield similar results and agree with the experiments. The main practical difference between the equal

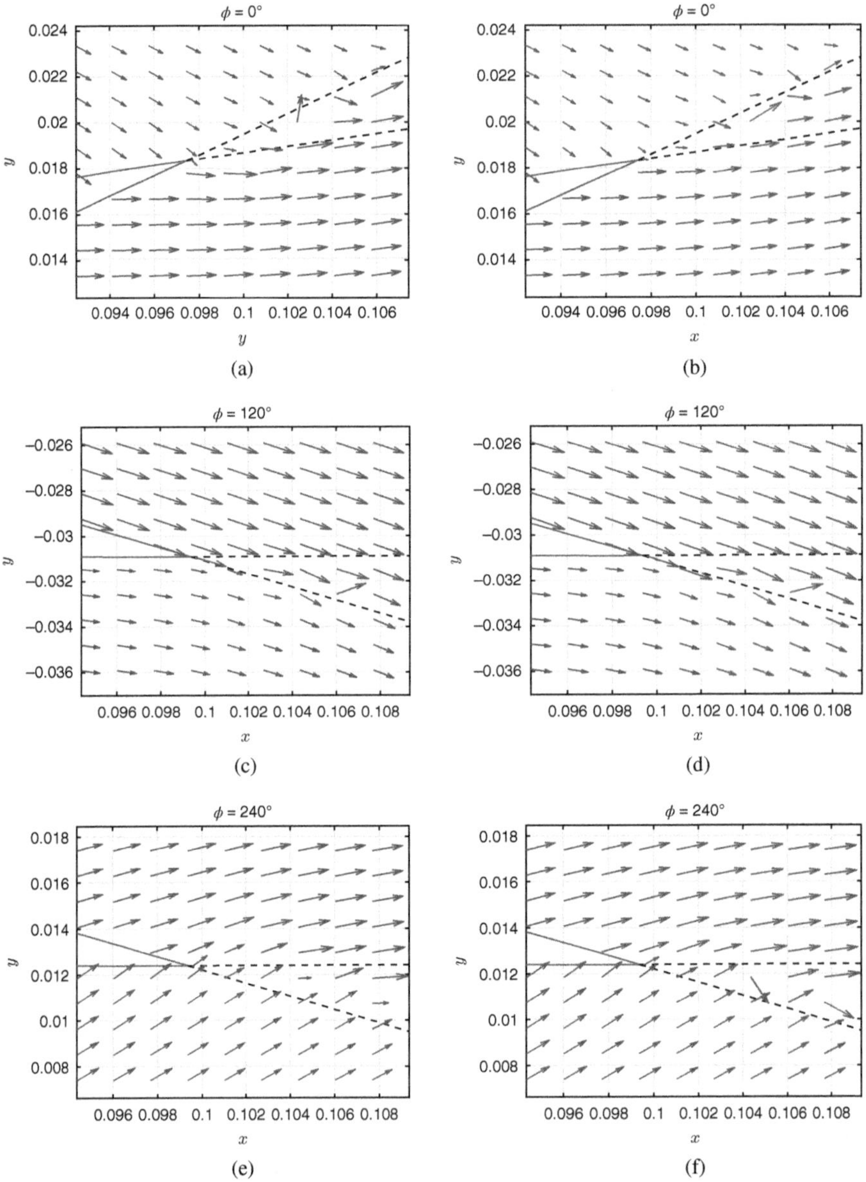

Figure 4.47 Flow around the trailing edge of a pitching and plunging airfoil for two different Kutta conditions. (a) Phase 0°, K1, (b) Phase 0°, K2, (c) Phase 120°, K1, (d) Phase 120°, K2, (e) Phase 240°, K1 and (f) Phase 240°, K2.

trailing edge pressure and zero trailing edge vorticity conditions is mathematical; the former is non-linear while the latter is linear.

As a final thought on the unsteady Kutta condition, we will look at the differences in the trailing edge flow angle obtained from the equal pressure and zero vorticity formulations. We repeat Example 4.14 for $k = 1.2$, $Str = 0.24$ using the two Kutta conditions, and we calculate the flow velocities in a grid around the trailing edge. Figure 4.47 shows three snapshots of the flowfields obtained by each of the Kutta conditions, at phase angles 0°, 120° and 240°. The plots on the left were obtained using the zero vorticity condition (K1) and those on the right using the equal pressure condition (K2). The extensions of the upper and lower trailing edge panels are also plotted on the graphs, in order to compare to the Giesing and Maskell condition. The flow velocity vector nearest to the trailing is calculated at a distance of $c/500$ downstream of that point. It can be seen that the two sets of flowfields look very similar, except for some fluctuations caused by the presence of the wake vortices. At a phase angle of 120°, the trailing edge flow vector is aligned with the upper trailing edge panel, but in the other two snapshots, this is not the case. In fact, the trailing edge flow angle can far exceed the angles of the two trailing edge panels. In general, the amplitude of the trailing edge flow angle obtained from the zero vorticity Kutta condition is slightly higher than that calculated using the equal pressure condition.

4.8 Concluding Remarks

The numerical methods presented in this chapter can give more information about 2D inviscid and incompressible flows than the analytical approaches of Chapter 3. However, Figure 4.46 has shown that there is a limited range of oscillation frequencies and amplitudes in which these methods can provide reliable estimates of unsteady aerodynamic loads. Nevertheless, when these estimates are valid, they can be used to investigate interesting physical phenomena such as the generation of thrust by pitching and plunging airfoils and the associated wake structure. Still, real flows are rarely two-dimensional. Chapter 5 will deal with unsteady flows around finite, three-dimensional wings.

4.9 Exercises

1 Throughout Section 4.2, the pitching moment is calculated using the distance between the bound vortices and a datum (quarter chord or pitching axis). What difference would it make if the pitching moment was calculated using the distance between the control points and the datum?

2 Repeat Example 4.8 with increasing amounts of camber. Is there a visible change in the frequency responses of the unsteady aerodynamic loads?

3 Develop frequency-domain versions of the SVPM and VPM techniques. Replace the wake model by the one used in Example 4.8. Compare the frequency responses of the aerodynamic loads of a pitching and plunging airfoil to those obtained from the time-domain versions.

4 Implement the non-circulatory lift calculation of Section 4.6 and use it to draw Figure 4.40.

References

Anderson, J.M., Streitlien, K., Barrett, D.S., and Triantafyllou, M.S. (1998). Oscillating foils of high propulsive efficiency. *Journal of Fluid Mechanics* 360: 41–72.

Basu, B.C. and Hancock, G.J. (1978). The unsteady motion of a two-dimensional aerofoil in incompressible flow. *Journal of Fluid Mechanics* 87 (1): 159–178.

Bisplinghoff, R.L., Ashley, H., and Halfman, R.L. (1996). *Aeroelasticity*. New York: Dover Publications, Inc.

Carr, L.W., McAlister, K.W., and McCroskey, W.J. (1977). Analysis of the development of dynamic stall based on oscillating airfoil experiments. Technical Note TN D-8382. NASA.

Cebeci, T., Platzer, M., Chen, H. et al. (2005). *Analysis of Low-Speed Unsteady Airfoil Flows*. Springer-Verlag.

Freymuth, P. (1988). Propulsive vortical signature of plunging and pitching airfoils. *AIAA Journal* 26 (7): 881–883.

Jones, K.D. and Platzer, M.F. (1996). Time-domain analysis of low-speed airfoil flutter. *AIAA Journal* 34 (5): 1027–1033.

Jones, K.D., Dohring, C.M., and Platzer, M.F. (1998). Experimental and computational investigation of the Knoller-Betz effect. *AIAA Journal* 36 (7): 1240–1246.

Jones, K.D., Lund, T.C., and Platzer, M.F. (2001). Experimental and computational investigation of flapping wing propulsion for micro air vehicles. In: *Fixed and Flapping Wing Aerodynamics for Micro Air Vehicle Applications, Progress in Astronautics and Aeronautics*, vol. 195 (ed. T.J. Mueller), 307–339. AIAA.

Katz, J. (1981). A discrete vortex method for the non-steady separated flow over an airfoil. *Journal of Fluid Mechanics* 102: 315–328.

Katz, J. and Plotkin, A. (2001). *Low Speed Aerodynamics*. Cambridge University Press.

Kim, M.J. and Mook, D.T. (1986). Application of continuous vorticity panels to general unsteady incompressible two-dimensional lifting flows. *Journal of Aircraft* 23 (6): 464–471.

Kuethe, A.M. and Chow, C.Y. (1986). *Foundations of Aerodynamics – Basis of Aerodynamic Design*. Wiley.

Küssner, H.G. (1936). Zusammenfassender bericht über den instationären auftrieb von flügeln. *Luftfahrtforschung* 13 (12): 410–424.

Loftin, L.K. Jr. and Cohen, K.S. (1948). Aerodynamic characteristics of a number of modified NACA four-digit-series airfoil sections. Technical Note TN-1591. NACA.

Poling, D.R. and Telionis, D.P. (1986). The response of airfoils to periodic disturbances – the unsteady Kutta condition. *AIAA Journal* 24 (2): 193–199.

Seetharam, H.C., Rodgers, E.J., and Wentz, W.H. Jr. (1977). Experimental studies of flow separation of the NACA 2412 airfoil at low speeds. Contractor Report CR-197497. NASA.

Simpson, R.J.S. and Palacios, R. (2013). Induced-drag calculations in the unsteady vortex lattice method. *AIAA Journal* 51 (7): 1775–1779.

Young, J. and Lai, J.C.S. (2007). Mechanisms influencing the efficiency of oscillating airfoil propulsion. *AIAA Journal* 45 (7): 1695–1702.

5

Finite Wings

5.1 Introduction

Up to this point we have only discussed 2D wing sections or, equivalently, 3D rectangular wings with infinite span. Furthermore, all the cases we treated concerned rigid wings whose geometry never changes. In other words, the only type of motion was rigid-body movements, such as plunging and pitching around a pitch axis. Real lifting surfaces can be subjected to both rigid-body and flexible motion. For example, traditional helicopter blades have flap, pitch and lead-lag degrees of freedom at their roots, but they can also bend and twist. Bird wings have the same three degrees of freedom at the elbow, as well as additional degrees of freedom at the wrist, and they can also bend and twist. Finally, aircraft wings have six degrees of freedom at the aircraft's centre of gravity, control surface deflection degrees of freedom and can undergo bending and torsion. Traditionally, the study of the rigid-body motion of aircraft is the objective of flight dynamics, while the study of flexible motion is the objective of aeroelasticity.

In this chapter and the next, we will study the attached flow aerodynamics of 3D wings that are forced to pitch, plunge, roll, yaw, deflect control surfaces and deform flexibly. Chapter 5 will focus on incompressible flows and Chapter 6 on compressible ones. First, we will define the various aspects of the geometry of a wing, as well as the non-dimensional parameters that can be used to describe them. Figure 5.1 plots the geometry of a typical swept and tapered trapezoidal wing. Looking at the top view of Figure 5.1a, the planform of the wing is defined by the span, b, root chord c_0, tip chord c_T and sweep angle at the quarter chord $\Lambda_{c/4}$. A wing can be swept back or swept forward, depending on the sign of the sweep angle. The taper ratio is defined as

$$\lambda = \frac{c_T}{c_0} \tag{5.1}$$

with limits $\lambda = 0$ for a wing with $c_T = 0$ and $\lambda = 1$ for an untapered wing. The planform area is given by

$$S = b\frac{c_T + c_0}{2}$$

Unsteady Aerodynamics: Potential and Vortex Methods, First Edition. Grigorios Dimitriadis.

Companion website: www.wiley.com/go/dimitriadis/unsteady_aerodynamics

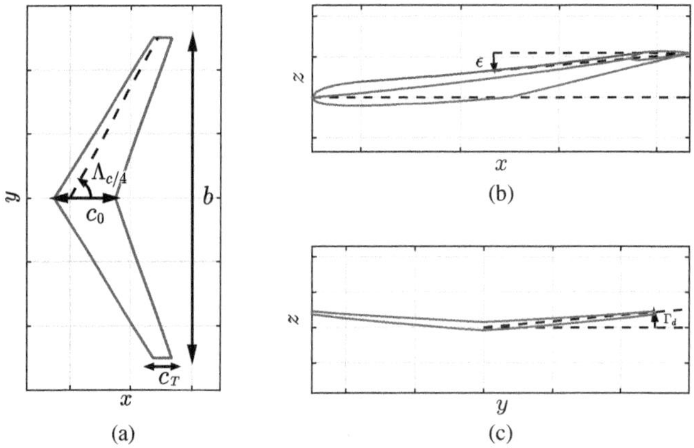

Figure 5.1 Swept and tapered trapezoidal wing geometry. (a) Top view, (b) side view, and (c) front view.

and the aspect ratio is defined as

$$\mathcal{R} = \frac{b^2}{S} \tag{5.2}$$

Looking at the side view of Figure 5.1b we can see that the angle of the root chord to the horizontal is zero, while the angle of the tip chord to the horizontal is negative. This characteristic of wings is known as twist and is denoted here by ϵ. Nose down (negative) twist is known as washout, while nose up (positive) twist is known as wash-in. For aircraft wings, the twist usually varies linearly from root to tip. Geometric twist can be complemented by aerodynamic twist, whereby the airfoil section changes along the span, such that the 2D zero-lift angle of the section also changes. Again, airfoil shape variations are usually linear along the span.

Finally, the front view of Figure 5.1c shows that the two half-wings are angled upwards by an angle Γ_d known as the dihedral. If the half-wings were angled downwards, Γ_d would be negative and known as the anhedral.

Many aircraft wings are made up of more than one trapezoidal section per half-wing, each with its own span, taper, sweep, twist, dihedral and airfoil section. A special wing geometry that is rarely used in practice but will be discussed in this chapter is the elliptical planform. This geometry features no sweep or twist (geometric or aerodynamic) and an elliptical chord-length variation along the span, such that $c_T = 0$, c_0 is the ellipse's minor axis, and the span line is the ellipse's major axis.

5.1.1 Rigid Wings and Flexible Wings

A wing can move rigidly so that all points of the wing move with the same translational and rotational velocity. This motion is usually defined by the motion of the centre of gravity of the aircraft and can be described in different frames of reference. These reference frames are detailed in numerous textbooks on flight mechanics; in this book, we will only use either an inertial frame or a body-fixed frame. The inertial frame of reference is used for motion cases such as those presented in Figures 2.1 and 2.2, while the body-fixed frame for the case depicted in Figure 2.3. The main characteristic of rigid-body motion is that the wing shape does not change; only the wing's position and orientation do. In the body-fixed frame of reference, the degrees of freedom of a wing are usually referred to as

- Surge. Displacement in the chordwise direction.
- Sideslip. Displacement in the spanwise direction.
- Plunge. Displacement in a direction vertical to the wing's surface.
- Pitch. Rotation around a spanwise axis.
- Roll. Rotation around a chordwise axis.
- Yaw. Rotation around an axis perpendicular to the wing's surface.

A wing can also deform flexibly. This kind of motion can be due to aeroelastic deflections, such as wing bending or wing torsion. Wings are usually said to deform with respect to their mode shapes, which are combinations of primarily three types of deflection:

- Spanwise bending. Spanwise sections of the wing move up or down.
- In-plane bending. Spanwise sections of the wing move forward or backward.
- Chordwise bending. Spanwise sections of the wing change camber. This type of deflection is generally considered negligible for wings of medium or high aspect ratios.
- Torsion. Spanwise sections of the wing rotate nose-up or nose-down.

Control surface deflections are rigid rotations around a hinge but can also be seen as flexible deflections since parts of the wing move with respect to other parts of the wing so that the wing shape is modified. Wing flapping can also be seen as flexible motion, as the two half-wings change position with respect to each other during the flapping cycle.

5.2 Finite Wings in Steady Flow

A lift-producing finite wing will necessarily be subjected to high pressure on the lower surface and low pressure on the upper surface. This pressure differential causes flow to move from the lower to the upper surface around the wingtips, as seen in Figure 5.2a in which the airflow behind a low-flying aircraft is visualised using smoke. Consequently, the flow behind the wingtip has two primary velocity components; a downstream translational velocity due to the free stream and a rotational velocity due to the motion from the lower to the upper surface. The rotational velocity component is strongest at the wingtips and decreases further inboard.

The flow topology seen in Figure 5.2a is often modelled as a thin vortex sheet that rolls up at the wingtips. Figure 5.2b plots a model of the wake behind a finite wing immersed in steady flow. The shade of the wake sheet denotes the local vortex strength, showing that this strength is stronger near the tips and weaker near the middle. Note that the wake sheet rolls up at the wingtips, but it can also be represented as flat in order to simplify the mathematical analysis.

In Figure 5.2b, both the wing and the wake have been represented by quadrilateral vortex rings; in the case of the wake, the vortex rings travel at the local flow velocity. We will discuss this type of modelling in Section 5.4. Here, we will start with a much simpler model known as Lifting Line Theory (see for example (Anderson Jr. 1985; Kuethe

Figure 5.2 Flow behind a lifting finite wing. (a) Visualisation of flow behind an aircraft, credit: NASA Langley Research Center and (b) Wake vortex sheet model behind a finite wing in steady flow. Source: NASA/Langley Research Center (NASA-LaRC)/Public Domain.

and Chow 1986)). The basis of the modelling is the horseshoe vortex, which is drawn in Figure A.9. As described in Appendix A.5, the horseshoe vortex is composed of three line segments, the bound segment that extends from the wingtip lying at $y = -b/2$ to the other wingtip at $y = b/2$ and the two trailing segments that extend downstream from the wingtips to infinity. All three vortex segments have equal and constant strength, Γ. The bound segment does not induce any velocity on itself, but by the Biot–Savart law of Eq. (2.227), the trailing segments induce a downward velocity along the bound segment given by Eq. (A.41)

$$w(y) = -\frac{\Gamma}{4\pi\,(y + b/2)} + \frac{\Gamma}{4\pi\,(y - b/2)} \tag{5.3}$$

Superimposing several horseshoe vortices, whose central segments all lie on the wing, leads to the complete lifting line wing model. Figure 5.3 plots a simple version of such a model for a rectangular wing of span $b = 2$ and constant chord $c = 0.4$, composed of three horseshoe vortices, Γ_1, Γ_2 and Γ_3. Note that the bound segment of Γ_1 extends along the entire span of the wing, that of Γ_2 along the central 3/5ths of the span and that of Γ_3 along the central 1/5th. As the bound segments of the vortices are collinear, their strengths are added such that the vortex strength along the quarter-chord line of the wing varies from Γ_1 at the wingtips to $\Gamma_1 + \Gamma_2 + \Gamma_3$ at midspan. The horseshoe vortices Γ_2 and Γ_3 induce upwash outboard of their trailing vortices and downwash inboard. Vortex Γ_1 induces downwash throughout the span. The wing is absent from Figure 5.3 as it is represented completely by the bound vortices.

The horseshoe vortex model of Figure 5.3 can become a continuous vorticity sheet if the number of vortices becomes infinite and their strength $d\Gamma$. This means that at every spanwise position y, there lies a trailing vortex of strength $d\Gamma(y)$, leading to a bound vorticity variation $\Gamma(y)$. The trailing vortices induce a downwash velocity distribution on the quarter

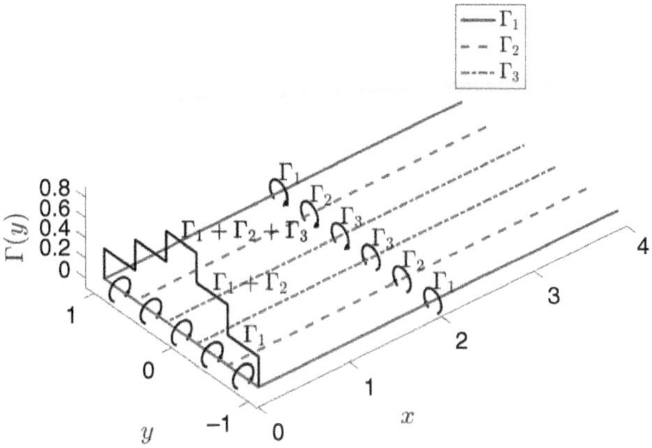

Figure 5.3 Finite wing modelling using horseshoe vortices.

chord line whose value is obtained by summing the contributions of all the horseshoe vortices. The strength of the trailing vortex lying at spanwise position y_0 can be written as

$$d\Gamma(y_0) = -\left.\frac{d\Gamma}{dy}\right|_{y_0} dy_0$$

where the negative sign is due to the fact that $d\Gamma/dy$ is positive for $y_0 < 0$ and negative for $y_0 > 0$, as seen in Figure 5.3. Consequently, for a continuous distribution of horseshoe vortices, Eq. (5.3) becomes

$$w_i(y) = -\frac{1}{4\pi}\int_{-b/2}^{b/2} \frac{\left.\frac{d\Gamma}{dy}\right|_{y_0}}{y - y_0} dy_0 \tag{5.4}$$

The continuous downwash distribution along the span caused by the trailing vortices changes the effective angle of attack seen by the wing at each spanwise position. Figure 5.4 plots a cross-section of the wing at spanwise station y. The wing is drawn parallel to the x axis and a free stream of magnitude Q_∞ is inclined to it by a geometric angle of attack $\alpha(y)$. Note that as the wing can be twisted, the geometric angle of attack can vary along the span. The downwash velocity induced by the trailing vortices at this spanwise station is denoted by $w_i(y)$; it is drawn downwards but is defined positive upwards. The effective (total) flow velocity seen by the cross-section has magnitude $\mathbf{Q}_{\text{eff}}(y) = (Q_\infty \cos\alpha(y), 0, Q_\infty \sin\alpha(y) + w_i(y))$ and angle of attack $\alpha_{\text{eff}}(y)$. This angle is given

$$\alpha_{\text{eff}}(y) = \tan^{-1}\left(\frac{Q_\infty \sin\alpha(y) + w_i(y)}{Q_\infty \cos\alpha(y)}\right) \approx \alpha(y) + \frac{w_i(y)}{Q_\infty}$$

assuming that $\alpha(y)$ is small and $w_i(y) \ll Q_\infty$. We define an induced angle of attack

$$\alpha_i(y) = -\frac{w_i(y)}{Q_\infty} \tag{5.5}$$

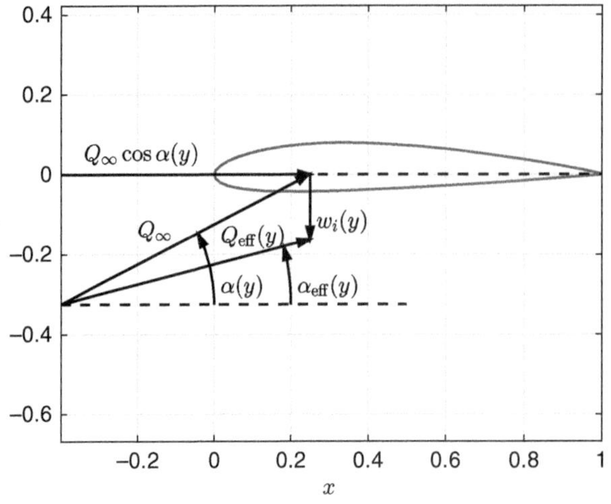

Figure 5.4 Spanwise induced and effective angles of attack.

so that the effective angle of attack is given by

$$\alpha_{\text{eff}}(y) = \alpha(y) - \alpha_i(y) \tag{5.6}$$

Lifting line theory (LLT) calculates the sectional lift coefficient $C_l(y)$ using the effective angle of attack and the result of 2D thin airfoil theory (see Eq. (3.24)):

$$C_l(y) = a_0(y)\left(\alpha_{\text{eff}}(y) - \alpha_{l=0}(y)\right) \tag{5.7}$$

where $a_0(y)$ is the 2D lift curve slope, usually approximated by 2π, and $\alpha_{l=0}(y)$ is the zero-lift angle for cambered airfoils, given by Eq. (3.25). Both a_0 and $\alpha_{l=0}$ are 2D sectional properties that can be obtained from experimental data (such as those presented in (Abbott and Von Doenhoff 1959)), thin airfoil theory, conformal transformation analysis or other approaches. They can vary in the spanwise direction if the geometry of the airfoil section is not constant; in this work, it will be assumed that the airfoil section does not change along the span for simplicity. Note that the notation $C_l(y)$ denotes the sectional lift coefficient of a 3D wing, while C_L denotes the total lift coefficient, calculated from

$$C_L = \frac{1}{S}\int_{-b/2}^{b/2} C_l(y)c(y)\mathrm{d}y \tag{5.8}$$

At this stage, we cannot use Eq. (5.7) since $\alpha_{\text{eff}}(y)$ is not yet known. We can of course combine Eqs. (5.4)–(5.6) such that

$$\alpha_{\text{eff}}(y) = \alpha(y) - \frac{1}{4\pi Q_\infty}\int_{-b/2}^{b/2} \frac{\left.\frac{\mathrm{d}\Gamma}{\mathrm{d}y}\right|_{y_0}}{y - y_0}\mathrm{d}y_0 \tag{5.9}$$

but we do not have a value for $\Gamma(y)$, so we can still not calculate $\alpha_{\text{eff}}(y)$. We need one more equation in order to close the problem; this equation is the Kutta–Joukowski theorem, again applied sectionally, which relates the sectional aerodynamic loads to the local circulation,

$$\mathbf{F}(y) = \rho\mathbf{Q}_{\text{eff}}(y) \times \mathbf{\Gamma}(y) \tag{5.10}$$

where the circulation vector is given by $\mathbf{\Gamma}(y) = (0, \Gamma(y), 0)$, the flow velocity vector by $\mathbf{Q}_{\text{eff}}(y) = (Q_\infty \cos\alpha(y), 0, Q_\infty \sin\alpha(y) + w_i(y))$ and the aerodynamic force is $\mathbf{F}(y) = (F_x(y), F_y(y), F_z(y))$. Substituting for $\mathbf{\Gamma}(y)$ and $\mathbf{Q}_{\text{eff}}(y)$ into Eq. (5.10), we obtain

$$F_x(y) = -\rho\left(Q_\infty \sin\alpha(y) + w_i(y)\right)\Gamma(y),\ F_y(y) = 0,\ F_z(y) = \rho Q_\infty \cos\alpha(y)\Gamma(y) \quad (5.11)$$

The sectional lift and drag are, respectively, parallel and perpendicular to the free stream, not the wing's surface, such that, after linearisation,

$$l(y) = F_z(y)\cos\alpha(y) - F_x(y)\sin\alpha(y) \approx \rho Q_\infty \Gamma(y) \quad (5.12)$$

$$d(y) = F_z(y)\sin\alpha(y) + F_x(y)\cos\alpha(y) \approx -\rho w_i(y)\Gamma(y) \quad (5.13)$$

Consequently, the sectional lift and drag coefficients are defined as

$$C_l(y) = \frac{l(y)}{\frac{1}{2}\rho Q_\infty^2 c(y)} = \frac{2\Gamma(y)}{Q_\infty c(y)} \quad (5.14)$$

$$C_d(y) = \frac{d(y)}{\frac{1}{2}\rho Q_\infty^2 c(y)} = -\frac{2w_i(y)\Gamma(y)}{Q_\infty^2 c(y)} = \alpha_i(y)C_l(y) \quad (5.15)$$

where we substituted from Eq. (5.5) in the expression for $C_d(y)$. Note that the drag is proportional to the lift coefficient and the induced angle of attack, which is why it is known as the induced drag. We can now combine Eqs. (5.7), (5.9) and (5.14) to obtain

$$a_0\left(\alpha(y) - \frac{1}{4\pi Q_\infty}\int_{-b/2}^{b/2}\frac{\left.\frac{d\Gamma}{dy}\right|_{y_0}}{y - y_0}dy_0 - \alpha_{l=0}\right) = \frac{2\Gamma(y)}{Q_\infty c(y)}$$

which can be re-arranged as

$$\alpha(y) = \frac{2\Gamma(y)}{a_0 Q_\infty c(y)} + \frac{1}{4\pi Q_\infty}\int_{-b/2}^{b/2}\frac{\left.\frac{d\Gamma}{dy}\right|_{y_0}}{y - y_0}dy_0 + \alpha_{l=0} \quad (5.16)$$

Equation (5.16) is the governing relation of lifting line theory; it is a single equation with a single unknown, $\Gamma(y)$. Once the circulation distribution along the span is calculated, it can be substituted into Eq. (5.14) in order to calculate the sectional lift coefficient and into (5.8) in order to evaluate the total lift. Unfortunately, Eq. (5.16) is integro-differential and does not have a general solution. It is usually solved by assuming a form for $\Gamma(y)$; two options are generally used:

- The elliptical lift distribution. The circulation is written as

$$\Gamma(y) = \Gamma_0\sqrt{1 - \left(\frac{2y}{b}\right)^2} \quad (5.17)$$

  and then substituted into Eq. (5.16) in order to solve for Γ_0. This type of distribution yields a simple analytical solution but is rarely encountered in real wings.
- The arbitrary lift distribution. The circulation is written as a Fourier series of the form

$$\Gamma(y) = \frac{a_0 c_0 Q_\infty}{2}\sum_{i=1}^{\infty} A_i \sin i\theta \quad (5.18)$$

where $y = (b/2)\cos\theta$, θ takes values from 0 to π and c_0 is the chord length at the centreline of the wing. Truncating and then substituting this circulation distribution into Eq. (5.16) lead to a linear systems of equations to be solved for the unknown coefficients A_i. This approach can be used for any unswept wing shape.

The application of LLT with these two types of distribution will be demonstrated in the following example.

Example 5.1 *Calculate the lift and drag of a wing with elliptical planform using both the elliptical and arbitrary lift distributions.*

The chord distribution of an elliptical wing is given by

$$c(y) = c_0\sqrt{1-\left(\frac{2y}{b}\right)^2} \tag{5.19}$$

The area of the wing is given by

$$S = \int_{-b/2}^{b/2} c(y)\mathrm{d}y = c_0 \int_{-b/2}^{b/2} \sqrt{1-\left(\frac{2y}{b}\right)^2}\,\mathrm{d}y$$

The integral in this expression can be calculated easily by substituting $y = (b/2)\cos\theta$ for θ taking values from 0 to π. Consequently,

$$S = -c_0 b/2 \int_{\pi}^{0} \sin^2\theta \mathrm{d}\theta = \frac{c_0 b\pi}{4}$$

so that the aspect ratio is

$$AR = \frac{b^2}{S} = \frac{4b}{c_0\pi} \tag{5.20}$$

Figure 5.5 plots the wing's planform for $c_0 = 2$ m and $b = 6$ m. The surface area is 9.4 m and the aspect ratio 3.8. The leading edge lies at $x = -c(y)/2$ and the trailing edge at $x = c(y)/2$. The wing is untwisted, has zero dihedral and its section is symmetric, so that $\alpha_{l=0} = 0$. We will also assume that $a_0 = 2\pi$.

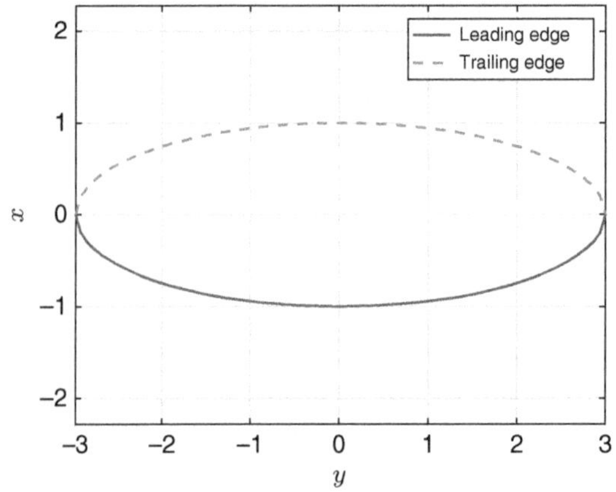

Figure 5.5 Elliptical wing planform.

We will first solve the lifting line problem using the elliptical circulation distribution. Substituting Eqs. (5.17) and (5.19) for the elliptical circulation and chord distributions, respectively, into Eq. (5.16), we obtain

$$\alpha(y) = \frac{2\Gamma_0}{a_0 Q_\infty c_0} - \frac{\Gamma_0}{\pi Q_\infty b^2} \int_{-b/2}^{b/2} \frac{y_0}{\sqrt{1-(2y_0/b)^2}(y-y_0)} dy_0$$

recalling that we have chosen $\alpha_{l=0} = 0$. *Carrying out the substitutions* $y = (b/2)\cos\theta$, $y_0 = (b/2)\cos\theta_0$, *we obtain*

$$\alpha(y) = \frac{2\Gamma_0}{a_0 Q_\infty c_0} + \frac{\Gamma_0}{2\pi Q_\infty b} \int_{\pi}^{0} \frac{\cos\theta_0}{\cos\theta - \cos\theta_0} d\theta_0$$

We now use the Glauert integral

$$\int_0^{\pi} \frac{\cos i\theta_0}{\cos\theta_0 - \cos\theta} d\theta_0 = \pi \frac{\sin i\theta}{\sin\theta} \tag{5.21}$$

for $i = 1$ *so that, finally,*

$$\alpha(y) = \frac{2\Gamma_0}{a_0 Q_\infty c_0} + \frac{\Gamma_0}{2Q_\infty b} \tag{5.22}$$

We can apply the same analysis to the definition of the downwash velocity of Eq. (5.4) in order to obtain

$$w_i(y) = -\frac{\Gamma_0}{2b}, \quad \alpha_i(y) = \frac{\Gamma_0}{2Q_\infty b} \tag{5.23}$$

which means that, for an elliptical lift distribution, the downwash velocity induced by the wake and the induced angle of attack are constant along the span.

We have chosen $\alpha(y) = \alpha$ *to be constant throughout the span so that solving Eq. (5.22) for* Γ_0 *yields*

$$\Gamma_0 = 2Q_\infty \alpha \frac{b a_0 c_0}{4b + a_0 c_0} \tag{5.24}$$

Substituting this result into Eq. (5.14), the sectional lift coefficient becomes

$$C_l(y) = \frac{4ba_0}{4b + a_0 c_0}\alpha \tag{5.25}$$

such that it is constant along the span. After integrating over the span using Eq. (5.8), we obtain the total lift coefficient

$$C_L = \frac{4ba_0}{4b + a_0 c_0}\alpha \tag{5.26}$$

and the 3D lift curve slope

$$C_{L_\alpha} = \frac{4ba_0}{4b + a_0 c_0} \tag{5.27}$$

Finally, we calculate the sectional drag from Eqs. (5.15), (5.25), (5.23) and (5.24), so that

$$C_d(y) = \alpha_i(y) C_l(y) = \frac{c_0}{4b}\left(\frac{4ba_0}{4b + a_0 c_0}\alpha\right)^2 = \frac{c_0}{4b} C_l(y)^2$$

The total induced drag is obtained by integrating $C_d(y)$ over the entire span, such that

$$C_D = \frac{1}{S}\int_{-b/2}^{b/2} c(y)C_d(y)\mathrm{d}y \tag{5.28}$$

Substituting from the equation for the aspect ratio (5.20) and evaluating the integral in Eq. (5.28), we obtain

$$C_D = \frac{C_L^2}{\pi Æ} \tag{5.29}$$

which states that the drag coefficient is proportional to the square of the lift coefficient.

Now, we will solve the same problem using the arbitrary circulation distribution of Eq. (5.18). Note that in this case, we do not make any assumptions about the shape of the circulation distribution, a Fourier series can represent an infinite number of shapes. Nevertheless, in order to solve the problem, we need to truncate the Fourier series to a finite number of terms so that the arbitrary circulation distribution becomes

$$\Gamma(y) = \Gamma(\theta) = \frac{a_0 c_0 Q_\infty}{2}\sum_{i=1}^{n_i} A_i \sin i\theta \tag{5.30}$$

where n_i is the chosen number of terms. The derivative of the circulation distribution can be calculated using the chain rule

$$\frac{\mathrm{d}\Gamma}{\mathrm{d}y} = \frac{\mathrm{d}\Gamma}{\mathrm{d}\theta}\frac{\mathrm{d}\theta}{\mathrm{d}y} = -\frac{a_0 c_0 Q_\infty}{b}\sum_{i=1}^{n_i} A_i \frac{i\cos i\theta}{\sin\theta}$$

Substituting this latest expression into Eq. (5.16), we obtain

$$\alpha(y) = \frac{c_0}{c(y)}\sum_{i=1}^{n_i} A_i \sin i\theta + \frac{a_0 c_0}{4\pi b}\int_\pi^0 \frac{\sum_{i=1}^{n_i} A_i i\cos i\theta_0}{\cos\theta - \cos\theta_0}\mathrm{d}\theta_0$$

Exchanging the order of the integral and the summation and applying the Glauert integral of Eq. (5.21) results in

$$\alpha(y) = \frac{c_0}{c(y)}\sum_{i=1}^{n_i} A_i \sin i\theta + \frac{a_0 c_0}{4b}\sum_{i=1}^{n_i} A_i i\frac{\sin i\theta}{\sin\theta} \tag{5.31}$$

Now, we recall that $\alpha(y)$ is constant and that $c(y) = c(\theta) = c_0 \sin\theta$, so that

$$\alpha = \sum_{i=1}^{n_i} A_i \frac{\sin i\theta}{\sin\theta}\left(1 + \frac{a_0 c_0 i}{4b}\right) \tag{5.32}$$

This is a single equation with n_i unknowns. However, it can be written out for $j = 1, 2, \dots, n_i$ different values of θ_j in order to set up the linear algebraic system

$$\begin{pmatrix} \left(1+\frac{a_0c_0}{4b}\right) & \frac{\sin 2\theta_1}{\sin\theta_1}\left(1+\frac{a_0c_0}{2b}\right) & \cdots & \frac{\sin n_i\theta_1}{\sin\theta_1}\left(1+\frac{a_0c_0n_i}{4b}\right) \\ \left(1+\frac{a_0c_0}{4b}\right) & \frac{\sin 2\theta_2}{\sin\theta_2}\left(1+\frac{a_0c_0}{2b}\right) & \cdots & \frac{\sin n_i\theta_2}{\sin\theta_2}\left(1+\frac{a_0c_0n_i}{4b}\right) \\ \vdots & \vdots & \ddots & \vdots \\ \left(1+\frac{a_0c_0}{4b}\right) & \frac{\sin 2\theta_{n_i}}{\sin\theta_{n_i}}\left(1+\frac{a_0c_0}{2b}\right) & \cdots & \frac{\sin n_i\theta_{n_i}}{\sin\theta_{n_i}}\left(1+\frac{a_0c_0n_i}{4b}\right) \end{pmatrix}\begin{pmatrix} A_1 \\ A_2 \\ \vdots \\ A_{n_i} \end{pmatrix} = \begin{pmatrix} \alpha \\ \alpha \\ \vdots \\ \alpha \end{pmatrix}$$

which can be easily solved for the unknown A_i values. As both the first column of the left-hand matrix and the right-hand vector are constant, the only possible solution is

$$A_1 = \frac{\alpha}{1 + \frac{a_0 c_0}{4b}}, \quad A_{2,\ldots,n_i} = 0 \tag{5.33}$$

Substituting back into Eq. (5.18), we obtain

$$\Gamma(y) = 2Q_\infty \alpha \frac{b a_0 c_0}{4b + a_0 c_0} \sin\theta = 2Q_\infty \alpha \frac{b a_0 c_0}{4b + a_0 c_0} \sqrt{1 - \left(\frac{2y}{b}\right)^2}$$

which is identical to the solution of Eqs. (5.17) and (5.24) that we obtained using the elliptical circulation distribution assumption. Then, the lift distribution is calculated using Eq. (5.14)

$$C_l(y) = \frac{a_0 c_0}{c(y)} \sum_{i=1}^{n_i} A_i \sin i\theta \tag{5.34}$$

Substituting into Eq. (5.8) yields the total lift coefficient as

$$C_L = \frac{1}{S} \int_{-b/2}^{b/2} c(y) C_l(y) \mathrm{d}y = \frac{a_0 c_0 b}{2S} \int_0^\pi \sum_{i=1}^{n_i} A_i \sin i\theta \sin\theta \mathrm{d}\theta$$

Noting that $\int_0^\pi \sin i\theta \sin\theta \mathrm{d}\theta = \pi/2$ *if* $i = 1$ *and zero otherwise, the lift coefficient becomes*

$$C_L = \frac{a_0 c_0 \pi b}{4S} A_1 = \frac{4 b a_0}{4b + a_0 c_0} \alpha \tag{5.35}$$

where A_1 is given from Eq. (5.33) and $S = c_0 \pi b/4$. Again, expression (5.35) is identical to Eq. (5.26) obtained earlier from the elliptical circulation distribution.

Finally, we calculate the sectional drag coefficient from Eq. (5.15). We will need the induced angle of attack for the arbitrary lift distribution, which is given by

$$\alpha_i(y) = \frac{a_0 c_0}{4b} \sum_{i=1}^{n_i} A_i i \frac{\sin i\theta}{\sin\theta}$$

Substituting this latest expression, as well as Eq. (5.34) into the definition of the sectional induced drag coefficient (5.15) yields

$$C_d(y) = \frac{a_0^2 c_0^2}{4bc(y)} \left(\sum_{i=1}^{n_i} A_i \sin i\theta \right) \left(\sum_{i=1}^{n_i} A_i i \frac{\sin i\theta}{\sin\theta} \right) \tag{5.36}$$

Substituting into Eq. (5.28) from expression (5.36) and carrying out the integration yields

$$C_{D_i} = \frac{a_0^2 c_0^2 \pi}{16S} \sum_{i=1}^{n_i} i A_i^2 \tag{5.37}$$

Finally, substituting from the lift coefficient Eq. (5.35), we can rewrite the drag coefficient as

$$C_{D_i} = \frac{C_L^2}{\pi Æ} \sum_{i=1}^{n_i} i \left(\frac{A_i}{A_1}\right)^2 = \frac{C_L^2}{\pi Æ}(1 + \sigma) \tag{5.38}$$

where $\sigma = \sum_{i=2}^{n_i} i (A_i/A_1)^2$. Recalling that $A_{2,\ldots,n_i} = 0$ for the wing with elliptical planform, it can be seen that $C_{D_i} = C_L^2/\pi Æ$, as already calculated from Eq. (5.29). This example is solved by Matlab code `wing_elliptic.m`.

The solutions obtained from the two distributions in the previous example were identical precisely because the planform of the wing was elliptical. For any other wing planform, only the arbitrary lift distribution can be used. Steady lifting line theory modifies steady 2D results by imposing a downwash velocity at each spanwise section. Consequently, it is possible to extend the theory to unsteady flows, if unsteady 2D results are modified in the same way.

5.3 The Impulsively Started Elliptical Wing

In both 2D and 3D, the most basic unsteady aerodynamic problem is impulsively started flow, whereby a symmetric wing lying still in a fluid accelerates impulsively to airspeed Q_∞ with a small angle of attack α. The problem is more complex in 3D because it depends not only on Q_∞ and α but also on the planform and aspect ratio. A standard solution exists for wings with elliptical planform so we will start by discussing this case.

5.3.1 The Solution by Jones

The most widely cited analytical solution for the aerodynamic loads acting on an impulsively started finite wing is the one developed by Jones (1939, 1940). The starting point of this analysis is Wagner's aerodynamic model of an impulsively started 2D flat plate, as detailed in Section 3.4 and Appendix C. Here, we repeat the fact that Wagner's theory models the flat plate as a circle in the complex plane $\zeta = \xi + i\eta$ and places pairs of vortices of strength $\pm\Delta\Gamma$, one lying at $(\Xi_0, 0)$ and one lying at $((c/2)^2/\Xi_0, 0)$, as shown in Figure 5.6a.

We have already shown that after applying the conformal transformation $z = \frac{1}{2}\left(\zeta + \frac{c^2}{4\zeta}\right)$ both vortices transform to point

$$X_0 = \left(\frac{1}{2}\left(\Xi_0 + \frac{c^2}{4\Xi_0}\right), 0\right)$$

in the flat plate plane. Jones's approach is slightly different; the vortex lying at Ξ_0 transforms to X_0 as before, but the vortex lying at $c^2/4\Xi_0$ remains there in the flat plate plane.

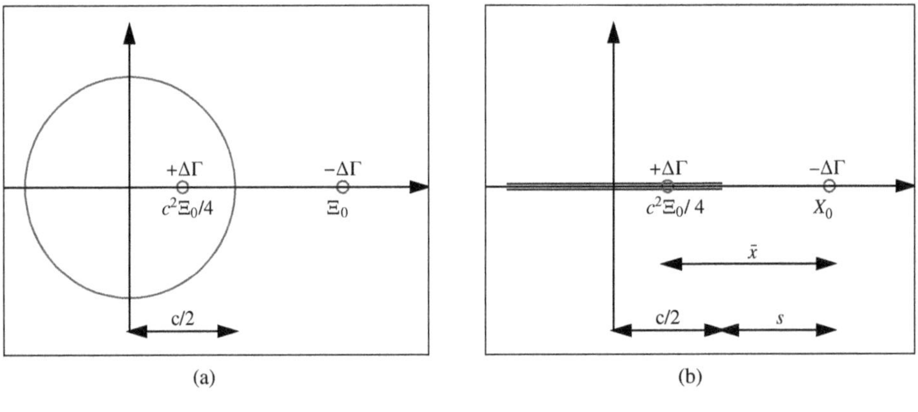

Figure 5.6 Sectional modelling of wing. (a) Circle plane and (b) Flat plate plane.

Furthermore, Jones defines the distance $X_0 - c/2$ as the flight path distance $s = Q_\infty t$. The objective is to retain two distinct vortices in the flat plate plane, one representing the bound vorticity and one representing the wake vorticity. Recall from Eq. (3.67) that

$$\Xi_0 = X_0 + \sqrt{X_0^2 - (c/2)^2} \tag{5.39}$$

As Jones defined the position of the wake vortex using $X_0 = c/2 + s$, its position in the circle plane is given by

$$\Xi_0 = c/2 + s + \sqrt{c^2/4 + cs + s^2 - c^2/4} = c/2 + s + \sqrt{s^2 + cs} \tag{5.40}$$

Hence, both X_0 and Ξ_0 are functions of s, that is functions of time. As $s \to \infty$, the bound vortex moves to

$$\lim_{s\to\infty} \frac{c^2}{4\Xi_0(s)} = 0$$

so it lies at the midchord point when the flow has reached steady-state conditions. Consequently, in the steady state, the wake vortex lies at an infinite distance behind the trailing edge, while the bound vortex lies at the half chord.

In three dimensions, the wake and bound vortices become vortex segments lying at $x = X_0$ and $x = c^2/4\Xi_0$, respectively, and extending from $y = -b/2$ to $y = b/2$. By Helmholtz's theorems, these vortices cannot end in space so they must be connected by two other vortices in order to form a closed rectangle, or vortex ring. The situation is illustrated in Figure 5.7 at a time t after the start of the motion. The wing is centred at the half-chord point and the wake vortex has reached position $s = Q_\infty t$ behind the trailing edge, while the starting vortex lies at position $c^2/4\Xi_0$. Both of these vortex segments have spanwise length b. They are connected by two trailing vortices lying at $\pm b/2$, starting at $c^2/4\Xi_0$ and ending at $s + c/2 = X_0$. The length of these trailing vortices is then $\bar{x} = X_0 - c^2/4\Xi_0$. This distance can be written as

$$\bar{x} = \frac{1}{2}\left(\Xi_0 + \frac{c^2}{4\Xi_0}\right) - \frac{c^2}{4\Xi_0} = \frac{1}{2}\left(\Xi_0 - \frac{c^2}{4\Xi_0}\right)$$

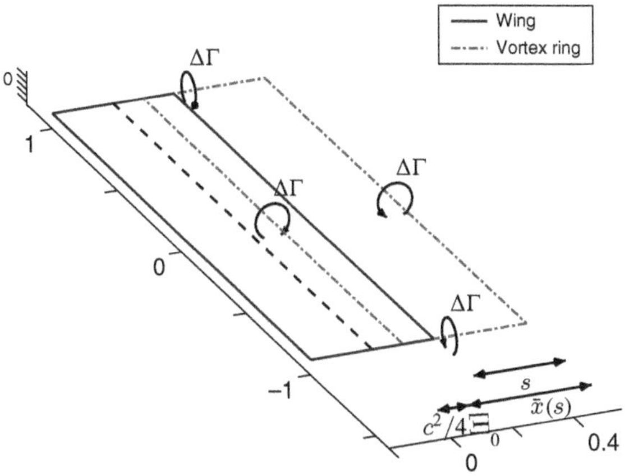

Figure 5.7 Single vortex ring placed on wing.

Substituting from Eq. (5.40), we obtain

$$\bar{x} = \frac{1}{2}\left(c/2 + s + \sqrt{s^2 + cs} - \frac{c^2}{4(c/2 + s + \sqrt{s^2 + cs})} \right) = \sqrt{s^2 + cs} \tag{5.41}$$

It must be stressed that the total circulation added to the flow by the vortex ring is zero. This total circulation is given by

$$\Gamma = \oint \Delta\Gamma \mathrm{d}s$$

which is a contour integral of a constant quantity and is therefore equal to zero. In fact, all the aerodynamic models we will present in this chapter and the next will make use of either vortex rings or doublet panels, which are equivalent. Consequently, Kelvin's theorem will always be automatically satisfied and will not be mentioned again until Chapter 7.

If only the vortex ring of Figure 5.7 is used in the aerodynamic modelling, the bound circulation will be constant in the spanwise direction. As we saw in Section 5.2, in steady flow (which corresponds to an impulsively started flow when $t \to \infty$), the bound circulation varies along the span. Therefore, the wing and wake cannot be modelled by a single vortex ring; we need to superimpose many vortex rings as we did for the steady lifting line method. Figure 5.8a demonstrates an example of this type of superposition. At the first time instance after the start of the motion, $t = \Delta t$, there are three concentric vortex rings with strengths $\Delta\Gamma(\bar{x}_1, y_1)$, $\Delta\Gamma(\bar{x}_1, y_2)$ and $\Delta\Gamma(\bar{x}_1, y_3)$. The lengths in the free stream direction of all three vortices is $\bar{x}_1$, while their lengths in the spanwise direction are $2y_1 = b/5$, $2y_2 = 3b/5$ and $2y_3 = b$. As in the case of the steady lifting line method, the strengths of the bound vortices are superimposed so that the total strength at the midspan point is $\Delta\Gamma(y_1) + \Delta\Gamma(y_2) + \Delta\Gamma(y_3)$, decreasing to $\Delta\Gamma(y_3)$ at the wingtips. If the number of vortex rings is increased to infinity, then the strength of each ring will be equal to $\mathrm{d}\Gamma$.

Figure 5.8b plots the wake at a time instance $t = 3\Delta t$ after the start of the motion. The three concentric vortex rings that were shed at $t = \Delta t$ now have length $\bar{x}_2$. Three more vortex rings were shed at $t = 2\Delta t$, whose length is now $\bar{x}_2$. Finally, three new vortex rings have just been shed with length $\bar{x}_1$. The bound vortices of all nine vortex rings are

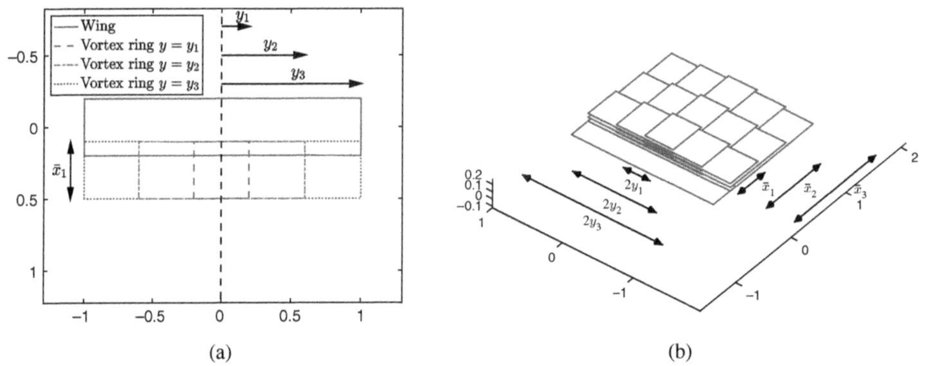

Figure 5.8 Jones' vortex ring model. (a) Superposition of vortex rings at $t = \Delta t$ and (b) Superposition of vortex rings at $t = 3\Delta t$.

coincident, which means that the bound circulation is the superposition of the strengths of all the vortex rings that exist at each time instance.

In order to proceed, we need to calculate the downwash induced by all the vortex rings. Recall that Eq. (5.23) shows that in steady conditions, the downwash is constant along the span for a wing with an elliptical lift distribution. We can assume that this is still true in the unsteady case so that we only need to calculate the downwash at one point along the span. Without loss of generality, we can choose the point $\bar{x} = 0, y = 0$. The task is then to calculate the downwash $dw(\bar{x}_0, y)$ induced by a vortex ring with strength $d\Gamma(y)$, span $2y$ and chord $\bar{x}_0$ at the midpoint of its bound vortex; then we can integrate this downwash between $y = (0, b/2)$ and $\bar{x}_0 = (0, \bar{x})$, that is

$$w(\bar{x}) = \int_0^{\bar{x}} \int_0^{b/2} dw(\bar{x}_0, y)$$

where $\bar{x}$ is the length of the longest vortex ring at time t. As $\bar{x}$ is a function of t, $w(\bar{x}) = w(t)$.

Figure 5.9 plots the ring at $(\bar{x}_0, y)$ and defines its vertices using numbers 1 to 4. As mentioned in Appendix A.7, the calculation of the velocity induced by a vortex ring must be carried out in the order 1–4. Note that the bound vortex segment (segment 41) does not induce any velocities on itself. The left and right trailing segments (segments 12 and 34), as well as the wake segment (segment 23), induce velocities on the midpoint of the bound vortex that can be calculated from Eq. (A.42), repeated here for convenience

$$\mathbf{u} = \frac{\Gamma}{4\pi} \frac{\mathbf{r}_{12} \times \mathbf{r}_1}{|\mathbf{r}_1 \times \mathbf{r}_2|^2} \mathbf{r}_{12} \cdot \left(\frac{\mathbf{r}_2}{|\mathbf{r}_2|} - \frac{\mathbf{r}_1}{|\mathbf{r}_1|} \right) \tag{5.42}$$

We will start with the right trailing edge segment 12, whose vertices are given by $\mathbf{x}_1 = (0, y, 0)$ and $\mathbf{x}_2 = (\bar{x}_0, y, 0)$. Then, since the point at which we want to calculate the induced velocity lies at the origin, $\mathbf{r}_1 = (0, y, 0)$, $\mathbf{r}_2 = (\bar{x}_0, y, 0)$ and $\mathbf{r}_{12} = \mathbf{r}_2 - \mathbf{r}_1 = (\bar{x}_0, 0, 0)$. Equation (5.42) becomes

$$dw_{12} = \frac{d\Gamma(y)}{4\pi} \frac{\bar{x}_0}{y\sqrt{\bar{x}_0^2 + y^2}}$$

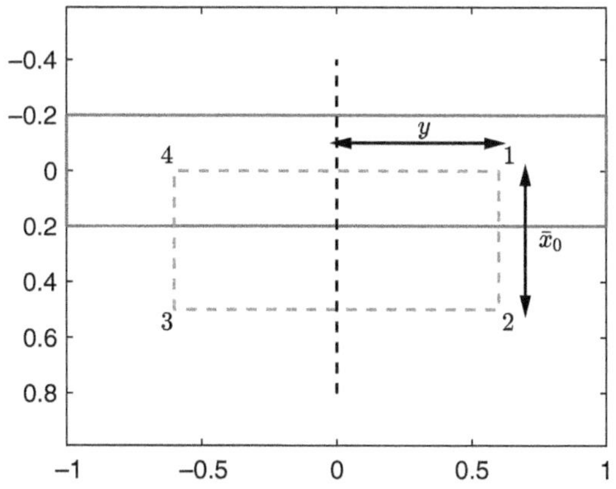

Figure 5.9 Vortex ring of span $2y$ and length $\bar{x}_0$.

since the u and v components of the induced velocity are zero. We now turn to the wake segment 23, for which $\mathbf{x}_1 = (\bar{x}_0, y, 0)$ and $\mathbf{x}_2 = (\bar{x}_0, -y, 0)$. Then, $\mathbf{r}_1 = (\bar{x}_0, y, 0)$, $\mathbf{r}_2 = (\bar{x}_0, -y, 0)$ and $\mathbf{r}_{12} = \mathbf{r}_2 - \mathbf{r}_1 = (0, -2y, 0)$. Hence,

$$dw_{23} = \frac{d\Gamma(y)}{2\pi} \frac{y}{\bar{x}_0 \sqrt{\bar{x}_0^2 + y^2}}$$

Finally, we treat the left vortex segment 34, for which $\mathbf{x}_1 = (\bar{x}_0, -y, 0)$ and $\mathbf{x}_2 = (0, -y, 0)$.

$$dw_{34} = \frac{d\Gamma(y)}{4\pi} \frac{\bar{x}_0}{y\sqrt{\bar{x}_0^2 + y^2}}$$

Assembling all the contributions yields a total downwash induced at the origin by this vortex ring equal to

$$dw(\bar{x}_0, y) = \frac{d\Gamma(y)}{2\pi} \left(\frac{\bar{x}_0}{y} + \frac{y}{\bar{x}_0} \right) \frac{1}{\sqrt{\bar{x}_0^2 + y^2}} \tag{5.43}$$

The elliptical lift distribution is given by Eq. (5.17), so that we can write

$$\Gamma(y) = \Gamma_0 \sin\theta$$

if we use the usual substitution $y = (b/2)\cos\theta$. Then

$$d\Gamma(y) = \Gamma_0 \cos\theta d\theta$$

and Eq. (5.43) becomes

$$dw(\bar{x}_0, y) = \frac{\Gamma_0}{2\pi} \left(\frac{\bar{x}_0}{(b/2)} + \frac{(b/2)\cos^2\theta}{\bar{x}_0} \right) \frac{1}{\sqrt{\bar{x}_0^2 + y^2}} d\theta \tag{5.44}$$

Jones states that $\bar{x}_0^2 + y^2$ can be written as

$$\bar{x}_0^2 + y^2 = \frac{(b/2)^2}{k_0^2}(1 - k_0^2\sin^2\theta) \tag{5.45}$$

where

$$k_0^2 = \frac{1}{1 + \left(\frac{2\bar{x}_0}{b}\right)^2} \tag{5.46}$$

Expression (5.45) can be easily proven by multiplying out the right-hand side; this proof is left as an exercise for the reader. Substituting the expression into Eq. (5.44) and integrating over θ from $\pi/2$ to 0 (corresponding to y values from 0 to $b/2$), we obtain

$$w(\bar{x}_0, \Gamma_0) = \frac{\Gamma_0}{\pi b} k_0 \int_0^{\pi/2} \left(\frac{\bar{x}_0}{(b/2)} + \frac{(b/2)\cos^2\theta}{\bar{x}_0} \right) \frac{1}{\sqrt{1 - k_0^2\sin^2\theta}} d\theta \tag{5.47}$$

In order to evaluate this integral, we will use the complete elliptic integrals of the first and second kind,

$$K(k_0) = \int_0^{\pi/2} \frac{d\theta}{\sqrt{1 - k_0^2 \sin^2\theta}} \tag{5.48}$$

$$E(k_0) = \int_0^{\pi/2} \sqrt{1 - k_0^2 \sin^2\theta}\, d\theta \tag{5.49}$$

Writing $\cos^2\theta = 1 - \sin^2\theta$ Eq. (5.47) becomes

$$w(\bar{x}_0, \Gamma_0) = \frac{\Gamma_0}{\pi b} k_0 \left[\frac{\bar{x}_0}{(b/2)} K(k_0) + \frac{(b/2)}{\bar{x}_0} \left(K(k_0) + \frac{1}{k_0^2} \int_0^{\pi/2} \frac{d\theta}{\sqrt{1 - k_0^2 \sin^2\theta}} \right.\right.$$

$$\left.\left. - \frac{1}{k_0^2} \int_0^{\pi/2} \frac{d\theta}{\sqrt{1 - k_0^2 \sin^2\theta}} - \frac{1}{k_0^2} \int_0^{\pi/2} \frac{k_0^2 \sin^2\theta}{\sqrt{1 - k_0^2 \sin^2\theta}} d\theta \right) \right]$$

where we have substituted from Eq. (5.48) and added and subtracted the integral $(1/k_0^2) \int_0^{\pi/2} d\theta/\sqrt{1 - k_0^2 \sin^2\theta}$ in the second term inside the square bracket. Simplifying and substituting from Eq. (5.49), we obtain the final value of the downwash induced by all the vortex rings with length $\bar{x}_0$ at the midpoint of the bound vortex ring

$$w(\bar{x}_0, \Gamma_0) = \frac{\Gamma_0}{\pi b} \left(\frac{2\bar{x}_0}{b} k_0 K(k_0) + \frac{b}{2\bar{x}_0} \left(K(k_0) \left(k_0 - \frac{1}{k_0} \right) + \frac{E(k_0)}{k_0} \right) \right) \tag{5.50}$$

Elliptic integrals can be calculated using Matlab function `ellipke.m` so that Eq. (5.50) is a complete analytical expression.

Equation (5.50) gives the downwash induced by all the vortex rings of length $\bar{x}_0$ on the midpoint of the bound vortex. However, Wagner theory calculates the 2D aerodynamic loads by summing the contributions of the wake vortices. Therefore, if 2D theory is used in order to calculate the 3D aerodynamic loads due to vortex rings whose wake vortices lie at $\bar{x}_0$, there will be an additional downwash contribution from the 2D vortex lying at the same position, given by

$$\frac{\Gamma_0}{2\pi \bar{x}_0}$$

As this downwash is already taken into account in the Wagner load calculation it must be subtracted from the total 3D downwash of Eq. (5.50), leading to

$$w(\bar{x}_0, \Gamma_0) = \frac{\Gamma_0}{\pi b} \left(\frac{2\bar{x}_0}{b} k_0 K(k_0) + \frac{b}{2\bar{x}_0} \left(K(k_0) \left(k_0 - \frac{1}{k_0} \right) + \frac{E(k_0)}{k_0} - 1 \right) \right) \tag{5.51}$$

Equation (5.51) gives the final value of downwash that we will be using for the 3D load calculation. Note that $w(0, \Gamma_0)$ is undefined, but it can be shown numerically that $\lim_{\bar{x}\to 0} w(\bar{x}, \Gamma_0) = 0$. Furthermore,

$$\lim_{\bar{x}\to\infty} w(\bar{x}, \Gamma_0) = \frac{\Gamma_0}{2b}$$

which is the steady lifting line result of Eq. (5.23). Figure 5.10 plots the variation of $w_\Gamma(\bar{x}) = w(\bar{x}, \Gamma_0)/\Gamma_0$ with $\bar{x}$ for elliptical wings with aspect ratios of 3 and 6. It can be seen that the smaller the aspect ratio, the faster the downwash tends towards the steady-state value.

At each time instance t_0, a row of vortex rings is shed with incremental strength $d\Gamma_0(t_0)$ and with initial trailing segment length $\bar{x}(t_0) = 0$. At a subsequent time instance t, the trailing segment length will have increased to $\bar{x}(t - t_0)$ and the row of vortex rings will induce an incremental downwash $dw_i(t, t_0)$ on the bound vortex whose strength is calculated from Eq. (5.51) as

$$dw_i(t, t_0) = w_\Gamma(t - t_0) d\Gamma_0(t_0)$$

where

$$w_\Gamma(t) = \frac{1}{\pi b}\left(\frac{2\bar{x}(t)}{b} k(t)K(k(t)) + \frac{b}{2\bar{x}(t)}\left(K(k(t))\left(k(t) - \frac{1}{k(t)}\right) + \frac{E(k(t))}{k(t)} - 1\right)\right) \tag{5.52}$$

noting that, from Eqs. (5.41) and (5.46),

$$\bar{x}(t) = \sqrt{(Q_\infty t)^2 + c_0 Q_\infty t},\ k(t)^2 = \frac{1}{1 + \left(\frac{2\bar{x}(t)}{b}\right)^2} \tag{5.53}$$

The total downwash acting at the mispan point of the wing at a general time t is then given by the sum of all the incremental downwash changes $dw_i(t, t_0)$ that occurred from time $t_0 = 0$ to time $t_0 = t$ and the downwash induced by the initial value of the circulation $\Gamma_0(0)$, that is

$$w_i(t) = w_\Gamma(t)\Gamma_0(0) + \int_0^t w_\Gamma(t - t_0) \left.\frac{\partial \Gamma_0}{\partial t}\right|_{t_0} dt_0 \tag{5.54}$$

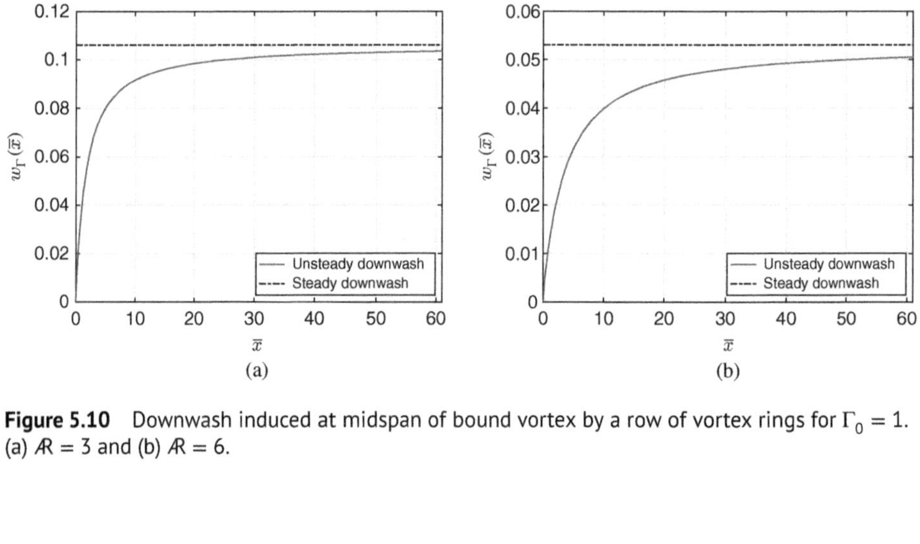

Figure 5.10 Downwash induced at midspan of bound vortex by a row of vortex rings for $\Gamma_0 = 1$. (a) $A\!\!R = 3$ and (b) $A\!\!R = 6$.

Equation (5.54) is not sufficient to calculate $w_i(t)$ since $\Gamma_0(t)$ is unknown. Jones calculated the latter using the vortex strength from 2D theory. Recall that the Kutta condition for 2D theory is given by Eq. (3.77), which can be solved for the unknown 2D incremental vortex strength $\bar{\Gamma}(X_0, t)\mathrm{d}X_0$. Then, the total circulation in the 2D wake at each time instance is given by

$$\Gamma_{2D}(t) = \int_{c_0/2}^{c_0/2+Q_\infty t} \bar{\Gamma}(X_0, t)\mathrm{d}X_0$$

The 2D circulation depends not only on time but also on α (see Eq. (3.77)), so that it must be denoted by $\Gamma_{2D}(\alpha, t)$. Jones combines this 2D circulation with the 3D effective angle of attack in order to determine the 3D circulation. By analogy to Eq. (5.9), we can define the 3D effective angle of attack as

$$\alpha_{\mathrm{eff}} = \alpha - \frac{w_i}{Q_\infty} \tag{5.55}$$

Note that this equation implies that the induced angle of attack is given by $\alpha_i = w_i/Q_\infty$, which is the opposite of the definition of Eq. (5.5). The difference is due to the fact that Jones defined w_i as positive downwards.

If a change in effective angle of attack $\mathrm{d}\alpha_{\mathrm{eff}}(t_0)$ occurs at time t_0, the incremental change in 3D circulation at all subsequent times is assumed by Jones to be given by

$$\mathrm{d}\Gamma_0(t, t_0) = \Gamma_{2D}(\mathrm{d}\alpha_{\mathrm{eff}}(t_0), t - t_0)$$

As Wagner theory is linear in α, this expression can be re-written as

$$\mathrm{d}\Gamma_0(t, t_0) = \frac{\Gamma_{2D}(t - t_0)}{\alpha}\mathrm{d}\alpha_{\mathrm{eff}}(t_0) = -\frac{1}{Q_\infty \alpha}\Gamma_{2D}(t - t_0)\mathrm{d}w_i(t_0)$$

where we have also substituted for $\mathrm{d}\alpha_{\mathrm{eff}}(t_0)$ from Eq. (5.55). The total 3D circulation at time t is then given by the sum of all the incremental changes $\mathrm{d}\Gamma_0(t, t_0)$ that occurred from $t_0 = 0$ to $t_0 = t$ and the circulation due to the initial value of the effective angle of attack, $\alpha_{\mathrm{eff}}(0) = \alpha - w_i(0)/Q_\infty$. Letting $\mathrm{d}w_i(t_0) = \partial w_i/\partial t|_{t_0}\mathrm{d}t_0$, we obtain

$$\Gamma_0(t) = \Gamma_{2D}(t)\left(1 - \frac{w_i(0)}{Q_\infty \alpha}\right) - \frac{1}{Q_\infty \alpha}\int_0^t \Gamma_{2D}(t - t_0)\left.\frac{\partial w_i}{\partial t}\right|_{t_0} \mathrm{d}t_0 \tag{5.56}$$

Equations (5.54) and (5.56) constitute a system of two equations with two unknowns, $w_i(t)$ and $\Gamma_0(t)$. Once $w_i(t)$ has been calculated, we can use again 2D theory in order to calculate the time response of the wing's lift coefficient. We repeat Eq. (3.114) for the circulatory lift obtained from Wagner theory for the general motion of a 2D airfoil

$$l_\Gamma(t) = \rho Q_\infty c\pi \left(w_{3/4}(0)\Phi(t) + \int_0^t \Phi(t - t_0)\dot{w}_{3/4}(t_0)\mathrm{d}t_0\right) \tag{5.57}$$

where we have substituted $c/2$ for the half-chord and changed the sign because here we define the lift as positive upwards. For an impulsively started motion, the circulatory lift is the only source of lift, $l(t) = l_\Gamma(t)$ and the upwash at the 3/4 chord point is simply $w_{3/4}(t) = Q_\infty \alpha$. However, for a section of a finite wing, the upwash must also include the downwash due to the trailing vortices, such that $w_{3/4}(y, t) = Q_\infty \alpha - w_i(y, t)$, where $w_i(y, t)$ is the downwash at spanwise station y. Then, the sectional lift acting on an impulsively started wing at spanwise position y is given by

$$l(y,t) = \rho Q_\infty^2 c_0 \pi \left(\left(\alpha - \frac{w_i(y,0)}{Q_\infty} \right) \Phi(t) - \frac{1}{Q_\infty} \int_0^t \Phi(t-t_0) \left. \frac{\partial w_i(y,t)}{\partial t} \right|_{t_0} \mathrm{d}t_0 \right)$$

Dividing by $\rho Q_\infty^2 c_0/2$ and setting $y = 0$, we obtain the sectional lift at the midspan point

$$C_l(0,t) = a_0 \left(\left(\alpha - \frac{w_i(0)}{Q_\infty} \right) \Phi(t) - \frac{1}{Q_\infty} \int_0^t \Phi(t-t_0) \left. \frac{\partial w_i(t)}{\partial t} \right|_{t_0} \mathrm{d}t_0 \right) \tag{5.58}$$

where $a_0 = 2\pi$ and $w_i(t) = w_i(0,t)$ is given by Eq. (5.54). Noting that the 2D lift coefficient for an impulsively started flat plate is $c_l(t) = a_0 \alpha \Phi(t)$, the equation above can be written as

$$C_l(0,t) = c_l(t) \left(1 - \frac{w_i(0)}{Q_\infty \alpha} \right) - \frac{1}{Q_\infty \alpha} \int_0^t c_l(t-t_0) \left. \frac{\partial w_i}{\partial t} \right|_{t_0} \mathrm{d}t_0 \tag{5.59}$$

As we are assuming that the downwash is constant along the span for an elliptical planform, the sectional lift will also be constant and the total lift acting on the wing is obtained from Eq. (5.8) as $C_L(t) = C_l(0,t)$. The problem with Eqs. (5.54) and (5.56) is that they are integral and therefore have no obvious analytical solution. Nevertheless, they can be solved numerically, as we will demonstrate in the next example.

Now, we will extend Jones' approach to the calculation of the drag response of an impulsively started elliptical wing. Recall that the drag response, $d(t)$, of a 2D flat plate in general motion is given by Eq. (3.123) as

$$d(t) = -\rho\pi \frac{c}{4} \left(-\frac{2 l_\Gamma(t)}{\rho Q_\infty c \pi} + \frac{c}{2} \dot{\alpha}(t) \right)^2 + \alpha(t) l(t)$$

where $l_\Gamma(t)$ is the 2D circulatory lift of Eq. (5.57) and we have again changed the sign of the lift as it is now defined positive upwards. Substituting $l_\Gamma(t) = l(t)$, $\alpha(t) = \alpha$ and $\dot{\alpha}(t) = 0$ for an impulsively started flat plate, we obtain

$$d(t) = -\rho\pi \frac{c}{4} \left(\frac{2l(t)}{\rho Q_\infty c \pi} \right)^2 + \alpha l(t)$$

Dividing throughout by $\rho Q_\infty^2 c/2$ yields the drag coefficient as

$$c_d(t) = -\frac{1}{2\pi} c_l^2(t) + \alpha c_l(t) \tag{5.60}$$

Finally, if instead of $c_l(t)$ for a 2D wing, we use the sectional lift coefficient $C_l(0,t)$ for a 3D wing, we obtain the sectional drag coefficient acting on the midspan of an impulsively started wing

$$C_d(0,t) = -\frac{1}{2\pi} C_l^2(0,t) + \alpha C_l(0,t) \tag{5.61}$$

As we argued earlier, since we assume that the downwash is constant along the span because of the elliptical planform, the sectional lift is also constant along the span. Consequently, the sectional drag is constant too and Eq. (5.28) gives $C_D(t) = C_d(0,t)$.

Example 5.2 *Calculate the unsteady lift acting on an impulsively started wing with elliptical planform*

The chord of the wing at its midspan is given by $c_0 = 2$ m and its aspect ratio is 6. From Eq. (5.20), its span is then $b = 9.43$ m. The free stream airspeed is set to $Q_\infty = 10$ m/s and the angle of attack to $\alpha = 5°$.

The first step is to calculate the circulation response of a 2D flat plate with the same chord using Wagner theory, exactly as was done in example 3.7. We set the maximum length of the wake to $30c_0$, and we calculate Γ_{2D} at $n = 3001$ time instances, such that the positions of the wake vortices are given by $X_{0_i} = i\Delta X_0$, where $\Delta X_0 = 30c_0/(n-1)$. The non-dimensional time step is $\Delta t = \Delta X_0/Q_\infty$, and the instantaneous wake length is $s_i = X_{0_i} - c0/2$. We calculate the strengths per unit length of the 2D vortices in the wake, $\bar{\Gamma}_i$, using Eq. (3.80) and then the dimensional strengths from $\Gamma_i = \bar{\Gamma}_i \Delta X_0$. The total circulation in the 2D wake at every time instance is given by

$$\Gamma_{2D}(t_i) = \sum_{j=1}^{i} \Gamma_j$$

Next, we deal with the flow around the 3D elliptical wing. We calculate the length of the trailing vortices, $\bar{x}(t_i)$, and the argument of the elliptical integrals $k(t_i)$ from Eq. (5.53). Jones recommends to adjust the starting value of $\bar{x}(t_i)$ so that

$$\bar{x}(t_i) = \frac{b}{\mathcal{R}} + \sqrt{(Q_\infty t_i)^2 + c_0 Q_\infty t_i}$$

where $b/\mathcal{R}$ is the mean chord. Then we apply Eq. (5.52) to calculate $w_\Gamma(t_i)$. At the first time instance, the convolutions of Eqs. (5.54) and (5.56) can be written as

$$w_i(t_1) = w_\Gamma(t_1)\Gamma_0(t_1)$$
$$\Gamma_0(t_1) = \Gamma_{2D}(t_1)\left(1 - \frac{w_i(t_1)}{Q_\infty \alpha}\right)$$

Solving this system of equations for $\Gamma_0(t_1)$, we obtain

$$\Gamma_0(t_1) = \frac{\Gamma_{2D}(t_1)}{1 + \frac{w_\Gamma(t_1)}{Q_\infty \alpha}},\ w_i(t_1) = w_\Gamma(t_1)\frac{\Gamma_{2D}(t_1)}{1 + \frac{w_\Gamma(t_1)}{Q_\infty \alpha}}$$

At the ith time instance Eqs. (5.54) and (5.56) become

$$w_i(t_i) = w_\Gamma(t_i)\Gamma_0(t_1) + w_\Gamma(t_1)\Delta\Gamma_0(t_i) + w_\Gamma(t_2)\Delta\Gamma_0(t_{i-1}) + \cdots + w_\Gamma(t_{i-1})\Delta\Gamma_0(t_2)$$
$$\Gamma_0(t_i) = \Gamma_{2D}(t_i)\left(1 - \frac{w_i(t_1)}{Q_\infty \alpha}\right) - \frac{\Gamma_{2D}(t_1)\Delta w_i(t_i)}{Q_\infty \alpha} - \frac{\Gamma_{2D}(t_2)\Delta w_i(t_{i-1})}{Q_\infty \alpha} - \cdots - \frac{\Gamma_{2D}(t_{i-1})\Delta w_i(t_2)}{Q_\infty \alpha}$$

where $\Delta\Gamma_0(t_i) = \Gamma_0(t_i) - \Gamma_0(t_{i-1})$ and $\Delta w_i(t_i) = w_i(t_i) - w_i(t_{i-1})$. We can re-write these expressions as the system of equations

$$\begin{pmatrix} 1 & 0 & 0 & -w_\Gamma(t_1) \\ 1 & -1 & 0 & 0 \\ 0 & \frac{\Gamma_{2D}(t_1)}{Q_\infty \alpha} & 1 & 0 \\ 0 & 0 & 1 & -1 \end{pmatrix} \begin{pmatrix} w_i(t_i) \\ \Delta w_i(t_i) \\ \Gamma_0(t_i) \\ \Delta\Gamma_0(t_i) \end{pmatrix} = \begin{pmatrix} w_\Gamma(t_i)\Gamma_0(t_1) + \sum_{j=2}^{i-1} w_\Gamma(t_j)\Delta\Gamma_0(t_{i-j+1}) \\ w_i(t_{i-1}) \\ \Gamma_{2D}(t_i)\left(1 - \frac{w_i(t_1)}{Q_\infty \alpha}\right) - \sum_{j=2}^{i-1} \frac{\Gamma_{2D}(t_j)\Delta w_i(t_{i-j+1})}{Q_\infty \alpha} \\ \Gamma_0(t_{i-1}) \end{pmatrix}$$

which can be solved easily for $w_i(t_i)$, $\Delta w_i(t_i)$, $\Gamma_0(t_i)$ and $\Delta\Gamma_0(t_i)$ at each time instance.

The lift coefficient of Eq. (5.59) can be written in discrete time as

$$C_l(0, t_i) = c_l(t_i)\left(1 - \frac{w_i(t_1)}{Q_\infty \alpha}\right) - \sum_{j=1}^{i-1} \frac{c_l(t_j)\Delta w_i(t_{i-j+1})}{Q_\infty \alpha}$$

whose evaluation is straightforward once $w_i(t_i)$ has been calculated. Then, the total drag is $C_L(t_i) = C_l(0, t_i)$. Finally, we calculate the sectional drag coefficient from Eq. (5.61) and set $C_D(t_i) = C_d(0, t_i)$.

Jones developed exponential approximations for $C_L(t)$ for wings with aspect ratios of 3 and 6, given by

$$C_{L_{\mathcal{R}=3}} = 1.2\pi\alpha\left(1 - 0.283e^{-0.540\tau}\right) \tag{5.62}$$

$$C_{L_{\mathcal{R}=6}} = 1.48\pi\alpha\left(1 - 0.361e^{-0.381\tau}\right) \tag{5.63}$$

where $\tau = 2Q_\infty t/c_0$ is the non-dimensional time.

Figure 5.11 plots the lift coefficient variation with time after the impulsive start for two different wings with elliptical planform and aspect ratios of 3 and 6. The Jones approximations of Eqs. (5.62) and (5.63) are also plotted, as well as the steady lift coefficients obtained from Eq. (5.26). Clearly, the agreement between the numerical solutions and Jones' approximations is not ideal, particularly for the $\mathcal{R} = 3$ case. This is due to the fact that Jones adapted his approximations in order to conform to an alternative value for $C_L(0)$, denoted by the dash-dot line in the figure. This value is given by

$$C_L(0) = \frac{\pi}{E(e)}\alpha \tag{5.64}$$

where $e = \sqrt{1 - (c_0/b)^2}$ is the eccentricity of the ellipse. Even after selecting the initial wake length as $b/\mathcal{R}$, the lift coefficient calculated from Eq. (5.59) cannot conform to this value; hence, Jones adapted the exponential approximations in order to do so.

Figure 5.12 plots the drag coefficient variation with time for the same two aspect ratios. The drag jumps initially to a value of around 0.012 and then decreases exponentially to the

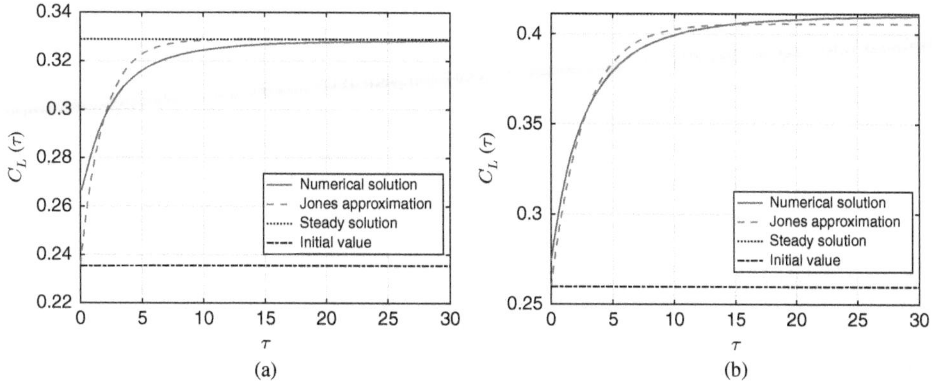

Figure 5.11 Lift coefficient response of impulsively started elliptical wings with two aspect ratios. (a) $\mathcal{R} = 3$ and (b) $\mathcal{R} = 6$.

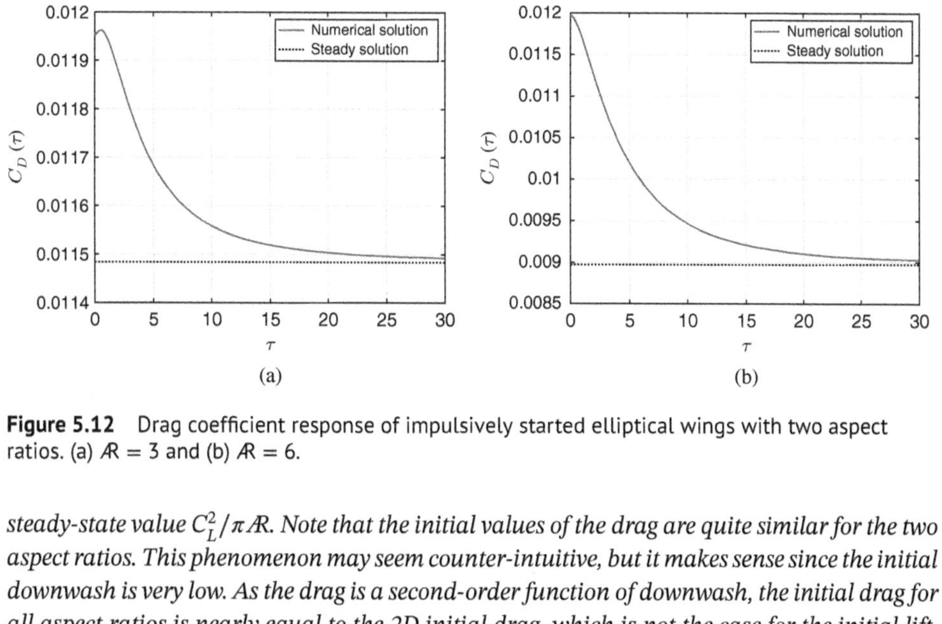

Figure 5.12 Drag coefficient response of impulsively started elliptical wings with two aspect ratios. (a) $\mathcal{R} = 3$ and (b) $\mathcal{R} = 6$.

steady-state value $C_L^2/\pi\mathcal{R}$. Note that the initial values of the drag are quite similar for the two aspect ratios. This phenomenon may seem counter-intuitive, but it makes sense since the initial downwash is very low. As the drag is a second-order function of downwash, the initial drag for all aspect ratios is nearly equal to the 2D initial drag, which is not the case for the initial lift. This example is solved by Matlab code `impulsive_Jones.m`*.*

In this example, the Jones solution was calculated numerically by discretising in time all the equations. Berci (2016) evaluated the same solution analytically, by approximating Eq. (5.52) as a sum of exponential terms of the form

$$w_\Gamma(t) = \frac{1}{\pi\mathcal{R}}\left(1 - \sum_{i=1}^{n} A_i e^{-B_i(\tau-\tau_0)/\mathcal{R}}\right) \tag{5.65}$$

where A_i and B_i are unknown coefficients to be determined by the solution. This approach takes advantage of the fact that Wagner's function $\Phi(t)$ can also be approximated as a sum of exponentials, as seen in Eq. (3.94). Note that expression (5.52) automatically satisfies the asymptotic conditions $w_\Gamma(0) = 0$ and $w_\Gamma(\infty)$. Finally, we will note that other authors have developed solutions for the impulsively started elliptical wing, such as Phlips et al. (1981). These approaches are generally referred to as unsteady lifting line methods.

5.3.2 Unsteady Lifting Line Solution

Lifting line theory makes use of thin airfoil theory and the Kutta–Joukowky theorem in order to determine the downwash caused by the trailing vortices. In other words, LLT assumes that the sectional lift of a 3D wing is given by 2D theory, but using an effective angle of attack due to the geometric angle and the downwash angle created by the trailing vortices. In order then to develop an unsteady version of lifting line theory, we need to adapt its main ingredients:

- Steady thin airfoil theory must be replaced by an unsteady alternative, Wagner theory (Section 3.4) or finite state theory (Section 3.6) for example.
- The steady Kutta–Joukowsky theorem must be replaced by an unsteady alternative.
- The trailing edge vortices cannot have constant strength. Their strength will be changing in time and space as a result of the instantaneous changes in bound vorticity. Furthermore, as the wake travels with a velocity comparable to the free stream velocity, the length of the trailing vortices will be finite at the start of the motion.

We have discussed unsteady 2D aerodynamic theories in detail in Chapter 3 and any one of them can be used to develop an unsteady lifting line methodology. An unsteady Kutta–Joukowski theorem can be derived from the linearized 2D Euler equations applied to the flow of Figure 5.4. From Eq. (3.204), the chordwise variation in pressure difference between the upper and lower surfaces of the wing section plotted in Figure 5.4 can be written as

$$\Delta p_b(x,y,t) = \rho\frac{\partial}{\partial t}\int_{-c(y)/2}^{x} \gamma_b(x,y,t)\mathrm{d}x + \rho Q_\infty \cos\alpha(y)\gamma_b(x,y,t) \tag{5.66}$$

where Q_∞ is the free stream airspeed coming at an angle of attack $\alpha(y)$, $\gamma_b(x,y,t)$ is the bound vorticity and $c(y)$ is the chord at spanwise station y. The aerodynamic force in the z direction is obtained by integrating this pressure difference over the chord, that is

$$F_z(y,t) = \int_{-c/2}^{c/2} \Delta p_b(x,y,t)\mathrm{d}x$$

In the case of lifting line theory, there is a single bound vortex with strength $\Gamma(y,t)$, so that

$$\int_{-c/2}^{c/2} \gamma_b(x,y,t)\mathrm{d}x = \Gamma(y,t)$$

Hence, after carrying out the integration, the unsteady vertical force becomes

$$F_z(y,t) = \rho\dot{\Gamma}(y,t)c(y) + \rho Q_\infty \cos\alpha(y)\Gamma(y,t) \tag{5.67}$$

The aerodynamic forces in the x and y directions are still given by Eq. (5.11) in the unsteady case, such that

$$F_x(y,t) = -\rho\left(Q_\infty \sin\alpha(y) + w_i(y)\right)\Gamma(y,t),\ F_y(y,t) = 0 \tag{5.68}$$

The sectional lift is given by Eq. (5.12) so that

$$l(y,t) = F_z(y,t)\cos\alpha(y) - F_x(y,t)\sin\alpha(y) \approx \rho Q_\infty \Gamma(y,t) + \rho\dot{\Gamma}(y,t)c(y)$$

after neglesting products of small quantities. The sectional lift coefficient is obtained after dividing by $1/2\rho Q_\infty^2 c(y)$,

$$C_l(y,t) = \frac{l(y)}{\frac{1}{2}\rho Q_\infty^2 c(y)} = \frac{2\dot{\Gamma}(y,t)}{Q_\infty^2} + \frac{2\Gamma(y,t)}{Q_\infty c(y)} \tag{5.69}$$

Equation (5.69) is an unsteady version of the Kutta–Joukowski theorem; it relates unsteady sectional lift to the unsteady circulation on the lifting line of a finite wing.

The elliptical lift distribution is given by Eq. (5.17), while the chord distribution by Eq. (5.19). Substituting we obtain

$$C_l(y,t) = \frac{2\dot{\Gamma}_0(t)\sqrt{1-\left(\frac{2y}{b}\right)^2}}{Q_\infty^2} + \frac{2\Gamma_0(t)}{Q_\infty c_0} \tag{5.70}$$

Hence, the unsteady lifting line problem is to find the time variation of $\Gamma_0(t)$ such that we can then calculate $C_l(y,t)$ from Eq. (5.70) and $C_L(t)$ from

$$C_L(t) = \frac{1}{S}\int_{-b/2}^{b/2} c(y)C_l(y,t)\mathrm{d}y = \frac{16\dot{\Gamma}_0(t)}{3\pi Q_\infty^2} + \frac{2\Gamma_0(t)}{Q_\infty c_0} \tag{5.71}$$

The value of $\Gamma_0(t)$ will depend on the strength and length of the trailing vortices, which depend in turn on the type of motion. Here, we will concentrate on impulsively started motion, whereby a 3D flat plate wing accelerates from rest impulsively to a constant airspeed Q_∞ and with constant geometric angle of attack α.

At the mid-span point, $y = 0$, Eq. (5.70) becomes

$$C_l(0,t) = \frac{2\dot{\Gamma}_0(t)}{Q_\infty^2} + \frac{2\Gamma_0(t)}{Q_\infty c_0} \tag{5.72}$$

We return to the sectional lift coefficient calculated from the convolution of the 2D lift and the 3D downwash (Eq. (5.59)) which we repeat here for convenience

$$C_l(0,t) = c_l(t)\left(1 - \frac{w_i(0)}{Q_\infty \alpha}\right) - \frac{1}{Q_\infty \alpha}\int_0^t c_l(t-t_0)\left.\frac{\partial w_i}{\partial t}\right|_{t_0}\mathrm{d}t_0$$

Using the lifting line theory logic, Eqs. (5.72) and (5.58) both give the sectional lift coefficient at the midspan point so they must be equal. Consequently,

$$c_l(t)\left(1 - \frac{w_i(0)}{Q_\infty \alpha}\right) - \frac{1}{Q_\infty \alpha}\int_0^t c_l(t-t_0)\left.\frac{\partial w_i}{\partial t}\right|_{t_0}\mathrm{d}t_0 = \frac{2\dot{\Gamma}_0(t)}{Q_\infty^2} + \frac{2\Gamma_0(t)}{Q_\infty c_0} \tag{5.73}$$

Expressions (5.54) and (5.73) form a system of two integro-differential equations for $\Gamma_0(t)$ and $w_i(t)$. Again, this system has no obvious analytical solution; we will demonstrate the numerical solution in the following example.

Once $\Gamma_0(t)$ and $w_i(t)$ have been evaluated, the sectional lift response can be obtained from Eq. (5.72) and the total lift coefficient response from (5.71). The sectional drag response can be obtained from expression (5.61). The total drag is given by

$$C_D(t) = \frac{1}{S}\int_{-b/2}^{b/2} c(y)C_d(y,t)\mathrm{d}y = \frac{1}{S}\int_{-b/2}^{b/2} c(y)\left(-\frac{1}{2\pi}C_l^2(y,t) + \alpha C_l(y,t)\right)\mathrm{d}y$$

Substituting for $C_l(y,t)$ from Eq. (5.70) and carrying out the integration, we obtain

$$C_D(t) = -\frac{12\pi Q_\infty^2\Gamma_0^2(t) + 64Q_\infty c_0\dot{\Gamma}_0(t)\Gamma_0(t) + 9\pi c_0^2\dot{\Gamma}_0^2(t)}{6\pi^2 Q_\infty^4 c_0^2} + \alpha C_L(t) \tag{5.74}$$

Example 5.3 *Repeat example 5.2 using unsteady lifting line theory*

We start by obtaining the 2D solution, in the same way as in Example 5.2. Once we have calculated $\bar{\Gamma}(X0)$, we evaluate the 2D lift using Eq. (3.84) and then we calculate $\bar{x}(t_i)$ and $w_\Gamma(t_i)$.

At the first time instance, we set $\dot{\Gamma}(t_1) = 0$ in order to avoid numerical discontinuities. Equations (5.54) and (5.73) become

$$w_i(t_1) = w_\Gamma(t_1)\Gamma_0(t_1)$$

$$\frac{2}{Q_\infty c_0}\Gamma_0(t_1) = c_l(t_1)\left(1 - \frac{w_i(t_1)}{Q_\infty\alpha}\right)$$

which can be readily solved for $\Gamma_0(t_1)$ and $w_i(t_1)$, that is

$$\Gamma_0(t_1) = \frac{1}{\frac{2}{c_l(t_1)Q_\infty c_0} + \frac{w_\Gamma(t_1)}{Q_\infty\alpha}},\quad w_i(t_1) = w_\Gamma(t_1)\Gamma_0(t_1)$$

At the ith time instance Eqs. (5.54) and (5.73) yield

$$w_i(t_i) = w_\Gamma(t_i)\Gamma_0(t_1) + w_\Gamma(t_1)\Delta\Gamma_0(t_i) + w_\Gamma(t_2)\Delta\Gamma_0(t_{i-1}) + \cdots + w_\Gamma(t_{i-1})\Delta\Gamma_0(t_2)$$

$$\frac{2\Delta\Gamma(t_i)}{Q_\infty^2\Delta t} + \frac{2\Gamma_0(t_i)}{Q_\infty c_0} + = c_l(t_i)\left(1 - \frac{w_i(t_1)}{Q_\infty\alpha}\right) - \frac{c_l(t_1)\Delta w_i(t_i)}{Q_\infty\alpha} - \frac{c_l(t_2)\Delta w_i(t_{i-1})}{Q_\infty\alpha} - \cdots - \frac{c_l(t_{i-1})\Delta w_i(t_2)}{Q_\infty\alpha}$$

where $\Delta\Gamma_0(t_i) = \Gamma_0(t_i) - \Gamma_0(t_{i-1})$ and $\Delta w_i(t_i) = w_i(t_i) - w_i(t_{i-1})$. We write these last two expressions as a system of algebraic equations, as we did in Example 5.2,

$$\begin{pmatrix} 1 & 0 & 0 & -w_\Gamma(t_1) \\ 1 & -1 & 0 & 0 \\ 0 & \frac{c_l(t_1)}{Q_\infty\alpha} & \frac{2}{Q_\infty c_0} & \frac{2}{Q_\infty^2\Delta t} \\ 0 & 0 & 1 & -1 \end{pmatrix}\begin{pmatrix} w_i(t_i) \\ \Delta w_i(t_i) \\ \Gamma_0(t_i) \\ \Delta\Gamma_0(t_i) \end{pmatrix} = \begin{pmatrix} w_\Gamma(t_i)\Gamma_0(t_1) + \sum_{j=2}^{i-1} w_\Gamma(t_j)\Delta\Gamma_0(t_{i-j+1}) \\ w_i(t_{i-1}) \\ c_l(t_i)\left(1 - \frac{w_i(t_1)}{Q_\infty\alpha}\right) - \sum_{j=2}^{i-1}\frac{c_l(t_j)\Delta w_i(t_{i-j+1})}{Q_\infty\alpha} \\ \Gamma_0(t_{i-1}) \end{pmatrix}$$

which can be solved for the values of $w_i(t_i)$, $\Delta w_i(t_i)$, $\Gamma_0(t_i)$ and $\Delta\Gamma_0(t_i)$ at each time instance. Finally, the sectional lift coefficient at the midspan point is given by the discrete version of Eq. (5.72),

$$C_l(0, t_i) = \frac{2\Delta\Gamma_0(t_i)}{Q_\infty^2 \Delta t} + \frac{2\Gamma_0(t_i)}{Q_\infty c_0}$$

while the total lift coefficient variation with time from the discrete version of Eq. (5.71) is

$$C_L(t_i) = \frac{16\Delta\Gamma_0(t_i)}{3\pi Q_\infty^2 \Delta t} + \frac{2\Gamma_0(t_i)}{Q_\infty c_0}$$

We also evaluate the sectional drag coefficient response from Eq. (5.61) and the total drag from (5.74).

Figure 5.13 plots the lift and drag responses obtained from the present method for wings with aspect ratios of 3 and 6, along with the responses estimated using Jones' method of example 5.2 as well as Jones' approximations of Eqs. (5.62) and (5.63). It can be seen that the unsteady lifting line result is in agreement with the value $C_L(0)$ obtained from expression (5.64) for both aspect ratios. Consequently, the unsteady lifting line technique does not suffer from

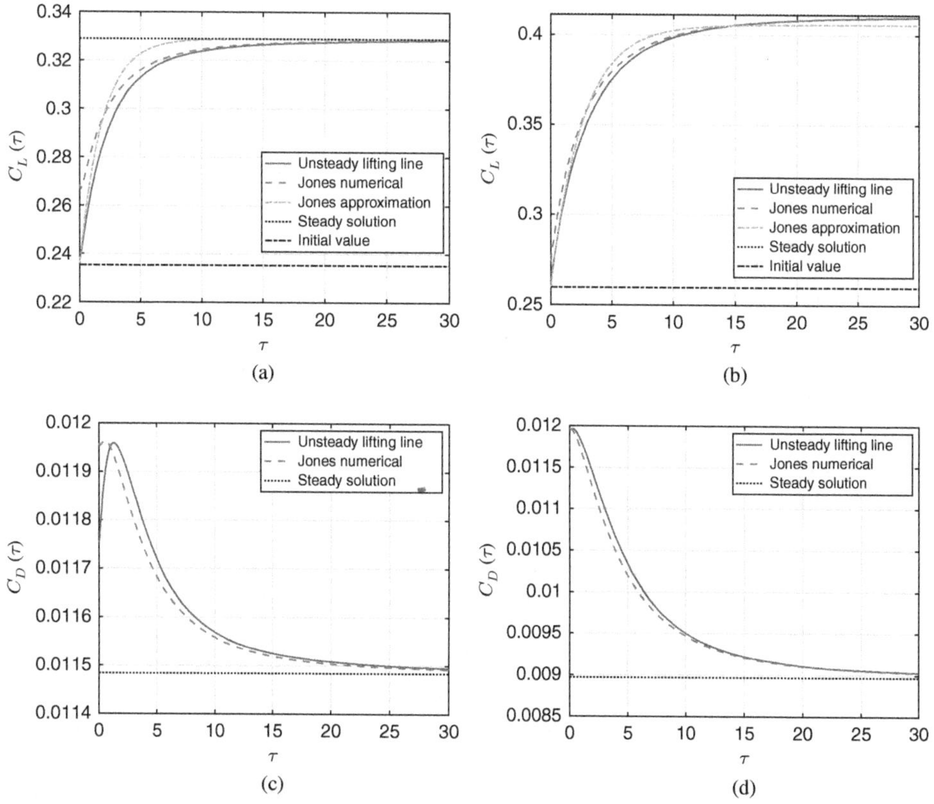

Figure 5.13 Lift coefficient response of impulsively started elliptical wings with two aspect ratios using unsteady lifting line theory. (a) $C_L(\tau)$ for $AR = 3$, (b) $C_L(\tau)$ for $AR = 6$, (c) $C_D(\tau)$ for $AR = 3$, and (d) $C_D(\tau)$ for $AR = 6$.

the inconsistency of Jones' approach at the start of the motion. For values of $\tau > 10$, the unsteady lifting line lift predictions are nearly identical to those of the numerical Jones solution. For values $\tau < 10$, there are some differences between the lift and drag responses obtained by the two methods. It should be noted though that these decrease as the aspect ratio increases. This example is solved by Matlab code `impulsive_LLT.m`.

Jones' method results in a simple quasi-analytical model of the impulsive start of an elliptical wing, but its usefulness is limited. Wings with elliptical planforms are not used in practice and impulsive starts are not often encountered either. Several unsteady lifting line theories can be applied to non-elliptical planforms and to general motion (see for Example (Boutet and Dimitriadis 2018; Izraelevitz et al. 2017; Jumper and Hugo 1994; Reissner 1947; Reissner and Stevens 1947)), but they all suffer from the fact that they cannot model swept wings. There exist correction techniques that can allow lifting line theory to represent sweep, but they are quite numerical in nature, thereby increasing the complexity and computational cost of LLT. We will therefore move on to panel methods, which are general and can deal with any planform and motion, as long as the flow remains attached and incompressible. As in the 2D case, there are two classes of panel methods:

- Lifting surface approaches that ignore the thickness of the wing
- Thickness modelling techniques

In this chapter, we will present one method from each class.

5.4 The Unsteady Vortex Lattice Method

The vortex lattice method (VLM) is a numerical approach for calculating the aerodynamic loads acting on a finite wing in both steady and unsteady flow conditions. It is described thoroughly by Katz and Plotkin (2001) and its popularity for unsteady aerodynamic and aeroelastic calculations has increased significantly over the last two decades, at least in the academic community.

The VLM solves the three-dimensional Laplace equation at discrete time instance and uses the unsteady Bernoulli equation to calculate the aerodynamic loads. It is a lifting surface method and therefore represents both the wing and wake as vortex sheets, modelled by means of a finite number of vortex rings. These rings can have any shape, but quadrilateral or triangular rings are most popular. The objective is then to calculate the strength of the vortex rings such that the boundary conditions are satisfied, namely impermeability and the Kutta condition.

The fact that wings are modelled as vorticity sheets means that wing thickness is neglected, and the wings are represented by their mean surface, often referred to as the camber surface. Figure 5.14 plots a cambered half-wing with sweep, taper and twist, along with its camber surface, which is the surface connecting all the sectional camber lines and is infinitely thin. The camber surface is then divided into m chordwise and n spanwise geometric panels, which are identified by $i = 1, \dots, m$, $j = 1, \dots, n$, i being the chordwise index and j being the spanwise index. Vortex rings are placed on the geometric panels, such that the leading edge of the i, j vortex ring coincides with the quarter of panel i, j while its

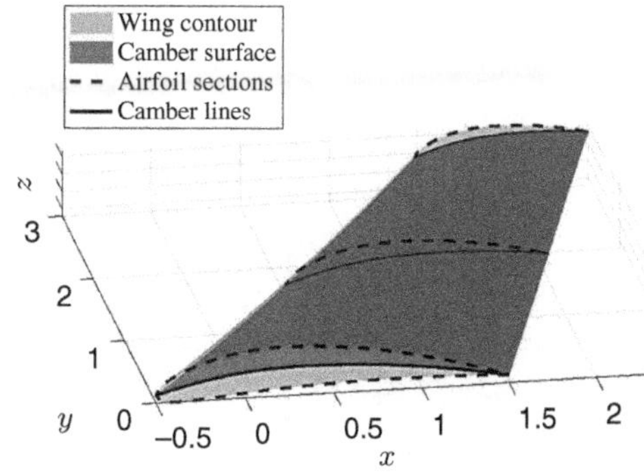

Figure 5.14 Camber surface of a half-wing.

trailing edge coincides with the quarter chord of panel $i+1, j$. It follows that the trailing edges of the last row of vortex rings lie in the wake. Each panel has a control point lying on its midspan 3/4 chord point. The wake is also made up of vortex rings, starting from the trailing edge of the last row of bound vortex rings and extending downstream. Wake vortex rings do not have control points.

The complete VLM panelling scheme is demonstrated in Figure 5.15 for a trapezoidal wing. Figure 5.15a draws all the geometric panels, bound vortex rings, control points and wake vortex rings for the entire wing, for $m = 3$ and $n = 8$, so that there are 24 panels in total. The following notation is used:

- The coordinates of the vertices of the geometric panels are denoted by $\mathbf{x}_{p_{i,j}} = (x_{p_{i,j}}, y_{p_{i,j}}, z_{p_{i,j}})$, for $i = 1, \dots, m+1, j = 1, \dots, n+1$.
- The coordinates of the vertices of the vortex rings are denoted by $\mathbf{x}_{v_{i,j}} = (x_{v_{i,j}}, y_{v_{i,j}}, z_{v_{i,j}})$, for $i = 1, \dots, m+1, j = 1, \dots, n+1$. The vortex rings and geometric panels have the same y coordinates, such that $y_{v_{i,j}} = y_{p_{i,j}}$.
- The coordinates of the control points are denoted by $\mathbf{x}_{c_{i,j}} = (x_{c_{i,j}}, y_{c_{i,j}}, z_{c_{i,j}})$, for $i = 1, \dots, m$, $j = 1, \dots, n$.
- The coordinates of the vertices of the wake panels are denoted by $\mathbf{x}_{w_{i,j}} = (x_{w_{i,j}}, y_{w_{i,j}}, z_{w_{i,j}})$, for $i = 1, \dots, m_w + 1$, $j = 1, \dots, n+1$, where m_w is the number of wake panels in the chordwise direction (the number of wake panels in the spanwise direction is n). A new row of wake panels is shed at the trailing edge at each time instance, such that at time $t_k = k\Delta t$, there are $m_w = k$ wake panels in the chordwise direction.

Figure 5.15b draws a single geometric panel, i, j, along with its corresponding vortex ring and control point. It can be seen that the two trailing segments of the vortex ring start at the quarter-chord points of the two trailing segments of the geometric panel. The control point lies at the panel's midspan point and $3(\Delta x_{p_{i,j}} + \Delta x_{p_{i,j+1}})/8$ behind the panel's leading edge, where $\Delta x_{p_{i,j}} = x_{p_{i+1,j}} - x_{p_{i,j}}$ and $\Delta x_{p_{i,j+1}} = x_{p_{i+1,j+1}} - x_{p_{i,j+1}}$.

Panel i, j can also be referred to using the single index $I = (i-1)n + j$, so that I takes values from 1 to mn. From this point onwards, we will use single indices I and J that both take values in the ranges $I, J = 1, \dots, mn$ or $I = 1, \dots, m_w n$, as well as double indices i, j, which take values in the ranges $i = 1, \dots, m$ or $i = 1, \dots, m_w$ and $j = 1, \dots, n$. The two numbering schemes are demonstrated in Figure 5.16 for a wing with three chordwise and eight spanwise panels. Both schemes will be used because each of them has advantages

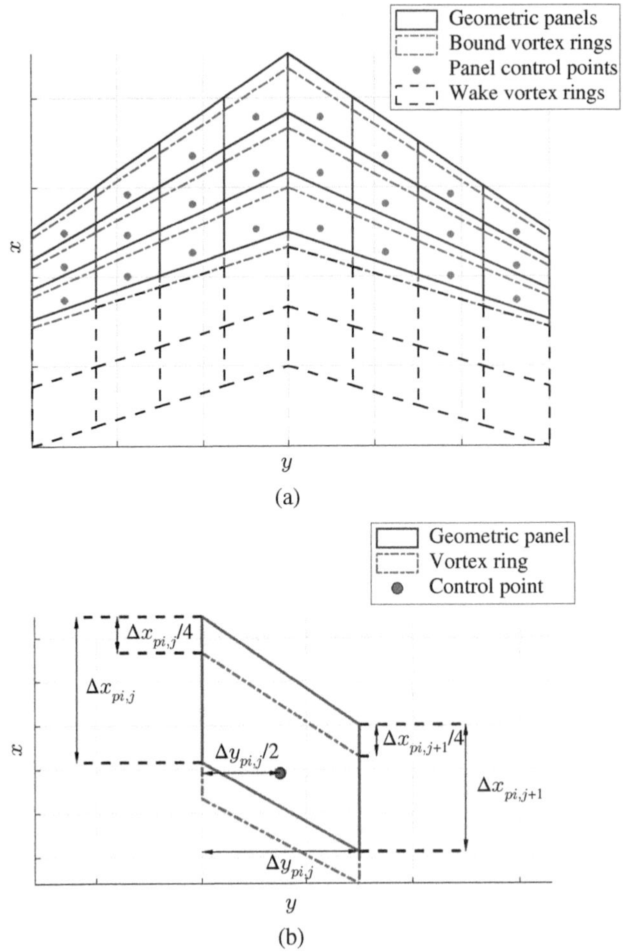

Figure 5.15 Vortex lattice panelling scheme. (a) Complete wing and (b) Single panel i, j.

and disadvantages. When calculating the influence of a panel on another it is easier to use single indices. The double index scheme is better for determining the chordwise neighbours of each panel. Using the double index scheme, the two chordwise neighbours of panel i, j are $i-1, j$ and $i+1, j$; using the single index scheme they are $I-n$ and $I+n$, which is less intuitive.

For an uncambered wing, the camber surface is flat and therefore the geometric panels and vortex rings are coplanar. This is not the case for cambered wings; Figure 5.17a plots two geometric panels on the surface of a cambered wing and the vortex ring they define. It is clear that the vortex ring only touches the panels at its leading and trailing edges, while the control point of panel i, j is not coplanar with the vortex ring. At the control point, we place the unit normal vector $\mathbf{n}_{i,j} = (n_{x_{i,j}}, n_{y_{i,j}}, n_{z_{i,j}})$. The vortex ring has strength $\Gamma_{i,j}$, and all of its segments have the same strength. An observer travelling along the ring in a clockwise direction will see circulation of the same strength and direction (anti-clockwise in this case) everywhere. An observer standing away from the ring will see circulation of equal strength but in opposite directions at opposite segments. This is a very important point because if

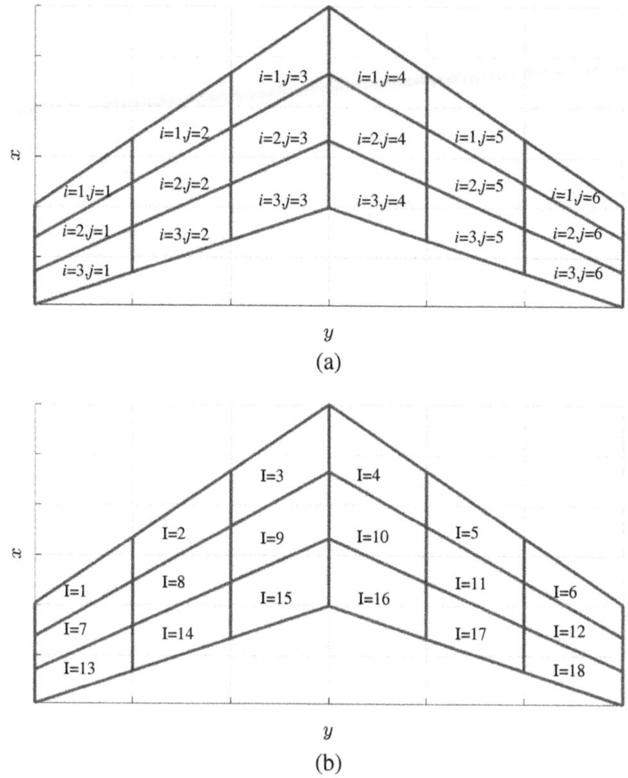

Figure 5.16 Two different panel numbering schemes for the VLM. (a) Double index numbering and (b) single index numbering.

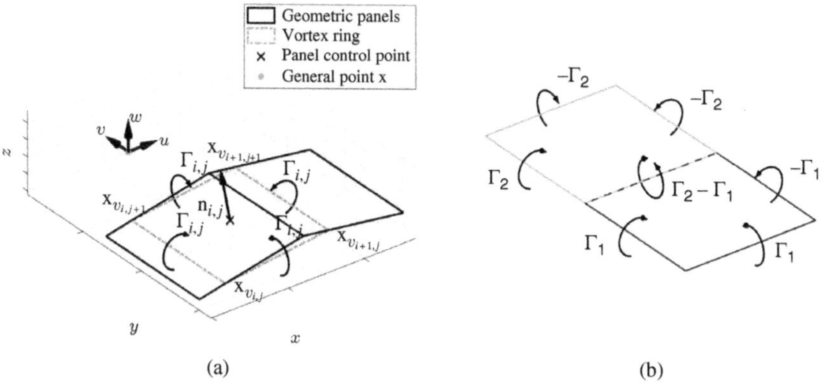

Figure 5.17 Vortex ring characteristics. (a) Vortex ring and geometric panels for a cambered wing and (b) direction of circulation.

we place two rings side-by-side so that they share one segment, the total circulation on this segment will be the difference between the strengths of the two rings. Figure 5.17b draws two vortex rings with strengths Γ_1 and Γ_2 that share a vortex segment. The direction of the circulation on each vortex ring is equal and opposite on opposite sides, so that the strength of the circulation on the shared segment is $\Gamma_2 - \Gamma_1$.

A vortex ring induces a velocity $\mathbf{u} = (u, v, w)$ at a general point located at $\mathbf{x}$, as seen in Figure 5.17a. This velocity can be calculated from Eq. (A.43). The velocity induced by the Jth bound vortex ring on the control point of the Ith panel is denoted by $\mathbf{u}_{I,J} = (u_{I,J}, v_{I,J}, w_{I,J})$. We can define the $mn \times mn$ influence coefficient matrices $\boldsymbol{A}_u(t_k), \boldsymbol{A}_v(t_k), \boldsymbol{A}_w(t_k)$, whose elements are given by

$$A_{u_{I,J}}(t_k) = \frac{u_{I,J}(t_k)}{\Gamma_J(t_k)},\ A_{v_{I,J}}(t_k) = \frac{v_{I,J}(t_k)}{\Gamma_J(t_k)},\ A_{w_{I,J}}(t_k) = \frac{w_{I,J}(t_k)}{\Gamma_J(t_k)}$$

for $I = 1, \dots, mn, J = 1, \dots, mn$. Note that the influence coefficient matrices are only functions of time if the motion is not modelled as quasi-fixed.

The Jth wake vortex ring also induces velocities $\mathbf{u}_{w_{I,J}} = (u_{w_{I,J}}, v_{w_{I,J}}, w_{w_{I,J}})$ on the Ith wing control panel; these velocities are again calculated using Eq. (A.43). We define the $mn \times m_w n$ wake influence coefficient matrices $\boldsymbol{B}_u(t_k)$, $\boldsymbol{B}_v(t_k)$, $\boldsymbol{B}_w(t_k)$, whose elements are given by

$$B_{u_{I,J}}(t_k) = \frac{u_{w_{I,J}}(t_k)}{\Gamma_{w_J}(t_k)},\ B_{v_{I,J}}(t_k) = \frac{v_{w_{I,J}}(t_k)}{\Gamma_{w_J}(t_k)},\ B_{w_{I,J}}(t_k) = \frac{w_{w_{I,J}}(t_k)}{\Gamma_{w_J}(t_k)}$$

for $I = 1, \dots, mn, J = 1, \dots, m_w n$ and where Γ_{w_J} is the strength of the Jth wake vortex ring. Adding up the velocities induced on the Ith control point by the bound vortices, the wake vortices, the free stream and the motion the impermeability condition of Eq. (2.43) becomes

$$\begin{aligned} &\sum_{J=1}^{mn} \left(A_{u_{I,J}}(t_k) n_{x_I}(t_k) + A_{v_{I,J}}(t_k) n_{y_I}(t_k) + A_{w_{I,J}}(t_k) n_{y_I}(t_k) \right) \Gamma_J(t_k) \\ &+ \sum_{J=1}^{m_w n} \left(B_{u_{I,J}}(t_k) n_{x_I}(t_k) + B_{v_{I,J}}(t_k) n_{y_I}(t_k) + B_{w_{I,J}}(t_k) n_{z_I}(t_k) \right) \Gamma_{w_J}(t_k) \\ &= -\left(\mathbf{Q}_\infty - \mathbf{v}_0(t_k) - \boldsymbol{\Omega}_0(t_k) \times (\mathbf{x}_{c_I}(t_k) - \mathbf{x}_f(t_k)) \right) \cdot \mathbf{n}_I(t_k) \end{aligned} \tag{5.75}$$

The relative velocity between the body and the fluid has u, v and w components at each control point so that it can be written in simpler form as

$$\mathbf{Q}_\infty - \mathbf{v}_0(t_k) - \boldsymbol{\Omega}_0(t_k) \times (\mathbf{x}_{c_I}(t_k) - \mathbf{x}_f(t_k)) = \left(u_{m_I}(t_k),\ v_{m_I}(t_k),\ w_{m_I}(t_k) \right)$$

Then, the impermeability boundary condition can be written in matrix form as

$$\boldsymbol{A}_n(t_k)\boldsymbol{\Gamma}(t_k) + \boldsymbol{B}_n(t_k)\boldsymbol{\Gamma}_w(t_k) = -\boldsymbol{u}_m(t_k) \circ \boldsymbol{n}_x(t_k) - \boldsymbol{v}_m(t_k) \circ \boldsymbol{n}_y(t_k) - \boldsymbol{w}_m(t_k) \circ \boldsymbol{n}_z(t_k) \tag{5.76}$$

where

$$A_{n_{I,J}} = A_{u_{I,J}} n_{x_I} + A_{v_{I,J}} n_{y_I} + A_{w_{I,J}} n_{y_I} \tag{5.77}$$

for $I = 1, \dots, mn, J = 1, \dots, mn$,

$$B_{n_{I,J}} = B_{u_{I,J}} n_{x_I} + B_{v_{I,J}} n_{y_I} + B_{w_{I,J}} n_{y_I}$$

for $I = 1, \dots, mn, J = 1, \dots, m_w n$ and $\boldsymbol{\Gamma}$, $\boldsymbol{\Gamma}_w$, $\boldsymbol{u}_m$, $\boldsymbol{v}_m$, $\boldsymbol{w}_m$, $\boldsymbol{n}_x$, $\boldsymbol{n}_y$, $\boldsymbol{n}_z$ are the $mn \times 1$ vectors whose elements are Γ_I, Γ_{w_I}, u_{m_I}, v_{m_I}, w_{m_I}, n_{x_I}, n_{y_I}, n_{z_I}, respectively. The $\circ$ operator in

Eq. (5.76) is the Hadamard product, denoting element-by-element multiplication of vectors of the same dimensions.

Equation (5.76) is a set of mn equations that can be solved directly for the mn unknown vortex strengths $\mathbf{\Gamma}(t_k)$ at each time instance. However, first we need to determine the strength of the wake vortices. At each time instance t_k, the old wake vortex rings move downstream, while a new row of vortex rings is shed from the trailing edge. As in the 2D case, once a vortex ring is shed into the wake, its strength remains constant at all times. The current second row of wake vortex rings was the first row at the previous time instance, and so on, so that $\Gamma_{w_{i,j}}(t_k) = \Gamma_{w_{i-1,j}}(t_{k-1})$ for $i > 1$ and $k > 1$. The strengths of the first row of wake vortex rings are calculated by different procedures, depending on whether the flow is steady or unsteady. We will demonstrate these procedures in the following steady and unsteady examples.

Once the strengths of the bound vortex rings are determined, we can calculate the aerodynamic loads acting on the wing using the Joukowski method (Simpson and Palacios 2013). There are two types of force:

- A quasi-steady force due to the circulation of each bound vortex segment. A general vortex ring will have four neighbours, such that each of its segments will coincide with a segment of one of the neighbouring vortex rings. Consider a segment going from point $\mathbf{x}_{v_{i,j}}$ to point $\mathbf{x}_{v_{i,j+1}}$ with vectorial length $\mathbf{x}_{v_{i,j+1}} - \mathbf{x}_{v_{i,j}}$ and midpoint $\mathbf{x}_{ms_{i,j}} = (\mathbf{x}_{i,j} + \mathbf{x}_{v_{i,j+1}})/2$, where the subscript ms stands for the midpoint of a spanwise vortex segment. The circulation of this segment is given by $\Gamma_{i,j} - \Gamma_{i-1,j}$. In order to apply the Kutta–Joukowski theorem, we also need the total flow velocity seen by this vortex segment due to the free stream, the motion and all the vortex rings. The usual choice is to calculate this flow velocity $\mathbf{u}_{ms_{i,j}}$ at the midpoint of the vortex segment. Then, the quasi-steady force acting on this vortex segment at its midpoint will be

$$\mathbf{F}_{s_{i,j}}(t_k) = \rho\left(\Gamma_{i,j}(t_k) - \Gamma_{i-1,j}(t_k)\right)\mathbf{u}_{ms_{i,j}}(t_k) \times \left(\mathbf{x}_{v_{i,j+1}}(t_k) - \mathbf{x}_{v_{i,j}}(t_k)\right) \tag{5.78}$$

  for $i = 1, \ldots, m+1, j = 1, \ldots, n$. Similarly, the quasi-steady force acting on the chordwise segment going from point $\mathbf{x}_{v_{i,j}}$ to point $\mathbf{x}_{v_{i+1,j}}$ will be equal to

$$\mathbf{F}_{c_{i,j}}(t_k) = \rho\left(\Gamma_{i,j}(t_k) - \Gamma_{i,j-1}(t_k)\right)\mathbf{u}_{mc_{i,j}}(t_k) \times \left(\mathbf{x}_{v_{i+1,j}}(t_k) - \mathbf{x}_{v_{i,j}}(t_k)\right) \tag{5.79}$$

  for $i = 1, \ldots, m, j = 1, \ldots, n+1$, where subscript mc stands for the midpoint of a chordwise segment and $\mathbf{u}_{mc_{i,j}}$ is the total flow velocity at the mid-chord point $\mathbf{x}_{mc_{i,j}} = (\mathbf{x}_{i,j} + \mathbf{x}_{v_{i+1,j}})/2$. It is important to keep in mind that forces $\mathbf{F}_{s_{i,j}}$ and $\mathbf{F}_{c_{i,j}}$ act on the midpoints of their corresponding vortex segments, not on the control points. Note that all the quantities in Eqs. (5.78) and (5.79) can be time-varying except for ρ. Vortex segments lying on the leading edge, trailing edge and wingtips do not have neighbours, and their strength is given by $\Gamma_{i,j}$ only.
- An unsteady force due to the change in strength of the vortex ring, given by

$$\mathbf{F}_{i,j}^{\text{unst}}(t_k) = \rho\frac{\partial\Gamma_{i,j}}{\partial t}s_{i,j}\mathbf{n}_{i,j}(t_k) \tag{5.80}$$

  where $s_{i,j}$ is the area of the corresponding geometric panel and the time derivative is calculated using first-order differences, such that

$$\frac{\partial\Gamma_{i,j}}{\partial t} = \frac{\Gamma_{i,j}(t_k) - \Gamma_{i,j}(t_{k-1})}{\Delta t} \tag{5.81}$$

  for $i = 1, \ldots, m, j = 1, \ldots, n$, noting that this force acts on the panel control points.

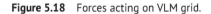

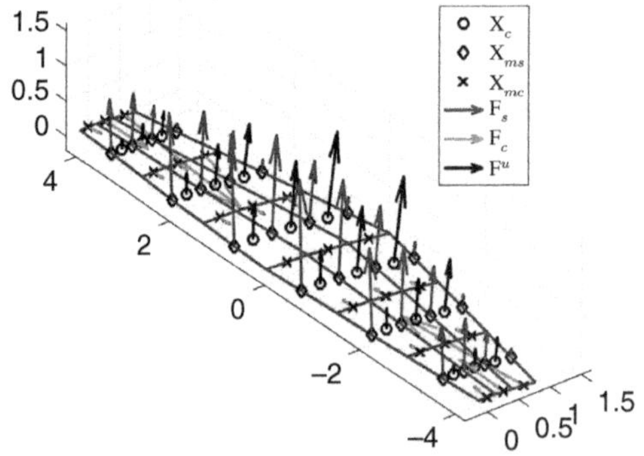

Figure 5.18 Forces acting on VLM grid.

The two force contributions of Eqs. (5.78)–(5.80) are in fact three-dimensional versions of the corresponding force contributions in Eq. (4.61), used in the load calculations for the unsteady 2D lumped vortex method. The total force acting on the entire wing is the sum of the quasi-steady loads from each vortex segment and the unsteady loads from each panel, that is

$$\mathbf{F}(t_k) = \sum_{i=1}^{m+1}\sum_{j=1}^{n}\mathbf{F}_{s_{i,j}}(t_k) + \sum_{i=1}^{m}\sum_{j=1}^{n+1}\mathbf{F}_{c_{i,j}}(t_k) + \sum_{i=1}^{m}\sum_{j=1}^{n}\mathbf{F}_{i,j}^{\text{unst}}(t_k) \tag{5.82}$$

Figure 5.18 draws the forces acting on the VLM grid and their points of application. It can be seen that the quasi-steady forces due to the spanwise vortex segments generate mainly lift, the quasi-steady forces due to the chordwise vortex segments generate mainly sideforce and the unsteady forces are perpendicular to the surface so they also generate mainly lift. We will first demonstrate the application of the VLM to a steady flow case.

Example 5.4 *Use the VLM to calculate the aerodynamic loads acting on a swept wing at several angles of attack and compare to experimental data.*

Kolbe and Boltz (1951) present a set of wind tunnel experiments on a swept trapezoidal wing of aspect ratio $\mathcal{R} = 3$*, taper ratio* $\lambda = 0.5$ *and leading edge sweep* $\Lambda_{LE} = 48.54°$*. The wing was tested at Mach numbers ranging from 0.08 to 0.96 and angles of attack from −2° to 30°. The data included pressure distributions along seven spanwise sections and total lift and drag forces measured using an aerodynamic balance. The half-wing model is drawn in Figure 5.19a, which also shows the positions of the pressure measurement taps. The wing section was a symmetric NACA 64A010 airfoil inclined by 45° to the plane of symmetry. As the VLM does not model thickness, we are only interested in the fact that the airfoil is uncambered and therefore the VLM panels will all lie on* $z = 0$*.*

The first task is to set up the geometric panels. The usual practice is to space the panels linearly in the chordwise direction and non-linearly in the spanwise direction. We set up non-dimensional coordinates, $\bar{x}_{p_i}$*,* $\bar{y}_{p_j}$ *and* $\bar{z}_{p_i}$ *such that*

$$\bar{x}_{p_i} = \frac{(i-1)}{m}$$

$$\bar{y}_{p_j} = \cos\theta_j \tag{5.83}$$

$$\bar{z}_{p_i} = 0 \tag{5.84}$$

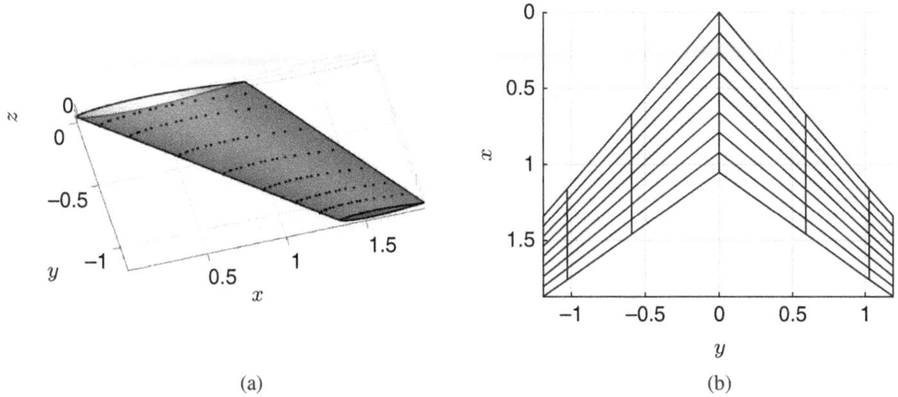

Figure 5.19 Wing tested by Kolbe and Boltz (1951) and its VLM model. (a) Actual geometry and (b) VLM model.

where $i = 1, \ldots, m+1, j = 1, \ldots, n+1, \theta_j = \pi - (j-1)\pi/n$. *The spanwise chord variation is given by*

$$c_j = (\lambda - 1)c_0|\bar{y}_{p_j}| + c_0 \tag{5.85}$$

while the leading edge chordwise position due to the sweep is

$$x_{LE_j} = \frac{b}{2}\tan\Lambda_{LE}|\bar{y}_{p_j}|$$

Consequently, the dimensional coordinates of the vertices of the geometric panels are calculated from

$$\begin{aligned} x_{p_{i,j}} &= x_{LE_j} + c_j\bar{x}_{p_i} \\ y_{p_{i,j}} &= \frac{b}{2}\bar{y}_{p_i} \\ z_{p_{i,j}} &= 0 \end{aligned} \tag{5.86}$$

Figure 5.19b plots the resulting planform for $m = 8$ *and* $n = 6$. *Even though the original wing is a half-wing, the VLM model is a full wing. In the experiments, half-wings are clamped to the vertical wall of the wind tunnel; the latter acts as a plane of symmetry that forces the air to flow in a direction parallel to the wing's centreline. In the numerical simulation, there is no such wall so that the complete wing must be modelled. There exist methods for imposing a symmetrical flow boundary condition on a half-wing, but they lie beyond the scope of this book.*

The next step is to calculate the coordinates of the vortex ring vertices. The x- and z-coordinates of the vortex ring points can be interpolated from those of the geometrical panels, except for the last row which lies behind the trailing edge. A good choice is to place this last row at a distance of $l_p = c_0/4m$ *behind the trailing edge, where* l_p *is known as the wake shedding distance and is constant along the span. Hence, the coordinates of the vortex ring vertices are given by*

$$x_{v_{i,j}} = \begin{cases} \frac{3x_{p_{i,j}} + x_{p_{i+1,j}}}{4} & \text{for } i \le m \\ \cos\left(\tan^{-1}\left(\frac{z_{p_{m+1,j}} - z_{p_{m,j}}}{x_{p_{m+1,j}} - x_{p_{m,j}}}\right)\right) l_p + x_{p_{m+1,j}} & \text{for } i = m+1 \end{cases} \tag{5.87}$$

$$y_{v_{i,j}} = y_{p_{i,j}}$$

$$z_{v_{i,j}} = \begin{cases} \frac{3z_{p_{i,j}}+z_{p_{i+1,j}}}{4} & \text{for } i \leq m \\ \sin\left(\tan^{-1}\left(\frac{z_{p_{m+1,j}}-z_{p_{m,j}}}{x_{p_{m+1,j}}-x_{p_{m,j}}}\right)\right) l_p + z_{p_{m+1,j}} & \text{for } i = m+1 \end{cases} \tag{5.88}$$

The coordinates of the control points can be obtained directly by interpolation in both the spanwise and chordwise directions, such as

$$x_{c_{i,j}} = \frac{1}{2}\left(\frac{x_{p_{i,j}} + 3x_{p_{i+1,j}}}{4} + \frac{x_{p_{i,j+1}} + 3x_{p_{i+1,j+1}}}{4}\right)$$

$$y_{c_{i,j}} = \frac{1}{2}\left(y_{p_{i,j}} + y_{p_{i,j+1}}\right) \tag{5.89}$$

$$z_{c_{i,j}} = \frac{1}{2}\left(\frac{z_{p_{i,j}} + 3z_{p_{i+1,j}}}{4} + \frac{z_{p_{i,j+1}} + 3z_{p_{i+1,j+1}}}{4}\right)$$

for $i = 1, \dots, m, j = 1, \dots, n$.

Next, we calculate unit vectors normal to the geometric panels. Consider the panel with vertices lying at $\mathbf{x}_{p_{i,j}}, \mathbf{x}_{p_{i,j+1}}, \mathbf{x}_{p_{i+1,j+1}}$ *and* $\mathbf{x}_{p_{i+1,j}}$. *Assuming that all the vertices lie on the same plane, a normal vector is given by the vector product of the two diagonals,* $\mathbf{r}_{13} = \mathbf{x}_{p_{i+1,j+1}} - \mathbf{x}_{p_{i,j}}$ *and* $\mathbf{r}_{42} = \mathbf{x}_{p_{i,j+1}} - \mathbf{x}_{p_{i+1,j}}$, *such that the unit normal vector is*

$$\mathbf{n}_{i,j} = \frac{\mathbf{r}_{13} \times \mathbf{r}_{42}}{|\mathbf{r}_{13} \times \mathbf{r}_{42}|} \tag{5.90}$$

If the wing is twisted, the panels are probably not flat, so that all four vertices will not be coplanar. This can be checked by creating a second normal vector from three of the four points, e.g.

$$\mathbf{n}_{2_{i,j}} = \frac{\mathbf{r}_{13} \times \mathbf{r}_{12}}{|\mathbf{r}_{13} \times \mathbf{r}_{12}|}$$

where $\mathbf{r}_{12} = \mathbf{x}_{p_{i,j+1}} - \mathbf{x}_{p_{i,j}}$ *and then calculating the vector product of the two vectors. If* $|\mathbf{n}_{i,j} \times \mathbf{n}_{2_{i,j}}| < \varepsilon$, *where* ε *is a small positive real number, then the panel is nearly planar and* $\mathbf{n}_{i,j}$ *is nearly normal to it. Here, we choose* $\varepsilon = 0.01$. *Note that using the vector product of the two diagonals gives a more representative normal vector for a non-planar panel than the vector product of any two sides.*

The area of the panel $s_{i,j}$ *can be calculated from the sum of the areas of the two triangles that make it up; these areas are equal to half the magnitude of the vector product of two of their sides. Consequently,*

$$s_{i,j} = \frac{1}{2}\left|\mathbf{r}_{13} \times \mathbf{r}_{12}\right| + \frac{1}{2}\left|\mathbf{r}_{13} \times \mathbf{r}_{14}\right| \tag{5.91}$$

where $\mathbf{r}_{14} = \mathbf{x}_{p_{i+1,j}} - \mathbf{x}_{p_{i,j}}$. *Finally, we define the wake vortex rings. As the flow is steady, only a single row of vortex rings is necessary, such that the leading segments of the wake vortex rings lie on the trailing segments of the corresponding bound vortex rings and the trailing segments of the wake vortex rings lie far downstream. In this case, we choose to place these trailing segments*

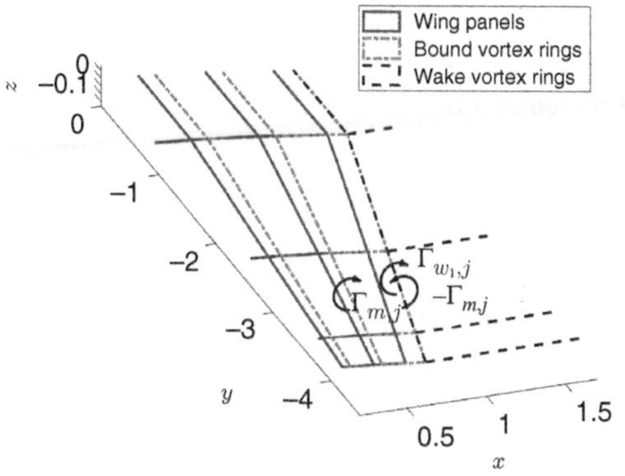

Figure 5.20 Vortex strength at the trailing edge for the steady VLM.

at a distance of $100c_0$ *downstream, such that the vertices of wake panel* $1, j$ *are given by*

$$x_{w_{1,j}} = x_{v_{m+1,j}},\ x_{w_{1,j+1}} = x_{v_{m+1,j+1}},$$

$$x_{w_{2,j+1}} = x_{v_{m+1,j+1}} + 100c_0,\ x_{w_{2,j}} = x_{v_{m+1,j}} + 100c_0$$

The other coordinates are all identical to those of the trailing segments of the bound vortex rings, that is $y_{w_{1,j}} = y_{w_{2,j}} = y_{v_{m+1,j}}$, $z_{w_{1,j}} = z_{w_{2,j}} = z_{v_{m+1,j}}$.

The strength of the wake vortex rings can be evaluated using the Kutta condition. Figure 5.20 plots the bound and wake vortex rings at the trailing edge of the wing. The trailing edge of bound ring m, j *coincides with the leading edge of wake vortex ring* $1, j$*, so that the total vortex strength of this vortex segment is equal to* $-\Gamma_{m,j} + \Gamma_{w_{1,j}}$. *Using the zero vorticity at the trailing edge form of the Kutta condition (see Section 4.7), we must set*

$$-\Gamma_{m,j} + \Gamma_{w_{1,j}} = 0$$

for $j = 1, \dots, n$, *or* $\Gamma_{w_{1,j}} = \Gamma_{m,j}$. *Using single-index numbering, bound panel* m, j *corresponds to panel* $(m-1)n + j$ *for* $j = 1, \dots, n$. *In matrix form, the Kutta condition can be written as*

$$\boldsymbol{\Gamma}_w = \boldsymbol{P}_c \boldsymbol{\Gamma} \tag{5.92}$$

where $\boldsymbol{P}_c = (\boldsymbol{0}_{n\times(m-1)n}, \boldsymbol{I}_n)$ *is a* $n \times mn$ *matrix that selects the last n elements of* $\boldsymbol{\Gamma}$, $\boldsymbol{0}_{n\times(m-1)n}$ *is a zero matrix with dimensions* $n \times (m-1)n$ *and* $\boldsymbol{I}_n$ *is a* $n \times n$ *unit matrix.*

The motion is steady so that the relative velocity between the wing and the flow is $\boldsymbol{Q}_\infty = (Q_\infty \cos\alpha,\ 0,\ Q_\infty \cos\alpha)$. *Substituting for* $\boldsymbol{\Gamma}_w$ *from expression (5.92) into the impermeability condition of Eq. (5.76), we obtain*

$$\left(\boldsymbol{A}_n \boldsymbol{\Gamma} + \boldsymbol{B}_n \boldsymbol{P}_c\right) \boldsymbol{\Gamma} = -Q_\infty \cos\alpha \boldsymbol{n}_x - Q_\infty \sin\alpha \boldsymbol{n}_z \tag{5.93}$$

The only unknowns in Eq. (5.93) are the mn bound vortex strengths, $\boldsymbol{\Gamma}$.

Now, we move to calculating the influence coefficient matrices $\boldsymbol{A}_n$ *and* $\boldsymbol{B}_n$ *in Eq. (5.93). Suppose we are calculating the influence of the Jth bound vortex ring on the Ith control point. We use Eq. (A.43), but we note that there are possible singularities if* $|\mathbf{r}_1| = 0$, $|\mathbf{r}_2| = 0$, *etc., or if*

$|\mathbf{r}_1 \times \mathbf{r}_2| = 0$, $|\mathbf{r}_2 \times \mathbf{r}_3| = 0$, etc. These singularities will occur if the control point lies on any of the vortex ring vertices or if it is collinear with a vortex segment. If any of these situations arise, we set the influence of the corresponding vortex segment(s) to zero. This should not be the case for the present calculation because the control points and bound and wake vortex rings are chosen such that none of these conditions can occur. However, it is good practice to check for singularities because the calculation of the aerodynamic loads will necessitate such checks. We carry out additional checks for $|\mathbf{r}_{12}| = 0$, $|\mathbf{r}_{23}| = 0$, etc., which can occur if the panel is triangular instead of quadrilateral. Once we have evaluated $\boldsymbol{A}_n$ and $\boldsymbol{B}_n$, we assemble Eq. (5.93) and solve it for the strengths of all the bound vortex rings. We can then calculate the strengths of the wake vortex rings from Eq. (5.92).

The final step is to calculate the aerodynamic loads acting on the wing using Eqs. (5.78)–(5.82) with $\mathbf{F}^{\text{unst}}_{i,j} = \mathbf{0}$ for all i and j since the flow is steady. First, we evaluate the midpoints of the spanwise and chordwise vortex segments, $\mathbf{x}_{ms_{i,j}} = (\mathbf{x}_{i,j} + \mathbf{x}_{v_{i,j+1}})/2$ and $\mathbf{x}_{mc_{i,j}} = (\mathbf{x}_{i,j} + \mathbf{x}_{v_{i+1,j}})/2$, respectively, and calculate the local flow velocities, $\mathbf{u}_{ms_{i,j}}$, $\mathbf{u}_{mc_{i,j}}$ using Eq. (A.43), before adding the free stream, such that

$$
\begin{aligned}
\boldsymbol{u}_{ms} &= Q_\infty \cos\alpha + \boldsymbol{A}_{u_{ms}}\boldsymbol{\Gamma} + \boldsymbol{B}_{u_{ms}}\boldsymbol{\Gamma}_w \\
\boldsymbol{v}_{ms} &= \boldsymbol{A}_{v_{ms}}\boldsymbol{\Gamma} + \boldsymbol{B}_{v_{ms}}\boldsymbol{\Gamma}_w \\
\boldsymbol{w}_{ms} &= Q_\infty \sin\alpha + \boldsymbol{A}_{w_{ms}}\boldsymbol{\Gamma} + \boldsymbol{B}_{w_{ms}}\boldsymbol{\Gamma}_w
\end{aligned}
\tag{5.94}
$$

where $\boldsymbol{u}_{ms}$, $\boldsymbol{v}_{ms}$, $\boldsymbol{w}_{ms}$, are the $(m+1)n$ column vectors with elements u_{ms_I}, v_{ms_I}, w_{ms_I}, respectively, $\boldsymbol{A}_{u_{ms}}$, $\boldsymbol{A}_{v_{ms}}$, $\boldsymbol{A}_{w_{ms}}$, are the $(m+1)n \times mn$ influence coefficient matrices of the bound vortex rings on the spanwise segment midpoints and $\boldsymbol{B}_{u_{ms}}$, $\boldsymbol{B}_{v_{ms}}$, $\boldsymbol{B}_{w_{ms}}$, are the $(m+1)n \times n$ influence coefficient matrices of the wake vortex rings on the spanwise segment midpoints. Similarly, the total flow velocities on the chordwise midpoints are given by

$$
\begin{aligned}
\boldsymbol{u}_{mc} &= Q_\infty \cos\alpha + \boldsymbol{A}_{u_{mc}}\boldsymbol{\Gamma} + \boldsymbol{B}_{u_{mc}}\boldsymbol{\Gamma}_w \\
\boldsymbol{v}_{mc} &= \boldsymbol{A}_{v_{mc}}\boldsymbol{\Gamma} + \boldsymbol{B}_{v_{mc}}\boldsymbol{\Gamma}_w \\
\boldsymbol{w}_{mc} &= Q_\infty \sin\alpha + \boldsymbol{A}_{w_{mc}}\boldsymbol{\Gamma} + \boldsymbol{B}_{w_{mc}}\boldsymbol{\Gamma}_w
\end{aligned}
\tag{5.95}
$$

As the points on which we calculate the velocities now lie on vortex segments, singularities will occur in the application of Eq. (A.43), and therefore, we must check for them and eliminate them, as discussed earlier. Finally, we apply Eqs. (5.78) and (5.79), noting that for $i = 1$

$$\mathbf{F}_{s_{1,j}} = \rho\Gamma_{1,j}\mathbf{u}_{ms_{1,j}} \times \left(\mathbf{x}_{v_{1,j+1}} - \mathbf{x}_{v_{1,j}}\right)$$

for $j = 1$

$$\mathbf{F}_{c_{i,1}} = \rho\Gamma_{i,1}\mathbf{u}_{mc_{i,1}} \times \left(\mathbf{x}_{v_{i+1,1}} - \mathbf{x}_{v_{i,1}}\right)$$

and for $j = n + 1$

$$\mathbf{F}_{c_{i,n+1}} = -\rho\Gamma_{i,n}\mathbf{u}_{mc_{i,n+1}} \times \left(\mathbf{x}_{v_{i+1,n+1}} - \mathbf{x}_{v_{i,n+1}}\right)$$

The trailing edge vortex segments ($i = m + 1$) have zero total strength due to the Kutta condition so they cause zero force. Finally, we apply Eq. (5.82), to obtain the total aerodynamic force

$$\mathbf{F} = (F_x, F_y, F_z)$$

where F_x is the chordwise force, F_y is the sideforce and F_z is the normal force. The total lift and drag coefficients are simply

$$C_L = \frac{F_z \cos\alpha - F_x \sin\alpha}{\frac{1}{2}\rho Q_\infty^2 S}, \quad C_D = \frac{F_z \sin\alpha + F_x \cos\alpha}{\frac{1}{2}\rho Q_\infty^2 S}$$

The spanwise variation of the sectional forces can be calculated by summing $\mathbf{F}_{s_{i,j}}$ and $\mathbf{F}_{c_{i,j}}$ in the chordwise direction only. Recall that $\mathbf{F}_{s_{i,j}}$ lies on the y-coordinates of the spanwise midpoints, which coincide with the y-coordinates of the control points, so that

$$f_{z_s}(y_{c_{1,j}}) = \sum_{i=1}^{m+1} F_{z_{s_{i,j}}}, \quad f_{x_s}(y_{c_{1,j}}) = \sum_{i=1}^{m+1} F_{z_{s_{i,j}}},$$

where f_z is the sectional normal force and f_x the sectional chordwise force. In contrast, $\mathbf{F}_{c_{i,j}}$ lies on the y-coordinates of the chordwise midpoints, which coincide with the y-coordinates of the vortex ring vertices, that is

$$f_{z_c}(y_{v_{1,j}}) = \sum_{i=1}^{m} F_{z_{c_{i,j}}}, \quad f_{x_c}(y_{v_{1,j}}) = \sum_{i=1}^{m} F_{z_{c_{i,j}}}$$

In order to add up these two force distributions, we need to interpolate $f_{z_c}(y_{v_{1,j}})$ and $f_{x_c}(y_{v_{1,j}})$ onto $y_{c_{1,j}}$, so that

$$f_{z_j} = f_{z_s}(y_{c_{1,j}}) + \frac{1}{2}\left(f_{z_c}(y_{v_{1,j}}) + f_{z_c}(y_{v_{1,j+1}})\right)$$

$$f_{x_j} = f_{x_s}(y_{c_{1,j}}) + \frac{1}{2}\left(f_{x_c}(y_{v_{1,j}}) + f_{x_c}(y_{v_{1,j+1}})\right)$$

for $j = 1, \ldots, n$. Then, the sectional normal and chordwise force distributions are given by

$$C_{x_j} = \frac{f_{x_j}}{1/2\rho U_\infty^2 \sum_{i=1}^{m} s_{i,j}}, \quad C_{z_j} = \frac{f_{z_j}}{1/2\rho U_\infty^2 \sum_{i=1}^{m} s_{i,j}}$$

again for $j = 1, \ldots, n$.

We carry out the VLM calculations using $m = 40$ and $n = 40$ and for angles of attack between $-2°$ and $24°$. Figure 5.21a plots the variation of the total lift coefficient against angle of attack and compares it to the experimental data obtained by Kolbe and Boltz (1951) for a Mach number of 0.08 and for two different values of the Reynolds number. It is clear that the VLM predicts very accurately the lift curve up to $\alpha = 13°$. In fact, if we compare to the data obtained for $Re = 6 \times 10^6$, the VLM lift estimates are good up to $\alpha = 14.5°$. Beyond that angle, viscous effects become important, and the experimental lift curve departs from linearity. Clearly, the VLM cannot predict the stall occurring at around $24.5°$. The drag polars of Figure 5.21b show that the VLM under-predicts the total drag coefficient because the drag is equal to zero at $C_L = 0$. For real flows, the drag polar can be represented by

$$C_D = C_{D_0} + i_d C_L^2$$

where C_{D_0} is the zero-lift drag and i_d is the induced drag factor. The values of C_{D_0} can be obtained from the experiments by curve-fitting the drag polar, leading to $C_{D_0} = 0.003$ for $Re = 4 \times 10^6$ and $C_{D_0} = 0.005$ for $Re = 6 \times 10^6$. If we add these values of C_{D_0} to the VLM drag polar, we obtain better agreement with the experimental data, at least for low values of C_L.

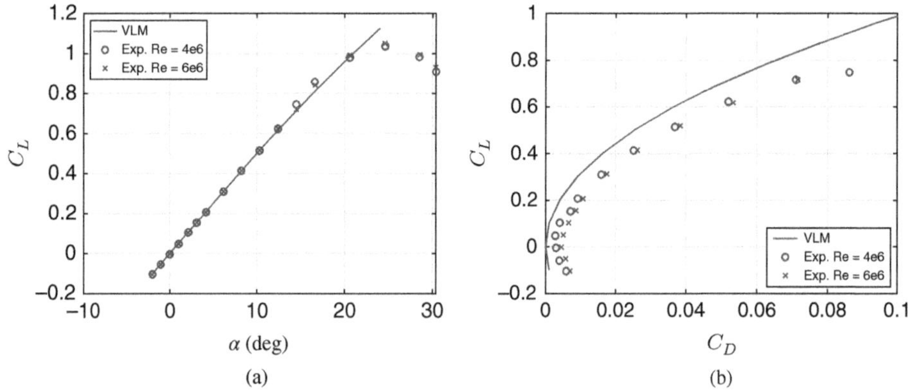

Figure 5.21 Lift and drag curves computed by the VLM. Experimental data by Kolbe and Boltz (1951). (a) C_L and (b) C_D. Source: Kolbe and Boltz (1951), credit: NACA.

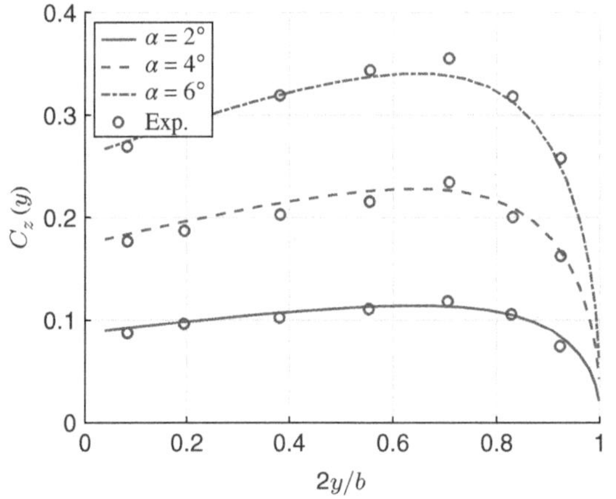

Figure 5.22 Spanwise normal force distribution for three angles of attack. Experimental data by Kolbe and Boltz (1951). Source: Kolbe and Boltz (1951), credit: NACA.

Figure 5.22 plots the normal force distribution along the span calculated by the VLM for angles of attack of 2°, 4° and 6°. The numerical estimates are compared to data presented in Kolbe and Boltz (1951), obtained at a Mach number of 0.25. This value of the Mach number is still low enough to assume that the flow is incompressible, particularly for the low angles of attack studied here. The agreement between the VLM normal force distributions and the experimental data is very good. This example is solved by Matlab code `steady_VLM.m`.

In this example, we imposed zero circulation on the last row of spanwise bound vortex segments, which lie behind the trailing edge. Therefore, the Kutta condition is numerically exact but spatially approximate, since the zero circulation condition is imposed at a distance of $l_p = c_0/4m$ behind the trailing edge. In fact, there are two aspects to the VLM Kutta condition:

- An implicit aspect due to the fact that the spanwise bound vortex segments are placed on the quarter chord of each panel (see Section 4.2).

- An explicit aspect due to the fact that the strength of the first row of wake vortex rings is equal to that of the last row of bound vortex rings, such that the strengths of the last bound spanwise vortex segments are equal to zero.

The combination of these two aspects of the Kutta condition gives satisfactory results, but there are more sophisticated explicit Kutta conditions, such as the one by Lee (1977), which lie beyond the scope of this book.

Before moving on to unsteady flows, we will first carry out a series of VLM calculations on untapered, unswept and untwisted rectangular wings with varying aspect ratios. The calculations are carried out exactly as in Example 5.4 but for only one angle of attack, $\alpha = 1°$. As the airfoil of all the wings is assumed to be symmetric, the 3D lift-curve slope of the wings is given by

$$C_{L_\alpha}(\mathcal{R}) = \frac{C_L(\mathcal{R})}{\alpha}$$

and the induced drag factor by

$$\frac{C_D(\mathcal{R})}{C_L^2(\mathcal{R})}$$

We also calculate the lift and drag for each of the wings using lifting line theory in order to compare them to the VLM estimates.

Figure 5.23a plots the variation of the lift-curve slopes of straight rectangular wings with aspect ratio values between 4 and 40. The results of the VLM and LLT are qualitatively similar but, quantitatively, the lifting line prediction is higher for all aspect ratios. The situation does not change if we change the values of the taper ratio, twist or camber; the lift-curve slope values predicted by LLT are always higher, particularly at low aspect ratios. In fact, the more accurate results are those obtained by the VLM; lifting line theory over-predicts the 3D lift-curve slope. Izraelevitz et al. (2017) also note this fact and make use of a correction technique by Hoerner and Borst (1985). The latter collated numerous experimental measurements of the 3D lift curve slopes of wings of different shapes and aspect ratios and showed that they all have lower lift curve slopes than those predicted by a simplified lifting line expression. Alternatively, the analytical lifting surface expressions in Ichikawa (2012) can be used to correct steady and unsteady lifting line theory results.

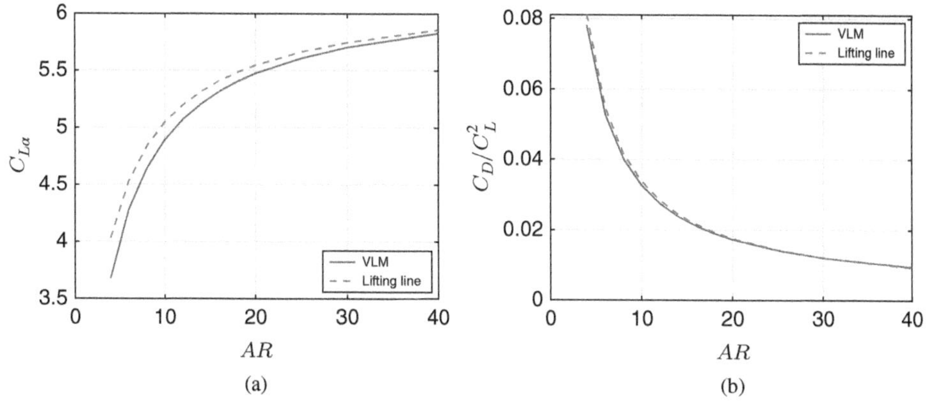

Figure 5.23 Lift-curve slopes and induced drag factors of straight rectangular wings with different aspect ratios. (a) C_{L_α} and (b) C_D/C_L^2.

In Section 5.4.1 we will calculate the impulsive start of a wing with elliptical planform and compare it to Jones' predictions. As the lift response after an impulsive start tends to the steady lift as time increases, we expect that there will be a steady-state disagreement between the unsteady VLM and unsteady lifting line approaches. Figure 5.23b shows that there is very good agreement between the two theories for the steady induced drag factor.

5.4.1 Impulsive Start of an Elliptical Wing

In the steady flow case, there was a single row of very long wake panels, whose strengths were unknown and calculated by imposing explicitly the Kutta condition. In the unsteady case, the VLM wake is made up of many rows of vortex rings with short streamwise lengths and the coordinates of the vertices change in time. Furthermore, the Kutta condition is no longer imposed explicitly; the wake is updated at the end of the calculations for each time instance.

As usual, there are two options for propagating the wake:

- Prescribed wake. The vertices of the wake panels are propagated at the free stream airspeed, that is

$$\mathbf{x}_{w_{i,j}}(t_k) = \mathbf{x}_{w_{i-1,j}}(t_{k-1}) + \mathbf{Q}_\infty \Delta t \tag{5.96}$$

- Free wake. The vertices of the wake panels are propagated at the local airspeed, that is

$$\mathbf{x}_{w_{i,j}}(t_k) = \mathbf{x}_{w_{i-1,j}}(t_{k-1}) + \left(\mathbf{Q}_\infty + \mathbf{u}_{w_{b_{i-1,j}}}(t_k) + \mathbf{u}_{ww_{i-1,j}}(t_k)\right) \Delta t \tag{5.97}$$

where $\mathbf{u}_{w_{b_{i,j}}}$ and $\mathbf{u}_{w_{w_{i,j}}}$ are the flow velocities induced on the vertices of the wake panels by the bound and wake vortex rings, respectively.

At the first time instance, there is no wake at all and the Kutta condition cannot be enforced; this makes sense in the context of an impulsive start. The flow visualisation of Figure 2.28b shows that immediately after the impulsive start, the flow around the trailing edge is strongly circulatory; forcing the total vortex strength at the trailing edge to be equal to zero would be unphysical. Within the context of an impulsive start, whereby the wing is rotated by an angle α and the free stream is $\mathbf{Q}_\infty = (U_\infty, 0, 0)$, the impermeability boundary condition (5.76) at the first time instance becomes

$$\boldsymbol{A}_n \boldsymbol{\Gamma}(t_1) = -U_\infty \boldsymbol{n}_x \tag{5.98}$$

There are n equations that can be solved for the $n \times 1$ vector $\boldsymbol{\Gamma}(t_1)$. The vortex strengths of the trailing edge panels are $\Gamma_{m,j}(t_1)$ for $j = 1, \dots, n$, and they are non-zero. We can now set up the wake as a single row of vortex rings with strengths

$$\Gamma_{w_{1,j}}(t_1) = \Gamma_{m,j}(t_1)$$

or in matrix form,

$$\boldsymbol{\Gamma}_w(t_1) = \boldsymbol{P}_c \boldsymbol{\Gamma}(t_1) \tag{5.99}$$

which is the unsteady form of Eq. (5.92). The leading edges of the wake vortex rings are coincident with the trailing edges of the corresponding bound vortex rings. The positions of the trailing vertices of the wake vortex rings are calculated from Eqs. (5.96) or (5.97), depending on the choice of wake model.

At the second time instance, the impermeability condition of expression (5.76) can then be written as

$$\boldsymbol{A}_n(t_2)\boldsymbol{\Gamma}(t_2) = -\boldsymbol{B}_n(t_2)\boldsymbol{\Gamma}_w(t_1) - U_\infty \boldsymbol{n}_x \tag{5.100}$$

where $\boldsymbol{B}_n(t_2)$ has dimensions $mn \times n$ and $\boldsymbol{\Gamma}_w(t_1)$ is a $n \times 1$ vector. Once this equation is solved for $\boldsymbol{\Gamma}(t_2)$, the total strength of the spanwise trailing vortex segments is $\Gamma_{m,j}(t_2) - \Gamma_{w_{1,j}}(t_1) = \Gamma_{m,j}(t_2) - \Gamma_{m,j}(t_1)$. Note that this vortex strength is not equal to zero and, hence, the Kutta condition is not satisfied at this stage. We have used the wake shape and strength of the previous time instance to calculate the current bound vortex strength. However, before we move on to the next time instance, we must first propagate the wake. At a general time instance t_k, the vortex rings already existing in the wake move downstream so that

$$\Gamma_{w_{i,j}}(t_k) = \Gamma_{w_{i-1,j}}(t_{k-1}) \tag{5.101}$$

for $i = 2, \dots, k$ and a new row of vortex rings is formed just behind the wing's trailing edge with strength

$$\Gamma_{w_{1,j}}(t_k) = \Gamma_{m,j}(t_k)$$

as shown in Figure 5.24. Now the total strength of the spanwise trailing vortex segments is $\Gamma_{m,j}(t_k) - \Gamma_{w_{1,j}}(t_k) = \Gamma_{m,j}(t_k) - \Gamma_{m,j}(t_k) = 0$ and the Kutta condition is satisfied. Therefore, the unsteady VLM solution is carried out in two sequential steps:

- First, we impose impermeability and solve for the current bound vortex strengths. At this stage, $\Gamma_{w_{1,j}}(t_k) = \Gamma_{m,j}(t_{k-1})$. We calculate the aerodynamic loads for this value of the wake strength, that is before enforcing the Kutta condition.
- Then we enforce the Kutta condition by propagating the wake so that $\Gamma_{w_{1,j}}(t_k) = \Gamma_{m,j}(t_k)$.

Also note that the wake becomes longer by one row of vortex rings at the end of each time instance. The positions of the wake vortex rings are propagated using Eqs. (5.96) or (5.97). If we choose to use a free wake model, then we need to calculate the $\mathbf{u}_{w_{b_{i,j}}}(t_k)$ and $\mathbf{u}_{w_{w_{i,j}}}(t_k)$ velocities before applying Eq. (5.97).

Example 5.5 *Use the VLM to calculate the aerodynamic load response of a wing with elliptical planform undergoing an impulsive start.*

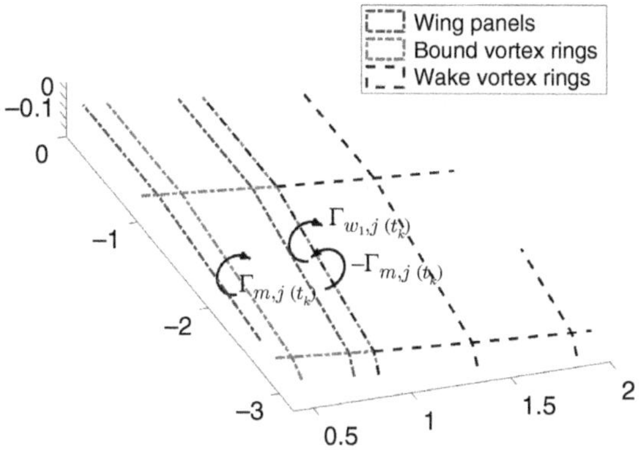

Figure 5.24 Vortex strength at the trailing edge for the unsteady VLM.

The characteristics of the wing are identical to those used in Example 5.2. As this is an unsteady simulation, we need to set the time step Δt and the duration of the simulation. The wake vortex rings must have similar size to the corresponding bound vortex rings (Simpson and Palacios 2013), so we choose a time step equal to

$$\Delta t = \frac{c_0}{mQ_\infty} \tag{5.102}$$

so that, if the wake vortices translate with the free stream airspeed, the distance between two successive spanwise vortex segments will be equal to the chordwise length of the midspan bound vortex rings. The simulation duration is chosen such that the total length of a prescribed wake will be $10c_0$ at the end of the simulation.

The panels and vortex rings are set up in the same way as in Example 5.4, but the spanwise chord variation is calculated from

$$c_j = c_0\sqrt{1-\bar{y}_{p_j}^2}$$

and the wing is uncambered so that $\bar{z}_{p_j} = 0$. A wing with elliptical planform is essentially a wing with a non-linear taper distribution; we set the non-dimensional taper axis $a_x = 1/2$, so that taper is defined around the half-chord point and the leading and trailing edges are mirror images of each other. The panel vertices are then given by

$$\begin{aligned} x_{p_{i,j},0} &= c_j(\bar{x}_{p_i} - a_x) \\ y_{p_{i,j},0} &= \frac{b}{2}\bar{y}_{p_i} \\ z_{p_{i,j},0} &= 0 \end{aligned} \tag{5.103}$$

so that the ellipse is symmetrical around the half-chord axis. The wing must then be rotated to the right angle of attack, such that

$$\begin{pmatrix} x_{p_{i,j}} \\ y_{p_{i,j}} \\ z_{p_{i,j}} \end{pmatrix} = \begin{pmatrix} \cos\alpha & 0 & \sin\alpha \\ 0 & 1 & 0 \\ -\sin\alpha & 0 & \cos\alpha \end{pmatrix} \begin{pmatrix} x_{p_{i,j},0} \\ y_{p_{i,j},0} \\ z_{p_{i,j},0} \end{pmatrix} \tag{5.104}$$

The vortex rings and panel control points are calculated using the procedures of Example 5.4, except that we have to force $x_{v_{i,1}} = x_{v_{i,n+1}} = z_{v_{i,1}} = z_{v_{i,n+1}} = 0$ for $i = 1, \dots, m$. The wake shedding distance is set to $l_p = 0.25U_\infty\Delta t$ (Katz and Plotkin 2001). Then we evaluate the aerodynamic influence coefficient matrix $A_{n_{I,J}}$ of Eq. (5.77). Note that the wingtip vertices for both the panels and vortex rings all lie on $x = 0, y = 0, z = 0$, since the chord-length at the tips is equal to zero. This is not a problem for the calculation of the influence coefficients, since the singularity checks we carry out will cause the velocities induced by any vortex segments with zero length to be equal to zero.

At the first time instance, the impermeability condition is given by Eq. (5.98), which can be readily solved for $\mathbf{\Gamma}(t_1)$. Next, we evaluate the velocities induced by the bound vortex rings on the midpoints of the spanwise and chordwise bound vortex segments, $\mathbf{u}_{ms_{i,j}}(t_1)$ and $\mathbf{u}_{mc_{i,j}}(t_1)$, respectively, which we use to calculate the aerodynamic loads from Eqs. (5.78)–(5.82). At the kth time instance, the time derivatives of the bound vortex strengths are given by Eq. (5.81), but at the first time instance, we set them to $\partial\Gamma_{i,j}/\partial t = 0$. Note that, as the free stream is aligned *with the x axis, the x and z components of $\mathbf{F}(t_k)$ are the drag and lift, respectively, that is $D(t_k) = F_x(t_k)$ and $L(t_k) = F_z(t_k)$.*

At this stage, the calculations for the first time instance are finished, but we need to create the wake before moving on to the second time instance. First, we calculate the velocities induced by the bound vortex rings on the vertices of the trailing edge vortex segments, $\mathbf{u}_{w_{b_{1,j}}}(t_1)$. Then, we set up the coordinates of the vertices of the first row of wake vortex rings

$$\mathbf{x}_{w_{1,j}}(t_1) = \mathbf{x}_{v_{m+1,j}}, \mathbf{x}_{w_{2,j}}(t_1) = \mathbf{x}_{v_{m+1,j}} + \left(\mathbf{Q}_\infty + \mathbf{u}_{w_{b_{1,j}}}(t_1)\right)\Delta t$$

so that the leading edges of the wake vortex rings lie on the trailing edges of the last row of bound vortex rings, while the trailing edges of the wake vortex rings lie at the position to which the leading edges have travelled after one time instance at the local flow velocity. Finally, we assign the strength of the first row of vortex rings, $\Gamma_{w_{1,j}}(t_2) = \Gamma_{m,j}(t_1)$.

At the second time instance, the impermeability condition is given by Eq. (5.100) and is solved for $\mathbf{\Gamma}(t_2)$. Note that the wake influence coefficient matrix, $\boldsymbol{B}_n(t_k)$, must be recalculated at each time instance since the wake changes size and shape. We evaluate the velocities induced by both the bound and wake vortex rings on the midpoints of the spanwise and chordwise bound vortex segments and calculate the aerodynamic loads, using Eq. (5.81) for $k = 2$ to estimate the time derivative of the bound vortex strengths. Next, we calculate the velocities induced by the bound and wake vortex rings on the vertices of the wake vortex rings, $\mathbf{u}_{w_{b_{i,j}}}(t_2)$ and $\mathbf{u}_{w_{w_{i,j}}}(t_2)$, respectively, for $i = 1, 2$ and $j = 1, \ldots, n+1$. Finally, we update the positions of the vertices of the wake vortex rings using Eq. (5.97), propagate the wake vortex strengths using Eq. (5.101) and shed the new row of wake vortex rings with strength $\Gamma_{w_{1,j}}(t_2) = \Gamma_{m,j}(t_2)$.

At the kth time instance, the procedure is as follows:

1. *Calculate the wake influence matrix $\boldsymbol{B}_n(t_k)$.*
2. *Set up and solve the impermeability condition.*

$$\boldsymbol{A}_n\mathbf{\Gamma}(t_k) = -\boldsymbol{B}_n(t_k)\mathbf{\Gamma}_w(t_k) - U_\infty \boldsymbol{n}_x \tag{5.105}$$

   for the unknown bound vortex strengths $\mathbf{\Gamma}(t_k)$.
3. *Calculate the velocities induced by the new bound and old wake vortex rings at the midpoints of the spanwise and chordwise vortex segments, $\mathbf{u}_{ms_{i,j}}(t_k)$ and $\mathbf{u}_{mc_{i,j}}(t_k)$.*
4. *Calculate the aerodynamic loads from Eqs. (5.78)–(5.82).*
5. *Calculate the velocities induced by the bound and wake vortex rings on the vertices of the wake vortex rings, $\mathbf{u}_{w_{b_{i,j}}}(t_k)$ and $\mathbf{u}_{w_{w_{i,j}}}(t_k)$, respectively, for $i = 1, \ldots, k$ and $j = 1, \ldots, n+1$.*
6. *Update the coordinates of the wake vertices using Eq. (5.97) for $i = 2, \ldots, k+1$ and $j = 1, \ldots, n+1$. The coordinates $\mathbf{x}_{w_{1,j}}$ are always equal to $\mathbf{x}_{v_{m+1,j}}$.*
7. *Propagate the strengths of the wake vortices so that $\Gamma_{w_{i,j}}(t_{k+1}) = \Gamma_{w_{i-1,j}}(t_k)$ for $i = 2, \ldots, k$ and $j = 1, \ldots, n$.*
8. *Set $\Gamma_{w_{1,j}}(t_{k+1}) = \Gamma_{m+1,j}(t_k)$ for $j = 1, \ldots, n$.*
9. *Increment k and go back to step 1.*

The calculation can be speeded up considerably if the wake is prescribed and propagated using Eq. (5.96). However, a free wake simulation can be used to visualize the wake's vortex system. Figure 5.25 plots a representative snapshot of the wing and wake. The wake vortex rings lying behind the wingtips actually roll up and down as they travel downstream, creating a realistic

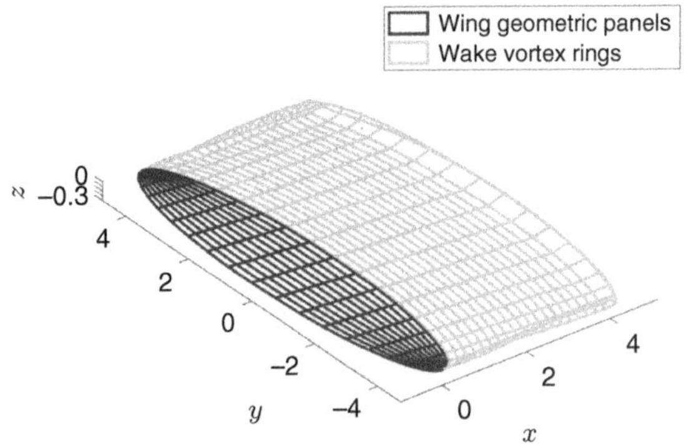

Figure 5.25 Wake shape behind impulsively started elliptical wing.

representation of the trailing vortices. Furthermore, the end of the wake rolls upwards and inwards, which is a representation of the starting vortex. The figure shows that the inviscid wake is a generally flat vorticity surface that rolls upwards at the tips and trailing edge, forming the trailing and starting vortices.

We carry out the VLM calculation with $m = 10$, $n = 20$ and compare the resulting lift and drag responses to the estimates obtained from Jones' method and the unsteady lifting line. Figure 5.26a plots this comparison against non-dimensional time $\tau = 2U_\infty t/c_0$. At the first time instance, the lift coefficient estimated by the VLM is nearly zero due to the fact that we have set $\partial\Gamma/\partial t = 0$. From the second time instance onwards, the VLM predictions become very similar to the results obtained from the unsteady lifting line technique, up to $\tau \approx 5$. The steady-state lift coefficient predicted by the VLM is around 5% lower than the one predicted by lifting line theory so that it is natural that there will be a divergence between the unsteady VLM lift predictions and those obtained from the analytical methods as τ increases.

The drag coefficient variation with time is plotted in Figure 5.26b and is compared to the drag predictions obtained from Jones' approach and the unsteady lifting line methodology. At the first time instance, the VLM predicts nearly zero drag, which jumps at the next time step to a value much closer to the analytical predictions at the second time instance. Then, the VLM

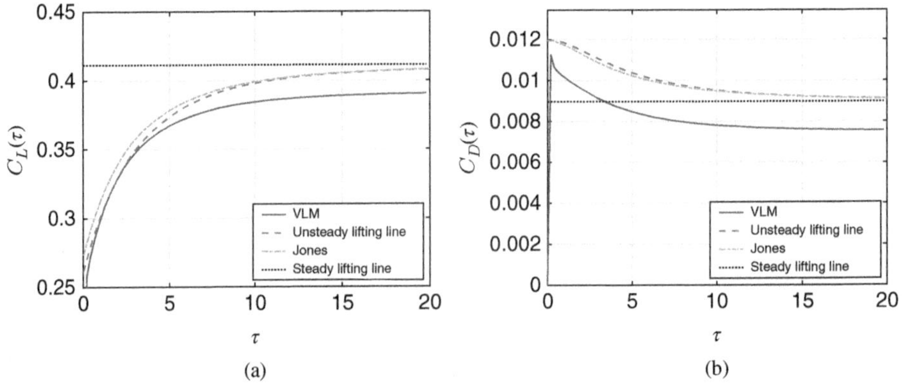

Figure 5.26 Aerodynamic load coefficient variation with time after impulsive start for an elliptical wing with $\alpha = 5°$ and $\mathcal{R} = 6$. (a) $C_L(\tau)$ and (b) $C_D(\tau)$.

drag estimates approach asymptotically a steady-state value that is lower than the lifting line steady-state C_D. This difference is again due to the over-prediction of the steady lift by lifting line theory, since Figure 5.23b shows that the steady induced drag factors predicted by the two methods are very similar.

Unsteady VLM calculations can become computationally expensive, particularly if the wake is modelled as free. The cost of the free wake calculation increases with the square of the number of time instances, while the cost of the prescribed wake calculation is proportional to the number of time instances. To improve the execution time, the influence coefficient matrix calculations are coded in C and must be compiled as a Matlab MEX-function. This example is solved by Matlab code `impulsive_VLM.m`.

The VLM technique is a less simplistic model of the wing than lifting line theory and is therefore said to be of higher fidelity. Calculating the flow over a lifting surface rather than a lifting line has a non-negligible effect on the predicted aerodynamic loads, but this effect is mainly visible as the flow reaches its steady state. Figure 5.26 shows that the initial load responses predicted by the Jones and unsteady lifting line techniques are in good agreement with the VLM results, at least for wings with an elliptical planform.

We now vary the aspect ratio and calculate the resulting lift and drag coefficient responses, again for $m = 10$, $n = 20$. Figure 5.27a plots $C_L(\tau)$ for $\alpha = 5°$ and five values of $\mathcal{R}$, as well as Wagner's solution for a 2D flat plate, denoted by $\mathcal{R} = \infty$. It is seen that the finite wings' lift responses approach that of the 2D plate as the aspect ratio increases from 4 to 40. Figure 5.27b plots $CD(\tau)$ for $\alpha = 5°$ and all aspect ratio values, along with the Wagner drag prediction for a 2D flat plate. For $\mathcal{R} \leq 6$, the drag of the finite wings at the start of the motion is lower than the 2D drag while at the end of the motion it is higher due to the effect of the steady-state induced drag. The cross over occurs somewhere between $\tau = 3$ and $\tau = 5$, depending on the aspect ratio. For $\mathcal{R} = 40$, the drag response of the finite wing lies very close to the 2D drag response.

The impulsive start results presented up to this point are interesting because they demonstrate the similarities and the differences between 2D and 3D wings. In both cases, the loads jump to an initial value at the start of the motion and then approach asymptotically their steady-state values. However, the steady-state drag of finite wings is not equal to zero, so

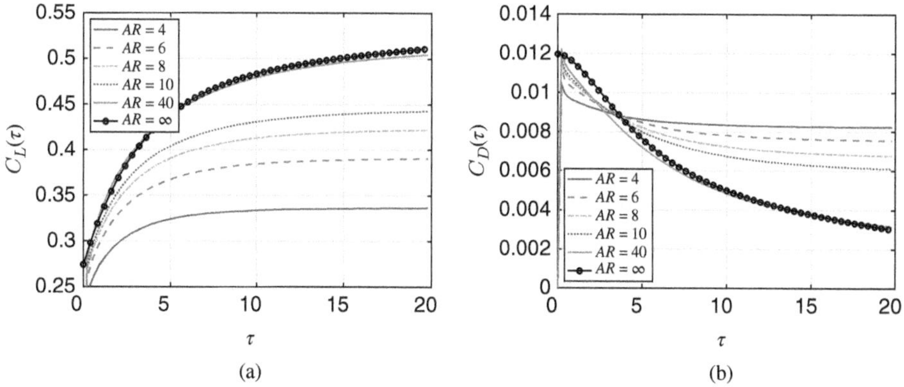

Figure 5.27 Aerodynamic load coefficient variation with time after impulsive start for elliptical wings with $\alpha = 5°$ for different aspect ratios. (a) $C_L(\tau)$ and (b) $C_D(\tau)$.

that the impulsive drag responses of 2D and 3D wings are fundamentally different. Another difference between 2D and 3D flow cases is the wake; in the 2D case, the wake is a 1D vorticity line, while in the 3D case, the wake is a 2D vorticity surface that rolls up at the wingtips. Nevertheless, this rollup has a very small effect on the aerodynamic loads for impulsively started motion. This is a very useful observation since prescribed wake VLM calculations are much faster than free wake calculations.

5.4.2 Other Planforms

Figure 5.27 shows how the lift and drag forces acting on a finite wing with elliptical planform vary with time for different aspect ratios. However, as finite wings can have many different planforms, it is interesting to observe how the lift and drag responses vary with planform for the same aspect ratio. For this, we will return to the rectangular wing implemented in Example 5.4, and we will add sweep and dihedral information to it. The objective is to explore the effect of planform on the impulsive load response of a wing with a given aspect ratio.

Example 5.6 *Calculate the aerodynamic loads acting on impulsively started rectangular wings with different planforms.*

We start the wing definition as we did in Example 5.4. We choose a baseline wing with $Æ = 6$, *taper* $\lambda = 1$ *and zero camber, sweep, twist and dihedral. We then vary the taper, sweep, twist and dihedral individually. The spanwise variation of the chord,* c_j, *is obtained from Eq. (5.85); those of the twist,* ε_j *and quarter-chord chordwise position due to sweep,* $x_{c/4_j}$ *are given by*

$$\begin{aligned} \varepsilon_j &= \varepsilon_t |\bar{y}_{p_j}| \\ x_{c/4_j} &= \frac{b}{2} \tan \Lambda_{c/4} |\bar{y}_{p_j}| \end{aligned} \tag{5.106}$$

where ε_t *is the twist angle at the wingtip,* $\Lambda_{c/4}$ *the sweep angle at the quarter-chord and* Γ_d *the dihedral angle. First, we set up the coordinates of the wing without twist, sweep or dihedral from Eq. (5.103), setting* $a_x = 1/4$. *Then we twist each chordwise section by the corresponding* ε_j *value, using*

$$\begin{pmatrix} x_{p_{i,j},1} \\ z_{p_{i,j},1} \end{pmatrix} = \begin{pmatrix} \cos \epsilon_j & \sin \epsilon_j \\ -\sin \epsilon_j & \cos \epsilon_j \end{pmatrix} \begin{pmatrix} x_{p_{i,j},0} \\ z_{p_{i,j},0} \end{pmatrix} \tag{5.107}$$

and $y_{p_{i,j},1} = y_{p_{i,j},0}$. *Next, we apply the sweep, such that*

$$x_{p_{i,j},2} = x_{p_{i,j},1} + x_{c/4_j}, \; y_{p_{i,j},2} = y_{p_{i,j},1}, \; z_{p_{i,j},2} = z_{p_{i,j},1}$$

and, finally, we apply the dihedral from

$$\begin{pmatrix} x_{p_{i,j}} \\ y_{p_{i,j}} \\ z_{p_{i,j}} \end{pmatrix} = \begin{pmatrix} 1 & 0 & 0 \\ 0 & \cos\Gamma_d & -\sin\Gamma_d \\ 0 & \sin\Gamma_d & \cos\Gamma_d \end{pmatrix} \begin{pmatrix} x_{p_{i,j},2} \\ y_{p_{i,j},2} \\ z_{p_{i,j},2} \end{pmatrix}$$

for $j = n/2 + 1, \dots, n + 1$; *for* $j = 1, \dots, n/2$, *we simply mirror across the* $y = 0$ *plane the points that we have just calculated. The wing's geometry is now correct, but it still needs to be rotated around the x axis by the angle of attack using Eq. (5.104). The bound vortex ring vertices and the control points are calculated exactly as was done in Example 5.4. We proceed by calculating the steady-state aerodynamic loads,* $C_L(\infty)$ *and* $C_D(\infty)$, *for an angle of attack of* $\alpha = 5^0$ *using the procedure of Example 5.4. The unsteady aerodynamic loads following an impulsive start are calculated using the procedure of Example 5.5.*

Figure 5.28 plots the ratios $C_L(\tau)/C_L(\infty)$ *and* $C_D(\tau)/C_L^2(\tau)$ *as functions of time for the baseline wing with aspect ratio values between 4 and 20. Recall that* $C_L(\tau)/C_L(\infty)$ *for finite wings is essentially a 3D version of the Wagner function so that the latter is also plotted in Figure 5.28a, labelled as* $Æ = \infty$. *As expected from Figure 5.27a, the ratio* $C_L(\tau)/C_L(\infty)$ *approaches the Wagner function as Æ increases and 3D aerodynamic effects become less important. Figure 5.28b shows that the time response of the induced drag factor* $C_D(\tau)/C_L^2(\tau)$ *also tends to its Wagner theory value as the aspect ratio increases.*

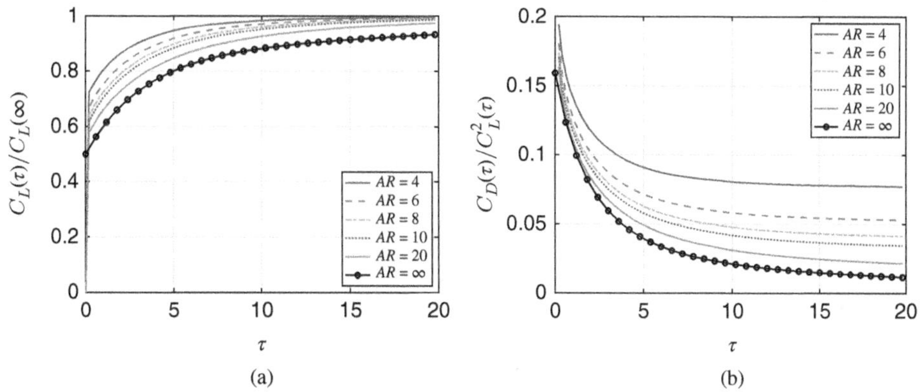

Figure 5.28 Effect of aspect ratio on the load responses after an impulsive start for rectangular wings. (a) $C_L(\tau)/C_L(\infty)$ and (b) $C_D(\tau)/C_L^2(\tau)$.

We will now determine the effect of taper on the impulsive load response. We set the aspect ratio back to 6 and impose taper values between 0.2 and 1. Figure 5.29a shows that as the taper ratio decreases, $C_L(\tau)/C_L(\infty)$ *tends faster towards its steady-state value, even though the initial values of the ratio are nearly identical. The induced drag factor plot of Figure 5.29b shows that* $C_D(\tau)/C_L^2(\tau)$ *also approaches its steady-state value faster as the taper ratio decreases. It follows that both the unsteady lift and drag responses are affected significantly by the taper ratio.*

We next set $\lambda = 1$ *and vary the sweep; we select five wings with sweep values of* $\Lambda_{c/4} = 0°$, $10°, 20°, 30°$ *and* $40°$ *and plot the resulting* $C_L(\tau)/C_L(\infty)$ *and* $C_D(\tau)/C_D(\infty)$ *responses in Figure 5.30. It can be seen that the effect of sweep on the lift response after an impulsive start is negligible. Its effect on the induced drag factor is slightly more important but still not as significant as the effect of the taper or aspect ratio. In fact, the effects of camber (assuming a*

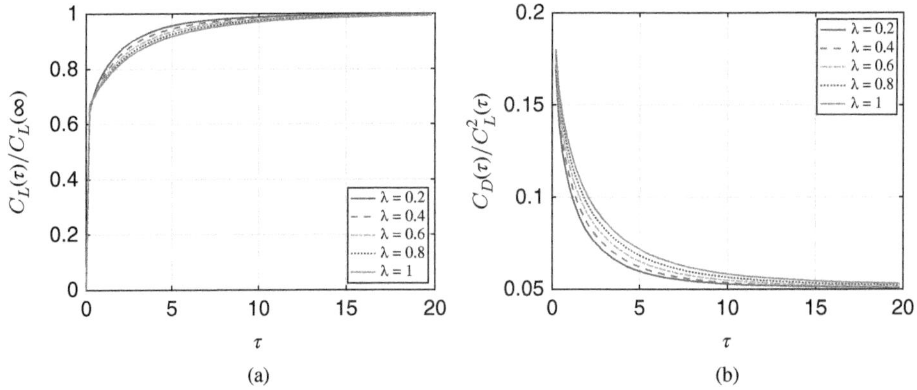

Figure 5.29 Effect of taper ratio on the load responses after an impulsive start for rectangular wings with $\mathcal{R} = 6$. (a) $C_L(\tau)/C_L(\infty)$ and (b) $C_D(\tau)/C_L^2(\tau)$.

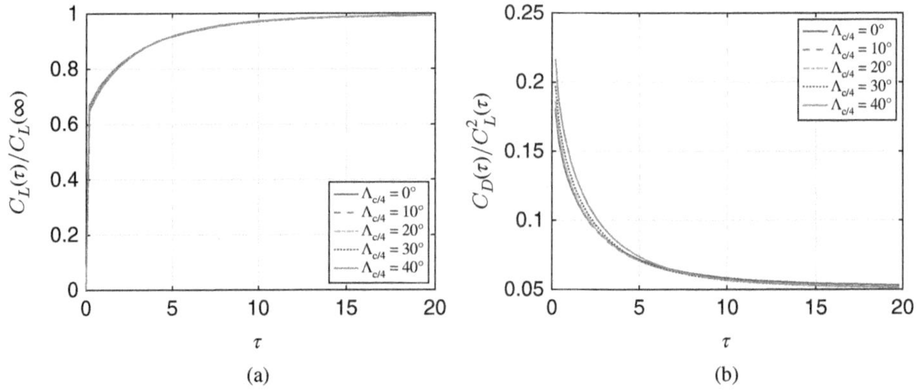

Figure 5.30 Effect of sweep on the load responses after an impulsive start for rectangular wings with $\mathcal{R} = 6$. (a) $C_L(\tau)/C_L(\infty)$ and (b) $C_D(\tau)/C_L^2(\tau)$.

NACA four digit camber line) and dihedral angle on the impulsive loads are also negligible; these calculations are left as an exercise for the reader.

Finally, we repeat the analysis for twist values between $\varepsilon_t = -4°$ and $1°$ and plot the resulting load responses in Figure 5.31. It is clear that twist has a measurable effect on $C_L(\tau)/C_L(\infty)$; increasing the washout decreases the value of the jump at $\tau = 0$. Figure 5.31b shows that twist has a major effect on the induced drag factor. Small amounts of washout cause a slight decrease in $C_D(\tau)/C_L^2(\tau)$, such that the minimum steady induced drag factor occurs at approximately $\varepsilon_t = -1.2°$. However, larger amounts of washout result in an enormous increase in both steady and unsteady C_D/C_L^2. It must be stressed that twist values used in practice usually lie between $\varepsilon_t = -3°$ and $-1°$ and are combined with taper. Furthermore, the effect of twist is sensitive to the rotation point. For example, if a_x in Eq. (5.103) is set to zero, the twist rotation occurs around the leading edge and the effect on $C_L(\tau)/C_L(\infty)$ is the opposite of what is observed in Figure 5.31a.

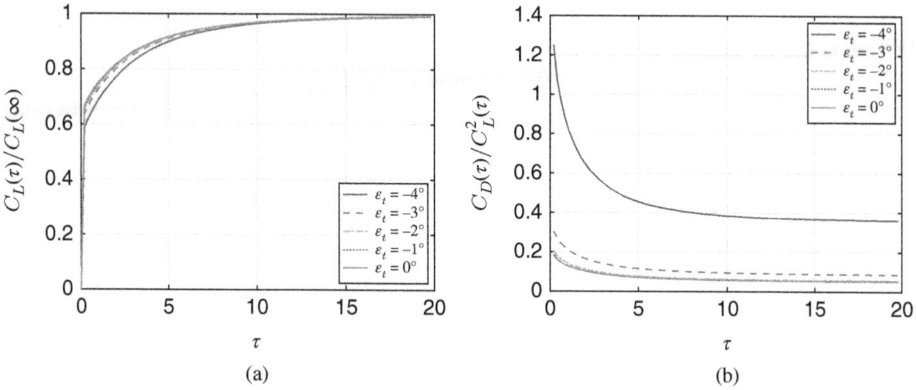

Figure 5.31 Effect of twist on the load responses after an impulsive start for rectangular wings with $\mathcal{R} = 6$. (a) $C_L(\tau)/C_L(\infty)$ and (b) $C_D(\tau)/C_L^2(\tau)$.

In conclusion, this example has shown that the most important parameter affecting the impulsive load response of a rectangular wing is the aspect ratio, followed by the taper ratio. The effects of sweep, dihedral and camber are negligible, while that of twist is only significant for unrealistic washout values. This example is solved by Matlab code `impulsive_VLM_rect.m`.

5.5 Rigid Harmonic Motion

In Chapters 3 and 4, we studied two types of unsteady motion for 2D airfoils, impulsive start and harmonic oscillations. We saw that analytical and numerical solutions exist for both these types of motion. In this chapter, we investigated the impulsive start of 3D wings and saw that there exists an analytical solution for elliptical planforms. For general planforms and for other types of motion, no such solutions are available. Therefore, in order to study the load response of finite wings undergoing general motion, we must resort to numerical techniques, such as the unsteady VLM. In this section, we will concentrate on rigid harmonic motion in both the longitudinal and lateral directions. Longitudinal motion involves pitch and plunge, while lateral motion involves roll, yaw and sideslip.

5.5.1 Longitudinal Harmonic Motion

We will start with longitudinal motion and consider a rigid finite wing that is forced to undergo harmonic pitching and plunging oscillations. The pitch and plunge motions are denoted by $\alpha(t)$ and $h(t)$; from this point onwards, the plunge is defined as positive upwards. The pitching occurs around an axis with direction (0, 1, 0) and passing through a point $(x_f, 0, 0)$. The oscillations are described by

$$h(t) = h_0 + h_1 \sin\left(\omega t + \phi_h\right),\ \alpha(t) = \alpha_0 + \alpha_1 \sin\left(\omega t + \phi_\alpha\right) \tag{5.108}$$

where h_1, α_1 are the plunge and pitch amplitudes, h_0, α_0 the mean values of the plunge and pitch, ω is the frequency and ϕ_h, ϕ_α are the plunge and pitch phases. The wing remains rigid so that its geometry does not change in time, but as it plunges and pitches, its position and angle of attack will change in time. Therefore, the VLM panel vertices, bound vortex ring vertices and control points will become functions of time. Equation (5.104) is written for panel vertex i, j at time t_k as

$$\begin{pmatrix} x_{p_{i,j}}(t_k) \\ y_{p_{i,j}}(t_k) \\ z_{p_{i,j}}(t_k) \end{pmatrix} = \begin{pmatrix} \cos\alpha(t_k) & 0 & \sin\alpha(t_k) \\ 0 & 1 & 0 \\ -\sin\alpha(t_k) & 0 & \cos\alpha(t_k) \end{pmatrix} \begin{pmatrix} x_{p_{i,j},0} - x_f \\ y_{p_{i,j},0} \\ z_{p_{i,j},0} \end{pmatrix} + \begin{pmatrix} x_f \\ 0 \\ h(t_k) \end{pmatrix} \tag{5.109}$$

for $i = 1, \dots, m, j = 1, \dots, n$. The coordinate of the panel control points, $\mathbf{x}_{c_{i,j}}(t_k)$, and of the vertices of the vortex rings, $\mathbf{x}_{v_{i,j}}(t_k)$, must be recalculated after this rotation.

The impermeability boundary condition is given by Eq. (5.75). Setting $\mathbf{Q}_\infty = (U_\infty, 0, 0)$, $\Omega = (0, \dot{\alpha}(t_k), 0)$ and $(\mathbf{x}_c(t_k) - \mathbf{x}_f(t_k)) = (x_{c_I}(t_k) - x_f, y_{c_I}(t_k), z_{c_I}(t_k) - h(t_k))$, the components of the relative velocity between the body and the fluid at the control points become

$$\begin{aligned} u_{m_I}(t_k) &= U_\infty - \dot{\alpha}(t_k)(z_{c_I}(t_k) - h(t_k)) \\ v_{m_I}(t_k) &= 0 \\ w_{m_I}(t_k) &= -\dot{h}(t_k) + \dot{\alpha}(t_k)(x_{c_I}(t_k) - x_f) \end{aligned} \tag{5.110}$$

Then, the matrix form of the impermeability boundary condition (5.76) can be used

$$\boldsymbol{A}_n(t_k)\boldsymbol{\Gamma}(t_k) + \boldsymbol{B}_n(t_k)\boldsymbol{\Gamma}_w(t_k) = -\boldsymbol{u}_m(t_k) \circ \boldsymbol{n}_x(t_k) - \boldsymbol{v}_m(t_k) \circ \boldsymbol{n}_y(t_k) - \boldsymbol{w}_m(t_k) \circ \boldsymbol{n}_z(t_k) \tag{5.111}$$

where $\boldsymbol{u}_m, \boldsymbol{v}_m, \boldsymbol{w}_m$ are the $mn \times 1$ vectors whose elements are the values of $u_{m_I}, v_{m_I}, w_{m_I}$ on all the control points. This equation can be solved for the unknown circulation strengths $\boldsymbol{\Gamma}(t_k)$, as usual.

The wake modelling choices are identical to the ones detailed in Section 4.3.1 for 2D airfoils. If prescribed wake modelling is chosen, then the wake will be propagated at the free stream airspeed and its shape will depend on the choice of wing motion. It can be completely flat if the wing is quasi-fixed and sinusoidal if the wing is moving. Flat and sinusoidal prescribed wake shapes are plotted in Figures 5.32a,b respectively. Finally, if the wake is modelled as free and the wing is moving, then the wake shape will resemble the one plotted in Figure 5.32c, featuring rollup at the wingtips.

Example 5.7 *Use the unsteady VLM to estimate the thrust produced by a harmonically plunging and pitching rectangular wing*

DeLaurier and Harris (1982) performed a series of experiments on a plunging and pitching finite wing in the wind tunnel and measured its thrust. The wing was rectangular with a NACA 0012 section, an aspect ratio of 4 and a chord of $c_0 = 0.051$ m. It was tested in a low speed wind tunnel at Reynolds numbers of 27800 and 43070 under harmonic plunge-only and combined pitch-plunge motion. Only the thrust was measured and plots of mean thrust coefficient against reduced frequency were presented. The mean thrust coefficient was defined as

$$C_T = -\frac{|D|}{1/2\rho U_\infty^2 S(2h_{\max}/c_0)^2} \tag{5.112}$$

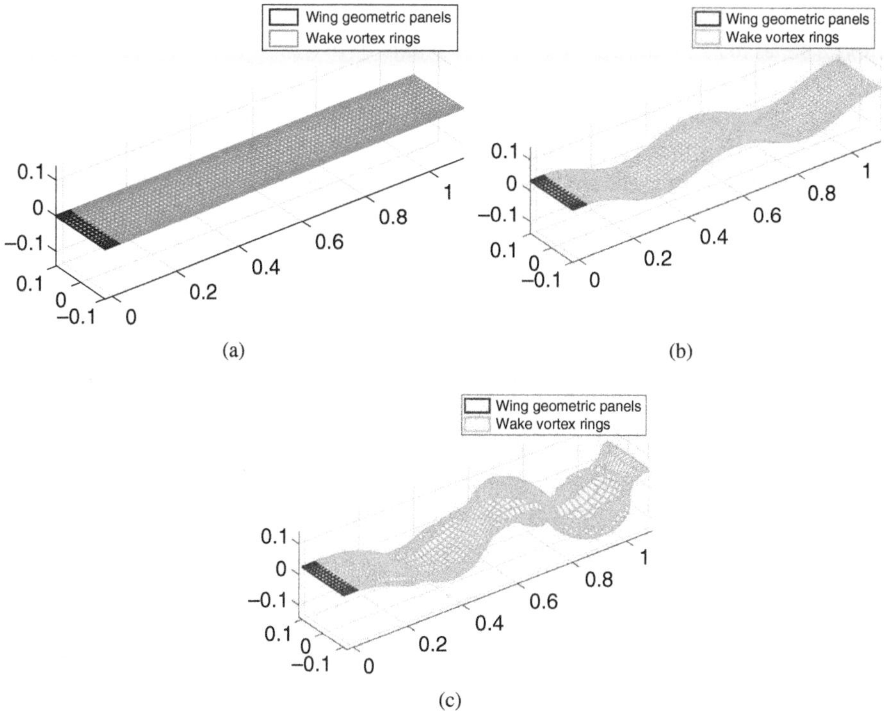

Figure 5.32 Types of wake modelling for a sinusoidally pitching and plunging finite wing. (a) Prescribed wake without wing motion (flat wake), (b) prescribed wake with wing motion and (c) free wake with wing motion.

where $|D|$ is the mean drag over a complete cycle of oscillation, $S = 0.01\ \mathrm{m}^2$ is the wing's surface area and $h_{\max}$ is the maximum vertical displacement of the leading or trailing edge, given by

$$h_{\max} = \begin{cases} \max\left(x_f \sin(\alpha(t)) + h(t)\right) & \text{if } \phi_\alpha \le 90^\circ \\ \max\left((x_f - c_0)\sin(\alpha(t)) + h(t)\right) & \text{if } \phi_\alpha > 90^\circ \end{cases}$$

In this example, we will first simulate a pure plunge motion case and compare the VLM load responses to the experimental measurements. The plunging motion is described by Eq. (5.108) with $h_1 = 0.625c_0$, $\phi_h = 0^\circ$, $h_0 = \alpha_0 = \alpha_1 = 0^\circ$. The reduced frequency $k = 2\omega U_\infty / c_0$ ranges between 0 and 0.16.

We start by defining the wing as in the baseline planform case of Example 5.6 and then we calculate the coordinates of the panel vertices. The numbers of chordwise and spanwise panels are denoted by m and n as usual. The time step is defined as $\Delta t = c_0/mU_\infty$ so that the number of time instance per cycle is

$$p_{pc} = \frac{T}{\Delta t} = \frac{\pi m}{k} \tag{5.113}$$

where $T = 2\pi/\omega$. As p_{pc} is unlikely to be an integer, we round it and adjust the time step such that there are $\lfloor p_{pc} \rceil$ time instances per period, that is

$$\Delta t = \frac{T}{\lfloor p_{pc} \rceil} \tag{5.114}$$

Consequently, the total number of time instances is equal to $n_c\lfloor p_{pc}\rceil$, where n_c is the number of complete cycles to be simulated. This number should be at least equal to 2 because the first cycle will be affected by the starting vortex.

As the position of the wing changes at each time instance, we calculate the time-varying positions of the panel corner-points using Eq. (5.109). We then calculate the coordinates of the panel control points and of the vertices of the bound vortex rings, as well as the unit vectors normal to the panels. The first row of the wake vortex rings is then shed from the trailing edge of the bound vortex rings, which is now time-varying,

$$\mathbf{x}_{w_{1,j}}(t_k) = \mathbf{x}_{v_{m+1,j}}(t_k)$$

Once the instantaneous geometry of the wing and wake has been set up, we can calculate all the influence coefficient matrices and then set up and solve the impermeability condition of Eq. (5.111) for the unknown bound vortex strengths. Finally, the quasi-steady aerodynamic force contributions are obtained from Eqs. (5.78) and (5.79) and the unsteady contribution from equation 5.80. In order to use these equations, we need the total flow velocities at the mid-points of the spanwise and chordwise bound vortex segments, which now include the relative motion between the wing and the flow. Combining Eqs. (5.94) and (5.95) with (5.110) leads to

$$\begin{aligned}
\boldsymbol{u}_{ms}(t_k) &= U_\infty - \dot{\alpha}(t_k)(\boldsymbol{z}_{ms}(t_k) - h(t_k)) + \boldsymbol{A}_{u_{ms}}(t_k)\boldsymbol{\Gamma}(t_k) + \boldsymbol{B}_{u_{ms}}(t_k)\boldsymbol{\Gamma}_w(t_k) \\
\boldsymbol{v}_{ms}(t_k) &= \boldsymbol{A}_{v_{ms}}(t_k)\boldsymbol{\Gamma}(t_k) + \boldsymbol{B}_{v_{ms}}(t_k)\boldsymbol{\Gamma}_w(t_k) \\
\boldsymbol{w}_{ms}(t_k) &= -\dot{h}(t_k) + \dot{\alpha}(t_k)(\boldsymbol{x}_{ms}(t_k) - x_f) + \boldsymbol{A}_{w_{ms}}(t_k)\boldsymbol{\Gamma}(t_k) + \boldsymbol{B}_{w_{ms}}(t_k)\boldsymbol{\Gamma}_w(t_k)
\end{aligned} \tag{5.115}$$

and

$$\begin{aligned}
\boldsymbol{u}_{mc}(t_k) &= U_\infty - \dot{\alpha}(t_k)(\boldsymbol{z}_{mc}(t_k) - h(t_k)) + \boldsymbol{A}_{u_{mc}}(t_k)\boldsymbol{\Gamma}(t_k) + \boldsymbol{B}_{u_{mc}}(t_k)\boldsymbol{\Gamma}_w(t_k) \\
\boldsymbol{v}_{mc}(t_k) &= \boldsymbol{A}_{v_{mc}}(t_k)\boldsymbol{\Gamma} + \boldsymbol{B}_{v_{mc}}(t_k)\boldsymbol{\Gamma}_w(t_k) \\
\boldsymbol{w}_{mc}(t_k) &= -\dot{h}(t_k) + \dot{\alpha}(t_k)(\boldsymbol{x}_{mc}(t_k) - x_f) + \boldsymbol{A}_{w_{mc}}(t_k)\boldsymbol{\Gamma}(t_k) + \boldsymbol{B}_{w_{mc}}(t_k)\boldsymbol{\Gamma}_w(t_k)
\end{aligned} \tag{5.116}$$

The drag is given by the x-component of

$$\mathbf{F}(t_k) = \sum_{i=1}^{m+1}\sum_{j=1}^{n}\mathbf{F}_{s_{i,j}}(t_k) + \sum_{i=1}^{m}\sum_{j=1}^{n+1}\mathbf{F}_{c_{i,j}}(t_k) + \sum_{i=1}^{m}\sum_{j=1}^{n}\mathbf{F}_{i,j}^{\text{unst}}(t_k) \tag{5.117}$$

If this drag is negative, then it is a thrust. We calculate the mean drag over the last cycle only, i.e. for $k = (n_c - 1)\lfloor p_{pc}\rceil + 1, \dots, n_c\lfloor p_{pc}\rceil + 1$ and the mean thrust coefficient from Eq. (5.112).

We choose $m = 4$, $n = 20$ and carry out simulations for k values between 0.04 and 0.16, calculating the mean thrust coefficient $C_T(k)$. Before comparing to the experimental measurements, we need to discuss an intricacy of unsteady wind tunnel experiments. When measuring the forces acting on a pitching and plunging wing with a load cell, we record not only the aerodynamic forces but also the inertial loads due to the motion, sometimes referred to as tare loads. Several solutions exist in order to identify the tare loads:

- *Test the wing forced to undergo the same motion in a vacuum. The measured loads will be purely inertial and can be subtracted from the wind-on measurements. This solution is ideal but can also be impractical, depending on the size of the model and its actuation mechanism.*

- *Replace the wing by a slender rod that has the same inertial characteristics and test it using the same kinematics in still air. Subtract the resulting loads from those measured using the wing at wind-on conditions. This solution is also good because a slender rod will generate negligible non-circulatory aerodynamic loads, at least at low to medium reduced frequencies. However, it works best for measuring the inertial loads due to plunging. It is difficult to build a slender rod system that has the same inertia in pitch as a wing and no aerodynamic influence.*
- *Carry out a test at a low frequency and subtract the mean of the resulting measurements from the mean of the higher-frequency data. This technique only works for mean loads and yields a relative estimate of the aerodynamic loads. Nevertheless, it is very cheap and can be easily replicated by numerical simulations.*

DeLaurier and Harris (1982) chose to estimate the tare drag due to the wing's support and actuation system by carrying out a test at $k = 0.034$ and subtracting it from the drag at higher reduced frequencies. We implement the same calculation for comparison purposes and plot the resulting $C_T(k)$ against k in Figure 5.33a, along with the experimental results by DeLaurier and Harris (1982) for the two tested Reynolds numbers. The VLM results (after tare removal) are in good agreement with the experimental data up to $k \approx 0.085$ for $Re = 43070$ and up to $k \approx 0.13$ for $Re = 278\,800$. The mean thrust follows a quadratic relationship with reduced frequency, tending to zero as k tends to zero. At higher values of k, the experimental curves flatten out due to flow separation. Carrying out a second-order curve fit of the VLM results with respect to k gives

$$C_T(k) = 1.37k^2 - 0.03k$$

which is in excellent agreement with the quasi-steady estimate $C_T(k)/k^2 = 1.4$ by Küchemann and Weber (1953) that is cited by DeLaurier and Harris (1982). Increasing the number of chordwise panels to $m = 10$ or using a free wake has very little effect on $C_T(k)$.

We now simulate a pitching and plunging case for which $h_1 = 0.625c_0$, $\phi_h = 0°$, $h_0 = \alpha_0 = 0°$, $\alpha_1 = 5.7°$ and $\phi_\alpha = 90°$. The pitching axis passes through the half-chord point, $x_f = c_0/2$.

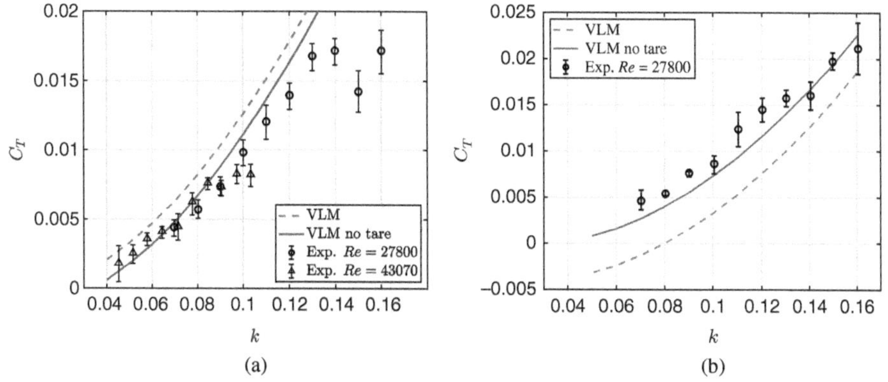

Figure 5.33 Mean thrust coefficient variation with reduced frequency for a pitching and plunging rectangular wing. (a) Plunge only and (b) pitch and plunge. Experimental results by DeLaurier and Harris (1982). Source: Adapted from DeLaurier and Harris (1982).

The other simulation parameters are identical to those of the plunge-only calculations and, again, we evaluate the tare drag at $k = 0.034$. Figure 5.33b plots the resulting $C_T(k)$ values and compares them to the experimental results. Clearly, the agreement between the numerical predictions and the experimental data is very good. However, the experimental results hide a very important element; the wing does not produce thrust at all values of k. In subtracting the tare drag calculated at $k = 0.034$, DeLaurier and Harris (1982) made the assumption that very little drag or thrust is produced at that value of the reduced frequency. This assumption is approximately correct for the plunge-only oscillations but not for the pitching and plunging motion. Figure 5.33b also plots the VLM $C_T(k)$ estimates without subtraction of the tare drag, showing that the wing produces drag for $k < 0.08$.

The pitch-plunge motion of this example is pitch-leading, which means that the pitch angle is positive as the wing is going up and negative as the wing is going down. Consequently, the motion-induced effective angle of attack is reduced so that the flow does not separate and the viscous drag remains low throughout the cycle. For small amplitudes, the effective angle of attack is given by

$$\alpha_{\text{eff}}(t) = \alpha(t) - \frac{\dot{h}(t)}{U_\infty}$$

Substituting for the sinusoidal forms of $\alpha(t)$ and $h(t)$, it is straightforward to show that the minimum magnitude of α_{eff} occurs when $\phi_\alpha = 90°$ and is given by

$$\min\,(|\alpha_{\text{eff}}(k)| = \alpha_1 - \frac{2kh_1}{c_0}$$

Clearly, as the frequency is reduced, the plunge contribution becomes small so that the major contributor to the effective angle of attack is the pitch angle. Furthermore, the unsteady aerodynamic load contribution also becomes small with decreasing k so that the wing produces mostly drag whether the effective angle of attack is positive or negative, as in the steady case.

We have only presented mean thrust results because this is the only type of unsteady data in DeLaurier and Harris (1982) that can be used for validation. However, it is interesting to look at both the lift and drag responses with respect to the phase of the effective angle of attack. Figure 5.34 plots $C_L(\tau)$ and $C_D(\tau)$ over the last cycle of oscillation for the pitching and

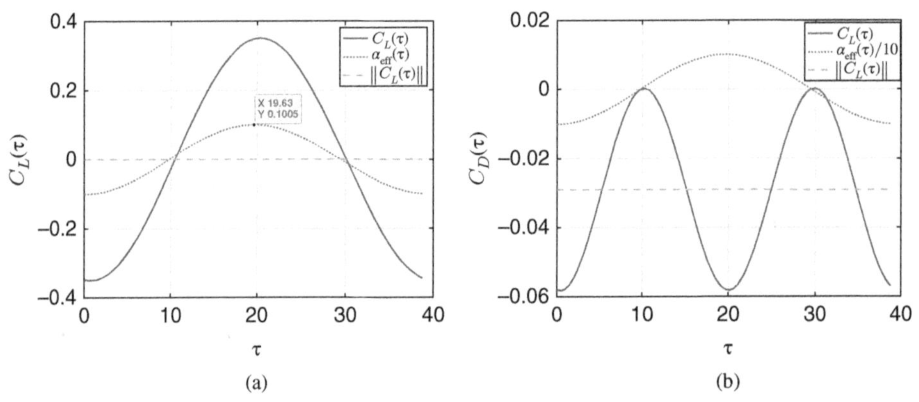

Figure 5.34 Lift and drag responses of the pitching and plunging wing of Figure 5.33b at $k = 0.16$. (a) $C_L(\tau)$ and (b) $C_D(\tau)$.

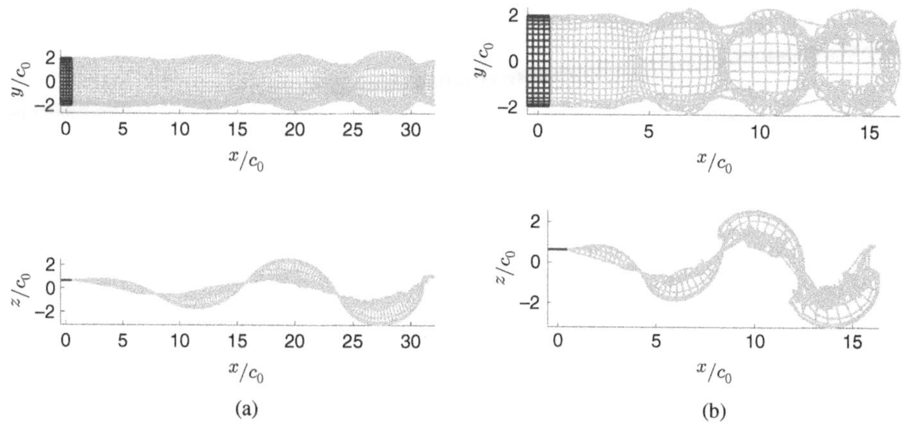

Figure 5.35 Free wake shape behind plunging rectangular wing with $\AE = 4$. (a) $k = 0.2$ and $k = 0.4$.

plunging case and for $k = 0.16$. The effective angle of attack $\alpha_{\text{eff}}(\tau)$ is also plotted on the same axes (scaled by a factor of 10 in Figure 5.34b). It can be seen that both the lift and the drag are nearly zero when the effective angle of attack is also zero; the wing is said to be aerodynamically neutral at these time instances. The lift goes through its maximum when $\alpha_{\text{eff}}(\tau)$ is also a maximum and vice versa. Conversely, the drag is minimised (or the thrust is maximised) when $\alpha_{\text{eff}}(\tau)$ is both maximum and minimum. This synchronisation is purely due to the 90° phase difference between the pitch and the plunge; changing ϕ_α introduces a phase shift between the minima of $C_D(\tau)$ and the extrema of $\alpha_{\text{eff}}(\tau)$. This example is solved by Matlab code `pitchplunge_VLM.m`.

Calculating the free wake behind an oscillating finite wing does not necessarily modify the aerodynamic load predictions but is of interest from the physical point of view. Figure 5.35 plots the free wake shape after two cycles of oscillation for the harmonically plunging $\AE = 4$ rectangular wing. For Example 5.7 Figure 5.35a shows that, for $k = 0.2$, the wake goes through narrower and wider sections, a phenomenon generally referred to as "concertina wake" in the flapping flight literature (Spedding et al. 1984). The wider sections of the wake were shed at times when the wing was producing significant lift or downforce; the narrow sections when the wing was aerodynamically neutral. As the reduced frequency increases, the narrow sections of the wake become even narrower, to the point that the wake separates into discrete rings, as seen in Figure 5.35b for $k = 0.4$. This separation phenomenon has also been observed experimentally behind a pitching and plunging rectangular wing with aspect ratio 3 (von Ellenrieder et al. 2003), albeit at higher reduced frequencies that also provoke leading edge separation.

5.5.2 Frequency Domain Load Calculations

The time domain calculations of Section 5.5.1 can be quite computationally expensive, particularly if we require to evaluate the load responses for different values of the frequency and amplitude of oscillation. Selecting a prescribed wake speeds up the calculations but separate simulations must still be carried out for different motion parameter values.

Furthermore, these calculations must last for several cycles in order to calculate the steady-state load response.

An alternative to repeated time domain simulations is to apply the Fourier Transform to the VLM and to carry out the calculations in the frequency domain. This approach has been used mainly for aeroelasticity (Dimitriadis 2017; Dimitriadis et al. 2018), but it can also be applied to purely aerodynamic calculations. The first step is to simplify the VLM modelling: the wing is quasi-fixed, and the wake is flat. The pitch and plunge motions are still given by Eq. (5.108), but since the wing is not moving, the pitch angle variation with time is modelled by rotating the free stream by $\alpha(t)$. Assuming that the pitch angle remains small at all times, Eq. (5.110) for the relative flow speeds seen by the Ith control point become

$$u_{m_I}(t) = U_\infty - \dot{\alpha}(t)z_{c_I},\ v_{m_I}(t) = 0,\ w_{m_I}(t) = U_\infty\alpha(t) + \dot{\alpha}(t)(x_{c_I} - x_f) - \dot{h} \tag{5.118}$$

where x_f is constant and $y_f = 0$ since the wing is not moving. The impermeability boundary condition of Eq. (5.111) is written at time t as

$$\boldsymbol{A}_n\boldsymbol{\Gamma}(t) + \boldsymbol{B}_n\boldsymbol{\Gamma}_w(t) = -\boldsymbol{u}_m(t) \circ \boldsymbol{n}_x - \boldsymbol{v}_m(t) \circ \boldsymbol{n}_y - \boldsymbol{w}_m(t) \circ \boldsymbol{n}_z \tag{5.119}$$

where all the influence coefficients and normal vectors are constant in time. Note that the multiplication by the constant n_{x_I} and n_{z_I} models the effect of the geometric camber. Substituting from Eq. (5.118), we obtain

$$\begin{aligned}\boldsymbol{A}_n\boldsymbol{\Gamma}(t) + \boldsymbol{B}_n\boldsymbol{\Gamma}_w(t) = &-\left(U_\infty - \dot{\alpha}(t)\boldsymbol{z}_c\right) \circ \boldsymbol{n}_x \\ &-\left(U_\infty\alpha(t) + \dot{\alpha}(t)(\boldsymbol{x}_c - x_f) - \dot{h}(t)\right) \circ \boldsymbol{n}_z\end{aligned} \tag{5.120}$$

Equation (5.120) can be combined with a prescribed wake propagation scheme and solved in the time domain as usual. The aerodynamic loads on the panels are calculated using the quasi-steady loads of Eqs. (5.78) and (5.79) and the unsteady loads of Eq. (5.80). The total aerodynamic loads are given by Eq. (5.117), but they are perpendicular and parallel to the chord, not to the free stream. In order to obtain the lift and drag, we must project $\mathbf{F}(t_k)$ in directions perpendicular and parallel to the free stream such that

$$L(t) = F_z(t) - F_x(t)\alpha(t),\ D(t) = F_z(t)\alpha(t) + F_x(t) \tag{5.121}$$

where $\mathbf{F}(t_k) = [F_x(t_k), F_y(t_k), F_z(t_k)]$.

The same VLM problem with a quasi-fixed wing and flat wake can be solved much more efficiently in the frequency domain, as was done in Section 4.4 for the 2D lumped vortex method. First, we need to recall that the wake vortex strengths are equal to the strengths of the bound trailing edge vortices at previous time steps such that

$$\boldsymbol{\Gamma}_w(t) = \begin{pmatrix} \Gamma_{(m-1)n+1}(t-\Delta t) \\ \vdots \\ \Gamma_{mn}(t-\Delta t) \\ \Gamma_{(m-1)n+1}(t-2\Delta t) \\ \vdots \\ \Gamma_{mn}(t-2\Delta t) \\ \vdots \\ \Gamma_{(m-1)n+1}(t-m_w\Delta t) \\ \vdots \\ \Gamma_{mn}(t-m_w\Delta t) \end{pmatrix} = \begin{pmatrix} \boldsymbol{P}_c\boldsymbol{\Gamma}(t-\Delta t) \\ \boldsymbol{P}_c\boldsymbol{\Gamma}(t-2\Delta t) \\ \vdots \\ \boldsymbol{P}_c\boldsymbol{\Gamma}(t-m_w\Delta t) \end{pmatrix} \tag{5.122}$$

where $\boldsymbol{P}_c = (\boldsymbol{0}_{n\times(m-1)n}, \boldsymbol{I}_n)$, $\boldsymbol{0}_{n\times(m-1)n}$ is the $n \times (m-1)n$ matrix whose elements are all zero and $\boldsymbol{I}_n$ is the $n \times n$ unit matrix.

The next step is to take the Fourier Transform of Eq. (5.120) such that

$$\boldsymbol{A}_n\boldsymbol{\Gamma}(\omega) + \boldsymbol{B}_n\boldsymbol{\Gamma}_w(\omega) = -\left(U_\infty\delta(\omega) - \mathrm{i}\omega\alpha(\omega)\boldsymbol{z}_c\right) \circ \boldsymbol{n}_x \\ - \left(U_\infty\alpha(\omega) + \mathrm{i}\omega\alpha(\omega)(\boldsymbol{x}_c - x_f) - \mathrm{i}\omega h(\omega)\right) \circ \boldsymbol{n}_z \tag{5.123}$$

where $\delta(\omega)$ is the Kronecker delta function of Eq. (3.145). Applying the Fourier Transform to the wake vortex strength vector results in (Morino and Gennaretti 1992)

$$\boldsymbol{\Gamma}_w(\omega) = \begin{pmatrix} \mathrm{e}^{-\mathrm{i}\omega\Delta t}\boldsymbol{P}_c\boldsymbol{\Gamma}(\omega) \\ \mathrm{e}^{-\mathrm{i}\omega 2\Delta t}\boldsymbol{P}_c\boldsymbol{\Gamma}(\omega) \\ \vdots \\ \mathrm{e}^{-\mathrm{i}\omega m_w\Delta t}\boldsymbol{P}_c\boldsymbol{\Gamma}(\omega) \end{pmatrix} = \boldsymbol{P}_e(\omega)\boldsymbol{P}_c\boldsymbol{\Gamma}(\omega) \tag{5.124}$$

with

$$\boldsymbol{P}_e(\omega) = \begin{pmatrix} \boldsymbol{I}_n\mathrm{e}^{-\mathrm{i}\omega\Delta t} \\ \boldsymbol{I}_n\mathrm{e}^{-\mathrm{i}\omega 2\Delta t} \\ \vdots \\ \boldsymbol{I}_n\mathrm{e}^{-\mathrm{i}\omega m_w\Delta t} \end{pmatrix}$$

Substituting into Eq. (5.123), we obtain

$$\left(\boldsymbol{A}_n + \boldsymbol{B}_n\boldsymbol{P}_e(\omega)\boldsymbol{P}_c\right)\boldsymbol{\Gamma}(\omega) = -\left(U_\infty\delta(\omega) - \mathrm{i}\omega\alpha(\omega)\boldsymbol{z}_c\right) \circ \boldsymbol{n}_x \\ - \left(U_\infty\alpha(\omega) + \mathrm{i}\omega\alpha(\omega)(\boldsymbol{x}_c - x_f) - \mathrm{i}\omega h(\omega)\right) \circ \boldsymbol{n}_z \tag{5.125}$$

which can be readily solved for $\boldsymbol{\Gamma}(\omega)$.

Once the bound vortex strengths have been calculated, we can proceed with the evaluation of the aerodynamic loads. First, we apply the Fourier Transform to the unsteady loads of Eq. (5.80) to obtain

$$\mathbf{F}^{\mathrm{unst}}_{i,j}(\omega) = \rho\mathrm{i}\omega s_{i,j}\Gamma_{i,j}(\omega)\mathbf{n}_{i,j} \tag{5.126}$$

We need to apply the Fourier Transform carefully to the quasi-steady loads of Eqs. (5.78) and (5.79). These equations contain products of time domain signals since $\Gamma_{i,j}(t)$ and $\mathbf{u}_{ms_{i,j}}(t)$ are both functions of time. Multiplication in the time domain becomes convolution in the frequency domain and vice-versa. Hence, Eqs. (5.78) and (5.79) become

$$\mathbf{F}_{s_{i,j}}(\omega) = \rho\left(\Gamma_{i,j}(\omega) - \Gamma_{i-1,j}(\omega)\right) * \mathbf{u}_{ms_{i,j}}(\omega) \times \left(\mathbf{x}_{v_{i,j+1}} - \mathbf{x}_{v_{i,j}}\right) \tag{5.127}$$

$$\mathbf{F}_{c_{i,j}}(\omega) = \rho\left(\Gamma_{i,j}(\omega) - \Gamma_{i,j-1}(\omega)\right) * \mathbf{u}_{mc_{i,j}}(\omega) \times \left(\mathbf{x}_{v_{i+1,j}} - \mathbf{x}_{v_{i,j}}\right) \tag{5.128}$$

where the $*$ symbol denotes convolution and $\mathbf{u}_{ms_{i,j}}(\omega)$, $\mathbf{u}_{mc_{i,j}}(\omega)$ are obtained by applying the Fourier transform to the time-varying versions of expressions (5.94) and (5.95) and setting $\cos\alpha(t) \approx 1$, $\sin\alpha(t) \approx \alpha(t)$, such that

$$\begin{aligned} \boldsymbol{u}_{ms}(\omega) &= U_\infty\delta(\omega) - \mathrm{i}\omega\alpha(\omega)\boldsymbol{z}_{ms} + \boldsymbol{A}_{u_{ms}}\boldsymbol{\Gamma}(\omega) + \boldsymbol{B}_{u_{ms}}\boldsymbol{\Gamma}_w(\omega) \\ \boldsymbol{v}_{ms}(\omega) &= \boldsymbol{A}_{v_{ms}}\boldsymbol{\Gamma}(\omega) + \boldsymbol{B}_{v_{ms}}\boldsymbol{\Gamma}_w(\omega) \\ \boldsymbol{w}_{ms}(\omega) &= U_\infty\alpha(\omega) - \mathrm{i}\omega h(\omega) + \mathrm{i}\omega\alpha(\omega)(\boldsymbol{x}_{ms} - x_f) + \boldsymbol{A}_{w_{ms}}\boldsymbol{\Gamma}(\omega) + \boldsymbol{B}_{w_{ms}}\boldsymbol{\Gamma}_w(\omega) \end{aligned} \tag{5.129}$$

and

$$\begin{aligned}
\boldsymbol{u}_{mc}(\omega) &= U_\infty \delta(\omega) - \mathrm{i}\omega\alpha(\omega)\boldsymbol{z}_{mc} + \boldsymbol{A}_{u_{mc}}\boldsymbol{\Gamma}(\omega) + \boldsymbol{B}_{u_{mc}}\boldsymbol{\Gamma}_w(\omega) \\
\boldsymbol{v}_{mc}(\omega) &= \boldsymbol{A}_{v_{mc}}\boldsymbol{\Gamma}(\omega) + \boldsymbol{B}_{v_{mc}}\boldsymbol{\Gamma}_w(\omega) \\
\boldsymbol{w}_{mc}(\omega) &= U_\infty \alpha(\omega) - \mathrm{i}\omega h(\omega) + \mathrm{i}\omega\alpha(\omega)(\boldsymbol{x}_{mc} - x_f) + \boldsymbol{A}_{w_{mc}}\boldsymbol{\Gamma}(\omega) + \boldsymbol{B}_{w_{mc}}\boldsymbol{\Gamma}_w(\omega)
\end{aligned} \tag{5.130}$$

Similarly, the lift and drag of Eq. (5.121) are transformed to

$$L(\omega) = F_z(\omega) - F_x(\omega) * \alpha(\omega),\; D(\omega) = F_z(\omega) * \alpha(\omega) + F_x(\omega) \tag{5.131}$$

The frequency domain formulation of the VLM becomes very computationally efficient if the excitation is harmonic; the impermeability Eq. (5.125) must only be solved once at the frequency of interest and once at $\omega = 0$. In contrast, the time domain solution requires one solution of the impermeability equation at each time instance and is therefore much more expensive.

Example 5.8 *Repeat example 5.7 using the frequency domain version of the VLM and study the effect of varying the phase on the thrust coefficient.*

We set up the wing exactly as in Example 5.7 but without rotating it to the physical angle of attack. The next step is to create the wake grid. We choose to use the same grid for all frequencies in order to avoid recalculating the wake influence coefficient matrices for each frequency. We select a total length of wake equal to $n_c = 30$ *chord-lengths so that the total length of the wake is* $n_c c_0$*. The time step is still equal to* $\Delta t = c_0/mU_\infty$ *so that the chordwise distance between successive wake panels is* $U_\infty \Delta t = c_0/m$*, starting from the trailing edges of the bound vortex rings. Consequently, the total chordwise number of wake panels is* $m_w = mn_c$*.*

We can now calculate the influence coefficients of the bound and wake vortex rings on the control points, $\boldsymbol{A}_n$ *and* $\boldsymbol{B}_n$*, using the procedure of Section 5.4. Next, we will set up and solve the impermeability condition (5.125), but first we need to define* $\alpha(\omega)$ *and* $h(\omega)$*. In the frequency domain, Eq. (5.108) become Eqs. (3.143) and (3.144)*

$$h(\omega) = h_0\delta(\omega) + \frac{h_1}{2\mathrm{i}}\mathrm{e}^{\mathrm{i}\phi_h}\delta(\omega - \omega_0) - \frac{h_1}{2\mathrm{i}}\mathrm{e}^{-\mathrm{i}\phi_h}\delta(\omega + \omega_0) \tag{5.132}$$

$$\alpha(\omega) = \alpha_0\delta(\omega) + \frac{\alpha_1}{2\mathrm{i}}\mathrm{e}^{\mathrm{i}\phi_\alpha}\delta(\omega - \omega_0) - \frac{\alpha_1}{2\mathrm{i}}\mathrm{e}^{-\mathrm{i}\phi_\alpha}\delta(\omega + \omega_0) \tag{5.133}$$

where ω_0 *is the frequency of oscillation and* ω *is any other frequency. As the values of* $h(\omega)$ *and* $\alpha(\omega)$ *at* $\omega = \omega_0$ *are the complex conjugates of their values at* $\omega = -\omega_0$*, we only need to solve Eq. (5.125) for two frequencies,* $\omega = 0$ *and* $\omega = \omega_0$*.*

We start with the zero frequency solution by substituting $\omega = 0$ *in Eq. (5.125) to obtain*

$$\left(\boldsymbol{A}_n + \boldsymbol{B}_n\boldsymbol{P}_e(0)\boldsymbol{P}_c\right)\boldsymbol{\Gamma}(0) = -U_\infty\boldsymbol{n}_x - U_\infty\alpha_0\boldsymbol{n}_z \tag{5.134}$$

where $\boldsymbol{P}_e(0) = (\boldsymbol{I}_n, \boldsymbol{I}_n, \ldots, \boldsymbol{I}_n)^T$*, so that* $\boldsymbol{P}_e(0)\boldsymbol{P}_c$ *becomes a selector matrix that is equal to 1 only for the elements of* $\boldsymbol{\Gamma}(0)$ *that lie on the trailing edge and zero for all other elements. Equation (5.134) can be easily solved for* $\boldsymbol{\Gamma}(0)$ *and then* $\boldsymbol{\Gamma}_w(0)$ *can be obtained from Eq. (5.124) as*

$$\boldsymbol{\Gamma}_w(0) = \boldsymbol{P}_e(0)\boldsymbol{P}_c\boldsymbol{\Gamma}(0)$$

There is no need to calculate the steady aerodynamic loads for $\omega = 0$, they are all equal to zero since the wing is flat ($\boldsymbol{n}_x = 0$) and the mean angle of attack is zero ($\alpha_0 = 0$), leading to $\boldsymbol{\Gamma}(0) = 0$. For a general wing geometry, we would need to use $\boldsymbol{\Gamma}(0)$ in order to calculate the flow speeds induced by the steady flow at the midpoints of the spanwise and chordwise bound vortex segments, $\boldsymbol{u}_{ms}(0)$ and $\boldsymbol{u}_{mc}(0)$, from expressions (5.129) and (5.130) and then apply Eqs. (5.127) and (5.128) for $\omega = 0$.

Next, we solve the imermeability Eq. (5.125) for $\omega = \omega_0$. Substituting from Eqs. (5.132) and (5.133), the impermeability condition becomes

$$\left(\boldsymbol{A}_n + \boldsymbol{B}_n \boldsymbol{P}_e(\omega_0)\boldsymbol{P}_c\right)\boldsymbol{\Gamma}(\omega_0) = -\left(-\mathrm{i}\omega_0 \frac{\alpha_1}{2\mathrm{i}} \mathrm{e}^{\mathrm{i}\phi_\alpha}\left(\boldsymbol{z}_c - z_f\right)\right) \circ \boldsymbol{n}_x - \left(\frac{\alpha_1}{2\mathrm{i}} \mathrm{e}^{\mathrm{i}\phi_\alpha}\left(U_\infty + \mathrm{i}\omega_0(\boldsymbol{x}_c - x_f)\right) - \mathrm{i}\omega_0 \frac{h_1}{2\mathrm{i}}\right) \circ \boldsymbol{n}_z \quad (5.135)$$

Again, this latest expression is solved for $\boldsymbol{\Gamma}(\omega_0)$ and then $\boldsymbol{\Gamma}_w(\omega_0)$ is obtained from Eq. (5.124). We also evaluate $\mathbf{u}_{ms_{i,j}}(\omega_0)$ and $\mathbf{u}_{mc_{i,j}}(\omega_0)$ by substituting expressions (5.132) and (5.133) into Eqs. (5.129) and (5.130).

The final step is to calculate the aerodynamic loads from Eqs. (5.126)–(5.131). The calculation of the unsteady term is straightforward. The quasi-steady loads on the spanwise segments of expression (5.127) are split into their three components

$$F_{xs_{i,j}}(\omega) = \rho\bar{\Gamma}_{i,j}(\omega) * \bar{u}_{ms_{i,j}}(\omega)$$
$$F_{ys_{i,j}}(\omega) = \rho\bar{\Gamma}_{i,j}(\omega) * \bar{v}_{ms_{i,j}}(\omega)$$
$$F_{zs_{i,j}}(\omega) = \rho\bar{\Gamma}_{i,j}(\omega) * \bar{w}_{ms_{i,j}}(\omega)$$

where

$$\bar{\Gamma}_{i,j}(\omega) = \Gamma_{i,j}(\omega) - \Gamma_{i-1,j}(\omega)$$
$$\bar{u}_{ms_{i,j}}(\omega) = u_{ms_{i,j}}(\omega) \times \left(x_{v_{i,j+1}} - x_{v_{i,j}}\right)$$
$$\bar{v}_{ms_{i,j}}(\omega) = v_{ms_{i,j}}(\omega) \times \left(y_{v_{i,j+1}} - x_{y_{i,j}}\right)$$
$$\bar{w}_{ms_{i,j}}(\omega) = w_{ms_{i,j}}(\omega) \times \left(z_{v_{i,j+1}} - z_{v_{i,j}}\right)$$

Our calculations up to now have been completely linear so that $\bar{\Gamma}_{i,j}(\omega)$, $\bar{u}_{ms_{i,j}}(\omega)$, etc., are sinusoidal and given by

$$\bar{\Gamma}_{i,j}(\omega) = \bar{\Gamma}_{i,j}(0)\delta(\omega) + \bar{\Gamma}_{i,j}(\omega_0)\delta(\omega - \omega_0) + \bar{\Gamma}^*_{i,j]}(\omega_0)\delta(\omega + \omega_0)$$
$$\bar{u}_{ms_{i,j}}(\omega) = \bar{u}_{ms_{i,j}}(0)\delta(\omega) + \bar{u}_{ms_{i,j}}(\omega_0)\delta(\omega - \omega_0) + \bar{u}^*_{ms_{i,j}}(\omega_0)\delta(\omega + \omega_0)$$
$$\bar{v}_{ms_{i,j}}(\omega) = \bar{v}_{ms_{i,j}}(0)\delta(\omega) + \bar{v}_{ms_{i,j}}(\omega_0)\delta(\omega - \omega_0) + \bar{v}^*_{ms_{i,j}}(\omega_0)\delta(\omega + \omega_0)$$
$$\bar{w}_{ms_{i,j}}(\omega) = \bar{w}_{ms_{i,j}}(0)\delta(\omega) + \bar{w}_{ms_{i,j}}(\omega_0)\delta(\omega - \omega_0) + \bar{w}^*_{ms_{i,j}}(\omega_0)\delta(\omega + \omega_0)$$

where the $*$ superscript denotes the complex conjugate. We calculate $\bar{\Gamma}_{i,j}(0)$, $\bar{u}_{ms_{i,j}}(0)$, $\bar{v}_{ms_{i,j}}(0)$, $\bar{w}_{ms_{i,j}}(0)$ from the steady solution and $\bar{\Gamma}_{i,j}(\omega_0)$, $\bar{u}_{ms_{i,j}}(\omega_0)$, $\bar{v}_{ms_{i,j}}(\omega_0)$, $\bar{w}_{ms_{i,j}}(\omega_0)$ from the oscillatory solution.

The convolution integrals are calculated by means of vector convolutions, as in Eq. (3.190). The vectors have three elements whose components correspond to frequencies $-\omega_0$, 0 and ω_0. For example

$$\bar{\Gamma}_{i,j}(\omega) * \bar{u}_{ms_{i,j}}(\omega) = \left(\bar{\Gamma}^*_{i,j}(\omega_0), \bar{\Gamma}_{i,j}(0), \bar{\Gamma}_{i,j}(\omega_0)\right) * \left(\bar{u}^*_{ms_{i,j}}(\omega_0), \bar{u}_{ms_{i,j}}(0), \bar{u}_{ms_{i,j}}(\omega_0)\right)$$

Similar calculations are carried out for the chordwise quasi-steady force contributions. The result is a vector with five elements, corresponding to frequencies $-2\omega_0$, $-\omega_0$, 0, ω_0 *and* $2\omega_0$. *Consequently, all the elements of* $\mathbf{F}_{s_{i,j}}(\omega)$ *and* $\mathbf{F}_{c_{i,j}}(\omega)$ *have the same five frequency components and the total aerodynamic load* $\mathbf{F}(\omega)$ *is of the form*

$$\mathbf{F}(\omega) = \mathbf{F}(0)\delta(\omega) + \mathbf{F}(\omega_0)\delta(\omega - \omega_0) + \mathbf{F}^*(\omega_0)\delta(\omega + \omega_0) + \mathbf{F}(2\omega_0)\delta(\omega - 2\omega_0) + \mathbf{F}^*(2\omega_0)\delta(\omega + 2\omega_0) \quad (5.136)$$

The convolution has doubled the frequency content of the aerodynamic loads with respect to that of the pitch and plunge oscillations. Evaluating the lift and drag from Eq. (5.131) further enriches the frequency content, as it adds components at $\pm 3\omega_0$, *which are however usually negligible. Therefore, we will assume that* $C_L(\omega)$ *and* $C_D(\omega)$ *have the same form as* $\mathbf{F}(\omega)$,

$$C_L(\omega) = C_L(0)\delta(\omega) + C_L(\omega_0)\delta(\omega - \omega_0) + C_L^*(\omega_0)\delta(\omega + \omega_0) + C_L(2\omega_0)\delta(\omega - 2\omega_0) + C_L^*(2\omega_0)\delta(\omega + 2\omega_0)$$
$$C_D(\omega) = C_D(0)(\omega) + C_D(\omega_0)\delta(\omega - \omega_0) + C_D^*(\omega_0)\delta(\omega + \omega_0) + C_D(2\omega_0)\delta(\omega - 2\omega_0) + C_D^*(2\omega_0)\delta(\omega + 2\omega_0)$$

We choose to apply the frequency domain version of the VLM for $\alpha_1 = 5.7°$, $h_1 = 0.625c_0$, *and a range of reduced frequencies between* $k = 0.034$ *(the tare frequency) and* $k = 1.6$. *For the pitch phase, we select values between* $\phi_\alpha = -180°$ *and* $180°$. *We calculate the vales of* $C_L(0)$, $C_L(\omega_0)$, $C_L(2\omega_0)$, $C_D(0)$, $C_D(\omega_0)$ *and* $C_D(2\omega_0)$ *and then calculate the mean thrust coefficient,* $C_T(k, \phi_\alpha)$ *using Eq. (5.112). In order to carry out this last calculation, we need to linearise* $h_{\max}$ *in the frequency domain such that*

$$h_{\max}(\phi_\alpha) = \begin{cases} x_f \frac{\alpha_1}{\mathrm{i}} e^{\mathrm{i}\phi_\alpha} + \frac{h_1}{\mathrm{i}} & \text{if } \phi_\alpha \leq 90° \\ (x_f - c_0)\frac{\alpha_1}{\mathrm{i}} e^{\mathrm{i}\phi_\alpha} + \frac{h_1}{\mathrm{i}} & \text{if } \phi_\alpha > 90° \end{cases}$$

Figure 5.36a plots the variation of C_T *with k and* ϕ_α, *after tare removal, along with all the experimental data published by DeLaurier and Harris (1982). For any frequency, the* C_T

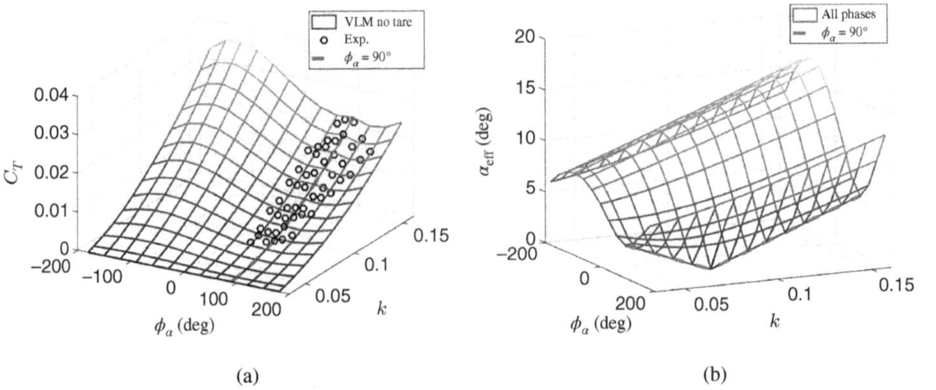

Figure 5.36 Mean thrust coefficient and effective angle of attack variation with reduced frequency and pitch phase angle. (a) $C_T(k, \phi_\alpha)$ and (b) $\alpha_{\text{eff}}(k, \phi_\alpha)$. Experimental results by DeLaurier and Harris (1982). Source: Adapted from DeLaurier and Harris (1982).

surface features a local maximum for $\phi_\alpha = 90°$, but the global maximum lies at $\phi_\alpha = -120°$. However, this is an inviscid calculation and, as discussed earlier, the effective angle of attack is minimum for $\phi_\alpha = 90°$ at all frequencies. Figure 5.36b plots the amplitude of α_{eff} against k and ϕ_α, showing that the effective angle of attack increases very quickly as ϕ_α moves away from 90°. At $\phi_\alpha = -120°$ the amplitude of α_{eff} reaches nearly 17°, which means that viscous effects are extremely important for this condition; significant flow separation is certain to occur over part of the cycle.

The agreement between the experimental data and the VLM estimates for the mean thrust coefficient is good at $\phi_\alpha = 90°$ but deteriorates at 60° and 105°, since these two phase angles result in much higher effective angles of attack. In reality, the best thrust is usually obtained for $\phi_\alpha = 90°$ and the global maximum predicted by the VLM for $\phi_\alpha = -120°$ does not exist. This example is solved by Matlab code `pitchplunge_VLM_freq.m`.

5.5.3 Lateral Harmonic Motion

As mentioned earlier, lateral motion involves the roll, yaw and sideslip degrees of freedom of an aircraft or wing so that the latter rotates around two axes simultaneously. Rotation around two or more axes must be handled carefully, starting with the determination of the axes themselves. The difference between the body-fixed and inertial frames of reference is demonstrated in Figure 5.37 for lateral motion. The wing is drawn in solid lines rotated in roll, pitch and yaw with respect to its initial position that is denoted by the dashed lines. The rotation axes are denoted by the unit vectors $\mathbf{n}_\phi$, $\mathbf{n}_\alpha$ and $\mathbf{n}_\psi$ for the roll, pitch and yaw axes respectively. In Figure 5.37a the rotation axes are body-fixed so that they have rotated with the wing. Conversely, in Figure 5.37b, the rotation axes are inertial so that they have remained parallel to the x, y and z axes, respectively.

When a wing rotates by an angle θ around axis $\mathbf{n}_\theta = (n_{\theta_1}, n_{\theta_2}, n_{\theta_3})^T$, its rotated coordinates can be calculated by

$$\mathbf{x}(\theta) = \boldsymbol{R}(\theta, \mathbf{n}_\theta)\mathbf{x}(0) \tag{5.137}$$

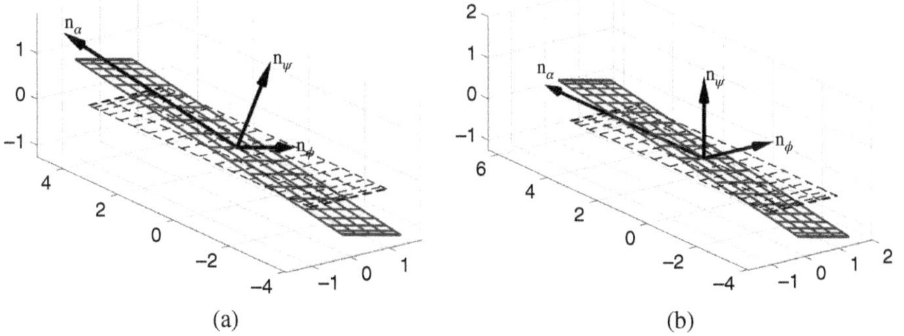

Figure 5.37 Body-fixed and inertial axes systems. (a) Body-fixed and (b) inertial.

where $\mathbf{x}(0) = (x(0), y(0), z(0))^T$ are the original coordinates, $\mathbf{x}(\theta)$ are the coordinates after the rotation and the elements of the 3×3 rotation matrix $\boldsymbol{R}(\theta, \mathbf{n}_\theta)$ are given by

$$\begin{aligned}
&R_{11} = \cos\theta + n_{\theta_1}^2(1-\cos\theta),\ R_{12} = n_{\theta_1}n_{\theta_2}(1-\cos\theta) - n_{\theta_3}\sin\theta \\
&R_{13} = n_{\theta_1}n_{\theta_3}(1-\cos\theta) + n_{\theta_2}\sin\theta,\ R_{21} = n_{\theta_1}n_{\theta_2}(1-\cos\theta) + n_{\theta_3}\sin\theta \\
&R_{22} = \cos\theta + n_{\theta_2}^2(1-\cos\theta),\ R_{23} = n_{\theta_2}n_{\theta_3}(1-\cos\theta) - n_{\theta_1}\sin\theta \\
&R_{31} = n_{\theta_1}n_{\theta_3}(1-\cos\theta) - n_{\theta_2}\sin\theta,\ R_{32} = n_{\theta_2}n_{\theta_3}(1-\cos\theta) + n_{\theta_1}\sin\theta \\
&R_{33} = \cos\theta + n_{\theta_3}^2(1-\cos\theta)
\end{aligned} \tag{5.138}$$

When the wing is undergoing a continuous rotation $\theta(t)$, Eq. (5.137) can be replaced by

$$\mathbf{x}(t) = \boldsymbol{R}(\Delta\theta, \mathbf{n}_\theta(t-\Delta t))\mathbf{x}(t-\Delta t) \tag{5.139}$$

$$\mathbf{n}_\theta(t) = \boldsymbol{R}(\Delta\theta, \mathbf{n}_\theta(t-\Delta t))\mathbf{n}_\theta(t-\Delta t) \tag{5.140}$$

so that the wing is rotated by $\Delta\theta = \theta(t) - \theta(t-\Delta t)$ at each time instance. If the rotation axis is body-fixed, then the rotation vector $\mathbf{n}_\theta(t)$ must also be rotated by the same amount. If $\Delta\theta$ is sufficiently small, the rotation matrix can be linearised to

$$\boldsymbol{R}(\Delta\theta, \mathbf{n}_\theta) \approx \begin{pmatrix} 1 & -n_{\theta_3}\Delta\theta & n_{\theta_2}\Delta\theta \\ n_{\theta_3}\Delta\theta & 1 & -n_{\theta_1}\Delta\theta \\ -n_{\theta_2}\Delta\theta & n_{\theta_1}\Delta\theta & 1 \end{pmatrix} \tag{5.141}$$

Finally, if a wing is rotating continuously around all three axes, its coordinates are given by

$$\mathbf{x}(t) = \boldsymbol{R}(\Delta\phi, \mathbf{n}_\phi(t-\Delta t))\boldsymbol{R}(\Delta\alpha, \mathbf{n}_\alpha(t-\Delta t))\boldsymbol{R}(\Delta\psi, \mathbf{n}_\psi(t-\Delta t))\mathbf{x}(t-\Delta t) \tag{5.142}$$

where $\boldsymbol{R}(\Delta\phi, \mathbf{n}_\phi(t-\Delta t))$, etc., are obtained by substituting the appropriate angle and rotation vector in Eqs. (5.138) or (5.141). The direction vectors must also be rotated, such that

$$\boldsymbol{N}_{\phi,\alpha,\psi}(t) = \boldsymbol{R}(\Delta\phi, \mathbf{n}_\phi(t-\Delta t))\boldsymbol{R}(\Delta\alpha, \mathbf{n}_\alpha(t-\Delta t))\boldsymbol{R}(\Delta\psi, \mathbf{n}_\psi(t-\Delta t))\boldsymbol{N}_{\phi,\alpha,\psi}(t-\Delta t) \tag{5.143}$$

where $\boldsymbol{N}_{\phi,\alpha,\psi}(t) = (\mathbf{n}_\phi(t), \mathbf{n}_\alpha(t), \mathbf{n}_\psi(t))$. Rotation is not a commutative operation, but we will assume that $\Delta\phi$, $\Delta\alpha$ and $\Delta\psi$ are so small that the order of the multiplication in Eqs. (5.142) and (5.143) is not important.

It should be recalled that the free stream is inertial so that it does not rotate with the wing and, hence, the angle of attack of the latter is calculated with respect to the direction of the former. Therefore, for a purely body-fixed axis system, rotation in yaw and roll will also result in a change in the angle of attack. Furthermore, if the wing is oscillating in yaw and roll, it will also oscillate in angle of attack. In order to avoid this phenomenon, we can set one of the axes to inertial and the others to body fixed. For example, the yaw axis can be inertial and always aligned with the z-axis, while the other two axes will rotate with the body. In this way, oscillations in angle of attack due to oscillations in roll and yaw will be minimized.

The relative flow velocity, $\mathbf{u}_m(t)$ induced by the free stream and the combined pitch, roll and yaw motion on any point $x(t), y(t), z(t)$ on the wing is given by

$$\mathbf{u}_m(t) = \mathbf{Q}_\infty - \left(\dot\phi\mathbf{n}_\phi(t)\times\mathbf{x}(t) + \dot\alpha\mathbf{n}_\alpha(t)\times\mathbf{x}(t) + \dot\psi\mathbf{n}_\psi(t)\times\mathbf{x}(t)\right)$$

assuming that the centre of rotation lies at the origin. After carrying out the vector products, we obtain

$$u_m(t) = U_\infty - (\dot{\phi} n_{\phi_2} + \dot{\alpha} n_{\alpha_2} + \dot{\psi} n_{\psi_2})z + (\dot{\phi} n_{\phi_3} + \dot{\alpha} n_{\alpha_3} + \dot{\psi} n_{\psi_3})y \tag{5.144}$$

$$v_m(t) = V_\infty + (\dot{\phi} n_{\phi_1} + \dot{\alpha} n_{\alpha_1} + \dot{\psi} n_{\psi_1})z - (\dot{\phi} n_{\phi_3} + \dot{\alpha} n_{\alpha_3} + \dot{\psi} n_{\psi_3})x \tag{5.145}$$

$$w_m(t) = W_\infty - (\dot{\phi} n_{\phi_1} + \dot{\alpha} n_{\alpha_1} + \dot{\psi} n_{\psi_1})y + (\dot{\phi} n_{\phi_2} + \dot{\alpha} n_{\alpha_2} + \dot{\psi} n_{\psi_2})x \tag{5.146}$$

where all the quantities on the right-hand sides of Eqs. (5.144) to (5.146) are functions of time. In the context of a VLM calculation, these equations can be used to calculate the velocities at the control points and at the mid-points of the spanwise and chordwise vortex segments, respectively, after substituting the appropriate coordinate values.

Example 5.9 *Determine if rolling or yawing oscillations can produce thrust*

We select a rectangular wing with $c_0 = 2$ m, $\mathcal{R} = 6$, immersed in an airflow with $U_\infty = 50$ m/s, $V_\infty = W_\infty = 0$. We set $m = 8$, $n = 20$ and calculate the coordinates of the geometric panel vertices, control points and wake vortex ring vertices in the unrotated position. The rotation centre is point $c_0/2, 0, 0$, so we move the wing forward by $c_0/2$ so that the rotation point lies at the origin. At the unrotated position, the body-fixed and inertial coordinate systems are aligned so that $\mathbf{n}_\phi = (1, 0, 0)$.

First, we will simulate a pure roll oscillation given by

$$\phi(t) = \phi_1 \cos(\omega t + \phi_\phi)$$

where $\phi_1 = 10°$ is the roll amplitude and $\phi_\phi = 90°$ is the roll phase; the pitch and yaw angles are zero at all times. At each time instance, we calculate $\Delta\phi = \phi(t_k) - \phi(t_{k-1})$ and then we use Eq. (5.142) three times to rotate the geometric panel vertices, control points and vortex rings vertices. As there is no motion in pitch or in yaw, we set $\Delta\alpha = 0$ and $\Delta\psi = 0$. At the first time instance, we set $\Delta\phi = \phi(t_1)$. Once all the coordinates have been rotated, we use Eq. (5.143) to rotate $\mathbf{n}_\phi$. Next, we calculate the coordinates of the midpoints of the spanwise and chordwise vortex segments, $\mathbf{x}_{ms}$ and $\mathbf{x}_{mc}$ and then we evaluate the relative flow velocities at these points, as well as the control points, using Eqs. (5.144)–(5.146). The rest of the simulation and load calculation procedure is identical to the one used in Example 5.7.

Figure 5.38 plots the results of the simulation for $k = 0.4$ after four oscillation cycles. The wake of Figure 5.38a is antisymmetric and increases slightly in width and drastically in height downstream of the wing. Figure 5.38b plots the drag coefficient during the last oscillation cycle; clearly, the wing is producing thrust for most of the cycle. The frequency of the drag signal is twice that of the motion, its mean is $C_D(0) = -0.043$ and its amplitude is $|C_D(2k)| = 0.045$. The lift is equal to zero at all times, since the motion is antisymmetric and positive lift on one half-wing is balanced by negative lift on the other at all times. The sideforce is also zero at all times due to the anti-symmetry.

The thrust production is due to leading edge suction; the unsteady and quasi-steady aerodynamic loads have very small components in the x direction, except for the quasi-steady forces acting on the midpoints of the leading edge spanwise midpoints, which are almost always directed forward. As the wing is rolling, one-half sees a positive angle of attack while the other half sees a negative angle of attack. As the aerodynamic force on the spanwise segments is the vector product between the local flow velocity and the vortex segments themselves, the force will

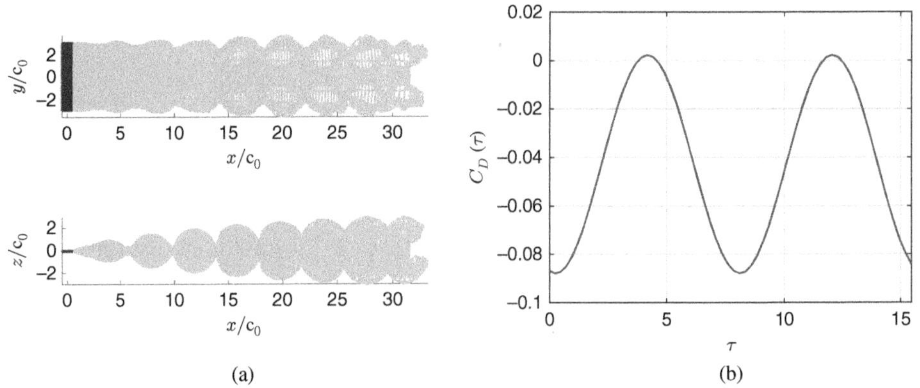

Figure 5.38 Simulation of rectangular wing with $\mathcal{R} = 6$ undergoing roll oscillations at $k = 0.4$. (a) Wake shape and (b) $C_D(\tau)$.

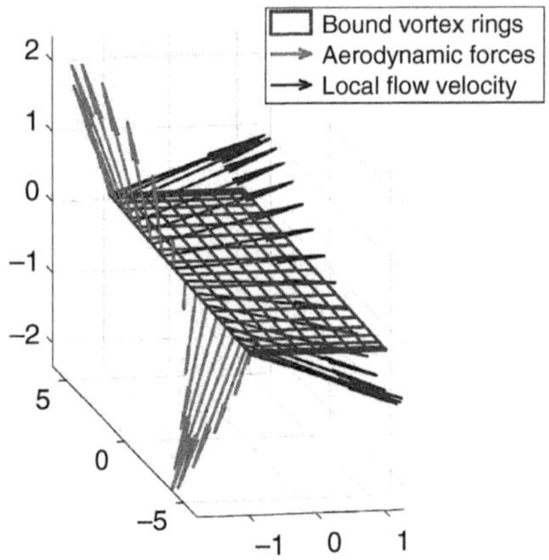

Figure 5.39 Aerodynamic loads acting on the leading edge spanwise segments of a rolling wing.

be leaning forward in both cases, as shown in Figure 5.39. Downstream of the leading edge, the local flow is aligned with the wing's surface because of the imposition of the impermeability condition so that the aerodynamic loads become nearly normal to the surface.[1] *This property of the VLM mirrors the physical phenomenon of leading edge suction force due to the acceleration of the flow around the leading edge.*

A pure yawing motion will not yield any aerodynamic loads at all if the pitch angle is zero. The yawing motion is parallel to the free stream so that impermeability is automatically satisfied and the strengths of the vortex rings are all equal to zero. This example is solved by Matlab code `lateral_VLM.m`.

1 The impermeability condition is applied at the control points, not at the midpoints of the vortex segments so that the local flow on the latter is never exactly parallel to the surface.

5.5.4 Aerodynamic Stability Derivatives

In flight dynamics, the unsteady aerodynamic loads are written as linear combinations of aerodynamic stability derivatives. The aircraft degrees of freedom are usually defined as perturbation quantities around an equilibrium flight condition with equilibrium velocity $\mathbf{Q}_e = (U_e, V_e, W_e)$. Using a body-fixed frame of reference, these degrees of freedom are three translations, surge u', sideslip v' and plunge w' and three rotations, roll ϕ, pitch θ and yaw ψ (see e.g.Example (Cook 1997)). If an air-trajectory frame of reference is used, the plunge is replaced by the perturbation angle of attack $\alpha \approx u'/W_e$ and the sideslip velocity by the perturbation sideslip angle $\beta \approx v'/Q_e$. All rotations are measured around a datum that is usually the centre of gravity of the aircraft. The discussion below will concern only the lateral degrees of freedom and loads, but similar analysis can be applied to longitudinal motion.

For lateral motion, the aerodynamic loads of interest are the sideforce Y, rolling moment l and yawing moment n, and they are also perturbations around their equilibrium values. The body-fixed degrees of freedom used to calculate these loads are the sideslip velocity v', the roll rate $p = \dot{\phi}$ and the yaw rate $r = \dot{\psi}$. The loads can be expanded as Taylor series around the equilibrium flight condition, such that

$$Y = Y(v', \dot{v}', p, \dot{p}, r, \dot{r}) = \frac{\partial Y}{\partial v'}v' + \frac{\partial Y}{\partial \dot{v}'}\dot{v}' + \frac{\partial Y}{\partial p}p + \frac{\partial Y}{\partial \dot{p}}\dot{p} + \frac{\partial Y}{\partial r}r + \frac{\partial Y}{\partial \dot{r}}\dot{r}$$

$$l = l(v', \dot{v}', p, \dot{p}, r, \dot{r}) = \frac{\partial l}{\partial v'}v' + \frac{\partial l}{\partial \dot{v}'}\dot{v}' + \frac{\partial l}{\partial p}p + \frac{\partial l}{\partial \dot{p}}\dot{p} + \frac{\partial l}{\partial r}r + \frac{\partial l}{\partial \dot{r}}\dot{r}$$

$$n = n(v', \dot{v}', p, \dot{p}, r, \dot{r}) = \frac{\partial n}{\partial v'}v' + \frac{\partial n}{\partial \dot{v}'}\dot{v}' + \frac{\partial n}{\partial p}p + \frac{\partial n}{\partial \dot{p}}\dot{p} + \frac{\partial n}{\partial r}r + \frac{\partial n}{\partial \dot{r}}\dot{r}$$

The derivatives in these equations are known as the aerodynamic stability derivatives. The basic assumption on which flight dynamic analysis is based is that these derivatives are constants in the frequency range of interest to aircraft motion. Their non-dimensional forms are defined as

$$C_{Y_{v'}} = Q_e \frac{\partial C_Y}{\partial v'}, C_{Y_{\dot{v}'}} = \frac{2Q_e^2}{b}\frac{\partial C_Y}{\partial \dot{v}'}, C_{Y_p} = \frac{2Q_e}{b}\frac{\partial C_Y}{\partial p}, C_{Y_{\dot{p}}} = \left(\frac{2Q_e}{b}\right)^2 \frac{\partial C_Y}{\partial \dot{p}}$$

and so on, for the l and n derivatives. The non-dimensional load coefficients are defined as

$$C_Y = \frac{Y}{\frac{1}{2}\rho Q_e^2 S}, \; C_l = \frac{l}{\frac{1}{2}\rho Q_e^2 Sb}, \; C_n = \frac{n}{\frac{1}{2}\rho Q_e^2 Sb} \tag{5.147}$$

If we non-dimensionalize the equation for Y, we obtain

$$C_Y(t) = \frac{1}{Q_e}C_{Y_{v'}}v'(t) + \frac{b}{2Q_e^2}C_{Y_{\dot{v}'}}\dot{v}'(t) + \frac{b}{2Q_e}C_{Y_p}p(t) + \left(\frac{b}{2Q_e}\right)^2 C_{Y_{\dot{p}}}\dot{p}(t)$$
$$+\frac{b}{2Q_e}C_{Y_r}r(t) + \left(\frac{b}{2Q_e}\right)^2 C_{Y_{\dot{r}}}\dot{r}(t)$$

Taking the Fourier Transform of this latest expression leads to

$$C_Y(\omega) = \frac{1}{Q_e} C_{Y_{v'}} v'(\omega) + \mathrm{i}\omega \frac{b}{2Q_e^2} C_{Y_{\dot{v}'}} v'(\omega) + \frac{b}{2Q_e} C_{Y_p} p(\omega) + \mathrm{i}\omega \left(\frac{b}{2Q_e}\right)^2 C_{Y_{\dot{p}}} p(\omega)$$
$$+ \frac{b}{2Q_e} C_{Y_r} r(\omega) + \mathrm{i}\omega \left(\frac{b}{2Q_e}\right)^2 C_{Y_{\dot{r}}} r(\omega) \tag{5.148}$$

Similarly, we can write frequency-domain equations for the rolling and yawing moment coefficients in terms of their respective aerodynamic stability derivatives such that

$$C_l(\omega) = \frac{1}{Q_e} C_{l_{v'}} v'(\omega) + \mathrm{i}\omega \frac{b}{2Q_e^2} C_{l_{\dot{v}'}} v'(\omega) + \frac{b}{2Q_e} C_{l_p} p(\omega) + \mathrm{i}\omega \left(\frac{b}{2Q_e}\right)^2 C_{l_{\dot{p}}} p(\omega)$$
$$+ \frac{b}{2Q_e} C_{l_r} r(\omega) + \mathrm{i}\omega \left(\frac{b}{2Q_e}\right)^2 C_{l_{\dot{r}}} r(\omega) \tag{5.149}$$

$$C_n(\omega) = \frac{1}{Q_e} C_{n_{v'}} v'(\omega) + \mathrm{i}\omega \frac{b}{2Q_e^2} C_{n_{\dot{v}'}} v'(\omega) + \frac{b}{2Q_e} C_{n_p} p(\omega) + \mathrm{i}\omega \left(\frac{b}{2Q_e}\right)^2 C_{n_{\dot{p}}} p(\omega)$$
$$+ \frac{b}{2Q_e} C_{n_r} r(\omega) + \mathrm{i}\omega \left(\frac{b}{2Q_e}\right)^2 C_{n_{\dot{r}}} r(\omega) \tag{5.150}$$

The aerodynamic stability derivatives can be evaluated from a frequency-domain VLM calculation that yields $C_Y(\omega)$, $C_l(\omega)$ and $C_n(\omega)$. Equations (5.148)–(5.150) are three equations with 18 unknown derivatives but they can be completed if the VLM calculation is carried out for at least six different nearby frequencies.

We have already written the VLM in the frequency domain for longitudinal motion; we can do the same for lateral motion. The wing is quasi-fixed, and the motion is modelled by rotating the free stream and imposing the relative flow velocities. The roll rotation vector is always aligned with the x axis, $\mathbf{n}_\phi = (1, 0, 0)$, and the yaw vector with the z axis, $\mathbf{n}_\psi = (0, 0, 1)$. Consequently, the relative flow velocities of Eqs. (5.144)–(5.146) become

$$\begin{aligned} u_m(t) &= U_e(t) + \dot{\psi}(t) y \\ v_m(t) &= V_e(t) + \dot{\phi}(t) z - \dot{\psi} x - v'(t) \\ w_m(t) &= W_e(t) - \dot{\phi}(t) y \end{aligned} \tag{5.151}$$

In the equilibrium flight condition, the wing lies at an equilibrium angle of attack α_e to the equilibrium free stream, whose velocity is $\mathbf{Q}_e = (U_e, 0, 0)$. As the wing is sideslipping, rolling and yawing, the free stream must be rotated with respect to the wing and in an opposite direction to the wing's motion. We can achieve this using the compound rotation of Eq. (5.142) such that

$$\mathbf{Q}_e(t) = \boldsymbol{R}(-\phi(t), \mathbf{n}_\phi) \boldsymbol{R}(-\alpha_e, \mathbf{n}_\alpha) \boldsymbol{R}(-\psi(t), \mathbf{n}_\psi) \mathbf{Q}_e$$

where the elements of the three rotation matrices are calculated from Eq. (5.138) for small angles. Then, the linearised compound rotation matrix becomes

$$\boldsymbol{R}(-\phi(t), -\alpha_e, -\psi(t)) = \begin{pmatrix} 1 & \psi(t) & -\alpha_e \\ -\psi(t) & 1 & \phi(t) \\ \alpha_e & -\phi(t) & 1 \end{pmatrix} \tag{5.152}$$

Multiplying this rotation matrix by the equilibrium free stream, we obtain the rotated free stream at each time instance

$$\mathbf{Q}_e(t) = \begin{pmatrix} 1 & \psi(t) & -\alpha_e \\ -\psi(t) & 1 & \phi(t) \\ \alpha_e & -\phi(t) & 1 \end{pmatrix} \begin{pmatrix} U_e \\ 0 \\ 0 \end{pmatrix} = U_e \begin{pmatrix} 1 \\ -\psi(t) \\ \alpha_e \end{pmatrix} \tag{5.153}$$

In the frequency domain, the lateral oscillations are described by

$$\begin{aligned} v'(\omega) &= \frac{v_1'}{2\mathrm{i}} \mathrm{e}^{\mathrm{i}\phi_{v'}} \delta(\omega - \omega_0) - \frac{v_1'}{2\mathrm{i}} \mathrm{e}^{-\mathrm{i}\phi_{v'}} \delta(\omega + \omega_0) \\ \phi(\omega) &= \frac{\phi_1}{2\mathrm{i}} \mathrm{e}^{\mathrm{i}\phi_\phi} \delta(\omega - \omega_0) - \frac{\phi_1}{2\mathrm{i}} \mathrm{e}^{-\mathrm{i}\phi_\phi} \delta(\omega + \omega_0) \\ \psi(\omega) &= \frac{\psi_1}{2\mathrm{i}} \mathrm{e}^{\mathrm{i}\phi_\psi} \delta(\omega - \omega_0) - \frac{\psi_1}{2\mathrm{i}} \mathrm{e}^{-\mathrm{i}\phi_\psi} \delta(\omega + \omega_0) \end{aligned} \tag{5.154}$$

and the zero frequency solution of Eq. (5.134) becomes

$$\left(\boldsymbol{A}_n + \boldsymbol{B}_n \boldsymbol{P}_e(0) \boldsymbol{P}_c\right) \boldsymbol{\Gamma}(0) = -U_e \boldsymbol{n}_x - U_e \alpha_e \boldsymbol{n}_z \tag{5.155}$$

Solving for $\boldsymbol{\Gamma}(0)$ leads to the steady bound vorticity on a wing lying at an angle of attack to a free stream. The impermeability boundary condition at $\omega = \omega_0$ becomes

$$\begin{aligned} \left(\boldsymbol{A}_n + \boldsymbol{B}_n \boldsymbol{P}_e(\omega_0) \boldsymbol{P}_c\right) \boldsymbol{\Gamma}(\omega_0) = &-\mathrm{i}\omega_0 \frac{\psi_1}{2\mathrm{i}} \mathrm{e}^{\mathrm{i}\phi_\psi} \boldsymbol{y}_c \circ \boldsymbol{n}_x \\ &- \left(-\frac{\psi_1}{2\mathrm{i}} \mathrm{e}^{\mathrm{i}\phi_\psi} \left(U_e + \mathrm{i}\omega_0 \boldsymbol{x}_c\right) + \mathrm{i}\omega_0 \frac{\phi_1}{2\mathrm{i}} \mathrm{e}^{\mathrm{i}\phi_\phi} \boldsymbol{z}_c - \frac{v_1'}{2\mathrm{i}} \mathrm{e}^{\mathrm{i}\phi_{v'}} \right) \circ \boldsymbol{n}_y \\ &- \left(-\mathrm{i}\omega_0 \frac{\phi_1}{2\mathrm{i}} \mathrm{e}^{\mathrm{i}\phi_\phi} \boldsymbol{y}_c \right) \circ \boldsymbol{n}_z \end{aligned} \tag{5.156}$$

again assuming that the rotation centre lies at the origin. Next, we calculate the relative flow velocities on the midpoints of the spanwise and chordwise vortex segments and follow the procedure of example 5.8 to calculate the total aerodynamic force $\mathbf{F}(\omega)$ in the body-fixed axis system in the form of Eq. (5.136). The lift, drag and sideforce coefficients are obtained from projecting $\mathbf{F}(\omega)$ onto the rotated free stream. Approximately unit vectors in the directions of the axes of the inertial system can be obtained from the columns of the rotation matrix of Eq. (5.152), so that the projection can be carried out by means of the convolution:

$$\begin{pmatrix} D(\omega) \\ Y(\omega) \\ L(\omega) \end{pmatrix} = \begin{pmatrix} 1 & \psi(\omega) & -\alpha(\omega) \\ -\psi(\omega) & 1 & \phi(\omega) \\ \alpha(\omega) & -\phi(\omega) & 1 \end{pmatrix}^T * \mathbf{F}(\omega) \tag{5.157}$$

where $\alpha(\omega) = \alpha_e \delta(\omega)$. The total rolling, pitching and yawing moments around the origin are calculated from the vector product of the force contributions in Eq. (5.117) by the appropriate vectorial distances,

$$\begin{aligned} \mathbf{M}(\omega) = &\sum_{i=1}^{m+1} \sum_{j=1}^{n} \mathbf{F}_{s_{i,j}}(\omega) \times \mathbf{x}_{ms_{i,j}} + \sum_{i=1}^{m} \sum_{j=1}^{n+1} \mathbf{F}_{c_{i,j}}(\omega) \times \mathbf{x}_{mc_{i,j}} \\ &+ \sum_{i=1}^{m} \sum_{j=1}^{n} \mathbf{F}_{i,j}^{\mathrm{unst}}(\omega) \times \mathbf{x}_{c_{i,j}} \end{aligned} \tag{5.158}$$

where $\mathbf{M}(\omega) = (l(\omega), m(\omega), n(\omega))$. The definition of the pitching moment coefficient in 3D is

$$C_m = \frac{m}{\frac{1}{2}\rho Q_e^2 S c_0}$$

Example 5.10 *Calculate the roll stability derivatives for a flat-plate wing with aspect ratio 6 and taper ratio 0.6 and compare to experimental measurements.*

Queijo et al. (1956) published results from wind tunnel tests on three wing geometries forced to oscillate in yaw. Here we will model the unswept wing, with $Æ = 6$, $c_0 = 0.29$ *m*, $b = 0.91$ *m and* $\lambda = 0.6$, *drawn in Figure 5.40a. The wing was forced to yaw sinusoidally around its midspan quarter chord at a reduced frequency based on the half-span*

$$k_b = \frac{\omega b}{2Q_e} = 0.23$$

for a range of mean angles of attack between $\alpha_e = 0°$ *and* $16°$. *The rolling and yawing moments were measured using strain gauges mounted on the wing's support strut. The actuation mechanism imposed yawing coupled with sideslip motion in the equal and opposite direction, as seen in Figure 5.40b. As the wing yaws to the right by* ψ, *it also translates to the left by a sideslip velocity* $v' = -Q_e \sin\psi$. *In this manner, pure yawing motion is obtained; we will discuss further this issue below.*

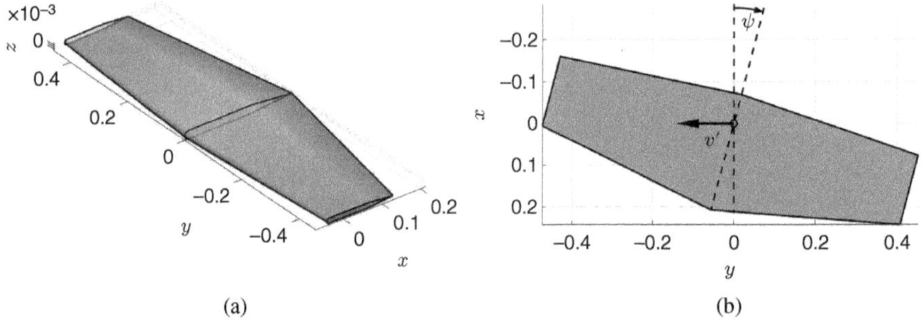

Figure 5.40 Wing model and kinematics tested by Queijo et al. (1956). (a) Wing model and (b) kinematics. Source: Adapted from Queijo et al. (1956).

We start by building the VLM model of the wing as in Example 5.6. The wing is tapered but has no sweep, twist, dihedral or camber. Once the panel vertices have been calculated, we rotate the wing by the mean angle of attack, α_e, *using Eq. (5.104). In general applications of the frequency-domain VLM approach, we rotate the free stream by the mean angle of attack instead. However, in these experiments, the x axis was aligned with the wind tunnel's free stream and the wing was pitched up by* α_e *with respect to it, so that yaw occurred around an axis perpendicular to the free stream and not to the wing. Consequently, we do not rotate the free stream by* α_e, *but we still rotate it by the mean and oscillatory yaw angles; Eq. (5.153) becomes*

$$\mathbf{Q}_e(0) = U_e \begin{pmatrix} 1 \\ -\psi_0 \\ 0 \end{pmatrix}, \quad \mathbf{Q}_e(\omega_0) = U_e \begin{pmatrix} 1 \\ -\frac{\psi_1}{2\mathrm{i}} \exp^{\mathrm{i}\phi_\psi} \\ 0 \end{pmatrix}$$

Transforming the relative veolocities of Eq. (5.151) to the frequency domain and applying sinusoidal motion in yaw and sideslip only leads to

$$u_m(0) = U_e,\ u_m(\omega_0) = \mathrm{i}\omega_0 U_e \frac{\psi_1}{2\mathrm{i}} \mathrm{e}^{\mathrm{i}\phi_\psi} y \tag{5.159}$$

$$v_m(0) = -U_e\psi_0 - v'(0),\ v_m(\omega_0) = -U_e \frac{\psi_1}{2\mathrm{i}} \mathrm{e}^{\mathrm{i}\phi_\psi} - v'(\omega_0) - \mathrm{i}\omega U_e \frac{\psi_1}{2\mathrm{i}} \mathrm{e}^{\mathrm{i}\phi_\psi} x$$

$$w_m(0) = 0,\ w_m(\omega_0) = 0 \tag{5.160}$$

We now recall that Queijo et al. (1956) imposed a sideslip velocity whose value can be linearised to $v'(t) \approx -U_e\psi(t)$, such that $v'(0) = -U_e\psi_0$ and $v'(\omega_0) = -U_e(\psi_1/2\mathrm{i})\mathrm{e}^{\mathrm{i}\phi_\psi}$. Substituting back into the equations for $v_m(0)$ and $v_m(\omega_0)$ above, we obtain

$$v_m(0) = 0,\ v_m(\omega_0) = -\mathrm{i}\omega U_e \frac{\psi_1}{2\mathrm{i}} \mathrm{e}^{\mathrm{i}\phi_\psi} x \tag{5.161}$$

so that there are no translational velocities and the relative motion between the wing and the flow is purely due to yaw rotation.

Equations (5.159)–(5.161) are used to calculate the relative flow velocities on the control points, spanwise vortex segment midpoints and chordwise vortex segment midpoints. The rest of the solution procedure is identical to that of Example 5.8, except that Eq. (5.157) must be used to calculate the drag, lift and sideforce. Finally, we evaluate the rolling, pitching and yawing moments from Eq. (5.158) and the respective moment coefficients from Eq. (5.147).

We now turn our attention to the sideforce coefficient $C_Y(\omega)$, which is a function of ψ only. Equation 5.148 becomes

$$C_Y(\omega) = \frac{b}{2Q_e} C_{Y_r} r(\omega) + \mathrm{i}\omega \left(\frac{b}{2Q_e}\right)^2 C_{Y_{\dot{r}}} r(\omega)$$

Substituting for $r(\omega) = \mathrm{i}\omega\psi(\omega)$ yields

$$C_Y(\omega_0) = \mathrm{i}k_b C_{Y_r} \psi(\omega) - k_b^2 C_{Y_{\dot{r}}} \psi(\omega)$$

Substituting for $\psi(\omega)$ from Eq. (5.154) yields

$$C_Y(\omega_0) = \mathrm{i}k_b C_{Y_r} \frac{\psi_1}{2\mathrm{i}} \mathrm{e}^{\mathrm{i}\phi_\psi} - k_b^2 C_{Y_{\dot{r}}} \frac{\psi_1}{2\mathrm{i}} \mathrm{e}^{\mathrm{i}\phi_\psi} \tag{5.162}$$

There are two unknowns to be evaluated, the aerodynamic stability derivatives C_{Y_r} and $C_{Y_{\dot{r}}}$, which are real. Equation (5.162) is complex, which means that both its real and imaginary parts must be satisfied. Consequently, we have two equations with two unknowns, to be solved for C_{Y_r} and $C_{Y_{\dot{r}}}$,

$$\Re(C_Y(\omega_0)) = k_b C_{Y_r} \frac{\psi_1}{2},\ \Im(C_l(\omega_0)) = k_b^2 C_{l_{\dot{r}}} \frac{\psi_1}{2}$$

since we can set $\phi_\psi = 0$ without loss of generality. Equations similar to (5.162) can be written for the rolling and yawing moment coefficients

$$C_l(\omega_0) = \mathrm{i}k_b C_{l_r} \frac{\psi_1}{2\mathrm{i}} \mathrm{e}^{\mathrm{i}\phi_\psi} - k_b^2 C_{l_{\dot{r}}} \frac{\psi_1}{2\mathrm{i}} \mathrm{e}^{\mathrm{i}\phi_\psi} \tag{5.163}$$

$$C_n(\omega_0) = \mathrm{i}k_b C_{n_r} \frac{\psi_1}{2\mathrm{i}} \mathrm{e}^{\mathrm{i}\phi_\psi} - k_b^2 C_{n_{\dot{r}}} \frac{\psi_1}{2\mathrm{i}} \mathrm{e}^{\mathrm{i}\phi_\psi} \tag{5.164}$$

We carry out VLM simulations with $m = 40$, $n = 40$, $psi_1 = 4°$, $\phi_\psi = 0°$, $k_b = 0.23$ and mean angles of attack between $-5°$ and $20°$. Figure 5.41 plots the variation of the sideforce, rolling

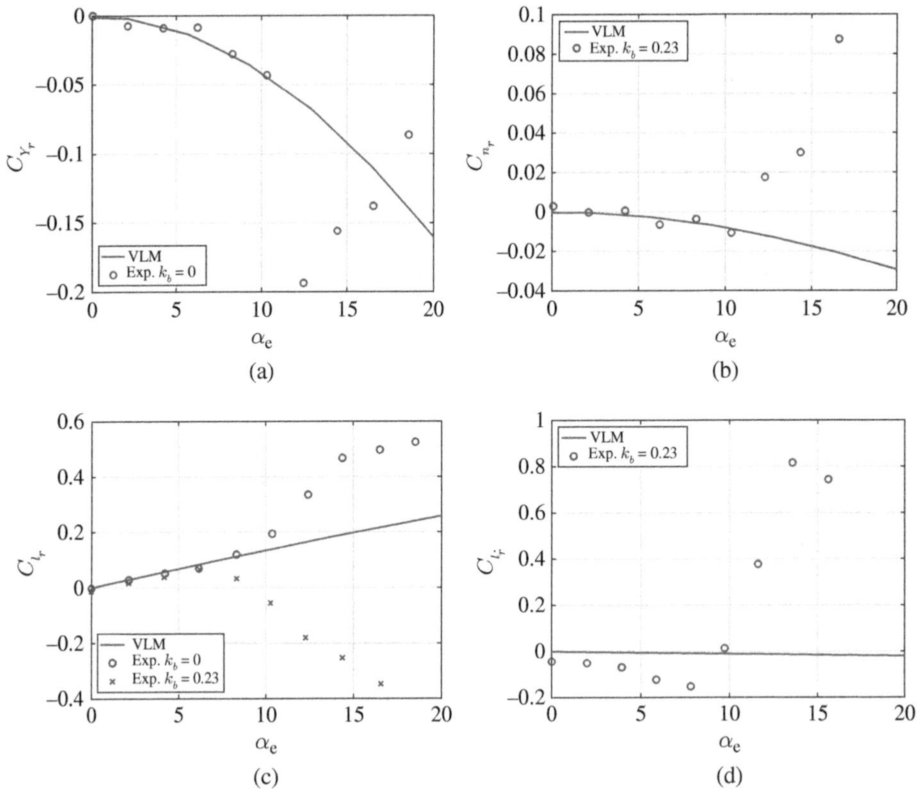

Figure 5.41 Sideforce and roll aerodynamic stability derivatives predicted by the VLM. Experimental data by Queijo et al. (1956). (a) C_{Y_r}, (b) C_{n_r}, (c) C_{l_r}, (d) $C_{l_{\dot{r}}}$. Source: Queijo et al. (1956), credit: NACA.

and yawing aerodynamic stability derivatives with α_e and compares them to the experimental measurements by Queijo et al. (1956). The experimental values for C_{Y_r} were obtained at steady conditions using a curved flow approach, and there are no measurements for $C_{Y_{\dot{r}}}$. Furthermore, the experimental definition of the positive direction for the rolling moment is opposite to the one used here so that the VLM results plotted in Figure 5.41 are $-C_{l_r}$ and $-C_{l_{\dot{r}}}$. The variation of C_{Y_r} with α_e is nonlinear and there is good agreement between the VLM predictions and the experimental data up to $\alpha_e = 10°$, as seen in Figure 5.41a. The variation of C_{l_r} with angle of attack is linear up to $\alpha_e = 6°$; beyond that value, the experimental data show that this derivative starts to vary non-linearly with α_e. Figures 5.41b and 5.41c show that the VLM predictions for C_{n_r} and C_{l_r} are in very good agreement with the experimental data inside the linear region. Theoretically, the value of $C_{l_{\dot{r}}}$ should be negligible for small angles of attack; the experimental measurements show that $C_{l_{\dot{r}}}$ is nearly constant for $\alpha_e < 5°$. At higher mean angles of attack, $C_{l_{\dot{r}}}$ starts to decrease and then increase due to the stalling characteristics of the wing. This example is solved by Matlab code `lateral_VLM_freq.m`.

The results of Figure 5.41 do not change significantly as the reduced frequency is increased. This is due to the linearity of the flow at low angles of attack, as well as the linearity of the frequency-domain VLM. Lichtenstein and Williams (1958) demonstrate experimentally that, for oscillations in sideslip, the frequency only has a significant effect at higher mean angles of attack, at which the flow starts to become non-linear. Consequently, the flight dynamic assumption that the aerodynamic stability derivatives are constant with frequency is valid for attached flow conditions and low frequencies.

5.6 The 3D Source and Doublet Panel Method

The 3D modelling methods presented up to this point treated the wing as either a lifting line or a lifting surface. In this section, we will extend the analysis to wings with thickness whose geometry is modelled in full. We will make use of Green's third identity introduced in Section 2.3.1; as we are interested in flow around wings, we must model both the wing and the wake so that Eq. (2.102) is extended to

$$\begin{aligned} E(\mathbf{x},t)\phi(\mathbf{x},t) = &-\frac{1}{4\pi}\int_{S(t)+S_w(t)} \mathbf{n}(\mathbf{x}_s,t)\cdot\nabla_s\phi(\mathbf{x}_s,t)\frac{1}{r(\mathbf{x},\mathbf{x}_s,t)}\mathrm{d}S \\ &+\frac{1}{4\pi}\int_{S(t)+S_w(t)} \phi(\mathbf{x}_s,t)\mathbf{n}(\mathbf{x}_s,t)\cdot\nabla_s\left(\frac{1}{r(\mathbf{x},\mathbf{x}_s,t)}\right)\mathrm{d}S \end{aligned} \quad (5.165)$$

where S is the total surface of the wing (that is, the total wetted area of both the upper and lower surfaces), S_w is the surface of the wake $\mathbf{x}$ is a point anywhere in the flow or on the surfaces, $\mathbf{x}_s$ are points on S or S_w and $E(\mathbf{x},t)$ is defined in Eq. (2.69). Recall that Green's third identity can be interpreted as the calculation of the potential induced at point $\mathbf{x}$ by a continuous surface distribution of points sources of strength $\mathbf{n}(\mathbf{x}_s,t)\cdot\nabla\phi(\mathbf{x}_s,t)$ and a continuous surface distribution of point doublets of strength $\phi(\mathbf{x}_s,t)$.

The wake contribution must be discussed further; Figure 5.42 plots the strict definition of the wing and wake surfaces, as seen already in Figure 2.29. The thickness of the wing's trailing edge has been exaggerated in order to show clearly the fact that the wake has an

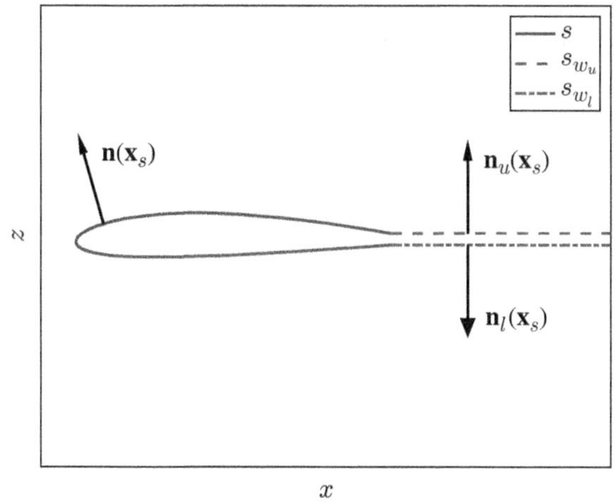

Figure 5.42 Wing and wake surfaces and unit normal vector definitions.

upper and a lower surface, S_{w_u} and S_{w_l}, respectively. Recall that the unit normal vector is always defined as pointing into the flow domain. This means that the normal vector on the upper wake surface, $\mathbf{n}_u$, points upwards while that on the lower surface, $\mathbf{n}_l$, points downwards. The wake's contribution to Green's third identity is then the sum of the contributions of the upper and lower surfaces,

$$-\frac{1}{4\pi}\int_{S_{w_u}(t)} \mathbf{n}_u(\mathbf{x}_s,t)\cdot\nabla_s\phi_u(\mathbf{x}_s,t)\frac{1}{r(\mathbf{x},\mathbf{x}_s,t)}\mathrm{d}S$$

$$+\frac{1}{4\pi}\int_{S_{w_u}(t)} \phi_u(\mathbf{x}_s,t)\mathbf{n}_u(\mathbf{x}_s,t)\cdot\nabla_s\left(\frac{1}{r(\mathbf{x},\mathbf{x}_s,t)}\right)\mathrm{d}S$$

$$-\frac{1}{4\pi}\int_{S_{w_l}(t)} \mathbf{n}_l(\mathbf{x}_s,t)\cdot\nabla_s\phi_l(\mathbf{x}_s,t)\frac{1}{r(\mathbf{x},\mathbf{x}_s,t)}\mathrm{d}S$$

$$+\frac{1}{4\pi}\int_{S_{w_l}(t)} \phi_l(\mathbf{x}_s,t)\mathbf{n}_l(\mathbf{x}_s,t)\cdot\nabla_s\left(\frac{1}{r(\mathbf{x},\mathbf{x}_s,t)}\right)\mathrm{d}S$$

where ϕ_u and ϕ_l are the values of the potential on the upper and lower surfaces of the wake, respectively. We can now assume that the two wake surfaces always lie infinitesimally close to each other so that they can be described by the single function $S_w(\mathbf{x}_s,t) = S_{w_u}(\mathbf{x}_s,t) = S_{w_l}(\mathbf{x}_s,t)$. The unit vector normal to the upper side of $S_{w_l}(t)$ is denoted by $\mathbf{n}(\mathbf{x}_s,t) = \mathbf{n}_u(\mathbf{x}_s,t)$ and is pointing upwards. The unit vector on the lower side of $S_{w_l}(t)$ must still point downwards so that

$$\mathbf{n}_l(\mathbf{x}_s,t) = -\mathbf{n}(\mathbf{x}_s,t)$$

Consequently, the wake's contribution simplifies to

$$-\frac{1}{4\pi}\int_{S_w(t)} \left(\mathbf{n}(\mathbf{x}_s,t)\cdot\nabla_s\phi_u(\mathbf{x}_s,t) - \mathbf{n}(\mathbf{x}_s,t)\cdot\nabla_s\phi_l(\mathbf{x}_s,t)\right)\frac{1}{r(\mathbf{x},\mathbf{x}_s,t)}\mathrm{d}S$$

$$\frac{1}{4\pi}\int_{S_w(t)} \left(\phi_u(\mathbf{x}_s,t) - \phi_l(\mathbf{x}_s,t)\right)\mathbf{n}_u(\mathbf{x}_s,t)\cdot\nabla_s\left(\frac{1}{r(\mathbf{x},\mathbf{x}_s,t)}\right)\mathrm{d}S$$

The usual approach to modelling the wake is to assume that it is an infinitely thin surface that causes a discontinuity in potential but no discontinuity in normal velocity. Consequently,

$$\mathbf{n}(\mathbf{x}_s,t)\cdot\nabla_s\phi_u(\mathbf{x}_s,t) - \mathbf{n}(\mathbf{x}_s,t)\cdot\nabla_s\phi_l(\mathbf{x}_s,t) = 0$$

so that the wake has no source contribution to Green's identity. It still has a doublet contribution, whose strength is denoted by

$$\mu_w(\mathbf{x}_s,t) = \phi_u(\mathbf{x}_s,t) - \phi_l(\mathbf{x}_s,t) \tag{5.166}$$

On the wing's surface, we will use the source and doublet strength definitions of Eqs. (2.98) and (2.99),

$$\sigma(\mathbf{x}_s,t) = \mathbf{n}(\mathbf{x}_s,t)\cdot\nabla_s\phi(\mathbf{x}_s,t) \tag{5.167}$$

$$\mu(\mathbf{x}_s,t) = \phi(\mathbf{x}_s,t) \tag{5.168}$$

so that Eq. (5.165) becomes

$$E(\mathbf{x},t)\phi(\mathbf{x},t) = -\frac{1}{4\pi}\int_{S(t)} \sigma(\mathbf{x}_s,t)\frac{1}{r(\mathbf{x},\mathbf{x}_s,t)}\mathrm{d}S + \frac{1}{4\pi}\int_{S(t)} \mu(\mathbf{x}_s,t)\mathbf{n}(\mathbf{x}_s,t)\cdot\nabla_s\left(\frac{1}{r(\mathbf{x},\mathbf{x}_s,t)}\right)\mathrm{d}S + \frac{1}{4\pi}\int_{S_w(t)} \mu_w(\mathbf{x}_s,t)\mathbf{n}(\mathbf{x}_s,t)\cdot\nabla_s\left(\frac{1}{r(\mathbf{x},\mathbf{x}_s,t)}\right)\mathrm{d}S \tag{5.169}$$

If the wing is static and immersed in a free stream of constant velocity $\mathbf{Q}_\infty = (U_\infty, V_\infty, W_\infty)$ the impermeability boundary condition of Eq. (2.43) becomes

$$\left(\nabla_s\phi(\mathbf{x}_s) + \mathbf{Q}_\infty\right)\cdot\mathbf{n}(\mathbf{x}_s) = 0, \text{ or } \mathbf{n}(\mathbf{x}_s)\cdot\nabla_s\phi(\mathbf{x}_s) = -\mathbf{Q}_\infty\cdot\mathbf{n}(\mathbf{x}_s) \tag{5.170}$$

Substituting into expression (5.167), we calculate the source strength as

$$\sigma(\mathbf{x}_s) = -\mathbf{Q}_\infty\cdot\mathbf{n}(\mathbf{x}_s) \tag{5.171}$$

For the moment we will assume that the wake propagates at the free stream airspeed, so that the surface S_w is prescribed and known. If we set $\mathbf{x} = \mathbf{x}_s$, then Eq. (5.169) becomes a single equation with two unknowns, $\mu(\mathbf{x}_s)$, $\mu_w(\mathbf{x}_s)$. The Kutta condition can be used to obtain a second relation between $\mu(\mathbf{x}_s)$ and $\mu_w(\mathbf{x}_s)$; the two equations can then be solved simultaneously over the entire wing surface S using the prescribed wake surface S_w and known free stream $\mathbf{Q}_\infty$. However, the integrals of expression (5.169) cannot be calculated analytically for every wing and wake geometry. The usual solution to this problem is to subdivide the surfaces S and S_w into surface panels on which the integrals can be calculated analytically, as shown in Eq. (2.94). Assuming that the surfaces of the wing and wake are discretized into N and N_w panels, respectively, see Figure 5.43, we can add the wake contribution to Eq. (2.94) to obtain

$$E(\mathbf{x},t)\phi(\mathbf{x},t) \approx -\frac{1}{4\pi}\sum_{J=1}^{N}\sigma(\mathbf{x}_{c_J},t)\int_{\Delta S_J}\frac{1}{r(\mathbf{x},\mathbf{x}_s,t)}\mathrm{d}S + \frac{1}{4\pi}\sum_{J=1}^{N}\mu(\mathbf{x}_{c_J},t)\int_{\Delta S_J}\mathbf{n}(\mathbf{x}_s,t)\cdot\nabla_s\left(\frac{1}{r(\mathbf{x},\mathbf{x}_s,t)}\right)\mathrm{d}S + \frac{1}{4\pi}\sum_{J=1}^{N_w}\mu_w(\mathbf{x}_{c_J},t)\int_{\Delta S_{w_J}}\mathbf{n}(\mathbf{x}_s,t)\cdot\nabla_s\left(\frac{1}{r(\mathbf{x},\mathbf{x}_s,t)}\right)\mathrm{d}S \tag{5.172}$$

where $\mathbf{x}_{c_J}$ is the control point of the Jth wing or wake panel, ΔS_J is the surface area of the Jth wing panel and ΔS_{w_J} that of the Jth wake panel. This aerodynamic modelling procedure is known as the source and doublet panel method (SDPM). The potential induced by quadrilateral source panels with constant strength is given in Appendix A.10 and that induced by doublet panels in Appendix A.11. Note that the panels considered in these appendices lie on the $z = 0$ plane; if a panel does not lie on this plane, its influence must be calculated after rotating the target point to the panel's local coordinates. Furthermore, the source strength is now calculated at the control points of the wing panels so that for steady flow, $\sigma(\mathbf{x}_{c_J}) = -\mathbf{Q}_\infty\cdot\mathbf{n}(\mathbf{x}_{c_J})$. Green's identity (5.169) is then used to evaluate the potential at the control points, $\phi(\mathbf{x}_{c_I}) = \mu(\mathbf{x}_{c_I})$, for $I = 1, \dots, N$, such that

$$\frac{1}{2}\mu(\mathbf{x}_{c_I}) = \sum_{J=1}^{N}A_{\phi_{I,J}}\sigma(\mathbf{x}_{c_J}) + \sum_{J=1}^{N}B_{\phi_{I,J}}\mu(\mathbf{x}_{c_J}) + \sum_{J=1}^{N_w}C_{\phi_{I,J}}\mu_w(\mathbf{x}_{c_J}) \tag{5.173}$$

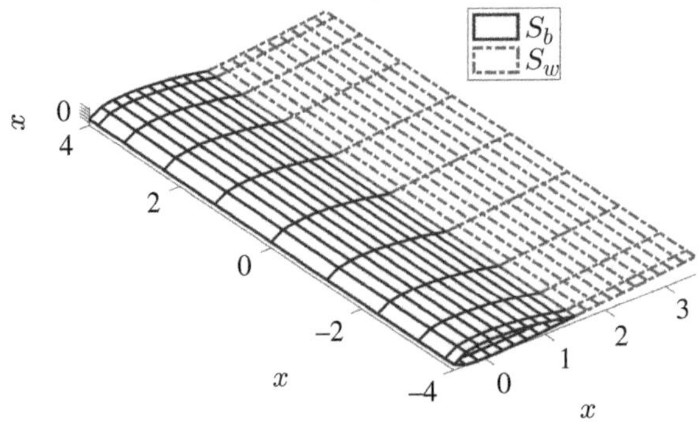

Figure 5.43 Wing and wake discretized into rectangular panels.

where

$$A_{\phi_{I,J}} = -\frac{1}{4\pi} \int_{\Delta S_J - \mathbf{x}_{c_I}} \frac{1}{r(\mathbf{x}_{c_I}, \mathbf{x}_s)} \mathrm{d}S$$

$$B_{\phi_{I,J}} = \frac{1}{4\pi} \int_{\Delta S_J - \mathbf{x}_{c_I}} \mathbf{n}(\mathbf{x}_{c_J}) \cdot \nabla_s \left(\frac{1}{r(\mathbf{x}_{c_I}, \mathbf{x}_s)} \right) \mathrm{d}S \tag{5.174}$$

$$C_{\phi_{I,J}} = \frac{1}{4\pi} \int_{\Delta S_{w_J}} \mathbf{n}(\mathbf{x}_{c_J}) \cdot \nabla_s \left(\frac{1}{r(\mathbf{x}_{c_I}, \mathbf{x}_s)} \right) \mathrm{d}S$$

are known as the influence coefficient matrices and the notation $\int_{\Delta S_J - \mathbf{x}_{c_I}}$ is used to state that if $I = J$, then the point $\mathbf{x}_{c_I}$ is excluded from the integral, as already discussed in the context of Eq. (2.66). There are N equations of the form of Eq. (5.173), which can be written in matrix form as

$$\boldsymbol{A}_\phi \boldsymbol{\sigma} + \left(\boldsymbol{B}_\phi - \frac{1}{2} \boldsymbol{I} \right) \boldsymbol{\mu} + \boldsymbol{C}_\phi \boldsymbol{\mu}_w = \mathbf{0} \tag{5.175}$$

where $\boldsymbol{A}_\phi$ is the $N \times N$ influence coefficient matrix of the wing source panels on the wing control points, $\boldsymbol{\sigma} = (\sigma(\mathbf{x}_{c_1}), \ldots, \sigma(\mathbf{x}_{c_N}))^T$ is the $N \times 1$ vector containing the source strengths of the wing panels, $\boldsymbol{B}_\phi$ is the $N \times N$ influence coefficient matrix of the wing doublet panels on the wing control points, $\boldsymbol{\mu} = (\mu(\mathbf{x})_{c_1}, \ldots, \mu(\mathbf{x}_{c_N}))^T$ is the $N \times 1$ vector containing the doublet strengths of the wing panels, $\boldsymbol{C}_\phi$ is the $N \times N_w$ influence coefficient matrix of the wake doublet panels on the wing control points and $\boldsymbol{\mu} = (\mu_w(\mathbf{x}_{c_1}), \ldots, \mu_w(\mathbf{x}_{c_{N_w}}))^T$ is the $N_w \times 1$ vector containing the doublet strengths of the wake panels. The source strengths are obtained from Eq. (5.171) as

$$\boldsymbol{\sigma} = -\left(U_\infty \boldsymbol{n}_x + V_\infty \boldsymbol{n}_y + W_\infty \boldsymbol{n}_z \right) \tag{5.176}$$

where $\boldsymbol{n}_x$, $\boldsymbol{n}_y$ and $\boldsymbol{n}_z$ are the $N \times 1$ vectors containing the x, y and z components, respectively, of the unit normal vectors of the control points of the wing panels, such that

$$\boldsymbol{n}_x = \left(n_x(\mathbf{x}_{c_1}), \ldots n_x(\mathbf{x}_{c_N}) \right)^T$$

$$\boldsymbol{n}_y = \left(n_y(\mathbf{x}_{c_1}), \ldots n_y(\mathbf{x}_{c_N}) \right)^T$$

$$\boldsymbol{n}_z = \left(n_z(\mathbf{x}_{c_1}), \ldots n_z(\mathbf{x}_{c_N}) \right)^T$$

for the wing normals; similar expressions are written for the wake normals. Equation (5.175) are a set of N equations with $N + N_w$ unknowns, $\boldsymbol{\mu}$ and $\boldsymbol{\mu}_w$. Closure is achieved using the Kutta condition and will be demonstrated in Example 5.11. The equations must be set up and solved at discrete time instances, unless the flow is steady, in which case none of the quantities in expressions (5.175) and (5.176) depends on time and only one solution is required.

Once we evaluate $\boldsymbol{\sigma}$, $\boldsymbol{\mu}$ and $\boldsymbol{\mu}_w$, we also know the potential on the surface, since from Eq. (5.168), $\phi(\mathbf{x}_s) = \mu(\mathbf{x}_s)$. The next step is to calculate the flow velocities on the surface of the wing. Off the surface, these velocities can be obtained from differentiating equations (5.172) with respect to x, y, z, or from Eq. (2.103) after discretisation into panels and addition of the wake panels. For example, the perturbation velocity in the x direction induced at point $\mathbf{x}$ by the panels on the wing and wake is given by

$$\begin{aligned}
u(\mathbf{x}) = &-\frac{1}{4\pi}\sum_{J=1}^{N}\sigma(\mathbf{x}_{c_J})\int_{\Delta S_J}\frac{\partial}{\partial x}\left(\frac{1}{r(\mathbf{x},\mathbf{x}_s)}\right)\mathrm{d}S \\
&+\frac{1}{4\pi}\sum_{J=1}^{N}\mu(\mathbf{x}_{c_J})\int_{\Delta S_J}\frac{\partial}{\partial x}\left(\mathbf{n}(\mathbf{x}_{c_J})\cdot\nabla_s\left(\frac{1}{r(\mathbf{x},\mathbf{x}_s)}\right)\right)\mathrm{d}S \\
&+\frac{1}{4\pi}\sum_{J=1}^{N_w}\mu_w(\mathbf{x}_{c_J})\int_{\Delta S_{w_J}}\frac{\partial}{\partial x}\left(\mathbf{n}(\mathbf{x}_{c_J})\cdot\nabla_s\left(\frac{1}{r(\mathbf{x},\mathbf{x}_s)}\right)\right)\mathrm{d}S
\end{aligned} \tag{5.177}$$

Similar expressions can be written for the perturbation velocities in the y and z directions. In Appendices A.10 and A.11, we show that it is possible to calculate analytically the velocities induced by a quadrilateral source or doublet panel on any point in the flow, but the resulting velocities lie in tangent and normal directions to the influencing panel. Therefore, we must first project them onto the x, y, z axes before substituting them into Eq. (5.177).

As mentioned at the end of Section 2.3.1, we cannot differentiate the left-hand side of Eq. (5.172) in directions x, y and z on the surface because $E(\mathbf{x}, t)$ is discontinuous there. Nevertheless, in directions parallel to the surface, E remains constant and equal to 1/2 and the differentiation can be carried out. The perturbation velocity on the surface along any tangential direction $\boldsymbol{\tau}(\mathbf{x}_s)$ is given by

$$u_\tau(\mathbf{x}_s) = \boldsymbol{\tau}(\mathbf{x}_s)\cdot\nabla\phi(\mathbf{x}_s) = \boldsymbol{\tau}(\mathbf{x}_s)\cdot\nabla\mu(\mathbf{x}_s)$$

where $u_\tau = \boldsymbol{\tau}\cdot(u, v, w)$ and u, v, w are the local surface velocities in the x, y and z directions, respectively. As the surface is a two-dimensional manifold in 3D space, it is sufficient to use two directions tangent to it, the chordwise and spanwise directions for instance. The corresponding unit vectors are $\boldsymbol{\tau}_m$ and $\boldsymbol{\tau}_n$ such that we obtain two equations for the three unknowns, u, v, w. The third equation comes from the boundary condition of expression (5.170). The three equations for the surface flow velocities become the system

$$\begin{aligned}
\boldsymbol{\tau}_m(\mathbf{x}_s)\cdot(u(\mathbf{x}_s), v(\mathbf{x}_s), w(\mathbf{x}_s)) &= \boldsymbol{\tau}_m(\mathbf{x}_s)\cdot\nabla\mu(\mathbf{x}_s) \\
\boldsymbol{\tau}_n(\mathbf{x}_s)\cdot(u(\mathbf{x}_s), v(\mathbf{x}_s), w(\mathbf{x}_s)) &= \boldsymbol{\tau}_n(\mathbf{x}_s)\cdot\nabla\mu(\mathbf{x}_s) \\
\mathbf{n}(\mathbf{x}_s)\cdot(u(\mathbf{x}_s), v(\mathbf{x}_s), w(\mathbf{x}_s)) &= -\mathbf{n}(\mathbf{x}_s)\cdot\mathbf{Q}_\infty = \sigma(\mathbf{x}_s)
\end{aligned} \tag{5.178}$$

where we have substituted for the source strength from expression (5.171). The tangential perturbation velocities $\boldsymbol{\tau}_m(\mathbf{x}_s)\cdot\nabla\mu(\mathbf{x}_s)$ and $\boldsymbol{\tau}_n(\mathbf{x}_s)\cdot\nabla\mu(\mathbf{x}_s)$ are usually calculated by differentiating numerically $\mu(\mathbf{x}_s)$ on the surface. This surface differentiation approach is the standard procedure for evaluating the SDPM surface velocities and is implemented in widely used software, such as PAN AIR (Epton and Magnus 1990)

Finally, once the perturbation flow velocities have been calculated, we can obtain the pressure coefficient from Eq. (2.31). For example, for steady flow,

$$c_p(\mathbf{x}_{c_I}) = 1 - \frac{\left(U_\infty + u(\mathbf{x}_{c_I})\right)^2 + \left(V_\infty + v(\mathbf{x}_{c_I})\right)^2 + \left(W_\infty + w(\mathbf{x}_{c_I})\right)^2}{Q_\infty^2} \tag{5.179}$$

where $Q_\infty = ||\mathbf{Q}_\infty||$. We will demonstrate the implementation of the SDPM approach in the following steady example.

Example 5.11 *Use the SDPM to compute the steady pressure distribution and aerodynamic loads acting on a rectangular wing with a NASA Advanced Laminar Flow Control airfoil section.*

In this example, we apply the SDPM to the experimental test case given in Applin and Gentry Jr. (1990). It concerns a rectangular untapered finite wing with an aspect ratio of 5.89 tested in a low-speed wind tunnel at several angles of attack (the maximum Mach number was 0.14; hence, the flow can be approximated as incompressible). The model was a semi-span wing, its full span was $b = 5.89$ m and its chord was $c_0 = 1$ m. There was a row of 55 pressure taps around the surface at a distance of 1.31 m from the centreline (that is 44% of the half-span). Pressure transducers were used to measure the pressure distribution at the tapped section and an aerodynamic balance to measure the total aerodynamic loads. The airfoil section was a modified version of the NASA Advanced Laminar Flow Control section. The exact shape of the airfoil is not given in the report but the coordinates of the pressure tapings are given so the shape can be reconstructed, albeit with a certain amount of uncertainty. Figure 5.44a plots the complete geometry of the half-wing wind tunnel model, including the positions of the pressure taps. Figure 5.44b draws the airfoil section, demonstrating that its shape is very unusual around the leading edge.

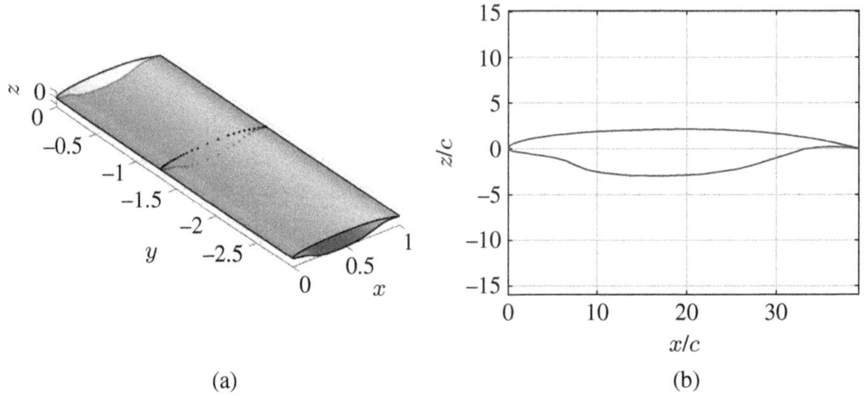

Figure 5.44 Wing tested by Applin and Gentry Jr. (1990) Source: Applin and Gentry Jr. (1990)/National Aeronautics and Space Administration/Public Domain. (a) Half-wing with pressure taps and (b) Airfoil section.

The first step is to create the panel grid, which must represent both the lower and the upper surfaces of the wing. The VLM panel numbering systems of Figure 5.16 are no longer adequate since the VLM only models the mean surface of the wing. For the SDPM, we will combine the VLM scheme with the 2D source and vortex panel method's numbering scheme of Figure 4.25, resulting in the schemes of Figure 5.45. The wing is split into 2m chordwise panels, m on the lower and m on the upper surface, and n spanwise panels. The leftmost lower trailing edge panel is the first and the chordwise index, $i = 1, \dots, 2m$, increases upstream on the lower surface and downstream on the upper surface. The spanwise index, $j = 1, \dots, n$, increases from left to right. The single indices $I, J = (i-1)n + j$ increase from left to right for each chordwise position and take values between 1 and $N = 2mn$.

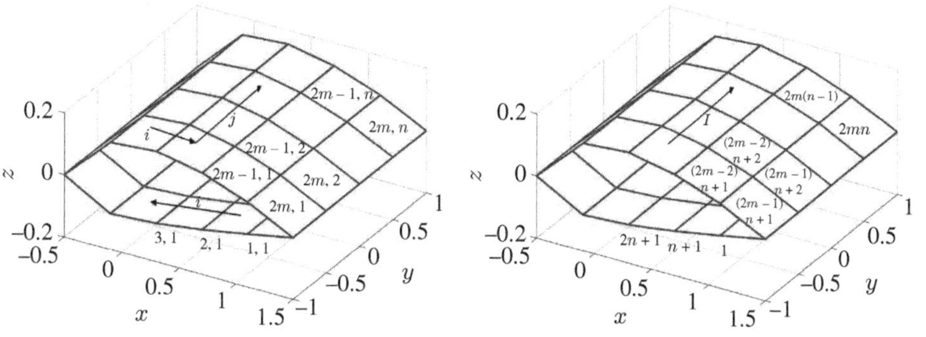

Figure 5.45 Two different panel numbering schemes for the SDPM. (a) Double index numbering and (b) single index numbering.

The panel discretisation used in the VLM was non-linear in the spanwise direction and linear in the chordwise direction. As the SDPM models the thickness of the wing, the leading edge must be reproduced as accurately as possible; hence, it is preferable to choose a non-linear chordwise panel spacing. We select the non-dimensional x coordinates of the panel vertices from

$$\bar{x}_{p_i} = 1 - \sin\theta_i \tag{5.180}$$

where $\theta_i = (i-1)\pi/(2m)$, for $i = 1, \dots, 2m+1$. The non-dimensional y coordinates of the panel vertices are identical to the VLM choice,

$$\bar{y}_{p_j} = \cos\theta_j \tag{5.181}$$

where $\theta_j = \pi - (j-1)\pi/n$ for $j = 1, \dots, n+1$. Then, the non-dimensional z coordinates of the panel vertices are obtained from the local airfoil geometry. For general wing geometries, the spanwise chord variation is given by Eq. (5.85), the dimensional coordinates of the geometric panel vertices by Eq. (5.86), and the twist is applied using Eq. (5.107). In the present case, there is no taper, twist, sweep or dihedral. The free stream velocities are then given by $U_\infty = Q_\infty \cos\alpha$, $V_\infty = 0$, $W_\infty = Q_\infty \sin\alpha$, where $Q_\infty = 50$ m/s is the magnitude of the free stream airspeed and $\alpha = 5°$ is the angle of attack.

The control points of the SDPM panels are the panel midpoints such that their coordinates are given by

$$x_{c_{i,j}} = \frac{1}{4}\left(x_{p_{i,j}} + x_{p_{i+1,j}} + x_{p_{i,j+1}} + x_{p_{i+1,j+1}}\right)$$
$$y_{c_{i,j}} = \frac{1}{4}\left(y_{p_{i,j}} + y_{p_{i+1,j}} + y_{p_{i,j+1}} + y_{p_{i+1,j+1}}\right)$$
$$z_{c_{i,j}} = \frac{1}{4}\left(z_{p_{i,j}} + z_{p_{i+1,j}} + z_{p_{i,j+1}} + z_{p_{i+1,j+1}}\right)$$

The unit vectors normal to the panels are obtained from Eq. (5.90) and the areas of the panels from Eq. (5.91). For the SDPM, we also need unit vectors tangent to the surface. There is an infinite number of such vectors, but we can select one that lies in the $x - z$ plane and a second that is perpendicular to both the normal and the first tangential vector. Tangent vectors are all perpendicular to the normal vector so they are defined by $(\tau_x, \tau_y, \tau_z) \cdot (n_x, n_y, n_z) = 0$. As we want our first tangent vector to lie in the $x - z$ plane, we must set its y component to zero, such that

$$[\tau_{xx}, 0, \tau_{xz}] \cdot [n_x, n_y, n_z] = 0$$

The tangential vector must also have unit length, that is $\tau_{xx}^2 + \tau_{xz}^2 = 1$. Solving the two equations together we have

$$\begin{aligned} \tau_{xx_{i,j}} &= \frac{1}{\sqrt{1 + (n_{x_{i,j}}/n_{z_{i,j}})^2}} \\ \tau_{xy_{i,j}} &= 0 \\ \tau_{xx_{i,j}} &= -\frac{n_{x_{i,j}}}{n_{z_{i,j}}}\tau_{xx_{i,j}} \end{aligned} \tag{5.182}$$

for $i = 1, \dots, 2m$, $j = 1, \dots, n$, so that our first tangential vector is $\boldsymbol{\tau}_{x_{i,j}} = (\tau_{xx_{i,j}}, \tau_{xy_{i,j}}, \tau_{xz_{i,j}})$. The second tangential vector is obtained simply from the vector product of the first and the normal vector

$$\boldsymbol{\tau}_{y_{i,j}} = \boldsymbol{\tau}_{x_{i,j}} \times \mathbf{n}_{i,j} \tag{5.183}$$

Finally, there is a single row of n wake panels whose leading edges lie on the trailing edges of the respective wing panels while their trailing edges lie far away, as shown in Figure 5.46. The coordinates of the wake panel vertices can then be set to

$$x_{w_{1,j}} = x_{p_{2m+1,j}},\ y_{w_{1,j}} = y_{p_{2m+1,j}},\ z_{w_{1,j}} = z_{p_{2m+1,j}},$$
$$x_{w_{2,j}} = x_{p_{2m+1,j}} + 100c_0,\ y_{w_{2,j}} = y_{p_{2m+1,j}},\ z_{w_{2,j}} = z_{p_{2m+1,j}}$$

for $j = 1, \dots, n+1$. The control points $x_{cw_{i,j}}, y_{cw_{i,j}}, z_{cw_{i,j}}$, normal unit vectors, $\mathbf{n}_{w_{i,j}}$ and tangential unit vectors, $\boldsymbol{\tau}_{xw_{i,j}}, \boldsymbol{\tau}_{yw_{i,j}}$ of the wake panels must also be calculated, for $i = 1, j = 1, \dots, n$.

The next step is to evaluate the source strength on each panel from Eq. (5.176), which can be rewritten as

$$\sigma_J = -\left(U_\infty n_{x_J} + V_\infty n_{y_J} + W_\infty n_{z_J}\right)$$

for $J = 1, \dots, 2mn$. Then, we calculate the source and doublet potential influence coefficient matrices appearing in Eq. (5.175) using Eqs. (A.74) and (A.84). We recall that these equations

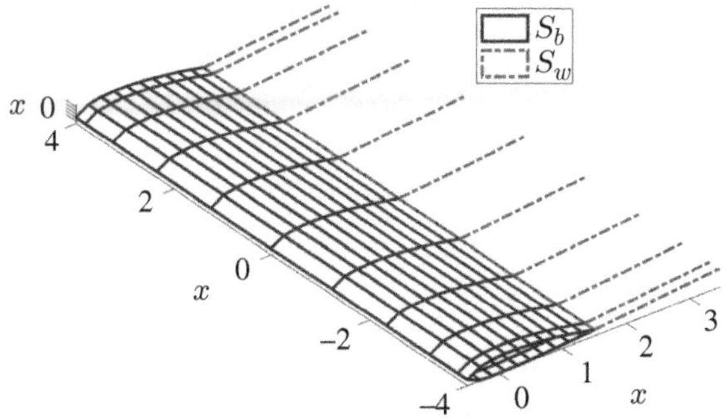

Figure 5.46 Wing and wake discretized into quadrilateral panels for steady flow.

are written for a panel that lies on the $z = 0$ plane, while the panels of the wing do not. Therefore, in order to use them, we need to express the point on which we seek to calculate the potential in the panel's local frame of reference. Suppose we need to calculate the potential induced by panel J on the control point of panel I. The Jth panel's local frame of reference is defined by the unit vectors $\boldsymbol{\tau}_{x_J}$, $\boldsymbol{\tau}_{y_J}$ and $\mathbf{n}_J$. We assemble this panel's vertices into vectors

$$\begin{aligned}
\mathbf{x} &= (x_{p_{i,j}}, x_{p_{i,j+1}}, x_{p_{i+1,j+1}}, x_{p_{i+1,j}}, x_{p_{i,j}}) \\
\mathbf{y} &= (y_{p_{i,j}}, y_{p_{i,j+1}}, y_{p_{i+1,j+1}}, y_{p_{i+1,j}}, y_{p_{i,j}}) \\
\mathbf{z} &= (z_{p_{i,j}}, z_{p_{i,j+1}}, z_{p_{i+1,j+1}}, z_{p_{i+1,j}}, z_{p_{i,j}})
\end{aligned} \tag{5.184}$$

where $J = (i-1)n + j$. We set up the rotation matrix

$$\mathbf{R} = \begin{pmatrix} \tau_{xx_J} & \tau_{xy_J} & \tau_{xz_J} \\ \tau_{yx_J} & \tau_{yy_J} & \tau_{yz_J} \\ n_{x_J} & n_{y_J} & n_{z_J} \end{pmatrix}$$

and then we calculate new coordinates for the panel vertices, rotated to the panel's local frame of reference from

$$\begin{pmatrix} x_{l_k} \\ y_{l_k} \\ z_{l_k} \end{pmatrix} = \mathbf{R} \begin{pmatrix} x_k - x_{c_J} \\ y_k - y_{c_J} \\ z_k - z_{c_J} \end{pmatrix}$$

where x_k, y_k, z_k are the kth elements of vectors $\mathbf{x}$, $\mathbf{y}$ and $\mathbf{z}$ for $k = 1, \dots, 4$. We also rotate the target control point (with index I) to the Jth panel's local frame of reference, such that

$$\begin{pmatrix} x_{l_c} \\ y_{l_c} \\ z_{l_c} \end{pmatrix} = \mathbf{R} \begin{pmatrix} x_{c_I} - x_{c_J} \\ y_{c_I} - y_{c_J} \\ z_{c_I} - z_{c_J} \end{pmatrix} \tag{5.185}$$

Finally, we compute the potential induced by the panel from Eqs. (A.74) and (A.84) for $\sigma = 1$ and $\mu = 1$, after substituting x_{l_k}, y_{l_k}, z_{l_k} for x_k, y_k, z_k and x_{l_c}, y_{l_c}, z_{l_c} for x, y, z and k for i in all relevant expressions. As the potential is a scalar quantity, it takes identical values in the local and global frames of reference so that we can finally assign

$$A_{\phi_{I,J}} = -\frac{1}{4\pi} \sum_{k=1}^{4} \left(f_k l_k - z_{l_c} t_k \right), B_{\phi_{I,J}} = \frac{1}{4\pi} \sum_{k=1}^{4} t_k \tag{5.186}$$

where f_k, l_k, t_l are obtained from Eqs. (A.72) and/or (A.73). However, we must recall that the integrals in Eq. (5.174) exclude the point $\mathbf{x}_{c_I}$ when $I = J$. The mathematical analysis in Appendices A.10 and A.11 shows that the self-influence of a doublet panel is equal to $-1/2$ if the control point is included in the integral (see Eq. (A.79)). However, if the control point is excluded, then the self-influence is equal to zero. Therefore, as we are excluding the control point here, we must set

$$B_{\phi_{I,I}} = 0$$

For the source panels, excluding the control point does not make any difference, since their self-influence is not zero away from that point. Hence,

$$A_{\phi_{I,I}} = -\frac{1}{4\pi}\sum_{k=1}^{4}\left(f_k l_k - z_{I_c} t_k\right)$$

The influence coefficient matrix of the wake $\mathbf{C}_\phi$ is obtained in the same way as that of the wing doublet panels, except that we do not need to exclude the control point since we do not need to calculate the influence of the wake panels on themselves. Matrices $\mathbf{A}_\phi$ and $\mathbf{B}_\phi$ are $2mn \times 2mn$, while matrix $\mathbf{C}_\phi$ is $2mn \times n$.

We are not yet in a position to solve Eq. (5.175) because there are $2mn$ equations for $(2m+1)n$ unknowns, the doublet strengths on the $2mn$ wing panels and the doublet strengths on the n wake panels. We make use of the Kutta condition in order to close the problem, remembering the vortex ring and doublet panel equivalence mentioned in Section A.11. This equivalence means that we can treat the edges of a doublet panels as vortex segments and therefore we can implement the Kutta condition by requiring that the total vortex strength on the trailing edge panels must be equal to zero. At each spanwise position, three panels have one edge lying on the trailing edge: wing panels $(1,j)$, $(2m,j)$ and wake panel $(1,j)$. The Kutta condition can therefore be written as

$$\mu_{1,j} - \mu_{2m,j} + \mu_{w_{1,j}} = 0 \tag{5.187}$$

for $j = 1, \ldots, n$, as seen in Figure 5.47. Recall that the vertices of the panels are arranged using the scheme of Eq. (5.184). If the panels are seen as equivalent vortex rings, then the trailing edge vortex rings are defined in the same direction for wing panel $1, j$ and wake panel $1, j$ and in the opposite direction for wing panel $2m, j$, as shown by the arrows in the figure. This is why $\mu_{2m,j}$ has a negative sign in Eq. (5.187).

The Kutta condition adds a further n equations so that we can now solve Eq. (5.175) in the form

$$\begin{pmatrix} \left(\mathbf{B}_\phi - \frac{1}{2}\mathbf{I}\right) & \mathbf{C}_\phi \\ \mathbf{K} & \mathbf{I}_n \end{pmatrix}\begin{pmatrix} \boldsymbol{\mu} \\ \boldsymbol{\mu}_w \end{pmatrix} = \begin{pmatrix} -\mathbf{A}_\phi\boldsymbol{\sigma} \\ \mathbf{0}_{n,1} \end{pmatrix} \tag{5.188}$$

where $\mathbf{K} = (\mathbf{I}_n, \mathbf{0}_{(2m-1)n,n}, -\mathbf{I}_n)$, $\mathbf{I}_n$ is the $n \times n$ unit matrix and $\mathbf{0}_{l,k}$ is a $l \times k$ matrix full of zeros.

The final step in the procedure is to calculate the velocities induced by the singularities on the control points of the panels by setting up Eq. (5.178). We will do this by means of numerical differentiation of the doublet strength on the surface. There are two possible directions in which we can calculate these derivatives, the chordwise direction i and the spanwise direction j.

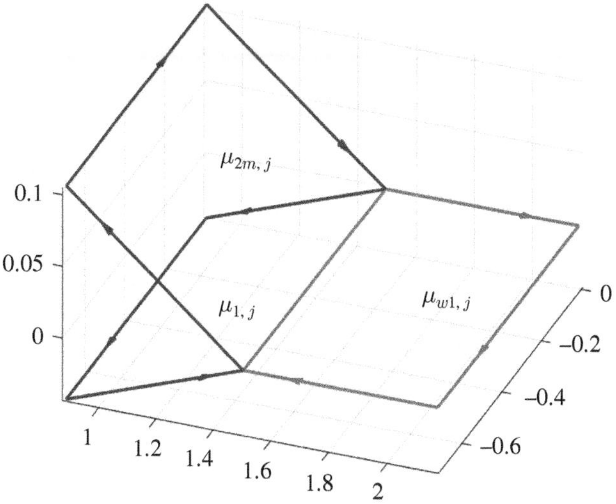

Figure 5.47 Kutta condition for the SDPM.

Any number of numerical differentiation schemes can be used; here we select a simple central difference approach. For the chordwise direction, we define

$$\Delta x_{i,j} = x_{c_{i+1,j}} - x_{c_{i-1,j}},\ \Delta y_{i,j} = y_{c_{i+1,j}} - y_{c_{i-1,j}},\ \Delta z_{i,j} = z_{c_{i+1,j}} - z_{c_{i-1,j}}$$
$$\Delta \mu_{i,j} = \mu_{i+1,j} - \mu_{i-1,j},\ \Delta s_{i,j} = \sqrt{\Delta x_{i,j}^2 + \Delta y_{i,j}^2 + \Delta z_{i,j}^2} \tag{5.189}$$
$$\tau_{mx_{i,j}} = \frac{\Delta x_{i,j}}{\Delta s_{i,j}},\ \tau_{my_{i,j}} = \frac{\Delta y_{i,j}}{\Delta s_{i,j}},\ \tau_{mz_{i,j}} = \frac{\Delta z_{i,j}}{\Delta s_{i,j}}$$

where $\tau_{mx_{i,j}}$, $\tau_{my_{i,j}}$ and $\tau_{mz_{i,j}}$ are the components of vector $\boldsymbol{\tau}_{m_{i,j}}$ in Eq. (5.178). Then the derivative of the doublet strength in the $\boldsymbol{\tau}_{m_{i,j}}$ direction is simply

$$\left.\frac{\partial \mu}{\partial \tau_m}\right|_{i,j} = \frac{\Delta \mu_{i,j}}{\Delta s_{i,j}} \tag{5.190}$$

For $i = 1$ and $i = 2m$, we use a first-order difference scheme. Similarly, for the spanwise direction, we define

$$\Delta x_{i,j} = x_{c_{i,j+1}} - x_{c_{i,j-1}},\ \Delta y_{i,j} = y_{c_{i,j+1}} - y_{c_{i,j-1}},\ \Delta z_{i,j} = z_{c_{i,j+1}} - z_{c_{i,j-1}}$$
$$\Delta \mu_{i,j} = \mu_{i,j+1} - \mu_{i,j-1},\ \Delta s_{i,j} = \sqrt{\Delta x_{i,j}^2 + \Delta y_{i,j}^2 + \Delta z_{i,j}^2} \tag{5.191}$$
$$\tau_{nx_{i,j}} = \frac{\Delta x_{i,j}}{\Delta s_{i,j}},\ \tau_{ny_{i,j}} = \frac{\Delta y_{i,j}}{\Delta s_{i,j}},\ \tau_{nz_{i,j}} = \frac{\Delta z_{i,j}}{\Delta s_{i,j}}$$

and the derivative of the doublet strength in the $\boldsymbol{\tau}_{n_{i,j}}$ direction is given by

$$\left.\frac{\partial \mu}{\partial \tau_n}\right|_{i,j} = \frac{\Delta \mu_{i,j}}{\Delta s_{i,j}} \tag{5.192}$$

Again, we use a first-order difference scheme for j = 1 and = n. Equations (5.178) on the i, jth panel become

$$\begin{aligned}\tau_{mx_{i,j}} u_{i,j} + \tau_{my_{i,j}} v_{i,j} + \tau_{mz_{i,j}} w_{i,j} &= \left.\frac{\partial \mu}{\partial \boldsymbol{\tau}_m}\right|_{i,j} \\ \tau_{nx_{i,j}} u_{i,j} + \tau_{ny_{i,j}} v_{i,j} + \tau_{nz_{i,j}} w_{i,j} &= \left.\frac{\partial \mu}{\partial \boldsymbol{\tau}_n}\right|_{i,j} \\ n_{x_{i,j}} u_{i,j} + n_{y_{i,j}} v_{i,j} + n_{z_{i,j}} w_{i,j} &= \sigma_{i,j}\end{aligned} \quad (5.193)$$

and can be solved for $u_{i,j}$, $v_{i,j}$ and $w_{i,j}$. Note that $\boldsymbol{\tau}_{m_{i,j}}$ and $\boldsymbol{\tau}_{n_{i,j}}$ are not necessarily orthogonal, and they are not identical to the tangent unit vectors $\boldsymbol{\tau}_{x_{i,j}}$ and $\boldsymbol{\tau}_{y_{i,j}}$ calculated from Eqs. (5.182) and (5.183). For example, $\boldsymbol{\tau}_{m_{i,j}}$ points upstream on the lower surface and downstream on the upper surface, while $\boldsymbol{\tau}_{x_{i,j}}$ points downstream on both surfaces. Nevertheless, $n_{x_{i,j}}$, $n_{y_{i,j}}$ and $n_{z_{i,j}}$ in Eq. (5.193) are the unit normal vector components already calculated from Eq. (5.90).

Once the flow velocities on the surface have been evaluated, we can calculate the pressure coefficient acting on each panel from Eq. (5.179). Then, the total aerodynamic force acting on each panel is obtained from

$$\mathbf{F}_{i,j} = -\frac{1}{2}\rho Q_\infty^2 c_{p_{i,j}} s_{i,j} \mathbf{n}_{i,j} \quad (5.194)$$

As we have rotated the free stream by α, the lift and drag are obtained from

$$L = -\sin\alpha \sum_{i=1}^{2m}\sum_{j=1}^{n} Fx_{i,j} + \cos\alpha \sum_{i=1}^{2m}\sum_{j=1}^{n} Fz_{i,j} \quad (5.195)$$

$$D = \cos\alpha \sum_{i=1}^{2m}\sum_{j=1}^{n} Fx_{i,j} + \sin\alpha \sum_{i=1}^{2m}\sum_{j=1}^{n} Fz_{i,j} \quad (5.196)$$

where $\mathbf{F}_{i,j} = (Fx_{i,j}, Fy_{i,j}, Fz_{i,j})$. As the wing is symmetric and the yaw angle is zero the total sideforce should be equal to zero.

We apply the SDPM simulation with $m = 50$, $n = 20$ and for angles of attack ranging from $-8.2°$ to $10°$, including the angles of attack for which experimental pressure distributions are presented in Applin and Gentry Jr. (1990). Figure 5.48 plots the pressure distribution around the tapped section for four values of the angle of attack, as calculated by the SDPM and measured experimentally. The spanwise position of the tapped section does not coincide with the spanwise position of any of the control points so the SDPM data plotted in Figure 5.48 are interpolated linearly between the neighbouring c_p values. The agreement between the numerical and experimental data is very good for $\alpha > -4.2°$. At this value of the angle of attack, the SDPM pressure distribution on the lower surface is inaccurate for $x/c < 0.1$. Normally, we would not expect viscous effects to be important at $\alpha = -4.2°$. However, Figure 5.44b shows that the airfoil section used to build this wing features an inflection point on the lower side, at around $x/c = 0.07$. This inflection point causes an adverse pressure gradient that can be significant at negative angles of attack and can cause localised flow separation.

Figure 5.49 plots the lift and drag curves calculated by the SDPM and compares them to the experimental measurements. Two sets of experimental data are plotted, one obtained from the aerodynamic balance and one obtained by integrating the measured pressure distribution.

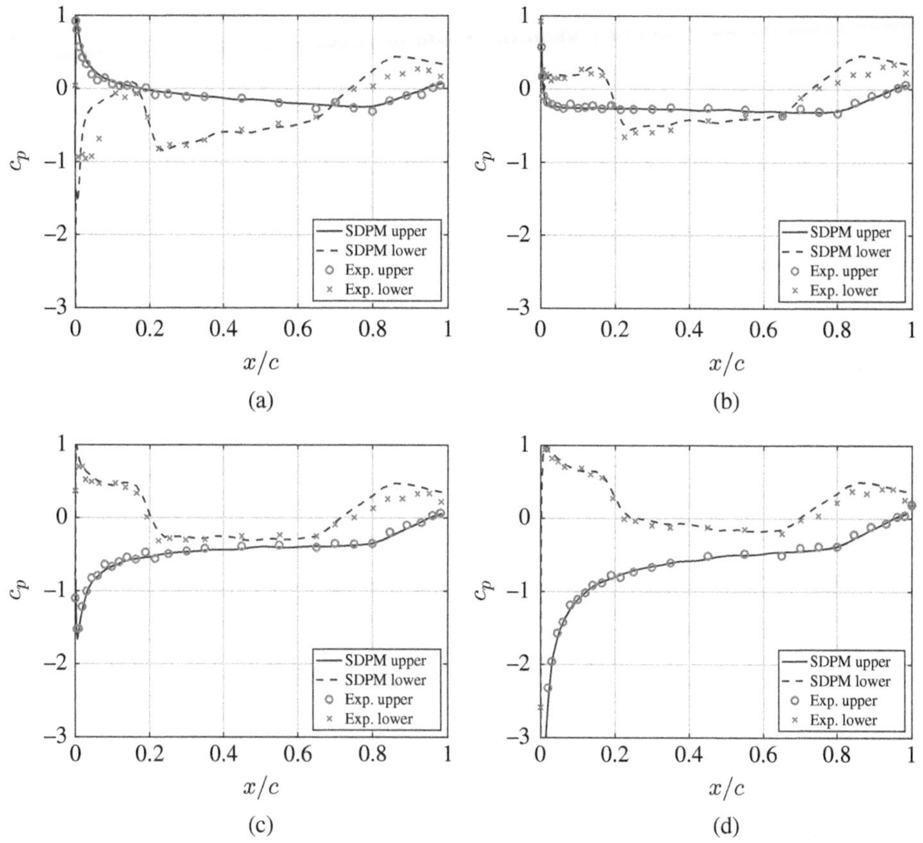

Figure 5.48 Steady pressure distribution around rectangular wing predicted by the SDPM. Experimental data by Applin and Gentry Jr. (1990). (a) $\alpha = -4.2°$, (b) (a) $\alpha = -0.2°$, (c) $\alpha = 3.8°$, and (d) $\alpha = 7.8°$. Source: Applin and Gentry Jr. (1990), credit: NASA.

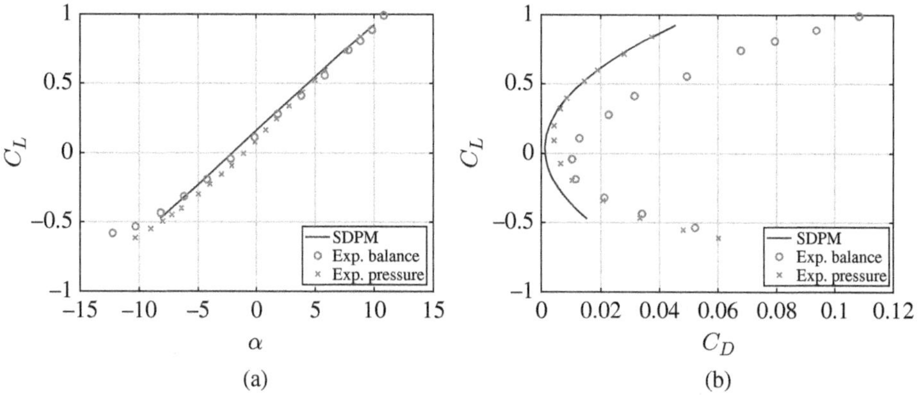

Figure 5.49 Total lift and drag coefficient predicted by the SDPM. Experimental data by Applin and Gentry Jr. (1990). (a) C_L vs α and (b) C_L vs C_D. Source: Applin and Gentry Jr. (1990), credit: NASA.

As there is only one spanwise pressure measurements station, Applin and Gentry Jr. (1990) assumed that the pressure distribution at 44% of the half-span was representative of the spanwise integral of the pressure distribution. Simulation results from another panel code were used to justify this assumption.

Figure 5.49a shows that the SDPM lift predictions are in good agreement with the experimental measurements (particularly the aerodynamic balance data) for all angles of attack higher than −8°. At lower angles, both sets of experimental data depart from linearity due to the viscous effects mentioned earlier. The drag polars of Figure 5.49b show that the SDPM severely underpredicts the drag coefficient compared to the balance measurements. On the other hand, the comparison with the pressure integration results is quite good for positive values of the lift. This means that the underprediction is due to the lack of modelling of skin friction phenomena that cannot be captured either by inviscid simulations or by integrating experimental pressures. This example is solved by Matlab code `steady_SDPM.m`.

The unsteady version of the SDPM is constructed in the same way as that of the VLM. The relative flow velocities on the control points of the wing panels are now functions of time, $\boldsymbol{u}_m(t_k)$, $\boldsymbol{v}_m(t_k)$, $\boldsymbol{w}_m(t_k)$, as is the source strength,

$$\sigma(t_k) = -\left(\boldsymbol{u}_m(t_k) \circ \boldsymbol{n}_x(t_k) + \boldsymbol{v}_m(t_k) \circ \boldsymbol{n}_y(t_k) + \boldsymbol{w}_m(t_k) \circ \boldsymbol{n}_z(t_k)\right) \tag{5.197}$$

Green's identity (5.188) becomes time-varying, such that

$$\left(\boldsymbol{B}_\phi(t_k) - \frac{1}{2}\boldsymbol{I}\right)\boldsymbol{\mu}(t_k) = -\boldsymbol{A}_\phi(t_k)\sigma(t_k) - \boldsymbol{C}_\phi(t_k)\boldsymbol{\mu}_w(t_k) \tag{5.198}$$

where the influence coefficient matrices are functions of time in the general case where the wing deforms. Once we have solved for $\boldsymbol{\mu}(t_k)$, we can calculate the perturbation velocities on the surface in exactly the same way as in the steady case, the pressure coefficient from the unsteady Bernoulli equation (2.31) and the aerodynamic loads. Finally, assuming a prescribed wake model, we propagate the positions of the wake panel vertices

$$\mathbf{x}_{w_{i,j}}(t_k) = \mathbf{x}_{w_{i-1,j}}(t_{k-1}) + \mathbf{Q}_\infty \Delta t$$

we propagate the wake doublet strength

$$\mu_{w_{i,j}}(t_k) = \mu_{w_{i-1,j}}(t_{k-1})$$

for $i = 2, \ldots, k$, and set the strength of the first row of wake vortex panels to

$$\mu_{w_{1,j}}(t_k) = \mu_{2m,j}(t_{k-1}) - \mu_{1,j}(t_{k-1}) \tag{5.199}$$

As in the case of the VLM, the Kutta condition is not enforced when solving for the doublet strengths of the wing panels but after calculating the aerodynamic loads.

5.7 Flexible Motion

All the motions considered up to this point have been rigid-body translations and rotations. In this section, we will introduce flexible motion, whereby the wing deforms with respect to its shape when at rest. Wing deflection is generally speaking an aeroelastic phenomenon, whereby inertial and aerodynamic loads cause wings to bend and twist. Aeroelasticity is not considered in this book, but we can apply prescribed deflections to wings in order to calculate the aerodynamic load responses. For small deflections, the time-varying shape of a wing can be expressed using a modal approximation of the form

$$\mathbf{x}_s(t) = \mathbf{x}_{s_0} + \sum_{l=1}^{N} \boldsymbol{\Theta}_l(\mathbf{x}_{s_0}) q_l(t) \tag{5.200}$$

where $\mathbf{x}_s(t) = (x_s(t), y_s(t), z_s(t))$ is the position of a point on a surface at time t, $\mathbf{x}_{s_0}$ is the position of the same point when the wing is undeformed, $\boldsymbol{\Theta}_l = (\Theta_{x_l}, \Theta_{y_l}, \Theta_{z_l})$ are spatial functions known as mode shapes and $q_l(t)$ are time functions referred to as generalized coordinates. In order to represent accurately any given deflection, the sum in Eq. (5.200) must be infinite, but in practice, it is truncated to N, the desired number of modes.

In writing Eq. (5.200), we have assumed that the wing's structure is linear and therefore the mode shapes only depend on the initial shape and do not change as the wing deforms. If we also assume that rotations are negligible, we can simplify Eq. (5.200) to

$$x_s(t) = x_{s_0} + \sum_{l=1}^{N_x} \Theta_{x,l}(x_{s_0}, y_{s_0}) q_{x_l}(t) \tag{5.201}$$

$$y_s(t) = y_{s_0} + \sum_{l=1}^{N_y} \Theta_{y,l}(x_{s_0}, y_{s_0}) q_{y_l}(t) \tag{5.202}$$

$$z_s(t) = z_{s_0} + \sum_{l=1}^{N_z} \Theta_{z,l}(x_{s_0}, y_{s_0}) q_{z_l}(t) \tag{5.203}$$

where $\Theta_{x,l}$ are chordwise in-plane mode shapes, $\Theta_{y,l}$ are spanwise in-plane modes, $\Theta_{z,l}$ are out-of-plane modes and $q_{x_l}, q_{y_l}, q_{z_l}$ are their respective generalised coordinates. We can further assume that in-plane deflections are negligible, such that only Eq. (5.203) is necessary to describe the wing's deflection.

As an example, Figure 5.50 plots four out-of-plane vibration mode shapes for a flat plate half-wing that is built-in at the root. The mode shapes were obtained by carrying out a finite element analysis on the structure of the wing, which was simply a solid flat plate made of aluminium and with thickness of 1 mm (De Oro Fernández et al. 2020). Mode 1 is a bending deflection although a small amount of torsion (twist) can be observed at the wingtip. Modes 2 and 3 are combinations of torsion and bending while mode 4 is mainly torsion. Superimposing different amounts of these four mode shapes using Eq. (5.203) can approximate the out-of-plane deflection of the real wing at any time instance.

We do not have to use finite element analysis in order to create mode shapes. Any set of physically plausible spatial functions can be used, as long as they satisfy boundary conditions. Determining the exact time history of the deflection of the wing using Eqs. (5.201)–(5.203) assumes that $q_{x_l}(t)$, $q_{y_l}(t)$ and $q_{z_l}(t)$ are known, for example because they were imposed or because they were measured experimentally. We will demonstrate this approach in the next example.

Example 5.12 *Use the SDPM to simulate the plunging flexible wing experiments presented by Heathcote et al. (2003)*

This test case consists in three different rectangular wings oscillated in plunge in a water tunnel. All wings were half-wings built in at the root and had an aspect ratio of 3 ($\mathcal{R} = 6$ for the

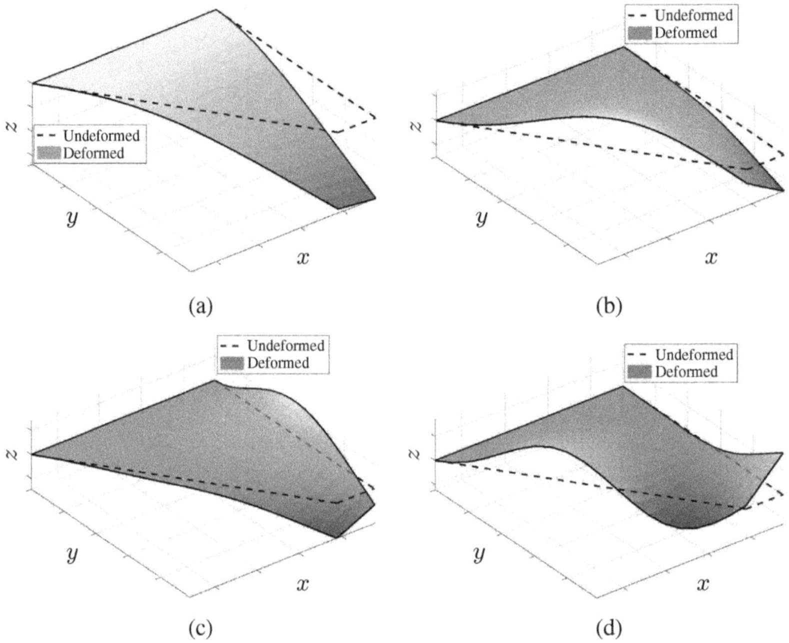

Figure 5.50 Out-of-plane mode shapes of a cantilevered flat plate wing. (a) Mode 1, (b) Mode 2, (c) Mode 3, and (d) Mode 4.

complete wing). The airfoil section was NACA 0012 in all three cases; the differences between the wings lay in the structure. One wing was made from nylon and was reinforced by two steel rods with 8 mm diameter; it is referred to as the 'inflexible' wing. The other two wings were made from polydimethylsiloxane rubber (PDMS) and were reinforced by a steel or aluminium plate with 1 mm thickness; they are referred to as the 'flexible' (steel plate) and 'very flexible' wing (aluminium plate).

The root of each wing was attached to a splitter plate and oscillated in plunge at reduced frequencies between $k = 0.4$ and $k = 1.8$. The splitter plate was used to ensure 2D flow at the root of the half-wing, as would have been the case if a full wing of aspect ratio 6 was oscillated in plunge. The imposed plunge time history was of the form

$$h(t) = h_1 \cos \omega t$$

where $h_1 = 0.175c_0$ is the plunge amplitude, c_0 the constant chord and ω the oscillation frequency, related to the reduced frequency by $\omega = 2U_\infty k/c_0$.

The bending deflection of the wings was observed using a high-speed camera. The authors state that the bending deflection was of order one, so that it increased monotonically from the root to the tip. This type of bending deflection can be described using the first cantilever beam mode shape equation (see for example Rao (2004))

$$\Theta(y) = \frac{1}{2}\left(\frac{\sinh\beta_1 - \sin\beta_1}{\cos\beta_1 + \cosh\beta_1}\left(\sin\beta_1\frac{2y}{b} - \sinh\beta_1\frac{2y}{b}\right) + \cosh\beta_1\frac{2y}{b} - \cos\beta_1\frac{2y}{b}\right) \tag{5.204}$$

where $\beta_1 = 1.875$ and y is distance from the root and takes values between 0 and $b/2$. Heathcote et al. (2003) do not give any information on torsional deflection so we will assume that it was *negligible. The same assumption was made by Stanford and Beran (2010) when they simulated this test case using the VLM. Furthermore, Heathcote et al. (2003) state that the time history of the generalised coordinate associated with this bending mode was sinusoidal of the form*

$$q(t) = q_1(k)\cos\left(\omega t + \phi_q(k)\right)$$

where $q_1(k)$ is the amplitude of the generalised coordinate and $\phi_q(k)$ is its phase lag with respect to the plunge motion. Consequently, the vertical displacement response of any part of the wing is given by

$$z(y,t) = \Theta(|y|)q_1(k)\cos\left(\omega t + \phi_q(k)\right) + h_1\cos\omega t \tag{5.205}$$

for y taking values between $-b/2$ and $b/2$. This displacement only depends on y because both the plunging and bending motions are constant in the chordwise direction. Heathcote et al. (2003) measured the displacement at the root and at the tip,

$$z(0,t) = h_1\cos\omega t,\ z(b/2,t) = q_1(k)\cos\left(\omega t - \phi_q(k)\right) + h_1\cos\omega t \tag{5.206}$$

since $\Theta(0) = 0$ and $\Theta(b/2) = 1$. They defined the tip bending ratio, $\bar{z}(k)$, as the tip bending amplitude divided by the root bending amplitude. They also defined the phase lag of the oscillations at the tip with respect to those at the root, $\phi_{b/2}(k)$. Figure 5.51 plots the variation of $\bar{z}(k)$ and $\phi_{b/2}(k)$ with reduced frequency for the three wings. Using the data of Figure 5.51 we can write the tip bending time response as

$$z(b/2,t) = h_1\bar{z}(k)\cos\left(\omega t + \phi_{b/2}(k)\right)$$

Equating this latest expression with the second of Eq. (5.206) and solving for $q_1(k)$ and $\phi_q(k)$, we obtain

$$q_1(k) = \sqrt{\left(h_1\bar{z}(k)\sin\phi_{b/2}(k)\right)^2 + \left(h_1\bar{z}(k)\cos\phi_{b/2}(k) - h_1\right)^2}$$
$$\phi_q(k) = \tan^{-1}\left(\frac{h_1\bar{z}(k)\sin\phi_{b/2}(k)}{h_1\bar{z}(k)\cos\phi_{b/2}(k) - h_1}\right) \tag{5.207}$$

where the inverse tangent is calculated use the four-quadrant arctangent function. Once we have evaluated $q_1(k)$ and $\phi_q(k)$, we can obtain the displacement anywhere along the span from

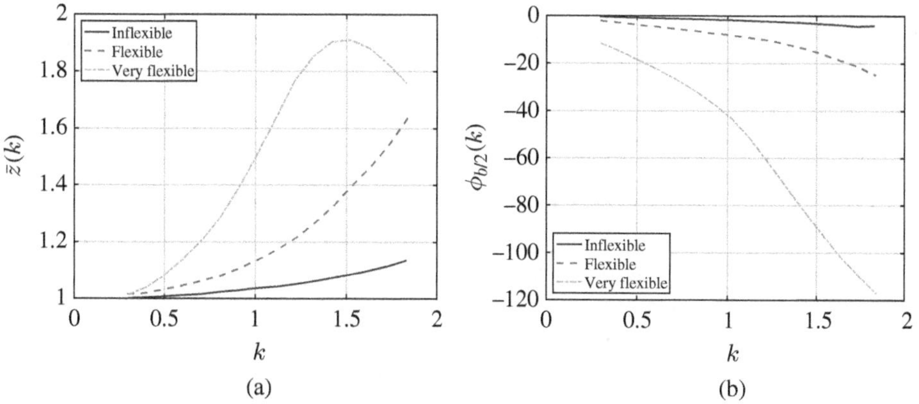

Figure 5.51 Tip bending amplitude and phase of plunging rectangular half-wing in Heathcote et al. (2003). (a) Tip bending amplitude and (b) bending phase lag. Source: Adapted from Heathcote et al. (2003).

Eq. (5.205). As we have neglected torsion, this displacement will be applied to all chordwise stations of the wing. Furthermore, the present SDPM calculation will not feature a splitter plate at the root so that a complete wing with aspect ratio 6 and extending from $-b/2$ to $+b/2$ will be simulated. We have taken the absolute value of y in $\Theta(|y|)$ because the displacement must be symmetric across the two half wings.

We start the SDPM calculation by defining the undeformed geometry of the wing and discretising it into 2m chordwise and n spanwise panels, as in example 5.11. The chord of the wing is $c_0 = 0.1$ m and its span $b = 0.6$ m. The resulting coordinates of the panel vertices are labeled as $x^0_{p_{i,j}}$, $z^0_{p_{i,j}}$ and $z^0_{p_{i,j}}$, where the 0 superscript denotes that the shape of the wing is undeformed. At each time instance, for a selected value of k, the deformed shape is given by

$$x_{p_{i,j}}(t_k) = x^0_{p_{i,j}},\ y_{p_{i,j}}(t_k) = y^0_{p_{i,j}},\ z_{p_{i,j}} = z^0_{p_{i,j}} + z(y^0_{p_{i,j}}, t_k)$$

where $z(y^0_{p_{i,j}}, t_k)$ is obtained from Eq. (5.205), noting that index k denotes the kth time instance while variable k denotes the reduced frequency. Figure 5.52 plots the deformed shape at time instances $t = T/4$ and $t = 3T/4$ for the very flexible wing at $k = 1.5$, where T is the oscillation period. As the phase lag is approximately 90° for this value of the reduced frequency, the wing is nearly undeformed at $t = 0$ and $t = T/2$ and the shapshots of Figure 5.52 show the two extremes of the deflection. The other two wings deform less for all values of k.

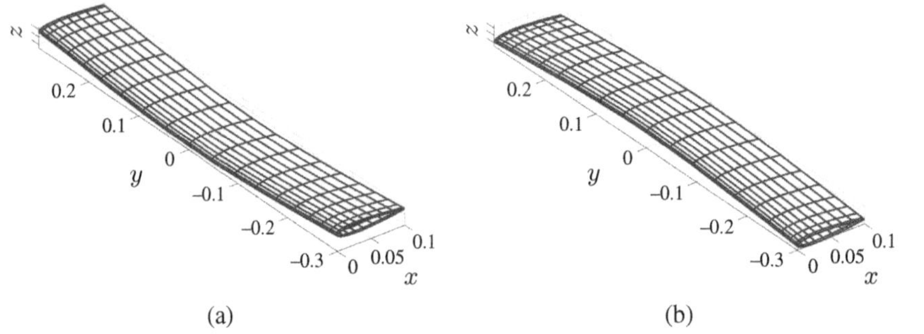

Figure 5.52 Two snapshots of the deformed shape of the very flexible wing for $k = 1.5$. (a) $t = T/4$ and (b) $t = 3T/4$.

Next, we calculate the positions of the control points, the panel areas and the normal and tangential vectors as in Example 5.11. The relative flow velocities on the control points are given by

$$u_{m_{i,j}}(t_k) = 0,\ v_{m_{i,j}}(t_k) = 0,\ w_{m_{i,j}}(t_k) = -\dot{z}(y^0_{c_{i,j}}, t_k)$$

As the shape of the wing changes due to the bending, we can choose to re-calculate the wing influence coefficient matrix at each time instance. Note that this re-calculation constitutes a significant additional computational cost.

The unsteady Green's identity at the kth time instance is given by Eq. (5.198), where the source strengths are obtained from expression (5.197) and the wake doublet strengths are given by Eq. (5.199). Since the doublet strengths of the bound trailing edge panels are related to those of the first row of wake panels, we must select the value of the time step such that the latter are of the same size as the former. We therefore use Eq. (5.102) to obtain

$$\Delta t = \frac{c_0}{mU_\infty}$$

as in the case of the VLM. This time step value is adjusted such that there is an integer number of time instances in each period of oscillation, p_{pc}, using Eq. (5.114). The number of cycles to be simulated is set to $n_c = 2$ and the number of time instances per cycle is given by Eq. (5.113).

At each time instance, the wake doublet strength is propagated using

$$\mu_{w_{i,j}}(t_k) = \mu_{w_{i-1,j}}(t_{k-1}) \tag{5.208}$$

for $i = 2, \ldots, k$ while $\mu_{w_{1,j}}(t_k)$ is given by Eq. (5.199). If the wake is prescribed, the positions of the wake panel vertices are updated using Eq. (5.96). If the wake is free, then we can use an adapted version of Eq. (5.97)

$$\mathbf{x}_{w_{i,j}}(t_k) = \mathbf{x}_{w_{i-1,j}}(t_{k-1}) + \left(\mathbf{Q}_\infty + \mathbf{u}_{w_{i-1,j}}(t_k)\right)\Delta t \tag{5.209}$$

where $\mathbf{Q}_\infty = (U_\infty, V_\infty, W_\infty)$ and $\mathbf{u}_{w_{i-1,j}}(t_k)$ are the velocities induced by the wing source, wing doublet and wake doublet panels on the wake panel vertices. These velocities can be obtained using expressions of the form of Eq. (5.177)

$$\begin{aligned}
\boldsymbol{u}_w(t_k) &= \boldsymbol{A}_u(\mathbf{x}_w, t_k)\boldsymbol{\sigma}(t_k) + \boldsymbol{B}_u(\mathbf{x}_w, t_k)\boldsymbol{\mu}(t_k) + \boldsymbol{C}_u(\mathbf{x}_w, t_k)\boldsymbol{\mu}_w(t_k) \\
\boldsymbol{v}_w(t_k) &= \boldsymbol{A}_v(\mathbf{x}_w, t_k)\boldsymbol{\sigma}(t_k) + \boldsymbol{B}_v(\mathbf{x}_w, t_k)\boldsymbol{\mu}(t_k) + \boldsymbol{C}_v(\mathbf{x}_w, t_k)\boldsymbol{\mu}_w(t_k) \\
\boldsymbol{w}_w(t_k) &= \boldsymbol{A}_w(\mathbf{x}_w, t_k)\boldsymbol{\sigma}(t_k) + \boldsymbol{B}_w(\mathbf{x}_w, t_k)\boldsymbol{\mu}(t_k) + \boldsymbol{C}_w(\mathbf{x}_w, t_k)\boldsymbol{\mu}_w(t_k)
\end{aligned} \tag{5.210}$$

where $\boldsymbol{u}_w(t_k)$, $\boldsymbol{v}_w(t_k)$, $\boldsymbol{w}_w(t_k)$ are the $(n+1)(k+1) \times 1$ vectors of velocities at the wake panel vertices, $\boldsymbol{A}_u(\mathbf{x}_w, t_k)$, $\boldsymbol{B}_u(\mathbf{x}_w, t_k)$, etc., are the $(n+1)(k+1) \times 2mn$ velocity influence coefficient matrices of the wing source and doublet panels on the wake panel vertices, while $\boldsymbol{C}_u(\mathbf{x}_w, t_k)$, $\boldsymbol{C}_v(\mathbf{x}_w, t_k)$, $\boldsymbol{C}_w(\mathbf{x}_w, t_k)$ are the $(n+1)(k+1) \times nk$ velocity influence coefficient matrices of the wake panels on the wake panel vertices. The velocity influence coefficient matrices of the source panels, $\boldsymbol{A}_u$, $\boldsymbol{A}_v$, $\boldsymbol{A}_v$, can be calculated using Eq. (A.75). The velocity influence coefficient matrices of the doublet panels should not be calculated using Eq. (A.85); the influence of a doublet panel on its own vertices is prone to numerical instability, especially for wake panels, which are generally not plane. It is best to make use of the mathematical analogy between the doublet panel and the vortex ring and apply Eq. (A.43) to calculate $\boldsymbol{B}_u$, $\boldsymbol{B}_v$, $\boldsymbol{B}_w$ and $\boldsymbol{C}_u$, $\boldsymbol{C}_v$, $\boldsymbol{C}_w$.

Once $\boldsymbol{\mu}(t_k)$ have been evaluated, the perturbation flow velocities on the control points, $u_{i,j}(t_k)$, $v_{i,j}(t_k)$, $w_{i,j}(t_k)$, are obtained from the unsteady version of Eq. (5.193)

$$\begin{aligned}
\tau_{mx_{i,j}}(t_k)u_{i,j}(t_k) + \tau_{my_{i,j}}(t_k)v_{i,j}(t_k) + \tau_{mz_{i,j}}(t_k)w_{i,j}(t_k) &= \left.\frac{\partial \mu(t_k)}{\partial \tau_m(t_k)}\right|_{i,j} \\
\tau_{nx_{i,j}}(t_k)u_{i,j}(t_k) + \tau_{ny_{i,j}}(t_k)v_{i,j}(t_k) + \tau_{nz_{i,j}}(t_k)w_{i,j}(t_k) &= \left.\frac{\partial \mu(t_k)}{\partial \tau_n(t_k)}\right|_{i,j} \\
n_{x_{i,j}}(t_k)u_{i,j}(t_k) + n_{y_{i,j}}(t_k)v_{i,j}(t_k) + n_{z_{i,j}}(t_k)w_{i,j}(t_k) &= \sigma_{i,j}
\end{aligned} \tag{5.211}$$

and the pressure coefficient distribution on the control panels is calculated using the unsteady Bernoulli equation (2.31)

$$c_{p_{i,j}}(t_k) = 1 - \frac{Q_{i,j}(t_k)^2}{Q_\infty^2} - \frac{2}{Q_\infty^2}\frac{\partial \phi_{i,j}(t_k)}{\partial t} \tag{5.212}$$

where

$$Q_{i,j}(t_k)^2 = \left(U_\infty + u_{m_{i,j}}(t_k) + u_{i,j}(t_k)\right)^2 + \left(V_\infty + v_{m_{i,j}}(t_k) + v_{i,j}(t_k)\right)^2 + \left(W_\infty + w_{m_{i,j}}(t_k) + w_{i,j}(t_k)\right)^2$$

and $\phi_{i,j}(t_k) = \mu_{i,j}(t_k)$ *thanks to Eq. (5.168). The time derivative of the latter can be approximated using the first-order backward difference*

$$\frac{\partial \mu_{i,j}(t_k)}{\partial t} \approx \frac{\mu_{i,j}(t_k) - \mu_{i,j}(t_{k-1})}{\Delta t}$$

Finally, the aerodynamic loads at each time instance, $C_L(t_k)$ *and* $C_D(t_k)$*, are obtained from Eqs. (5.194)–(5.196) with* $\alpha = 0$.

The mean thrust coefficient is defined here as

$$C_T = -\frac{1}{T}\int_0^T C_D(t)\mathrm{d}t$$

As the drag is only available at discrete time instances, the integration must be carried out numerically. Any number of numerical integration schemes can be used, but here we choose to apply the discrete Fourier Transform to the drag response from the last complete cycle of oscillation, such that we also estimate the amplitudes and phases of the higher frequency components. Therefore,

$$C_T = -\frac{1}{p_{pc}} \sum_{k=(n_c-1)p_{pc}+1}^{n_c p_{pc}} C_D(t_k)$$

The time-varying aerodynamic power is defined by Heathcote et al. (2003) as

$$C_P(t_k) = -C_L(t_k)\frac{\dot{h}(t_k)}{U_\infty}$$

which we will use here in order to compare to the experimental data. Then, the mean aerodynamic power is given by

$$C_P = \frac{1}{T}\int_0^T C_P(t)\mathrm{d}t = \frac{1}{p_{pc}} \sum_{k=(n_c-1)p_{pc}+1}^{n_c p_{pc}} C_P(t_k) \tag{5.213}$$

The propulsive efficiency of the wing is defined as

$$\eta = \frac{C_T}{C_P} \tag{5.214}$$

The complete SDPM calculations are carried out with $m = 20$, $n = 20$ *and repeated for* k *values between 0.4 and 1.8. The out-of-plane displacement amplitude and phase lag for these values of the reduced frequency were obtained by interpolating the data of Figure 5.51. The wake is modelled as prescribed (but not flat) in order to speed up the calculations.*

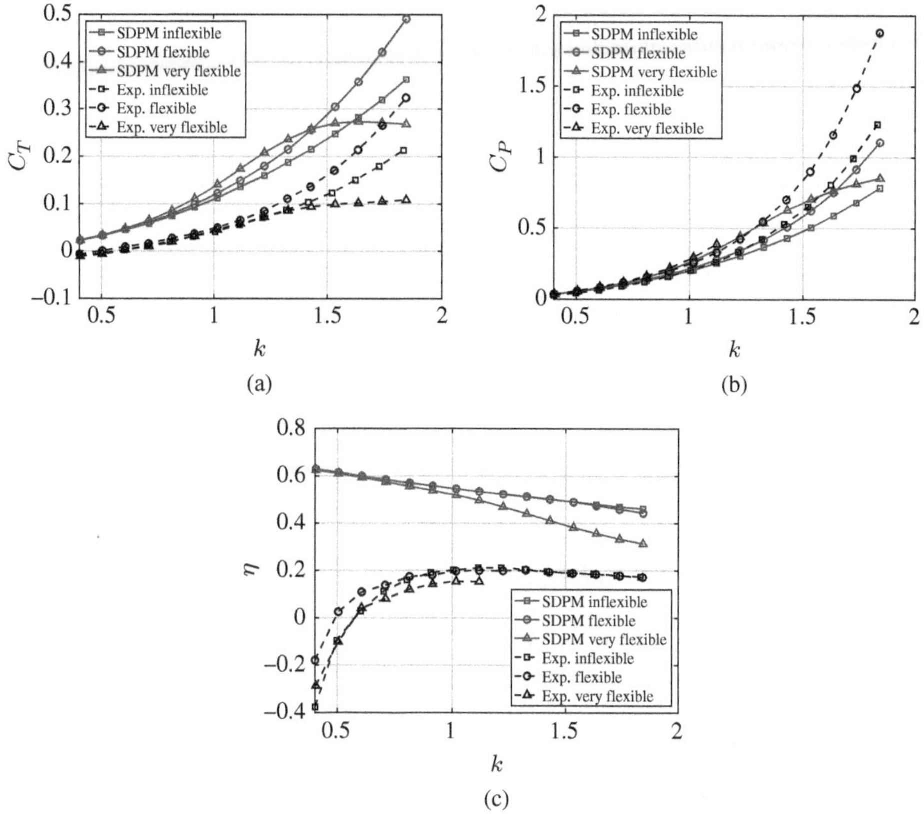

Figure 5.53 Mean thrust, mean propulsive power and propulsive efficiency variation with reduced frequency for the three wings, experimental data from Heathcote et al. (2003). (a) $C_T(k)$, (b) $C_P(k)$, and (c) $\eta(k)$. Source: Heathcote et al. (2003) reprinted with permission from Elsevier.

Figure 5.53 plots the resulting thrust, power and efficiency variation with reduced frequency for the three wings, along with the experimental data. The mean thrust values predicted by the SDPM are similar in form to the experimental data but higher in value; this is logical because the SDPM does not model viscous drag, which is present in the experimental measurements. As a consequence, the experimental thrust is negative for $k \leq 0.5$, which is not the case for the simulation results. Furthermore, the SDPM predictions for C_T increase faster with k than the measured data. Nevertheless, the same tendencies can be observed in the two sets of results: wing flexibility can have a beneficial or detrimental effect on mean thrust, depending on the value of the reduced frequency. The flexible wing produces more thrust than the inflexible one at all tested values of k. The very flexible wing produces more thrust for $k < 1.4$ and less thrust for higher k. Note that the experimental data show that the very flexible wing never produces more thrust than the other two.

The agreement between the simulation and the experiment is better for the mean propulsive power, as shown in Figure 5.53b. Note that, for the very flexible wing, Heathcote et al. (2003) only present C_P data for $k \leq 1.4$. The two sets of results are in acceptable agreement up to $k = 1.2$, but for higher frequencies the SDPM under-predicts the power. Cleaver et al. (2016)

show that significant flow separation phenomena occur at high plunging frequencies and amplitudes, significantly enhancing the lift, particularly for the flexible wings. As lift is the only component contributing to C_P, lift enhancement results in a significant increase in propulsive power. Nevertheless, even inviscid calculations show that, for the higher k values, flexibility has a significant effect on propulsive power, as C_P is higher for the flexible wing than the inflexible one, while the very flexible wing first increases the power and then decreases it.

Figure 5.53c plots the variation of propulsive efficiency with reduced frequency for the three wings. The efficiency is negative for all cases where $C_T < 0$. It can be seen that the η values predicted by the SDPM are much higher than the experimental measurements due to the overprediction of the thrust (and the underprediction of the power for $k > 1.2$). Nevertheless, both sets of results show that wing flexibility has a small effect on η. The experimental results show that the efficiency of the very flexible wing is lower than that of the other two wings for $k > 0.5$; the SDPM predicts the same phenomenon for $k > 1.2$. This example is solved by Matlab code `plungebend_SDPM.m`.

The data of Figure 5.53c could be taken as proof that wing flexibility is detrimental for propulsive efficiency. However, the wings tested by Heathcote et al. (2003) underwent very little torsion. There is a general consensus in the literature that wing flexibility has a positive effect on propulsive efficiency, but the wing must be able to undergo measurable torsion or chordwise bending. However, the exact way in which the structure deforms is very important in determining whether the propulsive efficiency will increase or decrease. Furthermore, optimal thrust and propulsive efficiency occur when the amplitude and frequency of the motion are sufficiently high to induce dynamic stall, releasing a strong vortex on the suction side of the wing that enhances both lift and thrust. There is evidence that optimal propulsive efficiency occurs at Strouhal numbers between 0.25 and 0.35 (Triantafyllou et al. 1993) or 0.2–0.4 (Taylor et al. 2003); many flying and swimming animals operate in these ranges. Dewey et al. (2013) found that optimal efficiency is obtained when the structural parameters are tuned to induce resonance inside the $Str = 0.25 - 0.35$ range. The study of the aerodynamic and hydrodynamic thrust production is still an active research area due to the complexity of the phenomena involved. Shyy et al. (2013) provide a very useful introduction to the subject.

The test case by Heathcote et al. (2003) is not ideal for validating potential flow methods because of the high amplitudes and frequencies involved that induced significant viscous effects. The amplitude of the plunge-induced effective angle of attack, $\dot{h}/U_\infty$, starts at 8° for $k = 0.4$ and increases to 37° for $k = 1.85$. Similarly, the Strouhal number values reported by Heathcote et al. (2003) range between 0.05 and 0.28, but they are based on the total amplitude of vertical oscillation midway between the root and the tip. At the wingtip itself, the Strouhal number reaches 0.36 at the highest reduced frequency applied to the flexible wing. The next test case we will present involves very low bending amplitudes and is better adapted to the validation of potential flow modelling techniques.

5.7.1 Source and Doublet Panel Method in the Frequency Domain

For motions of small amplitude, the SDPM can be transformed to the frequency domain, as was done for the VLM in section 5.5.2. We assume that flexibility can exist but its only non-negligible effect is on the relative velocity between the wing and the flow; the aerodynamic influence coefficient matrices are not affected significantly. We therefore set up the coordinates of the $N = 2mn$ airfoil panel vertices as usual and a flat wake of the desired length. The time step is set to $\Delta t = c_0/mU_\infty$ so that the chordwise length of the wake panels is $\Delta x = c_0/m$; both Δt and Δx are slightly adjusted so that there is an integer number of time instances per cycle of oscillation. Consequently, there are m_w chordwise and n spanwise wake panels. We will superimpose two types of motion:

- Horizontal free stream, U_∞, and rigid plunging and pitching around axis x_f, resulting in relative velocities

$$\boldsymbol{u}_m(t) = U_\infty - \dot{\alpha}(t)\left(\boldsymbol{z}_c - x_f\right)$$

$$\boldsymbol{v}_m(t) = 0$$

$$\boldsymbol{w}_m(t) = U_\infty \alpha(t) + \dot{\alpha}(t)(\boldsymbol{x}_c - x_f) - \dot{h}(t)$$

  where $\boldsymbol{u}_m, \boldsymbol{v}_m, \boldsymbol{w}_m$ are $N \times 1$ vectors of the velocities induced by the rigid-body motion on the control points of the wing panels, x_f is the position of the pitch axis and $\boldsymbol{x}_c$, $\boldsymbol{z}_c$ are the $N \times 1$ vectors of the x and z coordinates of the control points. Rolling and yawing degrees of freedom can be added from Eq. (5.151).
- Flexible motion of the form of Eq. (5.201)–(5.203), inducing velocities

$$\boldsymbol{u}_m(t) = -\sum_{l=1}^{N_x} \boldsymbol{\Theta}_{x,l}(\boldsymbol{x}_c, \boldsymbol{y}_c)\dot{q}_{x_l}(t) \tag{5.215}$$

$$\boldsymbol{v}_m(t) = -\sum_{l=1}^{N_y} \boldsymbol{\Theta}_{y,l}(\boldsymbol{x}_c, \boldsymbol{y}_c)\dot{q}_{y_l}(t) \tag{5.216}$$

$$\boldsymbol{w}_m(t) = -U_\infty \sum_{l=1}^{N_z} \frac{\partial \boldsymbol{\Theta}_{z,l}(\boldsymbol{x}_c, \boldsymbol{y}_c)}{\partial x} q_{z_l}(t) - \sum_{l=1}^{N_z} \boldsymbol{\Theta}_{z,l}(\boldsymbol{x}_c, \boldsymbol{y}_c)\dot{q}_{z_l}(t) \tag{5.217}$$

  where $\boldsymbol{\Theta}_{x,l}$, $\boldsymbol{\Theta}_{y,l}$, $\boldsymbol{\Theta}_{z,l}$ are $N \times 1$ vectors of mode shapes and q_{x_l}, q_{y_l}, q_{z_l} are generalized coordinates

The total relative flow velocities are obtained by adding the rigid and flexible contributions and the source strengths are calculated from Eq. (5.197). Transforming to the frequency domain, we obtain

$$\boldsymbol{\sigma}(\omega) = -\left(\boldsymbol{u}_m(\omega) \circ \boldsymbol{n}_x + \boldsymbol{v}_m(\omega) \circ \boldsymbol{n}_y + \boldsymbol{w}_m(\omega) \circ \boldsymbol{n}_z\right) \tag{5.218}$$

where $\boldsymbol{\sigma}(\omega)$ is a $N \times 1$ vector and now the normal vectors are constants. We also apply the Fourier transform to the unsteady version of Green's identity of Eq. (5.198), such that

$$\left(\boldsymbol{B}_\phi - \frac{1}{2}\boldsymbol{I}\right)\boldsymbol{\mu}(\omega) = -\boldsymbol{A}_\phi \boldsymbol{\sigma}(\omega) - \boldsymbol{C}_\phi \boldsymbol{\mu}_w(\omega) \tag{5.219}$$

where $\boldsymbol{A}_\phi$, $\boldsymbol{B}_\phi$ and $\boldsymbol{C}_\phi$ are constant influence coefficient matrices, calculated for the undeformed geometry. The wake is modelled as flat and the doublet strengths of its panels are obtained from Eqs. (5.199) and (5.208), such that

$$\boldsymbol{\mu}_w(t) = \begin{pmatrix} \mu_{(2m-1)n+1}(t-\Delta t) - \mu_1(t-\Delta t) \\ \vdots \\ \mu_{2mn}(t) - \mu_n(t) \\ \mu_{(2m-1)n+1}(t-2\Delta t) - \mu_1(t-2\Delta t) \\ \vdots \\ \mu_{2mn}(t-2\Delta t) - \mu_n(t-2\Delta t) \\ \vdots \\ \mu_{(2m-1)n+1}(t-m_w\Delta t) - \mu_1(t-m_w\Delta t) \\ \vdots \\ \mu_{2mn}(t-m_w\Delta t) - \mu_n(t-m_w\Delta t) \end{pmatrix} = \begin{pmatrix} \boldsymbol{P}_c\boldsymbol{\mu}(t-\Delta t) \\ \boldsymbol{P}_c\boldsymbol{\mu}(t-2\Delta t) \\ \vdots \\ \boldsymbol{P}_c\boldsymbol{\mu}(t-m_w\Delta t) \end{pmatrix} \tag{5.220}$$

where $\boldsymbol{P}_c = (\boldsymbol{0}_{n\times(2m-1)n}, \boldsymbol{I}_n) - (\boldsymbol{I}_n, \boldsymbol{0}_{n\times(2m-1)n})$, in direct analogy to the VLM wake Eq. (5.122).

Taking the Fourier transform of Eq. (5.220) and substituting it into Eq. (5.219), we obtain

$$\left(\boldsymbol{B}_\phi - \frac{1}{2}\boldsymbol{I} + \boldsymbol{C}_\phi\boldsymbol{P}_e(\omega)\boldsymbol{P}_c\right)\boldsymbol{\mu}(\omega) = -\boldsymbol{A}_\phi\boldsymbol{\sigma}(\omega) \tag{5.221}$$

where

$$\boldsymbol{P}_e(\omega) = \begin{pmatrix} \boldsymbol{I}_n \mathrm{e}^{-\mathrm{i}\omega\Delta t} \\ \boldsymbol{I}_n \mathrm{e}^{-\mathrm{i}2\omega\Delta t} \\ \vdots \\ \boldsymbol{I}_n \mathrm{e}^{-\mathrm{i}\omega m_w\Delta t} \end{pmatrix}$$

Equation 5.219 can be easily solved for the unknown bound doublet strengths $\boldsymbol{\mu}(\omega)$. The flow velocities on the control points are obtained from Eq. (5.211) such that

$$\begin{aligned} \boldsymbol{\tau}_{m_{i,j}} \cdot [u_{i,j}(\omega), v_{i,j}(\omega), w_{i,j}(\omega)] &= \frac{\partial \mu_{i,j}(\omega)}{\partial \boldsymbol{\tau}_m} \\ \boldsymbol{\tau}_{n_{i,j}} \cdot [u_{i,j}(\omega), v_{i,j}(\omega), w_{i,j}(\omega)] &= \frac{\partial \mu_{i,j}(\omega)}{\partial \boldsymbol{\tau}_n} \\ \mathbf{n}_{i,j} \cdot [u_{i,j}(\omega), v_{i,j}(\omega), w_{i,j}(\omega)] &= \sigma_{i,j}(\omega) \end{aligned} \tag{5.222}$$

and the pressure coefficient distribution is calculated from the Fourier transform of Eq. (5.212)

$$\begin{aligned} c_{p_{i,j}}(\omega) = \delta(\omega) - \frac{1}{U_\infty^2}\Big(&\left(u_{m_{i,j}}(\omega) + u_{i,j}(\omega)\right) * \left(u_{m_{i,j}}(\omega) + u_{i,j}(\omega)\right) \\ &+ \left(v_{m_{i,j}}(\omega) + v_{i,j}(\omega)\right) * \left(v_{m_{i,j}}(\omega) + v_{i,j}(\omega)\right) \\ &+ \left(w_{m_{i,j}}(\omega) + w_{i,j}(\omega)\right) * \left(w_{m_{i,j}}(\omega) + w_{i,j}(\omega)\right)\Big) - \frac{2}{U_\infty^2}\mathrm{i}\omega\mu_{i,j}(\omega) \end{aligned} \tag{5.223}$$

where the $*$ symbol denotes convolution and $\delta(\omega)$ is the Kronecker delta function, equal to 1 for $\omega = 0$ and zero otherwise. The convolutions are calculated using vector convolutions, as usual. Finally, the aerodynamic loads acting on the panels are obtained from Eq. 5.194

$$\mathbf{F}_{i,j}(\omega) = -\frac{1}{2}\rho U_\infty^2 c_{p_{i,j}}(\omega) s_{i,j}\mathbf{n}_{i,j} \tag{5.224}$$

while the lift and drag are calculated from Eq. (5.131).

Example 5.13 *Use the frequency domain version of the SDPM to calculate the unsteady pressure distribution on a wing oscillating in the first bending mode.*

Lessing et al. (1960) carried out a series of experiments on a rectangular wing that was forced to oscillate in bending. The wing had a constant chord $c_0 = 0.457$ m, span $b = 1.394$ m and a biconvex airfoil with 5% thickness. An internal mechanism forced it to deform in its first bending mode with a tip amplitude of around $z_0(b/2) = 5$ mm and a frequency of $\omega_0 = 26.5$ Hz. Pressure sensors were used to record the unsteady pressure at four spanwise stations,

$$\frac{2y}{b} = (0,\ 0.5,\ 0.7,\ 0.9) \tag{5.225}$$

Figure 5.54a plots the half-wing, showing the positions of the pressure taps. The pressure taps are not placed exactly at the spanwise stations mentioned earlier, but they are staggered by ± 1 cm around these stations. We will neglect this stagger in order to simplify the analysis.

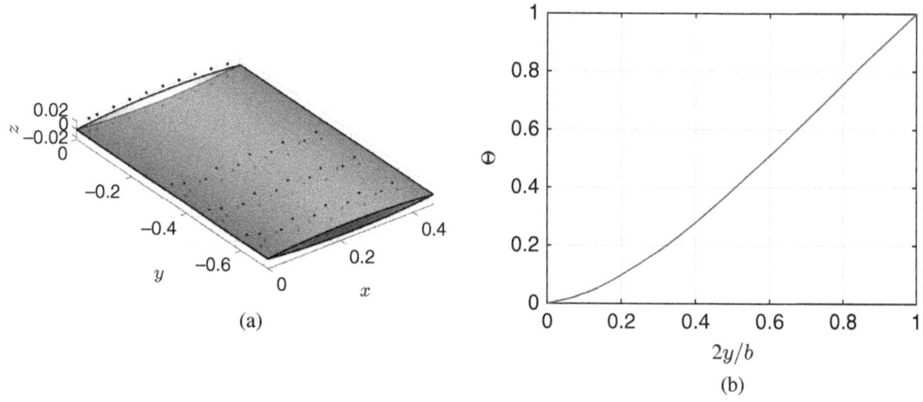

Figure 5.54 Wing and mode shape used by Lessing et al. (1960). (a) Half-wing and pressure taps and (b) First bending mode shape. Source: Lessing et al. (1960)/National Aeronautics and Space Administration/Public Domain.

The mode shape is given in graphical form by Lessing et al. (1960) and is reproduced here in Figure 5.54b, where Θ is the wing's deflection in the z direction. No deflections in the x or y directions are considered, as there was no twist and the tip amplitude was kept very small.

Lessing et al. (1960) carried out their experiments for a range of Mach numbers, from $M = 0.24$ to supersonic values. Here, we will only consider the $M = 0.24$ case since the SDPM used in this chapter is an incompressible approach. We start by setting up the grid in the same way we did for Example 5.12. The biconvex airfoil shape is given by

$$z_{p_i} = \pm 2(\bar{x}_{p_i} - \bar{x}_{p_i}^2)t_c$$

where $t_c = 0.05$ is the thickness of the airfoil and $\bar{x}_{p_i}$ are non-dimensional chordwise coordinates. The length of the wake is set to $10c_0$, the chordwise number of wake panels to $m_w = 10m$ and the chordwise spacing between the wake panels to c_0/m. We then calculate the influence coefficient matrices $\boldsymbol{A}_\phi$, $\boldsymbol{B}_\phi$ and $\boldsymbol{C}_\phi$.

The instantaneous position of the panel vertices is given by

$$z_{p_{i,j}}(t) = z_{p_{i,j}}^0 + \Theta(|y_{p_{i,j}}|)q(t) \tag{5.226}$$

where $z^0_{p_{i,j}}$ are the z coordinates of the wing's panel vertices in its undeformed shape, $\Theta(|y_{p_{i,j}}|)$ is the mode shape evaluated at the panel vertices by interpolating Figure 5.54b and $q(t)$ is the generalized coordinate. Since $\Theta(b/2) = 1$, the generalized coordinate is actually the tip displacement time history. The bending deflection is sinusoidal in time, such that

$$q(t) = q_1 \sin \omega_0 t$$

where $q_1 = 5$ mm is the amplitude of the tip displacement. The relative velocities between the wing and the fluid are given by

$$u_{m_{i,j}}(t) = U_\infty,\ v_{m_{i,j}}(t) = 0,\ w_{m_{i,j}}(t) = -\Theta(|y_{c_{i,j}}|)\dot{q}(t)$$

where $\Theta(|y_{c_{i,j}}|)$ is the mode shape evaluated at the control points. Taking the Fourier Transform of this latest expression, we obtain

$$\begin{aligned} &u_{m_{i,j}}(0) = U_\infty,\ v_{m_{i,j}}(0) = 0,\ w_{m_{i,j}}(0) = 0 \\ &u_{m_{i,j}}(\omega_0) = 0,\ v_{m_{i,j}}(\omega_0) = 0,\ w_{m_{i,j}}(\omega_0) = -\mathrm{i}\omega_0 \frac{q_1}{2\mathrm{i}}\Theta(|y_{c_{i,j}}|) \end{aligned} \tag{5.227}$$

since $w_{m_{i,j}}(t)$ is sinusoidal with zero mean and frequency ω_0.

First, we set $\omega = 0$ in Eq. (5.221) to obtain

$$\left(\boldsymbol{B}_\phi - \frac{1}{2}\boldsymbol{I} + \boldsymbol{C}_\phi \boldsymbol{P}_e(0)\boldsymbol{P}_c\right)\boldsymbol{\mu}(0) = U_\infty \boldsymbol{A}_\phi \boldsymbol{n}_z$$

which we solve for the steady doublet strength on the wing panels, $\boldsymbol{\mu}(0)$. We can then calculate the steady perturbation velocities $u_{i,j}(0)$, $v_{i,j}(0)$, $w_{i,j}(0)$ from Eq. (5.222). Next, we set $\omega = \omega_0$ in Eq. (5.221), such that

$$\left(\boldsymbol{B}_\phi - \frac{1}{2}\boldsymbol{I} + \boldsymbol{C}_\phi \boldsymbol{P}_e(\omega_0)\boldsymbol{P}_c\right)\boldsymbol{\mu}(\omega_0) = \boldsymbol{A}_\phi \boldsymbol{w}_m(\omega_0) \circ \boldsymbol{n}_z$$

from which we evaluate the oscillatory doublet strengths on the panels, $\boldsymbol{\mu}(\omega_0)$. Then, we calculate the oscillatory perturbation velocities $u_{i,j}(\omega_0)$, $v_{i,j}(\omega_0)$, $w_{i,j}(\omega_0)$ from Eq. (5.222). The complete spectrum of the doublet strength of panel i, j is given by

$$\mu_{i,j}(\omega) = \mu_{i,j}(0)\delta(\omega) + \mu_{i,j}(\omega_0)\delta(\omega_0 - \omega) + \mu^*_{i,j}(\omega_0)\delta(\omega_0 + \omega) \tag{5.228}$$

while the spectra of the perturbation velocities on the same panel are

$$\begin{aligned} u_{i,j}(\omega) &= u_{i,j}(0)\delta(\omega) + u_{i,j}(\omega_0)\delta(\omega - \omega_0) + u^*_{i,j}(\omega_0)\delta(\omega + \omega_0) \\ v_{i,j}(\omega) &= v_{i,j}(0)\delta(\omega) + v_{i,j}(\omega_0)\delta(\omega - \omega_0) + v^*_{i,j}(\omega_0)\delta(\omega + \omega_0) \\ w_{i,j}(\omega) &= w_{i,j}(0)\delta(\omega) + w_{i,j}(\omega_0)\delta(\omega - \omega_0) + w^*_{i,j}(\omega_0)\delta(\omega + \omega_0) \end{aligned} \tag{5.229}$$

We can substitute these spectra into Eq. (5.223) so that, after calculating the convolutions, the pressure coefficient distribution takes the form

$$\begin{aligned} c_{p_{i,j}}(\omega) = &\ c_{p_{i,j}}(0)\delta(\omega) + c_{p_{i,j}}(\omega_0)\delta(\omega - \omega_0) + c^*_{p_{i,j}}(\omega_0)\delta(\omega + \omega_0) \\ &+ c_{p_{i,j}}(2\omega_0)\delta(\omega - 2\omega_0) + c^*_{p_{i,j}}(2\omega_0)\delta(\omega + 2\omega_0) \end{aligned} \tag{5.230}$$

In order to compare the SDPM results with the measurements, we could ensure that the wing grid is such that there are control points on all the spanwise measurement positions but that would complicate the grid generation procedure. Alternatively, we can find the control points that lie closest to the spanwise positions of the measurement locations of expression (5.225) and interpolate linearly between them. Note that the SDPM does not place control points at midspan so that the pressures there are approximated by the pressures at the innermost control point location, $y_{c_{n/2+1}}$.

Figure 5.55 plots the mean pressure distribution around the wing at the four spanwise measurement locations, comparing the SDPM predictions to the experimental measurements. As the wing and motion are both symmetric, the pressure distributions are also symmetric across the upper and lower surfaces, except at the root where wall interference causes the experimental pressures to be slightly asymmetric. The SDPM underpredicts slightly the mean pressure distribution because the flow has been modelled as incompressible while the experimental Mach number was 0.24 and not zero.

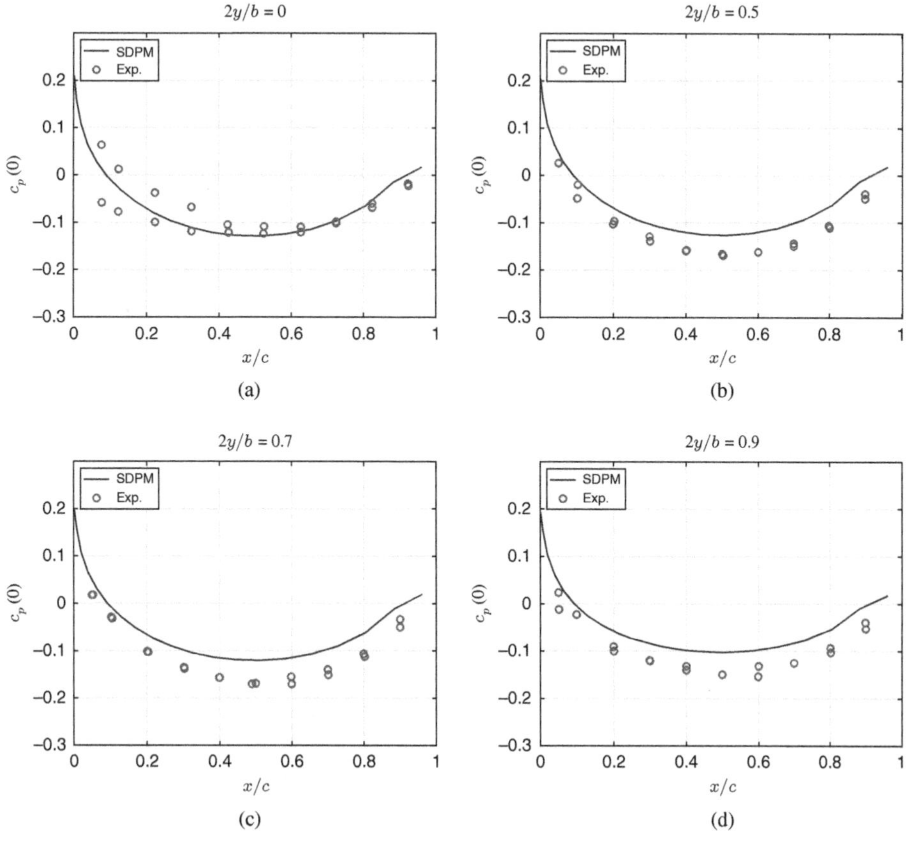

Figure 5.55 Mean pressure distribution predicted by the SDPM at $M_\infty = 0.24$. Experimental data by Lessing et al. (1960). (a) $2y/b = 0$, (b) $2y/b = 0.5$, (c) $2y/b = 0.7$, and (d) $2y/b = 0.9$. Source: Lessing et al. (1960), credit: NASA.

Lessing et al. (1960) give the oscillatory pressure distributions in the form of real and imaginary values of the pressure coefficient $c_p(\omega_0)$ *at the measurement positions. Figure (5.56) plots the real and imaginary parts of the pressure coefficient at the measurement locations,*

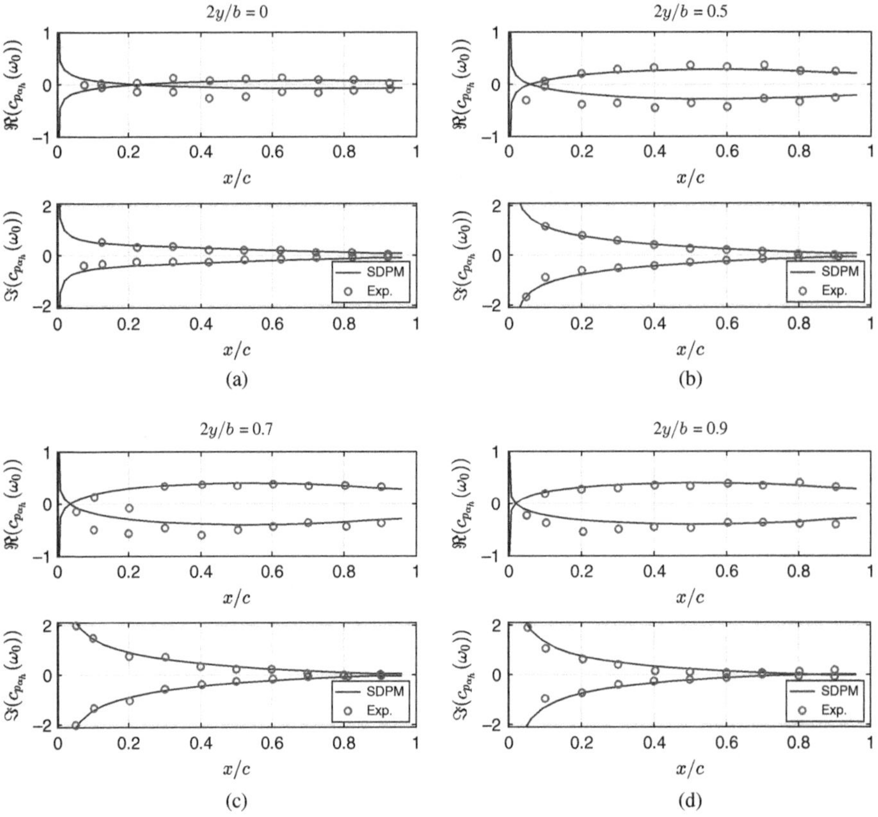

Figure 5.56 Real and imaginary parts of oscillatory pressure distribution predicted by the SDPM at $M_\infty = 0.24$. Experimental data by Lessing et al. (1960). (a) $2y/b = 0$, (b) $2y/b = 0.5$, (c) $2y/b = 0.7$, and (d) $2y/b = 0.9$. Source: Lessing et al. (1960), credit: NASA.

normalized by the motion-induced angle of attack at the tip

$$\alpha_h = \frac{\omega_0 q_1}{2\mathrm{i}U_\infty}$$

such that $c_{p_{\alpha_h}}(\omega_0) = c_p(\omega_0)/\alpha_h$. Only the frequency components at ω_0 are plotted since only these were reported in Lessing et al. (1960). It can be seen that there is very good agreement between the SDPM estimates and the experimental measurements. As the bending motion is symmetric with zero mean and the wing is not cambered, the pressure distributions are also symmetric. Furthermore, the magnitude of both the real and imaginary parts increases from root to tip, as the amplitude of the oscillation also increases monotonically with distance from the root. This example is solved by Matlab code `plungebend_SDPM_freq.m`.

5.8 Concluding Remarks

This chapter has shown that potential methods can be used to solve incompressible unsteady flows quickly and accurately, as long as their basic assumptions are respected.

Beyond their obvious benefits to aerodynamic analysis, such approaches can be used to set up flight dynamic and aeroelastic models, particularly when formulated in the frequency domain. Time-domain formulations are of interest when wing deflections are large, but the flow remains attached. This situation occurs in aircraft with very flexible wings with high aspect ratios, such as High Altitude Long Endurance aircraft. Both the VLM and the SDPM can be coupled with non-linear structural dynamic and flight dynamic models, leading to low-cost simulations that can represent complex aeroservoelastic behaviour (Murua et al. 2012). Potential flow techniques can also be used to predict the aerodynamic loads acting on rotating blades, such as helicopter or wind turbine blades (Hansen et al. 2006). In Chapter 6, we will extend the panel method methodology to compressible flow.

5.9 Exercises

1 Write a lifting line code to calculate the steady flow around a rectangular wing with any aspect ratio, taper twist and airfoil section.

2 Run the `impulsive_VLM_rect.m` code of example 5.6 for a flat rectangular wing with aspect ratio 6 and compare the resulting lift response to the experimental data published by Sawyer and Sullivan (1990). Note that there is significant uncertainty in this data set. The steady-state lift coefficient is too high for an $\mathcal{R} = 6$ wing at $\alpha = 10°$ and the lift responses have been filtered using a 15 Hz low-pass filter that has smoothed out their impulsive nature.

3 Repeat Example 5.10 for the other two wings tested by Queijo et al. (1956) and compare the VLM predictions to the experimental measurements.

4 Use the VLM and the analysis of Section 4.3 to model the encounter of a $\mathcal{R} = 4$ wing with a 1-cosine gust. Compare to the experimental results presented in Biler et al. (2019) for different gust ratios v_g/U_∞ and geometric angles of attack.

5 Repeat Example 5.4 using the SDPM and compare the resulting pressure distributions to the experimental pressures given in Kolbe and Boltz (1951).

6 Model the impulsive start of a wing with an elliptic planform using the SDPM and compare to the VLM and Jones/unsteady lifting line solutions.

7 Repeat Example 5.7 using the SDPM. Are the differences in the predictions of the two methods more important for the lift or for the thrust/drag? Do these differences reduce as you reduce the thickness of the airfoil?

8 Repeat Example 5.12 using the VLM. Does the VLM predict higher or lower aerodynamic loads than the SDPM?

References

Abbott, I.H. and Von Doenhoff, A.E. (1959). *Theory of Wing Sections: Including a Summary of Airfoil Data*. New York: Dover Publications, Inc.

Anderson, J.D. Jr. (1985). *Fundamentals of Aerodynamics*. McGraw-Hill International Editions.

Applin, Z.T. and Gentry, G.L. Jr. (1990). Experimental and theoretical aerodynamic characteristics of a high-lift semispan wing model. Technical Paper TP-2990. NASA.

Berci, M. (2016). Lift-deficiency functions of elliptical wings in incompressible potential flow: Jones' theory revisited. *Journal of Aircraft* 53 (2): 599–602.

Biler, H., Badrya, C., and Jones, A.R. (2019). Experimental and computational investigation of transverse gust encounters. *AIAA Journal* 57 (11): 4608–4622.

Boutet, J. and Dimitriadis, G. (2018). Unsteady lifting line theory using the Wagner function for the aerodynamic and aeroelastic modeling of 3D wings. *Aerospace* 5: 92.

Cleaver, D.J., Calderon, D.E., Wang, Z., and Gursul, I. (2016). Lift enhancement through flexibility of plunging wings at low Reynolds numbers. *Journal of Fluids and Structures* 64: 27–45.

Cook, M.V. (1997). *Flight Dynamics Principles*, 2e. Elsevier.

De Oro Fernández, E., Andrianne, T., and Dimitriadis, G. (2020). A database of flutter characteristics for simple low and medium aspect ratio wings at low speeds. *Proceedings of the AIAA SciTech 2020 Forum* number AIAA 2020-2172, Orlando, FL.

DeLaurier, J.D. and Harris, J.M. (1982). Experimental study of oscillating-wing propulsion. *Journal of Aircraft* 19 (5): 368–373.

Dewey, P.A., Boschitsch, B.M., Moored, K.W. et al. (2013). Scaling laws for the thrust production of flexible pitching panels. *Journal of Fluid Mechanics* 732: 29–46.

Dimitriadis, G. (2017). *Introduction to Nonlinear Aeroelasticity*. Chichester, West Sussex: Wiley.

Dimitriadis, G., Giannelis, N.F., and Vio, G.A. (2018). A modal frequency-domain generalised force matrix for the unsteady vortex lattice method. *Journal of Fluids and Structures* 76: 216–228.

Epton, M.A. and Magnus, A.E. (1990). PAN AIR- A Computer Program for Predicting Subsonic or Supersonic Linear Potential Flows About Arbitrary Configurations Using a Higher Order Panel Method. *Contractor Report CR-3251*. NASA.

Hansen, M., Sørensen, J., Voutsinas, S. et al. (2006). State of the art in wind turbine aerodynamics and aeroelasticity. *Progress in Aerospace Sciences* 42 (4): 285–330.

Heathcote, S., Wang, Z., and Gursul, I. (2003). Effect of spanwise flexibility on flapping wing propulsion. *Journal of Fluids and Structures* 24 (2): 183–199.

Hoerner, S.F. and Borst, H.V. (1985). *Fluid-Dynamic Lift: Practical Information on Aerodynamic and Hydrodynamic Lift*. Brick Town, NJ: Hoerner Fluid Dynamics.

Ichikawa, M. (2012). Analytical expression of induced drag for a finite elliptic wing. *Journal of Aircraft* 33 (3): 632–634.

Izraelevitz, J., Zhu, Q., and Triantafyllou, M.S. (2017). State-space adaptation of unsteady lifting line theory: twisting/flapping wings of finite span. *AIAA Journal* 55 (4): 1279–1294.

Jones, R.T. (1939). The Unsteady Lift of a Finite Wing. *Technical Report TN-682*. NACA.

Jones, R.T. (1940). The Unsteady Lift of a Wing of Finite Aspect Ratio. *Technical Report TR-681*. NACA.

Jumper, E.J. and Hugo, R.J. (1994). Loading characteristics of finite wings undergoing rapid unsteady motions: a theoretical treatment. *Journal of Aircraft* 31 (3): 495–502.

Katz, J. and Plotkin, A. (2001). *Low Speed Aerodynamics*. Cambridge University Press.

Kolbe, C.D. and Boltz, F.W. (1951). The forces and pressure distribution at subsonic speeds of a plane wing having 45° of sweepback, an aspect ratio of 3, and a taper ratio of 0.5. Research Memorandum RM A51G31. NACA.

Küchemann, D. and Weber, J. (1953). *Aerodynamics of Propulsion*. New York: McGraw-Hill.

Kuethe, A.M. and Chow, C.Y. (1986). *Foundations of Aerodynamics - Basis of Aerodynamic Design*. Wiley.

Lee, C.S. (1977). Prediction of steady and unsteady performance of marine propellers with or without cavitation be numerical lifting-surface theory. PhD thesis. Massachusetts Institute of Technology.

Lessing, H.C., Troutman, J.L., and Menees, G.P. (1960). Experimental determination of the pressure distribution on a rectangular wing oscillating in the first bending mode for Mach numbers from 0.24 to 1.30. Technical Note TN-D-344. NASA.

Lichtenstein, J.H. and Williams, J.L. (1958). Low speed investigation of the effects of frequency and amplitude of oscillation in sideslip on the lateral stability derivatives of 60° Delta wing, a 45° sweptback wing, and an unswept wing. Research Memorandum RM L58B26. NACA.

Morino, L. and Gennaretti, M. (1992). Boundary integral equation methods for aerodynamics. In: *Computational Nonlinear Mechanics in Aerospace Engineering, Progress in Astronautics and Aeronautics*, vol. 146 (ed. S. Atluri), 279–320. AIAA.

Murua, J., Palacios, R., and Graham, J.M.R. (2012). Applications of the unsteady vortex-lattice method in aircraft aeroelasticity and flight dynamics. *Progress in Aerospace Sciences* 55: 46–72.

Phlips, P.J., East, R.A., and Pratt, N.H. (1981). An unsteady lifting line theory of flapping wings with application to the forward flight of birds. *Journal of Fluid Mechanics* 112: 97–125.

Queijo, M.J., Fletcher, H.S., Marple, C.G., and Hughes, F.M. (1956). Preliminary measurements of the aerodynamic yawing derivatives of a triangular, a swept, and an unswept wing performing pure yawing oscillations, with a description of the instrumentation employed. Research Memorandum RM L55L14. NACA.

Rao, S.S. (2004). *Mechanical Vibrations*, 4e. Pearson Prentice Hall, International Edition.

Reissner, E. (1947). Effect of Finite Span on the Airload Distributions for Oscillating Wings: I - Aerodynamic Theory of Oscillating Wings of Finite Span. *Technical Report TN-1194*. NACA.

Reissner, E. and Stevens, J.E. (1947). Effect of Finite Span on the Airload Distributions for Oscillating Wings: II - Methods of Calculation and Examples of Application. *Technical Report TN-1195*. NACA.

Sawyer, R.S. and Sullivan, J.P. (1990). Lift response of a rectangular wing undergoing a step change in forward speed. *AIAA Journal* 28 (7): 1306–1307.

Shyy, W., Aono, H., Kang, C.-k., and Liu, H. (2013). *An Introduction to Flapping Wing Aerodynamics*. New York: Cambridge University Press.

Simpson, R.J.S. and Palacios, R. (2013). Induced-drag calculations in the unsteady vortex lattice method. *AIAA Journal* 51 (7): 1775–1779.

Spedding, G.R., Rayner, J.M.V., and Pennycuick, C.J. (1984). Momentum and energy in the wake of a pigeon (*Columba Livia*) in slow flight. *Journal of Experimental Biology* 111: 81–102.

Stanford, B.K. and Beran, P.S. (2010). Analytical sensitivity analysis of an unsteady vortex-lattice method for flapping-wing optimization. *Journal of Aircraft* 47 (2): 647–662.

Taylor, G.K., Nudds, R.L., and Thomas, A.L.R. (2003). Flying and swimming animals cruise at a Strouhal number tuned for high power efficiency. *Nature* 425: 707–711.

Triantafyllou, G., Triantafyllou, M., and Grosenbaugh, M. (1993). Optimal thrust development in oscillating foils with application to fish propulsion. *Journal of Fluids and Structures* 7 (2): 205–224.

von Ellenrieder, K.D., Parker, K., and Soria, J. (2003). Flow structures behind a heaving and pitching finite-span wing. *Journal of Fluid Mechanics* 490: 129–138.

6

Unsteady Compressible Flow

6.1 Introduction

Steady and unsteady compressible flow around wings and bodies is of paramount importance to aircraft design, since modern airliners cruise at high subsonic and transonic conditions. Potential flow methods can be used to obtain quick but reliable estimates of the steady and unsteady pressure distributions acting on such aircraft. Steady aerodynamic predictions are useful for preliminary performance calculations, while unsteady simulations are necessary for aeroelasticity, flight dynamics and gust response calculations. This chapter will concentrate on methods for finite wings but an example will also be given of how these techniques can be adapted to 2D airfoils. For subsonic flow, two main approaches will be presented: the Doublet Lattice Method (DLM), which is the industrial standard technique for aeroelasticity and flight dynamics, and the compressible Source and Doublet Panel Method (SDPM). For supersonic flow, the Mach box and Mach panel approaches will be discussed. Finally, the chapter will close with extensions and corrections of the DLM and SDPM for transonic flow. Only inviscid, irrotational and attached flow will be considered.

6.2 Steady Subsonic Potential Flow

In Section 2.4, it was shown that under inviscid, irrotational, isentropic and small disturbance assumptions, unsteady compressible flow is governed by the linearised small disturbance equation (2.143)

$$\beta^2\phi_{xx} + \phi_{yy} + \phi_{zz} - \frac{2M_\infty}{a_\infty}\phi_{xt} - \frac{1}{a_\infty^2}\phi_{tt} = 0 \tag{6.1}$$

where ϕ is the perturbation potential induced by the shape of the body and the motion, while the free stream flows along the x direction with speed U_∞, Mach number M_∞ and speed of sound a_∞. The compressibility factor β is defined as $\beta = \sqrt{1 - M_\infty^2}$. If we further assume that the flow is steady, then this equation simplifies to

$$\beta^2\phi_{xx} + \phi_{yy} + \phi_{zz} = 0 \tag{6.2}$$

Unsteady Aerodynamics: Potential and Vortex Methods, First Edition. Grigorios Dimitriadis.

Companion website: www.wiley.com/go/dimitriadis/unsteady_aerodynamics

which is very similar to Laplace's equation (2.22), except for the β coefficient. In fact, Eq. (6.2) can be transformed into Laplace's equation using the Prandtl–Glauert transformation (see for example (Anderson Jr. 1990))

$$\xi = \frac{x}{\beta}, \quad \eta = y, \quad \zeta = z \tag{6.3}$$

and then calculating the derivatives in Eq. (6.2) in terms of the new variables ξ, η, β. Using the chain rule, the first derivatives become

$$\phi_x = \frac{\partial \phi}{\partial \xi}\frac{\partial \xi}{\partial x} = \frac{1}{\beta}\phi_\xi, \quad \phi_y = \phi_\eta, \quad \phi_z = \phi_\zeta \tag{6.4}$$

and the second derivatives become

$$\phi_{xx} = \frac{\partial}{\partial \xi}\left(\frac{1}{\beta}\phi_\xi\right)\frac{\partial \xi}{\partial x} = \frac{1}{\beta^2}\phi_{\xi\xi}, \quad \phi_{yy} = \phi_{\eta\eta}, \quad \phi_{zz} = \phi_{\zeta\zeta} \tag{6.5}$$

Substituting back into the steady small disturbance equation (6.2), we obtain

$$\phi_{\xi\xi} + \phi_{\eta\eta} + \phi_{\zeta\zeta} = 0 \tag{6.6}$$

which is identical to Laplace's equation and can be solved using the methods of Chapter 5. In essence, we are transforming the compressible problem in x, y, z space to an incompressible problem in the scaled ξ, η, ζ space. Green's third identity in transformed space is obtained from Eq. (5.169), after replacing $\mathbf{x} = (x, y, z)$ by $\boldsymbol{\xi} = (\xi, \eta, \zeta)$, that is

$$\begin{aligned} E(\boldsymbol{\xi})\phi(\boldsymbol{\xi}) = &-\frac{1}{4\pi}\int_\Sigma \sigma(\boldsymbol{\xi}_s)\frac{1}{r(\boldsymbol{\xi}, \boldsymbol{\xi}_s)}\mathrm{d}\Sigma \\ &+\frac{1}{4\pi}\int_\Sigma \mu(\boldsymbol{\xi}_s)\mathbf{n}(\boldsymbol{\xi}_s)\cdot\bar{\nabla}_s\left(\frac{1}{r(\boldsymbol{\xi}, \boldsymbol{\xi}_s)}\right)\mathrm{d}\Sigma \\ &+\frac{1}{4\pi}\int_{\Sigma_w} \mu_w(\boldsymbol{\xi}_s)\mathbf{n}(\boldsymbol{\xi}_s)\cdot\bar{\nabla}_s\left(\frac{1}{r(\boldsymbol{\xi}, \boldsymbol{\xi}_s)}\right)\mathrm{d}\Sigma \end{aligned} \tag{6.7}$$

where $r(\boldsymbol{\xi}, \boldsymbol{\xi}_s) = \sqrt{(\xi - \xi_s)^2 + (\eta - \eta_s)^2 + (\zeta - \zeta_s)^2}$, $\Sigma(\boldsymbol{\xi}_s) = 0$ describes the transformed surface of the wing, $\Sigma_w(\boldsymbol{\xi}_s) = 0$ that of the wake, $\boldsymbol{\xi}_s$ is any point on the transformed wing or wake, $\mathbf{n}(\boldsymbol{\xi}_s)$ is a unit vector normal to $\Sigma(\boldsymbol{\xi}_s) = 0$ and $\bar{\nabla}_s = (\partial/\partial\xi_s, \partial/\partial\eta_s, \partial/\partial\zeta_s)$. The source and doublet strength are adapted in the same way from Eqs. (5.167) and (5.168) such that

$$\sigma(\boldsymbol{\xi}_s) = \mathbf{n}(\boldsymbol{\xi}_s)\cdot\bar{\nabla}_s\phi(\boldsymbol{\xi}_s) \tag{6.8}$$

$$\mu(\boldsymbol{\xi}_s) = \phi(\boldsymbol{\xi}_s) \tag{6.9}$$

In order to solve Eq. (6.7) for $\mu(\boldsymbol{\xi}_s)$, first, we need to calculate $\sigma(\boldsymbol{\xi}_s)$ from the impermeability boundary condition. As the wing is thick, we will use the zero mass flux condition of Eq. (2.157). Enforcing steady conditions, it can be written as

$$\rho_\infty\left((\beta^2\phi_x, \quad \phi_y, \quad \phi_z) + \mathbf{Q}_\infty\right)\cdot\mathbf{n}(\mathbf{x}_s) = 0 \tag{6.10}$$

We also recall from Section 2.2.3 that the definition of the normal to the surface is

$$\mathbf{n}(\mathbf{x}_s) = \frac{\nabla S(\mathbf{x}_s)}{||\nabla S(\mathbf{x}_s)||}$$

so that Eq. (6.10) can also be written as

$$\rho_\infty\left((\beta^2\phi_x, \phi_y, \phi_z) + \mathbf{Q}_\infty\right)\cdot\nabla S(\mathbf{x}_s) = 0 \tag{6.11}$$

after multiplying by $||\nabla S(\mathbf{x}_s)||$. This latest expression is written in compressible space x, y, z and must be transformed to incompressible space. First, we note that the equation of the surface in the transformed coordinates is

$$\Sigma(\xi, \eta, \zeta) = S(\beta\xi, \eta, \zeta) = 0$$

Using the fundamental properties of differentiation,

$$(\bar{\nabla}\Sigma)^T = \left(\frac{\partial S}{\partial(\beta\xi)}\frac{\partial(\beta\xi)}{\partial\xi}, \frac{\partial S}{\partial\eta}, \frac{\partial S}{\partial\zeta}\right)^T = \left(\beta\frac{\partial S}{\partial x}, \frac{\partial S}{\partial y}, \frac{\partial S}{\partial z}\right)^T = \boldsymbol{B}\nabla S^T$$

where

$$\boldsymbol{B} = \begin{pmatrix} \beta & 0 & 0 \\ 0 & 1 & 0 \\ 0 & 0 & 1 \end{pmatrix} \tag{6.12}$$

This equation can be solved for $\nabla S^T = \boldsymbol{B}^{-1}(\bar{\nabla}\Sigma)^T$, where

$$\boldsymbol{B}^{-1} = \begin{pmatrix} \frac{1}{\beta} & 0 & 0 \\ 0 & 1 & 0 \\ 0 & 0 & 1 \end{pmatrix}$$

Applying the Prandtl–Glauert transform to Eq. (6.11) leads to

$$\left(\left(\beta\phi_\xi, \phi_\eta, \phi_\zeta\right) + \mathbf{Q}_\infty\right)\boldsymbol{B}^{-1}(\bar{\nabla}\Sigma(\xi_s))^T = 0$$

or noting that $\left(\beta\phi_\xi, \phi_\eta, \phi_\zeta\right)\boldsymbol{B}^{-1} = \bar{\nabla}\phi$,

$$\bar{\nabla}\phi(\xi_s)(\bar{\nabla}\Sigma(\xi_s))^T = -\mathbf{Q}_\infty\boldsymbol{B}^{-1}(\bar{\nabla}\Sigma(\xi_s))^T$$

Finally, dividing throughout by $||\bar{\nabla}\Sigma(\xi_s)||$ results in

$$\bar{\nabla}\phi(\xi_s)\cdot\mathbf{n}(\xi_s) = -\mathbf{Q}_\infty\boldsymbol{B}^{-1}\mathbf{n}^T(\xi_s) \tag{6.13}$$

since $\mathbf{n}(\xi_s) = \bar{\nabla}\Sigma(\xi_s)/||\bar{\nabla}\Sigma(\xi_s)||$ is the unit vector normal to the transformed surface at point ξ_s. Substituting Eq. (6.13) into (6.8) gives the required source strength

$$\sigma(\xi_s) = -\mathbf{Q}_\infty\boldsymbol{B}^{-1}\mathbf{n}^T(\xi_s) = -(U_\infty, V_\infty, W_\infty)\cdot\left(\frac{n_\xi}{\beta}, n_\eta, n_\zeta\right) \tag{6.14}$$

Now that we have a value for $\sigma(\xi_s)$ and we can substitute it in Eq. (6.7) and solve the latter for the doublet strength $\mu(\xi_s)$, given that $\mu_w(\xi_s)$ is evaluated in terms of $\mu(\xi_s)$ using the Kutta condition. As we have not applied a transformation to the potential,

$$\mu(\xi_s) = \mu(\mathbf{x}_s) \tag{6.15}$$

This means that the potential on the transformed surface is equal to the potential on the corresponding points of the original geometry. We can then proceed to calculate the perturbation velocities; this can be done either in the original or the transformed coordinates. If we choose to work in the original coordinate system, we can use Eq. (5.178), but we have to adapt the third of these expressions to reflect the zero normal mass flux boundary condition (6.10), which can be written using the $\boldsymbol{B}$ matrix as

$$(\boldsymbol{B}\boldsymbol{B}\mathbf{n}^T(\mathbf{x}_s))^T\cdot(\phi_x(\mathbf{x}_s), \phi_y(\mathbf{x}_s), \phi_z(\mathbf{x}_s)) = -\mathbf{Q}_\infty\cdot\mathbf{n}(\mathbf{x}_s)$$

Then, Eq. (5.178) become

$$\begin{aligned}\boldsymbol{\tau}_m(\mathbf{x}_s)\cdot(\phi_x(\mathbf{x}_s),\,\phi_y(\mathbf{x}_s),\,\phi_z(\mathbf{x}_s)) &= \boldsymbol{\tau}_m(\mathbf{x}_s)\cdot\nabla\mu(\mathbf{x}_s)\\ \boldsymbol{\tau}_n(\mathbf{x}_s)\cdot(\phi_x(\mathbf{x}_s),\,\phi_y(\mathbf{x}_s),\,\phi_z(\mathbf{x}_s)) &= \boldsymbol{\tau}_n(\mathbf{x}_s)\cdot\nabla\mu(\mathbf{x}_s)\\ (\mathcal{B}\mathcal{B}\mathbf{n}^T(\mathbf{x}_s))^T\cdot(\phi_x(\mathbf{x}_s),\,\phi_y(\mathbf{x}_s),\,\phi_z(\mathbf{x}_s)) &= -\mathbf{Q}_\infty\cdot\mathbf{n}(\mathbf{x}_s)\end{aligned} \tag{6.16}$$

Alternatively, we can choose to work in transformed coordinates and calculate ϕ_ξ, ϕ_η, ϕ_ζ. Then we must apply the Prandtl–Glauert transformation to Eq. (5.178), such that

$$\begin{aligned}\boldsymbol{\tau}_m(\xi_s)\cdot(\phi_\xi(\xi_s),\,\phi_\eta(\xi_s),\,\phi_\zeta(\xi_s)) &= \boldsymbol{\tau}_m(\xi_s)\cdot\nabla\mu(\xi_s)\\ \boldsymbol{\tau}_n(\xi_s)\cdot(\phi_\xi(\xi_s),\,\phi_\eta(\xi_s),\,\phi_\zeta(\xi_s)) &= \boldsymbol{\tau}_n(\xi_s)\cdot\nabla\mu(\xi_s)\\ \mathbf{n}(\xi_s)\cdot(\phi_\xi(\xi_s),\,\phi_\eta(\xi_s),\,\phi_\zeta(\xi_s)) &= \sigma(\xi_s)\end{aligned} \tag{6.17}$$

We then need to calculate the perturbation velocities in the original coordinates from Eq. (6.4). The two approaches apply the same boundary condition and give identical results. Finally, we obtain the pressure distribution around the wing from the steady version of the second-order compressible pressure coefficient (2.153).

Example 6.1 *Use the Prandtl–Glauert transformation and the SDPM to calculate the pressure distribution around a swept untapered wing in steady subsonic compressible flow.*

Lockman and Seegmiller (1983) published a set of data from a wind tunnel experiment on a swept untapered and untwisted wing with aspect ratio 3 and with a NACA 0012 cross-section. The semi-span model had a leading edge sweep angle of $\Lambda = 20°$, was instrumented with pressure sensors on its upper surface only and was tested at Mach numbers ranging from 0.5 to 0.84. As the airfoil was symmetric and the wing was tested at both positive and negative angles of attack, the complete pressure distribution around it was obtained. For example, when testing the wing at an angle of attack $\alpha = 2°$, only the pressure distribution on the suction side was measured but, when testing it at $\alpha = -2°$, the pressure distribution on the pressure side of the $\alpha = 2°$ case was recorded. The pressure taps were placed at spanwise stations lying at $2y/b = 0.25$, 0.5, 0.75, 0.775, 0.8, 0.9, but only the stations at 0.25, 0.5 and 0.775 covered the entire chordline. Figure 6.1 draws the semi-span wing and the positions of the pressure taps.

This test case will be modelled here using the transformation of Eq. (6.3) and the SDPM for $2m$ chordwise and n spanwise panels. We start by calculating the wing panel vertices

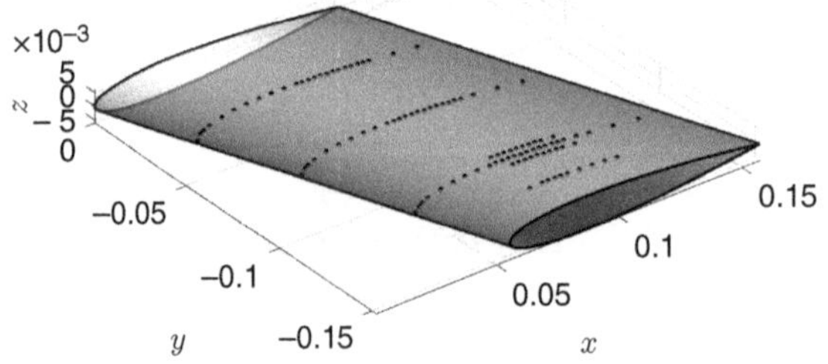

Figure 6.1 Half-wing tested by Lockman and Seegmiller (1983), showing the position of the pressure taps. Source: Adapted from Lockman and Seegmiller (1983).

$x_{p_{i,j}}, y_{p_{i,j}}, z_{p_{i,j}}$, *control points* $x_{c_{i,j}}, y_{c_{i,j}}, z_{c_{i,j}}$, *panel areas* $s_{i,j}$ *and panel normal vectors* $n_{x_{i,j}}, n_{y_{i,j}}, n_{z_{i,j}}$ *for* $i = 1, \dots, 2m$, $j = 1, \dots, n$, *exactly as in Example 5.11. Then, we set up the incompressible problem by defining*

$$\xi_{p_{i,j}} = \frac{x_{p_{i,j}}}{\beta}, \quad \eta_{p_{i,j}} = y_{p_{i,j}}, \quad \zeta_{p_{i,j}} = z_{p_{i,j}}$$

and calculate the control points $\xi_{c_{i,j}}, \eta_{c_{i,j}}, \zeta_{c_{i,j}}$ *and normal vectors* $n_{\xi_{i,j}}, n_{\eta_{i,j}}, n_{\zeta_{i,j}}$ *of the transformed panels. We also calculate the tangential vectors in the* ξ *direction* $\tau_{\xi\xi_{i,j}}, \tau_{\xi\eta_{i,j}}, \tau_{\xi\zeta_{i,j}}$ *from Eq. (5.182) and in the* η *direction* $\tau_{\eta\xi_{i,j}}, \tau_{\eta\eta_{i,j}}, \tau_{\eta\zeta_{i,j}}$ *from Eq. (5.183).*

The wake is also calculated in incompressible space such that the leading edge of the wake panels lies on $\xi_{w_{1,j}} = \xi_{p_{1,j}}, \eta_{w_{1,j}} = \eta_{p_{1,j}}, \zeta_{w_{1,j}} = \zeta_{p_{1,j}}$ *and the trailing edge on*

$$\xi_{w_{2,j}} = \xi_{p_{1,j}} + \frac{n_c c_0}{\beta}, \quad \eta_{w_{2,j}} = \eta_{w_{1,j}}, \quad \zeta_{w_{2,j}} = \zeta_{w_{1,j}}$$

for $j = 1, \dots, n$, *where* $n_c = 100$ *is the length of the wake in compressible chord lengths. As* $n_c c_0$ *is a very large distance, there is no need to transform it to incompressible space, we do this to be consistent. We also calculate the control points* $\xi_{cw_{i,j}}, \eta_{cw_{i,j}}, \zeta_{cw_{i,j}}$, *normal unit vectors,* $\mathbf{n}_{w_{i,j}}$ *and tangential unit vectors,* $\boldsymbol{\tau}_{\xi w_{i,j}}, \boldsymbol{\tau}_{\eta w_{i,j}}$ *of the wake panels, for* $i = 1, j = 1, \dots, n$.

As the wing can be placed at a small angle of attack, the free stream velocity is given by $\mathbf{Q}_\infty = Q_\infty(\cos\alpha, 0, \sin\alpha)$, *where* $Q_\infty = M_\infty a_\infty$ *is the magnitude of the free stream velocity and* $Q_\infty \sin\alpha \ll a_\infty$ *so that the flow in the z direction is not compressible. The values of the source strength on the control points that will ensure impermeability in compressible space are calculated from Eq. (6.14)*

$$\boldsymbol{\sigma} = -Q_\infty \left(\frac{\cos\alpha}{\beta} \mathbf{n}_\xi + \sin\alpha \mathbf{n}_\zeta \right) \tag{6.18}$$

where $\boldsymbol{\sigma}$ *is the* $2mn \times 1$ *column vector containing the source strengths of all the panels and* $\mathbf{n}_\xi$, $\mathbf{n}_\eta$ *and* $\mathbf{n}_\zeta$ *are the* $2mn \times 1$ *column vectors containing the* ξ, η *and* ζ *components of the unit normal vectors of the transformed wing panels at the control points. Then, we calculate the source* $\boldsymbol{A}_\phi$, *wing doublet* $\mathbf{B}_\phi$ *and wake doublet* $\mathbf{C}_\phi$ *influence coefficient matrices appearing in Eq. (5.175) using the procedure described in Example 5.11 and Eq. (5.186), recalling that* $B_{\phi_{I,I}} = 0$.

The Kutta condition is given by Eq. (5.187). Now, we can set up and solve Eq. (5.188)

$$\begin{pmatrix} \left(\boldsymbol{B}_\phi - \frac{1}{2}\boldsymbol{I}\right) & \boldsymbol{C}_\phi \\ \mathbf{K} & \boldsymbol{I}_n \end{pmatrix} \begin{pmatrix} \boldsymbol{\mu} \\ \boldsymbol{\mu}_w \end{pmatrix} = \begin{pmatrix} -\boldsymbol{A}_\phi \boldsymbol{\sigma} \\ \mathbf{0}_{n,1} \end{pmatrix} \tag{6.19}$$

in incompressible space for the values of the bound and wake doublet strengths, $\boldsymbol{\mu}$ *and* $\boldsymbol{\mu}_w$. *We can then calculate the velocities on the surface of the wing in incompressible space using Eq. (6.17) such that*

$$\begin{aligned} \tau_{m\xi_{i,j}} \phi_{\xi_{i,j}} + \tau_{m\eta_{i,j}} \phi_{\eta_{i,j}} + \tau_{m\zeta_{i,j}} \phi_{\zeta_{i,j}} &= \left.\frac{\partial \mu}{\partial \boldsymbol{\tau}_m}\right|_{i,j} \\ \tau_{n\xi_{i,j}} \phi_{\xi_{i,j}} + \tau_{n\eta_{i,j}} \phi_{\eta_{i,j}} + \tau_{n\zeta_{i,j}} \phi_{\zeta_{i,j}} &= \left.\frac{\partial \mu}{\partial \boldsymbol{\tau}_n}\right|_{i,j} \\ n_{\xi_{i,j}} \phi_{\xi_{i,j}} + n_{\eta_{i,j}} \phi_{\eta_{i,j}} + n_{\zeta_{i,j}} \phi_{\zeta_{i,j}} &= \sigma_{i,j} \end{aligned} \tag{6.20}$$

The compressible perturbation velocities are obtained by transforming back into compressible space using Eq. (6.4),

$$\phi_{x_{i,j}} = \frac{1}{\beta}\phi_{\xi_{i,j}}, \quad \phi_{y_{i,j}} = \phi_{\eta_{i,j}}, \quad \phi_{z_{i,j}} = \phi_{\zeta_{i,j}}$$

The pressure coefficient on the surface in compressible space can be calculated from the steady version of Eq. (2.153),

$$c_{p_{i,j}} = 1 - \frac{Q_{i,j}^2}{Q_\infty^2} + \frac{M_\infty^2}{Q_\infty^2}\phi_{x_{i,j}}^2 \tag{6.21}$$

for $i = 1, \dots, 2m$, $j = 1, \dots, n$, *where the total flow velocity on the control points is given simply by*

$$Q_{i,j} = \sqrt{(Q_\infty \cos\alpha + \phi_{x_{i,j}})^2 + \phi_{y_{i,j}}^2 + (Q_\infty \sin\alpha + \phi_{z_{i,j}})^2}$$

Finally, the aerodynamic forces acting on the panels can be evaluated from Eq. (5.194)

$$\mathbf{F}_{i,j} = -\frac{1}{2}\rho Q_\infty^2 c_{p_{i,j}} s_{i,j} \mathbf{n}_{i,j} \tag{6.22}$$

where $s_{i,j}$ *and* $\mathbf{n}_{i,j}$ *are the panel areas and unit normal vectors of the compressible, untransformed geometry. The lift and drag are evaluated from (5.195) and (5.196), respectively.*

As we noted at the end of Section 2.4.5, the range of validity of compressible potential solutions is restricted by the maximum local flow Mach number. For the present subsonic case, the solution is invalid if the local flow Mach number reaches or exceeds $M = 1$ *anywhere in the flow. It is therefore important to calculate this Mach number everywhere on the surface. The total flow velocity on the control points,* $Q_{i,j}$*, has already been calculated, but we also need to calculate the local speed of sound which, from adiabatic assumptions, is given by (Epton and Magnus 1990)*

$$a_{i,j} = a_\infty \left(1 - \frac{\gamma - 1}{2} M_\infty^2 \left(\frac{Q_{i,j}^2}{Q_\infty^2} - 1\right)\right) \tag{6.23}$$

Therefore, the local flow Mach number is

$$M_{i,j} = \frac{Q_{i,j}}{a_{i,j}}$$

At the end of the calculation, we check to see if any values of $M_{i,j}$ *have exceeded 1 in order to assess the validity of the solution and print out a warning if it is the case.*

We solve the problem using $m = 40$ *and* $n = 40$ *and interpolate linearly the resulting* $c_{p_{i,j}}$ *values in the spanwise direction in order to estimate the pressure coefficient at the measurement stations. Figure 6.2 plots the pressure coefficient at the measurement stations predicted by the compressible SDPM and compares it to the experimental data, for two values of the free stream Mach number. The pressure estimates obtained from the incompressible SDPM of the previous chapter are also plotted in order to demonstrate the need for compressible calculations. In both cases, the angle of attack is 2°. Figure 6.2a plots* c_p *for* $M_\infty = 0.601$*; it can be seen that the*

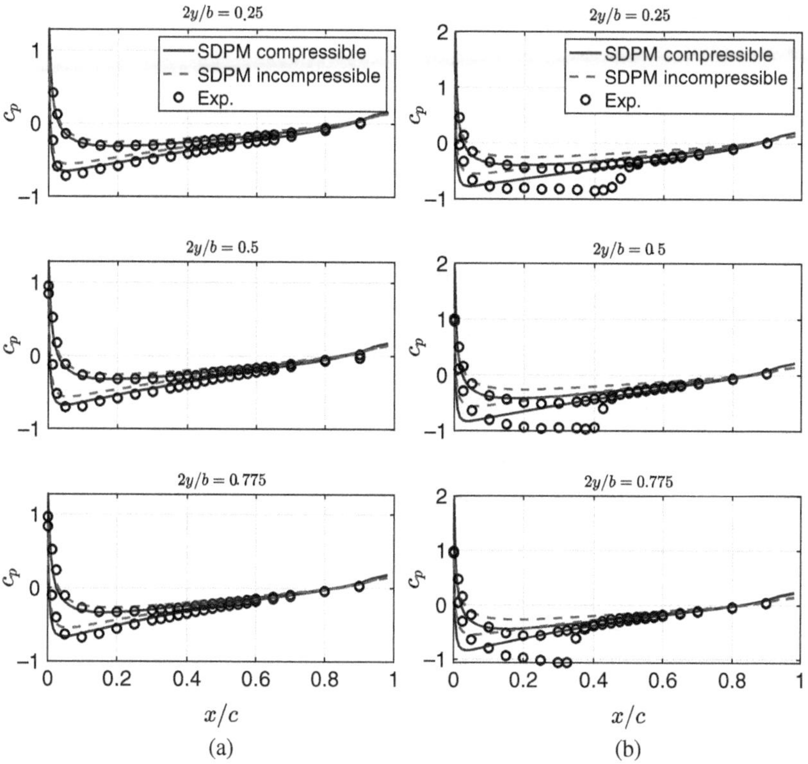

Figure 6.2 Pressure distribution predicted by the SDPM at three spanwise measurement stations for $\alpha = 2°$ and two values of the Mach number. Experimental data by Lockman and Seegmiller (1983). (a) $M_\infty = 0.601$. (b) $M_\infty = 0.794$. Source: Lockman and Seegmiller (1983), credit: NASA.

agreement between the pressures predicted by the compressible SDPM and the experimental measurements is quite good, although the suction peak is slightly underestimated. In this case, the local flow Mach number never exceeds unity. In Figure 6.2b, the free stream Mach number is $M_\infty = 0.794$ and the agreement between the SDPM and the experiment is very poor. In particular, the experiment shows that the pressure coefficient is quite flat over a large part of the suction side, before dropping abruptly. In fact, the flow is locally supersonic up to the location of the drop, where a shock wave forms. This phenomenon cannot be predicted by linear potential flow methods; it can only be predicted using non-linear approaches, such as solutions of the full potential equation (2.131) or of the transonic small disturbance equation (2.135). This example is solved by Matlab code `steady_subsonic_SDPM.m`.

Subsonic flow that features locally supersonic velocities is known as transonic or supercritical. Conversely, subcritical flow is purely subsonic and shock-free. For the moment, we will concentrate on the latter; transonic flows will be addressed in Section 6.5.

6.3 Unsteady Subsonic Potential Flow

Recall that unsteady subsonic potential flow is governed by the linearized small disturbance equation (6.1)

$$\beta^2\phi_{xx} + \phi_{yy} + \phi_{zz} - \frac{2M_\infty}{a_\infty}\phi_{xt} - \frac{1}{a_\infty^2}\phi_{tt} = 0 \tag{6.24}$$

where M_∞, a_∞ are the free stream Mach number and speed of sound, $\beta^2 = 1 - M_\infty^2$ and $\phi(x, y, z, t)$ is the perturbation potential. Once the potential distribution on the surface of the body is known, the pressure acting on it can be calculated from the second-order compressible unsteady Bernoulli equation (2.153)

$$c_p = 1 - \frac{Q^2}{Q_\infty^2} + \frac{M_\infty^2}{Q_\infty^2}\phi_x^2 - \frac{2}{Q_\infty^2}\phi_t + \frac{M_\infty^2}{Q_\infty^4}\phi_t^2 + \frac{2M_\infty^2}{Q_\infty^3}\phi_x\phi_t \tag{6.25}$$

In Section 2.5, we showed that the moving source and doublet are solutions to the linearised small disturbance equation. The perturbation potential induced at point $\mathbf{x} = (x, y, z)$ by a source or a doublet lying at $\mathbf{x}_0 = (x_0, y_0, z_0)$ is given by Eqs. (2.178) and (2.179) as

$$\phi_s(\mathbf{x}, \mathbf{x}_0, t) = -\frac{1}{4\pi}\frac{\sigma\left(t - \tau(\mathbf{x}, \mathbf{x}_0)\right)}{r_\beta(\mathbf{x}, \mathbf{x}_0)} \tag{6.26}$$

$$\phi_d(\mathbf{x}, \mathbf{x}_0, t) = \frac{1}{4\pi}\mathbf{n} \cdot \nabla\frac{\mu\left(t - \tau(\mathbf{x}, \mathbf{x}_0)\right)}{r_\beta(\mathbf{x}, \mathbf{x}_0)} \tag{6.27}$$

where $\sigma(x, y, z, t)$ is the source strength, $\mu(x, y, z, t)$ is the doublet strength and

$$r_\beta(\mathbf{x}, \mathbf{x}_0) = \sqrt{\left(x - x_0\right)^2 + \beta^2\left(y - y_0\right)^2 + \beta^2\left(z - z_0\right)^2} \tag{6.28}$$

$$\tau(\mathbf{x}, \mathbf{x}_0) = \frac{-M_\infty(x - x_0) + r_\beta(\mathbf{x}, \mathbf{x}_0)}{a_\infty\beta^2} \tag{6.29}$$

The linearised small disturbance equation is most often solved in the frequency domain. Taking the Fourier Transform of Eq. (6.24) leads to

$$\beta^2\phi_{xx}(\omega) + \phi_{yy}(\omega) + \phi_{zz}(\omega) - \frac{2\mathrm{i}\omega M_\infty}{a_\infty}\phi_x(\omega) + \frac{\omega^2}{a_\infty^2}\phi(\omega) = 0 \tag{6.30}$$

where ω is the frequency. Similarly, the potentials induced by the moving source and the moving doublet become

$$\phi_s(\mathbf{x}, \mathbf{x}_0, \omega) = -\frac{\sigma(\mathbf{x}_0, \omega)}{4\pi}\frac{\mathrm{e}^{-\mathrm{i}\omega\tau(\mathbf{x},\mathbf{x}_0)}}{r_\beta(\mathbf{x}, \mathbf{x}_0)} \tag{6.31}$$

$$\phi_d(\mathbf{x}, \mathbf{x}_0, \omega) = \frac{\mu(\mathbf{x}_0, \omega)}{4\pi}\mathbf{n} \cdot \nabla\frac{\mathrm{e}^{-\mathrm{i}\omega\tau(\mathbf{x},\mathbf{x}_0)}}{r_\beta(\mathbf{x}, \mathbf{x}_0)} \tag{6.32}$$

It can be verified that Eqs. (6.31) and (6.32) are indeed solutions to Eq. (6.30) by calculating the derivatives of the former and substituting them in the latter; this verification is left as an exercise for the reader. In the next two sections, we will demonstrate two numerical methods for solving the linearised small disturbance equation in the frequency domain, the DLM and the compressible SDPM.

6.3.1 The Doublet Lattice Method

The DLM (Albano and Rodden 1969; Kalman et al. 1971) is the industrial standard for unsteady aerodynamics at subsonic compressible airspeeds. It is a linearised, frequency-domain, quasi-fixed lifting surface method, like the frequency-domain version of the incompressible VLM so that it models the wing as a flat sheet with zero thickness. The impermeability boundary condition is applied by relating directly the relative velocity between the wing and the flow to the pressure jump across the sheet. In order to do this, the linearised small disturbance equation is re-written in terms of pressure instead of potential (Blair 1994; Gülçat 2016). We will carry out this derivation directly in the frequency domain. First, we multiply Eq. (6.30) by $\rho_\infty U_\infty$, differentiate it with respect to x and re-arrange the order of integration to obtain

$$\rho_\infty U_\infty \left(\beta^2 \frac{\partial^2}{\partial x^2}\phi_x(\omega) + \frac{\partial^2}{\partial y^2}\phi_x(\omega) + \frac{\partial^2}{\partial z^2}\phi_x(\omega) - \frac{2i\omega M_\infty}{a_\infty}\frac{\partial}{\partial x}\phi_x(\omega) + \frac{\omega^2}{a_\infty^2}\frac{\partial}{\partial x}\phi(\omega) \right) = 0 \tag{6.33}$$

Next, we multiply Eq. (6.30) by $\rho_\infty i\omega$

$$\rho_\infty i\omega \left(\beta^2 \frac{\partial^2}{\partial x^2}\phi(\omega) + \frac{\partial^2}{\partial y^2}\phi(\omega) + \frac{\partial^2}{\partial z^2}\phi(\omega) - \frac{2i\omega M_\infty}{a_\infty}\frac{\partial}{\partial x}\phi(\omega) + \frac{\omega^2}{a_\infty^2}\phi(\omega) \right) = 0$$

and we add this result to Eq. (6.33) such that

$$\begin{aligned}
&\beta^2 \frac{\partial^2}{\partial x^2}\left(\rho_\infty U_\infty \phi_x(\omega) + \rho_\infty i\omega\phi(\omega)\right) + \frac{\partial^2}{\partial y^2}\left(\rho_\infty U_\infty \phi_x(\omega) + \rho_\infty i\omega\phi(\omega)\right) \\
&+ \frac{\partial^2}{\partial z^2}\left(\rho_\infty U_\infty \phi_x(\omega) + \rho_\infty i\omega\phi(\omega)\right) - \frac{2M_\infty}{a_\infty} i\omega \frac{\partial}{\partial x}\left(\rho_\infty U_\infty \phi_x(\omega) + \rho_\infty i\omega\phi(\omega)\right) \\
&+ \frac{\omega^2}{a_\infty^2}\left(\rho_\infty U_\infty \phi_x(\omega) + \rho_\infty i\omega\phi(\omega)\right) = 0
\end{aligned} \tag{6.34}$$

Since the DLM is a linearised approach, we will use the linearised pressure coefficient of Eq. (2.154). In the frequency domain, this pressure coefficient becomes

$$c_p(\omega) = -\frac{2}{U_\infty}\phi_x(\omega) - \frac{2}{U_\infty^2} i\omega\phi(\omega) \tag{6.35}$$

We substitute for c_p from its definition (2.151) to obtain

$$p_\infty - p(\omega) = \rho_\infty U_\infty \phi_x(\omega) + \rho_\infty i\omega\phi(\omega) \tag{6.36}$$

where, in the present case, $Q_\infty = U_\infty$. Substituting this latest expression into Eq. (6.34) and defining $\bar{p}(\omega) = p_\infty - p(\omega)$ yields

$$\beta^2 \bar{p}_{xx}(\omega) + \bar{p}_{yy}(\omega) + \bar{p}_{zz}(\omega) - \frac{2M_\infty}{a_\infty} i\omega\bar{p}_x(\omega) + \frac{\omega^2}{a_\infty^2}\bar{p}(\omega) = 0 \tag{6.37}$$

which is the frequency-domain linearised small disturbance equation written in terms of the pressure difference $\bar{p}$. It is identical in form to Eq. (6.30) which means that it has source and doublet solutions of the form of expressions (6.31) and (6.32). These solutions are known as the pressure source and the pressure doublet, respectively.

The DLM aims to express directly the pressure difference to the relative velocity between the lifting surface and the flow. In order to do this, we return to the unsteady Bernoulli equation (6.36) and write it in the form

$$\phi_x(\omega) + \frac{\mathrm{i}\omega}{U_\infty}\phi(\omega) = \frac{\bar{p}(\omega)}{\rho_\infty U_\infty} \tag{6.38}$$

This expression is an inhomogeneous linear first-order ordinary differential equation. If we multiply both sides by $\mathrm{e}^{\mathrm{i}\omega x/U_\infty}$, we obtain

$$\frac{\partial}{\partial x}\left(\phi_x(\omega)\mathrm{e}^{\mathrm{i}\omega x/U_\infty}\right) = \frac{\bar{p}(\omega)}{\rho_\infty U_\infty}\mathrm{e}^{\mathrm{i}\omega x/U_\infty}$$

The potential can be obtained by integrating over x between $-\infty$ and x such that

$$\phi(\omega) = \frac{1}{\rho_\infty U_\infty}\int_{-\infty}^{x} \bar{p}(\omega)\mathrm{e}^{-\mathrm{i}\omega(x-\xi)/U_\infty}\,\mathrm{d}\xi \tag{6.39}$$

where ξ is an integration variable. The pressure $\bar{p}(\omega)$ in Eq. (6.39) is obtained from the solutions of Eq. (6.37), which are the pressure source and the pressure doublet. Since the wing has no thickness, it is modelled using only doublets; we therefore place a continuous distribution of doublets on the lifting surface, which lies on $z = 0$ and whose unit normal vector is $\mathbf{n} = (0, 0, 1)$. Consider a single pressure doublet placed on this surface at point $\mathbf{x}_s = (x_s, y_s, 0)$ and with strength $\psi(\mathbf{x}_s, \omega)$. By equivalence to Eq. (6.32), the pressure induced by this doublet at a general point $\mathbf{x}$ is given by

$$\bar{p}(\mathbf{x}, \mathbf{x}_s, \omega) = \frac{\psi(\mathbf{x}_s, \omega)}{4\pi}\frac{\partial}{\partial z}\left(\frac{\mathrm{e}^{-\mathrm{i}\omega\tau(\mathbf{x},\mathbf{x}_s)}}{r_\beta(\mathbf{x}, \mathbf{x}_s)}\right)$$

Substituting this latest expression into Eq. (6.39) leads to

$$\phi(\mathbf{x}, \mathbf{x}_s, \omega) = \frac{\psi(\mathbf{x}_s, \omega)}{4\pi\rho_\infty U_\infty}\int_{-\infty}^{x}\frac{\partial}{\partial z}\left(\frac{\mathrm{e}^{-\mathrm{i}\omega\tau(\xi,y,z,\mathbf{x}_s)}}{r_\beta(\xi, y, z, \mathbf{x}_s)}\right)\mathrm{e}^{-\mathrm{i}\omega(x-\xi)/U_\infty}\,\mathrm{d}\xi \tag{6.40}$$

Equation (6.40) gives the potential induced at any point $\mathbf{x}$ in the flow by a single pressure doublet lying at $\mathbf{x}_s$. If we place a continuous pressure doublet distribution on the wing's surface S, the total potential induced at point $\mathbf{x}$ will be

$$\phi(\mathbf{x}, \omega) = \frac{1}{4\pi\rho_\infty U_\infty}\int_S \psi(\mathbf{x}_s, \omega)\int_{-\infty}^{x}\frac{\partial}{\partial z}\left(\frac{\mathrm{e}^{-\mathrm{i}\omega\tau(\xi,y,z,\mathbf{x}_s)}}{r_\beta(\xi, y, z, \mathbf{x}_s)}\right)\mathrm{e}^{-\mathrm{i}\omega(x-\xi)/U_\infty}\,\mathrm{d}\xi\,\mathrm{d}S \tag{6.41}$$

If the point $\mathbf{x}$ lies on the surface, then, from Eq. (B.4)

$$\bar{p}(x_s, y_s, 0^\pm, \omega) = \mp\frac{\psi(x_s, y_s, \omega)}{2} \tag{6.42}$$

recalling that $\bar{p}(x, y, z) = p_\infty - p(x, y, z)$. The pressure jump across the surface is $\Delta p(x_s, y_s) = p(x_s, y_s, 0^-) - p(x_s, y_s, 0^+)$, or,

$$\begin{aligned}\Delta p(x_s, y_s) &= -\bar{p}(x_s, y_s, 0^-) + \bar{p}(x_s, y_s, 0^+)\\ &= -\frac{\psi(x_s, y_s, \omega)}{2} + \left(-\frac{\psi(x_s, y_s, \omega)}{2}\right) = -\psi(x_s, y_s, \omega)\end{aligned}$$

Substituting $\psi(\mathbf{x}_s, \omega) = -\Delta p(\mathbf{x}_s, \omega)$ into Eq. (6.41), we obtain

$$\phi(\mathbf{x}, \omega) = -\frac{1}{4\pi\rho_\infty U_\infty}\int_S \Delta p(\mathbf{x}_s, \omega)\int_{-\infty}^{x}\frac{\partial}{\partial z}\left(\frac{\mathrm{e}^{-\mathrm{i}\omega\tau(\xi,y,z,\mathbf{x}_s)}}{r_\beta(\xi, y, z, \mathbf{x}_s)}\right)\mathrm{e}^{-\mathrm{i}\omega(x-\xi)/U_\infty}\,\mathrm{d}\xi\,\mathrm{d}S \tag{6.43}$$

where $\mathbf{x}$ now lies on the surface. This latest equation relates directly the pressure jump across the surface to the potential at any point on the surface. It is to be solved for the unknown $\Delta p(\mathbf{x}_s, \omega)$ by imposing the impermeability boundary condition. Before we do that it is important to note that in all previous applications of 3D panel methods, we placed singularities not only on the wing but also in the wake. In the case of the DLM, we do not need to place doublets in the wake because, as stated in Kutta condition formulation 3 of Section 4.7, the pressure jump across the wake is equal to zero. Therefore, the wake does not contribute to Eq. (6.43). In essence, by not modelling the wake, we are imposing the Kutta condition implicitly. The practical consequence is a reduction in the computational cost of the DLM with respect to other panel techniques, although given the speed of modern computers, the difference in execution time is not necessarily significant.

In order to impose the impermeability boundary condition, we need to select the wing's motion. The flow equations are linearised and the wing is flat and lying on $z = 0$. Therefore, it makes sense to select linearised kinematics; if we limit the kinematics to pitching and plunging, the relative motion between the surface and the fluid is given by the Fourier transform of the linearised boundary condition of Eq. (2.46)

$$\phi_z(x_s, y_s, \omega) = -\left(U_\infty \alpha(\omega) - \mathrm{i}\omega h(\omega) + \mathrm{i}\omega\alpha(\omega)(x_s - x_f)\right) \tag{6.44}$$

where the plunge is defined as positive upwards here. Substituting for ϕ from Eq. (6.43), we obtain

$$w(\mathbf{x}, \omega) = -\frac{1}{4\pi\rho_\infty U_\infty}\int_S \Delta p(\mathbf{x}_s, \omega)\int_{-\infty}^{x}\frac{\partial^2}{\partial z^2}\left(\frac{\mathrm{e}^{-\mathrm{i}\omega\tau(\xi,y,z,\mathbf{x}_s)}}{r_\beta(\xi, y, z, \mathbf{x}_s)}\right)\mathrm{e}^{-\mathrm{i}\omega(x-\xi)/U_\infty}\,\mathrm{d}\xi\,\mathrm{d}S$$

where we have defined the upwash $w(\mathbf{x}, \omega) = \phi_z(x, y, \omega)$. The equation above is usually expressed in the form

$$w(\mathbf{x}, \omega) = -\frac{1}{4\pi\rho_\infty U_\infty}\int_S \Delta p(\mathbf{x}_s, \omega)K(\mathbf{x}, \mathbf{x}_s, \omega)\mathrm{d}S \tag{6.45}$$

where

$$K(\mathbf{x}, \mathbf{x}_s, \omega) = \int_{-\infty}^{x}\frac{\partial^2}{\partial z^2}\left(\frac{\mathrm{e}^{-\mathrm{i}\omega\tau(\xi,y,z,\mathbf{x}_s)}}{r_\beta(\xi, y, z, \mathbf{x}_s)}\right)\mathrm{e}^{-\mathrm{i}\omega(x-\xi)/U_\infty}\,\mathrm{d}\xi \tag{6.46}$$

is known as the Kernel function. Equation (6.45) is the fundamental equation of the DLM, as it relates directly the unknown pressure jump across the surface of the wing to the known motion-induced upwash. It is an integral equation that has no analytical solution so that the usual approach is to solve it numerically using a panelling scheme.

In the DLM method, the thin wing sheet is discretised into panels, as in the case of the VLM. The camber is ignored since the wing lies entirely on $z = 0$. On each panel, a pressure doublet line is placed on the quarter chord in the spanwise direction and the impermeability condition (6.45) is imposed at the 3/4 chord point of the panel. Figure 6.3a plots the discretisation of a complete wing, which is similar to that of Figure 5.15a for the VLM, except that there are only spanwise doublet lines on the 1/4 chord of each panel, there are no chordwise doublet lines and no wake. The panel numbering schemes are identical to those presented in Figure 5.16. Figure 6.3b plots a single panel, showing the positions of the doublet line and of the control point.

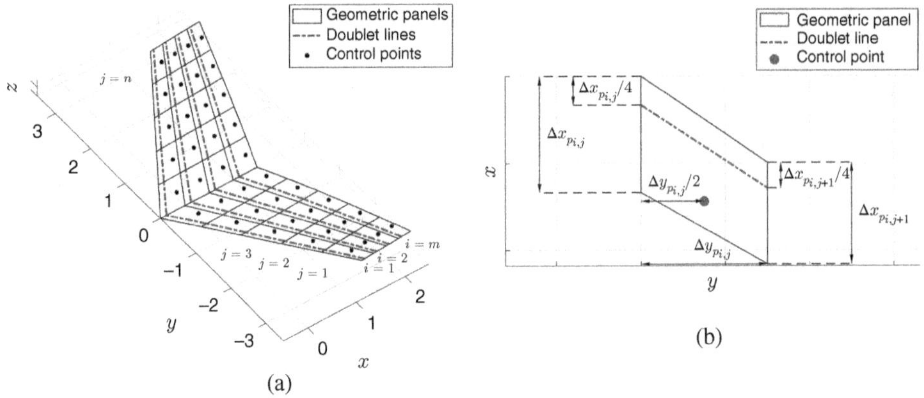

Figure 6.3 Discretization scheme for the Doublet Lattice Method. (a) Complete wing and (b) single panel.

Selecting m chordwise and n spanwise panels, the notation is similar to the one used for the VLM:

- The coordinates of the vertices of the geometric panels are denoted by $\mathbf{x}_{p_{i,j}} = (x_{p_{i,j}}, y_{p_{i,j}}, z_{p_{i,j}})$, for $i = 1, \dots, m+1, j = 1, \dots, n+1$.
- The coordinates of the vertices of the doublet lines are denoted by $\mathbf{x}_{d_{i,j}} = (x_{d_{i,j}}, y_{d_{i,j}}, z_{d_{i,j}})$, for $i = 1, \dots, m, j = 1, \dots, n+1$.
- The coordinates of the control points are denoted by $\mathbf{x}_{c_{i,j}} = (x_{c_{i,j}}, y_{c_{i,j}}, z_{c_{i,j}})$, for $i = 1, \dots, m$, $j = 1, \dots, n$.
- The coordinates of the midpoints of the doublet lines are denoted by $\mathbf{x}_{m_{i,j}} = (x_{m_{i,j}}, y_{m_{i,j}}, z_{m_{i,j}})$, for $i = 1, \dots, m, j = 1, \dots, n$.
- The mean panel chord lengths are denoted by $\Delta x_{i,j}$, for $i = 1, \dots, m, j = 1, \dots, n$.
- The panel half-spans are denoted by $\Delta y_{i,j}$ for $i = 1, \dots, m, j = 1, \dots, n$.

Using the DLM discretisation, the Kernel function integral of Eq. (6.46) can be evaluated approximately and the boundary condition of Eq. (6.45) becomes the discrete sum

$$w_I(\omega) = \sum_{J=1}^{mn} A_{I,J}(\omega)\Delta p_J(\omega) \tag{6.47}$$

for $I = 1, \dots, mn$, where $w_I(\omega)$ is the upwash at the control point of the Ith panel, $\Delta p_I(\omega)$ is the pressure jump at the control point of the Jth panel and $A_{I,J}(\omega)$ is the aerodynamic influence coefficient matrix. Equation (6.47) is usually recast in the non-dimensional form

$$\tilde{w}_I(k) = \sum_{J=1}^{mn} \tilde{A}_{I,J}(k)\Delta c_{p_J}(k) \tag{6.48}$$

where $k = \omega c_0/2U_\infty$ is the reduced frequency, $c_0/2$ is the root half-chord, $\tilde{w} = w/U_\infty$, $\Delta c_p = 2\Delta p/\rho_\infty U_\infty^2$. The influence coefficients $\tilde{A}_{I,J}(k)$ are calculated after non-dimensionalising x and y by $c_0/2$, such that $\tilde{x} = 2x/c_0, \tilde{y} = 2y/c_0$. Blair (1994) shows that the elements of $\tilde{A}_{I,J}(k)$ are given by

$$\tilde{A}_{I,J}(k) = -\frac{\Delta x_J}{8}\left(B_0\bar{K}_m + B_1\frac{\bar{K}_2 - \bar{K}_1}{2\Delta y_J} + B_2\frac{\bar{K}_1 - 2\bar{K}_m + \bar{K}_2}{2\Delta y_J^2}\right) \tag{6.49}$$

where

$$B_0 = \frac{2\Delta\tilde{y}_J}{(\tilde{y}_{c_I} - \tilde{y}_{m_J})^2 - \Delta\tilde{y}_J^2}$$

$$B_1 = \frac{1}{2}\ln\left(\frac{(\Delta\tilde{y}_J - \tilde{y}_{c_I} + \tilde{y}_{m_J})^2}{(\Delta\tilde{y}_J + \tilde{y}_{c_I} - \tilde{y}_{m_J})^2}\right) + \frac{2\Delta\tilde{y}_J(\tilde{y}_{c_I} - \tilde{y}_{m_J})}{(\tilde{y}_{c_I} - \tilde{y}_{m_J})^2 - \Delta\tilde{y}_J^2}$$

$$B_2 = 2\Delta\tilde{y}_J + (\tilde{y}_{c_I} - \tilde{y}_{m_J})\ln\left(\frac{(\Delta\tilde{y}_J - \tilde{y}_{c_I} + \tilde{y}_{m_J})^2}{(\Delta\tilde{y}_J + \tilde{y}_{c_I} - \tilde{y}_{m_J})^2}\right) + \frac{2\Delta\tilde{y}_J(\tilde{y}_{c_I} - \tilde{y}_{m_J})^2}{(\tilde{y}_{c_I} - \tilde{y}_{m_J})^2 - \Delta\tilde{y}_J^2}$$

$$\tilde{K}_1 = \tilde{K}(M_\infty, k, \tilde{x}_{c_I} - \tilde{x}_{d_J}, \tilde{y}_{c_I} - \tilde{y}_{d_J})$$

$$\tilde{K}_2 = \tilde{K}(M_\infty, k, \tilde{x}_{c_I} - \tilde{x}_{d_{J+1}}, \tilde{y}_{c_I} - \tilde{y}_{d_{J+1}})$$

$$\tilde{K}_m = \tilde{K}(M_\infty, k, \tilde{x}_{c_I} - \tilde{x}_{m_J}, \tilde{y}_{c_I} - \tilde{y}_{m_J})$$

The function $\tilde{K}(M_\infty, k, \Delta\tilde{x}, \Delta\tilde{y})$, for any $\Delta\tilde{x}$, $\Delta\tilde{y}$, is defined as

$$\tilde{K}(M_\infty, k, \Delta\tilde{x}, \Delta\tilde{y}) = -\left(I_1(u_1, k_1) + \frac{M_\infty|\Delta y|}{\sqrt{\Delta\tilde{x}^2 + \beta^2\Delta\tilde{y}^2}}\frac{e^{-ik_1u_1}}{\sqrt{1+u_1^2}}\right)e^{-ik\Delta\tilde{x}}$$

where

$$u_1 = \frac{M_\infty\sqrt{\Delta\tilde{x}^2 + \beta^2\Delta\tilde{y}^2} - \Delta\tilde{x}}{\beta^2|\Delta\tilde{y}|} \tag{6.50}$$

$$k_1 = k|\Delta\tilde{y}| \tag{6.51}$$

For $u_1 > 0$, the function $I_1(u_1, k_1)$ is given by

$$I_1(u_1, k_1) = e^{-ik_1u_1}\left(1 - \frac{u_1}{\sqrt{1+u_1^2}} - ik_1J_1(u_1, k_1)\right) \tag{6.52}$$

where

$$J_1(u_1, k_1) = \sum_{i=1}^{11} a_i \frac{e^{-idu_1}}{i^2d^2 + k_1^2}(id - ik_1) \tag{6.53}$$

while $d = 0.372$ and a_i are the elements of the vector

$$\mathbf{a} = \begin{pmatrix} 0.24186198 \\ -2.7918027 \\ 24.991079 \\ -111.59196 \\ 271.43549 \\ -305.75288 \\ -41.183630 \\ 545.98537 \\ -644.78155 \\ 328.72755 \\ -64.279511 \end{pmatrix}$$

Finally, for $u_1 < 0$, $I_1(u_1, k_1)$ is given by

$$I_1(u_1, k_1) = 2\Re\left(I_1(0, k_1)\right) - \Re\left(I_1(-u_1, k_1)\right) + \mathrm{i}\Im\left(I_1(-u_1, k_1)\right) \tag{6.54}$$

where $I_1(0, k_1)$ and $I_1(-u_1, k_1)$ are calculated from Eqs. (6.52) and (6.53). All the terms in Eqs. (6.49)–(6.54) are non-dimensional; all lengths have been non-dimensionalised by $c_0/2$. The DLM influence coefficients are the result of the approximation of the Kernel function by a second-degree polynomial function of distance along the doublet line and of the very careful integration of this polynomial in order to avoid singularities. Blair (1994) demonstrates the full derivation of Eqs. (6.49)–(6.54); see also Gülçat (2016).

Equation (6.48) is a set of mn algebraic equations that can be solved easily for the mn unknown pressure jumps $\Delta\tilde{p}_J(k)$. Once these have been calculated, the lift coefficient is obtained from

$$C_L(k) = \frac{1}{\tilde{S}} \sum_I^{mn} \Delta c_{p_I}(k) \tilde{s}_I \tag{6.55}$$

where $\tilde{S} = 4S/c_0^2$ is the non-dimensional wing area and $\tilde{s}_I = 4s_I/c_0^2$ is the non-dimensional area of the Ith panel. In fact, this lift is the normal force acting on the wing; as we are estimating pressure differences only and the wing is flat, no chordwise force can be predicted by the DLM.

Since the discretised wing is completely flat, its camber cannot be modelled directly but it can be represented using the upwash concept of thin airfoil theory, as described in Section 3.2. If the camber surface of the wing is described by $f_c(x, y)$, then Eq. (3.7) states that, for impermeability, the upwash due to the camber is given by

$$U_\infty \left.\frac{\mathrm{d}f_c}{\mathrm{d}x}\right|_{x,y}$$

Note that the $-U\alpha$ term in Eq. (3.7) is the upwash due to the angle of attack, only the term above is due to the camber. For example, for a rigid cambered wing that oscillates in pitch and plunge, the total upwash is obtained by superimposing the expression above with Eq. (6.44), that is

$$w(x, y, \omega) = -\left(U_\infty \alpha(\omega) - \mathrm{i}\omega h(\omega) + \mathrm{i}\omega\alpha(\omega)(x - x_f)\right) + U_\infty \left.\frac{\mathrm{d}f_c}{\mathrm{d}x}\right|_{x,y} \delta(\omega)$$

where $\delta(\omega)$ denotes that the upwash due to camber is only present for $\omega = 0$ since it is a constant term. For sinusoidal motion at frequency ω_0, the pitch and plunge oscillations are given by Eqs. (5.132) and (5.133)

$$h(\omega) = h_0\delta(\omega) + \frac{h_1}{2\mathrm{i}}\mathrm{e}^{\mathrm{i}\phi_h}\delta(\omega - \omega_0) - \frac{h_1}{2\mathrm{i}}\mathrm{e}^{-\mathrm{i}\phi_h}\delta(\omega + \omega_0)$$

$$\alpha(\omega) = \alpha_0\delta(\omega) + \frac{\alpha_1}{2\mathrm{i}}\mathrm{e}^{\mathrm{i}\phi_\alpha}\delta(\omega - \omega_0) - \frac{\alpha_1}{2\mathrm{i}}\mathrm{e}^{-\mathrm{i}\phi_\alpha}\delta(\omega + \omega_0)$$

Then the DLM Eq. (6.47) must be set up and solved twice, once for $\omega = 0$ and once for $\omega = \omega_0$,

$$-U_\infty\alpha_0 + U_\infty \left.\frac{\mathrm{d}f_c}{\mathrm{d}x}\right|_{x_{c_I}, y_{c_I}} = \sum_{J=1}^{mn} A_{I,J}(0)\Delta p_J(0) \tag{6.56}$$

$$-\left(U_\infty + \mathrm{i}\omega_0(x_{c_I} - x_f)\right)\frac{\alpha_1}{2\mathrm{i}}\mathrm{e}^{\mathrm{i}\phi_\alpha} + \mathrm{i}\omega_0\frac{h_1}{2\mathrm{i}}\mathrm{e}^{\mathrm{i}\phi_h} = \sum_{J=1}^{mn} A_{I,J}(\omega_0)\Delta p_J(\omega_0) \tag{6.57}$$

in order to evaluate the mean and oscillating pressure jump across the surface, $\Delta p_J(0)$ and $\Delta p_J(\omega_0)$, respectively.

Example 6.2 *Use the DLM to calculate the steady and unsteady pressure jumps across the surface of a swept and tapered wing oscillating in pitch.*

This test case (Zwaan 1985) consists in a swept tapered wing known as the Lockheed-Georgia, Air Force Flight Dynamics Laboratory, NASA Langley and NLR (LANN) wing. It was tested while undergoing pitching oscillations at the NLR High Speed Tunnel. The half-wing model had an aspect ratio of 7.92, taper ratio of 0.4, leading edge sweep of 27.493°, twist of −4.8°, root chord $c_0 = 0.3608$ *m, half-span* $b/2 = 1$ *m and half-area* $S = 0.2526$ *m*^{2}*. It featured a 12% thick supercritical airfoil whose ordinates at 8 spanwise stations are given in Zwaan (1985). The pitch axis lay at* $x_f = 0.224$ *m behind the root leading edge.*

Only one of the wind tunnel runs included in Zwaan (1985) featured subcritical flow. The free stream Mach number was $M_\infty = 0.621$*, the mean pitch angle* $\alpha_0 = 0.59°$*, the pitch amplitude* $\alpha_1 = 0.25°$*, the frequency 25 Hz and the reduced frequency based on the root half-chord* $k = 0.133$*. From this data, the free stream airspeed can be calculated as* $Q_\infty = 204.5$ *m/s. Mean and oscillating pressure measurements were taken at six spanwise stations* $2y/b = 0.2, 0.325, 0.475, 0.65, 0.825$ *and* 0.95*. There were up to 24 pressure taps on the upper surface and 16 on the lower surface, depending on the spanwise station. Figure 6.4 draws the half-wing with the pressure tap locations where mean pressures are available for the subcritical test run.*

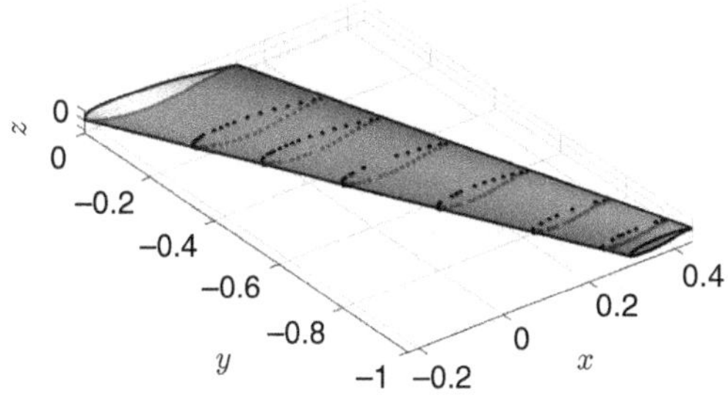

Figure 6.4 LANN half-wing with pressure taps. Source: Adapted from Zwaan (1985).

The first step is to set up the geometric panels, following the approach used in Examples 5.4 and 5.6 for the Vortex Lattice Method. The steady angle of attack of the wing is $\alpha_0 = 0.59°$*, but we do not rotate the wing; it will be modelled flat on the* $z = 0$ *plane and the mean angle of attack will be represented by the upwash it induces. We choose to discretise the wing into* $m = 40$ *chordwise and* $n = 40$ *spanwise linearly spaced panels. The non-dimensional coordinates of the panel vertices are calculated from*

$$\bar{x}_{p_i} = \frac{(i-1)}{m}, \quad \bar{y}_{p_i} = \frac{2(j-1)}{n} - 1, \quad \bar{z}_{p_i} = 0$$

and their dimensional versions are given by Eq. (5.86).

Once the coordinates of the geometric panel vertices have been determined, those of the doublet line vertices are given by

$$x_{d_{i,j}} = \frac{1}{4}\left(3x_{p_{i,j}} + x_{p_{i+1,j}}\right), \quad y_{d_{i,j}} = y_{p_{i,j}}, \quad z_{d_{i,j}} = 0$$

for $i = 1, \dots, m$, $j = 1, \dots, n+1$, *so that each doublet line starts at* $x_{d_{i,j}}, y_{d_{i,j}}, 0$ *and finishes at* $x_{d_{i,j+1}}, y_{d_{i,j+1}}, 0$. *The coordinates of the midpoints of the doublet lines are*

$$x_{m_{i,j}} = \frac{1}{2}\left(x_{d_{i,j}} + x_{d_{i,j+1}}\right), \quad y_{m_{i,j}} = \frac{1}{2}\left(y_{d_{i,j}} + y_{d_{i,j+1}}\right), \quad z_{m_{i,j}} = 0$$

for $i = 1, \dots, m$, $j = 1, \dots, n$. *The coordinates of the control points are calculated from Eq. (5.89), as for the VLM. The mean panel chord lengths are*

$$\Delta x_{i,j} = \frac{1}{2}\left(x_{p_{i+1,j}} - x_{p_{i,j}} + x_{p_{i+1,j+1}} - x_{p_{i,j+1}}\right)$$

for $i = 1, \dots, m$, $j = 1, \dots, n$ *while the panel half-spans are*

$$\Delta y_{i,j} = \frac{1}{2}|y_{p_{i,j+1}} - y_{p_{i,j}}|$$

for $i = 1, \dots, m$, $j = 1, \dots, n$.

In order to evaluate the steady pressure jump across the surface from Eq. (6.56), we also need to calculate the slope of the camber line at each control point. As mentioned earlier, Zwaan (1985) give the measured airfoil coordinates at 8 spanwise stations; these coordinates are non-dimensional and reflect not only the airfoil shape, which changes along the span but also the twist. We will denote these coordinates by $\bar{x}_{u_i}$, $\bar{x}_{l_k}$ *for the upper and lower surface chordwise positions and* $\bar{z}_{u_i}$, $\bar{z}_{l_k}$ *for the upper and lower surface heights. Counters i and k take different values depending on the spanwise station and* $\bar{x}_{u_i} \neq \bar{x}_{l_k}$. *We can calculate the camber line by taking the mean of the upper and lower side coordinates at each of the spanwise stations, but first we need to re-interpolate the upper and lower surface coordinates at the same* $m+1$ *non-dimensional chordwise positions,* $\bar{x}_{p_i}$, *leading to* $\bar{z}_{u_i}^{\text{int}}$ *and* $\bar{z}_{l_i}^{\text{int}}$. *Then, the camber line is simply*

$$\bar{f}_{c_i} = \frac{1}{2}\left(\bar{z}_{l_i}^{\text{int}} + \bar{z}_{u_i}^{\text{int}}\right)$$

for $i = 1, \dots, m+1$, *where* $\bar{f}_c = f_{c_i}/c$ *is the non-dimensional camber line. The original measured coordinates, interpolated coordinates and camber line at the root section of the wing are shown in Figure 6.5a.*

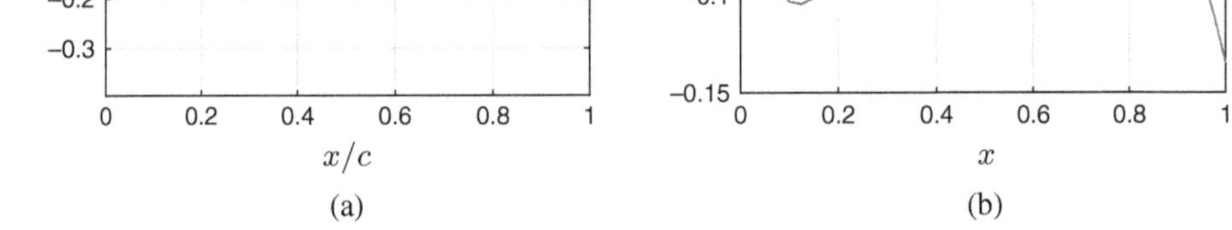

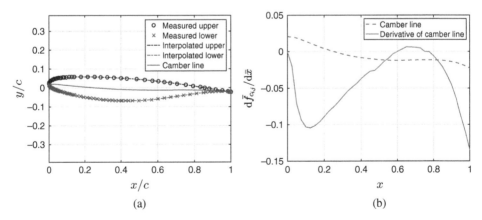

Figure 6.5 Calculation of camber line derivatives of the LANN wing. (a) Interpolation of airfoil shape and camber line, measured coordinates by Zwaan (1985). Source: Adapted from Zwaan (1985), credit: AGARD. (b) Derivative of camber line.

The values of $\bar{f}_{c_i}$ have now been obtained at the eight spanwise stations where the geometry was measured. The next step is to interpolate these values in the spanwise direction in order to obtain $\bar{f}_{c_{i,j}}$ at the spanwise coordinates $\bar{y}_{p_j}$ for $j = 1, \ldots, n+1$. The final step is to calculate the derivatives of the camber line in the chordwise direction. For each value of j, we curve fit the values $\bar{f}_{c_i}$ for $i = 1, \ldots, m+1$ using a spline, which we then differentiate to obtain the derivatives $\mathrm{d}\bar{f}_{c_{i,j}}/\mathrm{d}\tilde{x}$. Figure 6.5b plots the camber line and its derivative at $\bar{y}_{p_{n/2+1}} = 0$. It can be seen that while the camber line itself looks smooth, its derivative is noisy. The noise is due to the fact that differentiation amplifies errors in experimental data. Filtering could be applied to smooth this noise, but then the shape of the camber derivative would change, especially near the leading and trailing edges.

Next, we calculate the non-dimensional versions of all the panel vertices, doublet vertices, control points, mean panel chord lengths and panel half-spans (the derivative of the camber line is already non-dimensional)

$$\tilde{x}_{p_{i,j}} = \frac{2x_{p_{i,j}}}{c_0}, \quad \tilde{y}_{p_{i,j}} = \frac{2y_{p_{i,j}}}{c_0}, \quad \tilde{x}_{d_{i,j}} = \frac{2x_{d_{i,j}}}{c_0}, \quad \tilde{y}_{d_{i,j}} = \frac{2y_{d_{i,j}}}{c_0}, \quad \text{etc}$$

and then we evaluate the aerodynamic influence coefficient matrix from Eqs. (6.49)–(6.54). The quantity u_1 of Eq. (6.50) becomes singular when $\Delta\tilde{y} = 0$, that is when the spanwise distance between the influenced control point and the influencing doublet line midpoint is zero. This phenomenon occurs when the influencing doublet line lies in the same row of panels as the influenced control point. In order to resolve the singularity, we substitute $\Delta\tilde{y} = 10^{-12}$ or any other very small number when $\Delta\tilde{y} = 0$ (Blair 1994).

The last step is to calculate the values of the upwash induced on the control points by the pitching motion. These are obtained from Eqs. (6.56) and (6.57), after setting $h_1 = 0$. The non-dimensional forms of these equations are

$$-\alpha_0 + \left.\frac{\mathrm{d}\bar{f}_c}{\mathrm{d}\tilde{x}}\right|_{\tilde{x}_{c_I},\tilde{y}_{c_I}} = \sum_{J=1}^{mn} \tilde{A}_{I,J}(0)\Delta c_{p_J}(0) \tag{6.58}$$

$$-\left(1 + \mathrm{i}k_0(\tilde{x}_{c_I} - \tilde{x}_f)\right)\frac{\alpha_1}{2\mathrm{i}}\mathrm{e}^{\mathrm{i}\phi_\alpha} = \sum_{J=1}^{mn} \tilde{A}_{I,J}(k_0)\Delta c_{p_J}(k_0) \tag{6.59}$$

where $\tilde{x}_f = 2x_f/c_0$. Now, we can solve these two equations for the unknown pressure jumps at the control points, $\Delta c_{p_J}(0)$ and $\Delta c_{p_J}(k_0)$.

Figure 6.6 plots the mean pressure jump at four of the six spanwise measurement stations for $M_\infty = 0.62$ and compares it to the experimental measurements. The noise present in the camber derivatives causes noise in the steady pressure jump values. Nevertheless, the agreement between the predictions and the experimental data is adequate, except at the leading and trailing edges. At the leading edge, the DLM mean pressure undergoes a sharp increase that is not reflected by the measurements. At the trailing edge, the DLM consistently overestimates the mean pressure jump. For supercritical airfoils, the description of the geometry as a superposition of a thickness distribution and a camber line is not entirely representative. In NACA four-digit series airfoils, the camber derivative is positive from the leading edge to the maximum camber point and then negative all the way to the trailing edge. Figure 6.5b shows

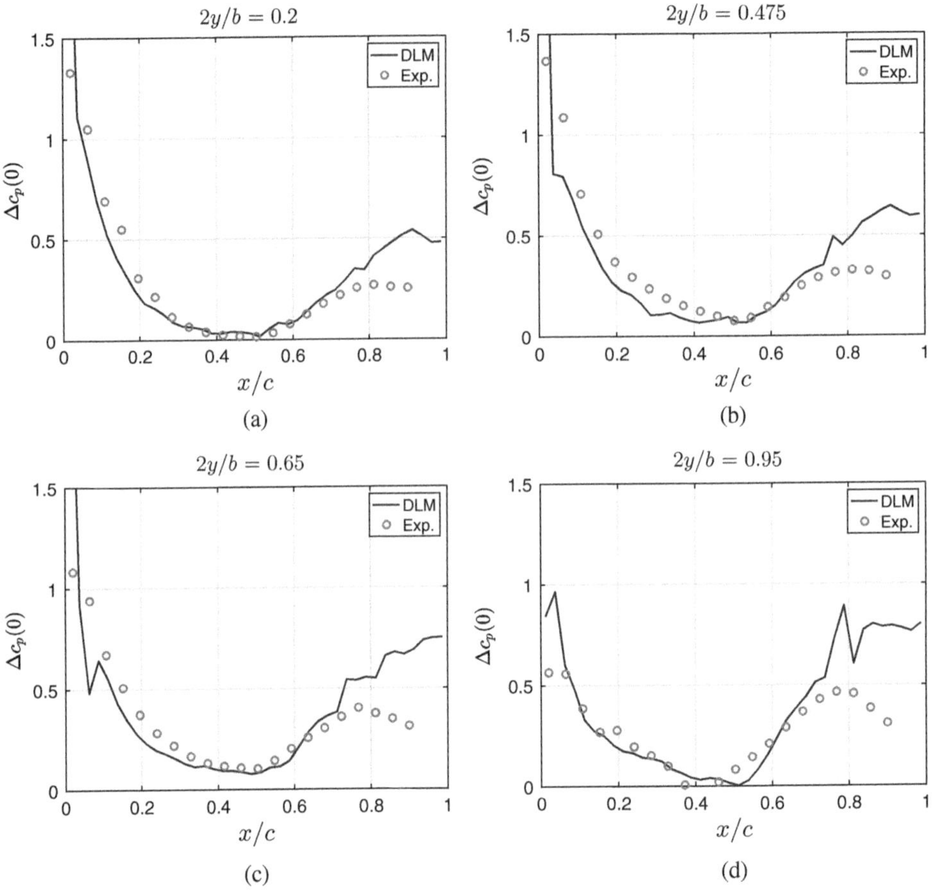

Figure 6.6 Amplitude and phase of mean pressure jump predicted by the DLM across LANN wing for $M_\infty = 0.62$. Experimental data by Zwaan (1985). (a) $2y/b = 0.2$, (b) $2y/b = 0.475$, (c) $2y/b = 0.65$ and (d) $2y/b = 0.95$. Source: Zwaan (1985), credit: AGARD.

that the camber derivative of the LANN wing's airfoil is significantly more complex.[1] *In fact, the definition of the camber line for modern airfoils is disputed. As an example Bagai (2006) states that the camber line should not be equidistant between the upper and lower surfaces in the vertical direction but in a direction normal to the camber line itself. Implementing such a definition on airfoils defined as experimentally measured points can be challenging.*

For the oscillatory pressure, we calculate the normalised pressure jumps

$$\Delta c_{p_\alpha}(k_0) = \frac{2\mathrm{i}\Delta c_p(k_0)}{\alpha_1}$$

in order to compare to the data plotted in Lessing et al. (1960). As the DLM does not necessarily calculate the pressure exactly at the spanwise tap locations, we interpolate $\Delta c_{p_{i,j}}$ *linearly*

1 The camber derivative of Figure 6.5b includes the effect of the local twist angle but this is constant over the chord.

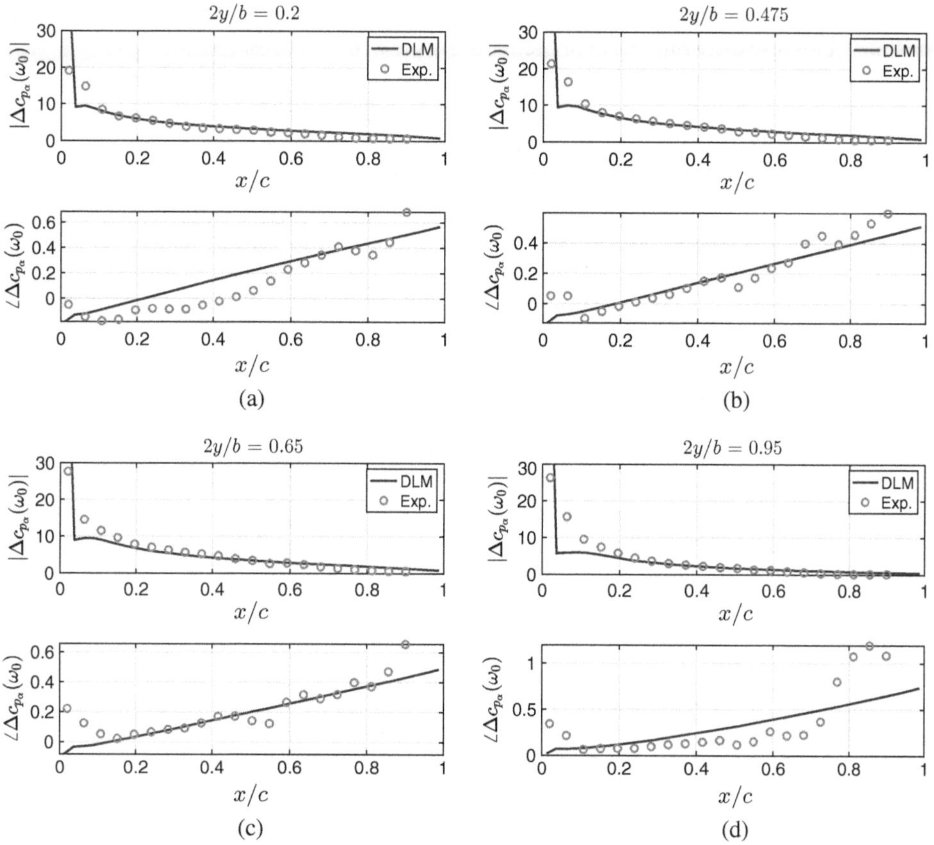

Figure 6.7 Amplitude and phase of oscillating pressure jump predicted by the DLM across LANN wing for $M_\infty = 0.62$. Experimental data by Zwaan (1985). (a) $2y/b = 0.2$, (b) $2y/b = 0.475$, (c) $2y/b = 0.65$ and (d) $2y/b = 0.95$. Source: Zwaan (1985), credit: AGARD.

between the two adjacent locations in order to compare to the experimental data. Figure 6.7 plots the magnitude and phase of Δc_{p_α} *at the same four spanwise measurement stations for* $M_\infty = 0.62$. *The figure shows that both the amplitude and phase of the pressure difference are estimated with satisfactory accuracy by the DLM. Nevertheless, the predicted amplitude still undergoes a sharp increase at the leading edge that is not representative of the experimental measurements.*

The total mean and oscillatory lift are calculated from Eq. (6.55). The mean lift calculated by the DLM is $C_L(0) = 0.375$ *while the experimental value is 0.285. The oscillatory lift predicted by the DLM is* $2\mathrm{i}C_L(k_0)/\pi\alpha_1 = 1.688 + 0.073\mathrm{i}$ *while the experimental measurements yield* $1.326 + 0.047\mathrm{i}$. *Therefore, the DLM overestimates both the mean and oscillatory lift, but it should be kept in mind that the experimental lift is the result of the summation of a limited number of pressure measurements and therefore does not reflect the pressure distribution over the entire wing very accurately. This example is solved by Matlab code* `pitchplunge_DLM.m`.

Example 6.2 has demonstrated the effectiveness of the DLM in calculating steady and unsteady pressure distributions over oscillating finite wings in subsonic flow. The approach is very fast since it ignores thickness and does not require a wake model. On the other hand, only pressure jumps across the surface can be calculated and camber must be modelled very carefully in order to obtain good estimates for the steady pressure jump. More importantly, as the pressure calculation is linearised, only aerodynamic loads normal to the surface can be predicted so that the DLM cannot estimate the drag. In the next section, we will describe a subsonic unsteady SDPM that can model the true geometry of a wing, including thickness and camber, and predict mean and oscillatory drag. As a final thought on the DLM, it should be mentioned that there are other approaches for calculating the influence coefficients, such as the Gaussian quadrature technique by Chen et al. (1993).

6.3.2 Unsteady 3D Subsonic Source and Doublet Panel Method

The unsteady subsonic SDPM presented here is based on the work by Morino (1980), Morino et al. (1975), Morino and Chen (1975). Its starting point is the compressible subsonic Green's theorem of Eq. (2.180), expressed in the frequency domain and on a quasi-fixed body geometry. First, we apply the Prandtl–Glauert transformation of Eq. (6.3) to the linearised small disturbance equation (2.143)

$$\beta^2\phi_{xx} + \phi_{yy} + \phi_{zz} - \frac{2M_\infty}{a_\infty}\phi_{xt} - \frac{1}{a_\infty^2}\phi_{tt} = 0 \tag{6.60}$$

Using the transformation procedure of Section 6.2, we obtain

$$\phi_{\xi\xi} + \phi_{\eta\eta} + \phi_{\zeta\zeta} - \frac{2M_\infty}{a_\infty\beta}\phi_{\xi t} - \frac{1}{a_\infty^2}\phi_{tt} = 0 \tag{6.61}$$

Applying the solution technique of Section 2.5 to Eq. (6.61), we guess a source solution of the form

$$\phi(\xi,\eta,\zeta,t) = \frac{F(t-\tau(\xi,\eta,\zeta))}{r(\xi,\eta,\zeta)} \tag{6.62}$$

where $r(\xi,\eta,\zeta) = \sqrt{(\xi-\xi_0)^2 + (\eta-\eta_0)^2 + (\zeta-\zeta_0)^2}$ and $\tau(\xi,\eta,\zeta)$ is a time delay. We define $\bar{t}(\xi,\eta,\zeta) = t - \tau(\xi,\eta,\zeta)$ so that $F(t-\tau(\xi,\eta,\zeta)) = F(\bar{t})$ and calculate all derivatives in Eq. (6.61) which we then proceed to substitute back into this equation to obtain

$$\begin{aligned}
&F\left(\frac{\partial^2}{\partial\xi^2}\left(\frac{1}{r}\right) + \frac{\partial^2}{\partial\eta^2}\left(\frac{1}{r}\right) + \frac{\partial^2}{\partial\zeta^2}\left(\frac{1}{r}\right)\right)\\
&+F'\left(-2\frac{\partial}{\partial\xi}\left(\frac{1}{r}\right)\frac{\partial\tau}{\partial\xi} - \frac{1}{r}\frac{\partial^2\tau}{\partial\xi^2} - 2\frac{\partial}{\partial\eta}\left(\frac{1}{r}\right)\frac{\partial\tau}{\partial\eta} - \frac{1}{r}\frac{\partial^2\tau}{\partial\eta^2}\right.\\
&\left.\quad -2\frac{\partial}{\partial\zeta}\left(\frac{1}{r}\right)\frac{\partial\tau}{\partial\zeta} - \frac{1}{r}\frac{\partial^2\tau}{\partial\zeta^2} - \frac{2M_\infty}{a_\infty\beta}\frac{\partial}{\partial\xi}\left(\frac{1}{r}\right)\right)\\
&+F''\left(\frac{1}{r}\left(\frac{\partial\tau}{\partial\xi}\right)^2 + \frac{1}{r}\left(\frac{\partial\tau}{\partial\eta}\right)^2 + \frac{1}{r}\left(\frac{\partial\tau}{\partial\zeta}\right)^2 + \frac{2M_\infty}{a_\infty\beta}\frac{1}{r}\frac{\partial\tau}{\partial\xi} - \frac{1}{a_\infty^2 r}\right) = 0
\end{aligned}$$

The coefficient of F in this latest expression is a steady source solution applied to Laplace's equation and is therefore equal to zero. If, by analogy to Eq. (2.175), we choose

$$\tau(\xi,\eta,\zeta) = \frac{-M_\infty(\xi-\xi_0)+r}{a_\infty\beta} \tag{6.63}$$

then the coefficients of F' and F'' also become equal to zero and the transformed linearised small disturbance equation is satisfied.

Next, we take the Fourier Transform of (6.61), leading to

$$\phi_{\xi\xi}(\omega)+\phi_{\eta\eta}(\omega)+\phi_{\zeta\zeta}(\omega)-\mathrm{i}\omega\frac{2M_\infty}{a_\infty\beta}\phi_\xi(\omega)+\omega^2\frac{1}{a_\infty^2}\phi(\omega)=0 \tag{6.64}$$

Taking also the Fourier Transform of the source solution of expression (6.62), we obtain

$$\phi(\xi,\eta,\zeta,\omega)=\frac{F(\omega)\mathrm{e}^{-\mathrm{i}\omega\tau(\xi,\eta,\zeta)}}{r}$$

In order to harmonize the notation with that of the incompressible case, we set $\sigma(\omega) = F(\omega)$. After substituting from Eq. (6.63), the source solution for the transformed unsteady linearised small perturbation equation in the frequency domain becomes

$$\phi(\xi,\eta,\zeta,\omega)=\sigma(\omega)\frac{\mathrm{e}^{-\mathrm{i}\Omega(-M_\infty(\xi-\xi_0)+r)}}{r} \tag{6.65}$$

where we have defined $\Omega = \omega/a_\infty\beta$.

The doublet solution is obtained by differentiating the source with respect to a unit vector, $\mathbf{n}$, that is

$$\phi(\xi,\eta,\zeta,\omega)=\mu(\omega)\mathbf{n}\cdot\nabla\left(\frac{\mathrm{e}^{-\mathrm{i}\Omega(-M_\infty(\xi-\xi_0)+r)}}{r}\right) \tag{6.66}$$

Consider the unsteady subsonic flow around a quasi-fixed closed surface whose coordinates, $\mathbf{x}_s$, are obtained from $S(\mathbf{x}_s) = 0$. The potential anywhere in this flow is given in terms of the potential on the surface by Green's theorem of Eq. (2.180); if we apply the Prandtl–Glauert transformation to the latter, we obtain

$$\begin{aligned}4\pi E(\mathbf{x},t)\phi(\xi,t) = &-\int_\Sigma \mathbf{n}(\xi_s)\cdot\bar{\nabla}_s\phi(\xi_s,t-\tau)\frac{1}{r(\xi,\xi_s)}\mathrm{d}\Sigma\\ &+\int_\Sigma \phi(\xi_s,t-\tau)\mathbf{n}(\xi_s)\cdot\bar{\nabla}_s\left(\frac{1}{r(\xi,\xi_s)}\right)\mathrm{d}\Sigma\\ &-\frac{\partial}{\partial t}\int_\Sigma\frac{\phi(\xi_s,t-\tau)}{r(\xi-\xi_s)}\mathbf{n}(\xi_s)\cdot\bar{\nabla}_s\tau(\xi,\xi_s)\mathrm{d}\Sigma\end{aligned} \tag{6.67}$$

where E is given by Eq. (2.95), $\tau(\xi,\xi_s)$ and $r(\xi,\xi_s)$ are given by

$$\tau(\xi,\xi_s)=\frac{-M_\infty(\xi-\xi_s)+r(\xi,\xi_s)}{a_\infty\beta}$$

$$r(\xi,\xi_s)=\sqrt{(\xi-\xi_s)^2+(\eta-\eta_s)^2+(\zeta-\zeta_s)^2}$$

while $\bar{\nabla}_s = (\partial/\partial\xi_s, \partial/\partial\eta_s, \partial/\partial\zeta_s)$, $\mathbf{n}(\xi_s)$ is the normal vector on the transformed geometry and Σ is the area of the transformed surface.

We now apply the Fourier transform to Eq. (6.67), such that $\phi(\xi, t)$ becomes $\phi(\xi, \omega)$, $\phi(\xi, t-\tau)$ becomes $\phi(\xi, \omega)e^{-i\omega\tau}$ and $\phi_t(\xi, t-\tau)$ becomes $i\omega\phi(\xi, \omega)e^{-i\omega\tau}$. Equation (6.67) is transformed to

$$\begin{aligned}4\pi E\phi(\xi, \omega) = &-\int_\Sigma \mathbf{n}(\xi_s) \cdot \bar{\nabla}_s\phi(\xi_s, \omega)e^{-i\omega\tau}\frac{1}{r(\xi, \xi_s)}d\Sigma \\ &+\int_\Sigma \phi(\xi_s, \omega)e^{-i\omega\tau}\bar{\nabla}_s\frac{1}{r(\xi, \xi_s)}d\Sigma \\ &-\int_\Sigma \frac{i\omega\phi(\xi_s, \omega)e^{-i\omega\tau}}{r(\xi, \xi_s)}\mathbf{n}(\xi_s) \cdot \bar{\nabla}_s\tau\, d\Sigma\end{aligned}$$

Noting that

$$\bar{\nabla}_s\left(\frac{e^{-i\omega\tau}}{r}\right) = e^{-i\omega\tau}\bar{\nabla}_s\left(\frac{1}{r}\right) - i\omega\bar{\nabla}_s\tau e^{-i\omega\tau}\frac{1}{r} \tag{6.68}$$

the potential anywhere in the flow is finally given by (Morino 1974)

$$\begin{aligned}E\phi(\xi, \omega) = &-\frac{1}{4\pi}\int_\Sigma \mathbf{n}(\xi_s) \cdot \left(\bar{\nabla}_s\phi(\xi_s, \omega)\frac{e^{-i\Omega(-M_\infty(\xi-\xi_s)+r(\xi,\xi_s))}}{r(\xi, \xi_s)}\right. \\ &\left.-\phi(\xi_s, \omega)\bar{\nabla}_s\left(\frac{e^{-i\Omega(-M_\infty(\xi-\xi_s)+r(\xi,\xi_s))}}{r(\xi, \xi_s)}\right)\right)d\Sigma\end{aligned} \tag{6.69}$$

where $\Omega = \omega/a_\infty\beta$. Note the similarity between Eq. (6.69) and the incompressible Green's identity of expression (2.95), which means that we can solve the two equations in essentially the same manner. Following the approach of Section 5.6, we add a wake contribution to Eq. (6.69) such that

$$\begin{aligned}E\phi(\xi, \omega) = &-\frac{1}{4\pi}\int_\Sigma \sigma(\xi_s, \omega)\frac{e^{-i\Omega(-M_\infty(\xi-\xi_s)+r(\xi,\xi_s))}}{r(\xi, \xi_s)}d\Sigma \\ &+\frac{1}{4\pi}\int_\Sigma \mu(\xi_s, \omega)\mathbf{n}(\xi_s) \cdot \bar{\nabla}_s\left(\frac{e^{-i\Omega(-M_\infty(\xi-\xi_s)+r(\xi,\xi_s))}}{r(\xi, \xi_s)}\right)d\Sigma \\ &+\frac{1}{4\pi}\int_{\Sigma_w} \mu_w(\xi_s, \omega)\mathbf{n}(\xi_s) \cdot \bar{\nabla}_s\left(\frac{e^{-i\Omega(-M_\infty(\xi-\xi_s)+r(\xi,\xi_s))}}{r(\xi, \xi_s)}\right)d\Sigma\end{aligned} \tag{6.70}$$

where Σ_w is the surface of the transformed wake, $\sigma(\xi_s, \omega) = \bar{\nabla}_s\phi(\xi_s, \omega) \cdot \mathbf{n}(\xi_s)$ is the source strength on the wing, $\mu(\xi_s, \omega) = \phi(\xi_s, \omega)$ is the doublet strength on the wing and $\mu(\xi_s, \omega) = \phi_u(\xi_s, \omega) - \phi_l(\xi_s, \omega)$ is the doublet strength on the wake.

The value of $\sigma(\xi_s, \omega)$ must be obtained from the impermeability boundary condition of Eq. (2.157)

$$\rho_\infty\left(\left(\beta^2\phi_x - \frac{M_\infty}{a_\infty^2}\phi_t, \phi_y, \phi_z\right) + \mathbf{Q}_\infty - \mathbf{v}_0(t) - \mathbf{\Omega}_0(t) \times (\mathbf{x}_s - \mathbf{x}_f)\right) \cdot \mathbf{n}(\mathbf{x}_s) = 0$$

having imposed the fact that the geometry is fixed in time. Dividing throughout by ρ_∞ and taking the Fourier transform of this condition leads to

$$\begin{aligned}&\left(\beta^2\phi_x(\omega) - \frac{i\omega M_\infty}{a_\infty^2}\phi(\omega), \phi_y(\omega), \phi_z(\omega)\right) \cdot \mathbf{n}(\mathbf{x}_s) \\ &\quad = -\left(\mathbf{Q}_\infty\delta(\omega) - \mathbf{v}_0(\omega) - \mathbf{\Omega}_0(\omega) \times (\mathbf{x}_s - \mathbf{x}_f)\right) \cdot \mathbf{n}(\mathbf{x}_s)\end{aligned}$$

Multiplying both sides by $||\nabla S(\mathbf{x}_s)||$ and then applying the Prandtl–Glauert transformation results in

$$\left(\beta\phi_{\xi}(\omega)-\frac{\mathrm{i}\omega M_\infty}{a_\infty^2}\phi(\omega),\ \phi_\eta(\omega),\ \phi_\zeta(\omega)\right)\boldsymbol{B}^{-1}(\bar{\nabla}\Sigma(\boldsymbol{\xi}_s))^T$$
$$=-\left(\mathbf{Q}_\infty\delta(\omega)-\mathbf{v}_0(\omega)-\boldsymbol{\Omega}_0(\omega)\times(\boldsymbol{\xi}_s-\boldsymbol{\xi}_f)\boldsymbol{B}\right)\boldsymbol{B}^{-1}(\bar{\nabla}\Sigma(\boldsymbol{\xi}_s))^T$$

where $\boldsymbol{B}$ is given by Eq. (6.12). Recalling that $\left(\beta\phi_\xi,\ \phi_\eta,\ \phi_\zeta\right)\boldsymbol{B}^{-1}=\bar{\nabla}\phi$ we can divide the expression above by $||\bar{\nabla}\Sigma(\boldsymbol{\xi}_s)||$ and re-arrange it into

$$\bar{\nabla}\phi(\boldsymbol{\xi}_s,\omega)\cdot\mathbf{n}(\boldsymbol{\xi}_s)=\frac{\mathrm{i}\omega M_\infty}{\beta a_\infty^2}\phi(\boldsymbol{\xi}_s,\omega)n_\xi(\boldsymbol{\xi}_s)-\left(\mathbf{Q}_\infty\delta(\omega)-\mathbf{v}_0(\omega)\right)\boldsymbol{B}^{-1}\mathbf{n}^T(\boldsymbol{\xi}_s)$$
$$+\boldsymbol{\Omega}_0(\omega)\times(\boldsymbol{\xi}_s-\boldsymbol{\xi}_f)\cdot\mathbf{n}(\boldsymbol{\xi}_s)$$

Finally, substituting for the source strength, doublet strength and Ω leads to

$$\sigma(\boldsymbol{\xi}_s,\omega)=\frac{\mathrm{i}\Omega M_\infty}{a_\infty}\mu(\boldsymbol{\xi}_s,\omega)n_\xi(\boldsymbol{\xi}_s)-\left(\mathbf{Q}_\infty\delta(\omega)-\mathbf{v}_0(\omega)\right)\boldsymbol{B}^{-1}\mathbf{n}^T(\boldsymbol{\xi}_s)$$
$$+\boldsymbol{\Omega}_0(\omega)\times(\boldsymbol{\xi}_s-\boldsymbol{\xi}_f)\cdot\mathbf{n}(\boldsymbol{\xi}_s) \tag{6.71}$$

which gives the source strength to be substituted into Eq. (6.70). For low values of the frequency, we can assume that $\mathrm{i}\Omega M_\infty/a_\infty \ll 1$ so that the source strength simplifies to

$$\sigma(\boldsymbol{\xi}_s,\omega)=-\left(\mathbf{Q}_\infty\delta(\omega)-\mathbf{v}_0(\omega)\right)\boldsymbol{B}^{-1}\mathbf{n}^T(\boldsymbol{\xi}_s)+\boldsymbol{\Omega}_0(\omega)\times(\boldsymbol{\xi}_s-\boldsymbol{\xi}_f)\cdot\mathbf{n}(\boldsymbol{\xi}_s) \tag{6.72}$$

Alternatively, Morino (1980) chose to approximate the normal velocity on the transformed geometry by that on the original geometry such that

$$\bar{\nabla}_s\phi(\boldsymbol{\xi}_s)\cdot\mathbf{n}(\boldsymbol{\xi}_s)\approx\nabla\phi(\boldsymbol{x}_s)\cdot\mathbf{n}(\boldsymbol{x}_s)$$
$$=-\left(\mathbf{Q}_\infty\delta(\omega)-\mathbf{v}_0(\omega)-\boldsymbol{\Omega}_0(\omega)\times(\mathbf{x}_s-\mathbf{x}_f)\right)\cdot\mathbf{n}(\boldsymbol{x}_s) \tag{6.73}$$

In the present work, we make use of Eq. (6.71), for compatibility with the steady case. The unsteady pressure distributions obtained using Eqs. (6.71) and (6.73) for the boundary condition are very similar.

As usual, calculating the integrals in Eq. (6.70) is impossible for general geometries so that a panelling approach must be used. Assuming that there are N wing and N_w wake panels, Eq. (6.70) can be approximated in the form of Eq. (5.173), that is

$$\frac{1}{2}\mu(\xi_{c_I},\omega)=\sum_{J=1}^{N}\bar{A}_{\phi_{I,J}}(\omega)\sigma(\xi_{c_J},\omega)+\sum_{J=1}^{N}\bar{B}_{\phi_{I,J}}(\omega)\mu(\xi_{c_J},\omega)+\sum_{J=1}^{N_w}\bar{C}_{\phi_{I,J}}(\omega)\mu_w(\xi_{c_J},\omega) \tag{6.74}$$

where

$$\bar{A}_{\phi_{I,J}}(\omega)=-\frac{1}{4\pi}\int_{\Delta\Sigma_J-\xi_{c_I}}\frac{\mathrm{e}^{-\mathrm{i}\Omega(-M_\infty(\xi_{c_I}-\xi_s)+r(\xi_{c_I},\xi_s))}}{r(\xi_{c_I},\xi_s)}\mathrm{d}\Sigma$$
$$\bar{B}_{\phi_{I,J}}(\omega)=\frac{1}{4\pi}\int_{\Delta\Sigma_J-\xi_{c_I}}\mathbf{n}(\xi_{c_J})\cdot\bar{\nabla}_s\left(\frac{\mathrm{e}^{-\mathrm{i}\Omega(-M_\infty(\xi_{c_I}-\xi_s)+r(\xi_{c_I},\xi_s))}}{r(\xi_{c_I},\xi_s)}\right)\mathrm{d}\Sigma \tag{6.75}$$
$$\bar{C}_{\phi_{I,J}}(\omega)=\frac{1}{4\pi}\int_{\Delta\Sigma_{w_J}}\mathbf{n}(\xi_{c_J})\cdot\bar{\nabla}_s\left(\frac{\mathrm{e}^{-\mathrm{i}\Omega(-M_\infty(\xi_{c_I}-\xi_s)+r(\xi_{c_I},\xi_s))}}{r(\xi_{c_I},\xi_s)}\right)\mathrm{d}\Sigma$$

In these expressions, ξ_{c_I}, ξ_{c_J} are the coordinates of the control point of the Ith and Jth panels, respectively, $\Delta\Sigma_J$ is the area of the Jth wing panel, $\Delta\Sigma_{w_J}$ is the area of the Jth wake panel, $\sigma(\xi_{c_J}, \omega)$ is the constant source strength on the Jth panel, $\mu(\xi_{c_J}, \omega)$ is the constant doublet strength on the Jth panel, $\mathbf{n}(\xi_{c_J})$ is the unit vector normal to the Jth panel, ξ_s are general points lying on the Jth panel and, for the case $I = J$, we exclude point ξ_{c_I} from the integrals.

Unfortunately, even for plane quadrilateral panels, there is no obvious way to calculate analytically the integrals in the expressions for $\bar{A}_{\phi_{I,J}}(\omega)$, $\bar{B}_{\phi_{I,J}}(\omega)$ and $\bar{C}_{\phi_{I,J}}(\omega)$, due to the presence of the exponential terms. Morino et al. (1975) sidestepped this difficulty by assuming that the exponential terms are constant over the panels and equal to their values at the control points, ξ_{c_J}, as is already done for the source and doublet strengths. Then, the source influence coefficients are approximated by

$$\bar{A}_{\phi_{I,J}}(\omega) \approx -\frac{e^{-i\Omega(-M_\infty(\xi_{c_I}-\xi_{c_J})+r(\xi_{c_I},\xi_{c_J}))}}{4\pi}\int_{\Delta\Sigma_J-\xi_{c_I}} \frac{1}{r(\xi_{c_I}, \xi_s)} d\Sigma \tag{6.76}$$

which is very useful since we have already calculated the integral of $1/r$ over a quadrilateral panel (see Section A.10). The doublet influence coefficients are given by

$$\bar{B}_{\phi_{I,J}}(\omega) = \frac{1}{4\pi}\int_{\Delta\Sigma_J-\xi_{c_I}} \mathbf{n}(\xi_{c_J}) \cdot \bar{\nabla}_s \left(\frac{e^{-i\Omega(-M_\infty(\xi_{c_I}-\xi_s)+r(\xi_{c_I},\xi_s))}}{r(\xi_{c_I}, \xi_s)} \right) d\Sigma \tag{6.77}$$

The derivative in this integral can be written as

$$\begin{aligned}
&\bar{\nabla}_s \left(\frac{e^{-i\Omega(-M_\infty(\xi_{c_I}-\xi_s)+r(\xi_{c_I},\xi_s))}}{r(\xi_{c_I}, \xi_s)} \right) \\
&= e^{-i\Omega(-M_\infty(\xi_{c_I}-\xi_s)+r)}\bar{\nabla}_s\left(\frac{1}{r}\right) - i\Omega e^{-i\Omega(-M_\infty(\xi_{c_I}-\xi_s)+r)}\left((M_\infty, 0, 0) + \bar{\nabla}_s r\right)\frac{1}{r} \\
&= (1 + i\Omega r)\, e^{-i\Omega(-M_\infty(\xi_{c_I}-\xi_s)+r)}\bar{\nabla}_s\left(\frac{1}{r}\right) - i\Omega e^{-i\Omega(-M_\infty(\xi_{c_I}-\xi_s)+r)}(M_\infty, 0, 0)\frac{1}{r}
\end{aligned} \tag{6.78}$$

since

$$\bar{\nabla}_s\left(\frac{1}{r}\right) = -\frac{1}{r^2}\bar{\nabla}_s r \text{ or } \frac{1}{r}\bar{\nabla}_s r = -r\bar{\nabla}_s\left(\frac{1}{r}\right)$$

We apply Morino's approximation by replacing every occurrence of ξ_s and ξ_s in expression (6.78) by ξ_{c_J} and ξ_{c_J}, except inside $1/r$. Consequently, the doublet influence coefficients of Eq. (6.77) become

$$\begin{aligned}
\bar{B}_{\phi_{I,J}}(\omega) \approx &\frac{1}{4\pi}\left(1 + i\Omega r(\xi_{c_I}, \xi_{c_J})\right) e^{-i\Omega(-M_\infty(\xi_{c_I}-\xi_{c_J})+r(\xi_{c_I},\xi_{c_J}))} \\
&\times \int_{\Sigma_J-\xi_{c_I}} \mathbf{n}(\xi_{c_J}) \cdot \bar{\nabla}_s\left(\frac{1}{r(\xi_{c_I}, \xi_s)}\right) d\Sigma \\
&-\frac{1}{4\pi} i\Omega e^{-i\Omega(-M_\infty(\xi_{c_I}-\xi_{c_J})+r(\xi_{c_I},\xi_{c_J}))} M_\infty n_\xi(\xi_{c_J}) \\
&\times \int_{\Sigma_J-\xi_{c_I}} \frac{1}{r(\xi_{c_I}, \xi_s)} d\Sigma
\end{aligned} \tag{6.79}$$

where $n_\xi(\xi_{c_J})$ is the ξ component of $\mathbf{n}(\xi_{c_J})$. Again, this is a very useful approximation since the integral of the doublet $\mathbf{n} \cdot \bar{\nabla}_s(1/r)$ over a quadrilateral panel has already been evaluated (see Section A.11).

Substituting the approximations of expressions (6.76) and (6.79) into Eq. (6.74) and writing the latter in matrix form, Green's theorem becomes

$$\left(\bar{\boldsymbol{B}}_{\phi}(\omega) - \frac{1}{2}\boldsymbol{I}\right)\boldsymbol{\mu}(\omega) + \bar{\boldsymbol{C}}_{\phi}(\omega)\boldsymbol{\mu}_{w}(\omega) = -\bar{\boldsymbol{A}}_{\phi}(\omega)\boldsymbol{\sigma}(\omega) \tag{6.80}$$

where the elements of matrices $\bar{\boldsymbol{A}}(\omega)$, $\bar{\boldsymbol{B}}_{\phi}(\omega)$ and $\bar{\boldsymbol{C}}_{\phi}(\omega)$ are given by

$$\bar{A}_{\phi_{I,J}}(\omega) = \mathrm{e}^{-\mathrm{i}\Omega(-M_{\infty}(\xi_{c_I}-\xi_{c_J})+r(\xi_{c_I},\xi_{c_J}))}A_{\phi_{I,J}} \tag{6.81}$$

$$\begin{aligned}\bar{B}_{\phi_{I,J}}(\omega) = {} & -\mathrm{i}\Omega \mathrm{e}^{-\mathrm{i}\Omega(-M_{\infty}(\xi_{c_I}-\xi_{c_J})+r(\xi_{c_I},\xi_{c_J}))}M_{\infty}n_{\xi}(\xi_{c_J})A_{\phi_{I,J}} \\ & + \left(1+\mathrm{i}\Omega r(\xi_{c_I},\xi_{c_J})\right)\mathrm{e}^{-\mathrm{i}\Omega(-M_{\infty}(\xi_{c_I}-\xi_{c_J})+r(\xi_{c_I},\xi_{c_J}))}B_{\phi_{I,J}}\end{aligned} \tag{6.82}$$

for $I = 1, \dots, N, J = 1, \dots, N$ and

$$\bar{C}_{\phi_{I,J}}(\omega) = \left(1+\mathrm{i}\Omega r(\xi_{c_I},\xi_{c_J})\right)\mathrm{e}^{-\mathrm{i}\Omega(-M_{\infty}(\xi_{c_I}-\xi_{c_J})+r(\xi_{c_I},\xi_{c_J}))}C_{\phi_{I,J}} \tag{6.83}$$

for $I = 1, \dots, N, J = 1, \dots, N_w$. In these last three expressions, the matrices $\boldsymbol{A}_{\phi}$, $\boldsymbol{B}_{\phi}$ and $\boldsymbol{C}_{\phi}$ are those obtained in Section 6.2 for steady subsonic flow, recalling that $B_{\phi_{II}} = 0$. Furthermore, we have made the assumption that the wake is flat and aligned with the x axis, so that $n_{\xi}(\xi_{c_J}) = 0$ for all the wake panels. Note that the unsteadiness of the motion only has an effect on the phase of the source contribution, not its amplitude. The doublet contributions, on the other hand, have both their phase and amplitude modified by unsteady effects.

Using Morino's approximation, extending the steady SDPM to unsteady flows becomes simply a matter of modifying the elements of the steady influence coefficient matrices. Note also that $\bar{\boldsymbol{A}}(\omega)$, $\bar{\boldsymbol{B}}_{\phi}(\omega)$ and $\bar{\boldsymbol{C}}_{\phi}(\omega)$ reduce to $\boldsymbol{A}_{\phi}$, $\boldsymbol{B}_{\phi}$ and $\boldsymbol{C}_{\phi}$ when $\Omega = 0$. Equation (6.80) can be solved using the frequency domain solution technique for the incompressible SDPM, as detailed in Section 5.7.1.

Morino's approximation is accurate when the influencing panel lies far from the influenced panel. For nearby panels, the approximation becomes increasingly accurate as the panel size is decreased. Nevertheless, as the exponential terms that are being approximated do not feature any singularities, the approximation is acceptable for panels of practical size. This means that the unsteady subsonic SDPM is computationally efficient, albeit more expensive than the DLM for the same grid size, since it requires the discretisation of three surfaces instead of one (the upper and lower surfaces of the wing and the wake).

Example 6.3 *Calculate the mean and oscillating pressure distributions around the LANN wing using the subsonic SDPM.*

Here we repeat Example 6.2 using the SDPM. The first step is to set up the wing geometry. The spanwise panel spacing is non-linear, so that the non-dimensional panel vertices are given by Eq. (5.181). The airfoil coordinates given in Zwaan (1985) are interpolated linearly to obtain the airfoil shape and local twist angle at the chosen spanwise coordinates. The chordwise panel spacing is also nonlinear, such that the non-dimensional spacing is given by Eq. (5.180). The airfoil coordinates are re-interpolated on the chosen chordwise coordinates. After applying the sweep, we obtain the dimensional coordinates of the panel vertices, $x_{p_{i,j}}$, $y_{p_{i,j}}$ *and* $z_{p_{i,j}}$, *for* $i = 1, \dots, 2m+1$ *and* $j = 1, \dots, n$. *The trailing edge thickness of the LANN wing is not zero, such that* $z_{p_{1,j}} \neq z_{p_{2m+1,j}}$. *Consequently, we have to decide on which side of the trailing edge to attach the wake. This choice does not have a significant influence on the solution; we chose to attach the wake to the upper side, that is*

$$x_{w_{1,j}} = x_{p_{2m+1,j}}, \quad y_{w_{1,j}} = y_{p_{2m+1,j}}, \quad z_{w_{1,j}} = z_{p_{2m+1,j}}$$

and set up the rest of the flat wake as in Example 5.13. This wake has a length of $10c_0$*, chordwise number of wake panels* $m_w = 10\,m$ *and chordwise spacing between panels of* $c_0/\beta m$.

The next step is to apply the Prandtl–Glauert transformation to the wing and wake. We then calculate the incompressible steady influence coefficient matrices $\mathbf{A}_\phi$, $\mathbf{B}_\phi$ *and* $\mathbf{C}_\phi$ *as in Example 6.1 and the compressible ones* $\bar{\mathbf{A}}_\phi$, $\bar{B}_\phi$ *and* $\bar{C}_\phi$ *from Eqs. (6.81)–(6.83). We calculate the wake doublet strength in terms of the bound doublet strength difference at the trailing edge using Eq. (5.220), as in the incompressible case. We calculate matrices*

$$\boldsymbol{P}_c = \left(\mathbf{0}_{n\times(2m-1)n}\, \boldsymbol{I}_n\right) - \left(\boldsymbol{I}_n\, \mathbf{0}_{n\times(2m-1)n}\right) \tag{6.84}$$

$$\boldsymbol{P}_e(\omega) = \begin{pmatrix} \boldsymbol{I}_n \mathrm{e}^{-\mathrm{i}\omega\Delta t} \\ \boldsymbol{I}_n \mathrm{e}^{-\mathrm{i}\omega 2\Delta t} \\ \vdots \\ \boldsymbol{I}_n \mathrm{e}^{-\mathrm{i}\omega m_w \Delta t} \end{pmatrix} \tag{6.85}$$

so that

$$\boldsymbol{\mu}_w = \boldsymbol{P}_e(\omega)\boldsymbol{P}_c\boldsymbol{\mu}$$

We then substitute this result for $\boldsymbol{\mu}_w$ *in Eq. (6.80) to obtain the compressible equivalent of Eq. (5.221)*

$$\left(\bar{\boldsymbol{B}}_\phi(\omega) - \frac{1}{2}\boldsymbol{I} + \bar{\boldsymbol{C}}_\phi(\omega)\boldsymbol{P}_e(\omega)\boldsymbol{P}_c\right)\boldsymbol{\mu}(\omega) = -\bar{\boldsymbol{A}}_\phi(\omega)\boldsymbol{\sigma}(\omega) \tag{6.86}$$

The source strength is calculated from Eq. (6.71), which can also be written as

$$\begin{aligned}\sigma(\xi_s,\omega) = {} & \frac{\mathrm{i}\Omega M_\infty}{a_\infty}\mu(\xi_s,\omega)n_\xi(\xi_s) \\ & - \left(\mathbf{Q}_\infty\delta(\omega) - \mathbf{v}_0(\omega) - \boldsymbol{\Omega}_0(\omega)\times(\mathbf{x}_s - \mathbf{x}_f)\right)\boldsymbol{B}^{-1}\mathbf{n}^T(\xi_s)\end{aligned} \tag{6.87}$$

On the control points, this expression becomes

$$\boldsymbol{\sigma}(\omega) = \frac{\mathrm{i}\Omega M_\infty}{a_\infty}\boldsymbol{n}_\xi \circ \boldsymbol{\mu}(\omega) - \left(\frac{1}{\beta}\boldsymbol{u}_m(\omega)\circ\boldsymbol{n}_\xi + \boldsymbol{v}_m(\omega)\circ\boldsymbol{n}_\eta + \boldsymbol{w}_m(\omega)\circ\boldsymbol{n}_\zeta\right)$$

where $\boldsymbol{u}_m$, $\boldsymbol{v}_m$, $\boldsymbol{w}_m$ *are the relative velocities on the control points of the original compressible geometry and* $\boldsymbol{n}_\xi$, $\boldsymbol{n}_\eta$, $\boldsymbol{n}_\zeta$ *are the components of the normal vectors on the transformed incompressible geometry. For sinusoidal pitching around point* $(x_f, 0, z_f)$ *with mean angle of attack* α_0*, amplitude* α_1*, phase* ϕ_α *and frequency* ω_0*, the relative velocities can be written as*

$$\begin{aligned}&\boldsymbol{u}_m(0) = Q_\infty\cos\alpha_0,\ \boldsymbol{u}_m(\omega_0) = -\mathrm{i}\omega_0(\boldsymbol{z}_c - z_f)\frac{\alpha_1}{2\mathrm{i}}\mathrm{e}^{\mathrm{i}\phi_\alpha} \\ &\boldsymbol{v}_m(0) = 0,\ \boldsymbol{v}_m(\omega_0) = 0 \\ &\boldsymbol{w}_m(0) = Q_\infty\sin\alpha_0,\ \boldsymbol{w}_m(\omega_0) = \left(Q_\infty + \mathrm{i}\omega_0(\boldsymbol{x}_c - x_f)\right)\frac{\alpha_1}{2\mathrm{i}}\mathrm{e}^{\mathrm{i}\phi_\alpha}\end{aligned} \tag{6.88}$$

We first solve the mean flow problem, for $\omega = 0$*. Equation (6.86) simplifies to*

$$\left(\boldsymbol{B}_\phi - \frac{1}{2}\boldsymbol{I} + \boldsymbol{C}_\phi\boldsymbol{P}_e(0)\boldsymbol{P}_c\right)\boldsymbol{\mu}(0) = -\boldsymbol{A}_\phi\boldsymbol{\sigma}(0) \tag{6.89}$$

and $\boldsymbol{\sigma}(0)$ becomes identical to Eq. (6.18) for fully steady flow such that

$$\boldsymbol{\sigma}(0) = -Q_\infty \left(\frac{\cos \alpha_0}{\beta} \mathbf{n}_\xi + \sin \alpha_0 \mathbf{n}_\zeta \right) \tag{6.90}$$

We solve for $\boldsymbol{\mu}(0)$ and calculate the steady incompressible velocities from Eq. (6.20) and the steady compressible velocities, $\phi_{x_{i,j}}(0)$, $\phi_{y_{i,j}}(0)$, $\phi_{z_{i,j}}(0)$, from (6.4).

For the unsteady problem, we solve Eq. (6.86) for $\omega = \omega_0$. The source strength is given by

$$\boldsymbol{\sigma}(\omega_0) = \frac{\mathrm{i}\Omega_0 M_\infty}{a_\infty} \boldsymbol{n}_\xi \circ \boldsymbol{\mu}(\omega_0) - \left(\frac{1}{\beta} \boldsymbol{u}_m(\omega_0) \circ \boldsymbol{n}_\xi + \boldsymbol{v}_m(\omega_0) \circ \boldsymbol{n}_\eta + \boldsymbol{w}_m(\omega_0) \circ \boldsymbol{n}_\zeta \right) \tag{6.91}$$

so that Green's identity (6.86) becomes

$$\left(\frac{\mathrm{i}\Omega_0 M_\infty}{a_\infty} \bar{\boldsymbol{A}}_\phi(\omega_0) \circ \boldsymbol{n}_\xi + \bar{\boldsymbol{B}}_\phi(\omega_0) - \frac{1}{2}\boldsymbol{I} + \bar{\boldsymbol{C}}_\phi(\omega_0) \boldsymbol{P}_e(\omega) \boldsymbol{P}_c \right) \boldsymbol{\mu}(\omega_0)$$
$$= \bar{\boldsymbol{A}}_\phi(\omega_0) \left(\frac{1}{\beta} \boldsymbol{u}_m(\omega_0) \circ \boldsymbol{n}_\xi + \boldsymbol{v}_m(\omega_0) \circ \boldsymbol{n}_\eta + \boldsymbol{w}_m(\omega_0) \circ \boldsymbol{n}_\zeta \right) \tag{6.92}$$

The next steps are to solve Eq. (6.92) for $\boldsymbol{\mu}(\omega_0)$ and to calculate the unsteady incompressible velocities from Eq. (6.20) and the unsteady compressible velocities, $\phi_{x_{i,j}}(\omega_0)$, $\phi_{y_{i,j}}(\omega_0)$, $\phi_{z_{i,j}}(\omega_0)$, from Eq. (6.4). Finally, we need to evaluate the mean and oscillating components of the pressure distribution. In Morino's original technique, this calculation is carried out by means of the linearised pressure coefficient of Eq. (2.154). Here, we use the non-linear equation (6.25) instead, in order to calculate more accurate pressure distributions (Martínez and Dimitriadis 2022). The procedure is similar to the one employed in Example 5.13 so that a multiplication in the time domain becomes a convolution in the frequency domain. The Fourier transform of Eq. (6.25) is

$$c_p(\omega) = \delta(\omega) - \frac{Q(\omega) * Q(\omega)}{Q_\infty^2} + \frac{M_\infty^2}{Q_\infty^2} \phi_x(\omega) * \phi_x(\omega) - \frac{2\mathrm{i}\omega}{Q_\infty^2} \phi(\omega)$$
$$+ \frac{M_\infty^2}{Q_\infty^4} (\mathrm{i}\omega\phi(\omega)) * (\mathrm{i}\omega\phi(\omega)) + \frac{2M_\infty^2}{Q_\infty^3} \phi_x(\omega) * (\mathrm{i}\omega\phi(\omega)) \tag{6.93}$$

where

$$\begin{aligned}
\phi_x(\omega) &= (\phi_x^*(\omega_0),\ \phi_x(0),\ \phi_x(\omega_0)) \\
\mathrm{i}\omega\phi(\omega) &= ((\mathrm{i}\omega_0\mu(\omega_0))^*,\ 0,\ \mathrm{i}\omega_0\mu(\omega_0)) \\
u(\omega) &= (u_m^*\left(\omega_0\right) + \phi_x^*(\omega_0),\ U_\infty + \phi_x(0), u_m\left(\omega_0\right) + \phi_x(\omega_0)) \\
v(\omega) &= (v_m^*\left(\omega_0\right) + \phi_y^*(\omega_0),\ V_\infty + \phi_y(0), v_m\left(\omega_0\right) + \phi_y(\omega_0)) \\
w(\omega) &= (w_m^*\left(\omega_0\right) + \phi_z^*(\omega_0),\ W_\infty + \phi_z(0), w_m\left(\omega_0\right) + \phi_z(\omega_0)) \\
Q(\omega) * Q(\omega) &= u(\omega) * u(\omega) + v(\omega) * v(\omega) + w(\omega) * w(\omega)
\end{aligned}$$

Since $\phi_x(\omega)$, $\mathrm{i}\omega\phi(\omega)$, $u(\omega)$, $v(\omega)$, $w(\omega)$ all have three frequency components, $c_p(\omega)$ will have five components due to the convolution. Consequently, the third component gives $c_p(0)$, the fourth the $c_p(\omega_0)$ and the fifth $c_p(2\omega_0)$. We can then calculate the total aerodynamic loads acting on the wing panels, $\mathbf{F}_{i,j}(0)$, $\mathbf{F}_{i,j}(\omega_0)$, $\mathbf{F}_{i,j}(2\omega_0)$, from Eq. (6.22) and the lift and drag coefficients from Eq. (5.131). The latter have frequency components up to $3\omega_0$ but only the components at 0, ω_0 and $2\omega_0$ take significant values.

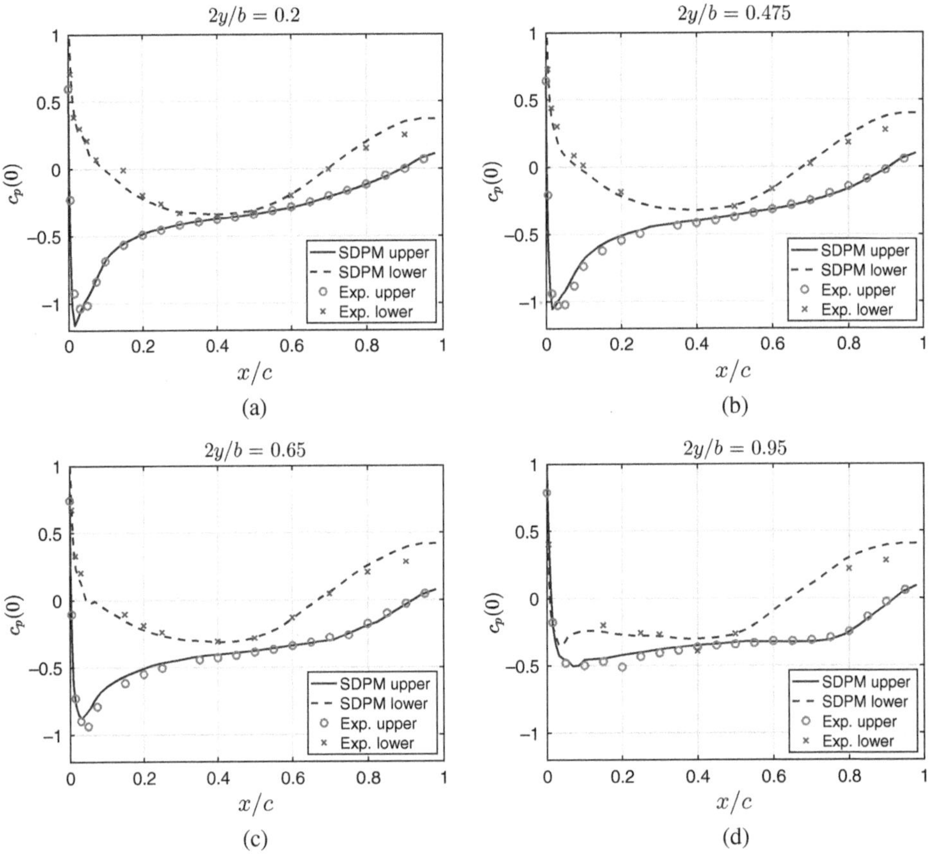

Figure 6.8 Mean pressure distribution predicted by the SDPM around LANN wing. Experimental data by Zwaan (1985). (a) $2y/b = 0.2$, (b) $2y/b = 0.475$, (c) $2y/b = 0.65$ and (d) $2y/b = 0.95$. Source: Zwaan (1985), credit: AGARD.

We carry out the SDPM calculation with $m = 40$ *chordwise and* $n = 40$ *spanwise panels. Figure 6.8 plots the mean pressure distribution* $c_p(0)$ *at four of the six measurement stations of the LANN wing. The pressures predicted by the SDPM are interpolated linearly in order to obtain the distribution at the measurement stations, as already carried out in Examples 5.13 and 6.2. It can be seen that there is very good agreement between the predicted and measured mean pressure distributions.*

Figure 6.9 plots the real and imaginary parts of the oscillating pressure distribution $c_{p_\alpha}(\omega_0) = 2ic_p(\omega_0)/\alpha_1$ *at the same four measurement stations. Again, the agreement between the numerical and experimental results is very good, except near the leading edge of the wingtip section, where the amplitude is underestimated. During the experiment, the pitching oscillations were causing slight bending and twisting of the wing, whose amplitude was maximum at the wingtip. Zwaan (1985) provide data for the bending and twisting mode shapes, which have not been implemented in the present calculation. This implementation is left as an exercise for the reader.*

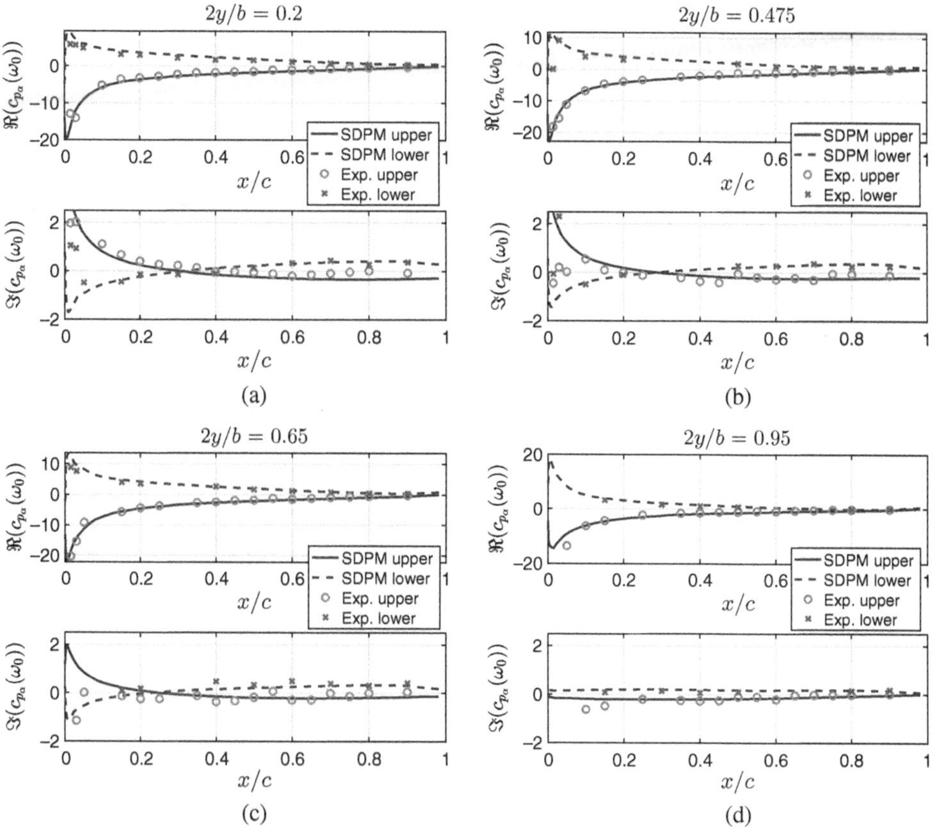

Figure 6.9 Real and imaginary parts of oscillatory pressure distribution predicted by the SDPM around LANN wing. Experimental data by Zwaan (1985). (a) $2y/b = 0.2$, (b) $2y/b = 0.475$, (c) $2y/b = 0.65$ and (d) $2y/b = 0.95$. Source: Zwaan (1985), credit: AGARD.

As the compressible SDPM represents the correct shape of the wing (including thickness, camber, twist, dihedral, etc.,) and as it uses a non-linear pressure calculation, it can calculate both the mean and oscillatory pressure distributions on the upper and lower surfaces with good accuracy. The fact that the exponential terms in Eq. (6.70) have been approximated by their values at the control points of the influencing panel is acceptable for the chosen number of panels but the choice $m = 20$, $n = 20$ *also gives reasonable pressure predictions. In fact, these exponential terms affect mainly the imaginary part of the pressure distribution; the real part is affected mostly by the steady influence coefficient matrices,* $\mathbf{A}_\phi$, $\mathbf{B}_\phi$ *and* $\mathbf{C}_\phi$. *The reader is invited to repeat the calculation of the present example after setting* $\Omega = 0$ *in Eqs. (6.81)–(6.83) and to compare the resulting oscillatory pressure distributions to the ones presented in Figure 6.9.*

The frequency components of the total aerodynamic lift are obtained by the SDPM is $C_L(0) = 0.299$ *and* $2\mathrm{i}C_L(\omega_0)/\pi\alpha_1 = 1.681 + 0.019\mathrm{i}$. *Recall from Example 6.2 that the corresponding experimental values are 0.285 and* $1.326 + 0.047\mathrm{i}$. *Therefore, the SDPM predicts a reasonably accurate value for the mean lift and overestimates the magnitude of the oscillatory lift although, as mentioned earlier, the experimental values are probably underestimated because of the limited number of pressure sensors. This example is solved by Matlab code* `pitchplunge_subsonic_SDPM.m`.

It is interesting to compare the predictions of the SDPM and DLM techniques. The pressure jump across the surface of the wing can be calculated from the SDPM results by subtracting the pressure distribution on the upper surface from that on the lower surface. Consequently, the pressure jumps

$$\Delta c_{p_{i,j}}(\omega) = c_{p_{m-i+1,j}}(\omega) - c_{p_{m+i,j}}(\omega) \tag{6.94}$$

correspond to non-dimensional chordwise control point coordinates $\bar{x}_{c_{m+i}}$ for $i = 1, \dots, m$. Figure 6.10 compares the mean pressure jump predictions across the LANN wing obtained by the DLM and SDPM techniques (Examples 6.2 and 6.3, respectively). It can be seen that the SDPM results are closer to the experimental measurements than the DLM predictions,

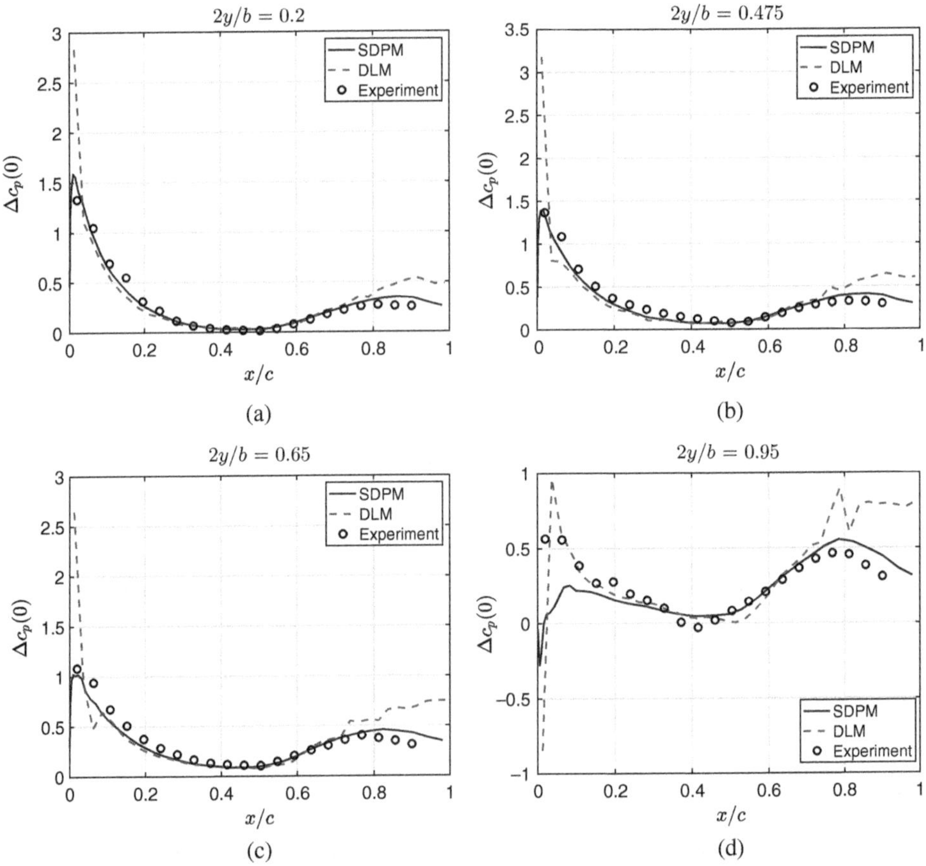

Figure 6.10 Comparison of mean pressure jump across the LANN wing predicted by the SDPM and the DLM. Experimental data by Zwaan (1985). (a) $2y/b = 0.2$, (b) $2y/b = 0.475$, (c) $2y/b = 0.65$ and ($2y/b = 0.95$). Source: Zwaan (1985), credit: AGARD.

particularly around the leading and trailing edges. The only exception is the wingtip (Figure 6.10d), where the DLM is more accurate around the leading edge. This exception may be an artefact due to the low number of pressure measurements near the leading edge at this spanwise position. The fact that the SDPM does not require the decomposition of the airfoil's geometry into a thickness distribution and camber line is a significant advantage; this decomposition leads the DLM to overestimate the pressure jump near the trailing edge, as if the local camber has been overestimated. Corrections to the estimated camber line can be applied to obtain better results from the DLM.

Figure 6.11 compares the predictions of the DLM and SDPM for the oscillating pressure jump across the LANN wing. It is clear that the SDPM predicts both the amplitude and phase of $\Delta c_{p_\alpha}(\omega_0)$ more accurately than the DLM. In particular, the SDPM better matches the amplitude of the pressure jump near the leading edge and the phase near the trailing edge. In this case, the improved accuracy of the SDPM is due to both the accurate geometric modelling and the nonlinear pressure calculation. The increased computational cost of the SDPM is therefore justified.

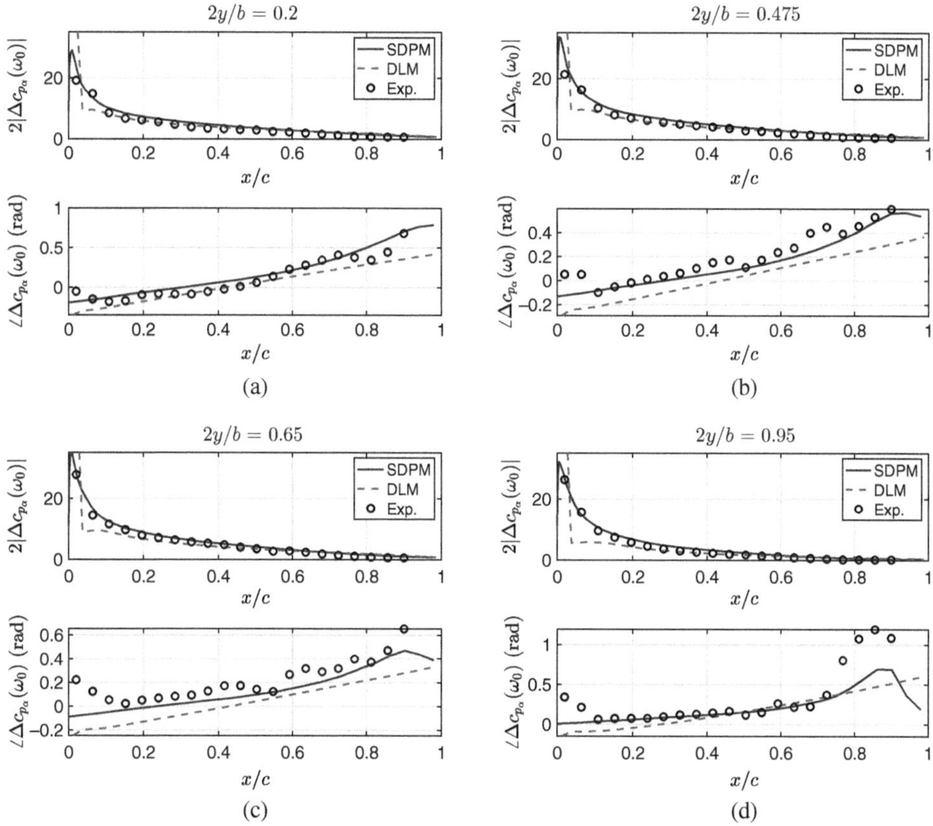

Figure 6.11 Comparison of oscillating pressure jump across the LANN wing predicted by the SDPM and the DLM. Experimental data by Zwaan (1985). (a) $2y/b = 0.2$, (b) $2y/b = 0.475$, (c) $2y/b = 0.65$ and $2y/b = 0.95$. Source: Zwaan (1985), credit: AGARD.

The convergence of the subsonic unsteady SDPM can be accelerated by improving on Morino's approximation (Martínez and Dimitriadis 2022). Instead of approximating the potential influence coefficient of a source panel as in Eq. (6.76), we can write

$$\bar{A}_{\phi_{I,J}}(\omega) = -\frac{e^{i\Omega M_\infty(\xi_{c_I}-\xi_{c_J})}}{4\pi}\int_{\Delta\Sigma_J-\xi_{c_I}} \frac{e^{-i\Omega r(\xi_{c_I},\xi_s)}}{r(\xi_{c_I},\xi_s)} d\Sigma \tag{6.95}$$

Then we can see that the function inside the integrand is identical to the potential induced by an acoustic subsonic source, as seen in Eq. (B.8) for $\Omega = \omega/\beta a_\infty$. This means that we can calculate the integral using the numerical approach of Eq. (B.10). Similarly, the approximation of the doublet influence coefficients given in Eq. (6.79) can be improved by writing

$$\begin{aligned}\bar{B}_{\phi_{I,J}}(\omega) = {} & \frac{1}{4\pi}e^{i\Omega M_\infty(\xi_{c_I}-\xi_{c_J})}\int_{\Delta\Sigma_J-\xi_{c_I}} \mathbf{n}(\xi_{c_J})\cdot\bar{\nabla}_s\left(\frac{e^{-i\Omega r(\xi_{c_I},\xi_s)}}{r(\xi_{c_I},\xi_s)}\right) d\Sigma \\ & + \frac{1}{4\pi}i\Omega e^{i\Omega M_\infty(\xi_{c_I}-\xi_{c_J})}M_\infty n_\xi(\xi_{c_J})\int_{\Delta\Sigma_J-\xi_{c_I}} \frac{e^{-i\Omega r(\xi_{c_I},\xi_s)}}{r(\xi_{c_I},\xi_s)} d\Sigma\end{aligned} \tag{6.96}$$

The integrand of the first integral is the acoustic subsonic doublet of Eq. (B.12) so that the integral can be evaluated from the numerical scheme of Eq. (B.14). The second integral contains the acoustic subsonic source, which has already been discussed.

6.3.3 Steady Correction of the Doublet Lattice Method

Figure 6.10 shows that the DLM predictions of the steady pressure jump across the surface can be inaccurate around the leading and trailing edges. As stated in Example 6.2, this inaccuracy is due to the difficulty in defining an adequate camber surface. An alternative to using an inadequate camber line is to correct the DLM equation by means of more accurate pressure data, either from another simulation method or from experimental measurements. Figure 6.10 demonstrates that the steady SDPM predictions are more accurate than the DLM results and, consequently, the former can be used to correct the latter.

We start with the steady DLM Eq. (6.58) and remove the contribution from the camber line to obtain

$$-\alpha_0 = \sum_{J=1}^{N} \tilde{A}_{I,J}(0)\Delta c_{p_J}(0)$$

or, in matrix form,

$$-\alpha_0 \boldsymbol{I}_{N,1} = \tilde{\boldsymbol{A}}(0)\Delta\boldsymbol{c}_p(0) \tag{6.97}$$

where $\boldsymbol{I}_{N,1}$ is a $N \times 1$ vector whose elements are all equal to 1. The steady pressure jump $\Delta\boldsymbol{c}_p(0, \alpha_0)$ that we will obtain by solving the equation above will be even worse than the DLM results of Figure 6.10 since the wing's camber is not represented at all. In fact, if we set $\alpha_0 = 0$ in Eq. (6.97), we will obtain $\Delta\boldsymbol{c}_p(0) = \boldsymbol{0}$. As the following discussion concerns only steady aerodynamics, we will drop the (0) and $_0$ notation that denotes that the frequency is equal to zero. Therefore, Eq. (6.97) becomes

$$-\alpha \boldsymbol{I}_{N,1} = \tilde{\boldsymbol{A}}\Delta\boldsymbol{c}_p \tag{6.98}$$

We will now assume that we know the correct pressure jump at a reference angle of attack α_{ref}, denoted by $\Delta \boldsymbol{c}_p(\alpha_{\text{ref}})$, and that we are interested in calculating the pressure jump at a slightly higher angle $\alpha_{\text{ref}} + \Delta\alpha$. Applying a first order Taylor expansion to $\Delta \boldsymbol{c}_p$ around α_{ref},

$$\Delta \boldsymbol{c}_p(\alpha_{\text{ref}} + \Delta\alpha) = \Delta \boldsymbol{c}_p(\alpha_{\text{ref}}) + \frac{\partial \Delta \boldsymbol{c}_p}{\partial \alpha}\Delta\alpha \tag{6.99}$$

The derivative $\partial \Delta \boldsymbol{c}_p / \partial \alpha$ can be calculated directly from Eq. (6.98)

$$\frac{\partial \Delta \boldsymbol{c}_p}{\partial \alpha} = -\tilde{\boldsymbol{A}}^{-1}\boldsymbol{I}_{N,1}$$

However, in writing this latest expression, we are assuming that the steady global influence coefficient matrix $\tilde{\boldsymbol{A}}$ calculated by the DLM is accurate, which is not necessarily the case. Therefore, we will correct $\tilde{\boldsymbol{A}}$ by a correction matrix $\boldsymbol{D}_{\text{corr}}$, such that

$$\Delta \boldsymbol{c}_p(\alpha_{\text{ref}} + \Delta\alpha) = \Delta \boldsymbol{c}_p(\alpha_{\text{ref}}) - \left(\boldsymbol{D}_{\text{corr}}\tilde{\boldsymbol{A}}\right)^{-1}\boldsymbol{I}_{N,1}\Delta\alpha \tag{6.100}$$

will yield an accurate pressure jump distribution $\Delta \boldsymbol{c}_p(\alpha_{\text{ref}} + \Delta\alpha)$, for every $\Delta\alpha$.

Values for the elements of $\boldsymbol{D}_{\text{corr}}$ can be obtained if accurate pressure jumps at both α_{ref} and $\alpha_{\text{ref}} + \Delta\alpha$ are calculated using a higher fidelity simulation method or measured experimentally. These accurate pressure values will be referred to as the reference pressures $\Delta \boldsymbol{c}_p^{\text{ref}}(\alpha_{\text{ref}})$ and $\Delta \boldsymbol{c}_p^{\text{ref}}(\alpha_{\text{ref}} + \Delta\alpha)$. Substituting them in Eq. (6.100) and rearranging leads to

$$\Delta \boldsymbol{c}_{p_\alpha}^{\text{ref}} = \frac{\Delta \boldsymbol{c}_p^{\text{ref}}(\alpha_{\text{ref}} + \Delta\alpha) - \Delta \boldsymbol{c}_p^{\text{ref}}(\alpha_{\text{ref}})}{\Delta\alpha} = -\left(\boldsymbol{D}_{\text{corr}}\tilde{\boldsymbol{A}}\right)^{-1}\boldsymbol{I}_{N,1} \tag{6.101}$$

where $\Delta \boldsymbol{c}_{p_\alpha}^{\text{ref}}$ is the derivative of the reference pressure jump with respect to the pitch angle. Every term in this expression is known except for $\boldsymbol{D}_{\text{corr}}$. However, the dimensions of $\boldsymbol{D}_{\text{corr}}$ are $N \times N$, where N is the number of panels, while there are only N equations. This problem is usually sidestepped by defining $\boldsymbol{D}_{\text{corr}}$ as a diagonal matrix so that only its N non-diagonal elements have non-zero values. Solving Eq. (6.101) for these elements yields (see for example (Friedewald et al. 2018))

$$D_{I,I}^{\text{corr}} = -\frac{1}{\sum_{J=1}^{N} \tilde{A}_{I,J}\Delta c_{p_\alpha,J}^{\text{ref}}} \tag{6.102}$$

for $I = 1, \dots, N$. Then, the corrected DLM pressure jump at any value of α is given by Eq. (6.100)

$$\Delta \boldsymbol{c}_p(\alpha) = -\left(\boldsymbol{D}_{\text{corr}}\tilde{\boldsymbol{A}}\right)^{-1}\boldsymbol{I}_{N,1}(\alpha - \alpha_{\text{ref}}) + \Delta \boldsymbol{c}_p^{\text{ref}}(\alpha_{\text{ref}}) \tag{6.103}$$

Example 6.4 *Correct the DLM predictions of the steady pressure jump across the LANN wing using results from the SDPM.*

We use the same geometry and parameter values for the LANN wing already mentioned in Examples 6.2 and 6.3. We set up matching grids for the DLM and SDPM; as the DLM grid is linear in both the chordwise and spanwise directions, we select the same setup for the SDPM. An alternative would be to evaluate the two solutions on different grids and then to interpolate the SDPM results onto the DLM grid. In order to reduce the complexity of the example, we choose the former approach, setting $m = 40$ *and* $n = 40$.

We calculate the steady solution from the SDPM using Eq. (6.89) for three values of the steady angle of attack, $\alpha_{\text{ref}} = 0°, 0.5°$ *and* $1°$. *We evaluate the reference pressure jumps from Eq. (6.94) for each of the* α_{ref} *values and then we use Eq. (6.101) to calculate the derivative of the reference pressure jump, selecting* $\alpha_{\text{ref}} = 0°$ *and* $\alpha_{\text{ref}} + \Delta\alpha = 1°$.

The next step is to set up the DLM equation (6.97), namely to calculate the $\tilde{\mathbf{A}}(0)$ *matrix, as detailed in Example 6.2. Equation (6.102) is then used to evaluate the correction matrix* $\mathbf{D}_{\text{corr}}$ *and finally we set up Eq. (6.103). We select* $\alpha = 0.5°$ *and evaluate the corrected DLM pressure jump distribution* $\Delta\mathbf{c}_{\text{p}}(\alpha)$. *Figure 6.12 compares the SDPM result for* $\alpha = 0.5°$ *to the prediction that we obtained from the corrected DLM method. It can be seen that the two sets of pressure jumps are nearly identical throughout the wing. Very good agreement can also be obtained at other values of* α, *but it must be kept in mind that the pressure calculation in the SDPM is non-linear; if* α *moves far from* α_{ref}, *the agreement between the corrected DLM and SDPM pressure values will start to degrade. This example is solved by Matlab code* `steady_DLM_corr.m`.

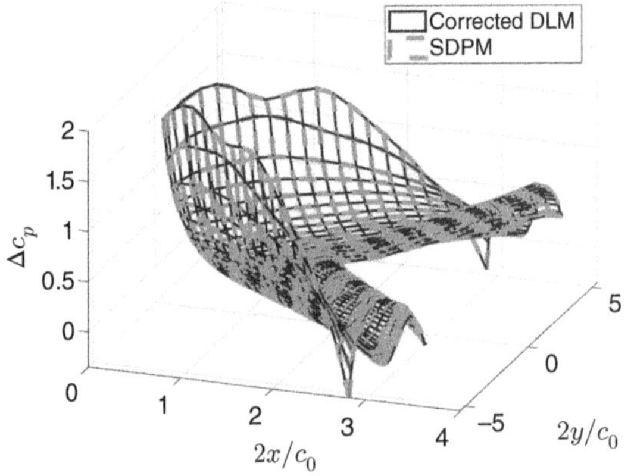

Figure 6.12 Comparison of pressure jump distribution on the LANN wing obtained from the corrected DLM and SDPM approaches.

Any trustworthy source can be used to obtain reference data, including experimental measurements or RANS, Euler, etc., simulations. The DLM correction method presented here is simple and effective, but several other approaches have been published. For example, the reference data can be integrated aerodynamic loads, such as the lift and pitching moment distribution along the span. Furthermore, even unsteady corrections can be applied to the DLM, for example in order to correct for unsteady transonic flow phenomena. One such approach will be presented later in this chapter. Palacios et al. (2001) provide a nice overview of DLM correction techniques.

6.3.4 Unsteady 2D Subsonic Source and Doublet Panel Method

Two-dimensional flow is a special case of three-dimensional flow so that the steady and unsteady pressure distributions around airfoil sections can be evaluated from finite wing methodologies. The 3D subsonic SDPM can be applied to unsteady flow around an airfoil section if the section is represented as a finite rectangular wing with very high aspect ratio. Only one panel is required in the spanwise direction, so that $n = 1$. The rest of the simulation procedure is identical to that of the 3D case.

Example 6.5 *Use the SDPM in order to predict the unsteady pressure distribution around a NLR 7301 airfoil with an oscillating flap.*

Zwaan (1982) published data from a wind tunnel experiment on a NLR 7301 supercritical airfoil undergoing oscillatory pitch and oscillatory flap motion. The shape of the 16.5% thick airfoil is given as a set of upper and lower surface coordinates. The airfoil, with $c = 0.18$ m, has a flap over the last 25% of its chord, the flap hinge lying at $x_h = 0.75c$, $z_h = 0$. Here we will model the oscillatory flap test case at $M_\infty = 0.502$; there is no pitching motion, just an oscillation of the flap angle, δ, given by

$$\delta(t) = \delta_0 + \delta_1 \sin\left(\omega_0 t + \phi_\delta\right)$$

where $\delta_0 = 0.02°$ is the mean angle, $\delta_1 = 0.97°$ is the amplitude and $\phi_\delta = 0°$ is the phase; $\delta(t)$ is defined as positive downwards (clockwise). The reduced frequency is $k = 0.098$ and the oscillation frequency is 30 Hz, from which we can deduce that the free stream airspeed is $U = 173.11$ m/s and the speed of sound $a_\infty = 344.84$ m/s. The pitch angle remains equal to zero at all times. Steady and unsteady pressure measurements are given at 23 chordwise positions on the upper and lower surfaces. Figure 6.13 plots the airfoil, the flap at three positions, the flap hinge and the pressure measurement points.

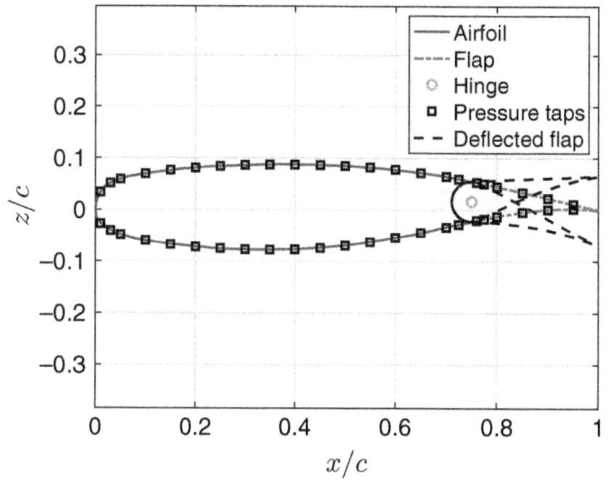

Figure 6.13 NLR 7301 supercritical airfoil with oscillating flap tested by Zwaan (1982). Source: Adapted from Zwaan (1982), credit: AGARD.

We model the airfoil as a 3D rectangular wing with span $b = 18$ m so that the aspect ratio is 100. There are 2m chordwise and $n = 1$ spanwise panels, with spanwise vertex coordinates $y_{p_{i,1}} = -b/2$, $y_{p_{i,2}} = b/2$ for $i = 1, \dots, 2m + 1$. This means that the 2m control points all lie on $y_{i,c} = 0$, the midspan point. The vertices of the chordwise panels are calculated by interpolating the measured airfoil geometry at x coordinates given by Eq. (5.180) using a cubic spline. The geometrical modelling is quasi-fixed, as usual, so that the flap does not deflect *but its motion is modelled by the relative flow velocity induced by the motion. For the general case where the wing is undergoing both pitching and flap oscillations, the mean relative flow velocities are*

$$\boldsymbol{u}_m(0) = Q_\infty \cos(\alpha_0)(1 - \boldsymbol{I}_f) + Q_\infty \cos(\alpha_0 + \delta_0)\boldsymbol{I}_f$$
$$\boldsymbol{v}_m(0) = 0$$
$$\boldsymbol{w}_m(0) = Q_\infty \sin(\alpha_0)(1 - \boldsymbol{I}_f) + Q_\infty \sin(\alpha_0 + \delta_0)\boldsymbol{I}_f$$

and the oscillatory ones

$$\boldsymbol{u}_m(\omega_0) = -\mathrm{i}\omega_0 \frac{\alpha_1}{2\mathrm{i}} \mathrm{e}^{\mathrm{i}\phi_\alpha}(\boldsymbol{z}_c - z_h) - \mathrm{i}\omega_0 \frac{\delta_1}{2\mathrm{i}} \mathrm{e}^{\mathrm{i}\phi_\delta}(\boldsymbol{z}_c - z_f) \circ \boldsymbol{I}_f$$
$$\boldsymbol{v}_m(\omega_0) = 0$$
$$\boldsymbol{w}_m(\omega_0) = \frac{\alpha_1}{2\mathrm{i}} \mathrm{e}^{\mathrm{i}\phi_\alpha} \left(Q_\infty + \mathrm{i}\omega_0(\boldsymbol{x}_c - x_f)\right) + \frac{\delta_1}{2\mathrm{i}} \mathrm{e}^{\mathrm{i}\phi_\delta} \left(Q_\infty + \mathrm{i}\omega_0(\boldsymbol{x}_c - x_h)\right) \circ \boldsymbol{I}_f$$

where $\boldsymbol{I}_f$ *is a* $2m \times 1$ *vector whose elements are equal to 1 if the corresponding control point belongs to the flap and 0 otherwise. The rest of the modelling procedure is identical to the one used in Example 6.3. The wake length is set to 10 chords and the chordwise length of the wake panels to* c_0/m.

Figure 6.14a plots the predicted mean pressure distribution around the airfoil for $m = 40$ *and compares it to the experimental measurements. The agreement between the SDPM and experimental mean pressure distributions is reasonably good, but the former is slightly fatter. Zwaan (1982) states that they observed a similar phenomenon when comparing steady theoretical predictions to the measurements. They attributed the difference to viscous effects and wind tunnel interference. Figure 6.14b plots the real and imaginary parts of the oscillatory pressure distribution,* $c_{p_\delta}(\omega_0) = 2\mathrm{i}c_p(\omega_0)/\delta_1$. *Even though the motion is restricted to the flap, the oscillatory pressure distribution over the entire airfoil is affected. There are two peaks in the real part of* $c_{p_\delta}(\omega_0)$, *around the leading edge and at the hinge axis; the imaginary*

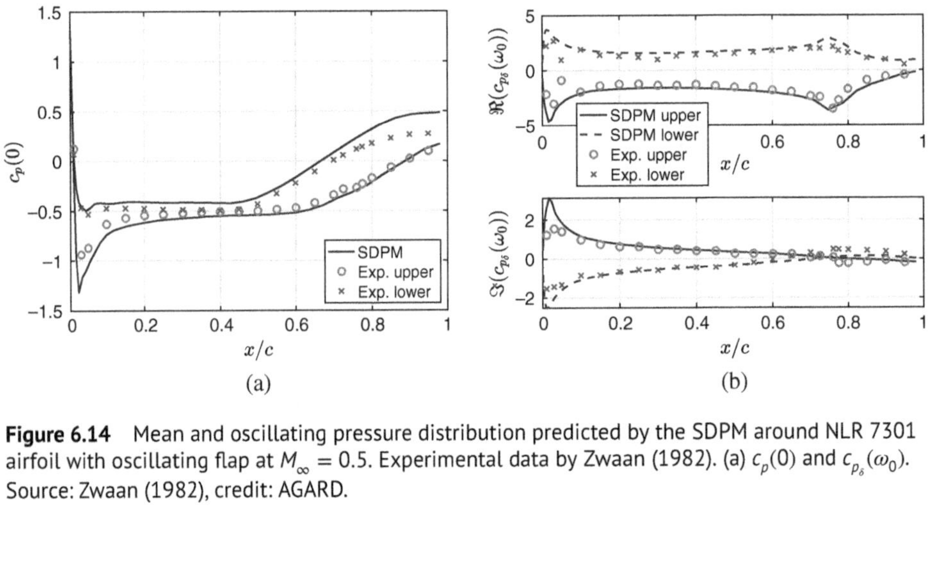

Figure 6.14 Mean and oscillating pressure distribution predicted by the SDPM around NLR 7301 airfoil with oscillating flap at $M_\infty = 0.5$. Experimental data by Zwaan (1982). (a) $c_p(0)$ and $c_{p_\delta}(\omega_0)$. Source: Zwaan (1982), credit: AGARD.

part also peaks around the leading edge but is equal to zero at the hinge axis. The imaginary distribution predicted by the SDPM is in generally good agreement with the experimental measurements, except around the leading edge and hinge axis where small differences can be observed. The SDPM real pressure distribution is accurate everywhere. As the airfoil coordinates are interpolated from measured data, increasing m will lead to noisy interpolations and, therefore, noisy pressure distributions. It is not recommended to use $m > 40$ for the present test case. This example is solved by Matlab code `pitchplunge_Machbox.m`.

6.4 Unsteady Supersonic Potential Flow

In Section 2.6, we showed that for supersonic flow, the linearised small perturbation equation is given by

$$-B^2\phi_{xx} + \phi_{yy} + \phi_{zz} - \frac{2M_\infty}{a_\infty}\phi_{xt} - \frac{1}{a_\infty^2}\phi_{tt} = 0 \tag{6.104}$$

where $B = \sqrt{M_\infty^2 - 1}$. A fundamental solution to this equation is the supersonic source of Eq. (2.194)

$$\phi(\mathbf{x}, \mathbf{x}_0, t) = \frac{F\left(t - \tau_+(\mathbf{x}, \mathbf{x}_0)\right) + F\left(t - \tau_-(\mathbf{x}, \mathbf{x}_0)\right)}{r_B(\mathbf{x}, \mathbf{x}_0)} \tag{6.105}$$

where $\phi(\mathbf{x}, t)$ is the potential induced at any point in space $\mathbf{x}$ by a source lying at $\mathbf{x}_0$ and

$$r_B(\mathbf{x}, \mathbf{x}_0) = \sqrt{\left(x - x_0\right)^2 - B^2\left(y - y_0\right)^2 - B^2\left(z - z_0\right)^2} \tag{6.106}$$

$$\tau_\pm(\mathbf{x}, \mathbf{x}_0) = \frac{M_\infty(x - x_0) \pm r_B(\mathbf{x}, \mathbf{x}_0)}{a_\infty B^2} \tag{6.107}$$

The supersonic doublet is also a solution to Eq. (6.104); it induces a potential given by

$$\phi(\mathbf{x}, \mathbf{x}_0, t) = \mathbf{n} \cdot \nabla \frac{F\left(t - \tau_+(\mathbf{x}, \mathbf{x}_0)\right) + F\left(t - \tau_-(\mathbf{x}, \mathbf{x}_0)\right)}{r_B(\mathbf{x}, \mathbf{x}_0)} \tag{6.108}$$

As usual, a flowfield can be built up from the superposition of sources and doublets lying on the surface of a body that is described by the function $S(\mathbf{x}_0, t) = 0$. For quasi-fixed geometries, the potential anywhere in this flow is given by Green's theorem of Eq. (2.196), which can be rewritten as

$$\begin{aligned}
4\pi E\phi(\mathbf{x}, t) = &-\int_S \mathbf{n}(\mathbf{x}_s) \cdot \nabla \left(\phi(\mathbf{x}_s, t - \tau_+) + \phi(\mathbf{x}_s, t - \tau_-)\right) \frac{1}{r_B(\mathbf{x}, \mathbf{x}_s)} \mathrm{d}S \\
&+ \int_S \left(\phi(\mathbf{x}_s, t - \tau_+) + \phi(\mathbf{x}_s, t - \tau_-)\right) \mathbf{n}(\mathbf{x}_s) \cdot \nabla \left(\frac{1}{r_B(\mathbf{x}, \mathbf{x}_s)}\right) \mathrm{d}S \\
&- \frac{\partial}{\partial t} \int_S \frac{\phi(\mathbf{x}_s, t - \tau_+)}{r_B(\mathbf{x}, \mathbf{x}_s)} \mathbf{n}(\mathbf{x}_s) \cdot \nabla \tau_+(\mathbf{x}, \mathbf{x}_s) \mathrm{d}S \\
&- \frac{\partial}{\partial t} \int_S \frac{\phi(\mathbf{x}_s, t - \tau_-)}{r_B(\mathbf{x}, \mathbf{x}_s)} \mathbf{n}(\mathbf{x}_s) \cdot \nabla \tau_-(\mathbf{x}, \mathbf{x}_s) \mathrm{d}S
\end{aligned} \tag{6.109}$$

Then, taking the Fourier transform of this latest expression and using Eq. (6.68), we obtain (Morino 1974)

$$4\pi E\phi(\mathbf{x},\omega) = -\int_S \mathbf{n}(\mathbf{x}_s)\cdot\nabla\phi(\mathbf{x}_s,\omega)\frac{e^{-i\omega\tau_+}+e^{-i\omega\tau_-}}{r_B(\mathbf{x},\mathbf{x}_s)}dS + \int_S \phi(\mathbf{x}_s,\omega)\mathbf{n}(\mathbf{x}_s)\cdot\nabla\left(\frac{e^{-i\omega\tau_+}+e^{-i\omega\tau_-}}{r_B(\mathbf{x},\mathbf{x}_s)}\right)dS$$

Substituting from Eq. (6.107), we see that the term $e^{-i\omega\tau_+}+e^{-i\omega\tau_-}$ can be written as

$$\begin{aligned} e^{-i\omega\tau_+}+e^{-i\omega\tau_-} &= e^{-i\omega M_\infty(x-x_s)/a_\infty B^2}\left(e^{-i\omega r_B/a_\infty B^2}+e^{i\omega r_B/a_\infty B^2}\right)\\ &= 2e^{-i\omega M_\infty(x-x_s)/a_\infty B^2}\cos\left(\omega r_B/a_\infty B^2\right)\end{aligned}$$

Therefore, for a fixed geometry and in the frequency domain, Green's theorem for supersonic flow becomes

$$2\pi E\phi(\mathbf{x},\omega) = -\int_S \mathbf{n}(\mathbf{x}_s)\cdot\nabla\phi(\mathbf{x}_s,\omega)\frac{e^{-i\omega M_\infty(x-x_s)/a_\infty B^2}\cos\left(\omega r_B/a_\infty B^2\right)}{r_B}dS + \int_S \phi(\mathbf{x}_s,\omega)\mathbf{n}(\mathbf{x}_s)\cdot\nabla\left(\frac{e^{-i\omega M_\infty(x-x_s)/a_\infty B^2}\cos\left(\omega r_B/a_\infty B^2\right)}{r_B}\right)dS \tag{6.110}$$

6.4.1 The Mach Box Method

The Mach box method (Moore and Andrew 1965; Pines et al. 1955) models wings as planar thin lifting surfaces, in the manner of the DLM, so that they lie completely on the $z = 0$ plane. The modelling approach is based on the analytical solutions for planar wings oscillating in supersonic flow developed by Garrick and Rubinow (1938). Two source distributions are placed on the wing, one on the upper side ($z = 0^+$) and one on the lower side ($z = 0^-$), such that the strengths of these distributions are equal and opposite at each point x_s, y_s. As the flow is supersonic, it is assumed that there is no circulation around the wing and therefore the upper and lower source distributions are isolated and do not communicate with each other; the strength of the source distribution on the upper surface is determined by the boundary condition on the upper surface and vice versa. The potential induced by each of the source distributions at any point in space is given by the first integral of Eq. (6.110), that is

$$2\pi E\phi(\mathbf{x},\omega) = -\int_S \mathbf{n}(\mathbf{x}_s)\cdot\nabla\phi(\mathbf{x}_s,\omega)\frac{e^{-i\omega M_\infty(x-x_s)/a_\infty B^2}\cos\left(\omega r_B/a_\infty B^2\right)}{r_B}dS \tag{6.111}$$

where $r_B = \sqrt{(x-x_s)^2 - B^2(y-y_s)^2 - B^2z^2}$, since the wing's surface lies on $z_s = 0$. We are interested in the potential induced on the surface itself, where $E = 1/2$ from Eq. (2.69). Hence, the potential induced at any point on the surface, $\mathbf{x} = (x, y, 0)$, by a source lying at another point on the surface $\mathbf{x}_s = (x_s, y_s, 0)$ is

$$\phi(\mathbf{x},\omega) = -\frac{1}{\pi}\int_{S_{MC}} \mathbf{n}(\mathbf{x}_s)\cdot\nabla\phi(\mathbf{x}_s,\omega)\frac{e^{-i\omega M_\infty(x-x_s)/a_\infty B^2}\cos\left(\omega r_B/a_\infty B^2\right)}{r_B}dS$$

where $r_B = \sqrt{(x-x_s)^2 - B^2(y-y_s)^2}$. Not all sources lying on S can induce potential at point $\mathbf{x}$; only those lying inside the upstream Mach cone of $\mathbf{x}$ can. Therefore, the integral is not evaluated over all of S, it is evaluated over S_{MC}, which is the area of the wing that lies inside the domain of dependence of point $\mathbf{x}$.

In order to ensure impermeability, the normal upwash $w(x_s, y_s)$ induced by the motion and the free stream on the surfacce must be equal to the normal velocity induced by the source distribution. On the upper surface,

$$w(x_s, y_s, \omega) = \mathbf{n}(\mathbf{x}_s) \cdot \nabla\phi(x_s, y_s, 0^+, \omega)$$

while on the lower surface,

$$-w(x_s, y_s, \omega) = \mathbf{n}(\mathbf{x}_s) \cdot \nabla\phi(x_s, y_s, 0^-, \omega)$$

Consequently, the potential on either of the surfaces becomes

$$\phi(x, y, 0^\pm, \omega) = \mp\frac{1}{\pi}\int_{S_{\mathrm{MC}}} w(x_s, y_s, \omega)\frac{\mathrm{e}^{-\mathrm{i}\omega M_\infty(x-x_s)/a_\infty B^2}\cos\left(\omega r_B/a_\infty B^2\right)}{r_B}\,\mathrm{d}S \tag{6.112}$$

From Eq. (2.154), the linearised pressure on the surface of the wing is given by

$$p(x_s, y_s, t) - p_\infty = -\rho_\infty\left(U_\infty\phi_x(x_s, y_s, t) + \phi_t(x_s, y, t)\right)$$

The potential takes equal and opposite values on the upper and lower sides of the surface such that $\phi(x_s, y_s, 0^-) = -\phi(x_s, y_s, 0^+)$. Hence, the pressure difference $\Delta p(x_s, y_s) = p(x_s, y_s, 0^+) - p(x_s, y_s, 0^-)$ is given by

$$\Delta p(x_s, y_s, t) = -2\rho_\infty\left(U_\infty\phi_x(x_s, y_s, t) + \phi_t(x_s, y_s, t)\right)$$

or after taking the Fourier Transform,

$$\begin{aligned}\Delta p(x_s, y_s, \omega) &= -2\rho_\infty\left(U_\infty\phi_x(x_s, y_s, \omega) + \mathrm{i}\omega\phi(x_s, y_s, \omega)\right)\\ &= -2\rho_\infty\left(\mathrm{i}\omega + U_\infty\frac{\partial}{\partial x}\right)\phi(x_s, y_s, \omega)\end{aligned} \tag{6.113}$$

Substituting from Eq. (6.112), we obtain the pressure difference

$$\begin{aligned}\Delta p(x_s, y_s, \omega) = &-\frac{2\rho_\infty}{\pi}\int_{S_{\mathrm{MC}}} w(x_s, y_s, \omega)\left(\mathrm{i}\omega + U_\infty\frac{\partial}{\partial x}\right)\\ &\times\frac{\mathrm{e}^{-\mathrm{i}\omega M_\infty(x-x_s)/a_\infty B^2}\cos\left(\omega r_B/a_\infty B^2\right)}{r_B}\,\mathrm{d}S\end{aligned} \tag{6.114}$$

This is the fundamental equation of the Mach box method; it expresses the pressure difference anywhere on the surface of a planar thin wing in terms of the known upwash induced by the motion. The equation is integral and therefore cannot be solved on general geometries, which means that a panelling scheme must be applied.

Following the approach by Pines et al. (1955), the wing's surface is split into $I = 1, \dots, M$ square boxes of side s so that the control point of each box lies on its geometric centre x_{c_I}, y_{c_I}. The transformation

$$\xi = \frac{x}{s}, \quad \eta = \frac{y}{s}$$

is applied to Eq. (6.114), such that the pressure induced on the Ith box by the Jth box is given by

$$\begin{aligned}\Delta p(\xi_{c_I}, \eta_{c_I}, \omega) = &-\frac{2\rho_\infty\omega s}{\pi}\int_{\Sigma_{\mathrm{MC}_J}} w_J(\omega)\left(\mathrm{i} + \frac{1}{k}\frac{\partial}{\partial\xi_{c_I}}\right)\\ &\times\frac{\mathrm{e}^{-\mathrm{i}k\gamma(\xi_{c_I}-\xi_J)/a_\infty B^2}\cos\left(k\gamma\rho_B/M_\infty\right)}{\rho_B}\,\mathrm{d}\Sigma\end{aligned} \tag{6.115}$$

where $\Sigma = S/s^2$ is the total transformed surface of the wing, Σ_{MC_J} is the area of panel J that lies within the upstream Mach cone of control point I, $\rho_B = \sqrt{\left(\xi_{c_I} - \xi_J\right)^2 - B^2\left(\eta_{c_I} - \eta_J\right)^2}$, ξ_J, η_J are the coordinates of any point lying on box J, the reduced frequency is defined as

$$k = \frac{\omega s}{U_\infty}$$

and $\gamma = M_\infty^2/B^2$. Pines et al. (1955) performed the integration in Eq. (6.115) by expanding the exponential and cosine terms as series in powers of k if boxes I and J lie nearby and by taking their mean values for far away boxes. If there are N boxes on the wing's surface, Eq. (6.115) becomes

$$\Delta p(\xi_{c_I}, \eta_{c_I}, \omega) = 2\rho_\infty \omega s \sum_{J=1}^{N} \left(R_{I,J} + \mathrm{i}I_{I,J}\right) w_J(\omega) \tag{6.116}$$

where $R_{I,J}$ and $I_{I,J}$ the real and imaginary parts of the influence coefficient of box J on the control point of box I, for $I = 1, \dots, N$. First we define the indicial distances between the control points of the influenced and influencing boxes, $n = (x_{c_I} - x_{c_J})/s$ and $l = |y_{c_I} - y_{c_J}|/s$. For example, if box J lies directly in front of box I, the distance between the control points of the two boxes will be described by $n = 1, l = 0$. If $n = 0$ and $l = 1$ the influencing box will lie directly to the left or right of the influenced control point. Then, for $n = 0$ and $n = 1$, the real and imaginary parts of the influence coefficients are given by

$$\begin{aligned} R_{I,J} = &-\frac{1}{kB\pi}\left(C_1 - C_2 - nC_3 + nC_4\right) \\ &+ \frac{3k\gamma}{4B^3\pi}\left((n+0.5)^2C_1 - (n+0.5)^2C_2 - n(n-0.5)^2C_3 + n(n-0.5)^2C_4\right) \\ &+ \frac{k\gamma(2M^2-3)}{4\pi B^2}\left((l-0.5)S_1 - (l+0.5)S_2 - n(l-0.5)S_3 + n(l+0.5)S_4\right) \end{aligned} \tag{6.117}$$

$$\begin{aligned} I_{I,J} = &\frac{1}{\pi}\left((l-0.5)H_1 - (l+0.5)H_2 - n(l-0.5)H_3 + n(l+0.5)H_4\right) \\ &+ \frac{1}{\pi B^3}\left((n+0.5)C_1 - (n+0.5)C_2 - n(n-0.5)C_3 + n(n-0.5)C_4\right) \end{aligned} \tag{6.118}$$

where

$$C_1 = \cos^{-1}\left(\frac{B(l-0.5)}{n+0.5}\right), \quad C_2 = \cos^{-1}\left(\frac{B(l+0.5)}{n+0.5}\right)$$

$$C_3 = \cos^{-1}\left(\frac{B(l-0.5)}{n-0.5}\right), \quad C_4 = \cos^{-1}\left(\frac{B(l+0.5)}{n-0.5}\right)$$

$$S_1 = \sqrt{(n+0.5)^2 - B^2(l-0.5^2)}, \quad S_2 = \sqrt{(n+0.5)^2 - B^2(l+0.5^2)}$$

$$S_3 = \sqrt{(n-0.5)^2 - B^2(l-0.5^2)}, \quad S_4 = \sqrt{(n-0.5)^2 - B^2(l+0.5^2)}$$

$$H_1 = \cosh^{-1}\left(\frac{n+0.5}{B(l-0.5)}\right), \quad H_2 = \cosh^{-1}\left(\frac{n+0.5}{B(l+0.5)}\right)$$

$$H_3 = \cosh^{-1}\left(\frac{n-0.5}{B(l-0.5)}\right), \quad H_4 = \cosh^{-1}\left(\frac{n-0.5}{B(l+0.5)}\right)$$

For $n > 1$, the following expressions are used for $R_{I,J}$ and $I_{I,J}$ instead of Eqs. (6.117) and (6.118),

$$R_{I,J} = -\frac{1}{\pi}\sin(nk\gamma)K_1 D - \frac{1}{kB\pi}\cos((n+0.5)k\gamma)K_2\left(C_1 - C_2\right) + \frac{1}{kB\pi}\cos((n-0.5)k\gamma)K_3\left(C_3 - C_4\right) \quad (6.119)$$

$$I_{I,J} = -\frac{1}{\pi}\cos(nk\gamma)K_1 D + \frac{1}{kB\pi}\sin((n+0.5)k\gamma)K_2\left(C_1 - C_2\right) - \frac{1}{kB\pi}\sin((n-0.5)k\gamma)K_3\left(C_3 - C_4\right) \quad (6.120)$$

where

$$K_1 = \cos\left(\frac{k\gamma}{M_\infty}\sqrt{n^2 - B^2 l^2}\right), \quad K_2 = \cos\left(\frac{k\gamma}{M_\infty}\sqrt{(n+0.5)^2 - B^2 l^2}\right)$$

$$K_3 = \cos\left(\frac{k\gamma}{M_\infty}\sqrt{(n-0.5)^2 - B^2 l^2}\right)$$

$$D = -(l-0.5)H_1 + (l+0.5)H_2 + (l-0.5)H_3 - (l+0.5)H_4 + \frac{n+0.5}{B}\left(C_1 - C_2\right) - \frac{n-0.5}{B}\left(C_3 - C_4\right)$$

Equations (6.17)–(6.20) were derived making the assumption that box J lies entirely within the upstream Mach cone of box I, which is not necessarily the case. In order to keep $R_{I,J}$ and $I_{I,J}$ real when box J lies partially within the upstream Mach cone, the values of the inverse cosine, square root and inverse hyperbolic cosine functions are adapted such that

$$\cos^{-1}(\alpha) = \begin{cases} 0 & \text{for} \quad |\alpha| > 1 \\ \cos^{-1}(\alpha) & \text{for} \quad |\alpha| \le 1 \end{cases}$$

$$\sqrt{\alpha} = \begin{cases} \sqrt{\alpha} & \text{for} \quad \alpha \ge 0 \\ 0 & \text{for} \quad \alpha < 0 \end{cases}$$

$$\cosh^{-1}(\alpha) = \begin{cases} \Re\left(\cosh^{-1}(\alpha)\right) & \text{for} \quad |\alpha| \ge 1 \\ 0 & \text{for} \quad |\alpha| < 1 \end{cases}$$

Equations (6.17)–(6.20) do not feature the coordinates of the control points or panel vertices; these are replaced by n and l, which is possible due to the fact that all the boxes are square and of the same size. Equation (6.116) can be written in more compact form as

$$\Delta p_I(\omega) = \sum_{J=1}^{N} P_{ww_{I,J}} w_J(\omega) \quad (6.121)$$

where the pressure influence coefficient $P_{ww_{I,J}}$ is given by

$$P_{ww_{I,J}} = 2\rho_\infty \omega s\left(R_{I,J} + \mathrm{i}I_{I,J}\right) \quad (6.122)$$

There is one more important issue to discuss about the Mach box method. The leading edge of supersonic wings is generally swept back and the sweep angle can be such that, depending on the free stream Mach number, the leading edge may lie inside or outside the downstream Mach cone starting at the apex of the wing. These two possibilities are depicted in Figure 6.15 where the planform of the wing of the F-5 aircraft is drawn, along with the Mach lines at two different Mach numbers. In Figure 6.15a, the free stream Mach number

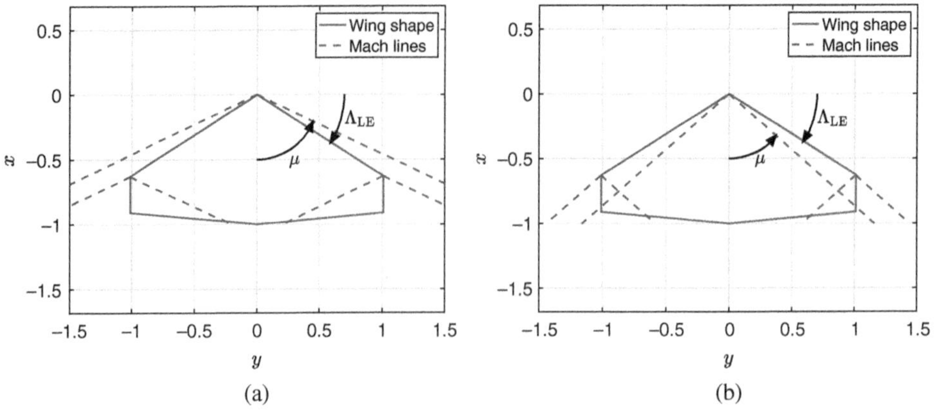

Figure 6.15 Subsonic (left) and supersonic (right) leading edges of the F-5 wing planform. (a) $M_\infty = 1.1$ and (b) $M_\infty = 1.32$.

is low and the leading edge lies entirely inside the downstream Mach cone coming from the wing's apex. This condition is known as a subsonic leading edge. Garrick and Rubinow (1938) refer to parts of the wing that lie outside the apex Mach cone as mixed supersonic regions. In Figure 6.15b, the free stream Mach number is higher and now the leading edge lies in front of the Mach cone. This situation is known as a supersonic leading edge or purely supersonic region. Subsonic leading edges require a different treatment to supersonic ones. For the former, the assumption that the flows on the upper and lower surfaces do not influence each other is not valid. Note that regions close to the two wingtips lie inside the downstream Mach cones coming from the wingtip leading edges. These regions are also subsonic (or mixed supersonic) and they too require the same special treatment.

The standard approach for subsonic regions is the diaphragm method, first proposed by Evvard (1947). Communication between the two sides of the wing in subsonic regions is allowed by adding boxes in the space between the leading edge or wingtip and the downstream Mach lines, known as the diaphragm. The upwash acting on the diaphragm is unknown but can be calculated from the condition that the pressure difference across a diaphragm box is equal to zero. The influence coefficients of the diaphragm on the wing, $P_{wd_{I,J}}$, of the wing on the diaphragm, $P_{dw_{I,J}}$ and of the diagram on itself, $P_{dd_{I,J}}$, are calculated using Eqs. (6.117)–(6.122). Then, the pressure difference on the diaphragm is given by

$$\Delta p_{d_I}(\omega) = \sum_{J=1}^{N_d} P_{dd_{I,J}} w_{D_J}(\omega) + \sum_{J=1}^{N} P_{dw_{I,J}} w_J(\omega) = 0 \tag{6.123}$$

for $I = 1, \dots, N_d$, where $w_{D_J}(\omega)$ is the unknown upwash acting on the diaphragm and N_d is the number of boxes in the diaphragm. Equation (6.123) can be solved for the unknown upwash $w_{D_J}(\omega)$ and then the pressure on the wing is obtained from

$$\Delta p_I(\omega) = \sum_{J=1}^{N} P_{ww_{I,J}} w_J(\omega) + \sum_{J=1}^{N_d} P_{wd_{I,J}} w_{D_J}(\omega) \tag{6.124}$$

for $I = 1, \dots, N$. Finally, the wake is not modelled at all by supersonic panel methods, since it lies behind the trailing edge and can therefore have no influence on the wing.

Example 6.6 *Calculate the oscillating pressure distribution on the wing of the F-5 aircraft pitching about the half-chord at $M_\infty = 1.32$.*

Tijdeman et al. (1979a,b) presented steady and unsteady pressure measurements from wind tunnel tests on a slightly modified model of the outer section of the wing of the F-5 aircraft at a range of Mach numbers. The half-wing model is depicted in Figure 6.16a, along with the positions of the pressure taps. The wing has leading edge sweep of 31.9°, taper ratio of 0.31 and aspect ratio of 2.98. The pressures are measured at eight spanwise stations, located at $2y/b = 0.18, 0.36, 0.51, 0.64, 0.72, 0.82, 0.88$ and 0.98. For each of the spanwise stations, there are 10 pressure taps in the chordwise direction, located at $x/c = 0.03, 0.1, 0.2, \dots, 0.9$, where c is the local chord. The airfoil section is a 4.8% thick modified NACA 65-A-004.8 with a droop nose, plotted in Figure 6.16b. This airfoil is only cambered over the first 40% of the chord; beyond this point it is symmetric, and the chord line is defined as the symmetry line of the aft section.

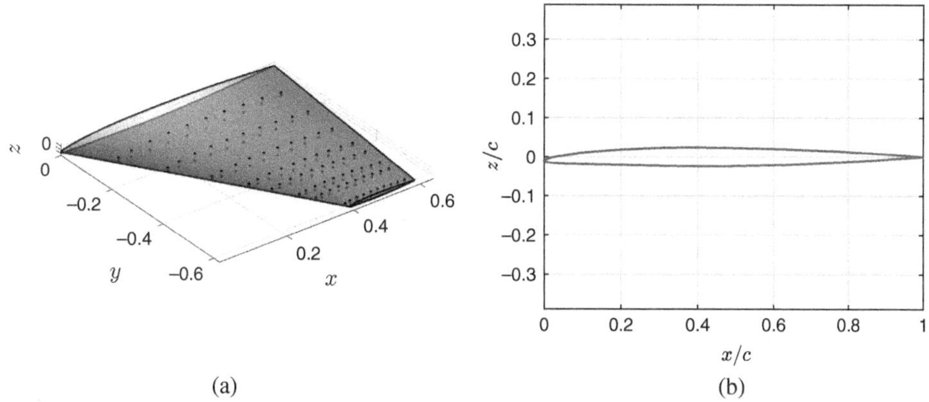

Figure 6.16 F-5 half-wing model tested by Tijdeman et al. (1979b). (a) Half-wing with pressure taps and (b) airfoil section. Source: Adapted from Tijdeman et al. (1979b), credit: AFFDL.

We select a test case whereby the wing is immersed in a free stream with $M_\infty = 1.35$ and is oscillating sinusoidally in pitch around the $x_f = 0.5c_0$ axis with amplitude $\alpha_1 = 0.222°$ and zero mean. The reduced frequency based on half the root chord is $k = 0.198$. The flow case is shown in Figure 6.15b; the leading edge lies outside the Mach cone coming from the apex of the wing and is therefore supersonic. Nevertheless, diaphragm boxes must be placed at the wingtips.

We begin by discretising the wing into square boxes. We choose to create m boxes in the chordwise direction at the root chord, such that $s = c_0/m$. The number of spanwise boxes is then $n = \lfloor b/s \rfloor$, where $\lfloor\rfloor$ denotes rounding down. We create a rectangular grid of mn boxes between $-b/2$ and $b/2$ and between 0 and c_0 and then we calculate the control points (mid-points) of all the boxes. If a control point lies inside the wing's planform, then its corresponding box is a valid wing box. If a control point lies inside the diaphragm, the corresponding box is a valid diaphragm box. The rest of the boxes are ignored. We denote the coordinates of the control points of the wing boxes by x_{c_l}, y_{c_l} and those of the diaphragm as x_{d_l}, y_{d_l}. Figure 6.17 plots the resulting wing and diaphragm grid for $m = 10$, along with the control points and Mach

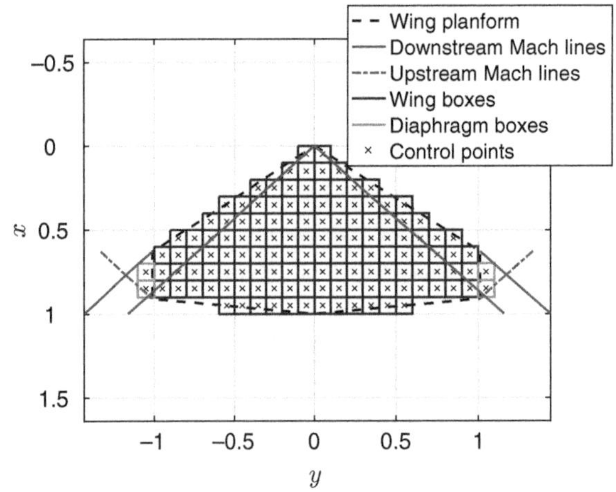

Figure 6.17 Mach box grid for F5 wing for $M_\infty = 1.32$ and $m = 10$.

lines. Note that the diaphragm grid extends from the downstream Mach lines coming from the wingtip leading edge to the upstream Mach lines coming from the wingtip trailing edge. Any boxes placed downstream of the latter would have no effect on the wing.

The next step is to calculate the pressure influence coefficients $P_{ww_{I,J}}$, $P_{wd_{I,J}}$, $P_{dw_{I,J}}$ *and* $P_{dd_{I,J}}$ *from Eqs. (6.117)–(6.122). The chordwise and spanwise indicial distances in these equations are calculated from the distances between the control points such that* $n = (x_{c_I} - x_{c_J})/s$, $l = |y_{c_I} - y_{c_J}|/s$. *Similar expressions are used for the distances between wing and diaphragm control points or diaphragm and diaphragm control points. The upwash acting on the Ith wing control point is due to the pitching motion and is therefore given by*

$$w_I = -\left(U_\infty + i\omega(x_{c_I} - x_f)\right)\frac{\alpha_1}{2i}e^{i\phi_\alpha} \tag{6.125}$$

(see Eq. (6.57) for example). Finally, we use Eq. (6.123) to calculate the upwash on the diaphragm control points, w_{d_I}, *and Eq. (6.124) to calculate the pressure difference on the wing. The pressure coefficient on the Ith wing control point is simply*

$$\Delta c_{p_I}(\omega) = \frac{\Delta p_I(\omega)}{\frac{1}{2}\rho_\infty U_\infty^2}$$

Figure 6.18 plots the amplitude and phase of $\Delta_{c_{p\alpha}} = 2i\Delta c_p(\omega)/\alpha_1$ *at four of the eight spanwise measurement stations, along with the experimental data. The Mach box results are obtained using* $m = 32$, *resulting in* $N = 1300$ *wing panels and* $N_d = 56$ *diaphragm panels. The calculated pressures must be interpolated in both the spanwise and chordwise directions in order to compare to the experimental data. It can be seen that the agreement between the predicted and measured pressure coefficient values is generally acceptable, although the amplitudes are in worse agreement for* $x/c < 0.5$. *Furthermore, the pressure values predicted by Mach box theory are noisy, which is uncharacteristic for numerical methods. The cause of this phenomenon is the fact that the wing is approximated by square boxes which all have the same shape and size. The only wings shapes that can be represented exactly using this*

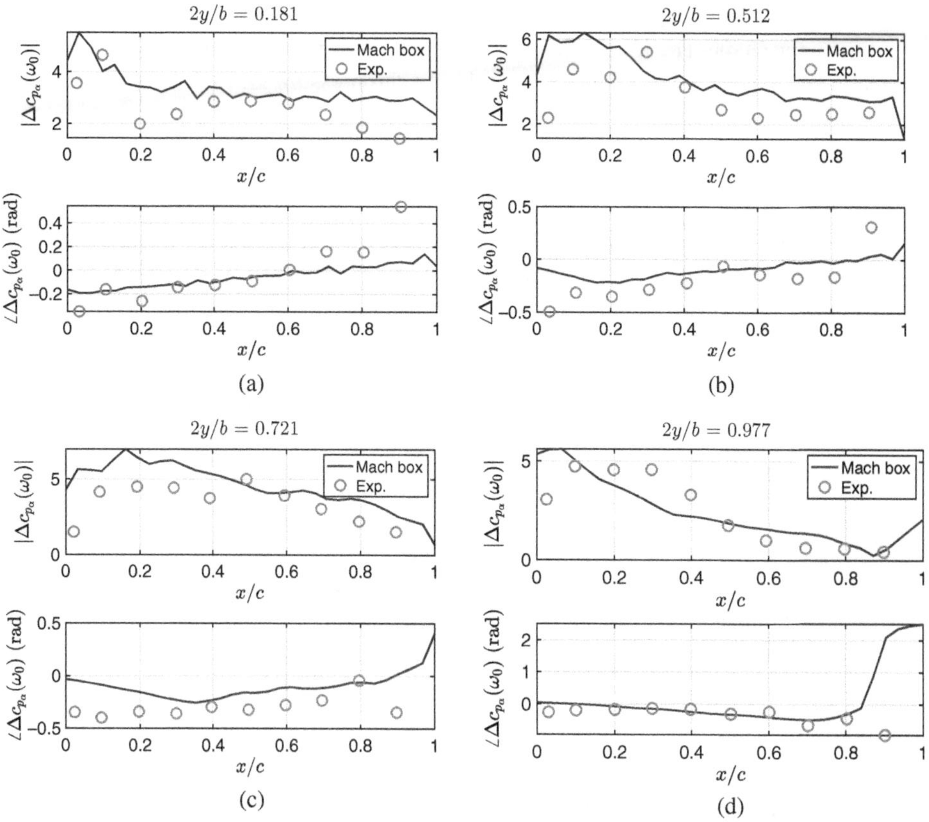

Figure 6.18 Oscillating pressure jump across the F-5 wing predicted by the Mach box method. Experimental data by Tijdeman et al. (1979b). (a) $2y/b = 0.181$, (b) $2y/b = 0.512$, (c) $2y/b = 0.721$ and (d) $2y/b = 0.977$. Source: Tijdeman et al. (1979b), credit: AFFDL.

approach are rectangular unswept and untapered wings with integer aspect ratio. All other shapes will have jagged leading and trailing edges, such that some boxes will protrude partially beyond the real leading and trailing edges, as seen in Figure 6.17. Furthermore, the approach assumes that boxes lie either entirely inside or entirely outside the domain of influence of the control points, even though many of them lie partially inside this domain. As a consequence, the influence of coefficient matrices are noisy and so are the resulting pressure distributions. In practice, Mach box pressure results are often smoothed before being plotted. This example is solved by Matlab code `pitchplunge_Machbox.m.m`*.*

The example showed that while the approach by Pines et al. (1955) is fast and gives reasonable results, the predicted pressure distributions are noisy. Other box methods have been published in the literature, but they generally suffer from the same issue, see for example Moore and Andrew (1965), Olsen (1969), Ueda and Dowell (1984). In order to avoid noisy pressure distributions, the aerodynamic modelling must be better adapted to the shape of the wing. This was done for example by Curtis and Lingard Jr. (1968) for Delta wings, using a variational principle to best select the positions of the control points that represent the wing shape. The Kernel function method (Cunningham 1966) models accurately the planform by expressing both the upwash and the pressure jump distributions on the wing as polynomial series.

6.4.2 The Mach Panel Method

Here we will present an alternative Mach panel method based on the panelling scheme we have been using throughout the book, quadrilateral panels of arbitrary shape. Going back to Eq. (6.113), we can calculate the pressure difference on the surface from

$$\Delta p(x_s, y_s, \omega) = -2\rho_\infty \left(U_\infty \phi_x(x_s, y_s, \omega) + \mathrm{i}\omega\phi(x_s, y_s, \omega)\right) \tag{6.126}$$

where $\phi(x_s, y_s, \omega)$ is the potential on one side of the surface, evaluated from Eq. (6.112)

$$\phi(x_s, y_s, \omega) = -\frac{1}{\pi}\int_{S_{\mathrm{MC}}} w(x_s, y_s, \omega)\frac{\mathrm{e}^{-\mathrm{i}\omega M_\infty (x-x_s)/a_\infty B^2}\cos\left(\omega r_B/a_\infty B^2\right)}{r_B}\,\mathrm{d}S$$

while $B^2 = M_\infty^2 - 1$ and $r_B = \sqrt{\left(x-x_0\right)^2 - B^2\left(y-y_s\right)^2}$. Now, we split the surface into $i = 1, \ldots, m$ chordwise and $j = 1, \ldots, n$ spanwise planar quadrilateral panels on which the upwash is assumed to be constant and equal to its value at the control points. The potential induced anywhere on the $z = 0$ plane by such a panel becomes

$$\phi(\mathbf{x}, \omega) = -\frac{w(\omega)}{\pi}\int_{S_{\mathrm{MC}}} \frac{\mathrm{e}^{-\mathrm{i}\omega M_\infty (x-x_s)/a_\infty B^2}\cos\left(\omega r_B/a_\infty B^2\right)}{r_B}\,\mathrm{d}S \tag{6.127}$$

The integral in this expression can be transformed from a surface integral to a line integral by applying the supersonic Prandtl–Glauert transformation

$$\xi = \frac{x}{B}, \quad \eta = y, \quad \zeta = z$$

so that Eq. (6.127) becomes

$$\phi(\xi, \omega) = -\frac{w(\omega)}{\pi}\int_{\Sigma_{\mathrm{MC}}} \frac{\mathrm{e}^{-\mathrm{i}\Omega M_\infty (\xi-\xi_s)}\cos(\Omega r)}{r}\,\mathrm{d}\Sigma \tag{6.128}$$

where $\Omega = \omega/a_\infty B$, $r = \sqrt{\left(\xi-\xi_0\right)^2 - \left(\eta-\eta_0\right)^2}$ and $\Sigma = S/B$ is the transformed area of the wing, while Σ_{MC} is the area of the influencing panel inside the domain of influence of point $\mathbf{x}$. In the transformed coordinates, r is real when $\left(\xi-\xi_0\right)^2 \geq \left(\eta-\eta_0\right)^2$ and therefore, the Mach cone is defined by

$$\left(\xi-\xi_0\right)^2 = \left(\eta-\eta_0\right)^2$$

so that the Mach angle is

$$\mu_{\mathrm{PG}} = \tan^{-1}1 = \frac{\pi}{4}$$

after applying the supersonic Prandtl–Glauert transformation. We now define

$$R = \sqrt{(\xi-\xi_s)^2 + (\eta-\eta_s)^2}$$

$$\theta = \tan^{-1}\left(\frac{\eta_s-\eta}{\xi_s-\xi}\right)$$

Writing r in terms of R, Appendix B.4 shows that

$$r = R\sqrt{\cos 2\theta}, \quad \xi - \xi_0 = -R\cos\theta, \quad \mathrm{d}\Sigma = R\,\mathrm{d}R\,\mathrm{d}\theta$$

Substituting these three expressions into Eq. (6.128) leads to

$$\phi(\xi,\omega) = -\frac{w(\omega)}{\pi}\oint \frac{1}{\sqrt{\cos 2\theta}}\int_0^R \mathrm{e}^{\mathrm{i}\Omega M_\infty R\sqrt{\cos 2\theta}}\cos\left(\Omega R\sqrt{\cos 2\theta}\right)\mathrm{d}R\,\mathrm{d}\theta$$

where $\oint$ denotes a contour integral around the perimeter of the section of the influencing panel that lies inside the upstream Mach cone of the influencing point. The integral over R in the latest expression is a standard integral, such that

$$\begin{aligned}\phi(\xi,\omega) = &-\frac{w(\omega)}{\pi}\oint \frac{1}{\Omega(M_\infty^2\cos^2\theta - \cos 2\theta)\sqrt{\cos 2\theta}}\\ &\times\left(\mathrm{e}^{\mathrm{i}\Omega M_\infty R\cos\theta}\left(\sqrt{\cos 2\theta}\sin(\Omega R\sqrt{\cos 2\theta}) + \mathrm{i}M_\infty\cos\theta\cos(\Omega R\sqrt{\cos 2\theta})\right)\right.\\ &\left.-\,\mathrm{i}M_\infty\cos\theta\right)\mathrm{d}\theta\end{aligned}$$

There is no obvious analytical expression for the remaining contour integral but it can be calculated numerically on quadrilateral panels, as shown in Appendix B.4. Note that the numerical evaluation of a line integral is much faster than that of a surface integral. Carrying out this calculation for the potential induced on all the control points by all the panels, Eq. (6.127) becomes

$$\phi(\xi_{c_I},\omega) = \sum_{J=1}^{N} A_{\phi_{I,J}}(\omega)w_J(\omega) \tag{6.129}$$

where $\phi(\xi_{c_I},\omega)$ is the potential induced on the Ith control point, $w_J(\omega)$ is the upwash acting on the Jth panel and $A_{\phi_{I,J}}(\omega)$ is the potential influence coefficient of the Jth panel on the Ith control point, for $I = 1,\dots,mn$ and $J = 1,\dots,mn$.

In applying the Prandtl–Glauert transformation, we did not scale ϕ, which means that $\phi(\mathbf{x}_{c_I},\omega) = \phi(\xi_{c_I},\omega)$. Consequently, we can substitute Eq. (6.129) directly into the pressure equation (6.126), except that we also need the horizontal velocity ϕ_x evaluated at the control points. This derivative can be obtained by means of a finite difference scheme, such that

$$\phi_x(\mathbf{x},\omega) = \frac{\phi(x+\delta x, y,\omega) - \phi(x,y,\omega)}{\delta x} \tag{6.130}$$

where $|\delta x| \ll 1$ is a small positive number. Then, $\phi_x(\mathbf{x}_{c_I},\omega)$ can be written as

$$\phi_x(\mathbf{x}_{c_{I,J}},\omega) = \sum_{J=1}^{N} A_{\phi_{x_{I,J}}}(\omega)w_J(\omega)$$

where

$$A_{\phi_{x_{I,J}}}(\omega) = \frac{A_\phi(x_{c_I}+\delta x,\omega) - A_{\phi_{I,J}}(\omega)}{\delta x}$$

Finally, the pressure jump induced on the Ith control point by all the panels becomes

$$\Delta p_I(\omega) = -2\rho_\infty\sum_{J=1}^{N}\left(U_\infty A_{\phi_{x_{I,J}}}(\omega) + \mathrm{i}\omega A_{\phi_{I,J}}(\omega)\right)w_J(\omega)$$

or, in non-dimensional form,

$$\Delta c_{p_I}(\omega) = -4\sum_{J=1}^{N}\left(A_{\phi_{x_{I,J}}}(\omega) + \frac{i\omega}{U_\infty}A_{\phi_{I,J}}(\omega)\right)\frac{w_J(\omega)}{U_\infty}$$
$$= \sum_{J=1}^{N} P_{ww_{I,J}}(\omega)\frac{w_J(\omega)}{U_\infty}$$

where $P_{ww_{I,J}}(\omega) = -4\left(A_{\phi_{x_{I,J}}}(\omega) + \frac{i\omega}{U_\infty}A_{\phi_{I,J}}(\omega)\right)$. Appendix B.4 also shows how to calculate the steady influence coefficients at $\omega = 0$, so that

$$P_{ww_{I,J}}(0) = -4A_{\phi_{x_{I,J}}}(0)$$

and

$$\Delta c_{p_I}(0) = \sum_{J=1}^{N} P_{ww_{I,J}}(0)\frac{w_J(0)}{U_\infty}$$

We also need to set up quadrilateral diaphragm panels and calculate their influences on the wing, $P_{wd_{I,J}}$, on themselves $P_{dd_{I,J}}$ as well as the influence on the wing on the diaphragm panels $P_{dw_{I,J}}$. We then solve Eq. (6.123) for the upwash on the diaphragm and Eq. (6.124) for the pressure on the wing, as in the case of the Mach box method.

Example 6.7 *Repeat Example 6.6 using the Mach panel method.*

The first step is to discretise the wing into quadrilateral panels, just as in the case of the VLM or DLM methods, see Example 5.4 or 6.2, respectively. The chordwise panel spacing is chosen to be linear while the spanwise spacing can be linear or non-linear. The discretisation results in panel vertices $x_{p_{i,j}}$ *and* $y_{p_{i,j}}$ *for* $i = 1, \dots, m$, $j = 1, \dots, n$ *(with* $z_{p_{i,j}} = 0$ *since all panels lie on the* $z = 0$ *plane). As the wing is quadrilateral, quadrilateral panels describe its exact shape.*

The next step is to set up the diaphragm, for which there are two possibilities. Figure 6.19a draws the subsonic leading edge case, for which $\mu > \pi/2 - \Lambda_{LE}$. *For this case, we need to place diaphragm panels in the area delimited by the downstream Mach lines coming from the nose*

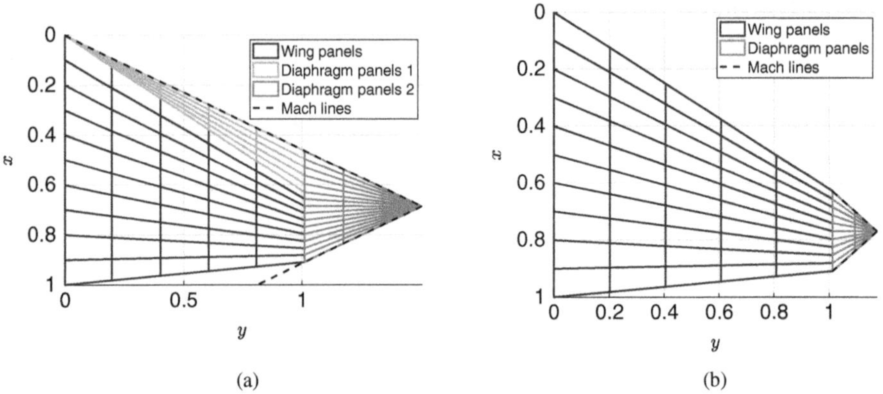

Figure 6.19 Wing and diaphragm panels for the F-5 wing at two Mach numbers. (a) $M_\infty = 1.1$ and (b) $M_\infty = 1.32$.

of the wing, the upstream Mach lines coming from the tip trailing edges and the wing contour. In order to discretise this area we define two diaphragm regions, one upstream of the wing from root to tip and one outboard of the tip. In the first area, the number of spanwise diaphragm panels is equal to $n/2$*, half the number of spanwise panels on the wing. The chordwise number of diaphragm panels can be chosen by the user or set to a number compatible with the chordwise spacing of the tip wing panels. In the second area, the chordwise number of diaphragm panels is equal to the sum of the chordwise numbers of panels on the wing and in the first diaphragm area. The spanwise number of diaphragm panels can be chosen by the user or set to a number compatible with the spanwise spacing of the tip wing panels. The central diaphragm panels in area 1 and the tip diaphragm panels in area 2 are triangular while all the other panels are quadrilateral. All diaphragm panels in Figure 6.19a must be mirrored onto the other half of the wing.*

Figure 6.19b plots the supersonic leading edge case. The diaphragm is much smaller and extends from the downstream Mach lines coming from the wingtip leading edges to the upstream Mach lines coming from the wingtip trailing edges. In this case, there are m *diaphragm panels in the chordwise direction, while the spanwise number of panels can be chosen by the user or set to a number compatible with the spanwise spacing of the tip wing panels. All panels are quadrilateral, except for those in the outermost row, which are triangular. Again, all diaphragm panels must be mirrored.*

Once the vertices of all wing and diaphragm panels have been calculated, we evaluate their control points, as the mean values of the coordinates of the four vertices, even for the triangular panels, for which two of the vertices are identical. Then we calculate the influence coefficients, $A_{\phi_{I,J}}(\omega)$ *of all panels (wing and diaphragm) on all control points (again wing and diaphragm) for* $\omega = 0$ *using Eq. (B.26) and* $\omega = \omega_0$ *using Eq. (B.21). The complete procedure of Appendix B.4 must be followed, including the determination of the panel segment lengths inside the upstream Mach cone of the control point and the approximation of the* $\sec 2\theta$ *function. We then repeat the calculation of the influence coefficients after augmenting the* x *coordinates of the control points by* $\delta x = 10^{-8}$ *in order to calculate the velocities of Eq. (6.130) and the influence coefficients* $A_{\phi_{x_{I,J}}}(0)$ *and* $A_{\phi_{x_{I,J}}}(\omega_0)$*.*

The steady upwash is given by

$$w_I(0) = -\alpha_0 + \left.\frac{\mathrm{d}\bar{f}_c}{\mathrm{d}\bar{x}}\right|_{\bar{x}_{c_I}, \bar{y}_{c_I}}$$

see Eq. (6.58) for the DLM. We calculate the camber of the airfoil of the F-5 wing and then its slope by interpolation, as we did in Example 6.2. Next, we set up the pressure influence coefficients $P_{ww_{I,J}}(0)$*,* $P_{wd_{I,J}}(0)$*,* $P_{dw_{I,J}}(0)$ *and* $P_{dd_{I,J}}(0)$*, solve Eq. (6.123) for* $w_{D_J}(0)$ *and Eq. (6.124) for* $\Delta p_I(0)$*.*

Figure 6.20 plots the mean pressure jump predicted by the Mach panel method at four spanwise stations and compares it to the experimental measurements. It can be seen that the pressure jump is only significant over the first 40% of the chord, where the nose droops and the airfoil is cambered. Downstream of this point the airfoil is symmetric and, as the angle of attack is zero, the local pressure jump is very low. Clearly, the agreement between the Mach panel predictions and the experimental data are very good. Unlike the LANN wing of Example 6.2, the camber line of the F-5 airfoil is clearly defined and, therefore, its effect on the pressure can be modelled adequately by a lifting surface technique.

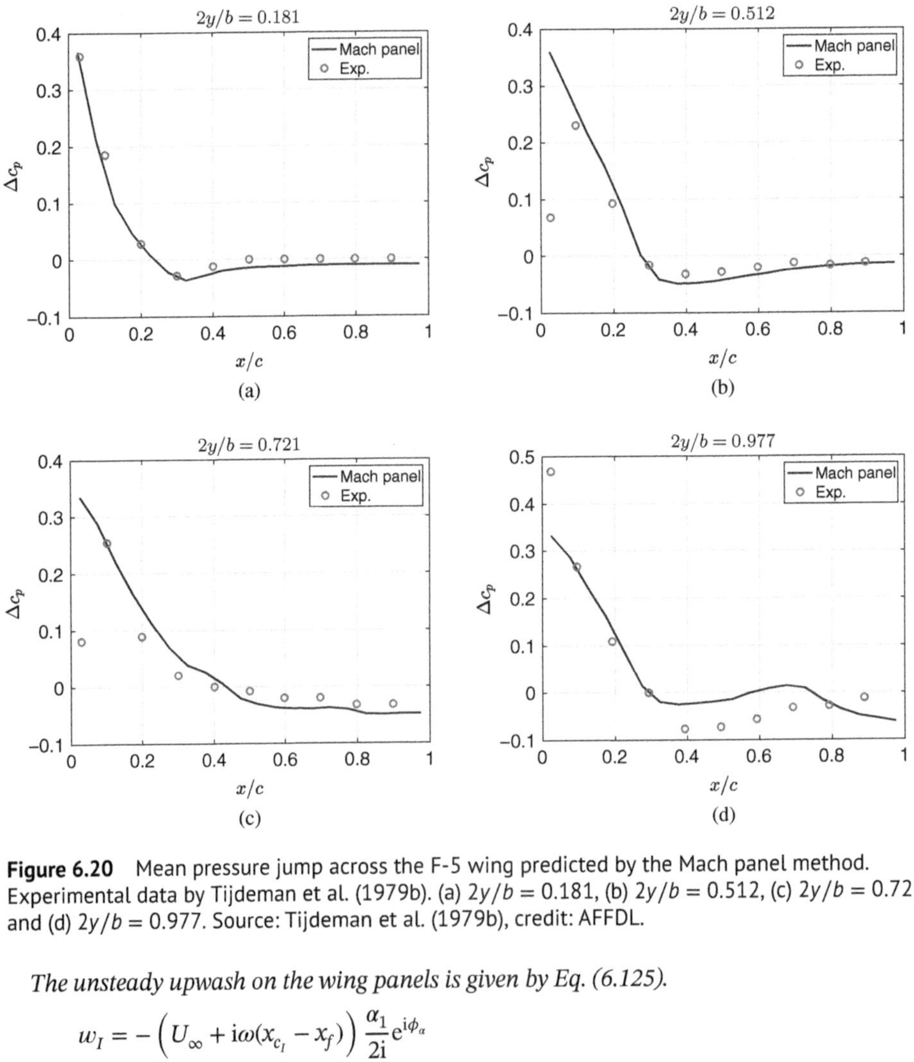

Figure 6.20 Mean pressure jump across the F-5 wing predicted by the Mach panel method. Experimental data by Tijdeman et al. (1979b). (a) $2y/b = 0.181$, (b) $2y/b = 0.512$, (c) $2y/b = 0.721$ and (d) $2y/b = 0.977$. Source: Tijdeman et al. (1979b), credit: AFFDL.

The unsteady upwash on the wing panels is given by Eq. (6.125).

$$w_I = -\left(U_\infty + i\omega(x_{c_I} - x_f)\right)\frac{\alpha_1}{2i}e^{i\phi_\alpha}$$

Again, we set up the pressure influence coefficients $P_{ww_{I,J}}(\omega_0)$, $P_{wd_{I,J}}(\omega_0)$, $P_{dw_{I,J}}(\omega_0)$ *and* $P_{dd_{I,J}}(\omega_0)$, *solve Eq. (6.123) for* $w_{D_J}(\omega_0)$ *and, finally, Eq. (6.124) for* $\Delta p_I(\omega_0)$.

Figure 6.21 plots the amplitude and phase of $\Delta c_{p_\alpha}(\omega_0) = 2i\Delta c_p(\omega_0)/\alpha_1$ *for* $m = 20$, $n = 20$, *for the case presented in Example 6.6. It can be seen that the amplitudes and phases predicted by the Mach panel method are smoother and closer to the experimental measurements than those evaluated using the Mach box approach. This example is solved by Matlab code* `pitch-plunge_Machpanel.m`.

Both the Mach box and Mach panel techniques depend on the inverse of the supersonic compressibility factor $1/B$. This means that, as $M_\infty \to 1$, the methods become numerically unstable. At sonic conditions, the linearised small disturbance equation (6.104) becomes

$$\phi_{yy} + \phi_{zz} - \frac{2}{a_\infty}\phi_{xt} - \frac{1}{a_\infty^2}\phi_{tt} = 0 \tag{6.131}$$

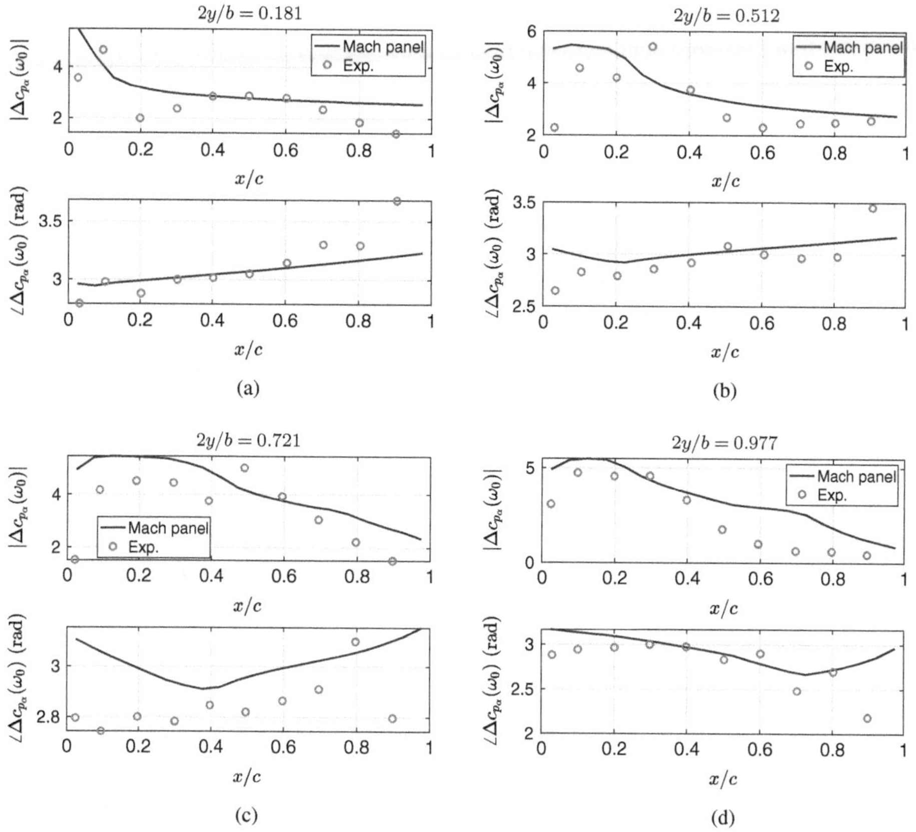

Figure 6.21 Oscillating pressure jump across the F-5 wing predicted by the Mach panel method. Experimental data by Tijdeman et al. (1979b). (a) $2y/b = 0.181$, (b) $2y/b = 0.512$, (c) $2y/b = 0.721$ and (d) $2y/b = 0.977$. Source: Tijdeman et al. (1979b), credit: AFFDL.

which is known as the linearised sonic equation; it has different solutions to the subsonic and supersonic small disturbance equations. Following Landahl (1961), we can apply the transformation

$$\xi = \frac{2x}{c_0}, \quad \eta = \frac{2y}{c_0}, \quad \zeta = \frac{2z}{c_0}$$

as well as the Fourier transform to Eq. (6.131) to obtain

$$\phi_{\eta\eta}(\omega) + \phi_{\zeta\zeta}(\omega) - \frac{\mathrm{i}\omega c_0}{a_\infty}\phi_\xi(\omega) + \frac{\omega^2 c_0^2}{4a_\infty^2}\phi(\omega) = 0$$

We can now introduce the reduced frequency $k = \omega c_0/2U_\infty$ such that

$$\phi_{\eta\eta}(\omega) + \phi_{\zeta\zeta}(\omega) - 2\mathrm{i}kM_\infty\phi_\xi(\omega) + k^2M_\infty^2\phi(\omega) = 0 \tag{6.132}$$

which is a linearised transonic equation that is valid for M_∞ very close or equal to 1, as long as k is high enough. Rodemich and Andrew (1965) developed a transonic box method for solving this equation on general planforms that is similar in spirit to the Mach box method; they even provided a Fortran code listing for their technique.[2] We will not discuss this approach in any more detail, we will instead move on to the analysis of transonic flow for subsonic free stream Mach numbers.

6.5 Transonic Flow

When the free stream Mach number is subsonic ($M_\infty < 1$) but the flow accelerates locally to supersonic speeds as it moves over a body, transonic conditions arise. Far downstream, the flow must return to the free stream Mach number, which means that it must decelerate from supersonic to subsonic speeds somewhere in the vicinity of the body. Such a deceleration generally occurs by means of a shock wave that lies on the surface of the body.

Figure 6.22a plots the Mach contours of the steady flow around a NASA SC(2)-0714 supercritical airfoil at a free stream Mach number of $M_\infty = 0.722$ and zero angle of attack. The solid black line is known as the sonic line, as it separates the subsonic from the supersonic region. The flow accelerates steadily around the leading edge until it becomes supersonic at around $x/c = 0.05$. The Mach number remains supersonic up to approximately $x/c = 0.58$ where it jumps abruptly to subsonic values by means of a shock wave. The resulting pressure distribution is plotted in Figure 6.22b; the pressure coefficient on the suction side is nearly constant throughout the supersonic region until it increases abruptly behind the shock wave.

The flow behaviour illustrated in Figure 6.22 is inherently non-linear and cannot be approximated by any linear approach. Transonic flow solutions of the highest fidelity can be obtained by solving numerically the full Navier–Stokes Eqs. (2.1)–(2.4) but such solutions are very computationally expensive for unsteady flow. Furthermore, for realistic aircraft Reynolds numbers, direct numerical simulations (DNS) of the Navier–Stokes equations are prohibitively expensive, both in terms of computational time and memory requirements.

Practical solutions of transonic problems are obtained by solving the Reynolds Averaged Navier–Stokes equations, in combination with a turbulence model for closure.

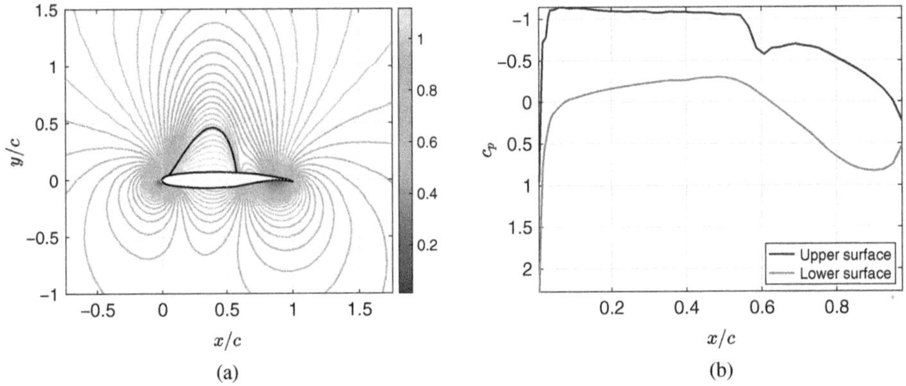

Figure 6.22 Mach contours and pressure distribution around NASA SC(2)-0714 airfoil at $M_\infty = 0.722$ and $\alpha = 0°$. (a) Mach contours and (b) c_p on surface.

2 See also Olsen (1966).

These schemes have much lower computational requirements than DNS, but they are still very expensive. Solutions of progressively lower fidelity can be obtained by solving the Euler equations ((2.5)–(2.7)), the full potential equation (2.131), or the simpler but still non-linear Transonic Small Disturbance equation (2.135). In all cases, the solutions are numerical, and they are evaluated not only on the surface of the body but also in a significant portion of the flowfield around it. The necessity to discretise the flowfield arises from the non-linearity of the equations and is the cause of the high computational cost of such techniques. In this section, we will focus on panel-like solutions of the TSD equation and on transonic correction techniques for subsonic panel methods.

6.5.1 Steady Transonic Flow

Under steady flow conditions, the transonic small disturbance equation (2.135) becomes

$$\left(1 - M_\infty^2 - \frac{\gamma+1}{U_\infty} M_\infty^2 \phi_x\right) \phi_{xx} + \phi_{yy} + \phi_{zz} = 0 \tag{6.133}$$

which is a non-linear partial differential equation. Recalling that $M_\infty < 1$, the role of the non-linear term is crucial in determining the character of the equation:

- If $\phi_x < U_\infty \left(1 - M_\infty^2\right) / (\gamma+1) M_\infty^2$, the sign of ϕ_{xx} is positive and the steady TSD is an elliptic equation, such as Laplace's equation.
- If $\phi_x > U_\infty \left(1 - M_\infty^2\right) / (\gamma+1) M_\infty^2$, the sign of ϕ_{xx} is negative and the steady TSD is a hyperbolic equation, such as the wave equation.

As ϕ_x varies in the flowfield, there can be areas where the equation is elliptic and others where it is hyperbolic. This is the mathematical definition of transonic flow, a flow that features both elliptic and hyperbolic behaviour.

Equation (6.133) can be solved numerically by discretizing the flowfield around the body and applying a finite difference, volume or element technique. Such approaches are beyond the scope of the present book; we will therefore look for an alternative that is more compatible with the panel methods already discussed up to now. The first steps are to write the transonic small perturbation equation (6.133) as

$$\beta^2 \phi_{xx} + \phi_{yy} + \phi_{zz} = \frac{\gamma+1}{U_\infty} M_\infty^2 \phi_x \phi_{xx}$$

where $\beta^2 = \sqrt{1 - M_\infty^2}$ and to apply the Prandtl–Glauert transformation of Eq. (6.3), so that

$$\phi_{\xi\xi} + \phi_{\eta\eta} + \phi_{\zeta\zeta} = \bar{\nabla}^2 \phi = \frac{\gamma+1}{U_\infty} \frac{M_\infty^2}{\beta^3} \phi_\xi \phi_{\xi\xi} \tag{6.134}$$

The equation is again non-linear and has no analytical solution, but we can still write Green's theorem for it. We return to the analysis presented in Section 2.3.1 and substitute the function of expression (2.62), written for the perturbation potential and in terms of transformed coordinates $\boldsymbol{\xi}$,

$$\mathbf{F}(\boldsymbol{\xi}) = \frac{1}{r(\boldsymbol{\xi}, \boldsymbol{\xi}_0)} \bar{\nabla} \phi(\boldsymbol{\xi}) - \phi(\boldsymbol{\xi}) \bar{\nabla} \left(\frac{1}{r(\boldsymbol{\xi}, \boldsymbol{\xi}_0)} \right)$$

in the divergence theorem (2.61)

$$\int_{\mathcal{V}} \bar{\nabla} \cdot \mathbf{F}(\xi) \mathrm{d}\mathcal{V} = \int_{\Sigma} \mathbf{n}(\xi_s) \cdot \mathbf{F}(\xi_s) \mathrm{d}\Sigma$$

where $\mathcal{V}$ is the transformed flow volume and Σ the transformed surface. We obtain

$$\int_{\mathcal{V}} \left(\frac{1}{r(\xi, \xi_0)} \bar{\nabla}^2 \phi(\xi) - \phi(\xi) \bar{\nabla}^2 \left(\frac{1}{r(\xi, \xi_0)} \right) \right) \mathrm{d}\mathcal{V}$$
$$= \int_{\Sigma} \mathbf{n}(\xi_s) \cdot \left(\frac{1}{r(\xi_s, \xi_0)} \bar{\nabla}\phi(\xi_s) - \phi(\xi_s) \bar{\nabla} \left(\frac{1}{r(\xi, \xi_0)} \right) \bigg|_{\xi_s} \right) \mathrm{d}\Sigma$$

Substituting for $\bar{\nabla}^2 \phi(\xi)$ from Eq. (6.134) yields

$$\int_{\mathcal{V}} -\phi(\xi) \bar{\nabla}^2 \left(\frac{1}{r(\xi, \xi_0)} \right) \mathrm{d}\mathcal{V}$$
$$= \int_{\Sigma} \mathbf{n}(\xi_s) \cdot \left(\frac{1}{r(\xi_s, \xi_0)} \bar{\nabla}\phi(\xi_s) - \phi(\xi_s) \bar{\nabla} \left(\frac{1}{r(\xi, \xi_0)} \right) \bigg|_{\xi_s} \right) \mathrm{d}\Sigma$$
$$- \frac{\gamma + 1}{U_\infty} \frac{M_\infty^2}{\beta^3} \int_{\mathcal{V}} \phi_\xi(\xi) \phi_{\xi\xi}(\xi) \frac{1}{r(\xi, \xi_0)} \mathrm{d}\mathcal{V}$$

Next, we substitute Eq. (2.64) into the left-hand side, such that

$$4\pi\phi(\xi_0) = \int_{\Sigma} \mathbf{n}(\xi_s) \cdot \left(\frac{1}{r(\xi_s, \xi_0)} \bar{\nabla}\phi(\xi_s) - \phi(\xi_s) \bar{\nabla} \left(\frac{1}{r(\xi, \xi_0)} \right) \bigg|_{\xi_s} \right) \mathrm{d}\Sigma$$
$$- \frac{\gamma + 1}{U_\infty} \frac{M_\infty^2}{\beta^3} \int_{\mathcal{V}} \phi_\xi(\xi) \phi_{\xi\xi}(\xi) \frac{1}{r(\xi, \xi_0)} \mathrm{d}\mathcal{V}$$

Finally, using the function $E(\xi_0)$ of Eq. (2.69) to generalise for the case where ξ_0 lies on the surface of the body, changing the direction of the normal vector, $\mathbf{n}(\xi_s)$, and dividing throughout by 4π leads to

$$E(\xi_0)\phi(\xi_0) = -\frac{1}{4\pi} \int_{\Sigma} \mathbf{n}(\xi_s) \cdot \bar{\nabla}\phi(\xi_s) \frac{1}{r(\xi_s, \xi_0)} \mathrm{d}\Sigma$$
$$+ \frac{1}{4\pi} \int_{\Sigma} \phi(\xi_s) \mathbf{n}(\xi_s) \cdot \bar{\nabla} \left(\frac{1}{r(\xi, \xi_0)} \right) \bigg|_{\xi_s} \mathrm{d}\Sigma$$
$$- \frac{1}{4\pi} \frac{\gamma + 1}{U_\infty} \frac{M_\infty^2}{\beta^3} \int_{\mathcal{V}} \phi_\xi(\xi) \phi_{\xi\xi}(\xi) \frac{1}{r(\xi, \xi_0)} \mathrm{d}\mathcal{V} \tag{6.135}$$

This latest relation is Green's theorem for steady transonic potential flow. It is written using the formulation of Eq. (2.82), whereby ξ_0 is the position of a single source lying somewhere on the surface or in the flowfield. For compatibility with the rest of this book, we will perform the change of variable that gave rise to Eq. (2.102), such that now there are continuous distributions of sources and doublets on the wing and sources in the flowfield. Furthermore, for a lifting wing we need to add a wake surface in the manner of Eq. (5.169) so that using the definitions of the wing source panel strength (5.167), wing doublet panel strength (5.168) and wake doublet panel strength (5.166), Eq. (6.135) becomes

$$E(\xi)\phi(\xi) = -\frac{1}{4\pi} \int_{\Sigma} \sigma(\xi_s) \frac{1}{r(\xi, \xi_s)} \mathrm{d}\Sigma$$

$$+\frac{1}{4\pi}\int_{\Sigma}\mu(\xi_s)\mathbf{n}(\xi_s)\cdot\bar{\nabla}\left(\frac{1}{r(\xi,\xi_s)}\right)\bigg|_{\xi_s}\mathrm{d}\Sigma$$
$$+\frac{1}{4\pi}\int_{\Sigma_w}\mu_w(\xi_s)\mathbf{n}(\xi_s)\cdot\bar{\nabla}\left(\frac{1}{r(\xi,\xi_s)}\right)\bigg|_{\xi_s}\mathrm{d}\Sigma$$
$$-\frac{1}{4\pi}\int_{\mathcal{V}}\sigma_V(\xi_V)\frac{1}{r(\xi,\xi_V)}\mathrm{d}\mathcal{V} \tag{6.136}$$

where ξ_V is the integration variable in the volume integral and denotes any point in the field, while $\sigma_V(\xi_V)$ is the source strength in the volume $\mathcal{V}$

$$\sigma_V(\xi_V)=\frac{\gamma+1}{U_\infty}\frac{M_\infty^2}{\beta^3}\phi_\xi(\xi_V)\phi_{\xi\xi}(\xi_V) \tag{6.137}$$

Equation (6.136) is the steady subsonic Green's identity of Eq. (6.7) with an additional volume integral that represents the non-linearity of the flow. It is implied that point ξ is excluded from all integrals.

No explicit modelling of shock waves appears in Eq. (6.136). A shock wave is an infinitely thin surface across which there is a discontinuous change in normal velocity but no change in potential. If $\mathbf{n}_{\text{shock}}$ is the vector normal to the shock's surface, and we use ϕ_1, ϕ_2 to denote the potential the potential upstream and downstream of the shock, respectively, then $\Delta\phi = \phi_2-\phi_1=0$ but $\Delta(\mathbf{n}_{\text{shock}}\cdot\nabla\phi)=\mathbf{n}_{\text{shock}}\cdot\nabla\phi_2-\mathbf{n}_{\text{shock}}\cdot\nabla\phi_1\neq 0$. In this sense, a shock is the opposite of a wake; across a wake there is a jump in potential, but the normal velocity is continuous. Building on the 2D analysis by Murman and Cole (1971), Tseng and Morino (1982) state that the jump in normal velocity across a 3D shock normal to the wing's surface is given by

$$\Delta(\mathbf{n}_{\text{shock}}\cdot\nabla\phi)-\frac{1}{2}\frac{\gamma+1}{U_\infty}\frac{M_\infty^2}{\beta^3}\Delta\left(\phi_\xi^2\right)n_{\xi_{\text{shock}}}=0$$

where $n_{\xi_{\text{shock}}}$ is the component of $\mathbf{n}_{\text{shock}}$ in the ξ direction. Using this value of the normal velocity jump, Tseng and Morino (1982) show that the shock is effectively already embedded in the volume integral of Eq. (6.136) and that there is no need to model it explicitly.

Techniques that solve equations similar to (6.136) are known as field-panel methods. The panel methods we have used so far are based on discretizing the surface of the wing and wake into flat panels and solving for the potential on the control points of the wing panels. Field panel methods also discretise a small volume around the wing, known as the field, into volume elements, such that the volume integral in Eq. (6.136) can be evaluated. Even after these discretisations, the equation cannot be solved directly because $\sigma_V(\xi)$ is a nonlinear function. Consequently, field-panel methods are iterative solutions.

The wing is discretised into $2m$ chordwise and n surface panels, for a total of $N=2mn$ wing panels, the wake into $N_w=n$ surface panels and the field into N_V volume elements. Equation (6.136) is written in discrete form on the control points of the wing panels as

$$\boldsymbol{A}_\phi\boldsymbol{\sigma}+\left(\boldsymbol{B}_\phi-\frac{1}{2}\boldsymbol{I}\right)\boldsymbol{\mu}+\boldsymbol{C}_\phi\boldsymbol{\mu}_w+\boldsymbol{A}_\phi^{Vb}\boldsymbol{\sigma}_V=\mathbf{0} \tag{6.138}$$

where $\boldsymbol{A}_\phi$, $\boldsymbol{B}_\phi$ are the $N\times N$ potential influence coefficient matrices of the wing's source and doublet panels on the wing control points, $\boldsymbol{C}_\phi$ is the $N\times N_w$ potential influence coefficient matrix of the wake's doublet panels on the wing control points and $\boldsymbol{A}_\phi^{Vb}$ is the

$N \times N_V$ potential influence coefficient matrix of the field elements on the wing panels. Furthermore, $\boldsymbol{\sigma}$ is the $N \times 1$ vector of the wing panel source strengths, $\boldsymbol{\mu}$ is the $N \times 1$ vector of the wing panel doublet strengths, $\boldsymbol{\mu}_w$ is the $N_w \times 1$ vector of the wake panel doublet strengths and $\boldsymbol{\sigma}_V$ is the $N_V \times 1$ vector of the field element source strengths.

On the control points of the field elements, Eq. (6.136) becomes

$$\boldsymbol{\phi} = \boldsymbol{A}_{\phi}^{bV}\boldsymbol{\sigma} + \boldsymbol{B}_{\phi}^{bV}\boldsymbol{\mu} + \boldsymbol{C}_{\phi}^{wV}\boldsymbol{\mu}_w + \boldsymbol{A}_{\phi}^{VV}\boldsymbol{\sigma}_V \tag{6.139}$$

where $\boldsymbol{\phi}$ is the $N_V \times 1$ vector of potential values at the control points of the field elements, $\boldsymbol{A}_{\phi}^{bV}$, $\boldsymbol{B}_{\phi}^{bV}$ are the $N_V \times N$ potential influence coefficient matrices of the body source and doublet panels on the field control points, $\boldsymbol{C}_{\phi}^{wV}$ is the $N_V \times N_w$ potential influence coefficient matrix of the wake doublet panels on the field control points and $\boldsymbol{A}_{\phi}^{VV}$ is the $N_V \times N_V$ potential influence coefficient matrix of the field elements on the field control points.

Equation (6.139) gives the potential in the field in terms of the unknown source strength $\boldsymbol{\sigma}_V$, which also depends on the potential in the field. Disretising Eq. (6.137) we obtain

$$\boldsymbol{\sigma}_V = \frac{\gamma + 1}{U_\infty}\frac{M_\infty^2}{\beta^3}\boldsymbol{\phi}_\xi \circ \boldsymbol{\phi}_{\xi\xi}$$

where $\circ$ denotes the Hadamard product. Equations (6.138) and (6.139) become

$$\boldsymbol{A}_\phi\boldsymbol{\sigma} + \left(\boldsymbol{B}_\phi - \frac{1}{2}\boldsymbol{I}\right)\boldsymbol{\mu} + \boldsymbol{C}_\phi\boldsymbol{\mu}_w + \boldsymbol{A}_\phi^{Vb}\frac{\gamma + 1}{U_\infty}\frac{M_\infty^2}{\beta^3}\boldsymbol{\phi}_\xi \circ \boldsymbol{\phi}_{\xi\xi} = \boldsymbol{0} \tag{6.140}$$

$$\boldsymbol{\phi} = \boldsymbol{A}_{\phi}^{bV}\boldsymbol{\sigma} + \boldsymbol{B}_{\phi}^{bV}\boldsymbol{\mu} + \boldsymbol{C}_{\phi}^{wV}\boldsymbol{\mu}_w + \boldsymbol{A}_{\phi}^{VV}\frac{\gamma + 1}{U_\infty}\frac{M_\infty^2}{\beta^3}\boldsymbol{\phi}_\xi \circ \boldsymbol{\phi}_{\xi\xi} \tag{6.141}$$

This is a set of two equations with three unknowns, the doublet strength on the surface $\boldsymbol{\mu}$, the doublet strength in the wake $\boldsymbol{\mu}_w$ and the potential in the field $\boldsymbol{\phi}$. They are completed by the Kutta condition of Eq. (6.19)

$$\boldsymbol{K}\boldsymbol{\mu} + \boldsymbol{\mu}_w = \boldsymbol{0} \tag{6.142}$$

recalling that $\boldsymbol{K} = (\boldsymbol{I}_n, \boldsymbol{0}_{(2m-1)n,n}, -\boldsymbol{I}_n)$, $\boldsymbol{I}_n$ is the $n \times n$ unit matrix and $\boldsymbol{0}_{l,k}$ is a $l \times k$ matrix full of zeros. We now have three equations with three unknowns, since $\boldsymbol{\sigma}$ is given by the impermeability boundary condition, but they are non-linear. They are therefore solved iteratively; at the first iteration, we set $\boldsymbol{\phi} = \boldsymbol{0}$ and solve Eqs. (6.140) and (6.142) as in the subsonic case. We then calculate the potential in the field from Eq. (6.141) as

$$\boldsymbol{\phi}_1 = \boldsymbol{A}_{\phi}^{bV}\boldsymbol{\sigma} + \boldsymbol{B}_{\phi}^{bV}\boldsymbol{\mu}_1 + \boldsymbol{C}_{\phi}^{wV}\boldsymbol{\mu}_{w_1}$$

At the kth iteration, Eqs. (6.140)–(6.142) become

$$\begin{aligned}
&\boldsymbol{A}_\phi\boldsymbol{\sigma} + \left(\boldsymbol{B}_\phi - \frac{1}{2}\boldsymbol{I}\right)\boldsymbol{\mu}_k + \boldsymbol{C}_\phi\boldsymbol{\mu}_{w_k} + \boldsymbol{A}_\phi^{Vb}\frac{\gamma + 1}{U_\infty}\frac{M_\infty^2}{\beta^3}\boldsymbol{\phi}_{\xi_{k-1}} \circ \boldsymbol{\phi}_{\xi\xi_{k-1}} = \boldsymbol{0} \\
&\boldsymbol{K}\boldsymbol{\mu}_k + \boldsymbol{\mu}_{w_k} = \boldsymbol{0} \\
&\boldsymbol{\phi}_k = \boldsymbol{A}_{\phi}^{bV}\boldsymbol{\sigma} + \boldsymbol{B}_{\phi}^{bV}\boldsymbol{\mu}_k + \boldsymbol{C}_{\phi}^{wV}\boldsymbol{\mu}_{w_k} + \boldsymbol{A}_{\phi}^{VV}\frac{\gamma + 1}{U_\infty}\frac{M_\infty^2}{\beta^3}\boldsymbol{\phi}_{\xi_{k-1}} \circ \boldsymbol{\phi}_{\xi\xi_{k-1}}
\end{aligned} \tag{6.143}$$

We stop the iterations when a convergence criterion has been met, such as $||\boldsymbol{\mu}_k - \boldsymbol{\mu}_{k-1}|| < \varepsilon$, where $\varepsilon \ll 1$ is a small positive real number.

In order to implement this iterative solution, we need to evaluate the derivatives $\boldsymbol{\phi}_{\xi}$ and $\boldsymbol{\phi}_{\xi\xi}$ from $\boldsymbol{\phi}$. The most popular method for doing this is the finite difference scheme by Murman and Cole (1971). The scheme is mixed hyperbolic and elliptic, that is it makes use of backwards or central differences, depending on whether the local flow Mach number is supersonic or subsonic, respectively. We also need to calculate the influence coefficients of the field sources on the wing and field control points, $\boldsymbol{A}_{\phi}^{Vb}$ and $\boldsymbol{A}_{\phi}^{VV}$, respectively. This calculation depends on the shape of the field volume elements; analytical expressions exist for orthogonal parallelepipeds, see for example Chu (1988).

Field-panel methods for steady flows have been investigated for many decades, since at least Ogana (1978), but the successes of such approaches are mitigated. The shock waves they predict are smoothed and attenuated compared to experimental results. As an example, Figure 6.23 plots the steady pressure distribution around the LANN wing of Examples 6.2 and 6.3 at a steady angle of attack $\alpha_0 = 2.6°$ and free stream Mach number $M_{\infty} = 0.77$. The figure compares the experimental results to the predictions of the linear SDPM and to those obtained from a field-panel method, labelled

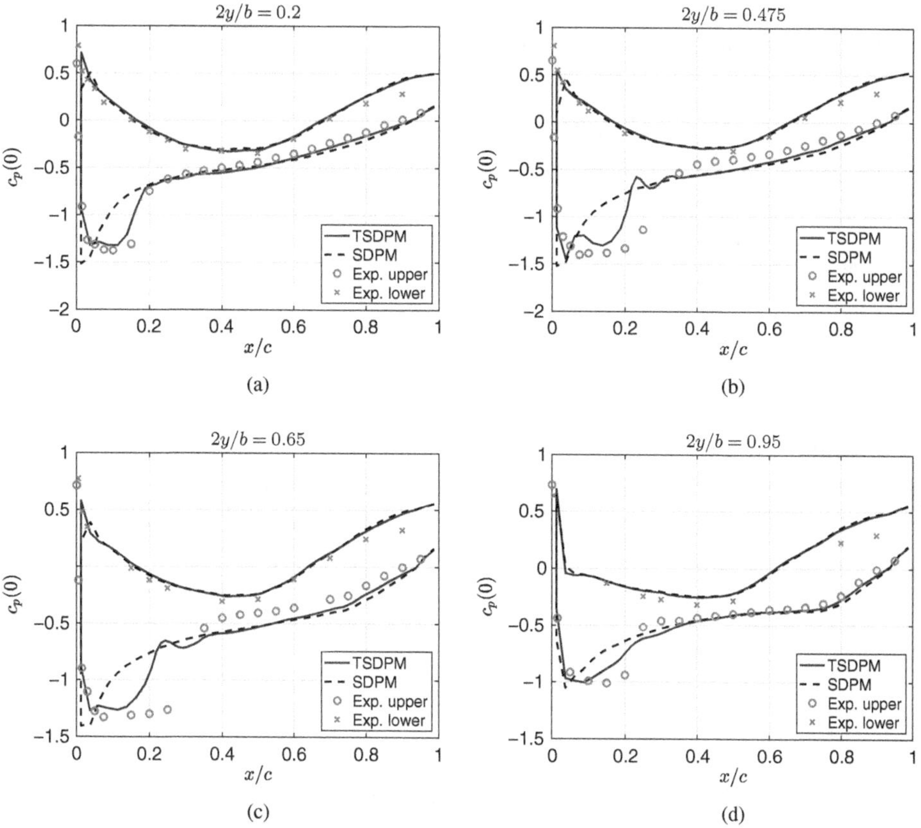

Figure 6.23 Steady pressure distribution predicted by the transonic SDPM around LANN wing for $M_{\infty} = 0.77$, $\alpha_0 = 2.6°$. Experimental data by Zwaan (1985). (a) $2y/b = 0.2$, (b) $2y/b = 0.475$, (c) $2y/b = 0.65$ and (d) $2y/b = 0.95$. Source: Zwaan (1985), credit: AGARD.

as TSDPM in the legend. Clearly, the field-panel results lie closer to the experimental data than the linear ones but the effect of the shock wave on the pressure distribution is smeared; the pressure recovers smoothly instead of jumping discontinuously. Furthermore, there are small-amplitude oscillations in the pressure distribution, which occur because a continuous method is used to model a discontinuous phenomenon. Increasing the number of panels, both on the surface and in the field, improves the overall match between simulation and experiment but also increases the amplitude of these oscillations.

More accurate predictions for steady transonic flows can be obtained by solving the RANS, Euler or full potential equations using finite volume or finite element discretisations of the flowfield. Such solutions can be obtained at reasonable computational cost for steady cases and can capture shock-induced discontinuities accurately, including multiple-shock cases. Crovato et al. (2020) present an extensive comparison of such solutions applied to the Onera M6 experimental test case. Nevertheless, field-panel methods can be applied to unsteady transonic flow problems, for which the computational cost of RANS/Euler/full potential solvers is still very high.

6.5.2 Time Linearised Transonic Small Perturbation Equation

Approximate treatments of unsteady transonic flows can be obtained by linearising the unsteady flow around a non-linear steady flow. This approach can be applied to the full potential equation, but, here, it will be demonstrated on the unsteady transonic small disturbance equation (2.135)

$$\phi_{tt} + 2U_\infty \phi_{xt} = a_\infty^2 \left(1 - M_\infty^2 - \frac{\gamma+1}{U_\infty} M_\infty^2 \phi_x\right) \phi_{xx} + a_\infty^2 \phi_{yy} + a_\infty^2 \phi_{zz} \tag{6.144}$$

to which we will apply directly the Prandtl–Glauert transformation of Eq. (6.3), to obtain

$$\phi_{\xi\xi} + \phi_{\eta\eta} + \phi_{\zeta\zeta} - \frac{2M_\infty}{a_\infty \beta} \phi_{\xi t} - \frac{1}{a_\infty^2} \phi_{tt} = \frac{\gamma+1}{U_\infty} \frac{M_\infty^2}{\beta^3} \phi_\xi \phi_{\xi\xi} \tag{6.145}$$

Once again, we model the body as quasi-fixed so that its motion is only represented by the relative flow velocity between the body and the fluid. In order to proceed, we define the potential $\phi(\xi, t)$ as the sum of a steady component $\bar{\phi}(\xi)$ and a small unsteady component $\phi'(\xi, t)$, such that

$$\phi(\xi, t) = \bar{\phi}(\xi) + \phi'(\xi, t) \tag{6.146}$$

The steady potential must satisfy the steady TSD Eq. (6.134)

$$\bar{\phi}_{\xi\xi} + \bar{\phi}_{\eta\eta} + \bar{\phi}_{\zeta\zeta} - \frac{\gamma+1}{U_\infty} \frac{M_\infty^2}{\beta^3} \bar{\phi}_\xi \bar{\phi}_{\xi\xi} = 0 \tag{6.147}$$

Equation (6.145) contains the non-linear term $\phi_\xi \phi_{\xi\xi}$; expressing this term using the definition of Eq. (6.146) we see that

$$\phi_\xi \phi_{\xi\xi} = \left(\bar{\phi}_\xi + \phi'_\xi\right) \left(\bar{\phi}_{\xi\xi} + \phi'_{\xi\xi}\right) \approx \bar{\phi}_\xi \bar{\phi}_{\xi\xi} + \bar{\phi}_\xi \phi'_{\xi\xi} + \phi'_\xi \bar{\phi}_{\xi\xi} = \bar{\phi}_\xi \bar{\phi}_{\xi\xi} + \frac{\partial}{\partial \xi}\left(\bar{\phi}_\xi \phi'_\xi\right)$$

since we assumed that ϕ' is small, $|\phi'| \ll |\bar{\phi}|$. Consequently, substituting from Eq. (6.146) into Eq. (6.144) yields

$$\phi'_{\xi\xi} + \phi'_{\eta\eta} + \phi'_{\zeta\zeta} - \frac{2M_\infty}{a_\infty \beta}\phi'_{\xi t} - \frac{1}{a_\infty^2}\phi'_{tt} - \frac{\gamma+1}{U_\infty}\frac{M_\infty^2}{\beta^3}\frac{\partial}{\partial \xi}\left(\bar{\phi}_\xi \phi'_\xi\right)$$
$$+ \bar{\phi}_{\xi\xi} + \bar{\phi}_{\eta\eta} + \bar{\phi}_{\zeta\zeta} - \frac{\gamma+1}{U_\infty}\frac{M_\infty^2}{\beta^3}\bar{\phi}_\xi \bar{\phi}_{\xi\xi} = 0 \tag{6.148}$$

The second line of this expression is the steady TSD equation and is equal to zero, so that

$$\phi'_{\xi\xi} + \phi'_{\eta\eta} + \phi'_{\zeta\zeta} - \frac{2M_\infty}{a_\infty \beta}\phi'_{\xi t} - \frac{1}{a_\infty^2}\phi'_{tt} = \frac{\gamma+1}{U_\infty}\frac{M_\infty^2}{\beta^3}\frac{\partial}{\partial \xi}\left(\bar{\phi}_\xi \phi'_\xi\right) \tag{6.149}$$

which is known as the Time-Linearised Transonic Small Disturbance (TLTSD) equation. Its left-hand side is identical to that of the linearized small disturbance equation (6.61). Its right-hand side is a linear term since $\bar{\phi}$ is a known function, obtained from the solution of Eq. (6.147).

Applying the Fourier transform to the time-linearised transonic small disturbance equation, we obtain

$$\phi'_{\xi\xi}(\omega) + \phi'_{\eta\eta}(\omega) + \phi'_{\zeta\zeta}(\omega) - \frac{2\mathrm{i}\omega M_\infty}{a_\infty \beta}\phi'_\xi(\omega) + \frac{\omega^2}{a_\infty^2}\phi'(\omega) = \frac{\gamma+1}{U_\infty \beta^3}M_\infty^2 \frac{\partial}{\partial \xi}\left(\bar{\phi}_\xi \phi'_\xi(\omega)\right) \tag{6.150}$$

Combining the Green's identities for unsteady subcritical flow (6.70) and steady supercritical flow (6.136), we obtain

$$\begin{aligned} E\phi'(\xi,\omega) = &-\frac{1}{4\pi}\int_\Sigma \sigma(\xi_s,\omega)\frac{\mathrm{e}^{-\mathrm{i}\Omega(-M_\infty(\xi-\xi_s)+r(\xi,\xi_s))}}{r(\xi,\xi_s)}\mathrm{d}\Sigma \\ &+\frac{1}{4\pi}\int_\Sigma \mu(\xi_s,\omega)\mathbf{n}(\xi_s)\cdot\bar{\nabla}_s\left(\frac{\mathrm{e}^{-\mathrm{i}\Omega(-M_\infty(\xi-\xi_s)+r(\xi,\xi_s))}}{r(\xi,\xi_s)}\right)\mathrm{d}\Sigma \\ &+\frac{1}{4\pi}\int_{\Sigma_w} \mu_w(\xi_s,\omega)\mathbf{n}(\xi_s)\cdot\bar{\nabla}_s\left(\frac{\mathrm{e}^{-\mathrm{i}\Omega(-M_\infty(\xi-\xi_s)+r(\xi,\xi_s))}}{r(\xi,\xi_s)}\right)\mathrm{d}\Sigma \\ &-\frac{1}{4\pi}\int_\mathcal{V} \sigma_V(\xi_V,\omega)\frac{\mathrm{e}^{-\mathrm{i}\Omega(-M_\infty(\xi-\xi_V)+r(\xi,\xi_V))}}{r(\xi,\xi_V)}\mathrm{d}\mathcal{V} \end{aligned} \tag{6.151}$$

where ξ_V is the integration variable in the volume integral, denoting any point in volume $\mathcal{V}$, and

$$\sigma_V(\xi_V,\omega) = \frac{\gamma+1}{U_\infty}\frac{M_\infty^2}{\beta^3}\frac{\partial}{\partial \xi}\left(\bar{\phi}_\xi \phi'_\xi\right)\bigg|_{\xi_V,\omega} \tag{6.152}$$

Note that the volume source term is multiplied by $\mathrm{e}^{-\mathrm{i}\Omega(-M_\infty(\xi-\xi_V)+r(\xi,\xi_V))}$, signifying the fact that, in the unsteady case, this term must also be delayed by the subsonic time delay $\tau(\xi,\xi_V)$. For a rigorous demonstration of this fact see Morino (1974).

As in the steady transonic case, we discretise the surface and field into N body panels, N_w wake panels and N_V field elements. Then, evaluating the discrete version of Eq. (6.151) on the surface control points and applying Morino's approximation of Eqs. (6.76) and (6.79), we obtain

$$\bar{\boldsymbol{A}}_\phi(\omega)\boldsymbol{\sigma}(\omega) + \left(\bar{\boldsymbol{B}}_\phi(\omega) - \frac{1}{2}\boldsymbol{I}\right)\boldsymbol{\mu}(\omega) + \bar{\boldsymbol{C}}_\phi(\omega)\boldsymbol{\mu}_w(\omega) + \bar{\boldsymbol{A}}_\phi^{Vb}(\omega)\boldsymbol{\sigma}_V(\omega) = \boldsymbol{0} \tag{6.153}$$

where the elements of matrices $\bar{\boldsymbol{A}}_\phi(\omega)$, $\bar{\boldsymbol{B}}_\phi(\omega)$, $\bar{\boldsymbol{C}}_\phi(\omega)$ are given by Eqs. (6.81)–(6.83) and the elements of $\bar{\boldsymbol{A}}_\phi^{Vb}(\omega)$ are given by

$$\bar{A}^{Vb}_{\phi_{IJ}}(\omega) = \mathrm{e}^{-\mathrm{i}\Omega(-M_\infty(\xi_{c_I}-\xi^V_{c_J})+r(\xi_{c_I},\xi^V_{c_J}))} A^{Vb}_{\phi_{I,J}} \tag{6.154}$$

for $I = 1,\dots,N, J = 1,\dots,N_V$, where $\xi^V_{c_J}$ are the coordinates of the control point of the Jth field element and $A^{Vb}_{\phi_{I,J}}$ the elements of the steady field influence coefficient matrix $\boldsymbol{A}^{Vb}_\phi$ in Eq. (6.138).

Evaluating the discrete version of Eq. (6.151) in the field leads to

$$\boldsymbol{\phi}'(\omega) = \bar{\boldsymbol{A}}^{bV}_\phi(\omega)\boldsymbol{\sigma}(\omega) + \bar{\boldsymbol{B}}^{bV}_\phi(\omega)\boldsymbol{\mu}(\boldsymbol{\omega}) + \bar{\boldsymbol{C}}^{wV}_\phi(\omega)\boldsymbol{\mu}_w(\boldsymbol{\omega}) + \bar{\boldsymbol{A}}^{VV}_\phi(\omega)\boldsymbol{\sigma}_V(\omega) \tag{6.155}$$

As in the steady case, the success of the scheme depends on the calculation of the source strength in the volume,

$$\boldsymbol{\sigma}_V(\omega) = \frac{\gamma+1}{U_\infty}\frac{M^2_\infty}{\beta^3}\left(\bar{\boldsymbol{\phi}}_{\xi\xi} \circ \boldsymbol{\phi}'_\xi(\omega) + \bar{\boldsymbol{\phi}}_\xi \circ \boldsymbol{\phi}'_{\xi\xi}(\omega)\right) \tag{6.156}$$

First, $\bar{\boldsymbol{\phi}}$ and its derivatives $\bar{\boldsymbol{\phi}}_\xi$, $\bar{\boldsymbol{\phi}}_{\xi\xi}$ must be calculated from a steady solution; as discussed in Section 6.5.1 such solutions are most accurate when obtained using a RANS, Euler or full potential equation solution based on finite element or finite volume discretization of the entire flowfield. The next step is to calculate $\boldsymbol{\phi}'_\xi(\omega)$ and $\boldsymbol{\phi}'_{\xi\xi}(\omega)$ from $\boldsymbol{\phi}'(\omega)$ using a finite difference scheme applied to the field grid. Such a scheme can be written in matrix form as

$$\boldsymbol{\phi}'_\xi(\omega) = \boldsymbol{T}_\xi\boldsymbol{\phi}'(\omega), \quad ,\boldsymbol{\phi}'_{\xi\xi}(\omega) = \boldsymbol{T}_{\xi\xi}\boldsymbol{\phi}'(\omega) \tag{6.157}$$

where $\boldsymbol{T}_\xi$ and $\boldsymbol{T}_{\xi\xi}$ are $N_V \times N_V$ sparse finite difference operator matrices. As Eq. (6.156) is linear, it can be written in the form

$$\boldsymbol{\sigma}_V(\omega) = \boldsymbol{D}\boldsymbol{\phi}'(\omega)$$

where

$$\boldsymbol{D} = \frac{\gamma+1}{U_\infty}\frac{M^2_\infty}{\beta^3}\left(\bar{\boldsymbol{\phi}}_{\xi\xi} \circ \boldsymbol{T}_\xi + \bar{\boldsymbol{\phi}}_\xi \circ \boldsymbol{T}_{\xi\xi}\right)$$

and the notation $\boldsymbol{x} \circ \boldsymbol{A}$ denotes the element-by-element multiplication of every column of matrix $\boldsymbol{A}$ by vector $\boldsymbol{x}$. Substituting back into Eq. (6.155) and solving for $\boldsymbol{\phi}'(\omega)$ yields

$$\boldsymbol{\phi}'(\omega) = \left(\boldsymbol{I} - \bar{\boldsymbol{A}}^{VV}_\phi(\omega)\boldsymbol{D}\right)^{-1}\left(\bar{\boldsymbol{A}}^{bV}_\phi(\omega)\boldsymbol{\sigma}(\omega) + \bar{\boldsymbol{B}}^{bV}_\phi(\omega)\boldsymbol{\mu}(\boldsymbol{\omega}) + \bar{\boldsymbol{C}}^{wV}_\phi(\omega)\boldsymbol{\mu}_w(\boldsymbol{\omega})\right)$$

and, hence,

$$\boldsymbol{\sigma}_V(\omega) = \boldsymbol{D}\left(\boldsymbol{I} - \bar{\boldsymbol{A}}^{VV}_\phi(\omega)\boldsymbol{D}\right)^{-1}\left(\bar{\boldsymbol{A}}^{bV}_\phi(\omega)\boldsymbol{\sigma}(\omega) + \bar{\boldsymbol{B}}^{bV}_\phi(\omega)\boldsymbol{\mu}(\omega) + \bar{\boldsymbol{C}}^{wV}_\phi(\omega)\boldsymbol{\mu}_w(\boldsymbol{\omega})\right)$$

Consequently, the source strength in the field becomes a linear combination of the source and doublet strength on the wing and the doublet strength in the wake. Substituting for $\boldsymbol{\sigma}_V(\omega)$ in Eq. (6.153) results in

$$\hat{\boldsymbol{A}}_\phi(\omega)\boldsymbol{\sigma}(\omega) + \hat{\boldsymbol{B}}_\phi(\omega)\boldsymbol{\mu}(\omega) + \hat{\boldsymbol{C}}_\phi(\omega)\boldsymbol{\mu}_w(\omega) = \boldsymbol{0}$$

where

$$\hat{\boldsymbol{A}}_\phi(\omega) = \bar{\boldsymbol{A}}_\phi(\omega) + \bar{\boldsymbol{A}}_\phi^{Vb}(\omega)\boldsymbol{D}\left(\boldsymbol{I} - \bar{\boldsymbol{A}}_\phi^{VV}(\omega)\boldsymbol{D}\right)^{-1}\bar{\boldsymbol{A}}_\phi^{bV}(\omega)$$

$$\hat{\boldsymbol{B}}_\phi(\omega) = \bar{\boldsymbol{B}}_\phi(\omega) - \frac{1}{2}\boldsymbol{I} + \bar{\boldsymbol{A}}_\phi^{Vb}(\omega)\boldsymbol{D}\left(\boldsymbol{I} - \bar{\boldsymbol{A}}_\phi^{VV}(\omega)\boldsymbol{D}\right)^{-1}\bar{\boldsymbol{B}}_\phi^{bV}(\omega)$$

$$\hat{\boldsymbol{C}}_\phi(\omega) = \bar{\boldsymbol{C}}_\phi(\omega) + \bar{\boldsymbol{A}}_\phi^{Vb}(\omega)\boldsymbol{D}\left(\boldsymbol{I} - \bar{\boldsymbol{A}}_\phi^{VV}(\omega)\boldsymbol{D}\right)^{-1}\bar{\boldsymbol{C}}_\phi^{wV}(\omega)$$

Finally, using Eqs. (6.84) and (6.85), we can write

$$\boldsymbol{\mu}_w(\omega) = \boldsymbol{P}_e(\omega)\boldsymbol{P}_c(\omega)\boldsymbol{\mu}(\omega)$$

so that the linearised unsteady transonic Green's identity becomes

$$\hat{\boldsymbol{A}}_\phi(\omega)\boldsymbol{\sigma}(\omega) + \left(\hat{\boldsymbol{B}}_\phi(\omega) + \hat{\boldsymbol{C}}_\phi(\omega)\boldsymbol{P}_e(\omega)\boldsymbol{P}_c(\omega)\right)\boldsymbol{\mu}(\omega) = \boldsymbol{0}$$

and can be solved easily for $\boldsymbol{\mu}(\omega)$, recalling that $\boldsymbol{\sigma}(\omega)$ is given by the boundary condition of Eq. (6.91). The flow velocity and pressure coefficient distributions on the surface are then obtained using the approach of Example 6.3.

Unlike the steady transonic case, the time linearisation of Eq. (6.144) has resulted in a linear unsteady problem whose solution is straightforward and non-iterative. Setting up the problem is less straightforward because it requires the calculation of the influence coefficients of the volume elements, $\bar{\boldsymbol{A}}_\phi^{Vb}(\omega)$ and $\bar{\boldsymbol{A}}_\phi^{VV}(\omega)$, as well as that of the finite difference operators $\boldsymbol{T}_\xi$ and $\boldsymbol{T}_{\xi\xi}$. Furthermore, matrix $\left(\boldsymbol{I} - \bar{\boldsymbol{A}}_\phi^{VV}(\omega)\boldsymbol{D}\right)$ is full and of size $N_V \times N_V$ so that its inversion can be computationally expensive, both in time and in memory requirements.

Arguably the first implementation of an unsteady field panel method was developed by Farr Jr. et al. (1975) and included in the TDSTRN and TDUTRN computer programmes. Perhaps the most successful applications of the TLTSD methodology have concerned the DLM; transonic variants of the DLM have been published by several authors, such as Lu and Voss (1992), Pi et al. (1979), Van Zyl (1997). Another lifting surface approach was developed by Hounjet (1971) and implemented into a software called FTRAN3. The ZAERO package includes the ZTAW solver, which is yet another transonic unsteady lifting surface method based on a the TLTSD equation (ZONA Technology Inc. 2004).

6.5.3 Unsteady Transonic Correction Methods

In Section 6.3.3, we demonstrated a technique for correcting the steady DLM predictions for camber effects. The exact same technique can be used for correcting for both camber and steady transonic effects. Furthermore, there are similar approaches for correcting the unsteady DLM predictions for transonic effects. Here we will demonstrate the successive kernel expansion technique by Silva et al. (2008).

The DLM equation (6.48) can be written in matrix form as

$$\mathbf{w}(k) = \tilde{\boldsymbol{A}}(k)\Delta\boldsymbol{c}_p(k) \tag{6.158}$$

where $\mathbf{w}(k)$ is the $N \times 1$ vector of upwash values acting at the control points of the N panels, $\tilde{\boldsymbol{A}}(k)$ is the $N \times N$ global influence coefficient matrix and $\Delta\boldsymbol{c}_p(k)$ is the pressure jump across the upper and lower surfaces of the wing and the panel control points. Equation (6.158) is valid at all reduced frequency values, down to $k = 0$. The basis of the successive kernel expansion technique is to expand $\mathbf{w}(k)$, $\tilde{\boldsymbol{A}}(k)$ and $\Delta\boldsymbol{c}_p(k)$ as power series in ik, such that

$$\mathbf{w}(k) = \mathbf{w}_0 + ik\mathbf{w}_1 + (ik)^2\mathbf{w}_2 + \cdots \tag{6.159}$$

$$\tilde{\boldsymbol{A}}(k) = \tilde{\boldsymbol{A}}_0 + ik\tilde{\boldsymbol{A}}_1 + (ik)^2\tilde{\boldsymbol{A}}_2 + \cdots \tag{6.160}$$

$$\Delta\boldsymbol{c}_p(k) = \Delta\boldsymbol{c}_{p_0} + ik\Delta\boldsymbol{c}_{p_1} + (ik)^2\Delta\boldsymbol{c}_{p_2} + \cdots \tag{6.161}$$

where $\tilde{\boldsymbol{A}}_0 = \tilde{\boldsymbol{A}}(0)$ is the steady global influence coefficient matrix. Substituting back into Eq. (6.158), we obtain

$$\mathbf{w}_0 + ik\mathbf{w}_1 = \left(\tilde{\boldsymbol{A}}_0 + ik\tilde{\boldsymbol{A}}_1 + (ik)^2\tilde{\boldsymbol{A}}_2\right)\left(\Delta\boldsymbol{c}_{p_0} + ik\Delta\boldsymbol{c}_{p_1} + (ik)^2\Delta\boldsymbol{c}_{p_2}\right)$$

noting that $\mathbf{w}_2 = 0$ in all the problems addressed in this book. Expanding the right-hand side and collecting orders of $(ik)^i$ leads to

$$\begin{aligned}\mathbf{w}_0 + ik\mathbf{w}_1 =& \tilde{\boldsymbol{A}}_0\Delta\boldsymbol{c}_{p_0} + ik\left(\tilde{\boldsymbol{A}}_0\Delta\boldsymbol{c}_{p_1} + \tilde{\boldsymbol{A}}_1\Delta\boldsymbol{c}_{p_0}\right)\\ &+ (ik)^2\left(\tilde{\boldsymbol{A}}_1\Delta\boldsymbol{c}_{p_1} + \tilde{\boldsymbol{A}}_0\Delta\boldsymbol{c}_{p_2} + \tilde{\boldsymbol{A}}_2\Delta\boldsymbol{c}_{p_0}\right) + \cdots\end{aligned}$$

Equating coefficients of $(ik)^i$ between the left- and right-hand sides yields

$$\begin{aligned}\mathbf{w}_0 &= \tilde{\boldsymbol{A}}_0\Delta\boldsymbol{c}_{p_0}\\ \mathbf{w}_1 &= \tilde{\boldsymbol{A}}_0\Delta\boldsymbol{c}_{p_1} + \tilde{\boldsymbol{A}}_1\Delta\boldsymbol{c}_{p_0}\\ 0 &= \tilde{\boldsymbol{A}}_1\Delta\boldsymbol{c}_{p_1} + \tilde{\boldsymbol{A}}_0\Delta\boldsymbol{c}_{p_2} + \tilde{\boldsymbol{A}}_2\Delta\boldsymbol{c}_{p_0}\end{aligned} \tag{6.162}$$

Solving for $\Delta\boldsymbol{c}_{p_0}$, $\Delta\boldsymbol{c}_{p_1}$ and $\Delta\boldsymbol{c}_{p_2}$ results in

$$\begin{aligned}\Delta\boldsymbol{c}_{p_0} &= \tilde{\boldsymbol{A}}_0^{-1}\mathbf{w}_0\\ \Delta\boldsymbol{c}_{p_1} &= \tilde{\boldsymbol{A}}_0^{-1}\left(\mathbf{w}_1 - \tilde{\boldsymbol{A}}_1\tilde{\boldsymbol{A}}_0^{-1}\mathbf{w}_0\right)\\ \Delta\boldsymbol{c}_{p_2} &= -\tilde{\boldsymbol{A}}_0^{-1}\left(\tilde{\boldsymbol{A}}_2\tilde{\boldsymbol{A}}_0^{-1}\mathbf{w}_0 + \tilde{\boldsymbol{A}}_1\tilde{\boldsymbol{A}}_0^{-1}\left(\mathbf{w}_1 - \tilde{\boldsymbol{A}}_1\tilde{\boldsymbol{A}}_0^{-1}\mathbf{w}_0\right)\right)\end{aligned}$$

The matrix $\tilde{\boldsymbol{A}}_0$ is the zero-frequency DLM influence coefficient matrix of Section 6.3.3 and can be corrected by multiplying it by the diagonal correction matrix $\boldsymbol{D}_{\text{corr}}$ whose elements are given by Eq. (6.102). Then, the corrected pressure jumps are obtained from

$$\begin{aligned}\Delta\boldsymbol{c}_{p_0} &= \left(\boldsymbol{D}_{\text{corr}}\tilde{\boldsymbol{A}}_0\right)^{-1}\mathbf{w}_0\\ \Delta\boldsymbol{c}_{p_1} &= \left(\boldsymbol{D}_{\text{corr}}\tilde{\boldsymbol{A}}_0\right)^{-1}\left(\mathbf{w}_1 - \tilde{\boldsymbol{A}}_1\tilde{\boldsymbol{A}}_0^{-1}\mathbf{w}_0\right)\\ \Delta\boldsymbol{c}_{p_2} &= -\left(\boldsymbol{D}_{\text{corr}}\tilde{\boldsymbol{A}}_0\right)^{-1}\left(\tilde{\boldsymbol{A}}_2\tilde{\boldsymbol{A}}_0^{-1}\mathbf{w}_0 + \tilde{\boldsymbol{A}}_1\tilde{\boldsymbol{A}}_0^{-1}\left(\mathbf{w}_1 - \tilde{\boldsymbol{A}}_1\tilde{\boldsymbol{A}}_0^{-1}\mathbf{w}_0\right)\right)\end{aligned} \tag{6.163}$$

noting that not all instances of $\tilde{\boldsymbol{A}}_0$ are corrected. Then, the corrected pressure jump distribution is obtained by substituting $\Delta\boldsymbol{c}_{p_0}$, $\Delta\boldsymbol{c}_{p_1}$ and $\Delta\boldsymbol{c}_{p_2}$ into Eq. (6.161). The coefficients $\tilde{\boldsymbol{A}}_0$, $\tilde{\boldsymbol{A}}_1$ and $\tilde{\boldsymbol{A}}_2$ in expression 6.160 can be evaluated by calculating $\tilde{\boldsymbol{A}}(k)$ from Eq. (6.49) at three values of the reduced frequency, including $k = 0$, and then curve-fitting the elements of the resulting matrices by a second-order polynomial in ik.

Example 6.8 *Use successive kernel expansion to approximate the unsteady pressure jump across the pitching LANN wing for $M_\infty = 0.77$, $\alpha_0 = 2.6°$, $\alpha_1 = 0.25°$ and $k = 0.108$.*

We start by repeating Example 6.2, changing the free stream Mach number to $M_\infty = 0.77$. We do not implement the camber line, as the steady pressure jump due to camber will be corrected using the experimental data. We evaluate the aerodynamic influence coefficient matrix using Eqs. (6.49)–(6.54) for $k_0 = 0$, $k_1 = 0.1$ and $k_2 = 0.2$, denoting the resulting matrices by $\tilde{\mathbf{A}}(k_0)$, $\tilde{\mathbf{A}}(k_1)$ and $\tilde{\mathbf{A}}(k_2)$. The next step is to curve fit these matrices using Eq. (6.160) in order to calculate $\tilde{\mathbf{A}}_0$, $\tilde{\mathbf{A}}_1$ and $\tilde{\mathbf{A}}_2$. We set up the system

$$\begin{pmatrix} \tilde{A}_{I,J}(k_0) \\ \tilde{A}_{I,J}(k_1) \\ \tilde{A}_{I,J}(k_2) \end{pmatrix} = \begin{pmatrix} 1 & ik_0 & (ik_0)^2 \\ 1 & ik_1 & (ik_1)^2 \\ 1 & ik_2 & (ik_2)^2 \end{pmatrix} \begin{pmatrix} \tilde{A}_{0_{I,J}} \\ \tilde{A}_{1_{I,J}} \\ \tilde{A}_{2_{I,J}} \end{pmatrix} \tag{6.164}$$

for $I = 1, \dots, N$, $J = 1, \dots, N$, where $N = mn$ is the total number of panels, and solve it for $\tilde{A}_{0_{I,J}}$, $\tilde{A}_{1_{I,J}}$ and $\tilde{A}_{2_{I,J}}$. Matrix $\tilde{\mathbf{A}}_0$ is real and equal to $\tilde{\mathbf{A}}(k_0)$ while $\tilde{\mathbf{A}}_1$ and $\tilde{\mathbf{A}}_2$ are complex.

The reference pressure jump is obtained from the experimental data. Zwaan (1985) carried out quasi-steady experiments for $\alpha_0 = 2.6°$, $\alpha_1 = 0.25°$ and $k = 0$ and tabulated the resulting mean pressure jump $\Delta c_p(0) = \Delta c_p(\alpha_{\text{ref}})$ and oscillatory pressure jump $\Delta c_{p_\alpha}(0) = \Delta c_{p_\alpha}^{\text{ref}}$, where $\alpha_{\text{ref}} = 2.6°$ and $\Delta\alpha = 0.25°$. Recall that pressures were measured on the LANN wing at 6 spanwise sections with up to 40 pressure taps each. Consequently, we must interpolate the experimentally measured pressure jumps onto the DLM grid. In order to avoid extrapolations, the experimental data are completed by setting the pressure jump equal to 0 throughout the wingtips and trailing edge. The interpolation is carried out by means of the built-in Matlab function `scatteredInterpolant.m`*. Figure 6.24 plots the resulting interpolated reference pressure jumps $\Delta c_p(\alpha_{\text{ref}})$ and $\Delta c_{p_\alpha}^{\text{ref}}$ at 2.6° and compares them to the experimental data. Note that the surface discretisation chosen here is identical to that of Example 6.2, that is $m = 40$, $n = 40$. The interpolated surface for $\Delta c_{p_\alpha}^{\text{ref}}$ is by no means smooth, demonstrating the disadvantage of using incomplete and noisy experimental data in order to correct the DLM predictions.*

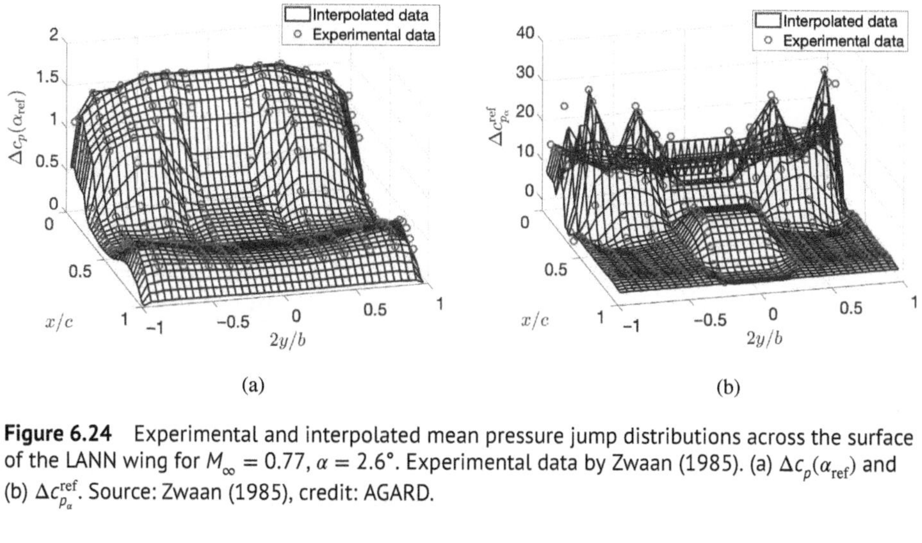

Figure 6.24 Experimental and interpolated mean pressure jump distributions across the surface of the LANN wing for $M_\infty = 0.77$, $\alpha = 2.6°$. Experimental data by Zwaan (1985). (a) $\Delta c_p(\alpha_{\text{ref}})$ and (b) $\Delta c_{p_\alpha}^{\text{ref}}$. Source: Zwaan (1985), credit: AGARD.

The next steps is to calculate the correction matrix $\mathbf{D}_{\mathrm{corr}}$ *from Eq. (6.102), before evaluating the steady corrected pressure jump at* $\alpha = 2.6°$ *from Eq. (6.103). In order to apply the unsteady transonic correction, first, we need to define the upwash vector* $\mathbf{w}(k)$ *in Eq. (6.159). Equation (6.59) gives this upwash as*

$$w_I(k) = -\left(1 + \mathrm{i}k(\tilde{x}_{c_I} - \tilde{x}_f)\right)\frac{\alpha_1}{2\mathrm{i}}\mathrm{e}^{\mathrm{i}\phi_\alpha}$$

for a general value of k and for $I = 1, \ldots, N$. *Comparing to Eq. (6.159), we can write*

$$\boldsymbol{w}_0 = -\frac{\alpha_1}{2\mathrm{i}}\mathrm{e}^{\mathrm{i}\phi_\alpha}\boldsymbol{I}_{mn\times 1}, \quad \boldsymbol{w}_1 = -(\tilde{\boldsymbol{x}}_c - \tilde{\boldsymbol{x}}_f)\frac{\alpha_1}{2\mathrm{i}}\mathrm{e}^{\mathrm{i}\phi_\alpha}, \quad \boldsymbol{w}(k) = \boldsymbol{w}_0 + \mathrm{i}k\boldsymbol{w}_1$$

so that $\mathbf{w}_2 = 0$ *indeed. Finally, we calculate* $\Delta\mathbf{c}_{p_0}$, $\Delta\mathbf{c}_{p_1}$ *and* $\Delta\mathbf{c}_{p_2}$ *from Eq. (6.163) and then evaluate the total corrected unsteady pressure jump,* $\Delta\mathbf{c}_p(k)$, *from Eq. (6.161) for* $k = 0.108$. *We may also calculate the uncorrected pressure jump from*

$$\mathbf{w}(k) = \left(\tilde{\boldsymbol{A}}_0 + \mathrm{i}k\tilde{\boldsymbol{A}}_1 + (\mathrm{i}k)^2\tilde{\boldsymbol{A}}_2\right)\Delta\mathbf{c}_p(k)$$

again for $k = 0.108$.

Figure 6.25 plots the chordwise variation of the corrected mean pressure jump and compares it to the experimental data. The agreement is very good, but this is to be expected since the DLM

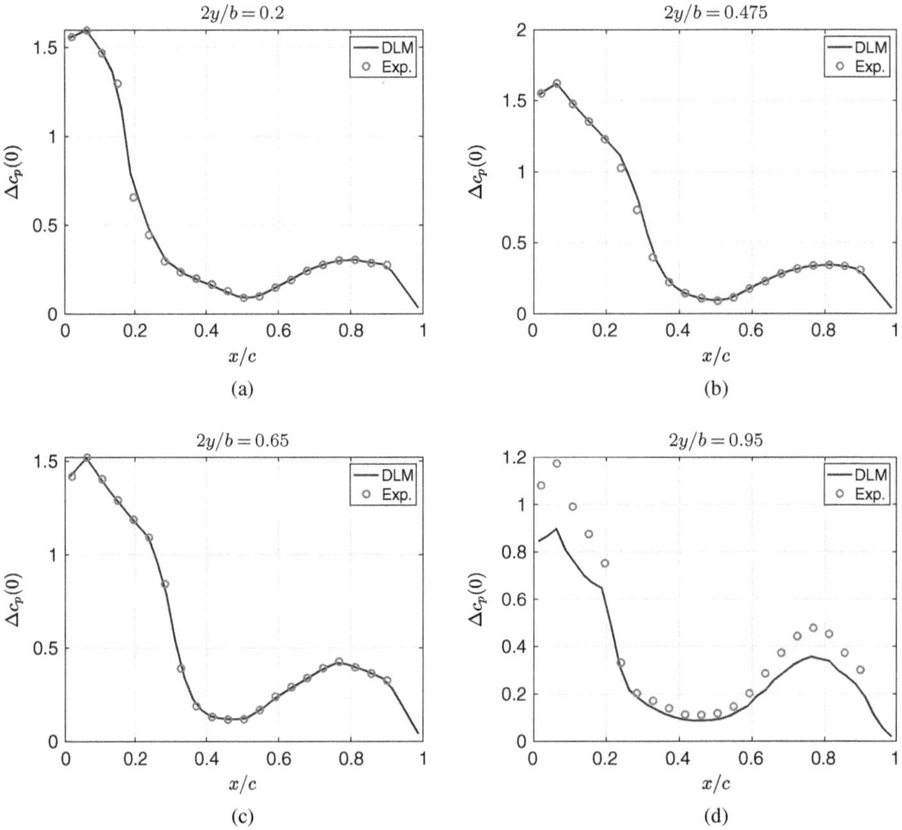

Figure 6.25 Corrected DLM mean pressure jump across LANN wing for $M_\infty = 0.77$, $\alpha_0 = 2.6°$. Experimental data by Zwaan (1985). (a) $2y/b = 0.2$, (b) $2y/b = 0.475$, (c) $2y/b = 0.65$ and (d) $2y/b = 0.95$. Source: Zwaan (1985), credit: AGARD.

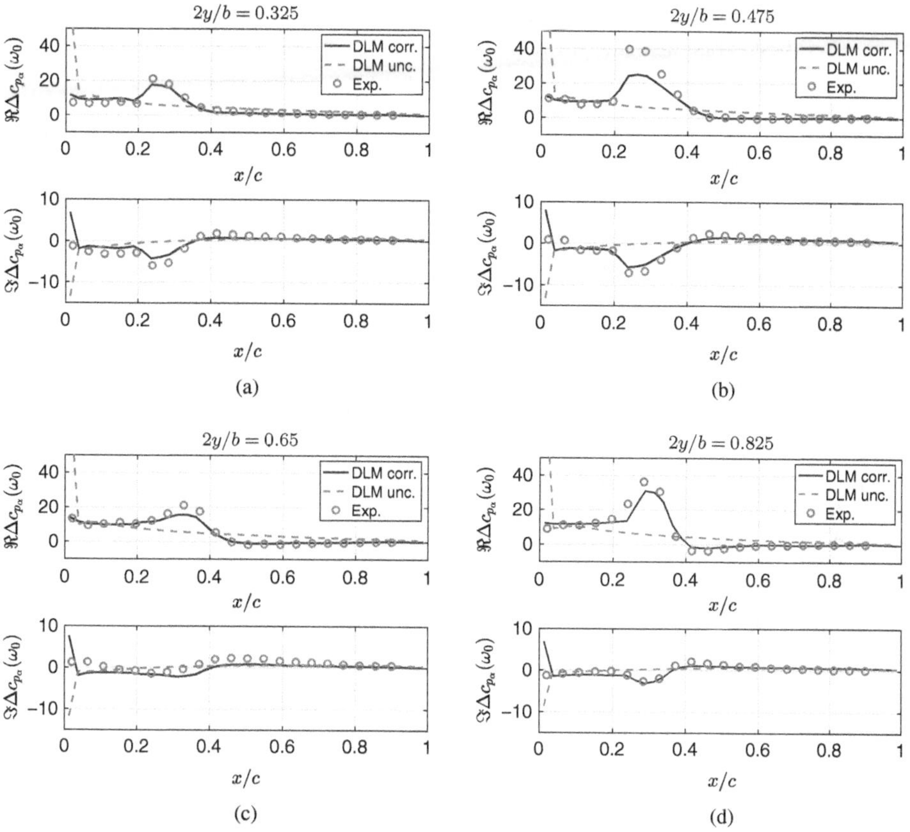

Figure 6.26 Corrected DLM oscillatory pressure jump across LANN wing for $M_\infty = 0.77$, $\alpha_0 = 2.6°$, $\alpha_1 = 0.25°$ and $k = 0.108$. Experimental data by Zwaan (1985). (a) $2y/b = 0.325$, (b) $2y/b = 0.475$, (c) $2y/b = 0.65$ and (d) $2y/b = 0.825$. Source: Zwaan (1985), credit: AGARD.

predictions were corrected using data obtained at nearly the same experimental conditions.[3] Figure 6.26 plots the real and imaginary parts of the corrected and uncorrected oscillatory pressure jumps predicted by the DLM and compares them again to the experimental measurements. The agreement between the corrected predictions and the experiment is reasonably good, except near the tip. The peak in the real part of the pressure jump occurring between $x/c = 0.2$ and 0.3 has been captured, even though its amplitude is not always identical to that observed experimentally. The corrected predictions are much more accurate than those obtained from the uncorrected DLM.

In assessing the results of Figure 6.26, it must be kept in mind that the reference pressure data was noisy and incomplete and had to be interpolated onto the DLM grid. DLM correction methods can yield very satisfactory results when the reference data are obtained from validated steady Euler or RANS simulations, if the latter are accurate. This example is solved by Matlab code `pitchplunge_transonic_DLM.m`.

3 The experimental data plotted in Figures 6.25 and 6.26 are taken from run 124 and those used for the correction from run 272; the only significant difference between the two runs is the value of the reduced frequency k, which is 0.108 for the former and 0 for the latter.

Transonic corrections can also be applied to the SDPM, but the pressure calculation must be linearised first. This means that the pressure coefficient is calculated from Eq. (2.154), after applying the Fourier and Prandtl–Glauert transforms

$$c_p(\omega) = -2\frac{\phi_\xi(\omega)}{\beta U_\infty} - \frac{2i\omega}{U_\infty^2}\phi(\omega)$$

Evaluating this equation at the control points of the surface panels and writing it in matrix form results in

$$\boldsymbol{c}_p(\omega) = -\frac{2}{\beta U_\infty}\boldsymbol{\phi}_\xi(\omega) - \frac{2i\omega}{U_\infty^2}\boldsymbol{\mu}(\omega) \tag{6.165}$$

The next step is to calculate $\boldsymbol{\phi}_\xi(\omega)$; first, we evaluate the tangential derivatives of the doublet strength on the surface from the Prandtl–Glauert transformation of Eqs. (5.189)–(5.192) and then ϕ_ξ, ϕ_η, ϕ_ζ from (6.20). In order to carry out the influence coefficient correction, we need to rewrite all these calculations as a single matrix operator. Equation (5.189) can be written in matrix–vector form as

$$\begin{aligned}
&\Delta\boldsymbol{\xi}_c = \boldsymbol{D}^m\boldsymbol{\xi}_c, \quad \Delta\boldsymbol{\eta}_c = \boldsymbol{D}^m\boldsymbol{\eta}_c, \quad \Delta\boldsymbol{\zeta}_c = \boldsymbol{D}^m\boldsymbol{\zeta}_c\\
&\Delta\boldsymbol{\mu} = \boldsymbol{D}^m\boldsymbol{\mu}, \quad \Delta\boldsymbol{s}_m = \sqrt{\Delta\boldsymbol{\xi}_c\circ\Delta\boldsymbol{\xi}_c + \Delta\boldsymbol{\eta}_c\circ\Delta\boldsymbol{\eta}_c + \Delta\boldsymbol{\zeta}_c\circ\Delta\boldsymbol{\zeta}_c}\\
&\boldsymbol{\tau}_{m_\xi} = \Delta\boldsymbol{\xi}_c\oslash\Delta\boldsymbol{s}_m, \quad \boldsymbol{\tau}_{m_\eta} = \Delta\boldsymbol{\eta}_c\oslash\Delta\boldsymbol{s}_m, \quad \boldsymbol{\tau}_{m_\zeta} = \Delta\boldsymbol{\zeta}_c\oslash\Delta\boldsymbol{s}_m
\end{aligned} \tag{6.166}$$

where the symbol $\oslash$ denotes element-by-element division, the square root is also applied from element to element and the chordwise finite difference matrix $\boldsymbol{D}^m$ is sparse with dimensions $N \times N$ and with the following non-zero elements

$$\begin{aligned}
&D^m_{i,i} = -1, \quad \text{for} \quad i = 1, \dots, n\\
&D^m_{i,n+i} = 1, \quad \text{for} \quad i = 1, \dots, n\\
&D^m_{(2m-1)n+i,(2m-2)n+i} = -1, \quad \text{for} \quad i = 1, \dots, n\\
&D^m_{(2m-1)n+i,(2m-1)n+i} = 1, \quad \text{for} \quad i = 1, \dots, n\\
&D^m_{n+i,i} = -1, \quad \text{for} \quad i = 1, \dots, 2(m-2)n\\
&D^m_{n+i,2n+i} = 1, \quad \text{for} \quad i = 1, \dots, 2(m-2)n
\end{aligned}$$

Consequently, Eq. (5.190) becomes

$$\boldsymbol{\mu}_{\tau_m} = \Delta\boldsymbol{\mu}\oslash\Delta\boldsymbol{s}_m = \left(\boldsymbol{D}^m\oslash\Delta\boldsymbol{s}_m\right)\boldsymbol{\mu} \tag{6.167}$$

where the notation $\boldsymbol{D}^m\oslash\Delta\boldsymbol{s}_m$ means that every column of $\boldsymbol{D}^m$ is divided in an element-by-element sense by the vector $\Delta\boldsymbol{s}_m$. Similarly, Eq. (5.191) are written as

$$\begin{aligned}
&\Delta\boldsymbol{\xi}_c = \boldsymbol{D}^n\boldsymbol{\xi}_c, \quad \Delta\boldsymbol{\eta}_c = \boldsymbol{D}^n\boldsymbol{\eta}_c, \quad \Delta\boldsymbol{\zeta}_c = \boldsymbol{D}^n\boldsymbol{\zeta}_c\\
&\Delta\boldsymbol{\mu} = \boldsymbol{D}^n\boldsymbol{\mu}, \quad \Delta\boldsymbol{s}_n = \sqrt{\Delta\boldsymbol{\xi}_c\circ\Delta\boldsymbol{\xi}_c + \Delta\boldsymbol{\eta}_c\circ\Delta\boldsymbol{\eta}_c + \Delta\boldsymbol{\zeta}_c\circ\Delta\boldsymbol{\zeta}_c}\\
&\boldsymbol{\tau}_{n_\xi} = \Delta\boldsymbol{\xi}_c\oslash\Delta\boldsymbol{s}_n, \quad \boldsymbol{\tau}_{n_\eta} = \Delta\boldsymbol{\eta}_c\oslash\Delta\boldsymbol{s}_n, \quad \boldsymbol{\tau}_{n_\zeta} = \Delta\boldsymbol{\zeta}_c\oslash\Delta\boldsymbol{s}_n
\end{aligned} \tag{6.168}$$

where the spanwise finite difference matrix $\boldsymbol{D}^n$ has non-zero elements

$$\begin{aligned}
&D^n_{1+(i-1)n,1+(i-1)n} = -1, \quad \text{for} \quad i = 1, \dots, 2m\\
&D^n_{1+(i-1)n,2+(i-1)n} = 1, \quad \text{for} \quad i = 1, \dots, 2m\\
&D^n_{in,in-1} = -1, \quad \text{for} \quad i = 1, \dots, 2m
\end{aligned}$$

$$D^n_{in,in} = 1, \quad \text{for} \quad i = 1, \dots, 2m$$
$$D^n_{(i-1)n+j,(i-1)n+j-1} = -1, \quad \text{for} \quad i = 1, \dots, 2m \quad \text{and} \quad j = 2, \dots, n-1$$
$$D^n_{(i-1)n+j,(i-1)n+j+1} = 1, \quad \text{for} \quad i = 1, \dots, 2m \quad \text{and} \quad j = 2, \dots, n-1$$

so that Eq. (5.192) becomes

$$\boldsymbol{\mu}_{\tau_n} = \Delta\boldsymbol{\mu} \oslash \Delta\boldsymbol{s}_n = \left(\boldsymbol{D}^n \oslash \Delta\boldsymbol{s}_n\right)\boldsymbol{\mu} \tag{6.169}$$

Next we must solve the system of Eq. (6.20) for the surface perturbation velocities in the ξ, η, ζ directions. Thanks to the linearisation of the pressure coefficient, we are only interested in $\boldsymbol{\phi}_\xi$. First, we calculate the vectors

$$\begin{aligned}
\boldsymbol{d} &= \mathbf{n}_\xi \circ \boldsymbol{\tau}_{m_\eta} \circ \boldsymbol{\tau}_{n_\zeta} - \mathbf{n}_\xi \circ \boldsymbol{\tau}_{m_\zeta} \circ \boldsymbol{\tau}_{n_\eta} - \mathbf{n}_\eta \circ \boldsymbol{\tau}_{m_\xi} \circ \boldsymbol{\tau}_{n_\zeta} + \mathbf{n}_\eta \circ \boldsymbol{\tau}_{m_\zeta} \circ \boldsymbol{\tau}_{n_\xi} \\
&\quad + \mathbf{n}_\zeta \circ \boldsymbol{\tau}_{m_\xi} \circ \boldsymbol{\tau}_{n_\eta} - \mathbf{n}_\zeta \circ \boldsymbol{\tau}_{m_\eta} \circ \boldsymbol{\tau}_{n_\xi} \\
\boldsymbol{a}_1 &= -(\mathbf{n}_\eta \boldsymbol{\tau}_{n_\zeta} - \mathbf{n}_\zeta \boldsymbol{\tau}_{n_\eta}) \oslash \boldsymbol{d} \\
\boldsymbol{a}_2 &= (\mathbf{n}_\eta \boldsymbol{\tau}_{m_\zeta} - \mathbf{n}_\zeta \boldsymbol{\tau}_{m_\eta}) \oslash \boldsymbol{d} \\
\boldsymbol{a}_3 &= (\boldsymbol{\tau}_{m_\eta} \boldsymbol{\tau}_{n_\zeta} - \mathbf{m}_\zeta \boldsymbol{\tau}_{n_\eta}) \oslash \boldsymbol{d}
\end{aligned} \tag{6.170}$$

Then, the perturbation velocity in the ξ direction is calculated from

$$\boldsymbol{\phi}_\xi = \boldsymbol{a}_1 \circ \boldsymbol{\mu}_{\tau_m} + \boldsymbol{a}_2 \circ \boldsymbol{\mu}_{\tau_n} + \boldsymbol{a}_3 \circ \boldsymbol{\sigma} \tag{6.171}$$

We will first correct the mean flow, for which $\boldsymbol{\mu}(0)$ is obtained from the solution of Eq. (6.89)

$$\boldsymbol{\mu}(0) = -\left(\boldsymbol{B}_\phi - \frac{1}{2}\boldsymbol{I} + \boldsymbol{C}_\phi \boldsymbol{P}_e(0)\boldsymbol{P}_c\right)^{-1} \boldsymbol{A}_\phi \boldsymbol{\sigma}(0) \tag{6.172}$$

Substituting from Eqs. (6.167), (6.169) and (6.172) into the expression for $\boldsymbol{\phi}_\xi$ (6.171) yields

$$\begin{aligned}
\boldsymbol{\phi}_\xi(0) &= -\left(\boldsymbol{a}_1 \circ \boldsymbol{D}^m \oslash \Delta\boldsymbol{s}_m + \boldsymbol{a}_2 \circ \boldsymbol{D}^n \oslash \Delta\boldsymbol{s}_n\right)\left(\boldsymbol{B}_\phi - \frac{1}{2}\boldsymbol{I} + \boldsymbol{C}_\phi \boldsymbol{P}_e(0)\boldsymbol{P}_c\right)^{-1} \boldsymbol{A}_\phi \boldsymbol{\sigma}(0) \\
&\quad + \boldsymbol{a}_3 \circ \boldsymbol{\sigma}(0)
\end{aligned}$$

Then, for $\omega = 0$, Eq. (6.165) becomes

$$\mathbf{c}_p(0) = \frac{1}{Q_\infty} \boldsymbol{K}(0)\boldsymbol{\sigma}(0) \tag{6.173}$$

where

$$\boldsymbol{K}(0) = \frac{2}{\beta}\left(\left(\boldsymbol{a}_1 \circ \boldsymbol{D}^m \oslash \Delta\boldsymbol{s}_m + \boldsymbol{a}_2 \circ \boldsymbol{D}^n \oslash \Delta\boldsymbol{s}_n\right)\left(\boldsymbol{B}_\phi - \frac{1}{2}\boldsymbol{I} + \boldsymbol{C}_\phi \boldsymbol{P}_e(0)\boldsymbol{P}_c\right)^{-1} \boldsymbol{A}_\phi - \boldsymbol{I}\boldsymbol{a}_3\right)$$

The source strength at zero frequency is given by Eq. (6.90), which can be linearized to

$$\boldsymbol{\sigma}(0) \approx -Q_\infty\left(\frac{1}{\beta}\boldsymbol{n}_\xi + \alpha_0 \boldsymbol{n}_\zeta\right)$$

so that the linearised pressure is given by

$$\mathbf{c}_p(0) = -\frac{1}{\beta}\boldsymbol{K}(0)\boldsymbol{n}_\xi - \boldsymbol{K}(0)\boldsymbol{n}_\zeta \alpha_0$$

Defining $\tilde{\boldsymbol{A}} = \boldsymbol{K}(0)^{-1}$, we obtain

$$\tilde{\boldsymbol{A}}\mathbf{c}_p(0) = -\frac{1}{\beta}\boldsymbol{n}_\xi - \boldsymbol{n}_\zeta \alpha_0$$

Finally, neglecting the constant term $-\boldsymbol{n}_\xi/\beta$ in the same way that we neglected the camber term in the DLM leads to

$$\tilde{\boldsymbol{A}}\boldsymbol{c}_p(0) = -\boldsymbol{n}_\zeta \alpha_0 \tag{6.174}$$

This latest expression is the SDPM equivalent of the steady DLM Eq. (6.97).

Equation (6.99) is written here in terms of the full pressure distribution around the wing's surface

$$\boldsymbol{c}_p(\alpha_{\text{ref}} + \Delta\alpha) = \boldsymbol{c}_p(\alpha_{\text{ref}}) + \frac{\partial \boldsymbol{c}_p}{\partial \alpha}\Delta\alpha$$

and the derivative $\partial \boldsymbol{c}_p/\partial\alpha$ is obtained from relation (6.174) as

$$\frac{\partial \boldsymbol{c}_p}{\partial \alpha} = -\tilde{\boldsymbol{A}}^{-1}\boldsymbol{n}_\zeta$$

The non-zero elements of the $N \times N$ correction matrix $\boldsymbol{D}_{\text{corr}}$ are given by a modified version of Eq. (6.102), that is

$$D_{I,I}^{\text{corr}} = -\frac{n_{\zeta_I}}{\sum_{J=1}^{N} \tilde{A}_{I,J} c_{p_\alpha,J}^{\text{ref}}} \tag{6.175}$$

where n_{ζ_I} is the ζ component of the normal vector on the Ith panel. Then, the corrected SDPM pressure at any angle α close to α_{ref} is given by

$$\boldsymbol{c}_p(\alpha) = -\left(\boldsymbol{D}_{\text{corr}}\tilde{\boldsymbol{A}}\right)^{-1}\boldsymbol{n}_\zeta(\alpha - \alpha_{\text{ref}}) + \boldsymbol{c}_p^{\text{ref}}(\alpha_{\text{ref}}) \tag{6.176}$$

In the unsteady case, $\boldsymbol{\mu}(\omega)$ and $\boldsymbol{\phi}_\xi(\omega)$ are given by

$$\boldsymbol{\mu}(\omega) = -\left(\bar{\boldsymbol{B}}_\phi(\omega) - \frac{1}{2}\boldsymbol{I} + \bar{\boldsymbol{C}}_\phi(\omega)\boldsymbol{P}_e(\omega)\boldsymbol{P}_c\right)^{-1}\bar{\boldsymbol{A}}_\phi(\omega)\boldsymbol{\sigma}(\omega)$$

$$\begin{aligned}\boldsymbol{\phi}_\xi(\omega) = &-\left(\boldsymbol{a}_1 \circ \boldsymbol{D}^m \oslash \Delta\boldsymbol{s}_m + \boldsymbol{a}_2 \circ \boldsymbol{D}^n \oslash \Delta\boldsymbol{s}_n\right)\\ &\times\left(\bar{\boldsymbol{B}}_\phi(\omega) - \frac{1}{2}\boldsymbol{I} + \bar{\boldsymbol{C}}_\phi(\omega)\boldsymbol{P}_e(\omega)\boldsymbol{P}_c\right)^{-1}\bar{\boldsymbol{A}}_\phi(\omega)\boldsymbol{\sigma}(\omega)\\ &+\boldsymbol{a}_3 \circ \boldsymbol{\sigma}(\omega)\end{aligned}$$

Then, Eq. (6.165) becomes

$$\begin{aligned}\boldsymbol{c}_p(\omega) &= \frac{1}{Q_\infty}\boldsymbol{K}_1(\omega)\boldsymbol{\sigma}(\omega) - \frac{2i\omega}{Q_\infty^2}\boldsymbol{\mu}(\omega) = \frac{1}{Q_\infty}\left(\boldsymbol{K}_1(\omega) + \frac{2i\omega}{Q_\infty}\boldsymbol{K}_2(\omega)\right)\boldsymbol{\sigma}(\omega)\\ &= \frac{1}{Q_\infty}\boldsymbol{K}(\omega)\boldsymbol{\sigma}(\omega)\end{aligned} \tag{6.177}$$

where

$$\begin{aligned}\boldsymbol{K}(\omega) &= \boldsymbol{K}_1(\omega) + \frac{2i\omega}{Q_\infty}\boldsymbol{K}_2(\omega)\\ \boldsymbol{K}_1(\omega) &= \frac{2}{\beta}\left(\left(\boldsymbol{a}_1 \circ \boldsymbol{D}^m \oslash \Delta\boldsymbol{s}_m + \boldsymbol{a}_2 \circ \boldsymbol{D}^n \oslash \Delta\boldsymbol{s}_n\right)\boldsymbol{K}_2(\omega) - \boldsymbol{I}\boldsymbol{a}_3\right)\\ \boldsymbol{K}_2(\omega) &= \left(\bar{\boldsymbol{B}}_\phi(\omega) - \frac{1}{2}\boldsymbol{I} + \bar{\boldsymbol{C}}_\phi(\omega)\boldsymbol{P}_e(\omega)\boldsymbol{P}_c\right)^{-1}\bar{\boldsymbol{A}}_\phi(\omega)\end{aligned} \tag{6.178}$$

and the elements of $\bar{\boldsymbol{A}}_\phi$, $\bar{\boldsymbol{B}}_\phi$, $\bar{\boldsymbol{C}}_\phi$ are given by Eqs. (6.81)–(6.83). Using the definition of the reduced frequency $k = \omega c_0/2Q_\infty$ and defining $\tilde{\boldsymbol{A}}(k) = \boldsymbol{K}^{-1}(k)$ Eq. (6.177) becomes

$$\tilde{\boldsymbol{A}}(k)\boldsymbol{c}_p(k) = \frac{1}{Q_\infty}\boldsymbol{\sigma}(k)$$

which is the SDPM equivalent of Eq. (6.158). Furthermore, $\boldsymbol{\sigma}(k)$ is the SDPM equivalent of the upwash $\boldsymbol{w}(k)$ in the DLM. Following the logic of the successive kernel expansion technique, we expand $\boldsymbol{\sigma}(k)$, $\tilde{\boldsymbol{A}}(k)$ and $\mathbf{c}_p(k)$ as power series in ik

$$\mathbf{c}_p(k) = \mathbf{c}_{p_0} + \mathrm{i}k\mathbf{c}_{p_1} + (\mathrm{i}k)^2\mathbf{c}_{p_2} \tag{6.179}$$

$$\tilde{\boldsymbol{A}}(k) = \tilde{\boldsymbol{A}}_0 + \mathrm{i}k\tilde{\boldsymbol{A}}_1 + (\mathrm{i}k)^2\tilde{\boldsymbol{A}}_2 \tag{6.180}$$

The source strength is given by Eq. (6.91) and the relative flow velocities by expressions (6.88). We make the hypothesis that $\Omega M_\infty/a_\infty \ll 1$ so that the first term in Eq. (6.91) becomes negligible and write $\boldsymbol{\sigma}(k)$ in the form

$$\begin{aligned}
\boldsymbol{\sigma}(k) &= Q_\infty(\boldsymbol{\sigma}_0 + \mathrm{i}k\boldsymbol{\sigma}_1)\\
\boldsymbol{\sigma}_0 &= -\frac{\alpha_1}{2\mathrm{i}}\mathrm{e}^{\mathrm{i}\phi_\alpha}\boldsymbol{n}_z\\
\boldsymbol{\sigma}_1 &= \frac{2}{c_0}\left(\frac{1}{\beta}\left(\mathbf{z}_c - z_f\right)\circ\boldsymbol{n}_x - \left(\mathbf{x}_c - x_f\right)\circ\boldsymbol{n}_z\right)\frac{\alpha_1}{2\mathrm{i}}\mathrm{e}^{\mathrm{i}\phi_\alpha}
\end{aligned} \tag{6.181}$$

Then, the SDPM equivalent of Eq. (6.163) becomes

$$\begin{aligned}
\mathbf{c}_{p_0} &= \left(\boldsymbol{D}_{\mathrm{corr}}\tilde{\boldsymbol{A}}_0\right)^{-1}\boldsymbol{\sigma}_0\\
\mathbf{c}_{p_1} &= \left(\boldsymbol{D}_{\mathrm{corr}}\tilde{\boldsymbol{A}}_0\right)^{-1}\left(\boldsymbol{\sigma}_1 - \tilde{\boldsymbol{A}}_1\tilde{\boldsymbol{A}}_0^{-1}\boldsymbol{\sigma}_0\right)\\
\mathbf{c}_{p_2} &= -\left(\boldsymbol{D}_{\mathrm{corr}}\tilde{\boldsymbol{A}}_0\right)^{-1}\left(\tilde{\boldsymbol{A}}_2\tilde{\boldsymbol{A}}_0^{-1}\boldsymbol{\sigma}_0 + \tilde{\boldsymbol{A}}_1\tilde{\boldsymbol{A}}_0^{-1}\left(\boldsymbol{\sigma}_1 - \tilde{\boldsymbol{A}}_1\tilde{\boldsymbol{A}}_0^{-1}\boldsymbol{\sigma}_0\right)\right)
\end{aligned} \tag{6.182}$$

Example 6.9 *Repeat Example 6.8 using the SDPM*

We start by setting up the SDPM simulation as in Example 6.3 and then calculate matrix $\mathbf{D}^m$ *in Eq. (6.167), vectors* $\Delta\boldsymbol{s}_m$, $\boldsymbol{\tau}_{m_\xi}$, $\boldsymbol{\tau}_{m_\eta}$ *and* $\boldsymbol{\tau}_{m_\zeta}$ *in (6.166), matrix* $\mathbf{D}^n$ *in (6.169), vectors* $\Delta\boldsymbol{s}_n$ $\boldsymbol{\tau}_{n_\xi}$, $\boldsymbol{\tau}_{n_\eta}$ *and* $\boldsymbol{\tau}_{n_\zeta}$ *in (6.168) and, hence, vectors* $\mathbf{d}$, $\boldsymbol{a}_1$, $\boldsymbol{a}_2$ *and* $\boldsymbol{a}_3$ *in Eq. (6.170). For reduced frequency values of* $k_0 = 0$, $k_1 = 0.1$ *and* $k_2 = 0.2$, *we evaluate matrices* $\bar{\mathbf{A}}_\phi(k)$, $\bar{\mathbf{B}}_\phi(k)$, $\bar{\mathbf{C}}_\phi(k)$ *from Eqs. (6.81)–(6.83),* $\mathbf{P}_c$ *from Eq. (6.84) and* $\mathbf{P}_e(k)$ *from Eq. (6.85). Now we can assemble matrix* $\mathbf{K}(k)$ *from Eq. (6.178), define* $\tilde{\mathbf{A}}(k) = \mathbf{K}(k)^{-1}$ *and solve the system of Eq. (6.164).*

The reference pressure distributions $\mathbf{c}_p(\alpha_{\mathrm{ref}})$ *and* $\mathbf{c}_{p_\alpha}^{\mathrm{ref}}$ *are obtained from the same quasi-steady experimental data used in Example 6.8. This time we interpolate the complete pressure distribution around the wing instead of the pressure jump. This means that we need to carry out two separate interpolations, one for the pressure distribution on the upper surface and one for that on the lower surface. We then assemble the complete* $\mathbf{c}_p(\alpha_{\mathrm{ref}})$ *and* $\mathbf{c}_{p_\alpha}^{\mathrm{ref}}$ *matrices.*

Next, we calculate the elements of the correction matrix $\mathbf{D}_{\mathrm{corr}}$ *from Eq. (6.175) and then evaluate the corrected mean pressure from expression (6.176). For the unsteady case, we calculate* $\boldsymbol{\sigma}_0$ *and* $\boldsymbol{\sigma}_1$ *from Eq. (6.181) before evaluating* $\mathbf{c}_{p_0}$, $\mathbf{c}_{p_1}$ *and* $\mathbf{c}_{p_2}$ *from expressions (6.182) and assembling the total corrected pressure coefficient from Eq. (6.179).*

Figure 6.27 plots the corrected mean pressure distribution around the wing and compares it to the uncorrected SDPM result and to the experimental measurements. As expected, the agreement between the corrected SDPM predictions and the experiment is excellent, since very similar experimental data were used to correct the former.

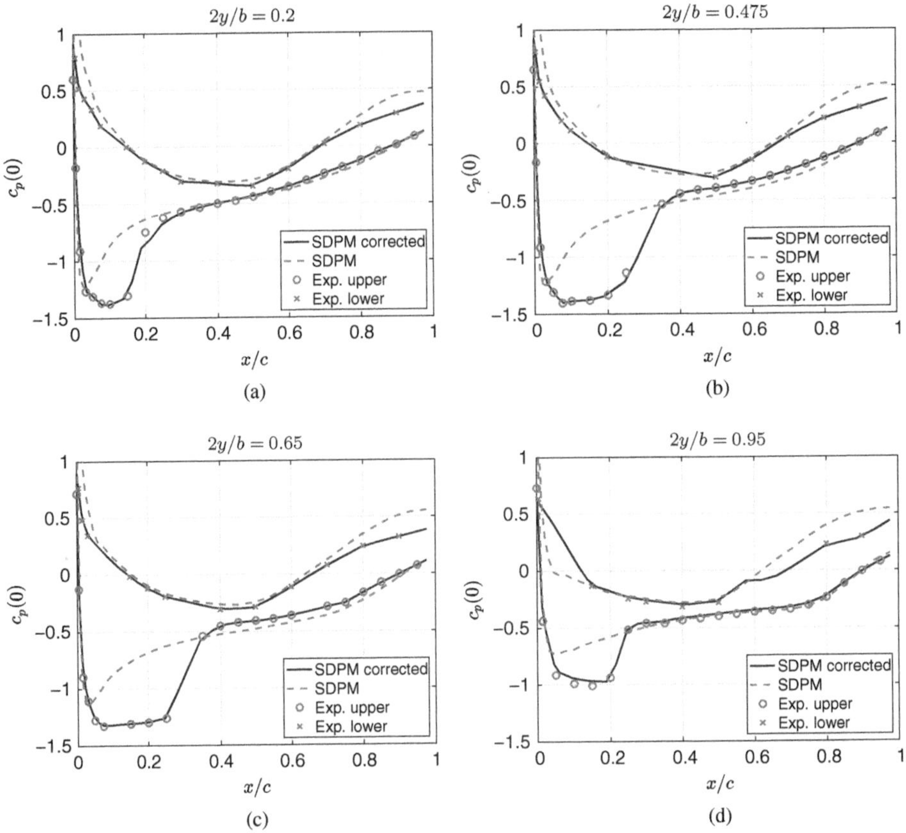

Figure 6.27 Corrected SDPM mean pressure jump across LANN wing for $M_\infty = 0.77$, $\alpha_0 = 2.6°$. Experimental data by Zwaan (1985). (a) $2y/b = 0.2$, (b) $2y/b = 0.475$, (c) $2y/b = 0.65$ and (d) $2y/b = 0.95$. Source: Zwaan (1985), credit: AGARD.

Figure 6.28 compares the corrected and uncorrected oscillatory pressure distributions obtained from the SDPM with the experimental measurements. The real part of the corrected pressure distribution is generally in good agreement with the experimental measurements, although the peaks at each spanwise position are underestimated. This underestimation is due to the interpolation procedure, since the highest possible value of $\Re(c_{p_\alpha}(\omega_0))$ at each spanwise position is the maximum experimental value of $\mathbf{c}^{\mathrm{ref}}_{p_\alpha}$. Since the SDPM grid does not include the locations on the wing where these maxima were measured, the peaks of $\Re(c_{p_\alpha}(\omega_0))$ are lower. The agreement between the SDPM and experimental imaginary parts of the pressure distribution is less good although still acceptable. It should be stated that the successive kernel expansion procedure for the DLM was inspired by a theoretical expansion of the DLM kernel by Watkins et al. (1955). No such expansion has been developed theoretically for the SDPM. The results plotted in Figure 6.28 were obtained for $m = 34$, $n = 34$. If the number of panels is increased beyond these values, the imaginary part of the oscillatory pressure distribution becomes increasingly inaccurate. This example is solved by Matlab code `pitchplunge_transonic_SDPM.m`.

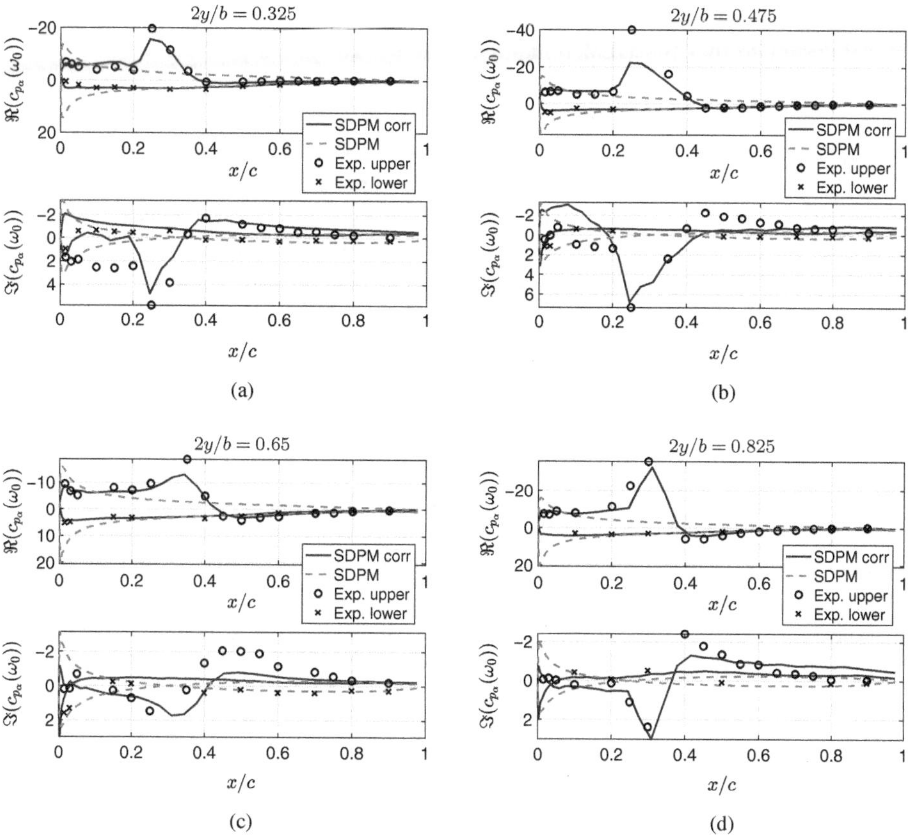

Figure 6.28 Corrected SDPM oscillatory pressure jump across LANN wing for $M_\infty = 0.77$, $\alpha_0 = 2.6°$. Experimental data by Zwaan (1985). (a) $2y/b = 0.325$, (b) $2y/b = 0.475$, (c) $2y/b = 0.65$ and (d) $2y/b = 0.825$. Source: Zwaan (1985), credit: AGARD.

The correction techniques presented in this section are only suitable for small-amplitude oscillations in the neighbourhood of the angles of attack at which reference pressure data are available. They cannot be used to correct predictions for high-amplitude oscillations, since they are based on a Taylor expansion of the pressure distribution around the reference angle of attack. In any case, potential flow is unsuitable for modelling high-amplitude oscillations since it cannot capture viscous phenomena.

6.6 Concluding Remarks

The methods presented in this chapter were shown to produce reliable aerodynamic predictions for compressible attached flow around 2D and 3D bodies. Consequently, they are ideal for use in early stages of aircraft design and in specialised applications, such as aeroelastic design. Higher fidelity approaches are increasingly being used in industry, but their higher computational cost makes them more suitable for later stages of aircraft design. Furthermore, the cost of unsteady high fidelity simulations is still prohibitive for practical design so that the DLM remains the industrial standard for aeroelasticity and flight dynamics. Nevertheless, it should be kept in mind that inviscid methods fail to predict important physical phenomena, such as viscous drag and shock-induced boundary layer separation. Effective aircraft performance calculations cannot be performed in the absence of viscous drag estimates; viscous corrections must therefore be applied to potential flow predictions using methods such as the flat plate analogy or viscous-inviscid interaction. Finally, shock-induced separation can induce transonic buffeting, which is an important aeroelastic instability and cannot always be predicted by inviscid methods.

6.7 Exercises

1 Repeat Example 5.13 for all Mach numbers tested by Lessing et al. (1960) using the incompressible SDPM code of the same example. How bad do the incompressible predictions get at Mach numbers of 0.7 and 0.9?

2 Implement the bending and twisting deformation of the LANN wing in Example 6.3. Does the agreement between the compressible SDPM predictions and the experimental measurements improve when including the flexible motion?

3 Repeat Example 6.3 using the approximations of Eqs. (6.95) and (6.96) for the calculation of the influence coefficient matrices. See also Martínez and Dimitriadis (2022).

4 Repeat Example 6.5 for the pitching oscillation test case at $k = 0.262$ and $M_\infty = 0.498$ reported in Zwaan (1982). How good is the agreement between the SDPM and experimental oscillatory pressures? Does the agreement improve if you increase the number of spanwise panels to, say, $n = 40$?

5 Repeat Examples 6.6 and 6.7 for the $M_\infty = 1.1$ test case reported in Tijdeman et al. (1979b). Are the Mach box and Mach panel method predictions as good as for the $M_\infty = 1.32$ case? Are there any numerical instabilities as $B \to 0$?

References

Albano, E. and Rodden, W.P. (1969). A doublet-lattice method for calculating lift distributions on oscillating surfaces in subsonic flows. *AIAA Journal* 7 (2): 279–285.

Anderson, J.D. Jr. (1990). *Modern Compressible Flow*, 2e. McGraw-Hill Publishing Company.

Bagai, A. (2006). Definition of the mean camber line from airfoil shapes. *Journal of the American Helicopter Society* 51 (4): 355–357.

Blair, M. (1994). A Compilation of the Mathematics Leading to the Doublet-Lattice Method. *Report WL-TR-95-3022*. Air Force Wright Laboratory.

Chen, P.C., Lee, H.W., and Liu, D.D. (1993). Unsteady subsonic aerodynamics for bodies and wings with external stores including wake effect. *Journal of Aircraft* 30 (5): 618–628.

Chu, L. (1988). Integral equation solution of the full potential equation for three-dimensional, steady, transonic wing flows. PhD thesis. Old Dominion University.

Crovato, A., Almeida, H.S., Vio, G. et al. (2020). Effect of levels of fidelity on steady aerodynamic and static aeroelastic computations. *Aerospace* 7: 42.

Cunningham, H.J. (1966). Improved numerical procedure for harmonically deforming lifting surfaces from the supersonic kernel function method. *AIAA Journal* 4 (11): 1961–1968.

Curtis, A.R. and Lingard, R.W. Jr. (1968). Unsteady aerodynamic distributions for harmonically deforming wings in supersonic flow. *Proceedings of the 6th AIAA Aerospace Sciences Meeting* number AIAA 68–74, New York.

Epton, M.A. and Magnus, A.E. (1990). PAN AIR- A Computer Program for Predicting Subsonic or Supersonic Linear Potential Flows About Arbitrary Configurations Using a Higher Order Panel Method. *Contractor Report CR-3251*. NASA.

Evvard, J.C. (1947). Distribution of wave drag and lift in the vicinity of wing tips at supersonic speeds. Technical Note TN-1382. NACA.

Farr, J.L. Jr., Traci, R.M., and Albano, E.D. (1975). Computer Programmes for Calculating Small Disturbance Transonic Flows About Oscillating Planar Wings. *Report AFFDL-TR-75-103*. Air Force Flight Dynamics Laboratory.

Friedewald, D., Thormann, R., Kaiser, C., and Nitzsche, J. (2018). Quasi-steady doublet-lattice correction for aerodynamic gust response prediction in attached and separated transonic flow. *CEAS Aeronautical Journal* 9: 53–66.

Garrick, I.E. and Rubinow, S.I. (1938). Theoretical Study of Air Forces on An Oscillating or Steady Thin Wing in a Supersonic Main Stream. *Technical Report TN-1383*. NACA.

Gülçat, U. (2016). *Fundamentals of Modern Unsteady Aerodynamics*, 2e. Springer.

Hounjet, M. (1971). A field panel/finite difference method for potential unsteady transonic flow. *AIAA Journal* 23 (4): 537–545.

Kalman, T.P., Rodden, W.P., and Giesing, J.P. (1971). Application of the doublet lattice method to nonplanar configurations in subsonic flow. *AIAA Journal* 8 (6): 406–413.

Landahl, M.T. (1961). *Unsteady Transonic Flow*. Dover Publications, Inc.

Lessing, H.C., Troutman, J.L., and Menees, G.P. (1960). Experimental determination of the pressure distribution on a rectangular wing oscillating in the first bending mode for Mach numbers from 0.24 to 1.30. Technical Note TN-D-344. NASA.

Lockman, W.K. and Seegmiller, H.L. (1983). An experimental investigation of the subcritical and supercritical flow about a swept semispan wing. Technical Memorandum TM-84367. NASA.

Lu, S. and Voss, R. (1992). A transonic doublet lattice method for 3D potential unsteady transonic flow calculation. *Report DLR-Forschungsbericht. 92-25*. DLR.

Sánchez Martínez, M. and Dimitriadis, G. (2022). Subsonic source and doublet panel methods. *Journal of Fluids and Structures* 113: 103–624.

Moore, M.T. and Andrew, L.V. (1965). Unsteady Aerodynamics for Advanced Configurations: Part IV - Application of the Supersonic Mach Box Method to Intersecting Planar Lifting Surfaces. *Report FDL TDR 64-152*. Air Force Flight Dynamics Laboratory.

Morino, L. (1974). A General Theory of Unsteady Compressible Potential Aerodynamics. *Contractor Report CR-2464*. NASA.

Morino, L. (1980). Steady, Oscillatory and Unsteady Subsonic and Supersonic Aerodynamic Forces - Production Version (SOUSSA-P 1.1) - Volume I - Theoretical Manual. *Contractor Report CR-159130*. NASA.

Morino, L. and Chen, L.T. (1975). Indicial compressible potential aerodynamics around complex aircraft configurations. *Aerodynamic Analyses Requiring Advanced Computers Conference, NASA SP-347*, 1067–1110. Hampton, VA, USA.

Morino, L., Chen, L., and Suciu, E.O. (1975). Steady and oscillatory subsonic and supersonic aerodynamics around complex configurations. *AIAA Journal* 13 (3): 368–374.

Murman, E.M. and Cole, J.D. (1971). Calculation of plane steady transonic flows. *AIAA Journal* 9 (1): 114–121.

Ogana, W. (1978). Solution of transonic flows by an integro-differential equation method. Technical Memorandum TM-78490. NASA.

Olsen, J.J. (1966). Demonstration of a Transonic Box Method for Unsteady Aerodynamics of Planar Wings. *Report AFFDL-TR-66-121*. Air Force Flight Dynamics Laboratory.

Olsen, J.J. (1969). Demonstration of a Supersonic Box Method for Unsteady Aerodynamics of Nonplanar Wings. *Report AFFDL-TR-67-104*. Air Force Flight Dynamics Laboratory.

Palacios, R., Climent, H., Karlsson, A., and Winzell, B. (2001). Assessment of strategies for correcting linear unsteady aerodynamics using CFD or test results. *Proceedings of the International Forum on Aeroelasticity and Structural Dynamics* number IFASD2001-017, Madrid, Spain.

Pi, W., Kelly, P., and Liu, D. (1979). A transonic doublet lattice method for unsteady flow calculations. *Proceedings of the 17th AIAA Aerospace Sciences Meeting* number AIAA 79-0078, New Orleans, LA, USA.

Pines, S., Dugundji, J., and Neuringer, J. (1955). Aerodynamic flutter derivatives for a flexible wing with supersonic and subsonic edges. *Journal of the Aeronautical Sciences* 22 (10): 693–700.

Rodemich, E.R. and Andrew, L.V. (1965). Unsteady Aerodynamics for Advanced Configurations. Part II - A Transonic Box Method for Planar Lifting Surfaces. *Report AFFDL-TDR-64-152-PT-2*. Air Force Flight Dynamics Laboratory.

Silva, R.G.A., Mello, O.A.F., ao Luiz, F. et al. (2008). Investigation on transonic correction methods for unsteady aerodynamics and aeroelastic analyses. *Journal of Aircraft* 45 (6): 1890–1903.

Tijdeman, H., Van Nunen, J.W.G., Kraan, A.N. et al. (1979a). Transonic Wind Tunnel Tests on An Oscillating Wing with external Stores. Part I: General Description. *Report AFFDL-TR-78-194-PT-1*. Air Force Flight Dynamics Laboratory.

Tijdeman, H., Van Nunen, J.W.G., Kraan, A.N. et al. (1979b). Transonic Wind Tunnel Tests on An Oscillating Wing with External Stores. Part II: The Clean Wing. *Report AFFDL-TR-78-194-PT-2*. Air Force Flight Dynamics Laboratory.

Tseng, K. and Morino, L. (1982). Nonlinear Green's function method for unsteady transonic flows. In: *Transonic Aerodynamics, Progress in Astronautics and Aeronautics*, vol. 81 (ed. D. Nixon), 565–603. AIAA.

Ueda, T. and Dowell, E.H. (1984). Doublet-point method for supersonic unsteady lifting surfaces. *AIAA Journal* 22 (2): 179–186.

Van Zyl, L.H. (1997). A transonic doublet lattice method for general configurations. *Proceedings of the International Forum on Aeroelasticity and Structural Dynamics*, 25–31. Rome, Italy.

Watkins, C.E., Runyan, H.L., and Woolston, D.S. (1955). On the Kernel Function of the Integral Equation Relating the Lift and Downwash Distributions of Oscillating Finite Wings in Subsonic Flow. *Report R-1234*. NACA.

ZONA Technology Inc. (2004). *ZAERO Version 7.2: User's Manual*, 13e. Scottsdale, AZ: ZONA Technology, Inc.

Zwaan, R.J. (1982). Data set 4: NLR 7301 supercritical airfoil. Oscillatory pitching and oscillating flap. In: *Compendium of Unsteady Aerodynamic Measurements* (ed. Various) number AGARD-R-702, 4:14–25. AGARD.

Zwaan, R.J. (1985). Data set 9: LANN wing pitching oscillation. In *Compendium of Unsteady Aerodynamic Measurements* (ed. Various) number AGARD-R-702, Addendum No. 1, 9:19–76. AGARD.

7

Viscous Flow

7.1 Introduction

Attached flow is particular to streamlined bodies lying at small angles of attack with respect to the oncoming free stream and with small motion amplitudes and frequencies. The flow over bluff bodies, such as circular, elliptical or rectangular cross sections, or over streamlined bodies at high angles of attack is at least partly separated. Such flows can be inherently unsteady, even if the body is not moving with respect to the free stream. As viscosity plays a crucial role in separated flows, the viscous flow equations must be used to model them. In this chapter, we will study some of these flows and detail some modelling approaches.

Prandtl was the first to state clearly that viscous phenomena are important only in specific parts of the flowfield (for an English translation of Prandtl's seminal 1904 paper see (Prandtl 2001)). In incompressible flows, the viscous regions are

- A thin layer of vorticity in contact with the surface of the body, known as the boundary layer. This layer can stay attached over the entire surface of the body and separates at the trailing edge. It can also separate somewhere else on the surface, becoming a free shear layer that also carries vorticity.
- Separated flow regions. These regions can be narrow, such as the wake behind a wing with attached flow, or wide, such as the flow around any body downstream of the boundary layer separation point.

In two dimensions, the incompressible viscous flow Eqs. (2.200)–(2.203) simplify to

$$\frac{\partial u}{\partial x} + \frac{\partial v}{\partial y} = 0$$

$$\frac{\partial u}{\partial t} + u\frac{\partial u}{\partial x} + v\frac{\partial u}{\partial y} = -\frac{1}{\rho}\frac{\partial p}{\partial x} + \nu\left(\frac{\partial^2 u}{\partial x^2} + \frac{\partial^2 u}{\partial y^2}\right)$$

$$\frac{\partial v}{\partial t} + u\frac{\partial v}{\partial x} + v\frac{\partial v}{\partial y} = -\frac{1}{\rho}\frac{\partial p}{\partial y} + \nu\left(\frac{\partial^2 v}{\partial x^2} + \frac{\partial^2 v}{\partial y^2}\right)$$

The thickness of the boundary layer, δ, is assumed to be small so that $\delta \ll c$. This means that, inside the boundary layer, y is of order δ, while x is of order c. The velocity u is of order

Unsteady Aerodynamics: Potential and Vortex Methods, First Edition. Grigorios Dimitriadis.

Companion website: www.wiley.com/go/dimitriadis/unsteady_aerodynamics

U_∞, but the velocity v must be of order $U_\infty \delta/c$ so that the two derivatives in the continuity equation are of the same order. Consequently, for a high enough Reynolds number, the momentum equations simplify to

$$\begin{aligned} \frac{\partial u}{\partial t} + u\frac{\partial u}{\partial x} + v\frac{\partial u}{\partial y} &= -\frac{1}{\rho}\frac{\partial p}{\partial x} + \nu\frac{\partial^2 u}{\partial y^2} \\ \frac{\partial p}{\partial y} &= 0 \end{aligned} \tag{7.1}$$

which are known as the boundary layer equations. Equation (7.1) is a non-linear partial differential equation so it is not easy to solve. Several semi-empirical solutions of the boundary layer equations exist, but they lie beyond the scope of this book.

A conceptual drawing of an attached boundary layer can be seen in Figure 7.1a. The shaded area is the body, while the dashed line denotes the edge of the boundary layer. At each point on the surface, the x coordinate it tangent to the surface while the y coordinate is normal to it; $u(x, y)$ is the flow velocity in the tangent direction and $v(x, y)$ in the normal direction. The thickness of the boundary layer is denoted by $\delta(x)$; beyond this thickness, the flow can be treated as inviscid. Exactly on the surface, the flow velocity is taken to be equal to zero, $u(x, 0) = 0$, $v(x, 0) = 0$. This condition is known as the no slip condition. The tangential velocity increases with y until it reaches the edge velocity $U_e(x)$, such that $u(x, y)/U_e(x)$ takes values between 0 and 1. Conversely, $v(x, y)/U_e(x) \ll 1$ for all values of y. Note that the slope

$$\left.\frac{\mathrm{d}u}{\mathrm{d}y}\right|_{x,y=0} > 0$$

wherever the boundary layer is attached. The vorticity in the boundary layer is given by Eq. (2.199) in 2D

$$\omega = \frac{\partial v}{\partial x} - \frac{\partial u}{\partial y}$$

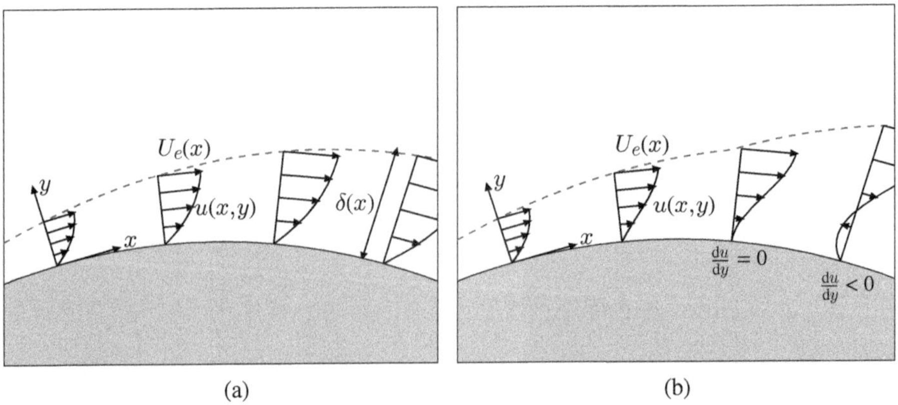

Figure 7.1 Conceptual view of attached and separating boundary layers. (a) Fully attached and (b) separating.

As the fist term is very small compared to the second,

$$\omega \approx -\frac{\partial u}{\partial y} \tag{7.2}$$

The velocity profiles in Figure 7.1a show that the vorticity is a maximum near the wall and becomes zero at the edge of the boundary layer. The direction of the vorticity is clockwise everywhere in this figure. If we assume that the velocity profile is triangular, then the vorticity can be approximated as constant over y, such that

$$\omega(x) \approx -\frac{U_e(x)}{\delta(x)}$$

For 2D flows, the definition of circulation (2.210) and Eq. (2.212) become

$$\Gamma = \int_S \omega \mathrm{d}S = \oint_C \mathbf{u} \cdot \boldsymbol{\tau} \mathrm{d}s \tag{7.3}$$

where S is the area of the flow in which we calculate the circulation and C the contour of this area. If S is chosen as the boundary layer from the stagnation point $x = 0$ up to any position $x = x_1$, its circulation can be obtained from the contour integral of Eq. (7.3) as

$$\Gamma(S) = \int_0^{x_1} u(x, \delta(x))\mathrm{d}x + \int_{\delta(x_1)}^{0} v(x_1, y)\mathrm{d}y + \int_{x_1}^{0} u(x, 0)\mathrm{d}x + \int_0^{\delta(0)} v(0, y)\mathrm{d}y$$

The height of the boundary layer at the stagnation point is zero so that $\delta(0) = 0$. Recalling that $u(x, \delta(x)) = U_e(x)$ and $u(x, 0) = 0$, we obtain

$$\Gamma(S) = \int_0^{x} U_e(x)\mathrm{d}x + \int_{\delta(x)}^{0} v(x, y)\mathrm{d}y \approx \int_0^{x} U_e(x)\mathrm{d}x \tag{7.4}$$

since $v(x, y) \ll U_e(x)$.

Figure 7.1b demonstrates the classical view of how a laminar boundary layer separates on a smooth surface. The slope of the tangential velocity is initially positive but decreases further along the surface until it becomes zero and then negative. Flow separation occurs where

$$\left.\frac{\mathrm{d}u}{\mathrm{d}y}\right|_{x_s, y=0} = 0 \tag{7.5}$$

and the point x_s is known as the separation point. This process is usually due to an adverse pressure gradient acting on the boundary layer, which occurs when the edge pressure $p_e(x)$ continuously increases. At the separation point, the vorticity is equal to zero at the wall, but it is still clockwise for $y > 0$. The point where the vorticity is a maximum has moved away from the wall and now lies at the inflexion point of the velocity profile. The separated boundary layer is a shear layer that carries the same clockwise vorticity. Note that, even downstream of the separation point, the flow velocity on the surface is equal to zero. However, just off the wall, the flow direction is reversed because the clockwise vorticity in the separated shear layer induces an upstream velocity under it.

7.1.1 Steady Flow Separation Mechanisms

Not all flow separation occurs by means of the mechanism depicted in Figure 7.1b. Separation can occur due to the presence of sharp corners in the geometry or of shock waves normal to the surface. Furthermore, a laminar boundary layer that separates can undergo turbulent transition and reattach, forming a separation bubble. Separated flow that does not reattach is unsteady and can affect conditions upstream, causing the position of the separation point to oscillate. Finally, flow separation can lead to the stall phenomenon, which is usually defined as the loss of lift due to flow separation. The terms 'flow separation' and 'stall' are often used interchangeably in the literature. For steady incompressible conditions, if the angle of attack of a 2D wing section increases from low to high values, three main flow separation mechanisms have been identified:

- Trailing edge stall. Under steady attached inviscid flow conditions, the pressure drops to its minimum value on the suction side, known as the suction peak, and then recovers downstream in order to reach the theoretical value of $c_p = 1$ at the trailing edge. This increase in pressure from the suction peak to the trailing edge is an adverse pressure gradient because it decelerates the flow in the boundary layer. Under attached viscous flow conditions, the boundary layer remains attached up to the trailing edge despite the adverse pressure gradient. As the angle of attack increases, so does the magnitude of the adverse pressure gradient, such that boundary layer separation starts to occur upstream of the trailing edge. The higher the angle of attack, the further upstream lies the separation point, until it reaches the leading edge. This flow separation mechanism is gradual and is demonstrated graphically in Figure 7.2a for a NACA 0018 airfoil at $Re = 5 \times 10^5$. The separated flow region travels upstream and becomes thicker as the angle of attack increases from 9° to 18°.
- Leading edge stall. This mechanism affects airfoils with moderate thickness and rounded leading edges. The flow starts separating near the leading edge; as the Reynolds number based on chordwise distance from the leading edge, Re_x, is very low at the separation point, the boundary layer is locally laminar. The suction peak is very high and narrow for this type of airfoil at moderate angles of attack and the laminar boundary layer cannot withstand a large amount of pressure recovery; it therefore separates. The separated shear layer is unstable and undergoes turbulent transition so that it reattaches to the surface a small distance downstream, forming a small laminar separation bubble. The boundary layer downstream of the bubble is turbulent. As the angle of attack increases, the separation bubble moves upstream because the suction peak also occurs further upstream; the size of the bubble remains mostly constant. At a critical value of the angle of attack, the separated shear layer can no longer reattach and suddenly the flow over the entire suction surface of the airfoil is separated. This flow separation mechanism is very abrupt and is often described as the bursting of the laminar separation bubble. It is depicted in Figure 7.2b, which is inspired by data from McCullough and Gault (1951) for the NACA $63_1 - 012$ airfoil at $Re = 5.8 \times 10^5$. For angles of attack between 4° and 12.8° flow separation is limited inside a small separation bubble near the leading edge, whose size is exaggerated for clarity. At 12.8° the entire flow over the suction side separates abruptly.
- Thin airfoil stall. This mechanism affects airfoils with sharp leading edges and thin airfoils with rounded leading edges. Even at small positive angles of attack, the stagnation point lies on the pressure side of the airfoil and the flow has to turn round the leading edge to reach the suction side. As the leading edge is sharp, the flow cannot follow its

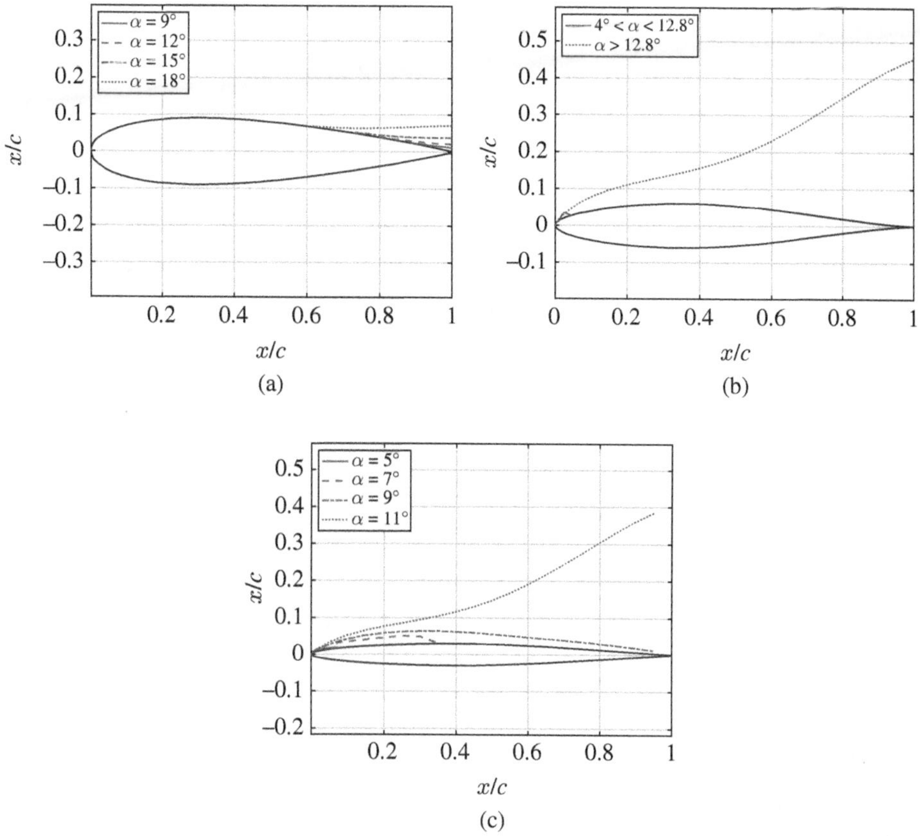

Figure 7.2 Three steady stall mechanisms. (a) Trailing edge stall, (b) leading edge stall, and (c) thin airfoil stall.

contour and separates. Again, the separated shear layer becomes turbulent and reattaches downstream, forming a separation bubble. As the angle of attack increases, the size of the separation bubble increases gradually until it spreads over the entire suction side. This flow separation mechanism is gradual and is demonstrated in Figure 7.2c, which is again inspired by data from McCullough and Gault (1951) for the NACA 64A006 airfoil at $Re = 5.8 \times 10^5$. At $\alpha = 5°$, there is a small separation bubble near the leading edge, whose size increases progressively at higher angles of attack. At $\alpha = 9°$, the separation bubble covers nearly the entire suction side. Complete flow separation is encountered at higher angles of attack.

Flow separation can lead to a loss of lift, whose magnitude depends on the type of stall. The maximum lift coefficient is denoted by $c_{l_{\max}}$ and the angle of attack at which it occurs by $\alpha_{c_{l_{\max}}}$. The term 'stall' refers to the decrease in c_l occurring at $\alpha > \alpha_{c_{l_{\max}}}$. Trailing edge and thin airfoil stall result in gradual loss of lift, while leading edge stall causes a significant and abrupt lift decrease once $\alpha_{c_{l_{\max}}}$ is exceeded. Depending on the geometry of the airfoil and the Reynolds number, leading edge stall can coexist with trailing edge stall.

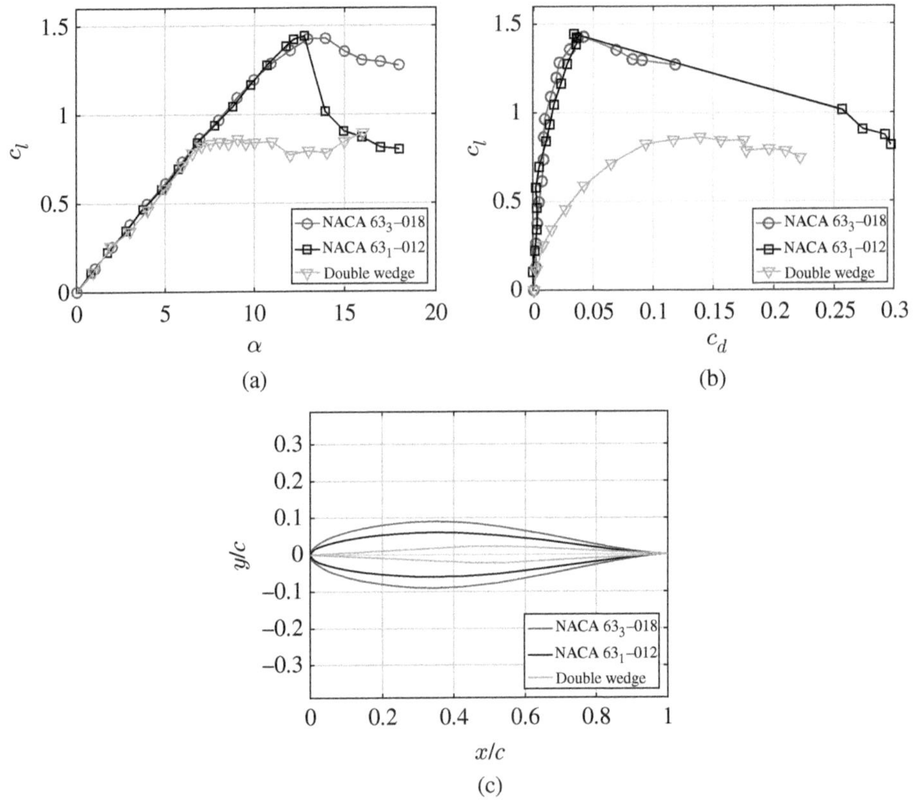

Figure 7.3 Lift curves and drag polars of three different airfoils at $Re = 5.8 \times 10^6$. Experimental data from McCullough and Gault (1951). (a) c_l vs α, (b) c_l vs c_d, and (c) airfoil shapes. Source: McCullough and Gault (1951), credit: NACA.

Figure 7.3 plots the variation of the lift coefficient with angle of attack and with drag coefficient for three symmetric airfoils, the 18% thick NACA $63_3 - 018$ undergoing trailing edge stall, the 12% thick NACA $63_1 - 012$ undergoing leading edge stall and a 9% thick double wedge airfoil undergoing thin airfoil stall. Initially, the lift of all three airfoils increases linearly with angle of attack, as predicted from ideal flow theory. The lift curve of the NACA $63_3 - 018$ departs from linearity for $\alpha > 10°$, but the lift coefficient keeps increasing up to $\alpha_{c_{l_{\max}}} = 14°$, where $c_{l_{\max}} = 1.43$; c_l starts decreasing gradually at higher angles of attack. This behaviour is typical of trailing edge stall. Initially, only a small region of the trailing edge undergoes flow separation; the effect of this separation is barely visible on the lift coefficient. As the separation point moves upstream, c_l starts increasing less quickly with α and, eventually, decreases. Figure 7.3b shows that once $c_{l_{\max}}$ is reached, the drag coefficient starts increasing discontinuously due to the significant flow separation.

The NACA $63_1 - 012$ is thinner than the NACA $63_3 - 018$ and has a smaller leading edge radius. Its lift curve departs from linearity for $\alpha > 11°$ and reaches $c_{l_{\max}} = 1.44$ at $\alpha_{c_{l_{\max}}} = 13°$. At $\alpha = 14°$, the lift coefficient drops abruptly to $c_l = 1.0$ and keeps dropping as the angle of attack increases further. The phenomenon observed here is leading edge stall. The laminar separation bubble at the leading edge has no significant effect on the lift coefficient until it bursts and the entire suction side of the airfoil experiences flow separation, leading to a significant loss of lift. Figure 7.3b shows that this loss of lift is accompanied by a tremendous increase in drag.

Finally, the double wedge airfoil is thinner than both NACA wings and has a sharp leading edge. Its lift curve departs from linearity for $\alpha > 7°$ and its c_l remains fairly constant thereafter, reaching $c_{l_{\max}} = 0.86$. This behaviour is typical of thin airfoil stall. The flow is fully attached only for $\alpha \approx 0°$. At all higher angles of attack, there is a separated flow region starting at the leading edge and ending at the reattachment point. The pressure distribution inside the separated flow region is flat, decreasing leading edge suction and causing higher drag values than those of the NACA airfoils for all $c_l > 0$. The reattachment point moves downstream as the angle of attack increases so that the entire suction surface is covered by the separation bubble at around $\alpha = 10°$; it is impossible to define $\alpha_{c_{l_{\max}}}$ since the lift coefficient is nearly constant from $\alpha = 8°$ to $11°$. At higher angles of attack, the lift coefficient first decreases and then increases.

It should be stated that the c_l and c_d data plotted in Figures 7.3a,b are mean values. In attached flow conditions, the variance of the instantaneous values is small and of the order of the experimental error. Separated flow is unsteady so that the variance of the instantaneous lift and drag starts to increase once significant sections of the surface undergo flow separation. Furthermore, the lift behaviour shown in Figure 7.3a is not necessarily complete. Lift curves can display considerable hysteresis when the angle of attack is first increased and then decreased, even at static conditions. Figure 7.4 plots the lift and drag coefficient variation with angle of attack for the NLR-7301 airfoil at $Re = 2.5 \times 10^6$ and $M = 0.183$ (McAlister et al. 1982). First, the angle of attack is increased in steps, the flow is allowed to stabilise at each step and then pressure measurements are taken, which are subsequently integrated to obtain the aerodynamic loads. After reaching $\alpha = 25°$, the angle of attack is decreased in steps and the measurements are repeated. The figure shows that the lift curve for decreasing α is considerably different to that for increasing α. Flow separation occurs at $\alpha = 20°$, but reattachment occurs at $\alpha = 12°$. This phenomenon is known as stall hysteresis (see for example (McCormick 1979)) and can be encountered frequently in non-linear systems. In essence, flow reattachment is not necessarily the inverse process of flow separation. Note that the same airfoil at different Reynolds and Mach numbers can

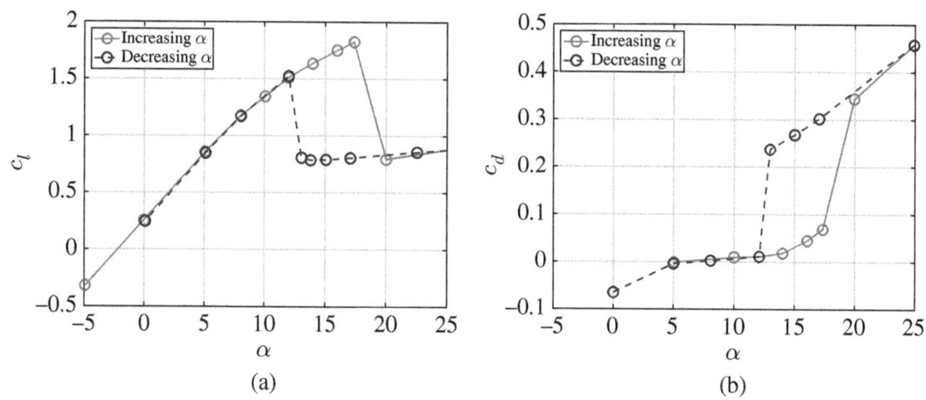

Figure 7.4 Lift and drag curves of the NLR-7301 airfoil at $Re = 2.5 \times 10^6$, $M = 0.183$. Experimental data from McAlister et al. (1982). (a) c_l vs α and (b) c_d vs α. Source: McAlister et al. (1982), credit: NASA.

display negligible hysteresis. Additionally, surface roughness and free stream turbulence can affect significantly flow separation and reattachment.

7.1.2 Dynamic Stall

In order to distinguish between static and moving wing cases, the term 'dynamic stall' is used to describe flow separation and re-attachment phenomena occurring over wings that move with respect to the free stream. Dynamic stall can also feature trailing edge separation and laminar separation bubbles but has three important differences to static stall, which are

- Stall delay. Stall on moving airfoils can occur at higher instantaneous angles of attack than on static airfoils.
- Delayed reattachment. Separated flow over a moving airfoil can reattach to the surface if the instantaneous angle of attack decreases sufficiently. The lift curve will nearly always display considerable hysteresis under dynamic stall conditions.
- Leading edge vortex formation and shedding: when a laminar separation bubble bursts, it can cause the creation of a free strong vortex that propagates downstream over the suction side of the airfoil.

We will demonstrate these phenomena by means of the following example.

Example 7.1 *Study the aerodynamic load responses of the NACA 0012 airfoil undergoing dynamic stall-inducing pitching oscillations, as reported in McAlister et al. (1982). Use the subsonic SDPM to determine the parts of the cycle during which the flow is attached.*

We start by repeating Example 3.15 for the same attached flow test case. The Mach number is $M_\infty = 0.3$, the mean pitch angle $\alpha_0 = 2.64°$, the amplitude $\alpha_1 = 10.16°$ and the reduced frequency $k_0 = 0.099$. The pitch phase angle is set to $\phi_\alpha = 0$ for all test cases. This test case is defined in McAlister et al. (1982) as frame 10309. The SDPM is implemented in 2D, as was done in Example 6.5, so that $b = 100c_0$. However, instead of using a single spanwise panel, we choose $n = 20$ because, in the present case, we will be calculating integrated aerodynamic loads and not just a sectional pressure distribution. The chordwise number of panels is set to $m = 40$. Furthermore, since the motion is pure pitching around the quarter chord, the motion-induced velocities are calculated from Eq. (6.88) with $x_f = c_0/4$, $z_f = 0$.

As the free stream Mach number is low and the instantaneous pitch angle takes significant values in the present test cases, we cannot assume that $Q_\infty \alpha(t) \ll Q_\infty$ so we cannot use Eq. (2.153) to calculate the pressure coefficient distribution on the surface. Instead, we will use the fourth-order equation (2.152) which, in the frequency domain, becomes

$$
\begin{aligned}
c_p(\omega) = &\left(\delta(\omega) - \frac{Q(\omega) * Q(\omega)}{Q_\infty^2} - \frac{2i\omega}{Q_\infty^2}\phi(\omega)\right) \\
&+ \frac{M_\infty^2}{4}\left(\delta(\omega) - \frac{Q(\omega) * Q(\omega)}{Q_\infty^2} - \frac{2i\omega}{Q_\infty^2}\phi(\omega)\right) \\
&* \left(\delta(\omega) - \frac{Q(\omega) * Q(\omega)}{Q_\infty^2} - \frac{2i\omega}{Q_\infty^2}\phi(\omega)\right)
\end{aligned}
\tag{7.6}
$$

We first calculate the incompressible pressure coefficient of Eq. (5.223), which we will call here $S_{i,j}(\omega)$,

$$
\begin{aligned}
S_{i,j}(\omega) = \delta(\omega) - \frac{1}{Q_\infty^2}\Big(&\left(u_{m_{i,j}}(\omega) + u_{i,j}(\omega)\right) * \left(u_{m_{i,j}}(\omega) + u_{i,j}(\omega)\right) \\
&+ \left(v_{m_{i,j}}(\omega) + v_{i,j}(\omega)\right) * \left(v_{m_{i,j}}(\omega) + v_{i,j}(\omega)\right) \\
&+ \left(w_{m_{i,j}}(\omega) + w_{i,j}(\omega)\right) * \left(w_{m_{i,j}}(\omega) + w_{i,j}(\omega)\right)\Big) - \frac{2}{Q_\infty^2} \mathrm{i}\omega\mu_{i,j}(\omega)
\end{aligned}
$$

which we then substitute into Eq. (7.6) to obtain

$$
c_{p_{i,j}}(\omega) = S_{i,j}(\omega) + \frac{M_\infty^2}{4} S_{i,j}(\omega) * S_{i,j}(\omega) \tag{7.7}
$$

All convolutions in the last two expressions are calculated using vector convolutions, as usual. There are five pressure components with non-negative frequencies, $c_{p_{i,j}}(0)$, $c_{p_{i,j}}(\omega_0)$, $c_{p_{i,j}}(2\omega_0)$, $c_{p_{i,j}}(3\omega_0)$ *and* $c_{p_{i,j}}(4\omega_0)$. *The corresponding aerodynamic forces,* $\mathbf{F}_{i,j}(\omega)$, *are calculated from Eq. (5.194) and also have five non-negative frequency components. The pitching moment around the pitch axis is calculated from the aerodynamic loads acting on the panels and from the distance between the control points and the pitch axis,* $\mathbf{r}_{i,j} = (\mathbf{x}_{c_{i,j}} - x_f, \mathbf{y}_{c_{i,j}} - y_f, \mathbf{z}_{c_{i,j}} - z_f)$,

$$
M_{i,j}(\omega) = \left(\mathbf{F}_{i,j}(\omega) \times \mathbf{r}_{i,j}\right) \cdot \mathbf{1}
$$

where $\mathbf{1} = (0, 1, 0)$ *is a unit vector in the direction of the pitch axis. The lift and drag coefficients are obtained from* $\mathbf{F}_{i,j}(\omega)$ *and Eq. (5.131); they have eight non-negative frequency components, up to* $7\omega_0$, *due to the convolution with* $\alpha(\omega)$. *In contrast, the sectional pitching moment coefficient at mid-span*

$$
c_m(\omega) = \sum_{i=1}^{2m} M_{i,n/2+1}(\omega) \frac{\Delta S_{i,n/2+1}}{c_0^2(y_{p_{1,n/2+2}} - y_{p_{1,n/2+1}})}
$$

has the same frequency components as the pressure distribution. For all the loads, the components occurring at frequencies above $3\omega_0$ *have small or negligible values. The time responses of the aerodynamic loads are obtained from the inverse Fourier transform of their frequency components such that*

$$
c_l(t) = C_l(0) + \sum_{j=1}^{7} \left(c_l(j\omega_0)e^{ji\omega_0 t} + c_l^*(j\omega_0)e^{-ji\omega_0 t}\right)
$$

$$
c_d(t) = C_d(0) + \sum_{j=1}^{7} \left(c_d(j\omega_0)e^{ji\omega_0 t} + c_d^*(j\omega_0)e^{-ji\omega_0 t}\right)
$$

$$
c_m(t) = C_m(0) + \sum_{j=1}^{4} \left(c_m(j\omega_0)e^{ji\omega_0 t} + c_m^*(j\omega_0)e^{-ji\omega_0 t}\right)
$$

Figure 7.5 plots $c_l(t)$, $c_d(t)$ *and* $c_m(t)$ *against* $\alpha(t)$ *and should be compared to Figure 3.31, calculated using Theodorsen theory. It can be seen that the compressible SDPM predictions are in better agreement with the experimental data than the Theodorsen results due to the fact that the SDPM neglects neither the thickness of the airfoil nor compressibility effects. Nevertheless,*

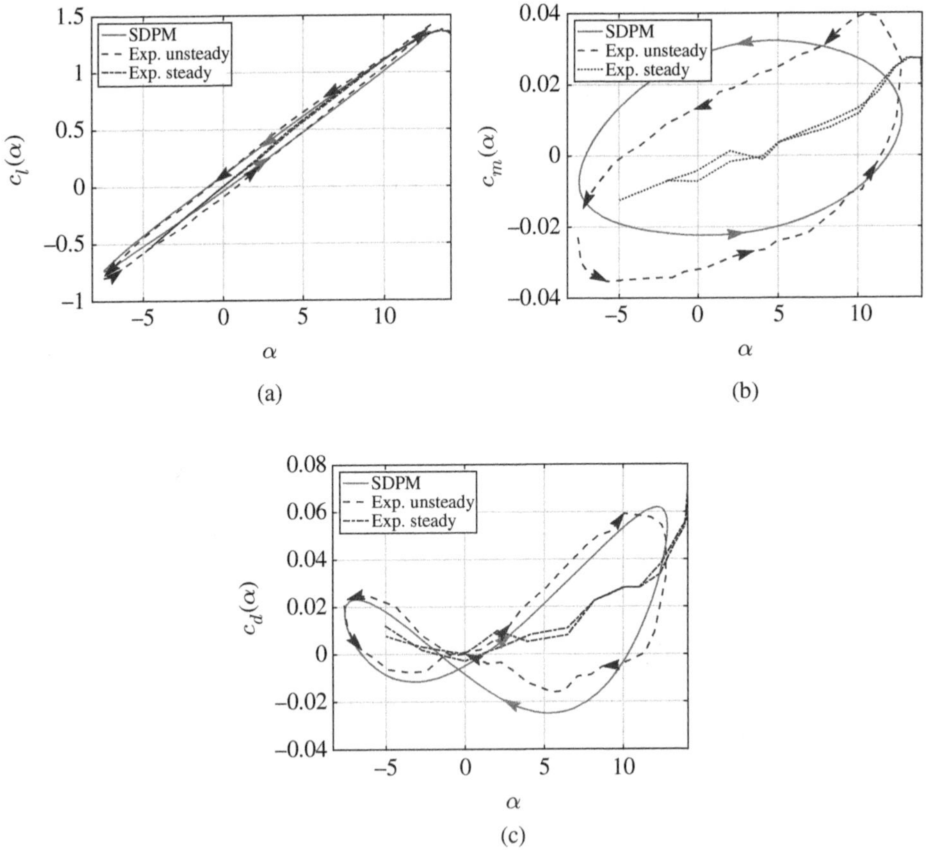

Figure 7.5 Aerodynamic load response predictions for sinusoidally pitching 2D NACA 0012 airfoil, frame 10309. Experimental data by McAlister et al. (1982). (a) $c_l(\alpha)$, (b) $c_{m_{xf}}(\alpha)$, and (c) $c_d(\alpha)$. Source: McAlister et al. (1982), credit: NASA.

this good agreement is due to the fact that the pitching motion has small mean and amplitude such that viscous effects are not very important and the flow remains mostly attached to the surface of the airfoil. Even so the maximum instantaneous pitch angle of 12.8° is quite high, and we can see some deformations of the experimental load responses around this angle; the pitch ellipse has a protuberance and is thinner than the inviscid result during the entire downstroke, while the drag response features a small deformation as the pitch angle decreases from $\alpha = 12.8°$ *to 10°. The arrows plotted on the SDPM and experimental load curves denote the direction of the variation of each load with pitch angle. For example, the lift increases on the upstroke and decreases on the downstroke, but the downstroke lift is always higher than the upstroke lift at the same pitch angle so that the lift curve is a counter-clockwise loop. The same is true for the moment, but the drag creates two loops, one clockwise and one anti-clockwise.*

We now repeat the calculations for a different test case, frame 7019, for which $\alpha_0 = 9°$, $\alpha_1 = 4.9°$ *and* $k = 0.05$. *The maximum pitch angle is 13.9°, slightly higher than the angle*

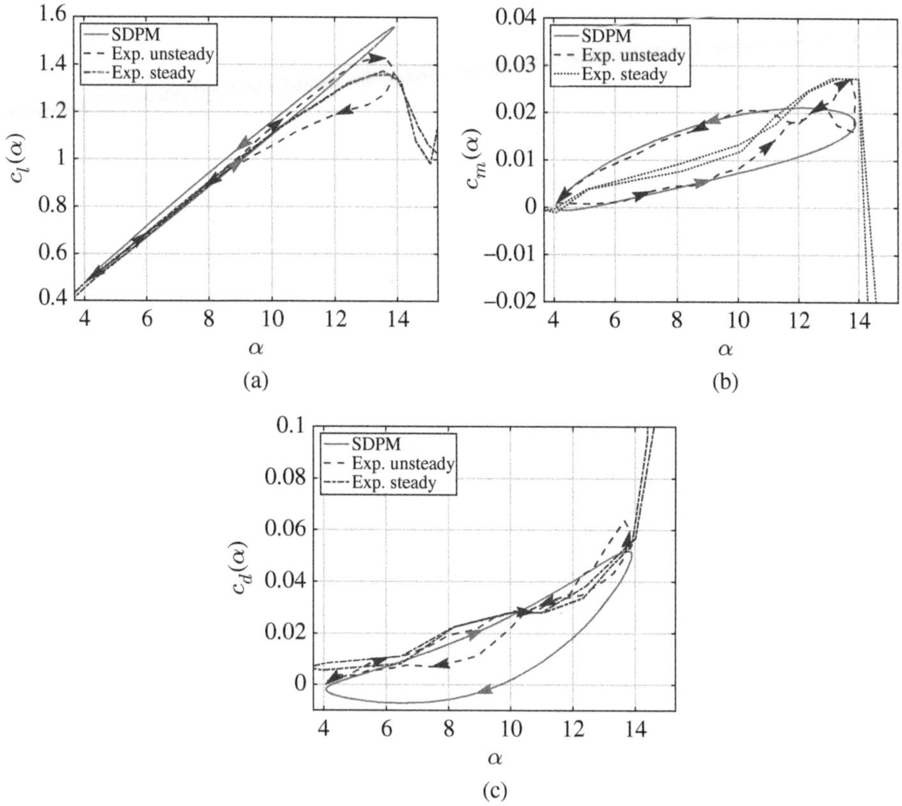

Figure 7.6 Aerodynamic load response predictions for sinusoidally pitching 2D NACA 0012 airfoil, frame 7019. Experimental data by McAlister et al. (1982). (a) $c_l(\alpha)$, (b) $c_{m_{xf}}(\alpha)$, and (c) $c_d(\alpha)$. Source: McAlister et al. (1982), credit: NASA.

of attack at which the steady lift takes its maximum value, $\alpha_{c_{l_{max}}} = 13.5°$. The value of the frequency is half that of the previous test case. Figure 7.6a plots the resulting lift coefficient against instantaneous pitch angle. The experimental and inviscid curves follow each other during the upstroke, up to $\alpha = 13°$. From then on, the experimental lift curve slopes to the left and forms a clockwise loop around the steady lift curve. The numerical lift prediction orbits in a counter-clockwise direction, just as in Figure 7.5a. It should also be noted that the maximum instantaneous value of the unsteady experimental lift is higher than the maximum steady lift.

McCroskey (1973) showed that inviscid flow can be used to predict a decrease in the adverse pressure gradient over the suction surface of an airfoil that is pitching upwards, compared to a static airfoil at the same angle of attack. This decrease in adverse pressure gradient is one of the mechanisms proposed to explain stall delay. Ericsson and Reding (1984) suggested an alternative, viscous mechanism, known as the leading edge jet effect. As the leading edge is pitching up, the flow in contact with the surface also pitches up due to the no slip condition. Consequently, the speed of the boundary layer at the wall is not zero but equal to the kinematic velocity of the surface and this additional momentum stabilises the boundary layer and delays stall.

Figure 7.6b plots the pitching moment response against α; the experimental data follow the inviscid prediction up to around 11° during the upstroke. From then on, the experimental unsteady response increases much more steeply than the linear result, in parallel with the steady moment curve. Near the maximum pitch angle, the viscous pitching moment drops significantly, only to recover quickly and eventually re-join the inviscid behaviour at around $\alpha = 10°$. As in the case of the lift, the pitching moment changes direction during stall, forming a clockwise loop. Finally, the drag plot of Figure 7.6c shows that the experimental c_d distances itself from the inviscid prediction at around $\alpha = 11°$, going through a higher maximum and not re-joining the SDPM results until the minimum pitch angle has been reached; the c_d values during the downstroke are quite similar to those observed during the upstroke.

For the next test case, the mean pitch angle and pitch amplitude are identical to those of the previous case ($\alpha_0 = 9°$, $\alpha_1 = 4.9°$), but the reduced frequency is twice as high, $k = 0.1$. The lift, drag and pitching moment responses are plotted against α in Figure 7.7. The experimental lift coefficient exceeds even further the maximum steady lift value, reaching a maximum almost as high as that of the inviscid prediction. The lift behaviour during the downstroke is

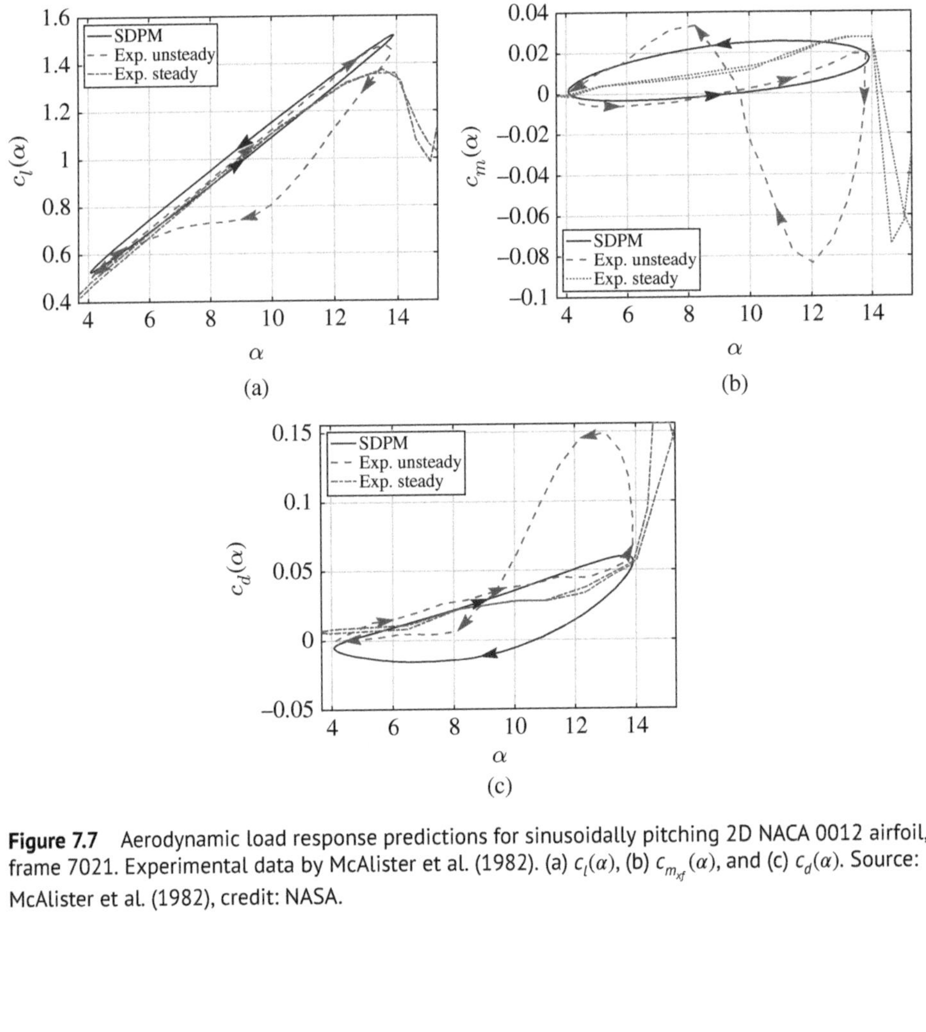

Figure 7.7 Aerodynamic load response predictions for sinusoidally pitching 2D NACA 0012 airfoil, frame 7021. Experimental data by McAlister et al. (1982). (a) $c_l(\alpha)$, (b) $c_{m_{xf}}(\alpha)$, and (c) $c_d(\alpha)$. Source: McAlister et al. (1982), credit: NASA.

very different to the SDPM predictions; the hysteresis is very high and the experimental curve loops in a clockwise direction. The pitching moment follows the inviscid behaviour up to nearly the maximum pitch angle but then drops significantly, reaching a minimum at $\alpha = 12°$ during the downstroke. It then recovers to re-join the SDPM curve at $\alpha = 6°$. The attached flow section of the experimental moment curve orbits counter-clockwise while the separated section clockwise. Finally, the experimental drag departs from the inviscid behaviour at the maximum pitch angle, forms a big counter-clockwise loop that reaches up to $c_d = 0.15$ and then decreases back to upstroke-like values at $\alpha = 8°$. Therefore, the higher reduced frequency of the present test case has provoked longer delays in both stall and reattachment than in the case of Figure 7.7. Furthermore, this test case has demonstrated more clearly that all three aerodynamic load curves reverse orbiting direction while dynamic stall is occurring.

Finally, we analyse a highly separated flow case, with $\alpha_0 = 11.8°$, $\alpha_1 = 9.9°$ and $k = 0.1$, plotted in Figure 7.8. The maximum unsteady lift coefficient reaches 1.9 and the lift curve always winds counter-clockwise, suggesting that the flow does not reattach during the downstroke. Furthermore, there is a discontinuity at around $\alpha = 16°$; the lift flattens out at this angle,

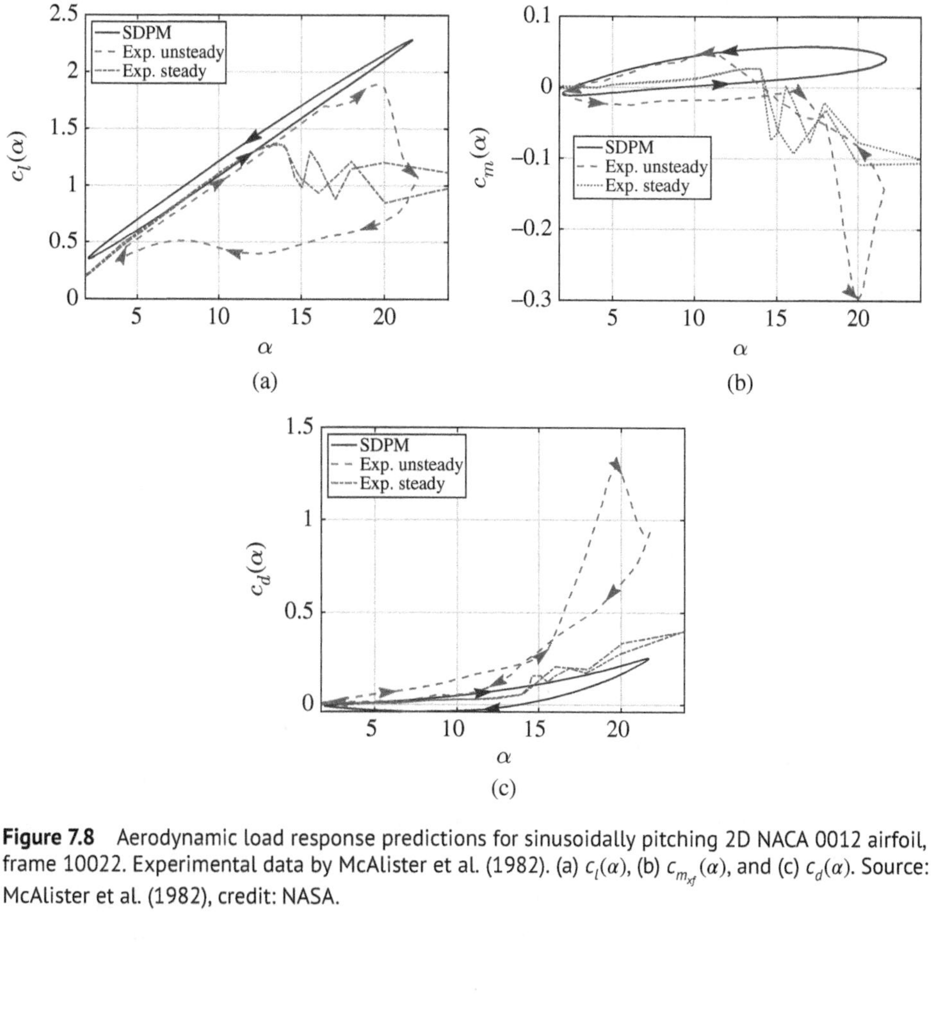

Figure 7.8 Aerodynamic load response predictions for sinusoidally pitching 2D NACA 0012 airfoil, frame 10022. Experimental data by McAlister et al. (1982). (a) $c_l(\alpha)$, (b) $c_{m_{xf}}(\alpha)$, and (c) $c_d(\alpha)$. Source: McAlister et al. (1982), credit: NASA.

but then starts to increase again to reach its global maximum at $\alpha = 19°$. Simultaneously, the pitching moment drops abruptly to −0.3 and the drag increases to nearly 1.3. These three phenomena are due to the shedding of a strong leading edge vortex that moves downstream with the local flow speed. The vortex induces very high speeds near its core and, therefore, a significant local drop in pressure. It translates over the suction side of the airfoil so that it causes an increase in lift up to the time instance when it clears the trailing edge. Furthermore, as it translates over the aft part of the wing, the low-pressure region lies behind the quarter chord so that it produces a very strong nose-down pitching moment. Finally, the vortex causes a significant increase in drag due to the low pressure over the aft section. This example is solved by Matlab function `pitchplunge_subsonic_SDPM_stall.m`.

7.2 Impulsively Started Flow around a 2D Flat Plate at High Angles of Attack

In chapters 3, 4 and 5 we modelled the aerodynamics of impulsively started 2D and 3D wings, assuming that the flow always remains attached over the entire surface of these wings and only separates at the trailing edge. If the angle of attack of the wing is high enough, experience shows that the flow will separate over the suction side of the wing sometime after the start of the motion and that it will remain separated at all subsequent time instances. The wing's motion may be steady after the impulsive start but that does not mean that the flow will also be steady. Separated flow can be highly unsteady due to the continuous generation and shedding of vorticity from the separated shear layer. Furthermore, the position of the separation point can also vary in time, causing further unsteadiness.

Flow separation and the shedding of a separated shear layer have already been modelled in this book. All attached flows have been forced to separate at the trailing edge, and the resulting shear layer was the trailing edge wake. The imposition of the Kutta condition and the shedding of a wake are in fact empirical modifications of potential flow theory that are implemented in order to represent physical flows that always involve viscosity. It stands to reason that additional empirical modifications could be used to represent flow separation at high angles of attack. In fact, numerous authors have developed potential flow models with separated shear layers. Here we will start with modifying the lumped vortex model of Section 4.2, which already sheds a trailing edge wake in the form of discrete vortices. The modification consists in shedding an additional set of discrete vortices to represent the separated shear layer, as was done for example by Katz (1981).

Recall that the lumped vortex method represents the flat plate as a series of panels that contain a discrete vortex on their quarter chord points. At each time instance, a discrete vortex is shed behind the trailing edge, whose strength is determined by Kelvin's theorem. The Kutta condition is implicitly satisfied by the placement of the bound vortices on the panels and the imposition of the impermeability boundary condition at the 3/4 chord point of each panel. Once shed, the wake vortices travel with the local flow velocity. In order to shed a second set of discrete vortices that represents flow separation, we need two pieces of information:

- The location of the separation point. For thick airfoils at moderate angles of attack, this point can lie anywhere on the surface and cannot be determined without performing viscous analysis. However, for a thin flat plate at a high angle of attack, it is logical to assume that the separation point lies at the leading edge.

- The strength of the first discrete vortices in the separated layer. This strength depends on the vorticity of the boundary layer at the separation point so that viscous methods are necessary to calculate it. A heuristic alternative is to impose the Kutta condition at the separation point. However, as the Kutta condition is implicit in the lumped vortex formulation, imposing it at the separation point does not give rise to an additional equation.

In order to adapt the lumped vortex method to separated flows, we will follow the implementation by Manar and Jones (2019). In the attached flow approach, there are n panels and n bound vortices on the wing. Here, the discretization is changed such that the bound vortices lie at the midpoints of the panels, while the impermeability boundary condition is imposed at the vertices of the panels. This means that there are $n+1$ equations for n bound vortices. The first trailing edge wake vortex, with strength Γ_{w_1} is placed at a distance $\Delta x_w, \Delta y_w$ from the trailing edge and the first separated layer vortex, with strength Γ_{s_1} is placed at a distance $\Delta x_w, \Delta y_w$ from the leading edge.

Figure 7.9 plots the modified lumped vortex scheme at the first time instance when there is only one separated and one wake vortex. Note that the flat plate is placed at a very high angle of attack. There are $n = 4$ panels whose vertices are also the control points x_{c_i}, y_{c_i} for $i = 1, \ldots, n+1$ and n bound vortices placed at the midpoints of the panels x_{b_i}, y_{b_i} for $i = 1, \ldots, n$. At the first time instance, Manar and Jones (2019) recommend to place both discrete vortices at a distance $\Delta x_w = 0.02c$, $\Delta y_w = 0$ from their respective shedding points so that

$$x_{s_1}(0) = x_{c_1} + 0.02c,\ y_{s_1}(0) = y_{c_1},\ x_{w_1}(0) = x_{c_{n+1}} + 0.02c,\ y_{w_1}(0) = y_{c_{n+1}} \tag{7.8}$$

where x_{s_i}, y_{s_i} are the coordinates of the separated layer vortices, while x_{w_i}, y_{w_i} are those of the wake vortices.

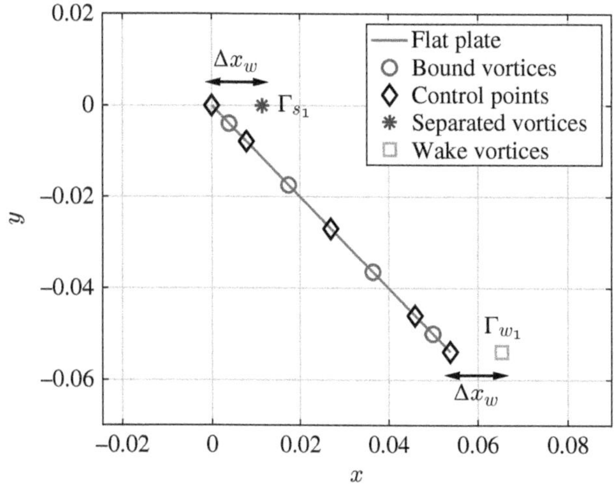

Figure 7.9 Modified lumped vortex scheme for separated flows.

At subsequent time instances, the shed vortices propagate at the local flow velocity. Generalising Eq. (4.56), the positions of the shed vortices become

$$\begin{aligned} x_{s_i}(t_k) &= x_{s_{i-1}}(t_{k-1}) + \left(U_\infty + u_{sb_i}(t_k) + u_{ss_i}(t_k) + u_{sw_i}(t_k)\right)\Delta t \\ y_{s_i}(t_k) &= y_{s_{i-1}}(t_{k-1}) + \left(v_{sb_i}(t_k) + v_{ss_i}(t_k) + v_{sw_i}(t_k)\right)\Delta t \\ x_{w_i}(t_k) &= x_{w_{i-1}}(t_{k-1}) + \left(U_\infty + u_{wb_i}(t_k) + u_{ws_i}(t_k) + u_{ww_i}(t_k)\right)\Delta t \\ y_{w_i}(t_k) &= y_{w_{i-1}}(t_{k-1}) + \left(v_{wb_i}(t_k) + v_{ws_i}(t_k) + v_{ww_i}(t_k)\right)\Delta t \end{aligned} \tag{7.9}$$

where k denotes the kth time instance, u_{sb_i}, v_{sb_i} are the velocities induced by the bound vortices on the separated vortices, u_{ss_i}, v_{ss_i} are those induced by the separated vortices on themselves, u_{sw_i}, v_{sw_i} those induced by the wake vortices on the separated vortices, u_{wb_i}, v_{wb_i} those induced by the bound vortices on the wake vortices, u_{ws_i}, v_{ws_i} those induced by the separated vortices on the wake vortices and u_{ww_i}, v_{ww_i} those induced by the wake vortices on themselves. At all time instances after $k = 1$, the first separated and wake vortices are placed at a third of the distance between the respective shedding point and the next vortex (Ansari et al. 2006), such that

$$\begin{aligned} x_{s_1}(t_k) &= x_{c_1} + (x_{s_2}(t_k) - x_{c_1})/3,\ y_{s_1}(t_k) = y_{c_1} + (y_{s_2}(t_k) - y_{c_1})/3 \\ x_{w_1}(t_k) &= x_{c_{n+1}} + (x_{w_2}(t_k) - x_{c_{n+1}})/3,\ y_{w_1}(t_k) = y_{c_{n+1}} + (y_{w_2}(t_k) - y_{c_{n+1}})/3 \end{aligned} \tag{7.10}$$

The influence coefficients of the bound vortices on the control points are calculated using Eqs. (4.20), (4.21) and (4.23), except that there are more control points than bound vortices so that the influence coefficient matrix $\boldsymbol{A}_n$ has dimensions $(n+1) \times n$. The velocities induced by all the free vortices on the control points are obtained by adding the influence of the separated vortices to Eqs. (4.36) and (4.37)

$$\mathbf{u}_w(t_k) = \boldsymbol{B}_u(t_k)\boldsymbol{\Gamma}_w(t_k) + \boldsymbol{C}_u(t_k)\boldsymbol{\Gamma}_s(t_k) \tag{7.11}$$

$$\mathbf{v}_w(t_k) = \boldsymbol{B}_v(t_k)\boldsymbol{\Gamma}_w(t_k) + \boldsymbol{C}_v(t_k)\boldsymbol{\Gamma}_s(t_k) \tag{7.12}$$

where $\boldsymbol{C}_u(t_k)$, $\boldsymbol{C}_v(t_k)$ are the $(n+1) \times k$ influence coefficient matrices of the separated shear layer and $\boldsymbol{\Gamma}_s(t_k)$ the $k \times 1$ vector containing the strengths of the separated vortices. Matrices $\boldsymbol{B}_u(t_k)$, $\boldsymbol{B}_v(t_k)$, $\boldsymbol{C}_u(t_k)$ and $\boldsymbol{C}_v(t_k)$ are all calculated using Eqs. (4.38) and (4.39) and the normal influence coefficient matrices, $\boldsymbol{B}_n(t_k)$ and $\boldsymbol{C}_n(t_k)$, are obtained from Eq. (4.41). Both of these matrices are separated into a $(n+1) \times 1$ column vector and a $(n+1) \times (k-1)$ matrix, such that

$$\begin{aligned} \boldsymbol{B}_n(t_k) &= \begin{pmatrix} \boldsymbol{b}_{n_1}(t_k) & \boldsymbol{B}_n^*(t_k) \end{pmatrix} \\ \boldsymbol{C}_n(t_k) &= \begin{pmatrix} \boldsymbol{c}_{n_1}(t_k) & \boldsymbol{C}_n^*(t_k) \end{pmatrix} \end{aligned}$$

At the kth time instance, Kelvin's theorem is obtained by adapting Eq. (4.35)

$$\sum_{j=1}^{n} \Gamma_{b_j}(t_k) + \Gamma_{w_1}(t_k) + \Gamma_{s_1}(t_k) = \sum_{j=1}^{n} \Gamma_{b_j}(t_{k-1}) \tag{7.13}$$

Consequently, the impermeability boundary condition of Eq. (4.42) is written as

$$\begin{pmatrix} \mathbf{A}_n & \mathbf{b}_{n_1}(t_k) & \mathbf{c}_{n_1}(t_k) \\ \mathbf{I}_{1\times n} & 1 & 1 \end{pmatrix} \begin{pmatrix} \boldsymbol{\Gamma}_b(t_k) \\ \Gamma_{w_1}(t_k) \\ \Gamma_{s_1}(t_k) \end{pmatrix} = \begin{pmatrix} \mathbf{b} - \mathbf{B}_n^*(t_k)\boldsymbol{\Gamma}_w^*(t_k) - \mathbf{C}_n^*(t_k)\boldsymbol{\Gamma}_s^*(t_k) \\ \sum_{j=1}^{n} \Gamma_{b_j}(t_{k-1}) \end{pmatrix} \tag{7.14}$$

where $\mathbf{b}$ is given by Eq. (4.24) and

$$\boldsymbol{\Gamma}_w(t_k) = \begin{pmatrix} \Gamma_{w_1}(t_k) \\ \boldsymbol{\Gamma}_w^*(t_k) \end{pmatrix}, \ \boldsymbol{\Gamma}_s(t_k) = \begin{pmatrix} \Gamma_{s_1}(t_k) \\ \boldsymbol{\Gamma}_s^*(t_k) \end{pmatrix}$$

Ejecting discrete vortices into the separated shear layer can cause significant numerical instability due to the singularity occurring when two vortices lie very close to each other, as discussed in Example 4.4. The wake vortices move away from the flat plate, but the separated layer vortices move along the wing, interacting very strongly with the bound vortices. It is therefore preferable to use a vortex core model for the calculation of the velocities induced by the vortices, such as the Vatistas model of Eq. (4.65). Consequently, all velocities and influence coefficients induced by all the vortices (both shed and bound) are evaluated using Eq. (4.66).

At each time instance, once the older vortices have been propagated and Eq. (7.14) has been solved for the strengths of the bound and newly shed vortices, we can calculate the aerodynamic force acting on the flat plate by means of the 2D version of the vorticity-momentum theorem of Eq. (2.232). As the flat plate is not moving and the free stream is steady, the second term in that equation is equal to zero. The first step is to calculate the first moment of vorticity from Eq. (2.230), which becomes

$$\alpha_x = \int_{S_\infty} y\omega \mathrm{d}S, \ \alpha_y = -\int_{S_\infty} x\omega \mathrm{d}S$$

for 2D flows, having set the datum $\mathbf{x}_0 = \mathbf{0}$. If the bound and shed vortices are considered to be 2D point vortices, then the vorticity is non-zero only at the locations of these vortices, as discussed in Section 2.7.1. This means that the second moment of vorticity contribution of the jth wake vortex will be given by

$$\alpha_{x_j} = y_{w_j} \int_{\delta S_j} \omega \mathrm{d}S, \ \alpha_{y_j} = -x_{w_j} \int_{\delta S_j} \omega \mathrm{d}S$$

where x_{w_j}, y_{w_j} are the coordinates of the vortex and δS_j is an infinitesimal area around the vortex, over which x and y can be assumed to be constant and equal to x_{w_j}, y_{w_j}. The integral in the two expressions above is the definition of the circulation of Eq. (2.210), written for 2D flows. Therefore, the first moment of vorticity of this vortex is simply $\alpha_{x_j} = y_{w_j}\Gamma_{w_j}$, $\alpha_{x_j} = -y_{w_j}\Gamma_{w_j}$. Summing over all the bound and shed vortices at the k time instance leads to

$$\alpha_x(t_k) = \sum_{j=1}^{n} y_{b_j}\Gamma_{b_j}(t_k) + \sum_{j=1}^{k} y_{s_j}\Gamma_{s_j} + \sum_{j=1}^{k} y_{w_j}\Gamma_{w_j}(t_k) \tag{7.15}$$

$$\alpha_y(t_k) = -\sum_{j=1}^{n} x_{b_j}\Gamma_{b_j}(t_k) - \sum_{j=1}^{k} x_{s_j}\Gamma_{s_j} - \sum_{j=1}^{k} x_{w_j}\Gamma_{w_j}(t_k) \tag{7.16}$$

Then, the aerodynamic loads can be evaluated from Eq. (2.232),

$$d(t_k) = \rho \left.\frac{d\alpha_x}{dt}\right|_{t_k}, \; l(t_k) = \rho \left.\frac{d\alpha_y}{dt}\right|_{t_k} \tag{7.17}$$

where $d(t_k)$ is the drag (parallel to the free stream) and $l(t_k)$ is the lift (perpendicular to the free stream). The time derivatives in these expressions can be calculated using a finite difference scheme. Note that the first moment of vorticity is taken around the origin in Eqs. (7.15) and (7.16). As the aerodynamic loads depend on the time derivative of this moment, changing the point around which they are calculated has a negligible effect on the loads. Potential approaches that shed a separated shear layer in the form of discrete vortices are known as Discrete Vortex Methods (DVM). The present DVM is based on a lumped vortex representation of the wing, but other representations, such as the source and vortex panel technique, are possible.

Example 7.2 *This example simulates a highly separated flow case that was studied experimentally and numerically by Manar and Jones (2019). It consists in a thin flat plate at an angle of attack $\alpha = 45°$ to the horizontal. The plate has an aspect ratio of eight, constant chord of $c = 0.0762$ m and is immersed in a towing tank filled with water. It is accelerated from rest to a constant final speed U_f; the speed of the plate is increased linearly from rest until it reaches U_f at time $\tau = 8$, where $\tau = tU_f/b$ is the non-dimensional time and $b = c/2$ is the half-chord. The speed of the plate is then given by*

$$U_\infty(t) = \begin{cases} \frac{U_f}{8}\tau & \text{if } 0 < \tau \leq 8 \\ U_f & \text{if } \tau > 8 \end{cases}$$

The final chord-based Reynolds number of the experiment is 12,500 so that the final flow speed is $U_f = 0.15$ m/s, given standard values for the density and viscosity of water. The loads acting on the plate were measured experimentally using a load cell while the circulation in the separated layer was estimated by means of Particle Image Velocimetry (PIV).

We start by discretizing the flat plate into $n = 100$ non-linearly spaced panels using Eq. (4.16). The reason for choosing a relatively high number of panels is that the plate must be reasonably solid so that it does not allow the vortices in the separated layer to flow through it. The panel vertices are also the control points, so that $x_{c_i} = \bar{x}_i c$, $y_{c_i} = 0$. We then rotate the control points around the leading edge by $\alpha = 45°$. The locations of the bound vortices are

$$x_{b_i} = (x_{c_{i+1}} + x_{c_i})/2, \; y_{b_i} = (y_{c_{i+1}} + y_{c_i})/2$$

for $i = 1, \dots, n$, while the components of the normal vectors at the control points are $n_{x_i} = \sin\alpha$, $n_{y_i} = \cos\alpha$, for $i = 1, \dots, n+1$.

The non-dimensional simulation end time is set to $\tau_f = 30$, which was also the end time of the experimentally reported data. The dimensional time step is set to $\Delta t = 3c/nU_f$, chosen by trial and error because it gives satisfactory results. At the first time instance, we shed discrete vortices at the locations given by Eq. (7.8). We then calculate the influence coefficients of the bound and shed vortices and set up Eq. (7.14) for $k = 1$

$$\begin{pmatrix} \mathbf{A}_n & \mathbf{b}_{n_1}(t_1) & \mathbf{c}_{n_1}(t_1) \\ \mathbf{I}_{1\times n} & 1 & 1 \end{pmatrix} \begin{pmatrix} \mathbf{\Gamma}_b(t_1) \\ \Gamma_{w_1}(t_1) \\ \Gamma_{s_1}(t_1) \end{pmatrix} = \begin{pmatrix} -U_\infty(t_1)\mathbf{n}_x \\ 0 \end{pmatrix} \tag{7.18}$$

assuming that the initial total circulation in the flow is equal to 0. At the first time step, there is only one shed vortex in each shed layer and their self-influence is zero. Note that, since $U_\infty(t_1) = 0$, all the vortex strengths will be equal to zero at the first time instance.

At the kth time instance, we propagate the shed vortices using Eq. (7.9) and their strengths using

$$\Gamma_{s_i}(t_k) = \Gamma_{s_{i-1}}(t_{k-1}),\ \Gamma_{w_i}(t_k) = \Gamma_{w_{i-1}}(t_{k-1})$$

and then we calculate the positions of the newly shed vortices using (7.10). Next, we evaluate the influence coefficients of all the free vortices on the control points. We solve Eq. (7.14) for the strengths of the bound and newly shed vortices and calculate the velocities induced by all the vortices on the two separated layers. The simulation ends when time τ_f is reached. The influences of the bound and shed vortices are calculated as follows:

- Influence of bound vortices on panel control points: inviscid 2D point vortex model
- Influence of bound vortices on all shed vortices: Vatistas vortex core model
- Influence of all shed vortices on panel control points: Vatistas vortex core model
- Influence of all shed vortices on all other shed vortices: Vatistas vortex core model

The vortex core radius, r_c, in the Vatistas model must be set to the smallest value that will ensure a stable simulation. This value depends on the chosen number of panels and time step and must also be selected by trial and error. It was found that $r_c = U_f \Delta t/4$ was an appropriate choice.

At each time instance, we also calculate the x and y components of the first moment of vorticity using Eqs. (7.15) and (7.16). We also store the instantaneous circulation in the separated shear layer

$$\Gamma_s(t_k) = \sum_{i=1}^{k} \Gamma_{s_i}(t_k)$$

for comparison with the experimental data. After the end of the simulation, we calculate the time derivatives of $\alpha_x(t_k)$, $\alpha_y(t_k)$ and $\Gamma_s(t_k)$ using a central finite difference scheme. Despite the use of the Vatistas vortex core model, there is still a certain amount of numerical instability in the calculation of the velocities induced by the vortices that, in turn, results in a certain amount of numerical noise in $\alpha_x(t_k)$, $\alpha_y(t_k)$ and $\Gamma_s(t_k)$, particularly at the latter stages of the simulation. Calculating numerical derivatives of noisy signals enhances the noise. In order to mitigate this problem, the time signals are first smoothed using a 20-point moving average filter. Finally, we calculate the lift and drag acting on the plate. The latter is accelerating between $\tau = 0$ and $\tau = 8$ so that the second term in Eq. (2.232) cannot be ignored. Adding this term to Eq. (7.17) leads to

$$d(t_k) = \rho \left.\frac{d\alpha_x}{dt}\right|_{t_k} + \rho \left.\frac{dU_\infty}{dt}\right|_{t_k} cd,\ l(t_k) = \rho \left.\frac{d\alpha_y}{dt}\right|_{t_k} \tag{7.19}$$

where $d = 0.003$ m is the thickness of the plate. The motion-induced term only affects the drag since there is no rotation and the acceleration is aligned with the x direction.

Figure 7.10 plots the positions of the shed vortices at two time instances. $\tau = 8$ and $\tau = 30$. At the end of the acceleration phase, Figure 7.10a shows that there are two counter-rotating

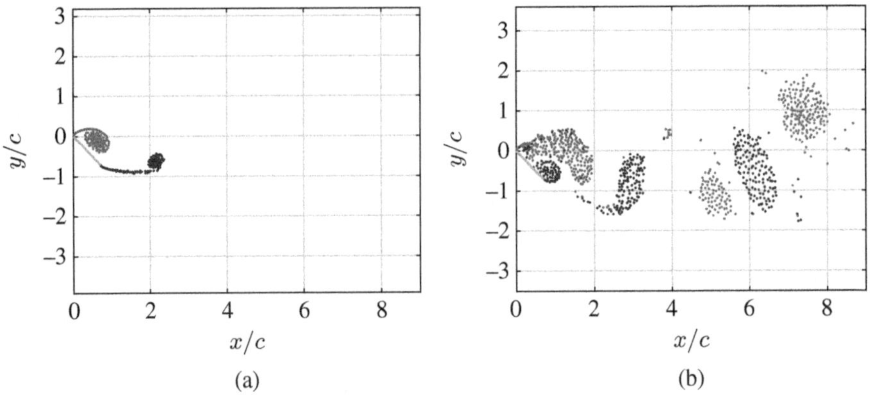

Figure 7.10 Vorticity around flat plate at $\alpha = 45°$ at two time instances after the start of the motion. (a) $\tau = 8$ and (b) $\tau = 20$.

vortical structures, one near the leading edge and one behind the trailing edge. At $\tau = 30$, the leading edge vortex has moved to $x/c = 8$ and two further leading edge vortices have been shed. The first trailing edge vortex has broken up and moved outside of the plotting window, but two more trailing edge vortices have been shed and a fourth is forming.

Figure 7.11 plots the time responses of the circulation in the separated shear layer, $\Gamma_s(t_k)$, and of its time derivative, $\dot{\Gamma}_s(t_k)$; the latter is referred to as the circulation production by Manar and Jones (2019). The values estimated by the Discrete Vortex Method are compared to those measured experimentally using PIV. The agreement is generally good, although $\Gamma_s(t_k)$ is slightly underestimated for $\tau > 25$ and $\dot{\Gamma}_s(t_k)$ is not always in phase with the experimental measurements. Finally, Figure 7.12 plots the time response of the normal and chordwise force coefficients, calculated from

$$c_n(t_k) = c_l(t_k)\cos\alpha + c_d(t_k)\sin\alpha,\ c_c(t_k) = -c_l(t_k)\sin\alpha + c_d(t_k)\cos\alpha$$

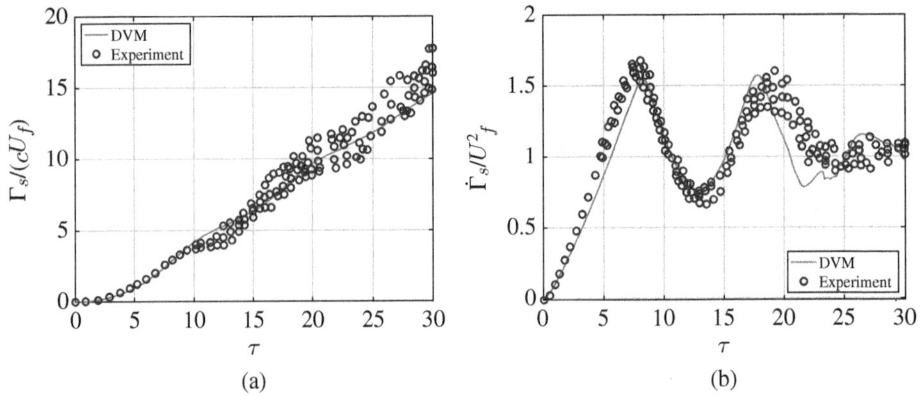

Figure 7.11 Time response of circulation in the separated shear layer. (a) $\Gamma_s(t_k)$ and (b) $\dot{\Gamma}_s(t_k)$. Source: Experimental data reprinted with permission from Manar and Jones (2019), copyright (2019) by the American Physical Society.

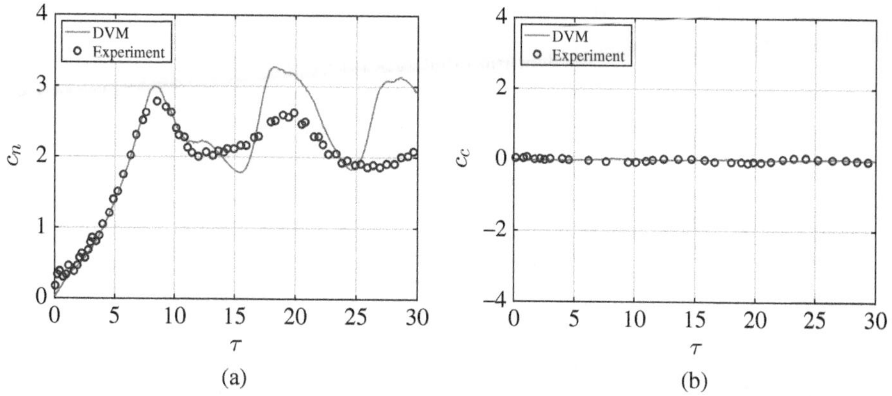

Figure 7.12 Time response of normal and chordwise force coefficient. (a) $c_n(t_k)$ and (b) $c_c(t_k)$. Source: Experimental data reprinted with permission from Manar and Jones (2019), copyright (2019) by the American Physical Society.

As the angle of attack is 45°, $\sin \alpha = \cos \alpha$ *and the lift and drag forces are also nearly equal. Consequently, the force in the chordwise direction is approximately zero at all times. The normal force coefficient predicted by the DVM is in good agreement with the experimental measurements up to around* $\tau = 16$. *At later times, the DVM over predicts the amplitude of the oscillation of the normal force coefficient by around 20%. There are several possible causes for this difference:*

- *The wing is treated as 2D in the simulation, but the experimental wing had an aspect ratio of eight and the experimentally measured aerodynamic loads reflect the three-dimensionality of the flow over it.*
- *There is no physical justification for applying an implicit Kutta condition at the leading edge. This modelling may be satisfactory over the first stages of the motion but less so at later stages.*
- *The Vatistas model implemented here does not feature any vortex dissipation. This is not likely to be the issue causing the overestimation of the normal force amplitude since the circulation in the shear layer appears to be well estimated at all times.*

Implementing vortex dissipation by means of the Vatistas–Leishman model of Eqs. (2.224) and (2.225) is straightforward and left as an exercise for the reader. This example is solved by Matlab code `impulsive_lv_sep.m`.

This example has shown that a quasi-viscous method such as the DVM can represent at least part of the time response of a highly separated flow with reasonable accuracy. The approach relies heavily on empirical information: the separation point is located at the leading edge, the nascent vortices are placed at a third of the distance between the separation point and the next vortex and all the free vortices are treated as vortex cores with an arbitrary radius. Note that the present implementation of the DVM is not suitable for simulating attached flows. The code runs more or less successfully for $\alpha = 5°$ if the number of panels is increased to around 400, but there will be a separated shear layer shed from the leading edge with significant vorticity, so that the lift coefficient will exceed $2\pi\alpha$, violating Wagner theory.

7.2.1 Flow Separation Criteria

In Example 7.2, it was known that leading edge separation occurs as soon as the motion starts and therefore we shed vortices of non-zero strength from the leading edge at each time instance. However, during unsteady motion, the flow may be attached over the leading edge at certain time instances, and we should not shed a separated layer when this happens. Furthermore, for both steady and unsteady conditions, the flow may not separate at the leading edge but further downstream. Several criteria have been proposed in order to determine if and where shedding must occur. We will start with the criterion for leading edge separation by Hammer et al. (2014), which states that the strength of vortex $\Gamma_{s_1}(t_k)$ is selected using

$$\Gamma_{s_1}(t_k) = \begin{cases} 0 & \text{if } \beta(t_k) < \beta_t \\ \frac{U_\infty \Delta t}{\Delta x}\Gamma_{b_1}(t_k) & \text{if } \beta(t_k) \geq \beta_t \end{cases}$$

where $\beta(t_k)$ is a flow angle at the leading edge, β_t is a threshold value of this flow angle and $\Delta x = c/n$ is the constant panel spacing on the wing. The flow angle can be the effective angle of attack, $\beta(t_k) = \alpha_{\text{eff}}(t_k) = \alpha(t_k) + \dot{h}(t_k)/U_\infty$ for a pitching and plunging airfoil, or it can be the angle that the local flow velocity forms with the surface at the leading edge,

$$\beta(t_k) = \tan^{-1}\frac{u_n(t_k)}{u_\tau(t_k)}$$

$u_n(t_k)$, $u_\tau(t_k)$ being the flow velocities normal and tangent to the leading edge, respectively, not including the effect of the motion. Selecting a value for β_t is not obvious; Chabalko et al. (2009) suggest $\beta_t = 30°$ for separation and a different value for reattachment. Note that, if the time step is chosen according to Eq. (4.55), then $\Gamma_{s_1}(t_k) = \Gamma_{b_1}(t_k)$ if $\beta(t_k) \geq \beta_t$.

Robertson et al. (2010) suggest an alternative approach based on Thwaites' laminar boundary layer solution. The momentum thickness of the boundary layer is defined as

$$\theta(x) = \int_0^\infty \left(1 - \frac{u}{U_e}\right)\frac{u}{U_e}\mathrm{d}y$$

Thwaites defined a boundary layer shape parameter

$$\lambda(x) = \frac{\theta^2}{\nu}\frac{\mathrm{d}U_e}{\mathrm{d}x} \tag{7.20}$$

where ν is the kinematic viscosity. Using known analytical solutions and experimental results he showed that for a laminar boundary layer, the momentum thickness is given by

$$\theta(x)^2 = \frac{0.45\nu}{U_e(x)^6}\int_0^x U_e(x)^5\mathrm{d}x$$

The edge velocity distribution $U_e(x)$ can be assumed to be equal to the inviscid tangential velocity on the surface, $\mathbf{u}(x) \cdot \boldsymbol{\tau}(x)$, so that the moment thickness and shape factor can be calculated. Then, separation occurs if $\lambda(x) < -0.09$ and the corresponding x coordinate is denoted by x_{sep}. It should be noted that the Thwaites solution is steady and is applied here at every time instance in a quasi-steady sense.

The vorticity of the separated shear layer can be set to $\Gamma_{s_1}(t_k) = \Gamma_b(x_s, t_k)$, where x_s is the separation point, but there are alternatives. Sarpkaya (1975) evaluated the circulation over a very short length of the shear layer near the separation point using an approach similar to the one that led to Eq. (7.4). Evaluating the contour integral from x_{sep} to $x_{\text{sep}} + \Delta x$, we obtain

$$\Delta\Gamma = \int_{x_{\text{sep}}}^{x_{\text{sep}}+\Delta x} u(x, \delta(x))\mathrm{d}x + \int_{\delta(x_{\text{sep}}+\Delta x)}^{0} v(x_{\text{sep}} + \Delta x, y)\mathrm{d}y$$
$$+ \int_{x_{\text{sep}}+\Delta x}^{x_{\text{sep}}} u(x, 0)\mathrm{d}x + \int_{0}^{\delta(x_{\text{sep}})} v(x_{\text{sep}}, y)\mathrm{d}y$$

Assuming that Δx is so small that $u(x_{\text{sep}} + \mathrm{d}x, y) = u(x_{\text{sep}}, y)$, $v(x_{\text{sep}} + \mathrm{d}x, y) = v(x_{\text{sep}}, y)$, the circulation simplifies to

$$\Delta\Gamma = \int_{x_{\text{sep}}}^{x_{\text{sep}}+\Delta x} u(x, \delta(x))\mathrm{d}x - \int_{x_{\text{sep}}}^{x_{\text{sep}}+\Delta x} u(x, 0)\mathrm{d}x$$
$$= u(x_{\text{sep}}, \delta(x))\Delta x - u(x_{\text{sep}}, 0)\Delta x$$

Then, dividing by Δt leads to

$$\frac{\Delta\Gamma}{\Delta t} = \left(u(x_{\text{sep}}, \delta(x)) - u(x_{\text{sep}}, 0)\right)\frac{\Delta x}{\Delta t}$$

For $\Delta x/\Delta t$, Katz and Plotkin (2001) propose the mean value between the upper and lower shear layer velocities such that

$$\frac{\Delta\Gamma}{\Delta t} = \left(u(x_{\text{sep}}, \delta(x)) - u(x_{\text{sep}}, 0)\right)\frac{\left(u(x_{\text{sep}}, \delta(x)) + u(x_{\text{sep}}, 0)\right)}{2}$$

Finally, taking the limit as $\Delta t \to 0$ results in the circulation rate shed into the flow

$$\frac{\partial\Gamma}{\partial t} = \frac{1}{2}u(x_{\text{sep}}, \delta(x))^2 - \frac{1}{2}u(x_{\text{sep}}, 0)^2$$

On a separated layer very near the separation point, the velocity $u(x_{\text{sep}}, 0)$ is small so that its square can be neglected. Consequently, the circulation rate can be written as

$$\frac{\partial\Gamma}{\partial t} \approx \frac{1}{2}u(x_{\text{sep}}, \delta)^2 \tag{7.21}$$

If we know the velocity $u(x_{\text{sep}}, \delta)$, we can set the change in circulation of the separated layer. Sarpkaya (1975) chose to approximate this velocity at the kth time instance as the mean of the velocities of the first four shed vortices, that is

$$u(x_{\text{sep}}, \delta, t_k) \approx \frac{1}{4}\sum_{i=1}^{4}\sqrt{u_{s_i}(t_k)^2 + v_{s_i}(t_k)^2}$$

where $u_{s_i}(t_k)$ and $v_{s_i}(t_k)$ are calculated from Eq. (7.9). Then, the strength of the nascent separated vortex at the kth time instance is approximated as

$$\Gamma_{s_1}(t_k) = \frac{\partial\Gamma}{\partial t}\Delta t = \frac{1}{2}u(x_{\text{sep}}, \delta, t_k)^2\Delta t$$

Alternatively, Katz (1981) chose $u(x_{\rm sep}, \delta)$ as the inviscid velocity tangent to the surface $U_e(x_{\rm sep})$ at the separation point (or at the nearest upstream control point). Then, he wrote the strength of the latest shed vortex as

$$\Gamma_{s_1}(t_k) = \frac{K}{2} U_e(x_{\rm sep}, t_k)^2 \Delta t \tag{7.22}$$

where K is a circulation reduction factor, with suggested values between 0.5 and 0.6.

Equations (7.20) and (7.21) are in fact simple versions of an approach known as viscous–inviscid interaction or viscous-inviscid matching. The potential flow equations are solved for the inviscid flow and some form of the boundary layer equations for the flow inside the boundary layer; the two solutions are matched at the exit of the boundary layer. The technique can be iterative, whereby, after the first matching, the two solutions are evaluated again and re-matched until convergence is achieved. Such unsteady approaches have been developed by Cebeci et al. (2005) and Riziotis and Voutsinas (2008); they both make use of a source and vortex panel method similar to the Basu & Hancock model for the inviscid flow. The separated shear layer is shed by means of a wake panel, just like the trailing edge wake. The boundary layer equations are solved using finite differences from the stagnation point to the point of separation on each of the surfaces of the airfoil. At the separation point, the strength, length and angle of the vortex-shedding panel is calculated. Boundary layer solutions lie beyond the scope of the present book; the interested reader is referred to Cebeci (1999) and Cebeci et al. (2005).

Ramesh et al. (2014) developed yet another leading edge separation criterion, based on the leading edge suction force of Section 3.4.2. Recall that the leading edge suction parameter (LESP), S, is given by Eq. (3.100)

$$u(x) = \frac{S(x)}{\sqrt{x+b}} \tag{7.23}$$

for a flat plate airfoil, if the origin of the coordinate system lies at the midchord point. Moving this origin to the leading edge, we obtain $S(0)$ as

$$S(0) = \lim_{x\to 0} u(x)\sqrt{x} \tag{7.24}$$

Ramesh et al. (2014) defined the non-dimensional LESP as

$$\text{LESP} = \frac{S(0)}{\sqrt{c}U_\infty}$$

They also expressed the unsteady bound vorticity using an unsteady version of the thin airfoil theory expansion of Eq. (3.11)

$$\gamma(\theta_0, t) = 2U_\infty \left(A_0(t) \frac{1+\cos\theta_0}{\sin\theta_0} + \sum_{n=1}^{\infty} A_n(t) \sin n\theta_0 \right) \tag{7.25}$$

Using Eq. (A.33) for a vortex panel of length c, Eq. (7.24) becomes

$$S(0) = \lim_{x\to 0} u(x)\sqrt{x} = \lim_{\theta\to 0} \frac{1}{2}\gamma(\theta_0, t)\sqrt{\frac{c}{2}\left(1-\cos\theta_0\right)} = \sqrt{c}U_\infty A_0(t)$$

so that, finally, the LESP simplifies to

$$\text{LESP} = A_0(t)$$

Generalising the fundamental equation of thin airfoil theory (3.10) for unsteady motion

$$-\frac{1}{2\pi}\int_0^\pi \frac{\gamma(\theta_0)\sin\theta_0}{\cos\theta-\cos\theta_0}\mathrm{d}\theta_0 = \frac{w(\theta,t)}{U_\infty} \tag{7.26}$$

where $w(\theta, t)$ is the velocity normal to the airfoil at point θ due to the camber, the motion, the wake and the separated shear layer, Eq. (3.15) becomes

$$\text{LESP}(t) = A_0(t) = -\frac{1}{\pi U_\infty}\int_0^\pi w(\theta_0,t)\mathrm{d}\theta_0$$

which can be calculated readily if the upwash is available. Ramesh et al. (2014) carried out several series of CFD simulations on pitching and plunging airfoils in order to show that the leading edge vortex is formed at a critical value of A_0 that is independent of the kinematics; it only depends on the airfoil shape and Reynolds number. They gave a table of these critical values ranging from 0.11 to 0.21 for three airfoils and several Reynolds numbers. Saini et al. (2021) further refined the LESP concept by using unsteady pressures around the leading edge in order to estimate LESP(t) and to detect vortex-related events, such as initiation, pinch-off and termination.

The most widely used dynamic stall modelling approach is the Leishman–Beddoes model (Beddoes 1976; Leishman 1988; Leishman and Beddoes 1989). It is an empirical technique that represents the complete dynamic stall phenomenon, including trailing edge separation, leading edge vortex formation and shedding and delayed reattachment. Its empirical nature means that it falls outside the scope of the present book, but it is worth mentioning that it also relies on a number of static and dynamic parameters that only depend on airfoil geometry, Reynolds and Mach number, not on kinematics. In particular, dynamic stall onset is defined as the condition:

$$x_9(t) > c_{n_1} \tag{7.27}$$

where c_{n_1} is an empirically determined value of the normal force coefficient and $x_9(t)$ is given by the differential equation

$$\dot{x}_9(t) = \frac{c_n^p(t) - x_9(t)}{T_p} \tag{7.28}$$

where $c_n^p(t)$ is the potential flow value of the instantaneous normal force coefficient and T_p is another empirically determined time delay parameter.

Example 7.3 *Use the Leishman–Beddoes criterion to pinpoint the onset of dynamic stall in the highly separated test case of Example 7.1.*

We start by repeating Example 7.1 for the case plotted in Figure 7.8. The kinematic parameters are $\alpha_0 = 11.8°$, $\alpha_1 = 9.9°$ and $k = 0.1$ and the Mach number $M_\infty = 0.3$. The dynamic stall parameters for the NACA 0012 airfoil at this Mach number are $c_{n_1} = 1.31$ and $T_p = 1.7c_0/2Q_\infty$. As we are solving for the aerodynamic loads in the frequency domain, we apply the Fourier transform to Eq. (7.28) such that

$$x_9(\omega) = \frac{c_n^p(\omega)}{\mathrm{i}\omega T_p + 1}$$

The aerodynamic load coefficients in the normal and chordwise directions are obtained from Eq. (6.22). We are interested in the normal force coefficient

$$c_n^p(\omega) = -\sum_{i=1}^{2m} c_{p_{i,n/2+1}}(\omega) s_{i,j} n_{z_{i,n/2+1}} \frac{1}{c_0(y_{p_{1,n/2+2}} - y_{p_{1,n/2+1}})}$$

We recall from Example 7.1 that $c_{p_{i,j}}(\omega)$ has frequency components at $\omega = 0$, ω_0, up to $4\omega_0$, so that the frequency components of $x_9(\omega)$ become

$$x_9(0) = c_n^p(0),\ x_9(\omega_0) = \frac{c_n^p(\omega)}{i\omega_0 T_p + 1},\ \ldots,\ x_9(4\omega_0) = \frac{c_n^p(4\omega)}{4i\omega_0 T_p + 1}$$

Applying the inverse Fourier transform, we obtain $x_9(t)$

$$x_9(t) = x_9(0) + \sum_{j=1}^{4} \left(x_9(j\omega_0) e^{ji\omega_0 t} + x_9^*(j\omega_0) e^{-ji\omega_0 t} \right)$$

Figure 7.13a plots $x_9(t)$, along with the SDPM and experimental $c_l(t)$ responses. It shows that $x_9(t)$ crosses the $c_{n_1} = 1.31$ critical value at a phase angle $\omega t = 14.5°$. The corresponding pitch

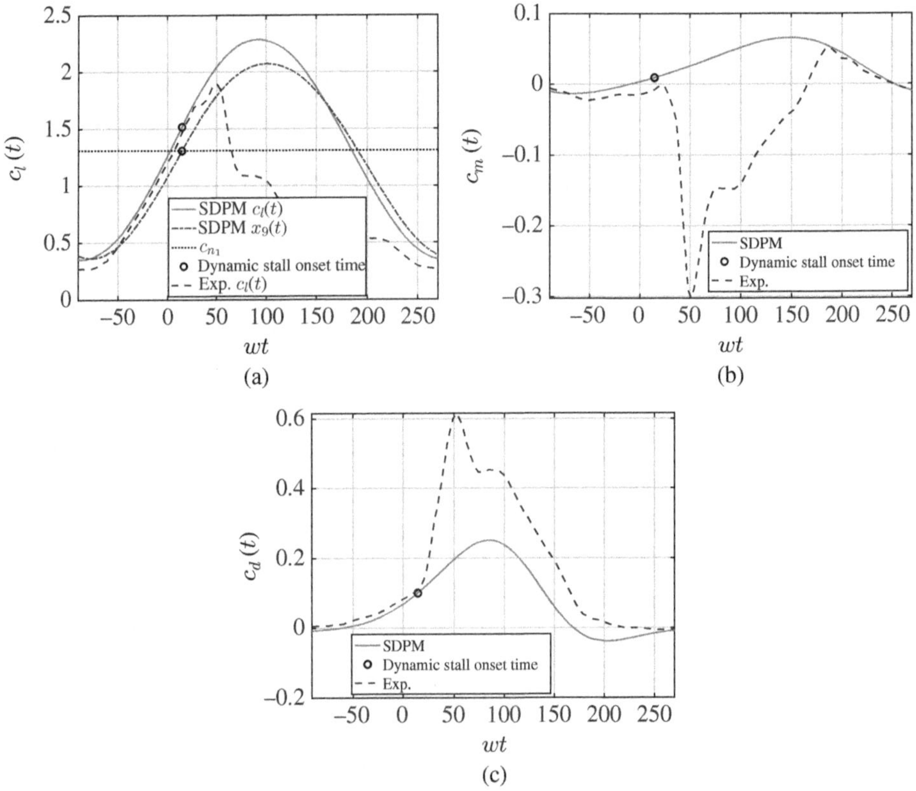

Figure 7.13 Aerodynamic load response predictions for sinusoidally pitching 2D pitching NACA 0012 airfoil and comparison to dynamic stall onset, frame 10022. Experimental data by McAlister et al. (1982). (a) $c_l(\alpha)$, (b) $c_{m_{xf}}(\alpha)$, and (c) $c_d(\alpha)$. Source: McAlister et al. (1982), credit: NASA.

angle is $\alpha_{ds} = 14.3°$. At this time instance, $c_l(t) = 1.52$ and continues to grow. The figure shows that dynamic stall onset does not have an immediate effect on the lift coefficient. Figure 7.13b plots the SDPM and experimental pitching moment responses along with the dynamic stall onset point. The experimental pitching moment response decreases abruptly after dynamic stall onset. Similarly, Figure 7.13c shows that the experimental drag curve increases abruptly just after the dynamic stall point. Similar behaviour is encountered for the other test cases of Example 7.1, except for the one plotted in Figure 7.5 where dynamic stall does not occur at all. This example is solved by Matlab code `pitchplunge_subsonic_SDPM_stall.m` *of Example 7.1.*

The Leishman–Beddoes dynamic stall onset criterion of Eq. (7.27) is effective precisely because the time delay T_p is applied to $c_n^p(t)$. Both T_p and c_{n_1} were determined for specific airfoils using experimental dynamic stall measurements. Nevertheless, the behaviours seen in Figures 7.13b,c suggest alternative phenomenological dynamic stall criteria. Sheng et al. (2006) postulated that dynamic stall onset involving a leading edge vortex can be identified from experimental load responses by one or more of the following criteria:

- A change in slope in the $c_n(\alpha)$ plot. The normal force coefficient is not given in McAlister et al. (1982) but a change in the slope of $c_l(\alpha)$ can be seen in Figure 7.8a, albeit at a higher angle than α_{ds}.
- A local maximum in the $c_c(\alpha)$ plot during the upstroke. Again, the chordwise force coefficient is not available McAlister et al. (1982) but such a local maximum can be seen in the $c_d(\alpha)$ plot of Figure 7.7c at around $\alpha = 11.5°$. No local maximum can be seen in Figure 7.8c before the end of the upstroke.
- An abrupt drop in $c_m(\alpha)$ of more than $\Delta c_m = 0.05$ during the upstroke. Such abrupt drops can be seen in Figures 7.7b and 7.8b, as well as in the time domain response plotted in 7.13b.

Applications of these criteria to three airfoils can be found in Boutet et al. (2020). We will conclude this section by observing that there is still no definitive method for determining dynamic stall onset. Even its definition is not entirely uniform across the literature. Dynamic stall can be defined as the apparition of any type of flow separation during unsteady motion of an airfoil. On the other hand, some authors associate dynamic stall with the formation and shedding of a leading edge vortex and therefore do not regard unsteady trailing edge separation as a dynamic stall phenomenon.

7.3 Flow Around a 2D Circular Cylinder

The flow around a circular cylinder is one of the most studied problems in fluid dynamics. In his review of Zdravkovich's seminal work on the subject (Zdravkovich 1997), P.W. Bearman argued that '... much of what we already know about fluid dynamics, and a great deal of what we still need to understand and predict, is present in the variety of phenomena generated by the flow around a circular cylinder' (Zdravkovich and Bearman 1998). Such phenomena include Reynolds-dependence, unsteady flow separation, transition in shear layers and boundary layers and co-existing quasi-stable flow states among others.

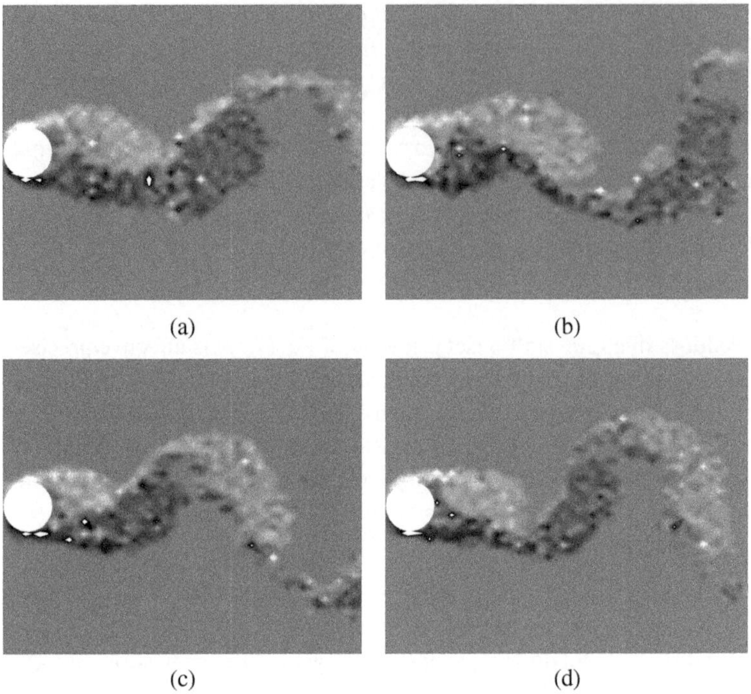

Figure 7.14 Snapshots of the vorticity field in the flow around a circular cylinder at four time instances. (a) Snapshot 1, (b) Snapshot 2, (c) Snapshot 3 and (a) Snapshot 4.

Figure 7.14 demonstrates qualitatively the behaviour of the vorticity field in the flow around a 2D circular cylinder at four time instances. Light gray blobs denote positive (clockwise) vorticity and dark gray negative (counter-clockwise) vorticity. The cylinder is static with respect to the free stream, yet the vorticity field varies in time both over the cylinder itself (especially its rear half) and in the wake. The vorticity is attached to the cylinder on the windward side but separates from the surface just after the maximum height, creating a separated shear layer. The separated layers from the two sides interact, forcing each other to translate upwards or downwards as a function of time. The separation points can also oscillate by a small amount. Finally, it can be seen that snapshot 4 is very similar to snapshot 1, which means that the time variation of the flow is nearly periodic. The periodic shedding of vorticity is a well-known phenomenon occurring in flows around bluff bodies and its dominant non-dimensional frequency is known as the Strouhal number, defined as

$$\mathrm{Str} = \frac{fH}{Q_\infty} \tag{7.29}$$

where f is the shedding frequency in Hz, H is the height of the body (in this case the circle's diameter D) and Q_∞ is the free stream velocity.

The Strouhal number for a given body depends strongly on the Reynolds number and can also be affected by the roughness of the surface and by the free stream turbulence. Figure 7.15a plots a set of data collected by Delany and Sorensen (1953) for

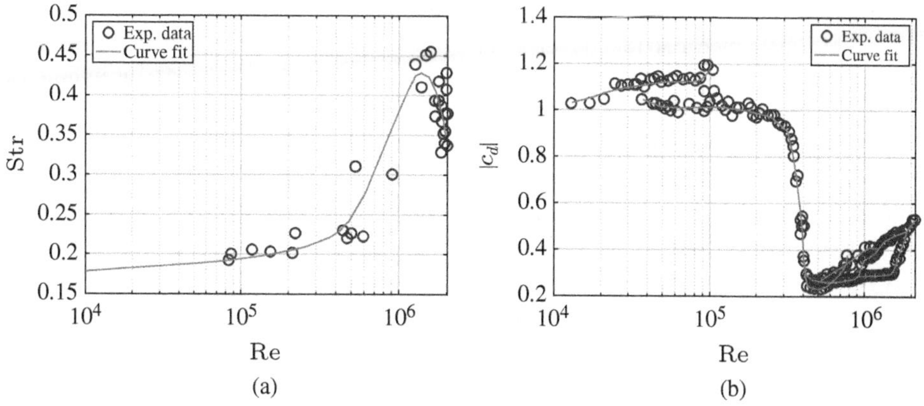

Figure 7.15 Variation of Strouhal number and mean drag coefficient with Reynolds number for circular cylinder. Experimental data by Delany and Sorensen (1953). (a) *Str* and (b) C_D. Source: Delany and Sorensen (1953), credit: NACA.

the Strouhal number of a circular cylinder in the Reynolds range 10^4 to 3×10^6. Up to $Re = 2 \times 10^5$ the Strouhal number lies between 0.18 and 0.20, but it increases to 0.45 in the range $Re = 2 \times 10^5 - 10^6$, only to drop again at higher Reynolds numbers. The range $Re = 2 \times 10^5 - 5 \times 10^5$ is referred to as the critical Reynolds number range because the separated shear layers near the cylinder start to become turbulent (Zdravkovich 1997). In the subcritical range, turbulence is confined to the wake region, but, as the Reynolds number increases, the separated layer near the surface starts to experience transition. A separated layer that becomes turbulent can reattach to the surface, forming a laminar separation bubble. Between $Re = 3 \times 10^5 - 4 \times 10^5$, known as the single-bubble range, only one side of the cylinder features a separation bubble. In the two-bubble range, $Re = 4 \times 10^5 - 5 \times 10^5$, separation bubbles exist on both sides of the cylinder. In the supercritical regime, the separation bubbles fragment in the spanwise direction such that the flow can no longer be considered as 2D. The separation line becomes irregular and the vortex shedding is no longer periodic. This lack of periodicity can be seen in Figure 7.15a, where there are two to three measured values of the Strouhal number for $Re > 10^6$.

The unsteadiness of the flow around the circular cylinder has a significant effect on the aerodynamic loads acting on it, which are also unsteady and oscillate at the Strouhal frequency and its harmonics. As the circular cylinder is symmetric, the mean lift is expected to be zero, which means that the mean drag is the major force acting on it. Figure 7.15b plots the variation of $|c_d|$ with the Reynolds number, showing that the mean drag coefficient lies in the 1–1.2 range, up to the onset of the critical Reynolds range. The drag decreases drastically in this range, since turbulent boundary layers can stay attached longer than laminar ones and the separated flow region on the surface of the cylinder decreases in size, thus reducing the pressure drag. This phenomenon is known as the drag crisis. Beyond $Re = 4 \times 10^5$, the mean drag starts to increase slowly. Note that different experiments give different results in the supercritical and subcritical Reynolds ranges, depending on the free stream turbulence and surface roughness.

It should also be mentioned that in the single-bubble range, the mean flow over the cylinder becomes asymmetric so that the mean lift is no longer equal to zero. Furthermore, there is no preferred side for the formation of the separation bubble on a perfectly symmetric cylinder so that it can lie on either of the sides. This means that there are two possible flow states, one where the separation bubble lies on the upper surface and one where it lies on the lower surface. Experiments have shown that the flow can switch between these two states at irregular intervals, a phenomenon known as flow bistability. Consequently, the lift can also jump from positive to negative mean values. Even small defects in the circularity of the cylinder can have a significant effect on this behaviour, as well as in the Reynolds number ranges in which the single-bubble and two-bubble regimes are found (see for example Benidir et al. (2018)).

In the present work, we will focus on the simulation of the flow past a circular cylinder in the subcritical Reynolds number range. A circular cylinder has infinite span and a circular cross-section so that it is often represented as a 2D flow around a circle, even though it has already been stated that in the supercritical Reynolds range, the flow is inherently 3D. The circle does not have leading or trailing edges such that the shedding strategy used in the previous section cannot be applied. An alternative is to apply the Discrete Vortex Method that will be described next.

7.3.1 The Discrete Vortex Method for Bluff Bodies

The incompressible 2D form of the flow equations (2.1)–(2.4) is given by

$$\frac{\partial u}{\partial x} + \frac{\partial v}{\partial y} = 0 \tag{7.30}$$

$$\frac{\partial u}{\partial t} + u\frac{\partial u}{\partial x} + v\frac{\partial u}{\partial y} = -\frac{1}{\rho}\frac{\partial p}{\partial x} + \nu\left(\frac{\partial^2 u}{\partial x^2} + \frac{\partial^2 u}{\partial y^2}\right) \tag{7.31}$$

$$\frac{\partial v}{\partial t} + u\frac{\partial v}{\partial x} + v\frac{\partial v}{\partial y} = -\frac{1}{\rho}\frac{\partial p}{\partial y} + \nu\left(\frac{\partial^2 v}{\partial x^2} + \frac{\partial^2 v}{\partial y^2}\right) \tag{7.32}$$

where $\nu = \mu/\rho$ is the constant kinematic viscosity. Equation (7.30) is the continuity equation and Eqs. (7.31)–(7.32) are the two momentum equations. They constitute a system of three equations with three unknowns, the velocities $u(x, y, t)$, $v(x, y, t)$ and the pressure $p(x, y, t)$ anywhere in the flow. They are non-linear and no general analytical solutions are known so that they must be solved numerically. One option is to discretise the complete flow around the body and to solve the equations using finite difference, finite element or finite volume approaches. Another option is to solve the vorticity-stream function formulation of the flow equations using a Lagrangian approach. The latter approach does not necessitate the discretisation of the flowfield (although it can still be used in order to accelerate the calculations) and has commonalities with the free wake modelling approach we have been using up to this point.

The vorticity-stream function formulation is obtained from the 2D version of the vorticity transport Eq. (2.208)

$$\frac{\partial \omega}{\partial t} + u\frac{\partial \omega}{\partial x} + v\frac{\partial \omega}{\partial y} = \nu\left(\frac{\partial^2 \omega}{\partial x^2} + \frac{\partial^2 \omega}{\partial y^2}\right) \tag{7.33}$$

where

$$\omega = \frac{\partial v}{\partial x} - \frac{\partial u}{\partial y} \tag{7.34}$$

is the vorticity of the flow, normal to the flow plane. For 2D flows, we can also define the scalar stream function Ψ such that

$$u = \frac{\partial \Psi}{\partial y},\ v = -\frac{\partial \Psi}{\partial x} \tag{7.35}$$

Note that the stream function automatically satisfies the continuity Eq. (7.30). Substituting Eq. (7.35) into the definition of the vorticity (7.34) yields

$$\omega = -\left(\frac{\partial^2 \Psi}{\partial x^2} + \frac{\partial^2 \Psi}{\partial y^2}\right) = -\nabla^2 \Psi \tag{7.36}$$

which is known as Poisson's equation. Clearly, if the flow is irrotational, then $\omega = 0$ everywhere and Poisson's equation becomes Laplace's equation written for the stream function. Separated flows cannot be assumed to be irrotational so that the vorticity cannot be zero everywhere in the flow.

Equations (7.30), (7.33) and (7.36) constitute the vorticity-stream function formulation of the flow equations. They are a system of three equations with three unknowns, the velocities $u(x, y, t)$, $v(x, y, t)$ and the vorticity $\omega(x, y, t)$ anywhere in the flow. Note that the pressure does not appear in this formulation, but it can be calculated a-posteriori once the solution has been obtained. An advantage of the vorticity–stream function approach is that Eq. (7.36) is linear and has known solutions.

We can make use of the linearity of Poisson's equation in order to derive its solution from the solution of its homogenous form

$$\nabla^2 \Psi = 0$$

As mentioned earlier, this is a Laplace equation that has known fundamental solutions. In 2D, the source solution is given by

$$G(\mathbf{x}, \mathbf{x}_0) = \frac{1}{2\pi} \ln r(\mathbf{x}, \mathbf{x}_0)$$

where $r(\mathbf{x}, \mathbf{x}_0) = \sqrt{(x - x_0)^2 + (y - y_0)^2}$, $\mathbf{x}$ is the point where we are evaluating the solution and $\mathbf{x}_0$ is the location of the source. Using $G(\mathbf{x}, \mathbf{x}_0)$ as a Green's function and applying Green's third identity to a rotational flow where the only boundary is the far-field (where Ψ is a constant and the vorticity is zero) it can be shown that

$$\Psi(\mathbf{x}, t) = \int_S \omega(\mathbf{x}_0, t) G(\mathbf{x}, \mathbf{x}_0) \mathrm{d}S_0 = \frac{1}{2\pi} \int_S \omega(\mathbf{x}_0, t) \ln r(\mathbf{x}, \mathbf{x}_0) \mathrm{d}S_0$$

where S is the entire flow domain and $\mathrm{d}S_0$ is an infinitesimal area around $\mathbf{x}_0$. Substituting into Eq. (7.35) and carrying out the differentiations, we obtain

$$u(\mathbf{x}, t) = \frac{1}{2\pi} \int_S \omega(\mathbf{x}_0, t) \frac{y - y_0}{r(\mathbf{x}, \mathbf{x}_0)^2} \mathrm{d}S_0 \tag{7.37}$$

$$v(\mathbf{x}, t) = -\frac{1}{2\pi} \int_S \omega(\mathbf{x}_0, t) \frac{x - x_0}{r(\mathbf{x}, \mathbf{x}_0)^2} \mathrm{d}S_0 \tag{7.38}$$

which is a form of the Biot–Savart law for distributed vorticity in a 2D flow and can be written in vector form as

$$\mathbf{u}(\mathbf{x},t) = \frac{1}{2\pi}\int_S \frac{\omega(\mathbf{x}_0,t)\mathbf{k}\times(\mathbf{x}-\mathbf{x}_0)}{r(\mathbf{x},\mathbf{x}_0)^2}\mathrm{d}S_0 \tag{7.39}$$

where $\mathbf{k} = (0,\ 0,\ 1)$ is the unit vector perpendicular to the flow plane. Substituting Eqs. (7.37) and (7.38) into the vorticity transport expression (7.33) leads to a single equation with a single unknown, $\omega(\mathbf{x},t)$. However, it is a non-linear partial differential equation with integral coefficients, which means that it cannot be solved for general flows.

The Discrete Vortex Method (DVM) constitutes an approximate methodology for solving the vorticity-stream function formulation of the flow equations. It relies on representing the continuous vorticity field in S by discrete vortex particles at positions $\mathbf{x}_{0_i}(t)$, for $i = 1,\ldots,m$, m being the total number of particles. The solution of Eq. (7.33) is carried out in two steps:

- The convection step where we solve the inviscid form of the equation:

$$\frac{\partial\omega}{\partial t} + u\frac{\partial\omega}{\partial x} + v\frac{\partial\omega}{\partial y} = \frac{\mathrm{D}\omega}{\mathrm{D}t} = 0 \tag{7.40}$$

- The diffusion step where we solve the linear diffusion equation:

$$\frac{\partial\omega}{\partial t} = \nu\left(\frac{\partial^2\omega}{\partial x^2} + \frac{\partial^2\omega}{\partial y^2}\right) \tag{7.41}$$

We start by discussing the convection step. Equation (7.40) implies that the vorticity in an infinitesimal area $\mathrm{d}S_0$ around each particle is conserved as it moves at the local flow velocity and can be represented by the circulation inside this area. From the definition of the circulation (2.210), we can write that the circulation of the ith vortex particle is given by

$$\mathrm{d}\Gamma_i = \omega(\mathbf{x}_{0_i})\mathrm{d}S_0$$

This circulation is constant in time but its position changes according to

$$\dot{\mathbf{x}}_{0_i} = \mathbf{u}(\mathbf{x}_{0_i},t) \tag{7.42}$$

The local flow velocity $\mathbf{u}(\mathbf{x}_{0_i},t)$ is calculated from the discrete version of Eq. (7.39):

$$\mathbf{u}(\mathbf{x}_{0_i}(t)) \approx \frac{1}{2\pi}\sum_{i=1}^{m}\frac{\Gamma_i\mathbf{k}\times(\mathbf{x}-\mathbf{x}_{0_i}(t))}{r(\mathbf{x},\mathbf{x}_0)^2} \tag{7.43}$$

where we have replaced $\mathrm{d}\Gamma_i$ by Γ_i to simplify the notation; it should always be kept in mind that Γ_i are small amounts of circulation. Consequently, if the circulation values and positions of all the vortex particles are known at a given time instance, their circulation values will be the same at all subsequent time instances and their positions can be evaluated from Eqs. (7.42) and (7.43). Exactly the same approach was taken in order to model the free wake behind an oscillating airfoil in Section 4.2.2. This step of the DVM is known as the convection step. The Biot–Savart law of Eq. (7.43) can induce numerical instabilities when the denominator is small so that it is generally replaced by a vortex core model, such as the Lamb–Oseen or Vatistas models of Section 2.7.1.

The next step in the DVM methodology is to solve the diffusion Eq. (7.41). Viscous diffusion can be approximated using the random walk technique (Chorin 1973). This approach is based on making an analogy between macroscopic diffusion in a fluid and the microscopic random motion of particles in that fluid. Therefore, if the vorticity in a flow is represented by a cloud of vortex particles, its diffusion can be modelled by a small-amplitude random motion of these particles. In the context of a discrete time simulation with time step Δt, the variance of this motion is given by $\sqrt{2\Delta t/Re}$ and its mean is zero. Convection and diffusion can be carried out in a single calculation. The position of the vortex particles at time instance $k+1$ is given by the discrete version of Eq. (7.42) and the random walk motion such that

$$\mathbf{x}_{0_i}(t_{k+1}) = \mathbf{x}_{0_i}(t_k) + \mathbf{u}(\mathbf{x}_{0_i}, t_k)\Delta t + \mathbf{Z}_i(t_k)\sqrt{2\Delta t/\mathrm{Re}} \tag{7.44}$$

where $\mathbf{Z}_i(t_k)$ is a 2×1 vector of random numbers selected from the normal distribution with zero mean and variance equal to 1.

From the discussion above, it becomes clear that the DVM is a vorticity propagation and diffusion technique but cannot directly determine the amount of vorticity that must be shed from a body. Once vorticity has been created and shed in the form of discrete vortices, the DVM can propagate it but other physical considerations must be used in order to calculate the amount of vorticity that is shed at each time instance. For example, in Chapter 4, the amount of vorticity to be shed from the trailing edge of a wing with attached flow was determined using the (implicit or explicit) Kutta condition. For separated flows, several different methods have been proposed in order to determine the shed vorticity, some of which will be presented in the next sections.

7.3.2 Modelling the Flow Past a Circular Cylinder Using the DVM

Chorin (1973) proposed a complete modelling method for the flow around a circular cylinder, based around the DVM. He determined the amount of vorticity to be released at each time instance by means of a very basic model of a boundary layer. Consider a surface that has been modelled using a potential method such that it is impermeable. This means that the flow velocity normal to the surface is zero while the tangential velocity is non-zero. However, boundary layer theory states that both the normal and tangential flow velocities on the surface must be equal to zero. This situation can be modelled by adding a point vortex at a small distance from the surface such that the velocity it induces on the surface is equal and opposite to that of the potential flow model.

Consider the impermeable surface in Figure 7.16 and the potential flow velocity $u_p(x_c, 0)$ at point $x_c, 0$ on this surface. We now add a vortex of clockwise strength Γ at point x_c, h. As this vortex lies directly above point $x_c, 0$, it cannot induce any normal velocity on it such that impermeability is not affected. The vortex can only induce a tangential velocity $u_\Gamma(x_c, 0)$. The total tangential velocity is then given by $u_p(x_c, 0) - u_\Gamma(x_c, 0)$ which can be equal to zero, depending on the strength of the vortex and the distance h. At point $x_c, 2h$, the total velocity in the direction tangent to the surface is $u_p(x_c, 2h) + u_\Gamma(x_c, 2h)$, which will be non-zero. Therefore, the addition of the vortex can force the total velocity to become zero on the surface and non-zero away from it, exactly as in the case of the boundary layer. The form of the velocity profile depends on the chosen vortex model; the inviscid vortex of Eq. (2.223) will induce a singular velocity at point x_c, h and the resulting velocity profile will be discontinuous. It is important then to choose a vortex core model instead, such as the Lamb–Oseen vortex of Eq. (2.221) or the Vatistas model (2.224). Chorin (1973) used a simpler model of the form:

$$u_\theta(r) = \begin{cases} \dfrac{\Gamma}{2\pi r} & \text{if } r \geq r_c \\ \dfrac{\Gamma}{2\pi r_c} & \text{if } r < r_c \end{cases} \tag{7.45}$$

where u_θ is the circumferential velocity induced by the vortex at r and at any angle θ. Looking at Figure 7.16, it is clear that $u_\Gamma(x_c, 0) = u_\theta(h)$.

It is clear that the vortex placement of Figure 7.16 will only produce the desired effect at point $x_c, 0$; at all other points on the surface, the tangential velocity will not be equal to zero. Furthermore, we have specified neither the distance h nor the strength Γ as yet. Consider a surface discretized into panels, three of which are plotted in Figure 7.17. The control points of the panels are denoted by $\mathbf{x}_{c_i} = (x_{c_i}, y_{c_i})$ and a source distribution is placed on the panels such that the surface is impermeable. The flow velocities on the control points are denoted by $\mathbf{u}_{c_i} = (u_{c_i}, v_{c_i})$.

We now place a vortex of the form of Eq. (7.45) near each panel, such that it lies at an infinitesimal distance, δr away from the control points in the local normal direction $\mathbf{n}_i$. Constant source distributions with strength σ_i are also placed on each of the panels in order to impose impermeability and calculate the potential flow velocities on the control points, $\mathbf{u}_{c_i}$. The strengths of the Γ_i vortices must represent the circulation in the boundary layer on the corresponding panels. If we assume that the velocity at the edge of the boundary

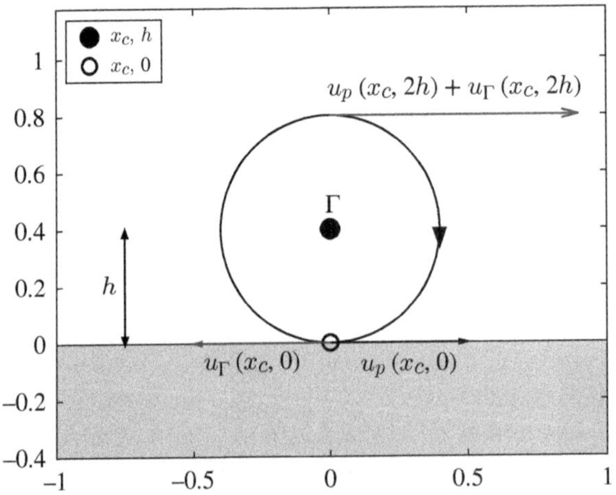

Figure 7.16 Approximate modelling of boundary layer.

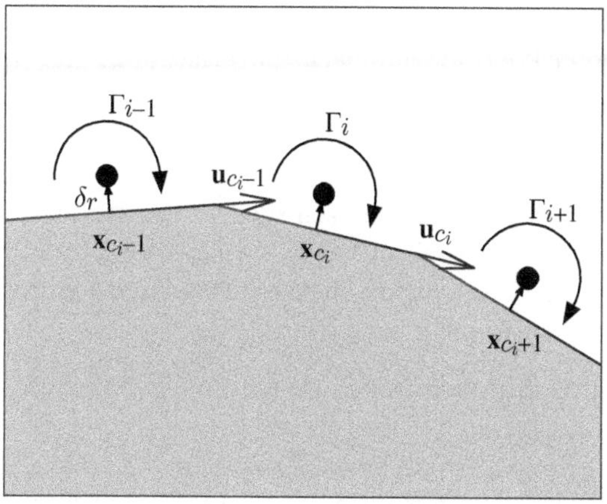

Figure 7.17 Vortex shedding strategy for circular cylinder.

layer on each panel is constant along the panel and equal to $\mathbf{u}_{c_i}$, then the circulation in the boundary layer over the ith panel is given from Eq. (7.4) as

$$\Gamma_i = \mathbf{u}_{c_i} \cdot \boldsymbol{\tau}_i s_i \tag{7.46}$$

where $\boldsymbol{\tau}_i$ is the unit vector tangent to the panel and s_i is the length of the panel. Assuming that the distance $\delta r < r_c$, Eq. (7.45) can be used to determine the velocity induced by each vortex on each panel, $\mathbf{u}_{\Gamma_i} = \Gamma_i/2\pi r_c$. In order to represent the boundary layer, the total velocity tangent to the ith panel must be equal to zero such that

$$-\frac{\Gamma_i}{2\pi r_c} + \mathbf{u}_{c_i} \cdot \boldsymbol{\tau}_i = 0$$

Substituting for Γ_i from Eq. (7.46) leads to

$$r_c = \frac{s_i}{2\pi}$$

such that the radius of the vortex core depends on the size of the vortex panel. For numerical stability, all the vortices must have the same vortex core radius, which we can choose as

$$r_c = \frac{1}{2\pi n} \sum_{i=1}^{n} s_i \tag{7.47}$$

where n is the total number of panels. Furthermore, the distance δr can be set to zero without affecting the model, since Eq. (7.45) dictates that the velocity $u_\theta(r) = \Gamma/2\pi r_c$ even when $r = \delta r = 0$.

At the next time instance, the shed vortices are propagated using Eq. (7.44), but first the local flow velocity must be calculated. Figure 7.16 shows that the shed vortex does not induce any normal velocity on the surface and that the total tangential velocity can be set to zero. Furthermore, Figure 7.17 shows that Γ_i does not induce normal velocity on the ith panel. However, it does induce normal velocities on all the other panels. Given that the source solution has been calculated in order to impose impermeability, the subsequent shedding of the vortices will induce additional velocities normal to the surface and render it permeable. This means that the source panels and vortices cannot coexist; the source panels are used only to calculate the tangential velocities $\mathbf{u}_{c_i}$ and hence Γ_i. Then, the source panels are replaced by the vortices that are placed on the panel control points. This means that when we start calculating the local flow velocities on each shed vortex, the source panels do not exist anymore, and these velocities are only due to the vortices, new and old[1].

A complete algorithm for viscous flow modelling around a body discretised into n source panels can then be summarised as

1. At the start of the kth time instance, there are n_w shed vortices in the flow, with strengths $\Gamma_i(t_{k-1})$ and positions $\mathbf{x}_{w_i}(t_{k-1})$ for $i = 1, n_w$.
2. Place source panels of constant strength $\sigma_i(t_k)$ on each panel. Calculate the velocities induced by the n_w vortices and n source panels on the panel control points in order to impose impermeability and to evaluate $\sigma_i(t_k)$ and the total flow velocities on the panels $\mathbf{u}_{c_i}(t_k)$.
3. Replace the source panels by new vortices lying on the control points whose strengths are given by Eq. (7.46). There are now $n + n_w$ vortices whose strengths are

$$\Gamma_i(t_k) = \begin{cases} \mathbf{u}_{c_i} \cdot \boldsymbol{\tau}_i s_i & \text{for } i \leq n \\ \Gamma_{i-n}(t_{k-1}) & \text{for } i > n \end{cases} \tag{7.48}$$

   so that the strengths of the older vortices are unaltered.
4. Calculate the local flow velocities $\mathbf{u}_{w_i}(t_k)$ on all the vortices, which lie at $\mathbf{x}_{c_i}$ for $i \leq n$ and $\mathbf{x}_{w_i}(t_{k-1})$ for $i > n$. The velocities at the newly created vortices are set equal to the tangential velocities calculated previously, $\mathbf{u}_{w_i}(t_k) = \mathbf{u}_{c_i}(t_k)$, for $i = 1, \ldots, n$. The local flow velocities at the positions of the older vortices, $i = n + 1, \ldots, n + n_w$, are calculated by summing the influence of all vortices on each other, as given by Eq. (7.45), and adding the free stream.
5. Propagate all vortices using Eq. (7.44)

$$\mathbf{x}_{w_i}(t_k) = \mathbf{x}_{w_i}(t_k - 1) + \mathbf{u}_{w_i}(t_k)\Delta t + \mathbf{Z}_i(t_k)\sqrt{2\Delta t/\mathrm{Re}} \tag{7.49}$$

   where $\mathbf{x}_{w_i}(t_k)$ are the positions of the vortex particles at time t_k, for $i = 1, \ldots, n + n_w$. Any vortices that cross the solid boundary are deleted.
6. Increment k and go back to step 2.

In order to accelerate the calculation, we can delete all vortices that lie further than a chosen distance from the body. At the first time instance, as the strength of the vortices depends on the source solution and the latter is irrotational, the sum of Γ_i will be equal to zero. This is not the case for subsequent time instances since vortices that cross the solid boundary or move too far from the body are deleted. Consequently, Kelvin's theorem is not imposed by this algorithm and, therefore, it is not possible to use the vorticity-momentum theorem to calculate the aerodynamic loads. However, we can calculate the pressure around

1 Re-calculating the total flow velocity on the control points using only the influences of the free stream, older vortices and newly shed vortices results in an approximately tangential velocity field. This result is logical because the strengths of the new vortices are calculated using an impermeable solution. The reader is invited to verity this statement.

the body from the Navier–Stokes equations at the wall. There the flow velocities u and v are both equal to zero at all times due to the no slip condition so that Eqs. (7.31) and (7.31) simplify to

$$\frac{1}{\rho}\frac{\partial p}{\partial x} = \nu\left(\frac{\partial^2 u}{\partial x^2} + \frac{\partial^2 u}{\partial y^2}\right) \tag{7.50}$$

$$\frac{1}{\rho}\frac{\partial p}{\partial y} = \nu\left(\frac{\partial^2 v}{\partial x^2} + \frac{\partial^2 v}{\partial y^2}\right) \tag{7.51}$$

Consider the unit vector normal to the wall, $\mathbf{n} = (n_x, n_y)$, multiply Eq. (7.50) by $-n_y$ and (7.51) by n_x and add them up to obtain

$$\frac{1}{\rho}\left(-n_y\frac{\partial p}{\partial x} + n_x\frac{\partial p}{\partial y}\right) = \nu\left(-n_y\frac{\partial^2 u}{\partial x^2} - n_y\frac{\partial^2 u}{\partial y^2} + n_x\frac{\partial^2 v}{\partial x^2} + n_x\frac{\partial^2 v}{\partial y^2}\right) \tag{7.52}$$

We can use the continuity Eq. (7.30) in the form

$$\frac{\partial u}{\partial x} = -\frac{\partial v}{\partial y}$$

to rewrite Eq. (7.52) as

$$\begin{aligned}\frac{1}{\rho}\left(-n_y\frac{\partial p}{\partial x} + n_x\frac{\partial p}{\partial y}\right) &= \nu\left(n_y\frac{\partial^2 v}{\partial x\partial y} - n_y\frac{\partial^2 u}{\partial y^2} + n_x\frac{\partial^2 v}{\partial x^2} - n_x\frac{\partial^2 u}{\partial x\partial y}\right)\\ &= \nu\left(n_x\frac{\partial}{\partial x}\left(\frac{\partial v}{\partial x} - \frac{\partial u}{\partial y}\right) + n_y\frac{\partial}{\partial y}\left(\frac{\partial v}{\partial x} - \frac{\partial u}{\partial y}\right)\right)\end{aligned}$$

Finally, recalling the definition of the 2D vorticity (7.34) and the fact that the unit vector tangent to the surface is given by $\boldsymbol{\tau} = (-n_y, n_x)$, the momentum equation at the wall becomes (Lin et al. 1997)

$$\frac{1}{\rho}\boldsymbol{\tau}\cdot\nabla p = \nu\mathbf{n}\cdot\nabla\omega$$

or using the notation $\mathbf{n}\cdot\nabla\omega = \partial\omega/\partial n$ and $\boldsymbol{\tau}\cdot\nabla p = \partial p/\partial\tau$,

$$\frac{1}{\rho}\frac{\partial p}{\partial \tau} = \nu\frac{\partial\omega}{\partial n} \tag{7.53}$$

This latest expression can be integrated over the surface in order to calculate the pressure but, first, we need to determine $\partial\omega/\partial n$. In order to do this, we consider the vorticity transport Eq. (7.33) which, on the wall, simplifies to

$$\frac{\partial\omega}{\partial t} = \nu\left(\frac{\partial^2\omega}{\partial x^2} + \frac{\partial^2\omega}{\partial y^2}\right)$$

As this equation is scalar, it should hold for any orthogonal coordinate system, e.g.

$$\frac{\partial\omega}{\partial t} = \nu\left(\frac{\partial^2\omega}{\partial \tau^2} + \frac{\partial^2\omega}{\partial n^2}\right)$$

It can be assumed that tangential diffusion is negligible compared to normal diffusion such that

$$\frac{\partial\omega}{\partial t} \approx \nu\frac{\partial^2\omega}{\partial n^2} \tag{7.54}$$

We assume that all the vorticity over a panel is concentrated in an infinitely thin region in which u and v are so small that we can still apply the wall forms of the Navier–Stokes and vorticity transport equations. Let us denote the height of this region by δ', where δ' is much smaller than the thickness of the boundary layer. We can now integrate Eq. (7.54) over this region to obtain

$$\int_0^{s_i}\int_0^{\delta'} \frac{\partial \omega}{\partial t} \mathrm{d}n\mathrm{d}\tau = \nu \int_0^{s_i}\int_0^{\delta'} \frac{\partial^2 \omega}{\partial n^2} \mathrm{d}n\mathrm{d}\tau$$

where s_i is the length of the ith panel. On the left-hand side, we take the differentiation with respect to time outside the integral, while on the right-hand side, we carry out the integration over n, leading to

$$\frac{\partial}{\partial t}\int_0^{s_i}\int_0^{\delta'} \omega \mathrm{d}n\mathrm{d}\tau = \nu \int_0^{s_i} \left(\left.\frac{\partial \omega}{\partial n}\right|_{\delta'} - \left.\frac{\partial \omega}{\partial n}\right|_0 \right) \mathrm{d}\tau$$

The integral on the left-hand side is the integral of vorticity in an area around the panel; therefore, it is the circulation of the panel, Γ_i. Furthermore, the vorticity at δ' was already assumed to be zero; as the vorticity at points outside δ' is also zero, it means that $\partial\omega/\partial n = 0$ at δ'. Therefore,

$$\frac{\partial \Gamma_i}{\partial t} = \nu \int_0^{s_i} \left(-\left.\frac{\partial \omega}{\partial n}\right|_0 \right) \mathrm{d}\tau$$

Finally, $\partial\omega/\partial n|_0$ is not zero, but we assume that it is constant over the panel such that

$$\frac{\partial \Gamma_i}{\partial t} = -\nu \frac{\partial \omega}{\partial n} s_i$$

or after re-arranging and converting to discrete time,

$$\nu \frac{\partial \omega}{\partial n} = -\frac{1}{s_i}\frac{\Delta \Gamma_i}{\Delta t}$$

We can substitute this latest expression into Eq. (7.53) and integrate it over the length of the panel to obtain

$$\Delta p_i = -\rho \frac{\Delta \Gamma_i}{\Delta t} \tag{7.55}$$

where Δp_i is the change in pressure between the previous panel and the next. The change in circulation $\Delta\Gamma_i$ is the circulation flow across the panel and can be calculated from the difference of the new vortex that was shed from the panel using Eq. (7.46) minus the circulation of any older vortices that crossed the ith panel and were deleted.

Example 7.4 *Simulate viscous flow around a circular cylinder for $Re = 10^5$ using the DVM and a source panel distribution on the cylinder's surface.*

We set the radius of the circle to $R = 1$ m, the free stream velocity to $U_\infty = 1$ m/s and the density of the fluid to $\rho = 1$ Kg/m^3. From Figure 7.15a, the Strouhal number for $Re = 10^5$ is around 0.2. The height, H, of the cylinder is equal to $2R$ so that Eq. (7.29) yields $f = 0.1$ Hz for the shedding frequency. We would like to simulate the unsteady flow with reasonable resolution at this frequency, so we choose

$$\Delta t = \frac{T}{40} = 0.25$$

where $T = 1/f$ is the shedding period and this value of Δt signifies that there will be 40 time instances per shedding cycle. As all the dimensions and flow parameters are fixed, changing Re is equivalent to changing the viscosity and the only effect it has on the calculation is to vary the variance of the random walk $\sqrt{2\Delta t/Re}$.

The surface of the cylinder is discretised into $n = 80$ panels with constant strength σ_i. The coordinates of the panel vertices are computed from

$$x_i = R\cos\left((1-i)\frac{2\pi}{n}\right) + R,\ y_i = R\sin\left((1-i)\frac{2\pi}{n}\right)$$

for $i = 1, \ldots, n+1$, so that they are numbered in a clockwise direction from point $2R, 0$ to $0,0$ and back to $2R, 0$. The control points, x_{c_i}, y_{c_i} lie at the midpoints of the panels and their coordinates are given by Eq. (4.103). The unit vectors tangent and normal to the panels are given by Eqs. (4.104) and (4.105), respectively, while the lengths of the panels are given by (4.106). The influence coefficients of the source panels on the control points in the x and y directions, $\mathbf{A}_u$, $\mathbf{A}_v$, are given by Eqs. (4.123) and (4.124).

At the start of the first time step, $k = 0$, there are no free vortices in the flow. The impermeability boundary condition is simply

$$\boldsymbol{A}_n\boldsymbol{\sigma}(t_0) = U_\infty \boldsymbol{n}_x$$

where the elements of the $n \times n$ matrix $\boldsymbol{A}_n$ are

$$A_{n_{i,j}} = A_{u_{i,j}} n_{x_i} + A_{v_{i,j}} n_{y_i}$$

while $\boldsymbol{\sigma} = (\sigma_1, \sigma_2, \ldots, \sigma_n)^T$ and $\mathbf{n}_x = (n_{x_1}, n_{x_2}, \ldots, n_{x_n})^T$. At a general time instance, k, the are n_w free vortices in the flowfield, lying at positions $x_{w_i}(t_{k-1})$, $y_{w_i}(t_{k-1})$ and with strengths $\Gamma_i(t_{k-1})$. Using Chorin's vortex core model, the x and y components of the velocities induced by the ith vortex at any point in the flow are given by

$$u_{\Gamma_i}(x, y) = u_\theta(r_i)\frac{y - y_{w_i}}{r_i}, \quad v_{\Gamma_i}(x) = -u_\theta(r_i)\frac{x - x_{w_i}}{r_i} \tag{7.56}$$

where $u_\theta(r_i)$ is obtained from Eq. (7.45) for $r_i = \sqrt{(x - x_{w_i})^2 + (y - y_{w_i})^2}$ and r_c is calculated from Eq. (7.47). The impermeability boundary condition becomes

$$\boldsymbol{A}_n\boldsymbol{\sigma}(t_k) = -(U_\infty + \boldsymbol{u}_\Gamma(\boldsymbol{x}_c, \boldsymbol{y}_c)) \circ \boldsymbol{n}_x - \boldsymbol{v}_\Gamma(\boldsymbol{x}_c, \boldsymbol{y}_c) \circ \boldsymbol{n}_y \tag{7.57}$$

where $\circ$ denotes the Hadamard product,

$$\boldsymbol{u}_\Gamma(\boldsymbol{x}_c, \boldsymbol{y}_c) = \sum_{i=1}^{n_w} \boldsymbol{u}_{\Gamma_i}(\boldsymbol{x}_c, \boldsymbol{y}_c),\ \boldsymbol{v}_\Gamma(\boldsymbol{x}_c, \boldsymbol{y}_c) = \sum_{i=1}^{n_w} \boldsymbol{v}_{\Gamma_i}(\boldsymbol{x}_c, \boldsymbol{y}_c)$$

and $\boldsymbol{u}_\Gamma = (u_{\Gamma_1}, u_{\Gamma_2}, \ldots, u_{\Gamma_n})^T$, $\boldsymbol{v}_\Gamma = (v_{\Gamma_1}, v_{\Gamma_2}, \ldots, v_{\Gamma_n})^T$. The n impermeability equations are solved for $\boldsymbol{\sigma}(t_k)$ and then the velocities on the control points are calculated from

$$\boldsymbol{u}_c(t_k) = U_\infty + \boldsymbol{A}_u\boldsymbol{\sigma}(t_k) + \boldsymbol{u}_\Gamma(\boldsymbol{x}_c, \boldsymbol{y}_c)),\ \boldsymbol{v}_c(t_0) = \boldsymbol{A}_v\boldsymbol{\sigma}(t_k) + \boldsymbol{v}_\Gamma(\boldsymbol{x}_c, \boldsymbol{y}_c))$$

where $\boldsymbol{u}_c = (u_{c_1}, u_{c_2}, \ldots, u_{c_n})^T$, $\boldsymbol{v}_c = (v_{c_1}, v_{c_2}, \ldots, v_{c_n})^T$. Next, we replace the source panels by n new vortices lying at the control points, such that their positions are

$$x_{w_i}(t_{k-1}) = x_{c_i},\ y_{w_i}(t_{k-1}) = y_{c_i}$$

and we calculate their strengths form Eq. (7.46),

$$\Gamma_i(t_k) = \left(u_{c_i}(t_k)\tau_{x_i} + v_{c_i}(t_k)\tau_{y_i}\right) s_i \tag{7.58}$$

for $i = 1, \dots, n$. *The total flow velocities at the newly created vortices are set to* $u_{w_i}(t_k) = u_{c_i}(t_k)$, $v_{w_i}(t_k) = v_{c_i}(t_k)$ *for* $i = 1, \dots, n$, *while the velocities at the older vortices are calculated from*

$$u_{w_i}(t_k) = \sum_{j=1, j\neq i}^{n+n_w} u_{\Gamma_j}(x_{w_i}(t_{k-1}), y_{w_i}(t_{k-1}))$$

$$v_{w_i}(t_k) = \sum_{j=1, j\neq i}^{n+n_w} v_{\Gamma_j}(x_{w_i}(t_{k-1}), y_{w_i}(t_{k-1}))$$

for $i = n + 1, \dots, n + n_w$. *The elements of the* Γ_i *vector are now given by Eq. (7.48).*

Next, we propagate all vortices using Eq. (7.49) and delete any vortices that have crossed inside the cylinder. We determine which vortices have done so using Matlab function `inpolygon`, *where the polygon is defined by the vertices of the panels,* x_i, y_i. *For each deleted vortex, we determine which panel they crossed by the minimum distance between the deleted vortex and the control points. Then we set up* $\Gamma_{\text{del}_i}(t_k)$, *the strength of the vortices that crossed the ith panel at the kth time instance, which may be equal to zero. Now, we can evaluate Eq. (7.55) in order to calculate the pressure on the ith panel*

$$p_i(t_k) = p_{i-1}(t_k) - \rho\frac{\Delta\Gamma_i(t_k)}{\Delta t} = p_{i-1}(t_k) - \rho\frac{\Gamma_i(t_k) - \Gamma_{\text{del}_i}(t_k)}{\Delta t} \tag{7.59}$$

for $i = 1, \dots, n$. *The starting pressure,* p_0, *can be set to any value since it will drop out in the calculation of the forces, which are circular integrals around the surface. On the other hand, if we want to compare the pressure distribution to experimental or theoretical values, we need to select* p_0 *more carefully. One option is to consider that panels* $n/2$ *and* $n/2 + 1$ *lie on either side of the front stagnation point, where the pressure must be equal to the dynamic pressure* $1/2\rho U_\infty^2 = 1/2$. *If we assume that the unsteady viscous flow around the upstream stagnation point can be adequately predicted by potential theory, then we can start the calculation of the pressure from panel* $n/2 + 1$ *and select* $p_0 = 1/2$. *The calculation of the pressure becomes*

$$p_i(t_k) = p_{i-1}(t_k) - \rho\frac{\Gamma_i(t_k) - \Gamma_{\text{del}_i}(t_k)}{\Delta t}$$

for $i = n/2 + 1, \dots, n$ *with* $p_{n/2} = 1/2$ *and*

$$p_i(t_k) = p_{i+1}(t_k) + 1/2 + \rho\frac{\Gamma_i(t_k) - \Gamma_{\text{del}_i}(t_k)}{\Delta t}$$

for $i = n/2, \dots, 1$ *with* $p_{n/2+1} = 1/2$. *Once the pressure distribution has been evaluated, the lift and drag can be calculated as usual,*

$$l(t_k) = \sum_{i=1}^{n} p_i(t_k)s_i n_{y_i}, \quad d(t_k) = \sum_{i=1}^{n} p_i(t_k)s_i n_{x_i}$$

and the lift and drag coefficients are obtained by dividing $l(t_k)$ *and* $d(t_k)$ *by* $1/2\rho U_\infty^2(2R)$. *In order to limit the cost of the simulation, we also delete any vortices that have moved to a position* $x_w > 24R$ *and therefore induce negligible velocities on the surface of the cylinder.*

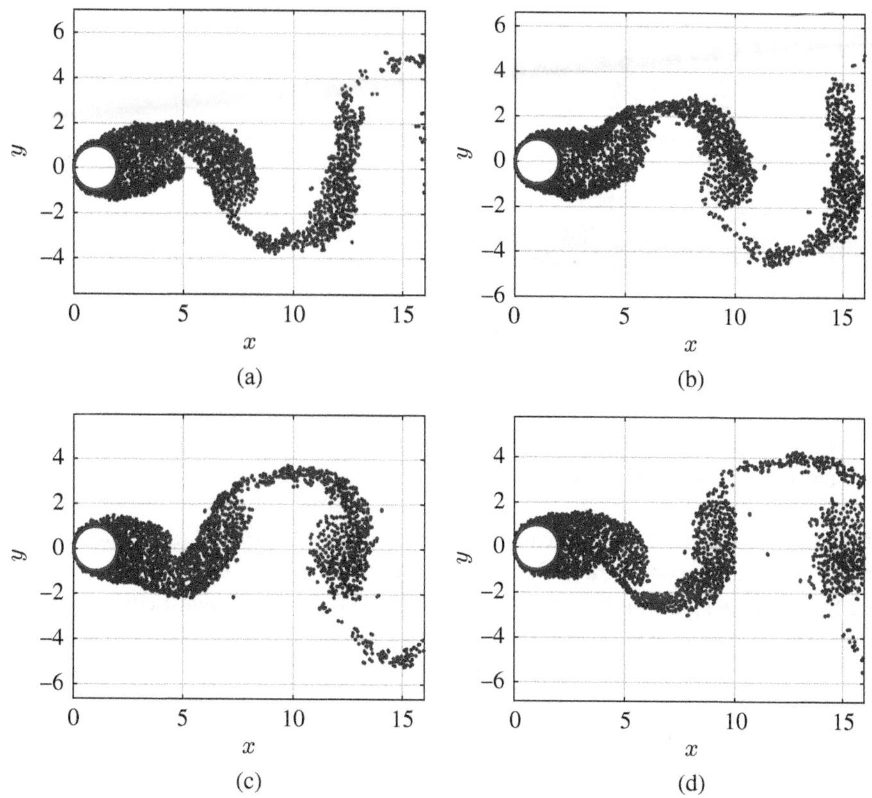

Figure 7.18 Snapshots of the discrete vortex positions at four time instances during the last Strouhal cycle. (a) $\varphi = 0$, (b) $\varphi = 120°$, (c) $\varphi = 228°$ and (d) $\varphi = 348°$.

The simulation is run for 24 shedding cycles, but only the last 20 cycles are used in order to display the computed aerodynamic loads and pressure distributions. Figure 7.18 plots the positions of the discrete vortices around the cylinder at four time instances during the last cycle, corresponding to phase angles of $0°$, $120°$, $228°$ *and* $348°$. *The vortices generated on the panels facing the wind will move nearly tangentially to the surface so that very few vortices lie ahead of the cylinder. This situation represents the attached flow we expect to encounter on the upstream face of the cylinder. Vortices shed from the top, bottom and rear of the cylinder move generally off the surface and propagate downstream, assembling into groups that represent physical vortical structures. Therefore, the DVM shedding process approximates the oscillatory shedding of vortices seen in Figure 7.14.*

Figure 7.19a plots the time response of the lift and drag coefficients. The lift is noisy but has a clear fundamental frequency. The drag coefficient is also very noisy but does not appear to have a fundamental frequency. Figure 7.19b plots the magnitudes of the discrete Fourier transforms of the time signals in Figure 7.19a. The steady-state components of these transforms are the mean values of the time signals; the mean lift is low, $c_l(0) = 0.05$, *while the mean drag is* $c_d(0) = 1.19$. *This mean drag is compatible with the value we would expect for the chosen Reynolds number; Figure 7.15b shows that the mean drag coefficient at* $Re = 10^5$ *should lie*

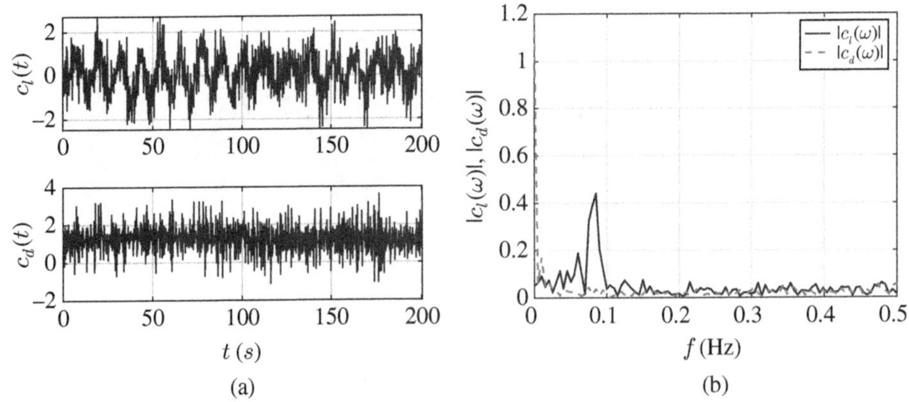

Figure 7.19 Time and frequency response of aerodynamic loads acting on circular cylinder at $Re = 10^5$. (a) Time response and (b) frequency response.

between 1 and 1.2. Note that due to the random walk, the results of the simulation are slightly different every time the code is run. The highest frequency component of the lift signal lies at $f = 0.85$ and its magnitude is 0.44. This means that the lift coefficient oscillates with an amplitude of 0.44 at a frequency of 0.85, which corresponds to a Strouhal number of 0.17. This value is lower than the expected Strouhal number of 0.2.

Figure 7.20 plots the simulated mean pressure coefficient distribution around the upper and lower surfaces and compares it to experimental results from Fage and Falkner (1931) and to the ideal flow distribution (see for example (Anderson Jr. 1985))

$$c_p(\theta) = 1 - 4\sin^2\theta$$

All pressure distributions are plotted such that $\theta = 0$ is the front stagnation point. The experimental data include measurements from different Reynolds numbers; the distributions for

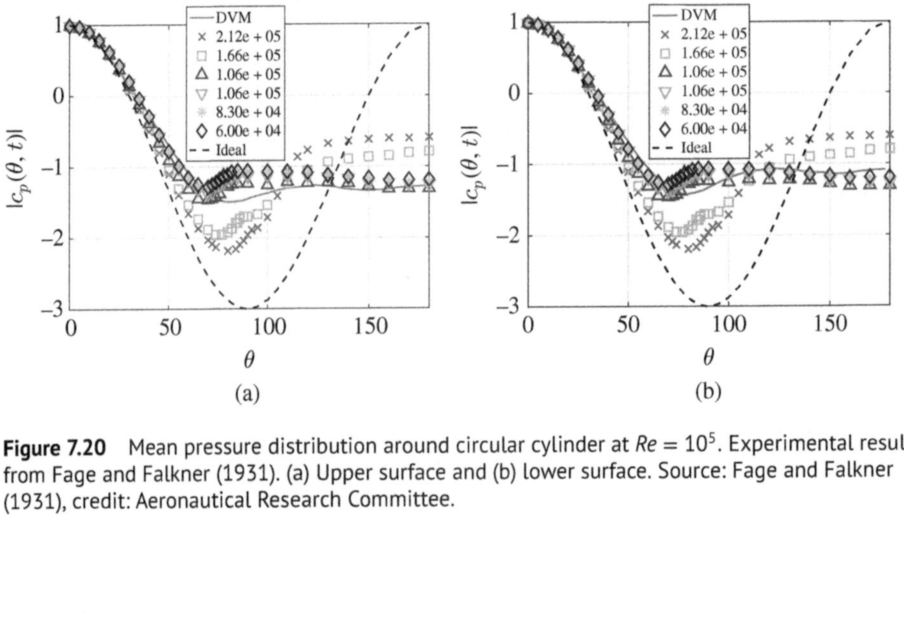

Figure 7.20 Mean pressure distribution around circular cylinder at $Re = 10^5$. Experimental results from Fage and Falkner (1931). (a) Upper surface and (b) lower surface. Source: Fage and Falkner (1931), credit: Aeronautical Research Committee.

$Re = 0.6 - 1.06 \times 10^5$ are very similar to each other. All experimental results lie close to the potential flow distribution for $\theta < 30°$, which means that the boundary layer is very thin in this region and therefore potential flow assumptions can be used to predict the pressure with reasonable accuracy. As the boundary layer thickens, the experimentally measured pressures move away from the potential flow predictions. For Reynolds numbers up to 10^5, boundary layer separation occurs at around $\theta = 70°$ so that the pressure increases first and then becomes nearly constant as θ increases. For the two higher Reynolds numbers, flow separation occurs near $\theta = 80°$ as the boundary layer is turbulent and separates less easily than laminar boundary layers. The mean pressure distribution predicted by the DVM agrees with the experimental measurements for $Re = 0.6 - 1.06 \times 10^5$ but is slightly asymmetric, which we would not expect at this Reynolds number range. It should be noted that the experimental data are only available on one of the surfaces. This example is solved by Matlab code `circ_DVM.m`.

The results presented in the previous example show that Chorin's implementation of the DVM can represent part of the physics of viscous flow around a circular cylinder. However, the Strouhal number is underestimated, and the mean pressure distribution is not symmetric. Changing the time step and number of panels does not improve the shortcomings of the method; decreasing the time step increases the Strouhal frequency to 0.2, but the mean pressure distribution becomes more asymmetric and moves closer to the experimental results for $Re = 1.66 - 2.12 \times 10^5$. The reader is invited to test different values for the time step and number of panel parameters. The main problem with the approach is the fact that Kelvin's theorem is not enforced so that the total circulation in the flow changes in time. In the next section, we will discuss an alternative approach that conserves circulation.

The Reynolds number affects only one parameter of the DVM, the variance of the random walk $\sqrt{2\Delta t/Re}$. This means that changing the Reynolds number only affects the dissipation of the discrete vortices but not their generation. In other words, the DVM cannot model transition in the boundary layer or turbulent re-attachment. As a consequence, the flow in the critical and supercritical Reynolds number ranges cannot be represented accurately; neither laminar separation bubbles, nor delayed boundary layer separation due to turbulence can be modelled by the DVM.

7.4 Flow Past 2D Rectangular Cylinders

Rectangular cylinders are canonical shapes that have been extensively studied in the fluid dynamics literature. Contrary to the circular cylinder, the geometry of a rectangular cylinder changes discontinuously at the corners, forcing local flow separation. Depending on the Reynolds number, face of the rectangle and angle of attack, flow that separates at a corner can remain separated or re-attach. In general, re-attachment occurs when a separated laminar boundary layer becomes turbulent. This behaviour can have a significant effect on the mean and oscillating aerodynamic loads acting on the cylinder. The other major difference between rectangular and circular cylinders is that the former are not axisymmetric and therefore changing their angle of attack alters the flow pattern.

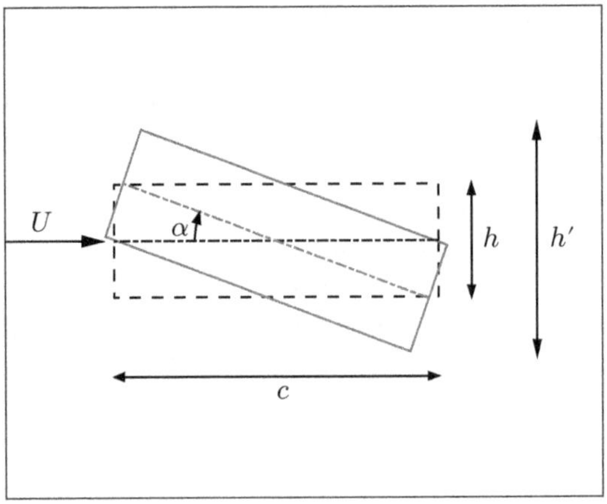

Figure 7.21 Rectangular cylinder of side ratio c/h at angle of attack α.

Figure 7.21 draws a rectangular cylinder of chord c and height h at an angle α to a free stream U_∞. The side ratio of the cylinder is usually defined as c/h, while the height of the inclined model h' is given by

$$h' = c \sin\alpha + h \cos\alpha$$

The study of the available literature must be undertaken with care because sometimes the side ratio is defined inversely, h/c, the aerodynamic load coefficients can be non-dimensionalised with respect to c, h or h' and the Strouhal number can be calculated with respect to h or h'. The Reynolds number is usually based on h.

7.4.1 Modelling the Flow Past Rectangular Cylinders Using the DVM

Chorin's method is based on imposing impermeability by means of a source panel distribution around the body. As this distribution cannot create circulation, it makes it impossible to enforce Kelvin's theorem. A popular alternative is to impose impermeability by means of vortex panels with linearly varying strength on the surface of the body, as demonstrated in Section 4.7. Recall that the strength of the ith vortex panel increases from γ_i at $\mathbf{x}_i$ to γ_{i+1} at $\mathbf{x}_{i+1}$, where $\mathbf{x}_i$ and $\mathbf{x}_{i+1}$ are the two ends of the panel. Then, if the surface is discretised into n panels, there will be $n+1$ unknown strengths to determine by applying impermeability at the n control points. The $n \times (n+1)$ influence coefficient matrix of the vortex panels on the control points, $\boldsymbol{A}_n$, is calculated using Eq. (4.197). At the kth time instance, there are n_w shed vortices so that the impermeability boundary condition becomes by analogy to Eq. (7.57)

$$\boldsymbol{A}_n\boldsymbol{\gamma} = -(U_\infty + \boldsymbol{u}_\Gamma(\boldsymbol{x}_c, \boldsymbol{y}_c)) \circ \boldsymbol{n}_x - \boldsymbol{v}_\Gamma(\boldsymbol{x}_c, \boldsymbol{y}_c) \circ \boldsymbol{n}_y$$

where $\boldsymbol{\gamma} = (\gamma_1, \gamma_2, \dots, \gamma_{n+1})$.

If the body is closed, points 1 and $n+1$ will coincide, and we do not want a discontinuous vortex distribution at this point. Therefore, we must select the vortex strength at $\mathbf{x}_1$ and $\mathbf{x}_{n+1}$ to be equal, $\gamma_{n+1} = \gamma_1$. This means that the impermeability condition reduces to

$$\boldsymbol{A}'_n\boldsymbol{\gamma}' = -(U_\infty + \mathbf{u}_\Gamma(\mathbf{x}_c, \mathbf{y}_c)) \circ \boldsymbol{n}_x - \mathbf{v}_\Gamma(\mathbf{x}_c, \mathbf{y}_c) \circ \boldsymbol{n}_y$$

where $\boldsymbol{\gamma}' = (\gamma_1 \, \gamma_2 \, \dots \gamma_n)$ and the elements of the $n \times n$ matrix $\boldsymbol{A}'_n$ are given by

$$A'_{n_{i,j}} = \begin{cases} A_{n_{i,1}} + A_{n_{i,n+1}} & \text{for } j = 1 \\ A_{n_{i,j}} & \text{for } j > 1 \end{cases}$$

for $i = 1, \dots, n$. Since the body is closed, matrix $\boldsymbol{A}'_n$ has rank $n-1$ because two of its rows are not linearly independent. In essence, as the body is closed and $\gamma_1 = \gamma_{n+1}$, if impermeability is imposed on $n-1$ panels, it will be automatically satisfied on the remaining panel (Lin et al. 1997). Consequently, the impermeability boundary condition is an underdetermined system and an additional equation is required to solve it, which is Kelvin's theorem. The circulation of the ith panel is given by

$$\Gamma_i = \begin{cases} (\gamma_i + \gamma_{i+1})\frac{s_i}{2} & \text{for } 1 \le i < n \\ (\gamma_n + \gamma_1)\frac{s_n}{2} & \text{for } i = n \end{cases} \tag{7.60}$$

Consequently, Kelvin's theorem becomes

$$\sum_{i=1}^{n-1} (\gamma_i + \gamma_{i+1})\frac{s_i}{2} + \frac{\gamma_n s_n}{2} + \frac{\gamma_1 s_n}{2} + \sum_{i=1}^{n_w} \Gamma_i = \Gamma_0 \tag{7.61}$$

where the first two terms one the left-hand side represent the total bound circulation, the third is the total circulation of the n_w vortices already shed into the wake and Γ_0 is the initial circulation in the flow. Adding Kelvin's theorem to the impermeability boundary condition results in

$$\begin{pmatrix} & \boldsymbol{A}'_n & \\ \frac{s_1+s_n}{2} & \dots & \frac{s_{n-1}+s_n}{2} \end{pmatrix} \boldsymbol{\gamma}'(t_k) = \begin{pmatrix} -(U_\infty + \boldsymbol{u}_\Gamma(\boldsymbol{x}_c, \boldsymbol{y}_c)) \circ \boldsymbol{n}_x - \boldsymbol{v}_\Gamma(\boldsymbol{x}_c, \boldsymbol{y}_c) \circ \boldsymbol{n}_y \\ \Gamma_0 - \sum_{i=1}^{n_w} \Gamma_i \end{pmatrix} \tag{7.62}$$

which is a set of $n+1$ equations with n unknowns, the elements of $\boldsymbol{\gamma}'(t_k)$.

Once the strengths $\gamma_i(t_k)$ have been determined, the vorticity of each of the panels is transformed into discrete vortices whose strengths are given by Eq. (7.60) and placed on the control points of the corresponding panels. The rest of the procedure is identical to Chorin's approach. The only other difference is that we are no longer constrained to use Chorin's vortex core model; we can use any other vortex model, such as the Lamb–Oseen or Vatistas models.

Example 7.5 *Calculate the unsteady aerodynamic loads acting on a 2D rectangular cylinder submerged in a steady free stream U_∞ at zero angle of attack for various values of the side ratio and compare to the results presented in Norberg (1993).*

We will simulate the flow around rectangular cylinders with chord $c = 1$ m and various heights, from $h = 5$ m to 0.24 m, which correspond to side ratios from 0.2 to 4, respectively. The experimental data summarized in Norberg (1993) were obtained for Reynolds numbers ranging from 0.5×10^3 to 4×10^4. Here we will select a value of $Re = 2 \times 10^4$. The fluid density is set to $\rho = 1$ kg/m^3 and the wind speed to $U_\infty = 1$ m/s.

The cylinders are discretised into n panels, whose mean length is

$$\bar{s} = (2c + 2h)/n$$

Therefore, the number of panels on the chordwise faces is $n_c = \lfloor c/\bar{s} \rceil$, where the notation $\lfloor\rceil$ denotes rounding to the nearest integer. We space the panels linearly, starting from the midpoint of the leeward face, winding round the bottom face to the midpoint of the windward *face and then round the top face back to the midpoint of the leeward face. In order to achieve this, we split the surface into six segments, with vertices*

$$(c, 0), (c, -h/2), (0, -h/2), (0,0), (0, h/2), (c, h/2), (c, 0)$$

The number of panels on each of these segments is $n_h/2$, n_c, $n_h/2$, $n_h/2$, n_c *and* $n_h/2$, *respectively, where* $n_h/2 = \lfloor h/2\bar{s} \rceil$. *The final number of panels is* $n = 2n_h + 2n_c$, *which can be different to the number specified initially. Figure 7.22 plots a rectangular cylinder with side ratio 0.7 immersed in a free stream of speed* U_∞ *and discretised into* $n = 20$ *panels. The figure details the numbering scheme for the panel vertices and defines the leeward, bottom, windward and top faces.*

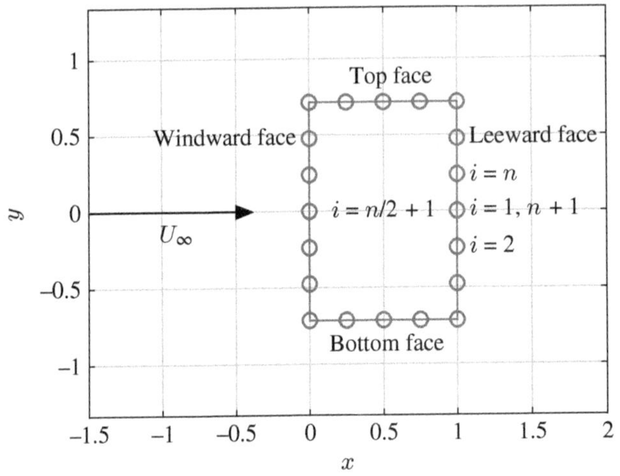

Figure 7.22 Definition of panel vertices on rectangular cylinder.

The coordinates of the control points, x_{c_i}, y_{c_i} *are obtained from Eq. (4.103), while the unit vectors tangent and normal to the panels are given by Eq. (4.104) and (4.105), respectively. We calculate the* $n \times (n+1)$ *influence coefficient matrix of the vortex panels on the control points,* $\mathbf{A}_n$ *using Eq. (4.197) and then we set up the boundary condition of Eq. (7.62) by neglecting one of the first n equations. Alternatively, several authors have chosen to keep all* $n + 1$ *equations for n unknowns and to solve Eq. (7.62) in a least squares sense.*

Once the strength of the vortex panels at the vertices, γ_i, *has been determined, the vortex panels are replaced by point vortices lying on the control points whose strengths are obtained from Eq. (7.60). The velocities induced by the vortices on each other can be calculated using the Chorin, Vatistas or Lamb–Oseen models. Here we select the Vatistas model of Eq. (2.224), such that the influence of the jth discrete vortex on the ith discrete vortex is given by*

$$u_{w_{i,j}} = \frac{\Gamma_j}{2\pi} \frac{(y_{w_i} - y_{w_j})}{(r_c^4 + r_{ij}^4)^{1/2}}$$

$$v_{w_{i,j}} = -\frac{\Gamma_j}{2\pi} \frac{(x_{w_i} - x_{w_j})}{(r_c^4 + r_{ij}^4)^{1/2}}$$

where $r_{ij} = \sqrt{(x_{w_i} - x_{w_j})^2 + (y_{w_i} - y_{w_j})^2}$ *and the vortex core radius is chosen as* $r_c = \bar{s}/2\pi$. *The wake is then propagated using Eq. (7.49) and any vortices that find themselves inside the body after the propagation are deleted, recording the strength of the vortices crossing the ith panel* Γ_{del_i}.

At the first time instance, there are no shed vortices, so that by Kelvin's theorem, the sum of the bound circulation is equal to Γ_0. *At the second time instance, n vortices with total circulation* Γ_0 *have been shed but some of them, say* n_{del}, *have been deleted. This means that, at the second time instance, Eq. (7.61) becomes*

$$\sum_{i=1}^{n-1} (\gamma_i + \gamma_{i+1}) \frac{s_i}{2} + \frac{\gamma_n s_n}{2} + \frac{\gamma_1 s_n}{2} + \Gamma_0 - \Gamma_{\mathrm{del}} = \Gamma_0$$

where $\Gamma_{\mathrm{del}} = \sum_{i=1}^{n_{\mathrm{del}}} \Gamma_{\mathrm{del}_i}$. *Consequently, at any general time instance, k, the sum of the bound circulation must be equal to the sum of the circulation that was deleted at the previous time instance,*

$$\sum_{i=1}^{n-1} (\gamma_i(t_k) + \gamma_{i+1}(t_k)) \frac{s_i}{2} + \frac{\gamma_n\left(t_k\right) s_n}{2} + \frac{\gamma_1\left(t_k\right) s_n}{2} = \Gamma_{\mathrm{del}}(t_{k-1})$$

Equation (7.62) becomes

$$\begin{pmatrix} & \boldsymbol{A}'_n & \\ \frac{s_1+s_n}{2} & \dots & \frac{s_{n-1}+s_n}{2} \end{pmatrix} \boldsymbol{\gamma}'(t_k) = \begin{pmatrix} -(U_\infty + \boldsymbol{u}_\Gamma(\boldsymbol{x}_c, \boldsymbol{y}_c, t_k)) \circ \boldsymbol{n}_x - \boldsymbol{v}_\Gamma(\boldsymbol{x}_c, \boldsymbol{y}_c, t_k) \circ \mathbf{n}_y \\ \Gamma_{\mathrm{del}}(t_{k-1}) \end{pmatrix} \tag{7.63}$$

The pressure on the ith panel is calculated from Eq. (7.59). As circulation is now conserved, we can also calculate the aerodynamic loads using the vorticity-momentum theorem by adapting Eqs. (7.15) and (7.16) to the current case, after the latest vortices have been shed and propagated, that is

$$\alpha_x(t_k) = \sum_{j=1}^{n_w(t_k)} y_{w_j}(t_k)\Gamma_j \tag{7.64}$$

$$\alpha_y(t_k) = -\sum_{j=1}^{n_w(t_k)} x_{w_j}(t_k)\Gamma_j \tag{7.65}$$

The lift and drag forces are then obtained from Eq. (7.17).

We simulate the flow around rectangles with side ratios from 0.2 to 4 using the same simulation parameters. The initial circulation is set to $\Gamma_0 = 0$. *The number of panels is set to* $n = 200$, *the number of time instances per shedding cycle to 60 and the shedding frequency to* $0.15U_\infty/h$, *where* 0.15 *is the expected Strouhal number for a rectangle with side ratio 4. The simulations are run for 12 shedding cycles but only the last six cycles are used for calculating flow quantities. The discrete Fourier transform is applied to the lift signals in order to identify the most important spectral components and the corresponding frequencies. For side ratios between 2.5 and 3.5, there are two important frequency components, resulting in two Strouhal numbers.*

Figure 7.23a plots the variation of Strouhal number with side ratio and compares them to the experimental values presented in Norberg (1993). It can be seen that the DVM results are in good agreement with the experimental data, except for side ratios less than 1, where

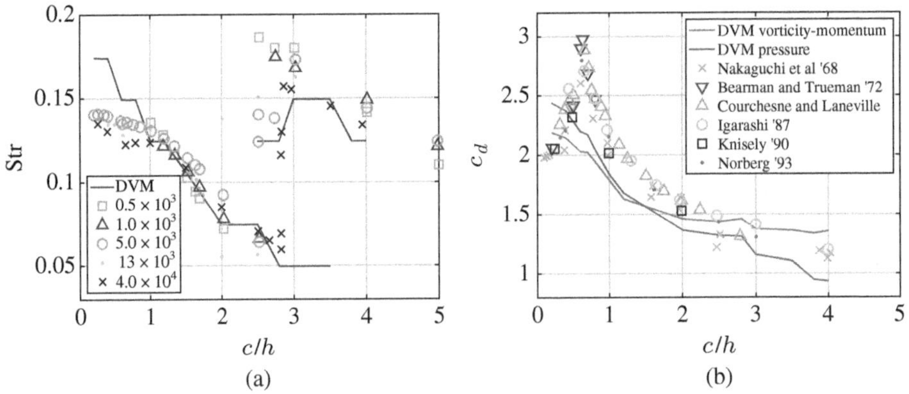

Figure 7.23 Strouhal number and mean drag coefficient of flow around rectangular cylinders of different side ratios. Experimental results from Norberg (1993). (a) *Str* and (b) c_d. Source: Norberg (1993)/with permission from Elsevier.

the Strouhal number is overestimated. In particular, the numerical results have replicated the discontinuous rise in Str between side ratios of 2.5 and 3. Experimental evidence shows that the vortex shedding becomes irregular in the $c/h = 2 - 3$ *range, and this phenomenon is captured by the DVM. This irregular shedding occurs because as the side ratio increases, flow that separates at the leading edge corners tends to reattach downstream on the bottom and top surfaces (Okajima 1982). For* $c/h = 2 - 3$*, re-attachment may or may not occur, depending on the value of the Reynolds number. For* $c/h > 3$ *separated flow always reattaches, such that the Reynolds-dependence of the Strouhal number is limited. Nakamura et al. (1991) show that such abrupt increases in Strouhal number also occur at* c/h *=6, 9 and 12.*

Figure 7.23b plots the variation of the mean drag coefficient with c/h*, comparing the DVM results to the experimental data from Norberg (1993). Drag coefficients obtained from both the vorticity-momentum equation and the pressure calculation are plotted. Both sets of results are underestimated, although the pressure-based* c_d *values are in reasonable agreement with the measurements by Nakaguchi et al. (1968), which were obtained for* $Re = 4 \times 10^4$*. Crucially, the simulated results completely miss the very high peak in mean drag coefficient occurring at* $c/h = 0.6$*. The drag is caused mainly by the difference in pressure between the windward and leeward sides. As the flow is attached to the leeward side, the pressure there is nearly independent of the side ratio, which means that variations in mean drag are due to variations in the pressure on the leeward side. At* $c/h = 0.62$*, the mean pressure on the leeward side is very low, causing a significant increase in drag. This high suction is probably caused by the formation of very strong vortices close to the body that induce high flow velocities tangential to the leeward face (Bearman and Trueman 1972). The fact that the DVM does not predict the high drag peak at* $c/h = 0.6$ *may mean that the strength of the vortices is underestimated or that they form too far from the body. This example is solved by Matlab code* `rect_vpDVM.m`.

It should be noted that the simulations in this latest example are very computationally expensive compared to any of the simulations presented so far in this book. The DVM works best when the vortex density is significant; at every time instance, 200 vortices are shed, and there are 720 time instances, resulting in a very high number of vortices towards the end of the simulation, despite the deletions. The most computationally expensive operation is the calculation of the velocities induced by the vortices on each other. Several methods can be used to decrease the cost of this calculation, such as the merging of faraway vortices and the application of the Fast Multipole Method (see for example (Rokhlin 1985)). The implementation of such approaches is beyond the scope of the present book.

7.5 Concluding Remarks

The methods presented in this chapter are the most heuristic and empirical of the entire book because potential flow methods cannot predict viscous phenomena by definition. Consequently, an array of empirical tools have been developed in order to adapt panel methods to the calculation of separated steady and unsteady flows. It should be noted that the separated flow examples given in this chapter work because the simulation strategy and parameters have been carefully fine-tuned for each application. For example, increasing the number of panels will not necessarily lead to better agreement with the experiment, but it may in fact decrease the accuracy of the simulation. It is generally hard to obtain robust panel-based methods for the calculation of separated flows; although a lot of research effort has been devoted to such techniques, they are seldom used in industrial practice, precisely because they require fine-tuning. Nevertheless, the research continues because the computational cost of potential flow techniques is significantly lower than that of RANS simulations.

7.6 Exercises

1 Repeat Example 7.2 using Eq. (7.22) to calculate the strength of the vortices shed from the separation point. Use the standard lumped vortex discretization of Example 4.4 and set to position of the separation point to $0.05c$, as suggested in Katz (1981).

2 Calculate the flow around the circular cylinder of Example 7.4 using the vortex panel method of Example 7.5.

References

Anderson, J.D. Jr. (1985). *Fundamentals of Aerodynamics*. McGraw-Hill International Editions.

Ansari, S.A., Zbikowski, R., and Knowles, K. (2006). Non-linear unsteady aerodynamic model for insect-like flapping wings in the hover. Part 1: Methodology and analysis. *Proceedings of the IMechE Part G: Journal of Aerospace Engineering* 220: 61–83.

Bearman, P.W. and Trueman, D.M. (1972). An investigation of the flow around rectangular cylinders. *Aeronautical Quarterly* 23: 229–237.

Beddoes, T.S. (1976). A synthesis of unsteady aerodynamic effects including stall hysteresis. *Vertica* 1: 113–123.

Benidir, A., Flamand, O., and Dimitriadis, G. (2018). The impact of circularity defects on bridge stay cable dry galloping stability. *Journal of Wind Engineering and Industrial Aerodynamics* 181: 14–26.

Boutet, J., Dimitriadis, G., and Amandolese, X. (2020). A modified Leishman–Beddoes model for airfoil sections undergoing dynamic stall at low Reynolds numbers. *Journal of Fluids and Structures* 93: 102852.

Cebeci, T. (1999). *An Engineering Approach to the Calculation of Aerodynamic Flows.* Springer-Verlag.

Cebeci, T., Platzer, M., Chen, H. et al. (2005). *Analysis of Low-Speed Unsteady Airfoil Flows.* Springer-Verlag.

Chabalko, C.C., Snyder, R.D., Beran, P.S., and Parker, G.H. (2009). The physics of an optimized flapping wing micro air vehicle. *Proceedings of the 47th AIAA Aerospace Sciences Meeting* number AIAA 2009-801, Orlando, FL, USA.

Chorin, A.J. (1973). Numerical study of slightly viscous flow. *Journal of Fluid Mechanics* 57 (4): 785–796.

Delany, N.K. and Sorensen, N.E. (1953). Low-speed drag of cylinders of various shapes. Technical Note TN-3038. NACA.

Ericsson, L.E. and Reding, J.P. (1984). Unsteady flow concepts for dynamic stall analysis. *Journal of Aircraft* 21 (8): 601–606.

Fage, A. and Falkner, V.M. (1931). Further Experiments on the Flow Around a Circular Cylinder. *Reports and Memoranda R. & M. No. 1369*. Aeronautical Research Committee.

Hammer, P., Altman, A., and Eastep, F. (2014). Validation of a discrete vortex method for low Reynolds number unsteady flows. *AIAA Journal* 52 (3): 643–649.

Katz, J. (1981). A discrete vortex method for the non-steady separated flow over an airfoil. *Journal of Fluid Mechanics* 102: 315–328.

Katz, J. and Plotkin, A. (2001). *Low Speed Aerodynamics*. Cambridge University Press.

Leishman, J.G. (1988). Validation of approximate indicial aerodynamic functions for two-dimensional subsonic flow. *Journal of Aircraft* 25 (10): 914–922.

Leishman, J.G. and Beddoes, T.S. (1989). A semi-empirical model for dynamic stall. *Journal of the American Helicopter Society* 34 (3): 3–17.

Lin, H., Vezza, M., and Galbraith, R.A.M.D. (1997). Discrete vortex method for simulating unsteady flow around pitching aerofoils. *AIAA Journal* 35 (3): 494–499.

Manar, F. and Jones, A.R. (2019). Evaluation of potential flow models for unsteady separated flow with respect to experimental data. *Physical Review Fluids* 4: 034702.

McAlister, K.W., Pucci, S.L., McCroskey, W.J., and Carr, L.W. (1982). An experimental study of dynamic stall on advanced airfoil sections, Volume 2. Pressure and force data. Technical Memorandum TM 84245. NASA.

McCormick, B.W. (1979). *Aerodynamics, Aeronautics and Flight Mechanics*. Wiley.

McCroskey, W.J. (1973). Inviscid flowfield of an unsteady airfoil. *AIAA Journal* 11 (8): 1130–1137.

McCullough, G.B. and Gault, D.E. (1951). Examples of three representative types of airfoil-section stall at low speed. Technical Note TN-2502. NACA.

Nakaguchi, H., Hashimoto, K., and Muto, S. (1968). An experimental study on aerodynamic drag of rectangular cylinders. *Journal of the Japan Society for Aeronautical and Space Sciences* 16: 1–5.

Nakamura, Y., Ohya, Y., and Tsuruta, H. (1991). Experiments on vortex shedding from flat plates with square leading and trailing edges. *Journal of Fluid Mechanics* 222: 437–447.

Norberg, C. (1993). Flow around rectangular cylinders: pressure forces and wake frequencies. *Journal of Wind Engineering and Industrial Aerodynamics* 49 (1–3): 187–196.

Okajima, A. (1982). Strouhal numbers of rectangular cylinders. *Journal of Fluid Mechanics* 123: 379–398.

Prandtl, L. (2001). On the motion of fluids with very little friction. In: *Early Developments of Modern Aerodynamics* (ed. J.A.D. Ackroyd, B.P. Axel, and A.I. Ruban). Auckland, Boston, MA, Johannesburg, Melbourne, New Delhi: Butterworh-Heinemann Oxford.

Ramesh, K., Gopalarathnam, A., Granlund, K. et al. (2014). An inviscid model of two-dimensional vortex shedding for transient and asymptotically steady separated flow over an inclined plate. *Journal of Fluid Mechanics* 751: 500–538.

Riziotis, V.A. and Voutsinas, S.G. (2008). Dynamic stall modelling on airfoils based on strong viscous-inviscid interaction coupling. *International Journal for Numerical methods in Fluids* 56: 185–208.

Robertson, D.K., Joo, J.J., and Reich, G.W. (2010). Vortex particle aerodynamic modelling of perching manoeuvres with micro air vehicles. *Proceedings of the 51st AIAA/ASME/ASCE/AHS/ASC Structures, Structural Dynamics, and Materials Conference* number AIAA2010-2825, Orlando, FL, USA.

Rokhlin, V. (1985). Rapid solution of integral equations of classical potential theory. *Journal of Computational Physics* 60 (2): 187–207.

Saini, A., Narsipur, S., and Gopalarathnam, A. (2021). Leading-edge flow sensing for detection of vortex shedding from airfoils in unsteady flows. *Physics of Fluids* 33: 087105.

Sarpkaya, T. (1975). An inviscid model of two-dimensional vortex shedding for transient and asymptotically steady separated flow over an inclined plate. *Journal of Fluid Mechanics* 68 (1): 109–128.

Sheng, W., Galbraith, R.A.M., and Coton, F.N. (2006). A new stall-onset criterion for low speed dynamic-stall. *Journal of Solar Energy Engineering* 128 (4): 461–471.

Zdravkovich, M.M. (1997). *Flow Around Circular Cylinders: Fundamentals*, vol. 1. Oxford University Press.

Zdravkovich, M.M. and Bearman, P.W. (1998). Flow around circular cylinders – volume 1: fundamentals. *ASME Journal of Fluids Engineering* 120: 216.

A

Fundamental Solutions of Laplace's Equation

This appendix presents some of the fundamental solutions of the two-dimensional Laplace equation (2.22). Only the solutions used in this book are discussed.

A.1 The 2D Point Source

The complex potential induced at a general point $z = x + iy$ by a source of strength σ lying on the origin is given by

$$F(z) = \frac{\sigma}{2\pi} \ln z \tag{A.1}$$

The complex velocity induced by the same source is

$$V(z) = \frac{dF}{dz} = \frac{\sigma}{2\pi} \frac{1}{z} \tag{A.2}$$

The real potential and the stream function are the real and imaginary parts of the complex potential, i.e. $F = \phi + i\psi$. Writing z in polar form as $z = re^{i\theta}$ and using the definition of the complex logarithm,

$$\phi = \frac{\sigma}{2\pi} \ln r = \frac{\sigma}{2\pi} \ln \sqrt{x^2 + y^2} \tag{A.3}$$

$$\psi = \frac{\sigma}{2\pi} (\theta + 2\pi k) = \frac{\sigma}{2\pi} \left(\tan^{-1} \frac{y}{x} + 2\pi k \right) \tag{A.4}$$

for any integer k. In practice, the potential and stream functions can be calculated either from the real and imaginary parts of Eq. (A.1) or from Eqs. (A.3) and (A.4).

Figure A.1a plots the real potential $\phi(x, y) = \Re(F(x, y))$ induced by a source of strength $\sigma = 1$ in cartesian coordinates. The equipotential lines are a set of concentric circles centred around the origin, while the potential itself increases with distance from the origin. Figure A.1b plots the stream function $\psi(x, y) = \Im(F(x, y))$ of the same flow. The streamlines are a set of semi-infinite straight lines extending from the origin in all directions. Note that the stream function is discontinuous across $y = 0$ for $x < 0$ ($\theta = \pi$ in polar coordinates) because Matlab calculates the complex logarithm in the same way as it does the four-quadrant inverse tangent function. We will discuss this issue in more detail in Section A2.

Unsteady Aerodynamics: Potential and Vortex Methods, First Edition. Grigorios Dimitriadis.

Companion website: www.wiley.com/go/dimitriadis/unsteady_aerodynamics

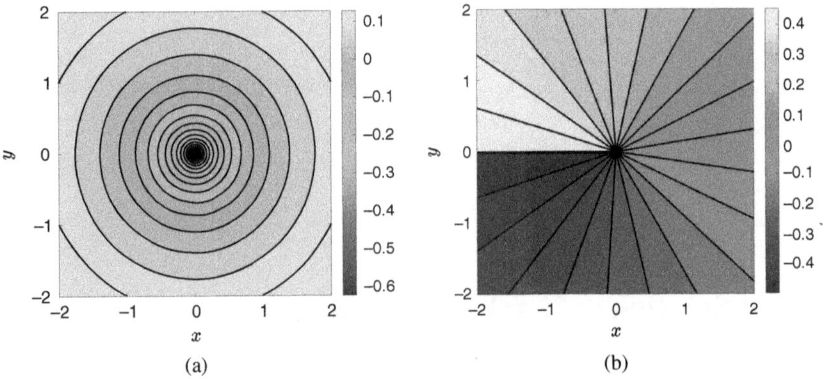

Figure A.1 Real potential and stream function induced by a source of strength $\sigma = 1$. (a) $\phi(x, y)$ and (b) $\psi(x, y)$.

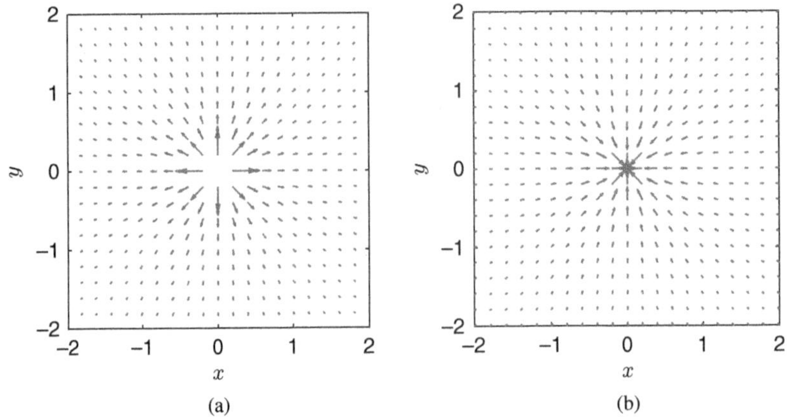

Figure A.2 Velocity field induced by a source and a sink. (a) $\sigma = 1$ and $\sigma = -1$.

Figure A.2a,b plot the velocity fields $u(x, y) = \Re(V(x, y))$, $v(x, y) = -\Im(V(x, y))$ for a source with strength $\sigma = 1$ and $\sigma = -1$, respectively. The source of Figure A.2a induces flow that moves from the origin outwards in all directions. In contrast, the source with negative strength (usually referred to as a sink) of Figure A.2b induces flow that moves towards the origin from all directions.

Finally, the complex potential and velocity induced by a source that lies on a general point z_0 instead of the origin are given by

$$F(z) = \frac{\sigma}{2\pi} \ln(z - z_0) \tag{A.5}$$

$$V(z) = \frac{\sigma}{2\pi} \frac{1}{z - z_0} \tag{A.6}$$

The associated real potential and velocities are

$$\phi(x, y) = \frac{\sigma}{2\pi} \ln \sqrt{(x - x_0)^2 + (y - y_0)^2} \tag{A.7}$$

$$u(x,y) = \frac{\sigma}{2\pi} \frac{x - x_0}{(x - x_0)^2 + (y - y_0)^2} \tag{A.8}$$

$$v(x,y) = \frac{\sigma}{2\pi} \frac{y - y_0}{(x - x_0)^2 + (y - y_0)^2} \tag{A.9}$$

A.2 The 2D Point Vortex

The complex potential induced at a general point $z = x + iy$ by a vortex of clockwise strength Γ lying on the origin is given by

$$F(z) = i\frac{\Gamma}{2\pi} \ln z \tag{A.10}$$

The complex velocity induced by the same vortex is given by

$$V(z) = \frac{dF}{dz} = i\frac{\Gamma}{2\pi}\frac{1}{z} \tag{A.11}$$

As in the case of the point source, writing z in polar form as $z = re^{i\theta}$ and using the definition of the complex logarithm, we obtain the real potential and stream function as

$$\phi = -\frac{\Gamma}{2\pi}(\theta + 2\pi k) = -\frac{\Gamma}{2\pi}\left(\tan^{-1}\frac{y}{x} + 2\pi k\right) \tag{A.12}$$

$$\psi = \frac{\Gamma}{2\pi} \ln r = \frac{\Gamma}{2\pi} \ln \sqrt{x^2 + y^2} \tag{A.13}$$

for any integer k.

Figure A.3 plots the real potential $\phi(x,y) = \Re(F(x,y))$ and stream function $\psi(x,y) = \Im(F(x,y))$ induced by a vortex of strength $\Gamma = 1$. In fact, the potential and stream function of the vortex correspond to the negative of the stream function and to the potential of the source, respectively. Hence, the streamlines are concentric circles centred around the origin, while the equipotential lines are semi-infinite straight lines extending from the origin in all directions. As in the case of the source's stream function, there is a discontinuity

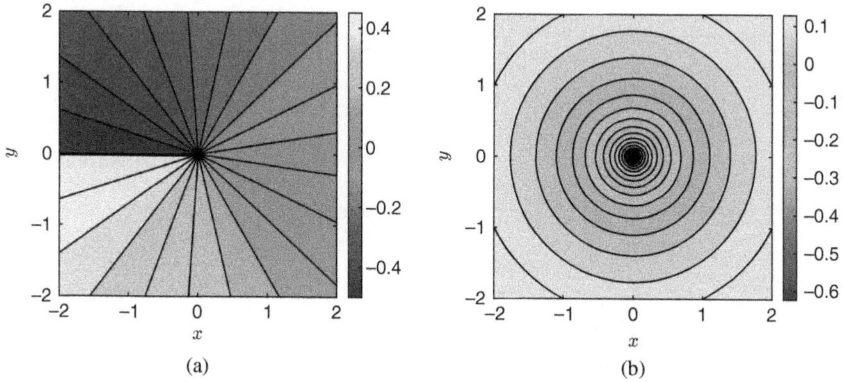

Figure A.3 Real potential and stream function induced by a vortex of strength $\Gamma = 1$. (a) $\phi(x,y)$ and (b) $\psi(x,y)$.

across $y = 0$ for $x < 0$ or $\theta = \pi$. Again, this discontinuity is caused by the definition of the four-quadrant inverse tangent function, `atan2(y,x)`, given by

$$\text{atan2}(y,x) = \begin{cases} \tan^{-1}(y/x) & \text{if } x > 0 \\ \tan^{-1}(y/x) + \pi & \text{if } x < 0 \text{ and } y \geq 0 \\ \tan^{-1}(y/x) - \pi & \text{if } x < 0 \text{ and } y < 0 \\ \pi/2 & \text{if } x = 0 \text{ and } y > 0 \\ -\pi/2 & \text{if } x = 0 \text{ and } y < 0 \end{cases} \tag{A.14}$$

so that the function takes values between $-\pi$ and π. Clearly, for $x < 0$, the atan2(y, x) function jumps from π to $-\pi$ as y crosses zero from positive to negative and so does the value of the potential induced by a point vortex. This discontinuity is not evident in the polar form of Eq. (A.12) since θ increases smoothly from 0 to 2π as we go around the four quadrants. However, the discontinuity is displaced to $\theta = 0$, where θ jumps down from 2π to 0 as we go around for another revolution. Consequently, there will always be a discontinuity in the value of the potential induced by a point vortex, if this potential is to remain single-valued.

The discontinuity in the potential induced by a vortex is not crucial, unless we want to calculate the induced velocity v across the jump by differentiating numerically the potential. Thankfully, the equation for the velocity (A.11) does not feature a discontinuity. Numerical methods that use vortices to solve for a flowfield generally do not require the explicit calculation of the potential.

Figures A.4a,b plot the velocity fields $u(x, y) = \Re(V(x, y))$, $v(x, y) = -\Im(V(x, y))$ for a vortex with strength $\Gamma = 1$ and $\Gamma = -1$, respectively. Both flows are rotating around the origin, with the one corresponding to $\Gamma = 1$ rotating in a clockwise direction, while the one corresponding to $\Gamma = -1$ rotates in a counter-clockwise direction.

Finally, the complex potential and velocity induced by a 2D point vortex that lies on a general point z_0 instead of the origin are given by

$$F(z) = \mathrm{i}\frac{\Gamma}{2\pi}\ln(z - z_0) \tag{A.15}$$

$$V(z) = \mathrm{i}\frac{\Gamma}{2\pi}\frac{1}{z - z_0} \tag{A.16}$$

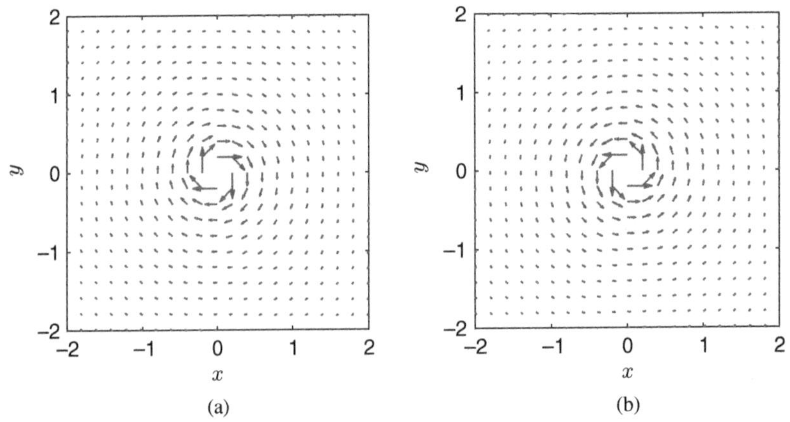

Figure A.4 Velocity field induced by vortices of opposite strength. (a) $\Gamma = 1$ and (b) $\Gamma = -1$.

and the corresponding real potential and velocities are

$$\phi(x,y) = -\frac{\Gamma}{2\pi}\tan^{-1}\frac{y-y_0}{x-x_0} \tag{A.17}$$

$$u(x,y) = \frac{\Gamma}{2\pi}\frac{y-y_0}{(x-x_0)^2+(y-y_0)^2} \tag{A.18}$$

$$v(x,y) = -\frac{\Gamma}{2\pi}\frac{x-x_0}{(x-x_0)^2+(y-y_0)^2} \tag{A.19}$$

A.3 The Source Line Panel

The source line panel is a straight or curved line segment on which we place a continuous distribution of sources. In this book, we will only consider straight panels. Consider the horizontal straight-line segment of Figure A.5, lying on the x axis and extending from x_1 to x_2, with point sources distributed along its length. For clarity, only 10 sources are plotted in the figure, but the source distribution is in fact continuous. The strength of the source distribution, $\sigma(x)\mathrm{d}x$, can be constant or can vary along the x direction.

Using Eq. (A.7) for the real potential induced by a single source, the potential induced at a general point x, y by the entire panel is given simply by

$$\phi(x,y) = \frac{1}{4\pi}\int_{x_1}^{x_2}\sigma(x_0)\ln\left((x-x_0)^2+y^2\right)\mathrm{d}x_0 \tag{A.20}$$

noting that all the sources lie on $y_0 = 0$. For constant source strength, $\sigma(x_0) = \sigma_0$, the integral can be evaluated using, for example Matlab's symbolic toolbox, such that

$$\int \ln\left((x-x_0)^2+y^2\right)\mathrm{d}x_0 = -2x_0 - 2y\tan^{-1}\left(\frac{x-x_0}{y}\right) - (x-x_0)\ln\left((x-x_0)^2+y^2\right)$$

Applying the limits of integration, we obtain

$$\begin{aligned}\phi(x,y) = \frac{\sigma_0}{4\pi}\Bigg(&2(x_1-x_2)+2y\left(\tan^{-1}\left(\frac{x-x_1}{y}\right)-\tan^{-1}\left(\frac{x-x_2}{y}\right)\right)\\ &+(x-x_1)\ln\left((x-x_1)^2+y^2\right)-(x-x_2)\ln\left((x-x_2)^2+y^2\right)\Bigg)\end{aligned} \tag{A.21}$$

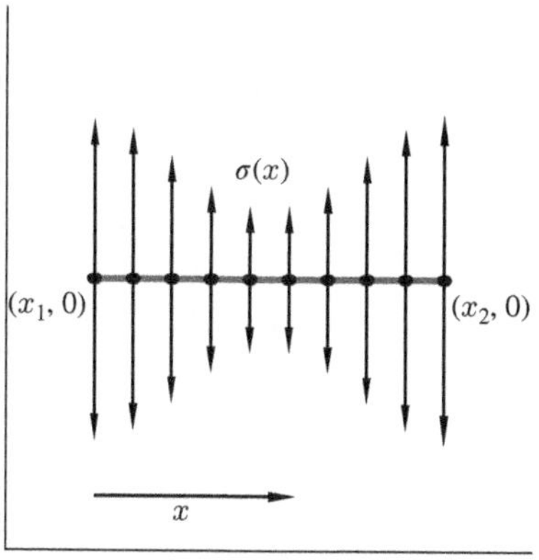

Figure A.5 The 1D source panel.

Note that, in the $x-y$ plane, the terms $\tan^{-1}((x-x_1)/y)$ and $\tan^{-1}((x-x_2)/y)$ are not the angles between the lines connecting point x, y and the vertices of the panel. These angles are given by $\tan^{-1}(y/(x-x_1))$ and $\tan^{-1}(y/(x-x_2))$. We can use the properties of the inverse tangent to show that

$$\tan^{-1} a - \tan^{-1} b = \tan^{-1}\frac{1}{b} - \tan^{-1}\frac{1}{a} \tag{A.22}$$

Applying this identity to expression (A.21), we obtain

$$\begin{aligned}\phi(x,y) = \frac{\sigma_0}{4\pi}\Bigg(2(x_1-x_2) + 2y\left(\tan^{-1}\left(\frac{y}{x-x_2}\right) - \tan^{-1}\left(\frac{y}{x-x_1}\right)\right)\\ + (x-x_1)\ln\left(\left(x-x_1\right)^2+y^2\right) - (x-x_2)\ln\left(\left(x-x_2\right)^2+y^2\right)\Bigg)\end{aligned} \tag{A.23}$$

The advantage of Eq. (A.23) over Eq. (A.21) is that the former allows the use of the four-quadrant inverse tangent function for numerical calculations.

From the definition of the potential, the horizontal velocity $u(x,y)$ induced by the source panel is given by

$$\begin{aligned}u(x,y) = \frac{\partial\phi}{\partial x} &= \frac{1}{4\pi}\frac{\partial}{\partial x}\int_{x_1}^{x_2}\sigma(x_0)\ln\left(\left(x-x_0\right)^2+y^2\right)\mathrm{d}x_0\\ &= \frac{1}{2\pi}\int_{x_1}^{x_2}\sigma(x_0)\frac{x-x_0}{\left(x-x_0\right)^2+y^2}\mathrm{d}x_0\end{aligned}$$

Again, for constant source strength, the integral in this latest expression can be evaluated using the standard integral

$$\int\frac{x}{ax^2+bx+c}\mathrm{d}x = \frac{1}{2a}\ln\left|ax^2+bx+c\right| - \frac{b}{a\sqrt{4ac-b^2}}\tan^{-1}\frac{2ax+b}{\sqrt{4ac-b^2}}$$

so that

$$u(x,y) = \frac{\sigma_0}{4\pi}\ln\left(\frac{(x-x_1)^2+y^2}{(x-x_2)^2+y^2}\right) \tag{A.24}$$

At the panel's midpoint, the horizontal velocity is obtained by setting $x = (x_1+x_2)/2, y = 0$ into Eq. (A.24), such that

$$u((x_1+x_2)/2, 0) = 0$$

At the two tips $u(x_{1,2}, 0) = \pm\infty$. This result is logical since the two tips behave as point sources in the outboard direction and therefore the velocity induced at a general point by each one of the sources tends to infinity as the point approaches the sources.

The vertical airspeed induced by the source panel at a general point x, y is

$$\begin{aligned}v(x,y) = \frac{\partial\phi}{\partial y} &= \frac{1}{4\pi}\frac{\partial}{\partial y}\int_{x_1}^{x_2}\sigma(x_0)\ln\left(\left(x-x_0\right)^2+y^2\right)\mathrm{d}x_0\\ &= \frac{1}{2\pi}y\int_{x_1}^{x_2}\frac{\sigma(x_0)}{\left(x-x_0\right)^2+y^2}\mathrm{d}x_0\end{aligned} \tag{A.25}$$

The vertical velocity induced by the panel on itself is obtained when $y = 0$. As the integral in this equation is multiplied by y on the numerator, this velocity is zero everywhere except at $x = x_0$, where it is undefined. In order to determine $v(x, 0)$, we can carry out the integral over a small region $x_0 - x = \varepsilon_0$ where $\varepsilon_0 \in [-\epsilon, \epsilon]$ and $\varepsilon \ll 1$ and then take the limit of the result as both y and ε tend to zero,

$$\lim_{y\to 0} v(x,y) = \frac{1}{2\pi} \lim_{y,\varepsilon\to 0} y \int_{-\varepsilon}^{\varepsilon} \frac{\sigma(x+\epsilon_0)}{\varepsilon_0^2 + y^2} d\varepsilon_0$$

We assume that $\sigma(x + \epsilon_0)$ is a smooth function whose variation in the region $x - \varepsilon$ to $x + \varepsilon$ is very small, so that $\sigma(x - \varepsilon) \approx \sigma(x + \varepsilon) \approx \sigma(x)$ is approximately constant in that region. We obtain

$$\lim_{y\to 0} v(x,y) = \frac{\sigma(x)}{2\pi} \lim_{y,\varepsilon\to 0} \int_{-\varepsilon}^{\varepsilon} \frac{y}{\varepsilon_0^2 + y^2} d\varepsilon_0$$

This is a standard integral, whose result is $\tan^{-1}(\varepsilon_0/y)$, so that by applying the limits, we obtain

$$\lim_{y\to 0} v(x,y) = \frac{\sigma(x)}{\pi} \lim_{y,\varepsilon\to 0} \left(\tan^{-1}\left(\frac{\varepsilon}{y}\right)\right)$$

We now let y and ε tend to zero in order to calculate $v(x, 0)$. However, we allow y to tend to zero faster than ε, such that the argument of the inverse tangent becomes infinite. Furthermore, if y is negative as it tends to 0^-, then $\tan^{-1} = -\pi/2$, while if y is positive as it tends to 0^+, $\tan^{-1} = +\pi/2$. Consequently,

$$v(x, 0^{\pm}) = \pm\frac{\sigma(x)}{2} \tag{A.26}$$

Therefore, a source panel induces everywhere on itself a normal velocity with magnitude equal to half the local strength.

If the source strength is constant over the entire panel, $\sigma(x_0) = \sigma_0$, expression (A.25) can be evaluated using the standard integral

$$\int \frac{1}{ax^2 + bx + c} dx = \frac{2}{\sqrt{4ac - b^2}} \tan^{-1} \frac{2ax + b}{\sqrt{4ac - b^2}}$$

The result is

$$v(x,y) = \frac{\sigma_0}{2\pi}\left(\tan^{-1}\left(\frac{x - x_1}{y}\right) - \tan^{-1}\left(\frac{x - x_2}{y}\right)\right) \tag{A.27}$$

Again, we can obtain a more practical expression for numerical calculations by applying the identity of Eq. (A.22), such that

$$v(x,y) = \frac{\sigma_0}{2\pi}\left(\tan^{-1}\left(\frac{y}{x - x_2}\right) - \tan^{-1}\left(\frac{y}{x - x_1}\right)\right) \tag{A.28}$$

Furthermore, from the result of Eq. (A.26),

$$v(x, 0^{\pm}) = \pm\frac{\sigma_0}{2} \tag{A.29}$$

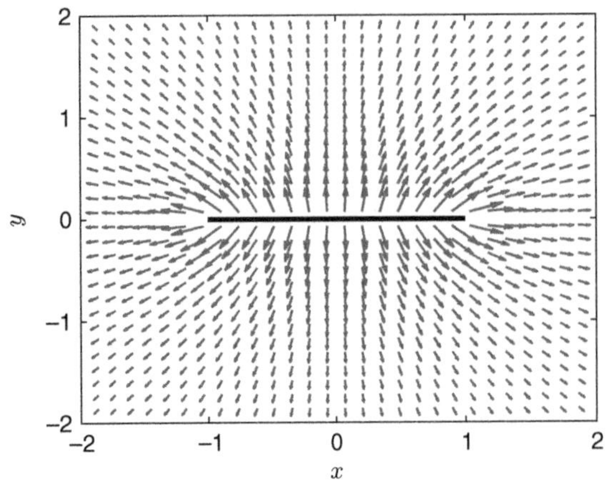

Figure A.6 Flow induced by a source panel with constant unit strength.

Figure A.6 plots the velocity vectors $u(x, y)$, $v(x, y)$ of the flow induced by a source panel with $x_1 = -1$, $x_2 = 1$ and $\sigma_0 = 1$ around itself. The flow is perfectly vertical at the centre of the panel ($x = 0$) but gets inclined horizontally towards the tips.

A.4 The Vortex Line Panel

Similarly to the source line panel, the vortex line panel is a straight or curved line segment on which we place a continuous distribution of vortices. Consider the horizontal straight-line segment of Figure A.7, lying on the x axis and extending from x_1 to x_2, with

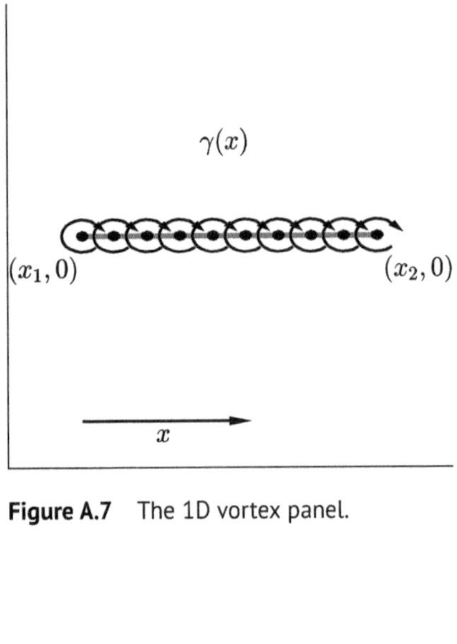

Figure A.7 The 1D vortex panel.

point vortices distributed along its length. The strength of the vortex distribution, $\gamma(x)\mathrm{d}x$, can be constant or can vary along the x direction.

Using Eq. (A.17) for the real potential induced by a single vortex, the potential induced at a general point x, y by the entire panel becomes

$$\phi(x,y) = -\frac{1}{2\pi}\int_{x_1}^{x_2} \gamma(x_0)\tan^{-1}\left(\frac{y}{x-x_0}\right)\mathrm{d}x_0 \tag{A.30}$$

noting that all the vortices lie on $y_0 = 0$. For constant vortex strength $\gamma(x_0) = \gamma_0$, Matlab's symbolic toolbox gives

$$\int \tan^{-1}\left(\frac{y}{x-x_0}\right)\mathrm{d}x_0 = -(x-x_0)\tan^{-1}\left(\frac{y}{x-x_0}\right) - \frac{y}{2}\ln\left((x-x_0)^2 + y^2\right)$$

Applying this outcome to Eq. (A.30) results in

$$\phi(x,y) = \frac{\gamma_0}{2\pi}\left((x-x_2)\tan^{-1}\left(\frac{y}{x-x_2}\right) - (x-x_1)\tan^{-1}\left(\frac{y}{x-x_1}\right)\right.$$
$$\left. + \frac{y}{2}\ln\left(\frac{(x-x_2)^2+y^2}{(x-x_1)^2+y^2}\right)\right) \tag{A.31}$$

The horizontal velocity $u(x, y)$ induced by the vortex panel is given by

$$u(x,y) = \frac{\partial\phi}{\partial x} = -\frac{1}{2\pi}\frac{\partial}{\partial x}\int_{x_1}^{x_2}\gamma(x_0)\tan^{-1}\left(\frac{y}{x-x_0}\right)\mathrm{d}x_0$$
$$= \frac{1}{2\pi}y\int_{x_1}^{x_2}\frac{\gamma(x_0)}{(x-x_0)^2+y^2}\mathrm{d}x_0 \tag{A.32}$$

Applying the arguments resulting in Eq. (A.26), the horizontal velocity induced at $y = 0$ by a vortex panel is

$$u(x, 0^{\pm}) = \pm\frac{\gamma(x)}{2} \tag{A.33}$$

everywhere on the panel. For constant vortex strength, we can use the standard integrals of Section A.3 to obtain

$$u(x,y) = \frac{\gamma_0}{2\pi}\left(\tan^{-1}\left(\frac{x-x_1}{y}\right) - \tan^{-1}\left(\frac{x-x_2}{y}\right)\right) \tag{A.34}$$

or after applying identity (A.22)

$$u(x,y) = \frac{\gamma_0}{2\pi}\left(\tan^{-1}\left(\frac{y}{x-x_2}\right) - \tan^{-1}\left(\frac{y}{x-x_1}\right)\right) \tag{A.35}$$

The vertical airspeed induced by the vortex panel at a general point x, y is

$$v(x,y) = \frac{\partial\phi}{\partial y} = -\frac{1}{2\pi}\frac{\partial}{\partial y}\int_{x_1}^{x_2}\gamma(x_0)\tan^{-1}\left(\frac{y}{x-x_0}\right)\mathrm{d}x_0$$
$$= -\frac{1}{2\pi}\int_{x_1}^{x_2}\gamma(x_0)\frac{x-x_0}{(x-x_0)^2+y^2}\mathrm{d}x_0 \tag{A.36}$$

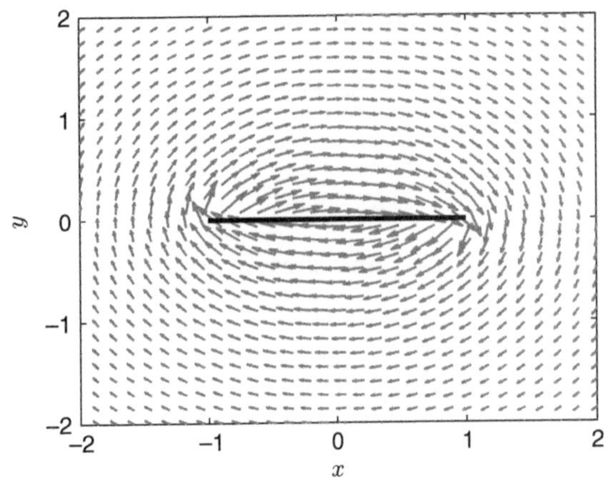

Figure A.8 Flow induced by a vortex panel with constant unit strength.

Again, using the standard integrals of Section A.3, the vertical airspeed induced by a vortex panel of constant strength becomes

$$v(x,y) = -\frac{\gamma_0}{4\pi} \ln\left(\frac{(x-x_1)^2+y^2}{(x-x_2)^2+y^2}\right) \tag{A.37}$$

Figure A.8 plots the velocity vectors $u(x,y)$, $v(x,y)$ of the flow induced by a vortex panel with $x_1 = -1, x_2 = 1$ and $\gamma_0 = 1$ around itself. The flow is tangent to the panel at its midpoint and recirculates in a clockwise direction around it.

If the strength of the vortex panel is not constant but varies linearly from γ_1 to γ_2 along the panel, then

$$\gamma(x_0) = \frac{\gamma_2-\gamma_1}{x_2-x_1}(x_0-x_1)+\gamma_1$$

where x_0 goes from x_1 to x_2. Substituting into Eqs. (A.32) and (A.36) and carrying out the integrations, we obtain the horizontal and vertical velocities induced by the panel as

$$\begin{aligned} u(x,y) &= u_1(x,y)\gamma_1 + u_2(x,y)\gamma_2 \\ v(x,y) &= v_1(x,y)\gamma_1 + v_2(x,y)\gamma_2 \end{aligned} \tag{A.38}$$

where

$$u_1(x,y) = \frac{1}{2\pi(x_1-x_2)}\left\{\left(\tan^{-1}\left(\frac{x-x_1}{y}\right) - \tan^{-1}\left(\frac{x-x_2}{y}\right)\right)(x-x_2) + \frac{y}{2}\ln\left(\frac{(x-x_2)^2+y^2}{(x-x_1)^2+y^2}\right)\right\}$$

$$u_2(x,y) = -\frac{1}{2\pi(x_1-x_2)}\left\{\left(\tan^{-1}\left(\frac{x-x_1}{y}\right) - \tan^{-1}\left(\frac{x-x_1}{y}\right)\right)(x-x_1) + \frac{y}{2}\ln\left(\frac{(x-x_2)^2+y^2}{(x-x_1)^2+y^2}\right)\right\}$$

$$v_1(x,y) = \frac{1}{2\pi(x_1-x_2)}\left\{(x_2-x_1) - y\left(\tan^{-1}\left(\frac{x-x_1}{y}\right) - \tan^{-1}\left(\frac{x-x_2}{y}\right)\right)\right.$$
$$\left. + \frac{(x-x_2)}{2}\ln\left(\frac{(x-x_2)^2+y^2}{(x-x_1)^2+y^2}\right)\right\}$$
$$v_2(x,y) = -\frac{1}{2\pi(x_1-x_2)}\left\{(x_2-x_1) - y\left(\tan^{-1}\left(\frac{x-x_1}{y}\right) - \tan^{-1}\left(\frac{x-x_2}{y}\right)\right)\right.$$
$$\left. + \frac{(x-x_1)}{2}\ln\left(\frac{(x-x_1)^2+y^2}{(x-x_1)^2+y^2}\right)\right\} \tag{A.39}$$

A.5 The Horseshoe Vortex

A horseshoe vortex is a vortex filament made up of two semi-infinite straight-line segments connected by one finite straight-line segment, as seen in Figure A.9. All three segments lie on the plane $z = 0$, the bound segment on the axis $x = 0$ and the trailing segments on the axes $y = \pm b/2$. In accordance with Helmholtz's theorems, the strength of the horseshoe vortex, Γds, is constant along its length, s. Such vortices are used in lifting line theory, whereby the bound segment lies on a wing with span b and the trailing segments in the wake. As a consequence, it is important to calculate the velocity induced by the horseshoe vortex on the bound segment.

The velocity induced at a general point (x, y, z) by a segment ds of the horseshoe vortex can be calculated from Eq. (2.229). First, we note that the coordinate along the vortex filaments, s, is parallel to x for the two trailing segments and parallel to y for the bound segment. Therefore, it is convenient to redefine s as

$$s = \begin{cases} x_0 & \text{for } y_0 = -b/2 \\ y_0 & \text{for } x_0 = 0 \\ x_0 & \text{for } y_0 = b/2 \end{cases}$$

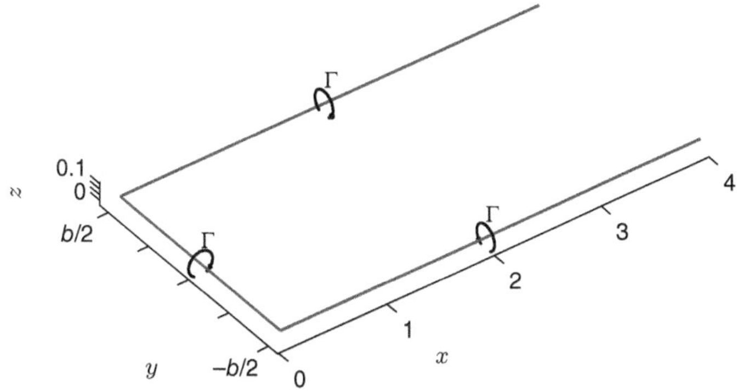

Figure A.9 Horseshoe vortex.

where the 0 subscript denotes a point (x_0, y_0, z_0) on the horseshoe vortex that induces a velocity on a general point (x, y, z) in the flowfield. Equation (2.229) can then be expressed as the sum of three integrals

$$\mathbf{u}(x,y,z) = \frac{\Gamma}{4\pi}\int_{-\infty}^{0}\frac{\boldsymbol{\tau}(x_0)\times\mathbf{r}(x_0)}{\left\|\mathbf{r}\left(x_0\right)\right\|^3}\mathrm{d}x_0\Bigg|_{y_0=-b/2} + \frac{\Gamma}{4\pi}\int_{-b/2}^{b/2}\frac{\boldsymbol{\tau}(y_0)\times\mathbf{r}(y_0)}{\left\|\mathbf{r}\left(y_0\right)\right\|^3}\mathrm{d}y_0\Bigg|_{x_0=0}$$

$$+\frac{\Gamma}{4\pi}\int_{0}^{\infty}\frac{\boldsymbol{\tau}(x_0)\times\mathbf{r}(x_0)}{\left\|\mathbf{r}\left(x_0\right)\right\|^3}\mathrm{d}x\Bigg|_{y_0=b/2} \tag{A.40}$$

where $y_0 = \pm b/2$ denotes the two trailing vortices and $x_0 = 0$ denotes the bound vortex. The unit tangent vector is given by

$$\boldsymbol{\tau}(x_0, y_0) = \begin{cases} (-1,\ 0,\ 0) & \text{for } y_0 = -b/2 \\ (0,\ 1,\ 0) & \text{for } x_0 = 0 \\ (1,\ 0,\ 0) & \text{for } y_0 = b/2 \end{cases}$$

Furthermore, the distance between any point $(x_0, y_0, 0)$ on the horseshoe vortex and any point $(0, y, 0)$ on the bound vortex is given by

$$\mathbf{r}(x_0, y_0) = \begin{cases} (-x_0,\ y + b/2,\ 0) & \text{for } y_0 = -b/2 \\ (0,\ y - y_0,\ 0) & \text{for } x_0 = 0 \\ (-x_0,\ y - b/2,\ 0) & \text{for } y_0 = b/2 \end{cases}$$

Substituting into Eq. (A.40), we obtain the velocity induced by the entire horseshoe vortex on the bound segment

$$\mathbf{u}(y) = \frac{\Gamma}{4\pi}\int_{-\infty}^{0}\frac{(0,0,-(y+b/2))}{\left(x_0^2+(y+b/2)^2\right)^{3/2}}\mathrm{d}x_0 + \frac{\Gamma}{4\pi}\int_{-b/2}^{b/2}\frac{(0,0,0)}{\left(y-y_0\right)^3}\mathrm{d}y_0$$

$$+\frac{\Gamma}{4\pi}\int_{0}^{\infty}\frac{(0,0,y-b/2)}{\left(x_0^2+(y-b/2)^2\right)^{3/2}}\mathrm{d}x_0$$

It is clear that the u and v components of $\mathbf{u} = (u,\ v,\ w)$ are both zero, while the velocity induced by the bound vortex on itself is also zero, except for $y = y_0$. We will exclude point $y = y_0$ from the calculation, such that the velocity induced by the bound vortex on any point lying on itself is always zero.

Applying the simplifications above, we obtain

$$w(y) = -\frac{\Gamma}{4\pi}\int_{-\infty}^{0}\frac{y+b/2}{\left(x_0^2+(y+b/2)^2\right)^{3/2}}\mathrm{d}x_0 + \frac{\Gamma}{4\pi}\int_{0}^{\infty}\frac{y-b/2}{\left(x_0^2+(y-b/2)^2\right)^{3/2}}\mathrm{d}x_0$$

We now make use of the general integral

$$\int\frac{a}{\left(x^2+a^2\right)^{3/2}}\mathrm{d}x = \frac{x}{a\left(x^2+a^2\right)^{1/2}}$$

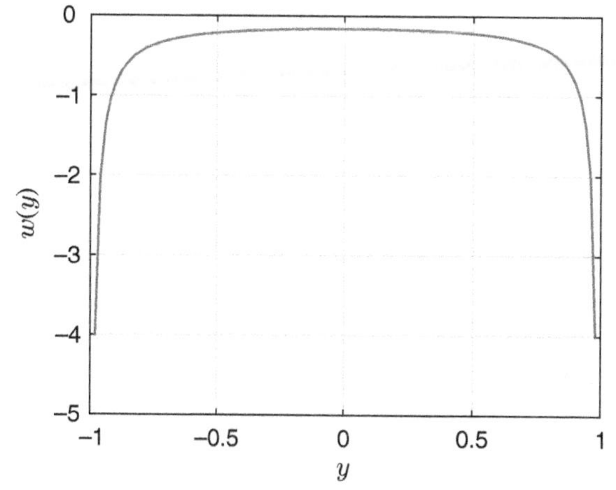

Figure A.10 Downwash induced by a horseshoe vortex.

for any real a, such that

$$w(y) = -\frac{\Gamma}{4\pi} \frac{x_0}{(y + b/2)\left(x_0^2 + (y + b/2)^2\right)^{1/2}}\Bigg|_{x_0=-\infty}^{x_0=0} + \frac{\Gamma}{4\pi} \frac{x_0}{(y - b/2)\left(x_0^2 + (y - b/2)^2\right)^{1/2}}\Bigg|_{x_0=0}^{x_0=\infty}$$

Finally, after calculating the general limit

$$\lim_{x\to\pm\infty} \frac{x}{a\left(x^2 + a^2\right)^{1/2}} = \pm\frac{1}{a}$$

and applying the limits of integration, we obtain

$$w(y) = -\frac{\Gamma}{4\pi\,(y + b/2)} + \frac{\Gamma}{4\pi\,(y - b/2)} = -\frac{\Gamma}{4\pi\,(b/2 + y)} - \frac{\Gamma}{4\pi\,(b/2 - y)} \tag{A.41}$$

This is the total velocity induced by the horseshoe vortex on its bound segment. As mentioned earlier, this induced velocity only has a vertical component. Furthermore, as y is bounded by $b/2$ and $-b/2$, this velocity will always be negative if the vortex strength is positive. Hence, the velocity induced by a horseshoe vortex on its bound segment is a downwash. Finally, it should be noted that the downwash becomes infinite at $y = \pm b/2$. Figure A.10 plots the downwash distribution induced by a horseshoe vortex with $b = 1$ and $\Gamma = 1$.

A.6 The Vortex Line Segment

The vortex line segment is a straight-line vortex filament of strength $\Gamma \mathrm{d}s$ and length l extending from point 1 to point 2, as shown in Figure A.11. The velocity it induces at a general

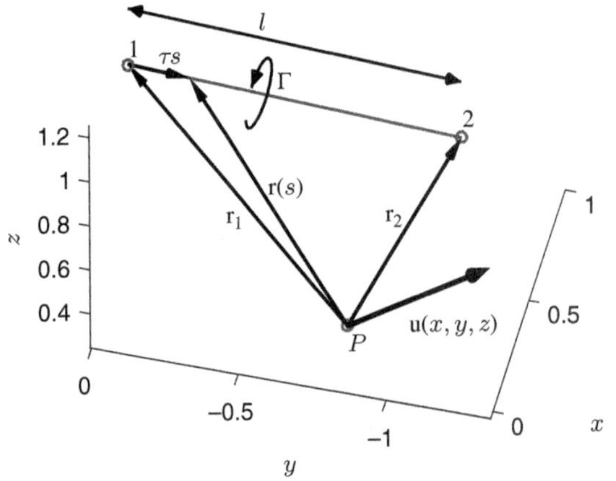

Figure A.11 Vortex line segment and the velocity it induces at point P.

point P lying at (x, y, z) is given by Eq. (2.229) and points in a direction perpendicular to the plane $P12$. The vectorial distances from P to point 1 and from P to point 2 are denoted by $\mathbf{r}_1$ and $\mathbf{r}_2$, respectively. Then, the distance from point 1 to point 2 is $\mathbf{r}_{12} = \mathbf{r}_2 - \mathbf{r}_1$.

As the filament is straight, its unit tangent vector is constant and given by

$$\tau = \frac{\mathbf{r}_{12}}{l}$$

The distance $\mathbf{r}(s)$ from point P to any general point on the filament is the equation of the straight line

$$\mathbf{r}(s) = \mathbf{r}_1 + \tau s = \mathbf{r}_1 + \frac{\mathbf{r}_{12}s}{l}$$

Substituting into Eq. (2.229), we obtain

$$\mathbf{u} = \frac{\Gamma}{4\pi}\int_0^l \frac{\frac{\mathbf{r}_{12}}{l} \times \left(\mathbf{r}_1 + \frac{\mathbf{r}_{12}s}{l}\right)}{\left\|\mathbf{r}_1 + \frac{\mathbf{r}_{12}s}{l}\right\|^3} ds = \frac{\Gamma}{4\pi l}\mathbf{r}_{12} \times \mathbf{r}_1 \int_0^l \frac{1}{\left\|\mathbf{r}_1 + \frac{\mathbf{r}_{12}s}{l}\right\|^3} ds$$

since $\mathbf{r}_{12} \times \mathbf{r}_{12} = 0$. The integral can be evaluated using Matlab's symbolic toolbox such that, after re-arranging, we obtain

$$\mathbf{u} = \frac{\Gamma}{4\pi}\frac{\mathbf{r}_{12} \times \mathbf{r}_1}{\left\|\mathbf{r}_1 \times \mathbf{r}_2\right\|^2}\mathbf{r}_{12} \cdot \left(\frac{\mathbf{r}_2}{\left\|\mathbf{r}_2\right\|} - \frac{\mathbf{r}_1}{\left\|\mathbf{r}_1\right\|}\right) \tag{A.42}$$

It should be kept in mind that in order to conform with Helmholtz's theorems, a vortex line segment cannot exist in isolation; it must be connected to other vortex filaments. Finally, if either $\left\|\mathbf{r}_1\right\| = 0$ or $\left\|\mathbf{r}_2\right\| = 0$, the induced velocity becomes singular. This means that the velocity induced by a vortex line segment on any point lying on the segment is singular. As we saw in Section A.5, if point P lies on a vortex segment, the velocity induced by every other point of the segment on P is equal to zero. It is only the velocity induced by point P on itself that is singular. Again, if we exclude this point from the calculation, we can state that the velocity induced by a vortex segment on itself is equal to zero.

A.7 The Vortex Ring

A vortex ring is a vortex filament that closes in on itself. The most widely used type of vortex ring is the quadrilateral, which is composed of four vortex line segments that connect nodes 1, 2, 3 and 4, as seen in Figure A.12. The four nodes are usually chosen to be co-planar. Non-coplanar nodes are not problematic for the calculation of the velocities induced by the vortex ring, but they make it difficult to define a unit vector normal to the plane of the vortex ring, which can have repercussions in the practical use of vortex rings.

The velocity that the quadrilateral vortex ring induces at a general point P lying at (x, y, z) is calculated by applying Eq. (A.42) four times, once for each vortex line segment. For the ring in Figure A.12, the velocity induced at P is

$$\mathbf{u} = -\mathbf{u}_{12} - \mathbf{u}_{23} - \mathbf{u}_{34} - \mathbf{u}_{41}$$

where $\mathbf{u}_{12}$, $\mathbf{u}_{23}$, etc., are the velocities induced by the segments connecting nodes 12, 23, etc., respectively, and the minus sign is used because Γ is defined positive using the right-hand rule. If the vectorial distances between point P and the four nodes are denoted by $\mathbf{r}_1$, $\mathbf{r}_2$, $\mathbf{r}_3$ and $\mathbf{r}_4$, then

$$\begin{aligned}\mathbf{u} = &-\frac{\Gamma}{4\pi}\frac{\mathbf{r}_{12}\times\mathbf{r}_1}{\|\mathbf{r}_1\times\mathbf{r}_2\|^2}\mathbf{r}_{12}\cdot\left(\frac{\mathbf{r}_2}{\|\mathbf{r}_2\|}-\frac{\mathbf{r}_1}{\|\mathbf{r}_1\|}\right)\\ &-\frac{\Gamma}{4\pi}\frac{\mathbf{r}_{23}\times\mathbf{r}_2}{\|\mathbf{r}_2\times\mathbf{r}_3\|^2}\mathbf{r}_{23}\cdot\left(\frac{\mathbf{r}_3}{\|\mathbf{r}_3\|}-\frac{\mathbf{r}_2}{\|\mathbf{r}_2\|}\right)\\ &-\frac{\Gamma}{4\pi}\frac{\mathbf{r}_{34}\times\mathbf{r}_3}{\|\mathbf{r}_3\times\mathbf{r}_4\|^2}\mathbf{r}_{34}\cdot\left(\frac{\mathbf{r}_4}{\|\mathbf{r}_4\|}-\frac{\mathbf{r}_3}{\|\mathbf{r}_3\|}\right)\\ &-\frac{\Gamma}{4\pi}\frac{\mathbf{r}_{41}\times\mathbf{r}_4}{\|\mathbf{r}_4\times\mathbf{r}_1\|^2}\mathbf{r}_{41}\cdot\left(\frac{\mathbf{r}_1}{\|\mathbf{r}_1\|}-\frac{\mathbf{r}_4}{\|\mathbf{r}_4\|}\right)\end{aligned} \tag{A.43}$$

where $\mathbf{r}_{12} = \mathbf{r}_2 - \mathbf{r}_1$, $\mathbf{r}_{23} = \mathbf{r}_3 - \mathbf{r}_2$, $\mathbf{r}_{34} = \mathbf{r}_4 - \mathbf{r}_3$ and $\mathbf{r}_{41} = \mathbf{r}_1 - \mathbf{r}_4$. It is important to carry out the calculation of the induced velocity in the same order for all the segments, as shown in Eq. (A.43).

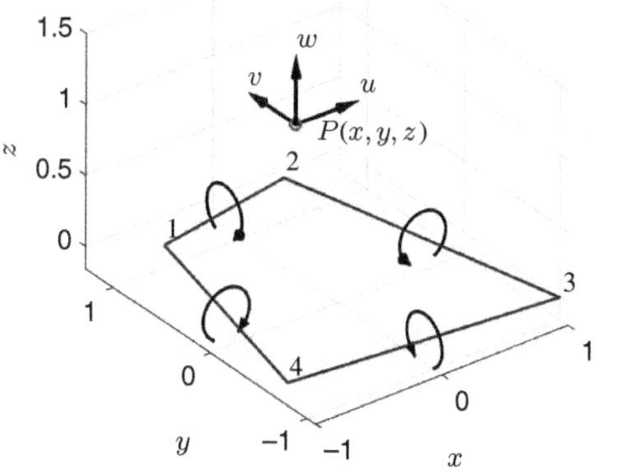

Figure A.12 Velocities induces at point P by a quadrilateral vortex ring.

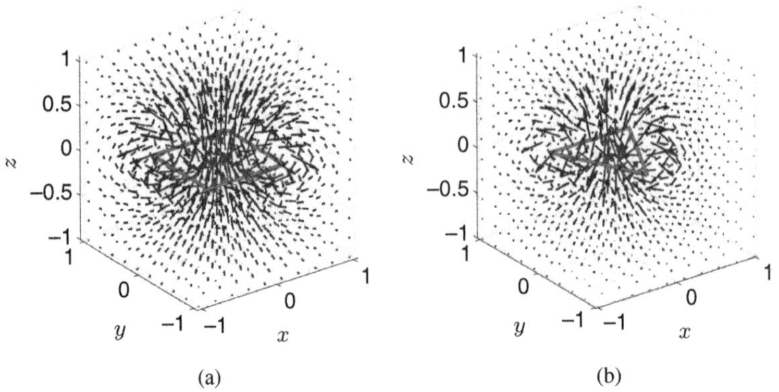

Figure A.13 Velocity field induced by vortex rings of constant unit strength. (a) Square vortex ring and (b) triangular vortex ring.

Triangular vortex rings are also popular; they can be used to model Delta wings and triangular wingtips. Strictly speaking, the velocity induced by the ring on any point P lying on itself is singular. As we stated before, we can bypass this singularity by excluding point P from the calculation of the induced velocity. For example, if point P lies on segment 23, the velocity induced at P is equal to the sum of the velocities induced by segments 12, 34 and 41 only; the velocity induced on P by segment 23 is set to zero. Figure A.13 plots the velocity field induced by two different ring geometries, a square and a triangle, for $\Gamma = 1$. The flow looks like a series of tubes rotating around each vortex segment, such that the fluid flows upwards through the ring.

A.8 The 3D Point Source

The potential induced at a general point $\mathbf{x} = (x, y, z)$ by a 3D point source of strength σ lying at $\mathbf{x}_0 = (x_0, y_0, z_0)$ is given by

$$\phi(\mathbf{x}, \mathbf{x}_0) = -\frac{\sigma}{4\pi}\frac{1}{r(\mathbf{x}, \mathbf{x}_0)} \tag{A.44}$$

where $r(\mathbf{x}, \mathbf{x}_0) = \sqrt{(x - x_0)^2 + (y - y_0)^2 + (z - z_0)^2}$, such that

$$\phi(\mathbf{x}, \mathbf{x}_0) = -\frac{\sigma}{4\pi}\frac{1}{\sqrt{(x - x_0)^2 + (y - y_0)^2 + (z - z_0)^2}} \tag{A.45}$$

while the velocities induced at the same point are given by

$$u(\mathbf{x}, \mathbf{x}_0) = \frac{\partial \phi}{\partial x} = \frac{\sigma}{4\pi}\frac{x - x_0}{\left((x - x_0)^2 + (y - y_0)^2 + (z - z_0)^2\right)^{3/2}}$$

$$v(\mathbf{x}, \mathbf{x}_0) = \frac{\partial \phi}{\partial y} = \frac{\sigma}{4\pi}\frac{y - y_0}{\left((x - x_0)^2 + (y - y_0)^2 + (z - z_0)^2\right)^{3/2}}$$

$$w(\mathbf{x}, \mathbf{x}_0) = \frac{\partial \phi}{\partial z} = \frac{\sigma}{4\pi}\frac{z - z_0}{\left((x - x_0)^2 + (y - y_0)^2 + (z - z_0)^2\right)^{3/2}}$$

The flowfield induced by a 3D point source and sink lying at the origin is plotted in Figure 2.4.

Differentiating u, v and w once more results in

$$\frac{\partial^2\phi}{\partial x^2} = \frac{\sigma}{4\pi}\frac{-2(x-x_0)^2+(y-y_0)^2+(z-z_0)^2}{\left((x-x_0)^2+(y-y_0)^2+(z-z_0)^2\right)^{5/2}}$$

$$\frac{\partial^2\phi}{\partial y^2} = \frac{\sigma}{4\pi}\frac{(x-x_0)^2-2(y-y_0)^2+(z-z_0)^2}{\left((x-x_0)^2+(y-y_0)^2+(z-z_0)^2\right)^{5/2}}$$

$$\frac{\partial^2\phi}{\partial z^2} = \frac{\sigma}{4\pi}\frac{(x-x_0)^2+(y-y_0)^2-2(z-z_0)^2}{\left((x-x_0)^2+(y-y_0)^2+(z-z_0)^2\right)^{5/2}}$$

The sum of these three expressions is equal to zero, so that the 3D point source is a solution of Laplace's equation $\nabla^2\phi = 0$, but only for $r \neq 0$. The second derivatives of ϕ are singular when $r \to 0$. In order to analyse the behaviour of $\nabla^2\phi$ around $r = 0$, we can study the non-singular function:

$$\phi(\mathbf{x}, \mathbf{x}_0, \varepsilon) = -\frac{1}{r(\mathbf{x}, \mathbf{x}_0) + \varepsilon} \tag{A.46}$$

where $\varepsilon \ll 1$ is a small positive real number. Taking the second derivatives of this function with respect to x, y, z and adding them up, we obtain

$$\nabla^2\phi(\mathbf{x}, \mathbf{x}_0, \varepsilon) = \frac{2\varepsilon}{r(r+\varepsilon)^3} \tag{A.47}$$

Clearly, this quantity is always positive and tends to $+\infty$ as $r \to 0$. As ε tends to zero, $\nabla^2\phi(\mathbf{x}, \mathbf{x}_0, \varepsilon)$ becomes zero for all r, except for $r = 0$, where we can apply L'Hôpital's rule to obtain

$$\lim_{\varepsilon,r\to 0}\frac{2\varepsilon}{r(r+\varepsilon)^3} = \lim_{\varepsilon,r\to 0}\frac{2}{3r(r+\varepsilon)^2} = \lim_{r\to 0}\frac{2}{3r^3} = +\infty$$

But $\lim_{\varepsilon\to 0}\nabla^2\phi(\mathbf{x}, \mathbf{x}_0, \varepsilon) = \nabla^2(-1/r)$, so that the expression above can be written as

$$\lim_{r\to 0}\nabla^2\left(\frac{-1}{r}\right) = +\infty$$

Consequently, the values of $\nabla^2(-1/r)$ for all r are

$$\nabla^2\left(\frac{-1}{r}\right) = \begin{cases} +\infty & \text{if } r = 0 \\ 0 & \text{if } r > 0 \end{cases} \tag{A.48}$$

which means that the 3D point source satisfies Laplace's equation everywhere except at $r = 0$, where the left-hand side of the equation becomes infinite.

Since $\nabla^2(-1/r)$ is not zero everywhere, it is interesting to calculate its volume integral. In order to do this, we will once again use the non-singular function (A.46) whose Laplacian derivative is given in Eq. (A.47). We calculate the integral of this Laplacian inside a sphere with centre $\mathbf{x}_0$ and radius R, using spherical coordinates defined as

$$x = r\cos\psi\sin\theta,\ y = r\sin\psi\sin\theta,\ z = r\cos\theta$$

where r takes values between 0 and R, ψ between 0 and 2π and θ between 0 and π. Then, $dV = r^2\sin\theta dr d\psi d\theta$ and the required integral becomes

$$I = \int_0^R\int_0^{2\pi}\int_0^{\pi}\frac{2\varepsilon}{r(r+\varepsilon)^3}r^2\sin\theta dr d\psi d\theta$$

First, we carry out the integrals of θ and ψ to obtain

$$I = \int_0^R \frac{8\pi\varepsilon r}{(r+\varepsilon)^3} \mathrm{d}r$$

The remaining integral can be evaluated using Matlab's symbolic toolbox as

$$I = \left. \frac{4\pi r^2}{(r+\varepsilon)^2} \right|_0^R = \frac{4\pi R^2}{(R+\varepsilon)^2}$$

Now, if we set $\varepsilon = 0$, the volume integral of $\nabla^2(-1/r)$ inside a sphere of any radius $R > 0$ is given by

$$\int_{V_R} \nabla^2 \left(-\frac{1}{r}\right) \mathrm{d}V = 4\pi \tag{A.49}$$

where V_R is the radius of the sphere. For the sink,

$$\int_{V_R} \nabla^2 \left(\frac{1}{r}\right) \mathrm{d}V = -4\pi \tag{A.50}$$

A.9 The 3D Point Doublet

The potential induced at a general point $\mathbf{x} = (x, y, z)$ by a 3D point doublet of strength μ lying at $\mathbf{x}_0 = (x_0, y_0, z_0)$ is given by

$$\phi(\mathbf{x}, \mathbf{x}_0) = \frac{\mu}{4\pi} \mathbf{n} \cdot \nabla \frac{1}{r(\mathbf{x}, \mathbf{x}_0)} \tag{A.51}$$

where $r(\mathbf{x}, \mathbf{x}_0) = \sqrt{(x-x_0)^2 + (y-y_0)^2 + (z-z_0)^2}$, $\nabla = (\partial/\partial x,\ \partial/\partial y,\ \partial/\partial z)$ and $\mathbf{n}$ is a unit vector pointing in a chosen direction. If $\mathbf{n}$ points in the z direction, i.e. $\mathbf{n} = (0,\ 0,\ 1)$, we obtain

$$\phi(\mathbf{x}, \mathbf{x}_0) = -\frac{\mu}{4\pi} \frac{z - z_0}{\left((x-x_0)^2 + (y-y_0)^2 + (z-z_0)^2\right)^{3/2}} \tag{A.52}$$

while the velocities induced at the same point are given by

$$u(\mathbf{x}, \mathbf{x}_0) = \frac{\partial \phi}{\partial x} = \frac{3\mu}{4\pi} \frac{(x-x_0)(z-z_0)}{\left((x-x_0)^2 + (y-y_0)^2 + (z-z_0)^2\right)^{5/2}}$$

$$v(\mathbf{x}, \mathbf{x}_0) = \frac{\partial \phi}{\partial y} = \frac{3\mu}{4\pi} \frac{(y-y_0)(z-z_0)}{\left((x-x_0)^2 + (y-y_0)^2 + (z-z_0)^2\right)^{5/2}}$$

$$w(\mathbf{x}, \mathbf{x}_0) = \frac{\partial \phi}{\partial z} = -\frac{\mu}{4\pi} \frac{(x-x_0)^2 + (y-y_0)^2 - 2(z-z_0)^2}{\left((x-x_0)^2 + (y-y_0)^2 + (z-z_0)^2\right)^{5/2}}$$

Figure 2.5 plots the flowfield induced by a 3D point source lying at the origin, for two values of $\mathbf{n}$.

A.10 The Source Surface Panel

Consider a flat surface panel lying on the $z = 0$ plane with area S. A continuous source distribution is placed on this panel such that the source strength of a small surface element dS lying at x_0, y_0 is denoted by $\sigma(x_0, y_0)\mathrm{d}S$. The incremental potential induced by this panel at any general point x, y and z by the element dS is obtained from Eq. (A.45)

$$\mathrm{d}\phi(x,y,z) = -\frac{1}{4\pi}\frac{\sigma(x_0,y_0)}{\sqrt{(x-x_0)^2+(y-y_0)^2+z^2}}\mathrm{d}S$$

and the total potential induced at the same point is the integral of $\mathrm{d}\phi(x, y, z)$ over S

$$\phi(x,y,z) = -\frac{1}{4\pi}\int_S \frac{\sigma(x_0,y_0)}{\sqrt{(x-x_0)^2+(y-y_0)^2+z^2}}\mathrm{d}S \tag{A.53}$$

The velocities induced by the panel at point x, y and z can be obtained by differentiating this equation such that

$$u(x,y,z) = \frac{\partial\phi}{\partial x} = \frac{1}{4\pi}\int_S \frac{x-x_0}{\left((x-x_0)^2+(y-y_0)^2+z^2\right)^{3/2}}\sigma(x_0,y_0)\mathrm{d}S$$

$$v(x,y,z) = \frac{\partial\phi}{\partial y} = \frac{1}{4\pi}\int_S \frac{y-y_0}{\left((x-x_0)^2+(y-y_0)^2+z^2\right)^{3/2}}\sigma(x_0,y_0)\mathrm{d}S$$

$$w(x,y,z) = \frac{\partial\phi}{\partial z} = \frac{1}{4\pi}\int_S \frac{z}{\left((x-x_0)^2+(y-y_0)^2+z^2\right)^{3/2}}\sigma(x_0,y_0)\mathrm{d}S \tag{A.54}$$

Figure A.14 plots the velocity field induced by two different panel geometries, a square and a triangle, for $\sigma(x_0, y_0) = 1$. All velocities point away from the panels in all directions. On the panel, the velocity is vertical around its midpoint, while it is horizontal and normal to the edges near the edge midpoints. If we look at the square panel along the x or y directions, we will see the velocity field of Figure A.6.

As $z \to 0$, the velocity $w(x, y, z)$ becomes zero everywhere except at the point $x = x_0$, $y = y_0$, where it is undefined. Therefore, for a point lying on $z = 0$ but outside the panel, $w(x, y, 0) = 0$ everywhere. For a point lying on the panel, the only panel element that has a non-zero contribution to w is the point $x_0 = x$, $y_0 = y$; all other contributions are equal to zero. Therefore, the integral of Eq. (A.54) only needs to be evaluated over a very small area delimited by $x_0 = x - \varepsilon$, $y_0 = y - \varepsilon$ and $x_0 = x + \varepsilon$, $y_0 = y + \varepsilon$, where $\varepsilon \ll 1$. Over this

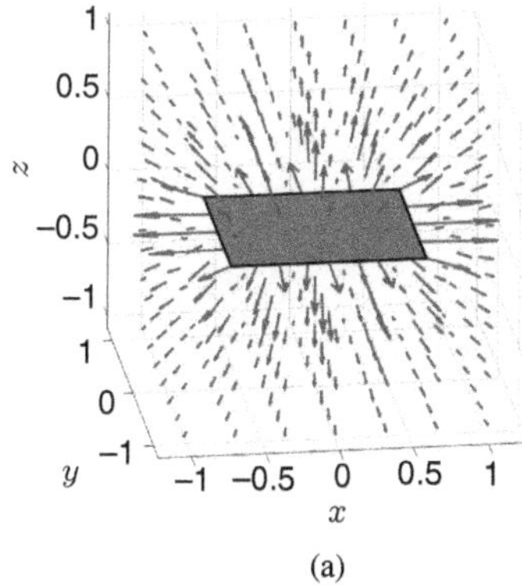

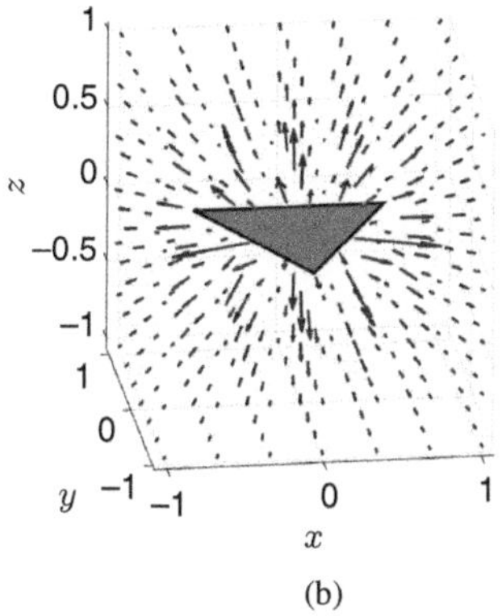

Figure A.14 Velocity field induced by source surface panels of constant unit strength. (a) Square panel and (b) triangular panel.

small area, it is assumed that $\sigma(x_0, y_0)$ will not change significantly, so we can replace it by $\sigma(x_0, y_0) \approx \sigma(x, y)$ and take it outside the integral. We define

$$x_0 = x + x', \ y_0 = y + y'$$

so that $dx_0 = dx'$ and $dy_0 = dy'$ and Eq. (A.54) becomes

$$w(x, y, 0^{\pm}) = \frac{\sigma(x, y)}{4\pi} \lim_{z,\varepsilon \to 0^{\pm}} \int_{-\varepsilon}^{\varepsilon} \left(\int_{-\varepsilon}^{\varepsilon} \frac{z}{\left(x'^2 + y'^2 + z^2\right)^{3/2}} dx' \right) dy' \tag{A.55}$$

We will first evaluate the integral over x',

$$\int_{-\varepsilon}^{\varepsilon} \frac{z}{\left(x'^2 + y'^2 + z^2\right)^{3/2}} dx' = \left. \frac{x' z}{(y'^2 + z^2)\sqrt{x'^2 + y'^2 + z^2}} \right|_{-\varepsilon}^{\varepsilon} = \frac{2\varepsilon z}{(y'^2 + z^2)\sqrt{\varepsilon^2 + y'^2 + z^2}}$$

Consequently, the integral over y' becomes

$$\int_{-\varepsilon}^{\varepsilon} \frac{2\varepsilon z}{(y'^2 + z^2)\sqrt{\varepsilon^2 + y'^2 + z^2}} dy' = 2\tan^{-1} \left. \frac{y' \varepsilon}{z\sqrt{\varepsilon^2 + y'^2 + z^2}} \right|_{-\varepsilon}^{\varepsilon} = 4\tan^{-1} \frac{\varepsilon^2}{z\sqrt{2\varepsilon^2 + z^2}}$$

Then, if we let z tend to $0^{\pm}$ faster than ε, the argument of the inverse tangent becomes $\pm\infty$ and the limit in expression (A.55) becomes $\pm 2\pi$. Substituting back into expression (A.55), we obtain simply

$$w(x, y, 0^{\pm}) = \pm \frac{\sigma(x, y)}{2} \tag{A.56}$$

so that the magnitude of the vertical velocity induced anywhere on the surface of the panel by the panel itself is equal to half of the local source strength.

For any general point x, y and z, the integral of Eq. (A.53) can be evaluated analytically for particular panel geometries. We will use the approach by Hess and Smith (1967) in order to calculate this integral. Consider Figure A.15a, which plots a quadrilateral panel lying on $z = 0$, with vertices x_{0_i} for $i = 1, \dots 4$ and a general point in space, lying at x, y, z. The objective is to calculate the potential induced by the panel at point x, y, z, with the

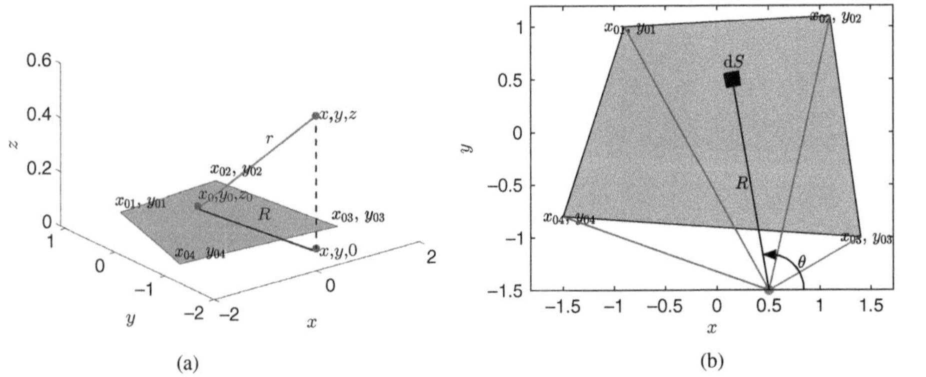

Figure A.15 Quadrilateral panel lying on $z = 0$ and general point in space, x, y, z. (a) 3D view and (b) view from the top.

assumption that the source strength is constant over the entire panel and equal to σ. The integral of Eq. (A.53) can be written as

$$\phi(x,y,z) = -\frac{\sigma}{4\pi}\int_S \frac{1}{r}\mathrm{d}S \tag{A.57}$$

where $r = \sqrt{(x-x_0)^2 + (y-y_0)^2 + z^2}$ and $x_0, y_0, 0$ is a general point on the panel, lying between the vertices. The projection of r on the z plane is denoted by R, such that

$$r^2 = R^2 + z^2 \tag{A.58}$$

Figure A.15b plots the panel and the point x, y, z from the top, so that they appear coplanar. The radius R is the distance between x, y and the point x_0, y_0 on the panel, while the angle θ is defined positive counter-clockwise so that the $\theta = 0$ half-axis lies directly to the right of point x, y. Therefore, R and θ are given by

$$R = \sqrt{\left(x-x_0\right)^2 + \left(y-y_0\right)^2} \tag{A.59}$$

$$\theta = \tan^{-1}\frac{y_0 - y}{x_0 - x} \tag{A.60}$$

The figure also plots a small area dS around point x_0, y_0. In cylindrical coordinates, the radial length of this area is dR, while its circumferential length is $R\mathrm{d}\theta$ so that

$$\mathrm{d}S = R\mathrm{d}R\mathrm{d}\theta$$

Using the cylindrical coordinates R, θ, z, Eq. (A.57) can be written as

$$\phi(x,y,z) = -\frac{\sigma}{4\pi}\oint\int_0^R \frac{1}{r}R\,\mathrm{d}R\,\mathrm{d}\theta \tag{A.61}$$

where $\oint$ denotes that the θ integral must be calculated for all the possible values of θ, from $\theta(x_{0_1}, y_{0_1})$ to $\theta(x_{0_4}, y_{0_4})$ and back to $\theta(x_{0_1}, y_{0_1})$. If x, y lies inside the rectangle, then θ will take values from $\theta = \theta(x_{0_1}, y_{0_1})$ to $\theta = \theta(x_{0_1}, y_{0_1}) + 2\pi$. In order to proceed, we need to change the variable of integration R to r, using the differential form of Eq. (A.58),

$$r\mathrm{d}r = R\mathrm{d}R \tag{A.62}$$

We also note that when $R = 0$, $r = |z|$. Equation (A.61) becomes

$$\phi(x,y,z) = -\frac{\sigma}{4\pi}\oint\int_{|z|}^{r}\frac{1}{r}r\mathrm{d}r\mathrm{d}\theta = -\frac{\sigma}{4\pi}\oint\int_{|z|}^{r}\mathrm{d}r\mathrm{d}\theta \tag{A.63}$$

After evaluating the inner integral, we have

$$\phi(x,y,z) = -\frac{\sigma}{4\pi}\oint(r - |z|)\mathrm{d}\theta$$

In order to evaluate this latest integral, we need to write r in terms of θ. Recall that the panel is defined by four straight-line segments and that the integration is carried out along these segments. Therefore, the potential is given by

$$\phi(x,y,z) = -\frac{\sigma}{4\pi}\sum_{i=1}^{4}\int_{\theta_i}^{\theta_{i+1}}(r - |z|)\mathrm{d}\theta = -\frac{\sigma}{4\pi}\sum_{i=1}^{4}\left(\int_{\theta_i}^{\theta_{i+1}} r\mathrm{d}\theta - |z|(\theta_{i+1} - \theta_i)\right) \tag{A.64}$$

where

$$\theta_i = \tan^{-1}\frac{y_{0_i} - y}{x_{0_i} - x} \tag{A.65}$$

while x_{0_i}, y_{0_i} is the starting point of the segment and x_{0_i+1}, y_{0_i+1} is the endpoint of the segment. If we can evaluate this integral over the ith segment, then we can evaluate it over all of them. The equation of the segment on the $z = 0$ plane is given by

$$y_0 = m_i\left(x_0 - x_{0_i}\right) + y_{0_i} = m_i x_0 + y_{0_i} - m_i x_{0_i} \tag{A.66}$$

where x_0, y_0 is any point on the segment and where the slope of the segment is given by

$$m_i = \frac{y_{0_i+1} - y_{0_i}}{x_{0_i+1} - x_{0_i}} \tag{A.67}$$

Figure A.15b shows that $x_0 = R\cos\theta + x$ and $y_0 = R\sin\theta + y$, where R is the distance between point x_0, y_0 and point x, y. Substituting into Eq. (A.66) and solving for R, we obtain

$$R = \frac{b_i}{\sin\theta - m_i\cos\theta}$$

where $b_i = y_{0_i} - y - m_i(x_{0_i} - x)$. We would like to express $\sin\theta - m_i\cos\theta$ as a single sine term, which can be done easily by writing it as $\sin\theta - m_i\cos\theta = A_i\sin(\theta + \phi_i)$, where $A_i = \sqrt{1 + m_i^2}$, $\phi_i = \sin^{-1}(-m_i/A_i)$. Substituting back into the expression for R, we obtain

$$R = \frac{b_i}{A_i\sin(\theta + \phi_i)} \tag{A.68}$$

We recall from Eq. (A.58) that $r = \sqrt{R^2 + z^2}$, so that

$$r = \sqrt{\frac{b_i^2}{A_i^2\sin^2(\theta + \phi_i)} + z^2} \tag{A.69}$$

and substitute back into Eq. (A.64) to obtain

$$\phi(x, y, z) = -\frac{\sigma}{4\pi}\sum_{i=1}^{4}\left(\int_{\theta_i}^{\theta_{i+1}}\sqrt{\frac{b_i^2}{A_i^2\sin^2(\theta + \phi_i)} + z^2 d\theta} - |z|(\theta_{i+1} - \theta_i)\right) \tag{A.70}$$

By trial and error, we determine the solution of the integral in this latest expression to be

$$\begin{aligned}\int\sqrt{\frac{b_i^2}{A_i^2\sin^2(\theta + \phi_i)} + z^2 d\theta} = &-\frac{\sin(\theta + \phi_i)}{\sqrt{b_i^2 + A_i^2 z^2\sin^2(\theta + \phi_i)}}\sqrt{z^2 + \frac{b_i^2}{A_i^2\sin^2(\theta + \phi_i)}}\\ &\times\left(\frac{b_i}{2}\ln\left(-\frac{b_i\cos(\theta + \phi_i) + \sqrt{b_i^2 + A_i^2 z^2\sin^2(\theta + \phi_i)}}{b_i\cos(\theta + \phi_i) - \sqrt{b_i^2 - A_i^2 z^2\sin^2(\theta + \phi_i)}}\right)\right.\\ &\left.-A_i z\tan^{-1}\left(\frac{\sqrt{b_i^2 - A_i^2 z^2\sin^2(\theta + \phi_i)}}{A_i z\cos(\theta + \phi_i)}\right)\right)\end{aligned}$$

This result can be verified by differentiating the right-hand side with respect to θ and comparing to the integrand of the left-hand side. This operation can be carried out using Matlab's symbolic toolbox. The integral can be simplified to

$$\int \sqrt{\frac{b_i^2}{A_i^2 \sin^2(\theta+\phi_i)} + z^2 \mathrm{d}\theta} = \mathrm{sgn}\left(\sin(\theta+\phi_i)\right)\left(z \tan^{-1}\left(\frac{\sqrt{b_i^2 - A_i^2 z^2 \sin^2(\theta+\phi_i)}}{A_i z \cos(\theta+\phi_i)}\right)\right.$$
$$\left. - \frac{b_i}{2A_i} \ln\left(-\frac{b_i \cos(\theta+\phi_i) + \sqrt{b_i^2 + A_i^2 z^2 \sin^2(\theta+\phi_i)}}{b_i \cos(\theta+\phi_i) - \sqrt{b_i^2 - A_i^2 z^2 \sin^2(\theta+\phi_i)}}\right)\right) \tag{A.71}$$

but now the right-hand side is not differentiable with respect to θ.

We can now apply the limits of integration for each of the segments and evaluate the potential from Eq. (A.70). However, there is a special case in which the calculation will fail; it can be seen from Eq. (A.67) that the slope of the ith segment becomes infinite if $x_{0_{i+1}} = x_{0_i}$. In this case, the equation of the segment is given simply by $x_0 = x_{0_i}$, where $x_0 = R\cos\theta + x$ as usual. Therefore, the radius R is obtained as

$$R = \frac{x_{0_i} - x}{\cos\theta} = \frac{x_{0_i} - x}{\sin(\theta + \pi/2)}$$

Comparing to Eq. (A.68), we can see that $A_i = 1$, $b_i = x_{0_i} - x$ and $\phi_i = \pi/2$. The potential calculation can then be carried out as before.

Katz and Plotkin (2001) give the potential induced by a source panel in a more compact form. First, we calculate the following quantities:

$$\begin{aligned}
d_i &= \sqrt{(x_{0_{i+1}} - x_{0_i})^2 + (y_{0_{i+1}} - y_{0_i})^2},\ m_i = \frac{y_{0_{i+1}} - y_{0_i}}{x_{0_{i+1}} - x_{0_i}} \\
r_i &= \sqrt{(x - x_{0_i})^2 + (y - y_{0_i})^2 + z^2},\ e_i = (x - x_{0_i})^2 + z^2 \\
h_i &= (x - x_{0_i})(y - y_{0_i}) \\
f_i &= \frac{(x - x_{0_i})(y_{0_{i+1}} - y_{0_i}) - (y - y_{0_i})(x_{0_{i+1}} - x_{0_i})}{d_i} \\
g_i &= r_i + r_{i+1} + d_i,\ \hat{g}_i = r_i + r_{i+1} - d_i,\ l_i = \ln\left(\frac{g_i}{\hat{g}_i}\right) \\
k_i &= \frac{m_i e_i - h_i}{r_i},\ \hat{k}_i = \frac{m_i e_{i+1} - h_{i+1}}{r_{i+1}} \\
t_i &= \tan^{-1}\frac{k_i}{z} - \tan^{-1}\frac{\hat{k}_i}{z}
\end{aligned} \tag{A.72}$$

for $i = 1, \ldots, 4$, with $x_{0_5} = x_{0_1}, y_{0_5} = y_{0_1}, r_5 = r_1, e_5 = e_1$ and $h_5 = h_1$. The term t_i can be problematic. First, we will set $t_i = 0$ if $x_{0_{i+1}} - x_{0_i} = 0$. Second, it can be combined into a single four-quadrant inverse tangent such that

$$t_i = \tan^{-1}\frac{k_i/z - \hat{k}_i/z}{1 + k_i\hat{k}_i/z^2} = \tan^{-1}\frac{z(k_i - \hat{k}_i)}{z^2 + k_i\hat{k}_i} \tag{A.73}$$

Then, the potential is given by[1]

$$\phi(x,y,z) = -\frac{\sigma}{4\pi}\sum_{i=1}^{4}(f_i l_i - z t_i) \tag{A.74}$$

while the velocities are given by

$$\begin{aligned} u(x,y,z) &= -\frac{\sigma}{4\pi}\sum_{i=1}^{4}\frac{y_{0_{i+1}} - y_{0_i}}{d_i} l_i \\ v(x,y,z) &= -\frac{\sigma}{4\pi}\sum_{i=1}^{4}\frac{x_{0_i} - x_{0_{i+1}}}{d_i} l_i \\ w(x,y,z) &= \frac{\sigma}{4\pi}\sum_{i=1}^{4} t_i \end{aligned} \tag{A.75}$$

For certain applications, particularly transonic flow corrections, we also need to calculate the second derivatives of the potential. These are obtained by differentiating again Eq. (A.75), such that

$$\begin{aligned} \phi_{xx}(x,y,z) &= -\frac{\sigma}{4\pi}\sum_{i=1}^{4}\frac{y_{0_{i+1}} - y_{0_i}}{d_i}\left(\frac{x - x_{0_i}}{r_i} + \frac{x - x_{0_{i+1}}}{r_{i+1}}\right)\left(\frac{1}{g_i} - \frac{1}{\hat{g}_i}\right) \\ \phi_{xy}(x,y,z) &= -\frac{\sigma}{4\pi}\sum_{i=1}^{4}\frac{y_{0_{i+1}} - y_{0_i}}{d_i}\left(\frac{y - y_{0_i}}{r_i} + \frac{y - y_{0_{i+1}}}{r_{i+1}}\right)\left(\frac{1}{g_i} - \frac{1}{\hat{g}_i}\right) \\ \phi_{xz}(x,y,z) &= -\frac{\sigma}{4\pi}\sum_{i=1}^{4}\frac{y_{0_{i+1}} - y_{0_i}}{d_i}\left(\frac{z}{r_i} + \frac{z}{r_{i+1}}\right)\left(\frac{1}{g_i} - \frac{1}{\hat{g}_i}\right) \\ \phi_{yy}(x,y,z) &= -\frac{\sigma}{4\pi}\sum_{i=1}^{4}\frac{x_{0_i} - x_{0_{i+1}}}{d_i}\left(\frac{y - y_{0_i}}{r_i} + \frac{y - y_{0_{i+1}}}{r_{i+1}}\right)\left(\frac{1}{g_i} - \frac{1}{\hat{g}_i}\right) \\ \phi_{yz}(x,y,z) &= -\frac{\sigma}{4\pi}\sum_{i=1}^{4}\frac{x_{0_i} - x_{0_{i+1}}}{d_i}\left(\frac{z}{r_i} + \frac{z}{r_{i+1}}\right)\left(\frac{1}{g_i} - \frac{1}{\hat{g}_i}\right) \\ \phi_{zz}(x,y,z) &= \frac{\sigma}{4\pi}\sum_{i=1}^{4}\left(\hat{k}_i + z^2\left(\frac{\hat{k}_i}{r_{i+1}^2} - \frac{2m_i}{r_{i+1}}\right)\right)\frac{1}{\hat{k}_i^2 + z^2} \\ &\quad -\left(k_i + z^2\left(\frac{k_i}{r_i^2} - \frac{2m_i}{r_i}\right)\right)\frac{1}{k_i^2 + z^2} \end{aligned} \tag{A.76}$$

A.11 The Doublet Surface Panel

We consider again a flat surface panel lying on the $z = 0$ plane with area S. A continuous doublet distribution is placed on this panel such that the source strength of a small surface element dS lying at x_0, y_0 is denoted by $\mu(x_0, y_0)dS$. The doublet distribution is aligned

1 The expression for the potential given by Katz and Plotkin (2001) is $\phi(x,y,z) = -\frac{\sigma}{4\pi}\sum_{i=1}^{4}(f_i l_i + |z| t_i)$. The form of the expression depends on the way the panel vertices and the normal and tangent vectors are defined. For the definitions used throughout this book, Eq. (A.74) is correct.

with the z axis ($\mathbf{n} = (0,\ 0,\ 1)$, such that the incremental potential induced by this panel at a general point x, y, z by the element $\mathrm{d}S$ is obtained from Eq. (A.52)

$$\mathrm{d}\phi(x,y,z) = -\frac{1}{4\pi}\frac{\mu(x_0,y_0)z}{\left((x-x_0)^2+(y-y_0)^2+(z-z_0)^2\right)^{3/2}}\mathrm{d}S$$

and the total potential induced at the same point is the integral of $\mathrm{d}\phi(x,y,z)$ over S

$$\phi(x,y,z) = -\frac{1}{4\pi}\int_S \frac{\mu(x_0,y_0)z}{\left((x-x_0)^2+(y-y_0)^2+(z-z_0)^2\right)^{3/2}}\mathrm{d}S \tag{A.77}$$

The velocities induced by the panel at point x, y, z can be obtained by differentiating this equation such that

$$\begin{aligned}
u(x,y,z) &= \frac{\partial\phi}{\partial x} = \frac{3}{4\pi}\int_S \frac{(x-x_0)z}{\left((x-x_0)^2+(y-y_0)^2+z^2\right)^{5/2}}\mu(x_0,y_0)\mathrm{d}S\\
v(x,y,z) &= \frac{\partial\phi}{\partial y} = \frac{3}{4\pi}\int_S \frac{(y-y_0)z}{\left((x-x_0)^2+(y-y_0)^2+z^2\right)^{5/2}}\mu(x_0,y_0)\mathrm{d}S\\
w(x,y,z) &= \frac{\partial\phi}{\partial z} = -\frac{1}{4\pi}\int_S \frac{(x-x_0)^2+(y-y_0)^2-2z^2}{\left((x-x_0)^2+(y-y_0)^2+z^2\right)^{5/2}}\mu(x_0,y_0)\mathrm{d}S
\end{aligned} \tag{A.78}$$

Figure A.16 plots the velocity field induced by two different panel geometries, a square and a triangle, for $\mu(x_0, y_0) = 1$. The flow fields are qualitatively identical to those produced by the vortex rings of Figure A.13. In fact, it can be shown that there is an equivalence between a vortex ring and a doublet panel (see Eq. (2.228) and Katz and Plotkin (2001)).

It should be noted that the form of the doublet potential of Eq. (A.77) is identical to that of the w velocity induced by a source panel (Eq. (A.54)). This means that we can apply the same arguments to the two equations for points lying on the panel. In other words, the potential induced by a doublet panel on a point lying on the panel is given by

$$\phi(x,y,0^{\pm}) = \mp\frac{\mu(x,y)}{2} \tag{A.79}$$

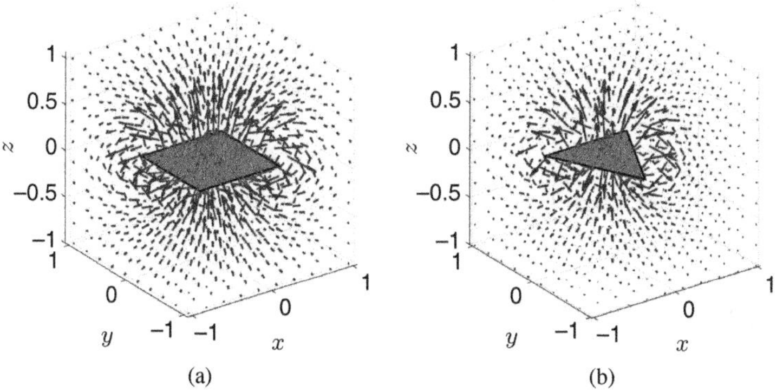

Figure A.16 Velocity field induced by doublet surface panels of constant unit strength. (a) Square panel and (b) triangular panel.

Since we have calculated the potential induced by the panel on itself, we can now calculate the u and v self-induced velocities,

$$u(x, y, 0^{\pm}) = \frac{\partial \phi(x, y, 0^{\pm})}{\partial x} = \mp \frac{1}{2} \frac{\partial \mu(x, y)}{\partial x} \tag{A.80}$$

$$v(x, y, 0^{\pm}) = \frac{\partial \phi(x, y, 0^{\pm})}{\partial y} = \mp \frac{1}{2} \frac{\partial \mu(x, y)}{\partial y} \tag{A.81}$$

If the doublet strength is constant over the entire panel, then both of these velocities will be equal to zero.

The potential induced by quadrilateral panels with constant doublet strength can be obtained from the derivative with respect to z of the potential induced by the same source panel. Differentiating Eq. (A.70) with respect to z and adapting it to the doublet panel leads to

$$\phi(x, y, z) = \frac{\mu}{4\pi} \sum_{i=1}^{4} \left(\frac{\partial}{\partial z} \left(\int_{\theta_i}^{\theta_{i+1}} \sqrt{\frac{b_i^2}{A_i^2 \sin^2(\theta + \phi_i)} + z^2 \mathrm{d}\theta} \right) - \mathrm{sgn}(z)(\theta_{i+1} - \theta_i) \right) \tag{A.82}$$

The derivative of the integral can be obtained by differentiating Eq. (A.71) with respect to z, leading to

$$\frac{\partial}{\partial z} \left(\int \sqrt{\frac{b_i^2}{A_i^2 \sin^2(\theta + \phi_i)} + z^2 \mathrm{d}\theta} \right) =$$

$$\mathrm{sgn}\left(\sin(\theta + \phi_i)\right) \tan^{-1} \left(\frac{\sqrt{b_i^2 - A_i^2 z^2 \sin^2(\theta + \phi_i)}}{A_i z \cos(\theta + \phi_i)} \right) \tag{A.83}$$

As with the source panel, more compact results are given in Katz and Plotkin (2001) for the doublet panel. The potential is obtained from

$$\phi(x, y, z) = \frac{\mu}{4\pi} \sum_{i=1}^{4} t_i \tag{A.84}$$

where t_i is given by Eq. (A.72) or by the four-quadrant alternative of expression (A.73). The velocities are

$$\begin{aligned}
u(x, y, z) &= -\frac{\mu}{4\pi} \sum_{i=1}^{4} z(y_{0_{i+1}} - y_{0_i}) \frac{p_i}{\hat{p}_i} \\
v(x, y, z) &= -\frac{\mu}{4\pi} \sum_{i=1}^{4} z(x_{0_i} - x_{0_{i+1}}) \frac{p_i}{\hat{p}_i} \\
w(x, y, z) &= \frac{\mu}{4\pi} \sum_{i=1}^{4} b_i \frac{p_i}{\hat{p}_i}
\end{aligned} \tag{A.85}$$

where

$$b_i = (x - x_{0_{i+1}})(y - y_{0_i}) - (x - x_{0_i})(y - y_{0_{i+1}})$$
$$p_i = r_i + r_{i+1}$$

$$\hat{p}_i = r_i r_{i+1} q_i$$
$$q_i = r_i r_{i+1} + \left((x - x_{0_i})(x - x_{0_{i+1}}) + (y - y_{0_i})(y - y_{0_{i+1}}) + z^2\right)$$

for $i = 1, \dots, 4$. The second derivatives of the potential are obtained by differentiating equations (A.85), such that

$$\begin{aligned}
\phi_{xx}(x,y,z) = \frac{\mu}{4\pi}\sum_{i=1}^{4} z\frac{(y_{0_i} - y_{0_{i+1}})}{\hat{p}_i} & \left\{ \frac{(x - x_{0_i})}{r_i} + \frac{(x - x_{0_{i+1}})}{r_{i+1}} \right. \\
& - \frac{p_i}{q_i}\left(2x - x_{0_i} - x_{0_{i+1}} + \frac{(x - x_{0_i})r_{i+1}}{r_i} + \frac{(x - x_{0_{i+1}})r_i}{r_{i+1}}\right) \\
& \left. - \frac{p_i(x - x_{0_i})}{r_i^2} - \frac{p_i(x - x_{0_{i+1}})}{r_{i+1}^2} \right\} \\
\phi_{xy}(x,y,z) = \frac{\mu}{4\pi}\sum_{i=1}^{4} z\frac{(y_{0_i} - y_{0_{i+1}})}{\hat{p}_i} & \left\{ \frac{(y - y_{0_i})}{r_i} + \frac{(y - y_{0_{i+1}})}{r_{i+1}} \right. \\
& - \frac{p_i}{q_i}\left(2y - y_{0_i} - y_{0_{i+1}} + \frac{(y - y_{0_i})r_{i+1}}{r_i} + \frac{(y - y_{0_{i+1}})r_i}{r_{i+1}}\right) \\
& \left. - \frac{p_i(y - y_{0_i})}{r_i^2} - \frac{p_i(y - y_{0_{i+1}})}{r_{i+1}^2} \right\} \\
\phi_{xz}(x,y,z) = \frac{\mu}{4\pi}\sum_{i=1}^{4} (y_{0_i} - y_{0_{i+1}})\frac{p_i}{\hat{p}_i} & \left\{ 1 - \frac{z^2}{r_{i+1}^2} - \frac{z^2}{r_i^2} + \frac{z}{p_i}\left(\frac{z}{r_i} + \frac{z}{r_{i+1}}\right) \right. \\
& \left. - \frac{z}{q_i}\left(2z + z\frac{r_{i+1}}{r_i} + \frac{z r_i}{r_{i+1}}\right) \right\} \\
\phi_{yy}(x,y,z) = \frac{\mu}{4\pi}\sum_{i=1}^{4} z\frac{(x_{0_i} - x_{0_{i+1}})}{\hat{p}_i} & \left\{ -\frac{(y - y_{0_i})}{r_i} - \frac{(y - y_{0_{i+1}})}{r_{i+1}} \right. \\
& + \frac{p_i}{q_i}\left(2y - y_{0_i} - y_{0_{i+1}} + \frac{(y - y_{0_i})r_{i+1}}{r_i} + \frac{(y - y_{0_{i+1}})r_i}{r_{i+1}}\right) \\
& \left. + \frac{p_i(y - y_{0_i})}{r_i^2} + \frac{p_i(y - y_{0_{i+1}})}{r_{i+1}^2} \right\} \\
\phi_{yz}(x,y,z) = \frac{\mu}{4\pi}\sum_{i=1}^{4} (x_{0_i} - x_{0_{i+1}})\frac{p_i}{\hat{p}_i} & \left\{ -1 + \frac{z^2}{r_{i+1}^2} + \frac{z^2}{r_i^2} - \frac{z}{p_i}\left(\frac{z}{r_i} + \frac{z}{r_{i+1}}\right) \right. \\
& \left. + \frac{z}{q_i}\left(2z + z\frac{r_{i+1}}{r_i} + \frac{z r_i}{r_{i+1}}\right) \right\} \\
\phi_{zz}(x,y,z) = \frac{\mu}{4\pi}\sum_{i=1}^{4} z\frac{b_i}{\hat{p}_i} & \left\{ \frac{1}{r_i} + \frac{1}{r_{i+1}} - \frac{p_i}{q_i}\left(2 + \frac{r_{i+1}}{r_i} + \frac{r_i}{r_{i+1}}\right) - \frac{p_i}{r_{i+1}^2} - \frac{p_i}{r_i^2} \right\}
\end{aligned} \tag{A.86}$$

References

Hess, J.L. and Smith, A.M.O. (1967). Calculation of potential flow about arbitrary bodies. *Progress in Aerospace Sciences* 8 (1): 1–138.

Katz, J. and Plotkin, A. (2001). *Low Speed Aerodynamics*. Cambridge University Press.

B

Fundamental Solutions of the Linearized Small Disturbance Equation

This appendix presents some of the fundamental solutions of the linearized small disturbance equation for compressible potential flow (2.143) that are used in this book.

B.1 The Subsonic Doublet Surface Panel

Consider a flat surface lying on the $z = 0$ plane with area S, exposed to a subsonic free stream of speed U_∞ and Mach number M_∞. A continuous doublet distribution is placed on this panel such that the doublet strength of a small surface element $\mathrm{d}S$ lying at x_0, y_0 is denoted by $\mu(x_0, y_0, t)\mathrm{d}S$. The incremental potential induced by this panel at any of the general points x, y, z by the element $\mathrm{d}S$ is obtained in the frequency domain from Eq. (6.32)

$$\begin{aligned}\mathrm{d}\phi(\mathbf{x}, \omega) &= \frac{\mu(x_0, y_0, \omega)}{4\pi}\frac{\partial}{\partial z}\left(\frac{\mathrm{e}^{-\mathrm{i}\omega\tau(\mathbf{x},\mathbf{x}_0)}}{r_\beta(\mathbf{x}, \mathbf{x}_0)}\right)\mathrm{d}S \\ &= -\frac{\mu(\omega)}{4\pi}\frac{z\mathrm{e}^{-\mathrm{i}\omega\tau}}{r_\beta^2}\left(\frac{\beta^2}{r_\beta} + \frac{\mathrm{i}\omega}{a_\infty}\right)\mathrm{d}S\end{aligned}$$

where

$$r_\beta(\mathbf{x}, \mathbf{x}_0) = \sqrt{\left(x - x_0\right)^2 + \beta^2\left(y - y_0\right)^2 + \beta^2 z^2} \tag{B.1}$$

$$\tau(\mathbf{x}, \mathbf{x}_0) = \frac{-M_\infty(x - x_0) + r_\beta(\mathbf{x}, \mathbf{x}_0)}{a_\infty\beta^2} \tag{B.2}$$

The total potential induced at the same point is the integral of $\mathrm{d}\phi(x, y, z)$ over S

$$\phi(\mathbf{x}, \omega) = -\frac{1}{4\pi}\int_S \mu(x_0, y_0, \omega)\frac{z\mathrm{e}^{-\mathrm{i}\omega\tau}}{r_\beta^2}\left(\frac{\beta^2}{r_\beta} + \frac{\mathrm{i}\omega}{a_\infty}\right)\mathrm{d}S \tag{B.3}$$

Of particular interest is the case where the point $\mathbf{x}$ lies on the panel itself so that $z = 0$. As the integrand in Eq. (B.3) is multiplied by z, if the point at which we evaluate the potential lies on the panel, then the integrand will be equal to zero everywhere except at $x_0 = x$, $y_0 = 0$,

Unsteady Aerodynamics: Potential and Vortex Methods, First Edition. Grigorios Dimitriadis.

Companion website: www.wiley.com/go/dimitriadis/unsteady_aerodynamics

where it will be undefined. We can therefore proceed to calculate the potential induced by the panel on itself in the same way we did in Section A.10 and define

$$x_0 = x + x', \quad y_0 = y + y'$$

so that $dx_0 = dx'$ and $dy_0 = dy'$. We will calculate the integral of Eq. (B.3) for x' and y' values between $-\varepsilon$ and ε, where $\varepsilon \ll 1$. Over such a small surface, we assume that the strength of the doublet is constant and equal to $\mu(\omega)$. Next, we need to think of the value of τ, which becomes

$$\tau = \frac{-M_\infty x' + \sqrt{x'^2 + \beta^2 y'^2 + \beta^2 z^2}}{a_\infty \beta^2}$$

As x' and y' take values between $-\varepsilon$ and ε and z approaches zero, τ is of order ε / a_∞ and

$$\omega\tau = O\left(\frac{\omega\varepsilon}{a_\infty}\right)$$

This means that we can approximate the exponential term in the integrand of Eq. (B.3) by

$$e^{-i\omega\tau} = 1$$

as long as $\omega \ll a_\infty / \varepsilon$. Since ε is arbitrarily small, this condition implies that ω should be finite. After substituting this result in Eq. (B.3), we obtain

$$\phi(\mathbf{x}, \omega) = -\frac{\mu(\omega)}{4\pi} \int_{-\varepsilon}^{+\varepsilon} \int_{-\varepsilon}^{+\varepsilon} \left(\frac{\beta^2 z}{r_\beta(x', y', z)^3} + \frac{i\omega}{a_\infty} \frac{z}{r_\beta(x', y', z)^2} \right) dx' \, dy'$$

Blair (1994) observed that, since x', y' and z are small,

$$\frac{z}{r_\beta(x', y', z)^2} \ll \frac{z}{r_\beta(x', y', z)^3}$$

We can therefore neglect the small term in the integral, such that

$$\phi(\mathbf{x}, \omega) = -\frac{\mu(\omega)}{4\pi} \int_{-\varepsilon}^{+\varepsilon} \int_{-\varepsilon}^{+\varepsilon} \frac{\beta^2 z}{r_\beta(x', y', z)^3} dx' \, dy'$$

First, we evaluate the integral in x', which becomes

$$\int_{-\varepsilon}^{+\varepsilon} \frac{\beta^2 z}{r_\beta(x', y', z)^3} dx' = \left. \frac{x' z}{(y'^2 + z^2) r_\beta} \right|_{-\varepsilon}^{+\varepsilon} = \frac{2\varepsilon z}{(y'^2 + z^2)\sqrt{\varepsilon^2 + \beta^2 y'^2 + \beta^2 z^2}}$$

and then we integrate over y' to obtain

$$\phi(\mathbf{x}, \omega) = -\frac{\mu(\omega)}{\pi} \tan^{-1} \frac{\varepsilon^2}{z\sqrt{(1 + \beta^2)\varepsilon^2 + \beta^2 z^2}}$$

As usual, we let z tend to $0^\pm$ faster than ε so that the argument in the inverse tangent becomes $\pm\infty$. Hence, taking the limit of $\varphi(x, y, z, \omega)$ as $z \to 0^\pm$ gives

$$\phi(x, y, 0^\pm, \omega) = \mp \frac{\mu(\omega)}{2} \tag{B.4}$$

This result makes sense because as only the doublet lying at x, y has non-zero influence at point $x, y, 0$, the distance from the point to the doublet is zero, and hence, there is no delay between the time a change occurs in the strength of the doublet and the time it is felt at point $x, y, 0$. As a consequence, the self-influence of a subsonic doublet panel is equal to that of an incompressible doublet panel.

B.2 The Acoustic Source Surface Panel

Consider the linearized small perturbation equation (2.143)

$$\left(1 - M_\infty^2\right)\phi_{xx} + \phi_{yy} + \phi_{zz} - \frac{2M_\infty}{a_\infty}\phi_{xt} - \frac{1}{a_\infty^2}\phi_{tt} = 0 \tag{B.5}$$

and set $M_\infty = 0$, so that there is no free stream. The result is an acoustic equation of the form

$$\phi_{xx} + \phi_{yy} + \phi_{zz} - \frac{1}{a_\infty^2}\phi_{tt} = 0 \tag{B.6}$$

Applying the Fourier Transform to this equation, we obtain its frequency domain version

$$\phi_{xx}(\omega) + \phi_{yy}(\omega) + \phi_{zz}(\omega) + \frac{\omega^2}{a_\infty^2}\phi(\omega) = 0 \tag{B.7}$$

where ω is the frequency of oscillation of $\phi(\omega)$. The acoustic equation has a fundamental source solution such that the potential induced at a general point x, y, z by an acoustic source of strength $\sigma(\omega)$ lying at x_0, y_0, z_0 is given by

$$\phi\left(\mathbf{x}, \mathbf{x}_0, \omega\right) = -\frac{\sigma(\omega)}{4\pi}\frac{\mathrm{e}^{-\mathrm{i}\omega r(\mathbf{x},\mathbf{x}_0)/a_\infty}}{r(\mathbf{x}, \mathbf{x}_0)} \tag{B.8}$$

where

$$r(\mathbf{x}, \mathbf{x}_0) = \sqrt{\left(x - x_0\right)^2 + \left(y - y_0\right)^2 + \left(z - z_0\right)^2}$$

This acoustic source solution can be obtained from Eq. (6.31) after setting $M_\infty = 0$.

We now consider a flat quadrilateral panel lying on the $z = 0$ plane with area S and apply a constant source distribution on it, such that the source strength of any small surface element $\mathrm{d}S$ lying at x_0, y_0 is $\sigma(\omega)\mathrm{d}S$. The total potential induced by this acoustic source panel at a general point $\mathbf{x}$ is given by

$$\phi(\mathbf{x}, \omega) = -\frac{\sigma(\omega)}{4\pi}\int_S \frac{\mathrm{e}^{-\mathrm{i}\Omega r(\mathbf{x},\mathbf{x}_0)}}{r(\mathbf{x}, \mathbf{x}_0)}\mathrm{d}S \tag{B.9}$$

where we have defined $\Omega = \omega/a_\infty$ and $r(\mathbf{x}, \mathbf{x}_0) = \sqrt{\left(x - x_0\right)^2 + \left(y - y_0\right)^2 + z^2}$. Figure A.15 shows two views of the quadrilateral panel and the point $\mathbf{x}$. We will use the Hess and Smith (1967) method of Section A.10 in order to evaluate the potential of Eq. (B.9). Recalling that distance R and angle θ are given by Eqs. (A.59) and (A.60) and substituting $\mathrm{d}S = R\,\mathrm{d}R\,\mathrm{d}\theta$, Eq. (B.9) becomes

$$\phi(\mathbf{x}, \omega) = -\frac{\sigma(\omega)}{4\pi}\oint\int_0^R \frac{\mathrm{e}^{-\mathrm{i}\Omega r}}{r}R\,\mathrm{d}R\,\mathrm{d}\theta$$

so that after substituting for $R\,\mathrm{d}R = r\,\mathrm{d}r$ (expression (A.62)), we obtain

$$\phi(\mathbf{x}, \omega) = -\frac{\sigma(\omega)}{4\pi}\oint\int_{|z|}^r \mathrm{e}^{-\mathrm{i}\Omega r}\,\mathrm{d}r\,\mathrm{d}\theta$$

The inner integral can be evaluated easily such that

$$\phi(\mathbf{x}, \omega) = \frac{\mathrm{i}\sigma(\omega)}{4\pi\Omega}\oint\left(\mathrm{e}^{-\mathrm{i}\Omega r} - \mathrm{e}^{-\mathrm{i}\Omega|z|}\right)\mathrm{d}\theta$$

In order to calculate this last remaining integral, we split it into the four segments of the quadrilateral, as we did in Section A.10. Then,

$$\phi(\mathbf{x},\omega) = \frac{\mathrm{i}\sigma(\omega)}{4\pi\Omega}\sum_{i=1}^{4}\int_{\theta_i}^{\theta_{i+1}}\left(\mathrm{e}^{-\mathrm{i}\Omega r} - \mathrm{e}^{-\mathrm{i}\Omega|z|}\right)\mathrm{d}\theta$$

where θ_i is given by Eq. (A.65). We also substitute for r in terms of θ from Eq. (A.69) to obtain

$$\phi(\mathbf{x},\omega) = \frac{\mathrm{i}\sigma(\omega)}{4\pi\Omega}\sum_{i=1}^{4}\int_{\theta_i}^{\theta_{i+1}}\left(\mathrm{e}^{-\mathrm{i}\Omega\sqrt{\frac{b_i^2}{A_i^2\sin^2(\theta+\phi_i)}+z^2}} - \mathrm{e}^{-\mathrm{i}\Omega|z|}\right)\mathrm{d}\theta$$

Unfortunately, there is no obvious analytical solution for this integral, but it can be evaluated numerically. The calculation starts with creating n_k values of $\theta_{i,k}$, linearly spaced between θ_i and θ_{i+1}. Then, we calculate $r_{i,k}(\theta_{i,k})$ from Eq. (A.69) and finally, we evaluate the potential from

$$\phi(\mathbf{x},\omega) = \frac{\mathrm{i}\sigma(\omega)}{4\pi\Omega}\sum_{i=1}^{4}\sum_{k=1}^{n_k}\left(\mathrm{e}^{-\mathrm{i}\Omega r_{i,k}} - \mathrm{e}^{-\mathrm{i}\Omega|z|}\right)\Delta\theta \tag{B.10}$$

where $\Delta\theta = \theta_{i,2} - \theta_{i,1}$.

B.3 The Acoustic Doublet Surface Panel

The acoustic doublet is obtained from the subsonic doublet if we set $U_\infty = M_\infty = 0$, such that τ in Eq. (B.1) becomes

$$\tau(\mathbf{x},\mathbf{x}_0) = \frac{r(\mathbf{x},\mathbf{x}_0)}{a_\infty} \tag{B.11}$$

Then, the potential induced at a point $\mathbf{x}$ by an acoustic doublet of strength $\mu(\omega)\mathrm{d}S$ lying at $\mathbf{x}_0$ and pointing in the direction $\mathbf{n}$ is given in the frequency domain by

$$\phi = -\frac{\mu(\omega)}{4\pi}\mathbf{n}\cdot\nabla\left(\frac{\mathrm{e}^{-\mathrm{i}\Omega r(\mathbf{x},\mathbf{x}_0)}}{r(\mathbf{x},\mathbf{x}_0)}\right) \tag{B.12}$$

where, again, $\Omega = \omega/a_\infty$.

If we consider a flat surface lying on the $z = 0$ plane with area S and apply a constant doublet distribution on it such that the doublet strength of a small surface element $\mathrm{d}S$ lying at x_0, y_0 is denoted by $\mu(\omega)\mathrm{d}S$, the total potential induced at any general point x, y, z is given from Eq. (B.3) as

$$\phi(\mathbf{x},\omega) = -\frac{\mu(\omega)}{4\pi}\int_S \frac{z\mathrm{e}^{-\mathrm{i}\Omega r}}{r^2}\left(\frac{1}{r} + \mathrm{i}\Omega\right)\mathrm{d}S \tag{B.13}$$

where $\Omega = \omega/a_\infty$. We will evaluate this integral using the approach that was applied to the acoustic source panel in Section B.2. Setting $\mathrm{d}S = R\,\mathrm{d}R\,\mathrm{d}\theta = r\,\mathrm{d}r\,\mathrm{d}\theta$, where R and θ are given by Eqs. (A.59) and (A.60), we obtain

$$\phi(\mathbf{x},\omega) = -\frac{\mu(\omega)z}{4\pi}\oint\int_{|z|}^{r}\frac{\mathrm{e}^{-\mathrm{i}\Omega r}}{r}\left(\frac{1}{r} + \mathrm{i}\Omega\right)\mathrm{d}r\,\mathrm{d}\theta$$

The integral over r is easy because

$$\frac{\mathrm{d}}{\mathrm{d}r}\left(\frac{\mathrm{e}^{-\mathrm{i}\Omega r}}{r}\right) = -\frac{\mathrm{e}^{-\mathrm{i}\Omega r}}{r}\left(\frac{1}{r} + \mathrm{i}\Omega\right)$$

Applying the limits of integration, we obtain

$$\phi(\mathbf{x}, \omega) = \frac{\mu(\omega)z}{4\pi} \oint \left(\frac{\mathrm{e}^{-\mathrm{i}\Omega r}}{r} - \frac{\mathrm{e}^{-\mathrm{i}\Omega|z|}}{|z|}\right) \mathrm{d}\theta$$

The final calculation of this integral is numerical, as in the case of Eq. (B.10). We split the integral into four sums, one for each line segment. For the ith segment, we create n_k values of $\theta_{i,k}$, linearly spaced between θ_i and θ_{i+1}, and calculate $r_{i,k}(\theta_{i,k})$ from Eq. (A.69). Then, the potential of a quadrilateral acoustic doublet panel is given by

$$\phi(\mathbf{x}, \omega) = \frac{\mu(\omega)z}{4\pi} \sum_{i=1}^{4} \sum_{k=1}^{n_k} \left(\frac{\mathrm{e}^{-\mathrm{i}\Omega r_{i,k}}}{r_{i,k}} - \frac{\mathrm{e}^{-\mathrm{i}\Omega|z|}}{|z|}\right) \Delta\theta \tag{B.14}$$

where $\Delta\theta = \theta_{i,2} - \theta_{i,1}$.

B.4 The Supersonic Source Surface Panel

The potential induced by a supersonic source panel of constant strength $\sigma(\omega)\mathrm{d}S$ lying on the plane $z = 0$ on a general point lying at $\mathbf{x} = (x, y, 0)$ is given by

$$\phi(\mathbf{x}, \omega) = -\frac{\sigma(\omega)}{\pi} \int_{S_{\mathrm{MC}}} \frac{\mathrm{e}^{-\mathrm{i}\omega M_\infty (x-x_0)/a_\infty B^2} \cos\left(\omega r_B / a_\infty B^2\right)}{r_B} \mathrm{d}S$$

where $B^2 = M_\infty^2 - 1$ and $r_B = \sqrt{\left(x - x_0\right)^2 - B^2\left(y - y_0\right)^2}$. The integral over S_{MC} denotes that only points on the panel that lie within the upstream Mach cone of point $x, y, 0$ are included. The Mach angle is given by $\mu = \sin^{-1}(1/M_\infty)$. After applying the supersonic Prandtl–Glauert transformation

$$\xi = \frac{x}{B}, \quad \eta = y, \quad \zeta = z \tag{B.15}$$

the potential becomes

$$\phi(\mathbf{x}, \omega) = -\frac{\sigma(\omega)}{\pi} \int_{\Sigma_{\mathrm{MC}}} \frac{\mathrm{e}^{-\mathrm{i}\Omega M_\infty (\xi - \xi_0)} \cos(\Omega r)}{r} \mathrm{d}\Sigma \tag{B.16}$$

where $\Omega = \omega / a_\infty B$ and $r = \sqrt{\left(\xi - \xi_0\right)^2 - \left(\eta - \eta_0\right)^2}$. Note that $\Sigma = S/B$ is the area of the transformed panel and Σ_{MC} is the area of the transformed panel lying inside the transformed upstream Mach cone, whose Mach angle is 45°.

We will start by assuming that the entire panel lies inside the upstream Mach cone of the point, as shown in Figure B.1. Using polar coordinates, we can see that

$$\xi - \xi_0 = -R\cos\theta, \quad \eta - \eta_0 = -R\sin\theta, \quad \mathrm{d}\Sigma = R\,\mathrm{d}R\,\mathrm{d}\theta$$

where $R = \sqrt{(\xi - \xi_0)^2 + (\eta - \eta_0)^2}$ and $\theta = \tan^{-1}((\eta_0 - \eta)/(\xi_0 - \xi))$. Furthermore, r can be written as

$$r^2 = (\xi - \xi_0)^2 - (\eta - \eta_0)^2 = R^2 - 2(\eta - \eta_0)^2 = R^2(1 - 2\sin^2\theta) = R^2 \cos 2\theta$$

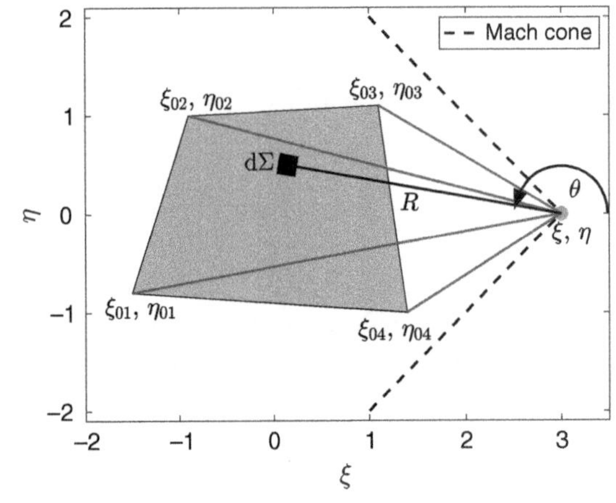

Figure B.1 Quadrilateral panel lying on $\zeta = 0$ and general point in space, ξ, η, ζ.

or

$$r = R\sqrt{\cos 2\theta}$$

Substituting for r, $\xi - \xi_0$ and $\mathrm{d}\Sigma$ in Eq. (B.16), we obtain

$$\phi(\mathbf{x}, \omega) = -\frac{\sigma(\omega)}{\pi} \oint \int_0^R \frac{\mathrm{e}^{\mathrm{i}\Omega M_\infty R\cos\theta} \cos(\Omega R\sqrt{\cos 2\theta})}{\sqrt{\cos 2\theta}} \mathrm{d}R\,\mathrm{d}\theta \tag{B.17}$$

where the integral over θ is calculated for all possible θ values, starting at θ_1, going to θ_4 and back to θ_1. In this way, the area of the triangle formed by points (ξ, η), (ξ_{0_3}, η_{0_3}) and (ξ_{0_4}, η_{0_4}) is subtracted from the area of the pentagon formed by the four vertices of the panel and point (ξ, η).

The inner integral in expression (B.17) can be evaluated from

$$\int \mathrm{e}^{\mathrm{i}aR} \cos(bR)\mathrm{d}R = -\frac{\mathrm{e}^{\mathrm{i}aR}\,(b\sin(bR) + \mathrm{i}a\cos(bR))}{a^2 - b^2} \tag{B.18}$$

where $a = \Omega M_\infty \cos\theta$ and $b = \Omega\sqrt{\cos 2\theta}$. Substituting and applying the limits leads to

$$\begin{aligned}\phi(\mathbf{x}, \omega) = &-\frac{\sigma(\omega)}{\pi} \oint \frac{1}{\Omega(M_\infty^2 \cos^2\theta - \cos 2\theta)\sqrt{\cos 2\theta}} \\ &\times (\mathrm{e}^{\mathrm{i}\Omega M_\infty R\cos\theta} \left(\sqrt{\cos 2\theta}\sin(\Omega R\sqrt{\cos 2\theta}) + \mathrm{i}M_\infty \cos\theta\cos(\Omega R\sqrt{\cos 2\theta})\right) \\ &- \mathrm{i}M_\infty \cos\theta)\mathrm{d}\theta\end{aligned} \tag{B.19}$$

There is no obvious analytical result for the line integral in Eq. (B.19), but it can be evaluated numerically, as in the case of Eq. (B.10). First, we need to determine if any part of the panel lies in the upstream Mach cone of the point ξ, η. For each one of the segments of the panel, we calculate the angle θ_i between the point ξ, η and the vertices of the segment using Eq. (A.66). It is very important to use the four-quadrant inverse tangent function for this calculation, such that the resulting angles will lie either between 0 and π or between 0 and $-\pi$.

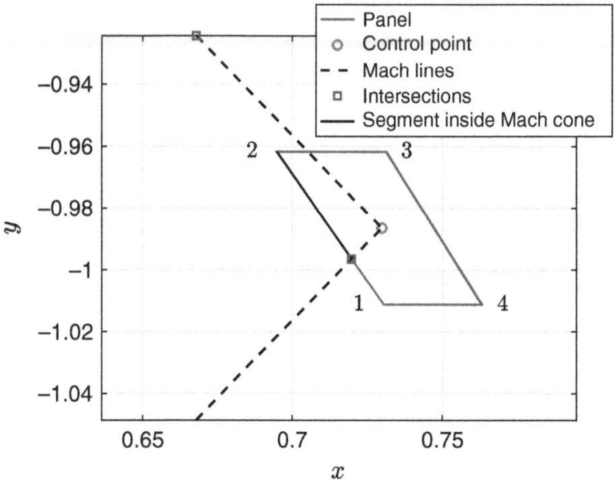

Figure B.2 Quadrilateral supersonic source panel, control panel, Mach lines and intersections.

This means that if $|\theta_i| > 3\pi/4$, then the ith vertex of the panel will lie in the upstream Mach cone of point ξ, η. However, if none of the vertices lies in this cone, a region of the panel may still lie inside. One way to check if this is the case is to calculate the intersections between the panel segments and the upstream Mach lines, as seen in Figure B.2. In this figure, we are determining which segment of side 12 of the panel lies inside the forward Mach cone of the control point of the same panel. We denote the co-ordinates of the vertices of segment 12 by $\boldsymbol{\xi}_0 = (\xi_{0_1}, \xi_{0_2})$, $\boldsymbol{\eta}_0 = (\eta_{0_1}, \eta_{0_2})$ and the coordinates of the two intersections by $\hat{\boldsymbol{\xi}} = (\hat{\xi}_1, \hat{\xi}_2)$, $\hat{\boldsymbol{\eta}} = (\hat{\eta}_1, \hat{\eta}_2)$.

- If point 1 lies outside the Mach cone, then
  - If point 2 lies inside the Mach cone, the side will surely intersect one of the Mach lines. Find which intersection point lies closest to point 1 and set $\xi_{0_1} = \hat{\xi}_k$, $\eta_{0_1} = \hat{\eta}_k$, where k is the index of the relevant intersection point. Point ξ_{0_2}, η_{0_2} remains unchanged.
  - If point 2 also lies outside the Mach cone, there will be either 0 or two intersections. If there are two intersections, then
    - If $\eta_{0_1} - \eta \leq 0$, set $\boldsymbol{\xi}_0 = \hat{\boldsymbol{\xi}}$, $\boldsymbol{\eta}_0 = \hat{\boldsymbol{\eta}}$
    - If $\eta_{0_1} - \eta > 0$, set $\boldsymbol{\xi}_0 = (\hat{\xi}_2, \hat{\xi}_1)$, $\boldsymbol{\eta}_0 = (\hat{\eta}_2, \hat{\eta}_1)$
- If point 1 lies inside the Mach cone, then
  - If point 2 lies inside the Mach cone, keep $\boldsymbol{\xi}_0$ unchanged.
  - If point 2 lies outside the Mach cone, the side will surely intersect one of the Mach lines. Find which intersection point lies closest to point 2 and set $\xi_{0_2} = \hat{\xi}_k$, $\eta_{0_2} = \hat{\eta}_k$, where k is the index of the relevant intersection point. Point ξ_{0_1}, η_{0_1} remains unchanged.
- In all other cases, the side 12 does not lie inside the Mach cone at all and the side has no influence on the potential induced at point ξ, η.

If a part of a segment lies inside the Mach cone, the vertices of this part will be given by $\boldsymbol{\xi}_0$, $\boldsymbol{\eta}_0$. We recalculate the angles $\theta_{1,2}$ from Eq. (A.65) and then we unwrap $\theta_{1,2}$, so that the jump between the angles of the vertices of the segment cannot be greater than π. We create

n_k values of $\theta_{i,k}$ linearly spaced between θ_1 and θ_2 and then we calculate the corresponding $R_{i,k}$ values from Eq. (A.68). Next, we calculate $a_{i,k} = \Omega M_\infty \cos\theta_{i,k}$, $b_{i,k} = \Omega\sqrt{\cos 2\theta_{i,k}}$ and the integrand of Eq. (B.19)

$$I_{i,k}(\omega) = \left(\frac{e^{ia_{i,k}R_{i,k}}\left(b_{i,k}\sin(b_{i,k}R_{i,k}) + ia_{i,k}(\cos(b_{i,k}R_{i,k}) - 1)\right)}{a_{i,k}^2 - b_{i,k}^2} \right) \sqrt{\sec 2\theta_{i,k}} \tag{B.20}$$

where $i = 1, \dots, 4$ denotes the ith segment and $k = 1, \dots, n_k$ denotes the kth value of $\theta_{i,k}$ on the ith segment. We can now evaluate the integral

$$\phi(\xi, \omega) = -\frac{\sigma(\omega)}{\pi} \sum_{i=1}^{4} \sum_{k=1}^{n_k} I_{i,k}(\omega)\Delta\theta \tag{B.21}$$

if $\sec 2\theta_{i,k}$ remains finite throughout the segment. However, if one of the two vertices of the segment is an intersection with a Mach line, then $\sec 2\theta_{i,k}$ will become infinite at that vertex and the integral cannot be evaluated. One way to overcome this difficulty is to replace the function $\sec 2\theta$ by another function that is not singular at the Mach lines. Moore and Andrew (1965) replaced the entire term $\cos \Omega r/r$ in the integrand of Eq. (B.16) by an infinite series, which they then truncated in order to evaluate it numerically. Here, we will do something similar, we will replace $\sec 2\theta$ by its Taylor expansion around $\theta = 0$,

$$\sec 2\theta = \sum_{l=0}^{\infty} d_{2l}(2\theta)^{2l} = 1 + \frac{1}{2}(2\theta)^2 + \frac{5}{24}(2\theta)^4 + \frac{61}{720}(2\theta)^6 + \frac{277}{8064}(2\theta)^8 \dots$$

The series is infinite, but the values of the coefficients become smaller than machine precision after the 40th term. Therefore, $\sec 2\theta$ can be written as

$$\sec 2\theta \approx \sum_{l=0}^{l_{max}} d_{2l}(2\theta)^{2l} \tag{B.22}$$

where $l_{max} \le 40$, depending on the desired accuracy. In this expansion θ must take values between $-\pi/4$ and $\pi/4$, while the values of $\theta_{i,k}$ in Eq. (B.20) lie in the intervals $\pm[3\pi/4, 5\pi/4]$. This means that integer multiples of π must be added to or subtracted from $\theta_{i,k}$ before using Eq. (B.22).

In the steady case, $\omega = \Omega = 0$, the potential of Eq. (B.16) simplifies to

$$\phi(\mathbf{x}, 0) = -\frac{\sigma(0)}{\pi} \int_{\Sigma_{MC}} \frac{1}{r} \mathrm{d}\Sigma \tag{B.23}$$

Using polar coordinates, as in the unsteady case, $r = R\sqrt{\cos 2\theta}$ and $\mathrm{d}\Sigma = R\,\mathrm{d}R\,\mathrm{d}\theta$. Substituting into Eq. (B.23), we obtain

$$\phi(\mathbf{x}, 0) = -\frac{\sigma(0)}{\pi} \oint \int_0^R \frac{1}{\sqrt{\cos 2\theta}} \mathrm{d}R\,\mathrm{d}\theta$$

The integral over R is trivial, so that

$$\phi(\mathbf{x}, 0) = -\frac{\sigma(0)}{\pi} \oint \frac{R}{\sqrt{\cos 2\theta}} \mathrm{d}\theta \tag{B.24}$$

The value of R must be substituted from Eq. (A.68) for the ith segment, but then there is no obvious analytical form for the resulting integral. We therefore follow the numerical integration procedure outlined for the unsteady case. The steady version of the integrand of Eq. (B.20) is simply

$$I_{i,k}(0) = R_k \sqrt{\sec 2\theta_k} \tag{B.25}$$

and the total potential becomes

$$\phi(\xi, 0) = -\frac{\sigma(0)}{\pi} \sum_{i=1}^{4} \sum_{k=1}^{n_k} I_k(0) \Delta\theta \tag{B.26}$$

References

Blair, M. (1994). A Compilation of the Mathematics Leading to the Doublet-Lattice Method. *Report WL-TR-95-3022*. Air Force Wright Laboratory.

Hess, J.L. and Smith, A.M.O. (1967). Calculation of potential flow about arbitrary bodies. *Progress in Aerospace Sciences* 8 (1): 1–138.

Moore, M.T. and Andrew, L.V. (1965). Unsteady Aerodynamics for Advanced Configurations: Part IV – Application of the Supersonic Mach Box Method to Intersecting Planar Lifting Surfaces. *Report FDL TDR 64-152*. Air Force Flight Dynamics Laboratory.

C

Wagner's Derivation of the Kutta Condition

This appendix presents the derivation of the Kutta condition developed by Wagner (1925). The derivation of Eq. (3.76) given in Chapter 3 is the one by Theodorsen. As Wagner did not include any sources in his modelling and only considered impulsively started flows, he followed a simpler approach.

Wagner's theory models the flat plate as a circle in the complex $\zeta = \xi + i\eta$ plane and places pairs of vortices of strength $\pm\Gamma$, one lying at $(\Xi_0, 0)$ and one lying at $(b^2/\Xi_0, 0)$. Figure C.1 is a reproduction of Figure 3.8 but without the sources, showing the circle of radius b in the ζ plane and the placement of a pair of vortices.

After applying the conformal transformation

$$z = \frac{1}{2}\left(\zeta + \frac{b^2}{\zeta}\right)$$

the circle in the ζ plane is transformed into a flat plate of chord c in the $z = x + iy$ plane. The vortices lying at Ξ_0 and b^2/Ξ_0 both transform to point

$$X_0 = \left(\frac{1}{2}\left(\Xi_0 + \frac{b^2}{\Xi_0}\right), 0\right)$$

in the flat plate plane (see Section 3.3.5). Wagner then calculated the vertical velocity at the trailing edge $\xi = b$, $\eta = 0$ induced by the two vortices using Eq. (A.19), such that

$$dv(b, 0) = -\frac{\Gamma_0}{2\pi}\left(\frac{1}{b - b^2/\Xi_0} - \frac{1}{b - \Xi_0}\right) = -\frac{\Gamma_0}{2\pi b}\left(\frac{\Xi_0 + b}{\Xi_0 - b}\right)$$

Finally, after substituting from Eq. (3.67) and simplifying, the vertical velocity induced at the trailing edge by a single pair of vortices becomes

$$dv(b, 0) = -\frac{\Gamma_0}{2\pi b}\sqrt{\frac{X_0 + b}{X_0 - b}}$$

The total vertical velocity induced by all the vortices is given by

$$v(b, 0) = -\frac{1}{2\pi b}\int_b^\infty \sqrt{\frac{X_0 + b}{X_0 - b}}\bar{\Gamma}(X_0, t)dX_0$$

Unsteady Aerodynamics: Potential and Vortex Methods, First Edition. Grigorios Dimitriadis.

Companion website: www.wiley.com/go/dimitriadis/unsteady_aerodynamics

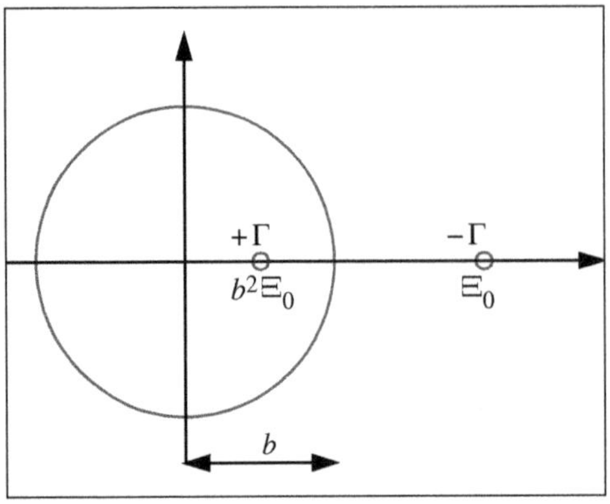

Figure C.1 Wagner's flow model.

where we have used the definition $\Gamma(X_0, t) = \bar{\Gamma}(X_0, t)\mathrm{d}X_0$ of Eq. (3.72). Wagner enforced the Kutta condition by requiring that the total vertical flow velocity at the trailing edge is equal to zero, that is $U\alpha + v(b, 0) = 0$. Consequently,

$$\frac{1}{2\pi b}\int_b^\infty \sqrt{\frac{X_0 + b}{X_0 - b}}\bar{\Gamma}(X_0, t)\mathrm{d}X_0 = U\alpha$$

which is identical to Eq. (3.76) for an impulsively started plate, i.e. if $\dot{h} = \dot{\alpha} = 0$.

Reference

Wagner, H. (1925). Über die entstehung des dynamischen auftriebes von tragflügeln. *Zeitschrift für Angewandte Mathematik und Mechanik* 5 (1): 17–35.

Index

Unsteady Aerodynamics: Potential and Vortex Methods, First Edition. Grigorios Dimitriadis.

Companion website: www.wiley.com/go/dimitriadis/unsteady_aerodynamics

k

l

m

n

p

r

s